Vorticity and Incompressible Flow

This book is a comprehensive introduction to the mathematical theory of vorticity and incompressible flow ranging from elementary introductory material to current research topics. Although the contents center on mathematical theory, many parts of the book showcase the interactions among rigorous mathematical theory, numerical, asymptotic, and qualitative simplified modeling, and physical phenomena. The first half forms an introductory graduate course on vorticity and incompressible flow. The second half comprises a modern applied mathematics graduate course on the weak solution theory for incompressible flow.

Andrew J. Majda is the Samuel Morse Professor of Arts and Sciences at the Courant Institute of Mathematical Sciences of New York University. He is a member of the National Academy of Sciences and has received numerous honors and awards including the National Academy of Science Prize in Applied Mathematics, the John von Neumann Prize of the American Mathematical Society and an honorary Ph.D. degree from Purdue University. Majda is well known for both his theoretical contributions to partial differential equations and his applied contributions to diverse areas besides incompressible flow such as scattering theory, shock waves, combustion, vortex motion and turbulent diffusion. His current applied research interests are centered around Atmosphere/Ocean science.

Andrea L. Bertozzi is Professor of Mathematics and Physics at Duke University. She has received several honors including a Sloan Research Fellowship (1995) and the Presidential Early Career Award for Scientists and Engineers (PECASE). Her research accomplishments in addition to incompressible flow include both theoretical and applied contributions to the understanding of thin liquid films and moving contact lines.

Cambridge Texts in Applied Mathematics

Maximum and Minimum Principles
M. J. SEWELL

Solitons
P. G. DRAZIN AND R. S. JOHNSON

The Kinematics of Mixing
J. M. OTTINO

Introduction to Numerical Linear Algebra and Optimisation
PHILIPPE G. CIARLET

Integral Equations
DAVID PORTER AND DAVID S. G. STIRLING

Perturbation Methods
E. J. HINCH

The Thermomechanics of Plasticity and Fracture
GERARD A. MAUGIN

Boundary Integral and Singularity Methods for Linearized Viscous Flow
C. POZRIKIDIS

Nonlinear Wave Processes in Acoustics
K. NAUGOLNYKH AND L. OSTROVSKY

Nonlinear Systems
P. G. DRAZIN

Stability, Instability and Chaos
PAUL GLENDINNING

Applied Analysis of the Navier–Stokes Equations
C. R. DOERING AND J. D. GIBBON

Viscous Flow
H. OCKENDON AND J. R. OCKENDON

Scaling, Self-Similarity, and Intermediate Asymptotics
G. I. BARENBLATT

A First Course in the Numerical Analysis of Differential Equations
ARIEH ISERLES

Complex Variables: Introduction and Applications
MARK J. ABLOWITZ AND ATHANASSIOS S. FOKAS

Mathematical Models in the Applied Sciences
A. C. FOWLER

Thinking About Ordinary Differential Equations
ROBERT E. O'MALLEY

A Modern Introduction to the Mathematical Theory of Water Waves
R. S. JOHNSON

Rarefied Gas Dynamics
CARLO CERCIGNANI

Symmetry Methods for Differential Equations
PETER E. HYDON

High Speed Flow
C. J. CHAPMAN

Wave Motion
J. BILLINGHAM AND A. C. KING

An Introduction to Magnetohydrodynamics
P. A. DAVIDSON

Linear Elastic Waves
JOHN G. HARRIS

Introduction to Symmetry Analysis
BRIAN J. CANTWELL

Vorticity and Incompressible Flow

ANDREW J. MAJDA
New York University

ANDREA L. BERTOZZI
Duke University

CAMBRIDGE
UNIVERSITY PRESS

PUBLISHED BY THE PRESS SYNDICATE OF THE UNIVERSITY OF CAMBRIDGE
The Pitt Building, Trumpington Street, Cambridge, United Kingdom

CAMBRIDGE UNIVERSITY PRESS
The Edinburgh Building, Cambridge CB2 2RU, UK
40 West 20th Street, New York, NY 10011-4211, USA
10 Stamford Road, Oakleigh, VIC 3166, Australia
Ruiz de Alarcón 13, 28014 Madrid, Spain
Dock House, The Waterfront, Cape Town 8001, South Africa

http://www.cambridge.org

First published 2002

Printed in the United Kingdom at the University Press, Cambridge

Typeface Times Roman 10/13 pt. *System* LaTeX 2_ε [TB]

A catalog record for this book is available from the British Library.

Library of Congress Cataloging in Publication Data
Majda, Andrew, 1949–
Vorticity and incompressible flow / Andrew J. Majda, Andrea L. Bertozzi.
p. cm. – (Cambridge texts in applied mathematics ; 27)
Includes bibliographical references and index.
ISBN 0-521-63057-6 (hb) – ISBN 0-521-63948-4 (pb)
1. Vortex-motion. 2. Non-Newtonian fluids. I. Bertozzi, Andrea L. II. Title. III. Series.
QA925 .M35 2001
532′.059 – dc21 00-046776

ISBN 0 521 63057 6 hardback
ISBN 0 521 63948 4 paperback

Contents

Preface

Vorticity is perhaps the most important facet of turbulent fluid flows. This book is intended to be a comprehensive introduction to the mathematical theory of vorticity and incompressible flow ranging from elementary introductory material to current research topics. Although the contents center on mathematical theory, many parts of the book showcase a modern applied mathematics interaction among rigorous mathematical theory, numerical, asymptotic, and qualitative simplified modeling, and physical phenomena. The interested reader can see many examples of this symbiotic interaction throughout the book, especially in Chaps. 4–9 and 13. The authors hope that this point of view will be interesting to mathematicians as well as other scientists and engineers with interest in the mathematical theory of incompressible flows.

The first seven chapters comprise material for an introductory graduate course on vorticity and incompressible flow. Chapters 1 and 2 contain elementary material on incompressible flow, emphasizing the role of vorticity and vortex dynamics together with a review of concepts from partial differential equations that are useful elsewhere in the book. These formulations of the equations of motion for incompressible flow are utilized in Chaps. 3 and 4 to study the existence of solutions, accumulation of vorticity, and convergence of numerical approximations through a variety of flexible mathematical techniques. Chapter 5 involves the interplay between mathematical theory and numerical or quantitative modeling in the search for singular solutions to the Euler equations. In Chap. 6, the authors discuss vortex methods as numerical procedures for incompressible flows; here some of the exact solutions from Chaps. 1 and 2 are utilized as simplified models to study numerical methods and their performance on unambiguous test problems. Chapter 7 is an introduction to the novel equations for interacting vortex filaments that emerge from careful asymptotic analysis.

The material in the second part of the book can be used for a graduate course on the theory for weak solutions for incompressible flow with an emphasis on modern applied mathematics. Chapter 8 is an introduction to the mildest weak solutions such as patches of vorticity in which there is a complete and elegant mathematical theory. In contrast, Chap. 9 involves a discussion of subtle theoretical and computational issues involved with vortex sheets as the most singular weak solutions in two-space dimensions with practical significance. This chapter also provides a pedagogical introduction to the mathematical material on weak solutions presented in Chaps. 10–12.

Chapter 13 involves a theoretical and computational study of the one-dimensional Vlasov–Poisson equations, which serves as a simplified model in which many of the unresolved issues for weak solutions of the Euler equations can be answered in an explicit and unambiguous fashion.

This book is a direct outgrowth of several extensive lecture courses by Majda on these topics at Princeton University during 1985, 1988, 1990, and 1993, and at the Courant Institute in 1995. This material has been supplemented by research expository contributions based on both the authors' work and on other current research.

Andrew Majda would like to thank many former students in these courses who contributed to the write-up of earlier versions of the notes, especially Dongho Chae, Richard Dziurzynski, Richard McLaughlin, David Stuart, and Enrique Thomann. In addition, many friends and scientific collaborators have made explicit or implicit contributions to the material in this book. They include Tom Beale, Alexandre Chorin, Peter Constantin, Rupert Klein, and George Majda. Ron DiPerna was a truly brilliant mathematician and wonderful collaborator who passed away far too early in his life; it is a privilege to give an exposition of aspects of our joint work in the later chapters of this book.

We would also like to thank the following people for their contributions to the development of the manuscript through proofreading and help with the figures and typesetting: Michael Brenner, Richard Clelland, Diego Cordoba, Weinan E, Pedro Embid, Andrew Ferrari, Judy Horowitz, Benjamin Jones, Phyllis Kronhaus, Monika Nitsche, Mary Pugh, Philip Riley, Thomas Witelski, and Yuxi Zheng. We thank Robert Krasny for providing us with Figures 9.4 and 9.5 in Chap. 9.

1

An Introduction to Vortex Dynamics for Incompressible Fluid Flows

In this book we study incompressible high Reynolds numbers and incompressible inviscid flows. An important aspect of such fluids is that of *vortex dynamics*, which in lay terms refers to the interaction of local swirls or eddies in the fluid. Mathematically we analyze this behavior by studying the rotation or *curl* of the velocity field, called the *vorticity*. In this chapter we introduce the Euler and the Navier–Stokes equations for incompressible fluids and present elementary properties of the equations. We also introduce some elementary examples that both illustrate the kind of phenomena observed in hydrodynamics and function as building blocks for more complicated solutions studied in later chapters of this book.

This chapter is organized as follows. In Section 1.1 we introduce the equations, relevant physical quantities, and notation. Section 1.2 presents basic symmetry groups of the Euler and the Navier–Stokes equations. In Section 1.3 we discuss the motion of a particle that is carried with the fluid. We show that the particle-trajectory map leads to a natural formulation of how quantities evolve with the fluid. Section 1.4 shows how locally an incompressible field can be approximately decomposed into translation, rotation, and deformation components. By means of exact solutions, we show how these simple motions interact in solutions to the Euler or the Navier–Stokes equations. Continuing in this fashion, Section 1.5 examines exact solutions with shear, vorticity, convection, and diffusion. We show that although deformation can increase vorticity, diffusion can balance this effect. Inviscid fluids have the remarkable property that vorticity is transported (and sometimes stretched) along streamlines. We discuss this in detail in Section 1.6, including the fact that vortex lines move with the fluid and circulation over a closed curve is conserved. This is an example of a quantity that is locally conserved. In Section 1.7 we present a number of global quantities, involving spatial integrals of functions of the solution, such as the kinetic energy, velocity, and vorticity flux, that are conserved for the Euler equation. In the case of Navier–Stokes equations, diffusion causes some of these quantities to dissipate. Finally, in Section 1.8, we show that the incompressibility condition leads to a natural reformulation of the equations (which are due to Leray) in which the pressure term can be replaced with a nonlocal bilinear function of the velocity field. This is the sense in which the pressure plays the role of a Lagrange multiplier in the evolution equation. The appendix of this chapter reviews the Fourier series and the Fourier transform

(Subsection 1.9.1), elementary properties of the Poisson equation (Subsection 1.9.2), and elementary properties of the heat equation (Subsection 1.9.3).

1.1. The Euler and the Navier–Stokes Equations

Incompressible flows of homogeneous fluids in all of space $\mathbb{R}^N$, $N = 2, 3$, are solutions of the system of equations

$$\frac{Dv}{Dt} = -\nabla p + \nu \Delta v, \tag{1.1}$$

$$\operatorname{div} v = 0, \qquad (x, t) \in \mathbb{R}^N \times [0, \infty), \tag{1.2}$$

$$v|_{t=0} = v_0, \qquad x \in \mathbb{R}^N, \tag{1.3}$$

where $v(x, t) \equiv (v^1, v^2, \ldots, v^N)^t$ is the fluid velocity, $p(x, t)$ is the scalar pressure, D/Dt is the convective derivative (i.e., the derivative along particle trajectories),

$$\frac{D}{Dt} = \frac{\partial}{\partial t} + \sum_{j=1}^{N} v^j \frac{\partial}{\partial x_j}, \tag{1.4}$$

and div is the divergence of a vector field,

$$\operatorname{div} v = \sum_{j=1}^{N} \frac{\partial v^j}{\partial x_j}. \tag{1.5}$$

The gradient operator ∇ is

$$\nabla = \left(\frac{\partial}{\partial x_1}, \frac{\partial}{\partial x_2}, \ldots, \frac{\partial}{\partial x_N} \right)^t, \tag{1.6}$$

and the Laplace operator Δ is

$$\Delta = \sum_{j=1}^{N} \frac{\partial^2}{\partial x_j^2}. \tag{1.7}$$

A given kinematic constant viscosity $\nu \geq 0$ can be viewed as the reciprocal of the Reynolds number R_e. For $\nu > 0$, Eq. (1.1) is called the *Navier–Stokes* equation; for $\nu = 0$ it reduces to the *Euler equation*. These equations follow from the conservation of momentum for a continuum (see, e.g., Chorin and Marsden, 1993). Equation (1.2) expresses the incompressibility of the fluid (see Proposition 1.4). The initial value problem [Eqs. (1.1)–(1.3)] is unusual because it contains the time derivatives of only three out of the four unknown functions. In Section 1.8 we show that the pressure $p(x, t)$ plays the role of a Lagrange multiplier and that a nonlocal operator in $\mathbb{R}^N$ determines the pressure from the velocity $v(x, t)$.

This book often considers examples of incompressible fluid flows in the *periodic* case, i.e.,

$$v(x + e_i, t) = v(x, t), \qquad i = 1, 2, \ldots, N, \tag{1.8}$$

for all x and $t \geq 0$, where e_i are the standard basis vectors in $\mathbb{R}^N$, $e_1 = (1, 0, \ldots,)^t$, etc. Periodic flows provide prototypical examples for fluid flows in bounded domains $\Omega \subset \mathbb{R}^N$. In this case the bounded domain Ω is the N-dimensional torus T^N. Flows on the torus serve as especially good elementary examples because we have Fourier series techniques (see Subsection 1.9.1) for computing explicit solutions. We make use of these methods, e.g., in Proposition 1.18 (the Hodge decomposition of T^N) in this chapter and repeatedly throughout this book.

In many applications, e.g., predicting hurricane paths or controlling large vortices shed by jumbo jets, the viscosity ν is very small: $\nu \sim 10^{-6} - 10^{-3}$. Thus we might anticipate that the behavior of inviscid solutions (with $\nu = 0$) would give a lot of insight into the behavior of viscous solutions for a small viscosity $\nu \ll 1$. In this chapter and Chap. 2 we show this to be true for explicit examples. In Chap. 3 we prove this result for general solutions to the Navier–Stokes equation in $\mathbb{R}^N$ (see Proposition (3.2).

1.2. Symmetry Groups for the Euler and the Navier–Stokes Equations

Here we list some elementary symmetry groups for solutions to the Euler and the Navier–Stokes equations. By straightforward inspection we get the following proposition.

Proposition 1.1. *Symmetry Groups of the Euler and the Navier–Stokes Equations. Let v, p be a solution to the Euler or the Navier–Stokes equations. Then the following transformations also yield solutions:*

(i) Galilean invariance: For any constant-velocity vector $\mathbf{c} \in \mathbb{R}^N$,

$$\begin{aligned} v_{\mathbf{c}}(x, t) &= v(x - \mathbf{c}t, t) + \mathbf{c}, \\ p_{\mathbf{c}}(x, t) &= p(x - \mathbf{c}t, t) \end{aligned} \tag{1.9}$$

is also a solution pair.

(ii) Rotation symmetry: for any rotation matrix $\mathcal{Q}$ ($\mathcal{Q}^t = \mathcal{Q}^{-1}$),

$$\begin{aligned} v_{\mathcal{Q}}(x, t) &= \mathcal{Q}^t v(\mathcal{Q}x, t), \\ p_{\mathcal{Q}}(x, t) &= p(\mathcal{Q}x, t) \end{aligned} \tag{1.10}$$

is also a solution pair.

(iii) Scale invariance: for any $\lambda, \tau \in \mathbb{R}$,

$$v_{\lambda,\tau}(x, t) = \frac{\lambda}{\tau} v\left(\frac{x}{\lambda}, \frac{t}{\tau}\right), \qquad p_{\lambda,\tau}(x, t) = \frac{\lambda^2}{\tau^2} p\left(\frac{x}{\lambda}, \frac{t}{\tau}\right), \tag{1.11}$$

is a solution pair to the Euler equation, and for any $\tau \in \mathbb{R}^+$,

$$v_\tau(x, t) = \tau^{-1/2} v\left(\frac{x}{\tau^{1/2}}, \frac{t}{\tau}\right), \qquad p_\tau(x, t) = \tau^{-1} p\left(\frac{x}{\tau^{1/2}}, \frac{t}{\tau}\right), \tag{1.12}$$

is a solution pair to the Navier–Stokes equation.

We note that scaling transformations determine the two-parameter symmetry group given in Eqs. (1.11) for the Euler equation. The introduction of viscosity $\nu > 0$, however, restricts this symmetry group to the one-parameter group given in Eqs. (1.12) for the Navier–Stokes equation.

1.3. Particle Trajectories

An important construction used throughout this book is the *particle-trajectory mapping* $X(\cdot, t)\colon \alpha \in \mathbb{R}^N \to X(\alpha, t) \in \mathbb{R}^N$. Given a fluid velocity $v(x, t)$, $X(\alpha, t) = (X_1, X_2, \ldots, X_N)^t$ is the location at time t of a fluid particle initially placed at the point $\alpha = (\alpha_1, \alpha_2, \ldots, \alpha_N)^t$ at time $t = 0$. The following nonlinear ordinary differential equation (ODE) defines particle-trajectory mapping:

$$\frac{dX}{dt}(\alpha, t) = v(X(\alpha, t), t), \qquad X(\alpha, 0) = \alpha. \tag{1.13}$$

The parameter α is called the Lagrangian particle marker. The particle-trajectory mapping X has a useful interpretation: An initial domain $\Omega \subset \mathbb{R}^N$ in a fluid evolves in time to $X(\Omega, t) = \{X(\alpha, t)\colon \alpha \in \Omega\}$, with the vector v tangent to the particle trajectory (see Fig. 1.1).

Next we review some elementary properties of $X(\cdot, t)$. We define the Jacobian of this transformation by

$$J(\alpha, t) = \det(\nabla_\alpha X(\alpha, t)). \tag{1.14}$$

We use subscripts to denote partial derivatives and variables of differential operators, e.g., $f_t = \partial/\partial t f$, $\nabla_\alpha = [(\partial/\partial\alpha_1), \ldots, (\partial/\partial\alpha_N)]$. The time evolution of the Jacobian J satisfies the following proposition.

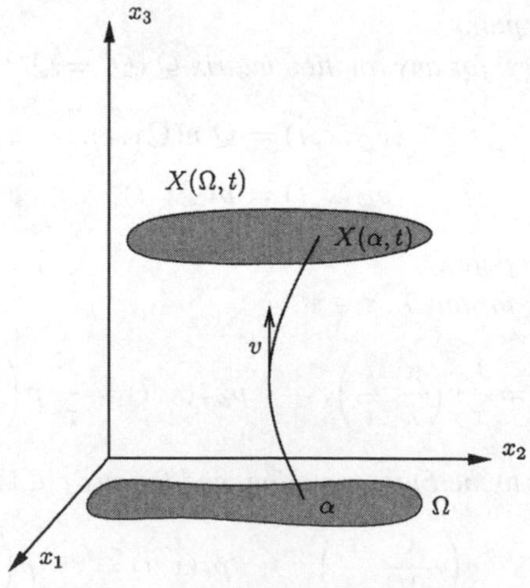

Figure 1.1. The particle-trajectory map.

Proposition 1.2. *Let $X(\cdot, t)$ be a particle-trajectory mapping of a smooth velocity field $v \in \mathbb{R}^N$. Then*

$$\frac{\partial J}{\partial t} = (\mathrm{div}_x\, v)|_{(X(\alpha,t),t)} J(\alpha, t). \tag{1.15}$$

We also frequently need a formula to determine the rate of change of a given function $f(x, t)$ in a domain $X(\Omega, t)$ moving with the fluid. This calculus formula, called the transport formula, is the following proposition.

Proposition 1.3. *(The Transport Formula). Let $\Omega \subset \mathbb{R}^N$ be an open, bounded domain with a smooth boundary, and let X be a given particle-trajectory mapping of a smooth velocity field v. Then for any smooth function $f(x, t)$,*

$$\frac{d}{dt}\int_{X(\Omega,t)} f\, dx = \int_{X(\Omega,t)} [f_t + \mathrm{div}_x(f v)] dx. \tag{1.16}$$

We give the proofs of Propositions 1.2 and 1.3 below. As an immediate application of these results, we note that either $J(\alpha, t) = 1$ or $\mathrm{div}\, v = 0$ implies incompressibility.

Definition 1.1. *A flow $X(\cdot, t)$ is incompressible if for all subdomains Ω with smooth boundaries and any $t > 0$ the flow is volume preserving:*

$$\mathrm{vol}\, X(\Omega, t) = \mathrm{vol}\, \Omega.$$

Applying the transport formula in Eq. (1.16) for $f \equiv 1$, we get $\mathrm{div}\, v = 0$. Moreover, then Eq. (1.15) yields $J(\alpha, t) = J(\alpha, 0) = 1$. We state this as a proposition below.

Proposition 1.4. *For smooth flows the following three conditions are equivalent:*

(i) a flow is incompressible, i.e., $\forall \Omega \subset \mathbb{R}^N$, $t \geq 0$ $\mathrm{vol}\, X(\Omega, t) = \mathrm{vol}\, \Omega$,
(ii) $\mathrm{div}\, v = 0$,
(iii) $J(\alpha, t) = 1$.

Now we give the proof of Proposition 1.2.

Proof of Proposition 1.2. Because the determinant is multilinear in columns (rows), we compute the time derivative

$$\frac{\partial J}{\partial t} = \frac{\partial}{\partial t} \det\left[\frac{\partial X^i}{\partial \alpha_j}(\alpha, t)\right] = \sum_{i,j} A_i^j \frac{\partial}{\partial t}\frac{\partial X^i}{\partial \alpha_j}(\alpha, t),$$

where A_i^j is the minor of the element $\partial X^i / \partial \alpha_j$ of the matrix $\nabla_\alpha X$. The minors satisfy the well-known identity

$$\sum_j \frac{\partial X^k}{\partial \alpha_j} A_i^j = \delta_i^k J, \qquad \text{where} \quad \delta_i^k = \begin{cases} 1, & k = i \\ 0, & k \neq i \end{cases}.$$

The definition of the particle trajectories in Eq. (1.13) then gives

$$\frac{\partial J}{\partial t} = \sum_{i,j,k} A_i^j \frac{\partial X^k}{\partial \alpha_j} v_{x_k}^i = \sum_{i,k} v_{x_k}^i \delta_i^k J = J \text{ div } v. \qquad \Box$$

Finally we give the proof of Proposition 1.3.

Proof of Proposition 1.3. By the change of variables $\alpha \to X(\alpha, t)$, we reduce the integration over the moving domain $X(\Omega, t)$ to the integration over the fixed domain Ω:

$$\int_{X(\Omega,t)} f(x,t)dx = \int_\Omega f(X(\alpha,t),t)J(\alpha,t)d\alpha.$$

The definition in Eq. (1.13) implies that

$$\begin{aligned}\frac{d}{dt}\int_{X(\Omega,t)} f\,dx &= \int_\Omega \left[\left(\frac{\partial f}{\partial t} + \frac{dX}{dt}\cdot\nabla f\right)J + f\frac{\partial J}{\partial t}\right]d\alpha \\ &= \int_\Omega \left(\frac{\partial f}{\partial t} + v_x\cdot\nabla f + f\,\text{div}_x v\right)J\,d\alpha \\ &= \int_{X(\Omega,t)} \left[\frac{\partial f}{\partial t} + \text{div}_x(fv)\right]dx. \qquad \Box\end{aligned}$$

1.4. The Vorticity, a Deformation Matrix, and Some Elementary Exact Solutions

First we determine a simple local description for an incompressible fluid flow. Every smooth velocity field $v(x,t)$ has a Taylor series expansion at a fixed point (x_0, t_0):

$$v(x_0 + h, t_0) = v(x_0, t_0) + (\nabla v)(x_0, t_0)h + \mathcal{O}(h^2), \qquad h \in \mathbb{R}^3. \tag{1.17}$$

The 3×3 matrix $\nabla v = (v_{x_j}^i)$ has a *symmetric* part $\mathcal{D}$ and an *antisymmetric* part Ω:

$$\mathcal{D} = \frac{1}{2}(\nabla v + \nabla v^t), \tag{1.18}$$

$$\Omega = \frac{1}{2}(\nabla v - \nabla v^t). \tag{1.19}$$

$\mathcal{D}$ is called the deformation or rate-of-strain matrix, and Ω is called the rotation matrix. If the flow is incompressible, div $v = 0$, then the trace tr $\mathcal{D} = \sum_i d_{ii} = 0$. Moreover, the vorticity ω of the vector field v,

$$\omega = \text{curl } v \equiv \left(\frac{\partial v^3}{\partial x_2} - \frac{\partial v^2}{\partial x_3}, \frac{\partial v^1}{\partial x_3} - \frac{\partial v^3}{\partial x_1}, \frac{\partial v^2}{\partial x_1} - \frac{\partial v^1}{\partial x_2}\right)^t, \tag{1.20}$$

satisfies

$$\Omega h = \frac{1}{2}\omega \times h, \qquad \forall h \in \mathbb{R}^3. \tag{1.21}$$

Using the Taylor series expression (1.17) and the new definitions, we have Lemma 1.1

Lemma 1.1. *To linear order in* $(|x - x_0|)$*, every smooth incompressible velocity field* $v(x, t)$ *is the (unique) sum of three terms:*

$$v(x, t_0) = v(x_0, t_0) + \frac{1}{2}\omega \times (x - x_0) + \mathcal{D}(x - x_0), \tag{1.22}$$

where $\mathcal{D}$ *is the (symmetric) deformation matrix with* $\operatorname{tr} \mathcal{D} = 0$ *and* ω *is the vorticity.*

The successive terms in Eq. (1.22) have a natural physical interpretation in terms of translation, rotation, and deformation. Retaining only the term $v(x_0, t_0)$ in Eq. (1.22) gives

$$X(\alpha, t) = \alpha + v(x_0, t_0)(t - t_0),$$

which describes an *infinitesimal translation.*

By a rotation of axes, without loss of generality, we can assume that $\omega = (0, 0, \omega)^t$, so

$$\omega \times (x - x_0) = \begin{bmatrix} 0 & -\omega & 0 \\ \omega & 0 & 0 \\ 0 & 0 & 0 \end{bmatrix} (x - x_0).$$

Thus retaining only this term in the velocity for the particle-trajectory equation gives the particle trajectories $X = (X', X_3)$ as

$$X'(\alpha, t) = x_0' + \mathcal{Q}\left(\frac{1}{2}\omega t\right)(x' - x_0'), \qquad X_3(\alpha, t) = x_0^3,$$

where $\mathcal{Q}$ is the rotation matrix in the $x_1 - x_2$ plane:

$$\mathcal{Q}(\varphi) = \begin{bmatrix} \cos\varphi & -\sin\varphi \\ \sin\varphi & \cos\varphi \end{bmatrix}.$$

These trajectories are circles on the $x_1 - x_2$ plane, so the second term $\frac{1}{2}\omega \times (x - x_0)$ in Eq. (1.22) is an infinitesimal rotation in the direction of ω with angular velocity $\frac{1}{2}|\omega|$.

Finally, because $\mathcal{D}$ is a symmetric matrix, there is a rotation matrix $\mathcal{Q}$ so that $\mathcal{Q}\mathcal{D}\mathcal{Q}^t = \operatorname{diag}(\gamma_1, \gamma_2, \gamma_3)$. Moreover, traces are invariant under similarity transformations so that $\gamma_1 + \gamma_2 + \gamma_3 = 0$. Thus, without loss of generality, assume that

$$\mathcal{D} = \operatorname{diag}[\gamma_1, \gamma_2, -(\gamma_1 + \gamma_2)].$$

Retaining only the term $\mathcal{D}(x - x_0)$ from Eq. (1.22) in the particle-trajectory equation yields

$$X(\alpha, t) = x_0 + \begin{bmatrix} e^{\gamma_1(t-t_0)} & 0 & 0 \\ 0 & e^{\gamma_2(t-t_0)} & 0 \\ 0 & 0 & e^{-(\gamma_1+\gamma_2)(t-t_0)} \end{bmatrix} (\alpha - x_0).$$

For example, if we set $\gamma_1, \gamma_2 > 0, x_0 = 0$, the fluid is compressed along the x_1–x_2 plane but stretched along the x_3 axis, creating a jet. This corresponds to a sharp deformation of the fluid. Thus the third term in Eq. (1.22) represents an infinitesimal deformation velocity in the direction $(x - x_0)$.

We have just proved the following corollary.

Corollary 1.1. *To linear order in* $(|x-x_0|)$, *every incompressible velocity field* $v(x,t)$ *is the sum of infinitesimal translation, rotation, and deformation velocities.*

A large part of this book addresses the interactions among these three contributions to the velocity field. To illustrate the interaction between a vorticity and a deformation, we now derive a large class of exact solutions for both the Euler and the Navier–Stokes equations.

Proposition 1.5. *Let* $\mathcal{D}(t)$ *be a real, symmetric,* 3×3 *matrix with* $\operatorname{tr} \mathcal{D}(t) = 0$. *Determine the vorticity* $\omega(t)$ *from the ODE on* $\mathbb{R}^3$,

$$\frac{d\omega}{dt} = \mathcal{D}(t)\omega, \qquad \omega|_{t=0} = \omega_0 \in \mathbb{R}^3, \tag{1.23}$$

and the antisymmetric matrix Ω *by means of the formula* $\Omega h = \frac{1}{2}\omega \times h$. *Then*

$$\begin{aligned} v(x,t) &= \frac{1}{2}\omega(t) \times x + \mathcal{D}(t)x, \\ p(x,t) &= -\frac{1}{2}[\mathcal{D}_t(t) + \mathcal{D}^2(t) + \Omega^2(t)]x \cdot x \end{aligned} \tag{1.24}$$

are exact solutions to the three-dimensional (3D) Euler and the Navier–Stokes equations.

The solutions in Eqs. (1.24) can be trivially generalized by use of the Galilean invariance (see Proposition 1.1). Because the pressure p has a quadratic behavior in x, these solutions have a direct physical meaning only locally in space and time. Also, because the velocity is linear in x, the effects of viscosity do not alter these solutions. Nevertheless, these solutions model the typical local behavior of incompressible flows.

Before proving this proposition, first we give some examples of the exact solutions in Eqs. (1.24) that illustrate the interactions between a rotation and a deformation.

Example 1.1. Jet Flows. Taking $\omega_0 = 0$ and $\mathcal{D} = \operatorname{diag}(-\gamma_1, -\gamma_2, \gamma_1 + \gamma_2)$, $\gamma_j > 0$, from Eqs. (1.23 and 1.24) we get

$$v(x,t) = [-\gamma_1 x_1, -\gamma_2 x_2, (\gamma_1 + \gamma_2)x_3]^t. \tag{1.25}$$

This flow is irrotational, $\omega = 0$, and forms two jets along the positive and the negative

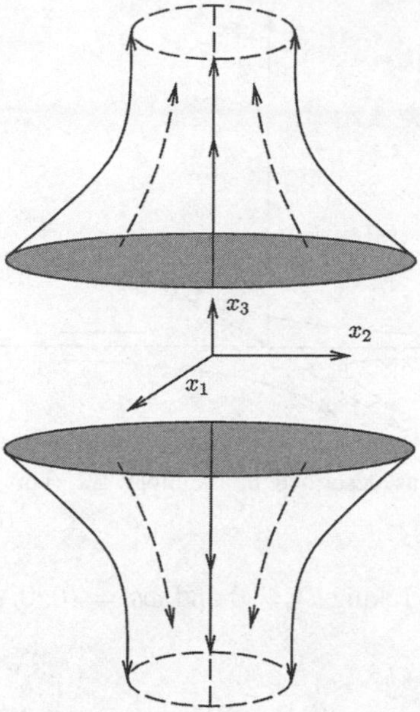

Figure 1.2. A jet flow as described in Example 1.1.

directions of the x_3 axis, with particle trajectories (see Fig. 1.2)

$$X(\alpha, t) = \begin{bmatrix} e^{-\gamma_1 t} & 0 & 0 \\ 0 & e^{-\gamma_2 t} & 0 \\ 0 & 0 & e^{(\gamma_1+\gamma_2)t} \end{bmatrix} \alpha.$$

Observe that

$$\left(X_1^2 + X_2^2\right)(\alpha, t) = e^{-2(\gamma_1+\gamma_2)t}\left(\alpha_1^2 + \alpha_2^2\right),$$

so the distance of a given fluid particle to the x_3 axis decreases exponentially in time. A jet flow is one type of axisymmetric flow without swirl, which will be discussed in Subsection 2.3.3 of Chap. 2.

Example 1.2. Strain Flows. Now taking $\omega_0 = 0$ and $\mathcal{D} = \text{diag}(-\gamma, \gamma, 0)$, $\gamma > 0$, from Eq. (1.25) we get

$$v(x, t) = (-\gamma x_1, \gamma x_2, 0)^t. \tag{1.26}$$

This flow is irrotational $\omega = 0$ and forms a strain flow (independent of x_3) with the particle trajectories $X = (X', X_3)$ (see Fig. 1.3):

$$X'(\alpha, t) = \begin{bmatrix} e^{-\gamma t} & 0 \\ 0 & e^{\gamma t} \end{bmatrix} \alpha', \qquad X_3(\alpha, t) = \alpha_3.$$

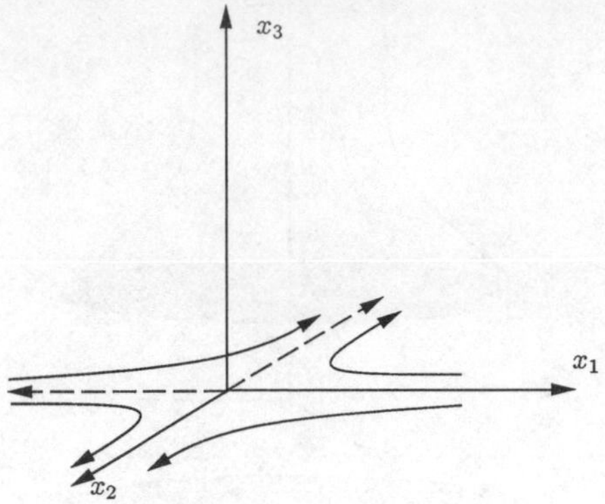

Figure 1.3. A strain flow as described in Example 1.2. This flow is independent of the variable x_3.

Example 1.3. A Vortex. Taking $\mathcal{D} = 0$ and $\omega_0 = (0, 0, \omega)^t$, from Eqs. (1.23) and (1.24) we get

$$v(x, t) = \left(-\frac{1}{2}\omega_0 x_2, \frac{1}{2}\omega_0 x_1, 0 \right)^t. \tag{1.27}$$

This flow is a rigid rotation motion in the $x_1 - x_2$ plane, with the angular velocity $\frac{1}{2}\omega_0$ and the particle trajectories $X = (X', X_3)$, $X' = (X^1, X^2)$ (see Fig. 1.4):

$$X'(\alpha, t) = \mathcal{Q}\left(\frac{1}{2}\omega_0 t\right)\alpha', \qquad X_3(\alpha, t) = \alpha_3,$$

where $\mathcal{Q}$ is the 2×2 rotation matrix

$$\mathcal{Q}(\varphi) = \begin{bmatrix} \cos\varphi & -\sin\varphi \\ \sin\varphi & \cos\varphi \end{bmatrix}.$$

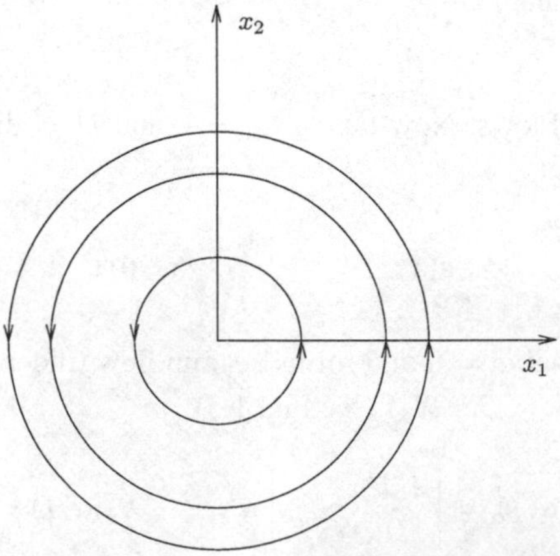

Figure 1.4. A two-dimensional vortex as described in Example 1.3.

Example 1.4. A Rotating Jet. Now we take the superposition of a jet and a vortex, with $\mathcal{D} = \text{diag}(-\gamma_1, -\gamma_2, \gamma_1 + \gamma_2)$, $\gamma_j > 0$, and $\omega_0 = (0, 0, \omega_0)^t$. Then Eq. (1.23) reduces to the scalar vorticity equation, and

$$\omega(t) = \omega_0 e^{(\gamma_1+\gamma_2)t}.$$

Observe that in this case the vorticity ω aligns with the eigenvector $e_3 = (0, 0, 1)^t$ corresponding to the positive eigenvalue $\lambda_3 = \gamma_1 + \gamma_2$ of $\mathcal{D}$ and that the vorticity $\omega(t)$ increases exponentially in time.

The corresponding velocity v given by the first of Eqs. (1.24) is

$$v(x, t) = \left[-\gamma_1 x_1 - \frac{1}{2}\omega(t)x_2, \frac{1}{2}\omega(t)x_1 - \gamma_2 x_2, (\gamma_1 + \gamma_2)x_3\right]^t. \tag{1.28}$$

Now the $X' = (X_1, X_2)$ coordinates of particle trajectories satisfy the coupled ODE,

$$\frac{dX'}{dt} = \begin{bmatrix} \gamma_1 & -\frac{1}{2}\omega(t) \\ \frac{1}{2}\omega(t) & -\gamma_2 \end{bmatrix} X',$$

and $X_3(\alpha, t) = e^{(\gamma_1+\gamma_2)t}\alpha_3$. From the above equation we get

$$\frac{1}{2}\frac{d}{dt}\left(X_1^2 + X_2^2\right) = -\gamma_1 X_1^2 - \gamma_2 X_2^2,$$

so

$$e^{-2\max(\gamma_1,\gamma_2)t}\left(\alpha_1^2 + \alpha_2^2\right) \le \left(X_1^2 + X_2^2\right)(\alpha, t) \le e^{-2\min(\gamma_1,\gamma_2)t}\left(\alpha_1^2 + \alpha_2^2\right).$$

Thus, although the particle trajectories spiral around the x_3 axis with increasing angular velocity $\frac{1}{2}\omega(t)$ (see Fig. 1.5), the minimal and the maximal distances of a given fluid particle to the x_3 axis are the same as those for the jet without rotation (see Example 1.1). A rotating jet is a type of axisymmetric flow with swirl, discussed in detail in Subsection 2.3.3.

Finally, we give the proof of Proposition 1.5.

Proof of Proposition 1.5. Computing the ∂_{x_k} derivative of the Navier-Stokes equation, we get componentwise

$$\left(v^i_{x_k}\right)_t + v^j\left(v^i_{x_k}\right)_{x_j} + v^j_{x_k}v^i_{x_j} = -p_{x_i x_k} + \nu\left(v^i_{x_k}\right)_{x_j x_j}.$$

Thus, introducing the notation $V \equiv (v^i_{x_k})$ and $P \equiv (p_{x_i x_k})$ for the Hessian matrix of the pressure p, we get the matrix equation for V:

$$\frac{DV}{Dt} + V^2 = -P + \nu\Delta V. \tag{1.29}$$

If we want to see how the rotation and the deformation interact, it is natural to decompose V into its symmetric part $\mathcal{D} = \frac{1}{2}(V + V^t)$ and antisymmetric part

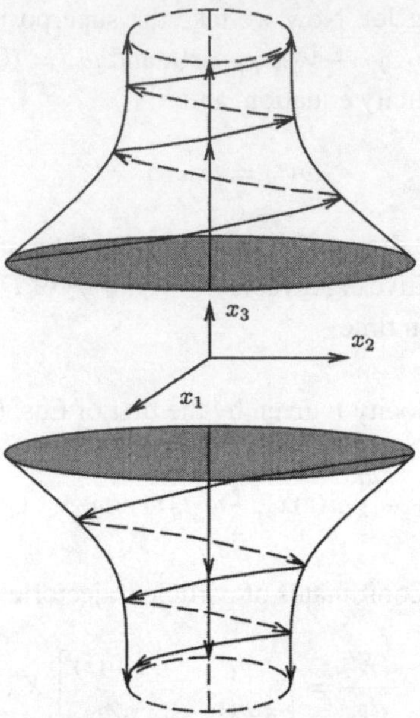

Figure 1.5. A rotating jet as described in Example 1.4. The particle trajectories spiral around the x_3 axis with increasing angular velocity $\frac{1}{2}\omega(t)$. However, the distance to the x_3 axis remains the same as in the case of the nonrotational jet.

$\Omega = \frac{1}{2}(V - V^t)$. Then $\mathcal{D}$ and V satisfy the formula

$$V^2 = (\Omega^2 + \mathcal{D}^2) + (\Omega\mathcal{D} + \mathcal{D}\Omega),$$

where the first term is symmetric and the second one is antisymmetric. The symmetric part of Eq. (1.29) thus is

$$\frac{D\mathcal{D}}{Dt} + \mathcal{D}^2 + \Omega^2 = -P + \nu\Delta\mathcal{D}, \tag{1.30}$$

where D/Dt is a convective derivative, as given in Eq. (1.4), and the antisymmetric part of Eq. (1.29) is

$$\frac{D\Omega}{Dt} + \Omega\mathcal{D} + \mathcal{D}\Omega = \nu\Delta\Omega. \tag{1.31}$$

Because the antisymmetric matrix Ω is defined in terms of the vorticity ω by Eq. (1.21), $\Omega h \equiv \frac{1}{2}\omega \times h$, after some straightforward calculations Eq. (1.31) gives the following *vorticity equation* for the dynamics of ω:

$$\frac{D\omega}{Dt} = \mathcal{D}\omega + \nu\Delta\omega. \tag{1.32}$$

General equations (1.30) and (1.32) are fundamental for the developments presented later in this book. Vorticity equation (1.32) is derived directly from the Navier–Stokes equation in Section 2.1 of Chap. 2. It is a key fact in the study of continuation of solutions in Chaps. 3 and 4, the design of vortex methods in Chap. 6, and the notion of vortex patches in Chap. 8.

To continue the proof, now we postulate the velocity $v(x,t)$ as in Eqs. (1.24):

$$v(x,t) = \frac{1}{2}\omega(t) \times x + \mathcal{D}(t)x.$$

Because curl $v = \omega(t)$, ω does not depend on the spatial variable x so that $\Delta\omega$ and $v \cdot \nabla\omega$ both vanish. Vorticity equation (1.32) reduces to the scalar ODE:

$$\frac{d\omega}{dt} = \mathcal{D}\omega.$$

Thus, given $\mathcal{D}(t)$, we can solve this equation for $\omega(t)$.

Now we show that the symmetric part in Eq. (1.30) of the Navier–Stokes equation determines the pressure p. Because ω determines Ω by $\Omega h = \frac{1}{2}\omega \times h$, Eq. (1.30) gives

$$-P = \frac{d\mathcal{D}}{dt} + \mathcal{D}^2 + \Omega^2,$$

where the right-hand side is a known symmetric matrix. Because $P(t)$ is a symmetric matrix, it is the Hessian of a scalar function; in fact, $p(x,t)$ is the explicit function

$$p(x,t) = \frac{1}{2}P(t)x \cdot x.$$

By the above construction, v and p satisfy the Navier–Stokes equation. □

1.5. Simple Exact Solutions with Convection, Vortex Stretching, and Diffusion

Vorticity equation (1.32) for the dynamics of ω,

$$\frac{D\omega}{Dt} = \mathcal{D}\omega + \nu\Delta\omega, \tag{1.33}$$

plays a crucial role in our further analysis of the Euler and the Navier–Stokes equations. The viscous term $D\omega/Dt$ represents a convection of ω along the particle trajectories. In Example 1.4 we saw that the term $\mathcal{D}\omega$ is responsible for vortex stretching: The vorticity ω increases (decreases) when ω roughly aligns with eigenvectors corresponding to positive (negative) eigenvalues of the matrix $\mathcal{D}$. The viscous term $\nu\Delta\omega$ leads to a diffusion of ω.

To illustrate the competing effects of convection, vortex stretching, and diffusion, we construct a large class of exact *shear-layer solutions* to the Navier–Stokes equation. We take the irrotational strain flow from Eq. (1.26),

$$v(x,t) = (-\gamma x_1, \gamma x_2, 0)^t, \qquad \gamma > 0,$$

with the corresponding pressure $p(x,t) = \frac{1}{2}\gamma(x_1^2 + x_2^2)$. Now we want to admit also a nonzero third component $v^3(x_1, t)$ of the velocity that depends on only time and the x_1 variable. Thus we seek a solution to the Navier–Stokes equation with the velocity

$$v(x,t) = [-\gamma x_1, \gamma x_2, v^3(x_1,t)]^t.$$

This velocity field solves the Navier–Stokes equation provided that v^3 satisfies the much simpler linear diffusion equation

$$\begin{aligned} v_t^3 - \gamma x_1 \partial_{x_1} v^3 &= \nu \frac{\partial^2 v^3}{\partial x_1^2}, \\ v^3|_{t=0} &= v_0^3(x_1). \end{aligned} \tag{1.34}$$

The vorticity corresponding to this velocity is $\omega(x,t) = [0, -v_{x_1}^3(x_1,t), 0]^t$. We differentiate Eq. (1.34) to obtain the equation for the second component $\omega^2(x_1,t)$ of vorticity:

$$\begin{aligned} \omega_t^2 - \gamma x_1 \omega_{x_1}^2 &= \gamma\omega^2 + \nu \frac{\partial^2 \omega^2}{\partial x_1^2}, \\ \omega^2|_{t=0} &= \omega_0^2(x_1) \equiv -\frac{\partial v_0^3}{\partial x_1}(x_1). \end{aligned} \tag{1.35}$$

This equation is the second component of general vorticity equation (1.33) written for our velocity v, and it illustrates the competing effects from Eq. (1.33) in a simpler context: There is the convection velocity $-\gamma x_1$, the stretching of vorticity by $\gamma\omega^2$, and the diffusion through the term $\nu\partial^2\omega^2/\partial x_1^2$. Moreover, because the vorticity ω aligns with the eigenvector $e_2 = (0,1,0)^t$ corresponding to the positive eigenvalue γ of the matrix $\mathcal{D}$, the vorticity ω increases in time.

Observe that the velocity v^3 is determined from the vorticity ω^2 by the nonlocal integral operator

$$v^3(x_1,t) = -\int_{-\infty}^{x_1} \omega^2(\xi_1,\tau)d\xi_1. \tag{1.36}$$

In Section 2.1 we show that for any solution to the Euler or the Navier–Stokes equation the velocity v can be recovered from the vorticity ω by a certain nonlocal operator, and Eq. (1.36) provides a simple illustration of this fact.

Using Eqs. (1.34)–(1.36), first we illustrate the effects of convection. To simplify notation we suppress superscripts in the scalar velocity v^3 and the vorticity ω^2 and subscripts in the space variable x_1. We have Example 1.5.

Example 1.5. A Basic Shear-Layer Flow. Taking $\gamma = \nu = 0$, from Eq. (1.35) we get the vorticity (denote $x = x_1$)

$$\omega(x,t) = [0, \omega_0(x), 0]^t, \qquad x \in \mathbb{R}. \tag{1.37}$$

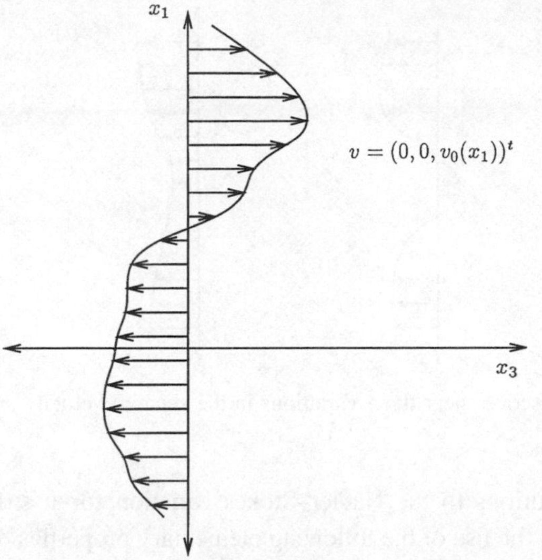

Figure 1.6. A basic shear flow (Example 1.5).

Moreover, by Eq. (1.34) [or Eq. (1.36)] the velocity is (see Fig. 1.6)

$$v(x, t) = [0, 0, v_0(x)]^t, \qquad x \in \mathbb{R}. \tag{1.38}$$

Next we incorporate diffusion effects into this solution. We have Example 1.6.

Example 1.6. A Viscous Shear-Layer Solution. With $\gamma = 0$ and $\nu > 0$, Eq. (1.35) reduces to the heat equation on the scalar vorticity $\omega(x, t)$:

$$\omega_t = \nu\omega_{xx},$$
$$\omega|_{t=0} = \omega_0(x).$$

The solution ω is (see Lemma 1.13 in Subsection 1.9.3).

$$\omega(x, t) = \int_{\mathbb{R}} H(x - y, \nu t)\omega_0(y)dy, \qquad x \in \mathbb{R}, \tag{1.39}$$

where H is the one-dimensional (1D) Gaussian heat kernel

$$H(y, t) = (4\pi t)^{-1/2} e^{-|y|^2/4t}. \tag{1.40}$$

The x_3 component of the velocity is determined by Eq. (1.34) or Eq. (1.36) as

$$v(x, t) = \int_{\mathbb{R}} H(x - y, \nu t)v_0(y)dy, \qquad x \in \mathbb{R}. \tag{1.41}$$

Thus the diffusion spreads the solution ω and v (see Fig. 1.7).

The above examples of the inviscid and the viscous shear-layer solutions are also good illustrations of the general problem: How well do solutions to the Euler equation

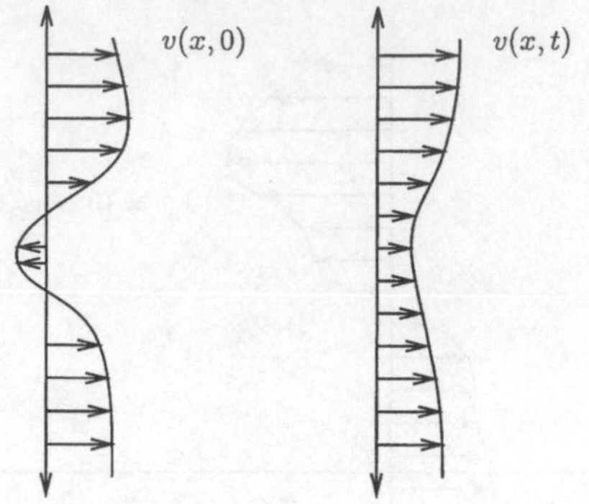

Figure 1.7. The viscous shear flow. Variations in the sheared velocity are smoothed out over time.

approximate solutions to the Navier–Stokes equation for $\nu \ll 1$? We answer this question through the use of the following elementary properties of the heat equation (their proofs are given in Subsection 1.9.3).

Lemma 1.2. *Let $u(x,t)$ be a solution to the heat equation in $\mathbb{R}^N$:*

$$u_t = \nu \Delta u, \qquad u|_{t=0} = u_0.$$

If the initial condition satisfies $|u_0| + |\nabla u_0| \le M$, then

$$|u(x,t) - u_0(x)| \le cM(\nu t)^{1/2}.$$

Lemma 1.3. *Let $u(x,t)$ be a solution to the heat equation in $\mathbb{R}^N$:*

$$u_t = \nu \Delta u, \qquad u|_{t=0} = u_0.$$

Then for a fixed x and ν,

$$u(x,t) \sim (4\pi \nu t)^{-N/2} \int_{\mathbb{R}^N} u_0(y) dy + \mathcal{O}\big[(\nu t)^{-N/2}\big], \qquad t \gg 1.$$

Now we need only to quote these results to get Proposition 1.6.

Proposition 1.6. *Let $v_0(x_1)$ be a rapidly decreasing function such that $|v_0| + |\nabla v_0| \le M$. Let $v^E(x_1,t)$ be the basic Euler ($\nu = 0$) shear-layer solution in Eq. (1.38), and let $v^\nu(x_1,t)$ be the viscous shear-layer solution in Eq. (1.40). Then*

$$|v^E(x_1,t) - v^\nu(x_1,t)| \le c(\nu t)^{1/2}, \qquad x_1 \in \mathbb{R}, \tag{1.42}$$

$$v^\nu(x_1,t) \sim (4\pi \nu t)^{-1/2} \int_{\mathbb{R}} v_0(y) dy, \qquad x_1 \in \mathbb{R}, \quad t \gg 1. \tag{1.43}$$

The estimate in relation (1.42) supports our intuition that for high Reynolds numbers ($\nu \ll 1$) the inviscid solution v^E approximates the viscous solution v^ν. For a fixed ν

this approximation deteriorates as $t \nearrow \infty$. Asymptotic formula (1.43) indicates that for large time $t \gg 1$ the viscous solution v^ν "remembers" only the mean value of the initial condition. This is due to the diffusion that spreads gradients in the initial condition. In Section 3.1 we show the convergence $v^\nu \to v^E$ for arbitrary smooth solutions to the Navier–Stokes and the Euler equations for any fixed time interval.

Now we look for exact solutions that exhibit the stretching of vorticity. Using formula (1.35) first we discuss Example 1.7.

Example 1.7. Inviscid Strained Shear-Layer Solutions. With $\gamma > 0$ and $\nu = 0$, Eq. (1.35) reduces to the linear ODE, on the scalar vorticity $\omega(x, t)$:

$$\omega_t - \gamma x \omega_x = \gamma \omega, \qquad x \in \mathbb{R},$$
$$\omega|_{t=0} = \omega_0(x).$$

In contrast to the basic shear-layer solution in Eq. (1.37), this solution is time dependent. When the characteristic curves $X(\alpha, t) = \alpha e^{-\gamma t}$ are used, the above equation reduces to an ODE, yielding the solution

$$\omega(x, t) = \omega_0(x e^{\gamma t}) e^{\gamma t}, \qquad x \in \mathbb{R}. \tag{1.44}$$

The corresponding x_3 component of velocity is

$$v(x, t) = v_0(x e^{\gamma t}), \qquad x \in \mathbb{R}.$$

Thus, because of the convection, the x_2 component of vorticity is compressed in the x_1 direction and is stretched in the x_3 direction (see Fig. 1.8).

Finally we incorporate the effects of diffusion into this exact solution. We have Example 1.8.

Example 1.8. Viscous Strained Shear-Layer Solutions. With $\gamma, \nu > 0$, the dynamics of the scalar vorticity $\omega(x, t)$ is given by the linear convection–diffusion

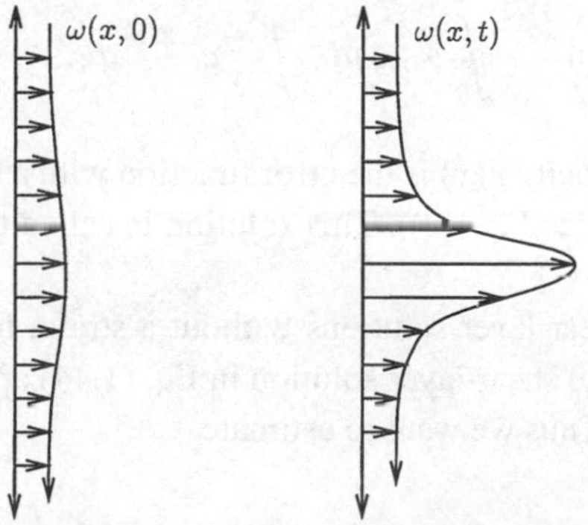

Figure 1.8. Inviscid strained shear layer (Example 1.7).

equation (1.35):

$$\omega_t - \gamma x \omega_x = \gamma \omega + \nu \omega_{xx}, \qquad x \in \mathbb{R},$$
$$\omega|_{t=0} = \omega_0(x). \tag{1.45}$$

We can solve this equation explicitly, namely, first we observe that by taking $\omega = e^{\gamma t}\tilde{\omega}$ we reduce this problem to

$$\tilde{\omega}_t - \gamma x \tilde{\omega}_x = \nu \tilde{\omega}_{xx}, \qquad \tilde{\omega}|_{t=0} = \omega_0.$$

Next, by the change of variables $\xi = xe^{\gamma t}$ and $\zeta = (\nu/2\gamma)(e^{2\gamma t} - 1)$ we reduce it to the heat equation

$$\tilde{\omega}_\zeta = \tilde{\omega}_{\xi\xi}, \qquad \tilde{\omega}|_{\zeta=0} = \omega_0, \qquad x \in \mathbb{R}.$$

Thus the solution $\omega(x,t)$ to our problem is

$$\omega(x,t) = e^{\gamma t} \int_{\mathbb{R}} H\left[xe^{\gamma t} - \xi, \frac{\nu}{2\gamma}(e^{2\gamma t} - 1)\right] \omega_0(\xi) d\xi, \qquad x \in \mathbb{R}, \tag{1.46}$$

where H is the 1D Gaussian kernel in (1.40). Note that the above vorticity equation may admit nontrivial steady-state solutions, for the spreading from the diffusion term $\nu\omega_{xx}$ may balance the compression from the convection term $-\gamma x \omega_x$ and the stretching of vorticity from $\gamma\omega$. Indeed, by the dominated convergence theorem we get

$$\lim_{t\to\infty} \omega(x,t) = \omega(x) \equiv \left(\frac{2\pi\nu}{\gamma}\right)^{-1/2} e^{-\frac{\gamma}{2\nu}x^2} \int_{\mathbb{R}} \omega_0(\xi) d\xi, \qquad x \in \mathbb{R}. \tag{1.47}$$

By a direct substitution the reader may check that this steady-state vorticity satisfies Eq. (1.35). The solution $\omega(x)$ has a sharp transition layer of the thickness $\mathcal{O}(\nu/\gamma)$.

The corresponding x_3 component of velocity is given by Eq. (1.36) as

$$v(x) = \left(\frac{2\pi\nu}{\gamma}\right)^{-1/2} \int_{\mathbb{R}} \omega_0(\xi) d\xi \int_{-\infty}^{x} e^{-\frac{\gamma}{2\nu}\xi^2} d\xi, \qquad x \in \mathbb{R}, \tag{1.48}$$

i.e., the steady-state velocity $v(x)$ is the error function with a sharp transition layer of the thickness $\mathcal{O}(\nu/\gamma)$ (see Fig. 1.9). This solution is called the *Burgers shear-layer solution*.

As before for the shear-layer solutions without a strain, finally we examine how well the inviscid strained shear-layer solution in Eq. (1.44) approximates the viscous solution in Eq. (1.46). Thus we want to estimate

$$|\omega^0(x,t) - \omega^\nu(x,t)| = e^{\nu t} \left| \omega_0(xe^{\gamma t}) - (4\pi a)^{1/2} \int_{\mathbb{R}} e^{-\frac{|xe^{\gamma t} - \xi|^2}{4\pi}} \omega_0(\xi) d\xi \right|,$$

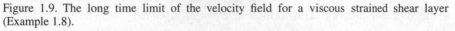

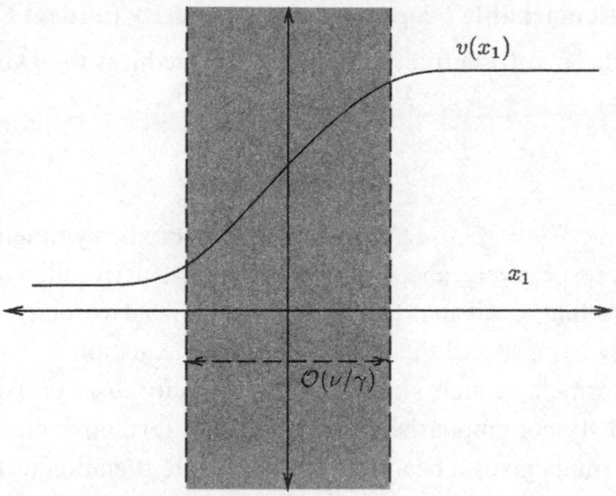

Figure 1.9. The long time limit of the velocity field for a viscous strained shear layer (Example 1.8).

where $a = (\nu/2\gamma)(e^{2\gamma t} - 1)$. By changing the variables $\eta = (\xi - xe^{\gamma t})/\sqrt{a}$ and using the mean-value theorem we get

$$\begin{aligned} |\omega^0(x,t) - \omega^\nu(x,t)| &\le e^{\nu t}(4\pi)^{1/2} \int_{\mathbb{R}} e^{-\frac{|\eta|^2}{4}} |\omega_0(xe^{\gamma t}) - \omega_0(xe^{\gamma t} + \sqrt{a}\eta)| d\eta \\ &\le C \sup|\nabla\omega_0| \int_{\mathbb{R}} |\eta| e^{-\frac{|\eta|^2}{4}} d\eta e^{\nu t} \sqrt{a} \\ &\le C(\omega_0) e^{\nu t} \left[\frac{\nu}{2\gamma}(e^{2\gamma t} - 1)\right]^{1/2}, \qquad x \in \mathbb{R}. \end{aligned}$$

Thus we have proved the following proposition.

Proposition 1.7. *Let $\omega_0(x)$ be a rapidly decreasing function such that $|\omega_0| + |\nabla\omega_0| \le M$. Let $\omega^0(x,t)$ be the inviscid and $\omega^\nu(x,t)$ the viscous strained shear-layer solutions in Eqs. (1.44) and (1.46), respectively. Then*

$$|\omega^0(x,t) - \omega^\nu(x,t)| \le c(M) e^{\nu t} \left[\frac{\nu}{2\gamma}(e^{2\gamma t} - 1)\right]^{1/2}, \qquad x \in \mathbb{R}. \tag{1.49}$$

Estimates for the velocity $v^3(x_1, t)$ follow directly from relation (1.49) by an integration.

We see that for any fixed t and $\gamma > 0$ the inviscid solution ω^0 approximates the viscous solution ω^ν as $\nu \searrow 0$. This approximation, however, deteriorates not only in time but also with the thickness $\mathcal{O}(\nu/\gamma)$ of the sharp transition layer in the Burger's shear-layer solution. For large γ, the estimate in relation (1.49) is misleading because it involves maximum norms – using a norm like the L^1 norm, which is not sensitive to sharp transition layers, gives better convergence estimates.

1.6. Some Remarkable Properties of the Vorticity in Ideal Fluid Flows

For inviscid fluids ($\nu = 0$) vorticity equation (1.33) reduces to $(D\omega/Dt) = \mathcal{D}\omega$ or, equivalently, to

$$\frac{D\omega}{Dt} = \omega \cdot \nabla v \tag{1.50}$$

because the matrix $\nabla v = \mathcal{D} + \Omega$, where $\mathcal{D}$ and Ω are the symmetric and the antisymmetric parts, respectively, and $\omega \times \omega \equiv 0$. First, we derive the vorticity-transport formula, an important result that proves that the inviscid vorticity equation can be integrated exactly by means of the particle-trajectory equation.

This simple formula, which states that the vorticity ω is propagated by inviscid flows, is usually not emphasized in the standard textbooks on fluid mechanics. However, this formula gives a beautiful geometric interpretation of the stretching of vorticity. In Chap. 4 we use this formula in the study of properties of general solutions to the 3D Euler equation. This formula also has an important role in designing vortex methods for the numerical solution to the 3D Euler equation: At the end of Chap. 4 this is discussed briefly and Chap. 6 is devoted to this topic.

Proposition 1.8. *Vorticity-Transport Formula. Let $X(\alpha, t)$ be the smooth particle trajectories corresponding to a divergence-free velocity field $v(x, t)$. Then the solution ω to the inviscid vorticity equation (1.50) is*

$$\omega(X(\alpha, t), t) = \nabla_\alpha X(\alpha, t)\omega_0(\alpha). \tag{1.51}$$

Before giving the proof of Proposition 1.8, we discuss the significance formula (1.51). First, we show that Eq. (1.51) leads to an interpretation of vorticity amplification (or decay). Recall that, because the fluid is incompressible, by Proposition 1.2 $\det(\nabla_\alpha X(\alpha, t)) = 1$. Thus for any fixed α and t, $\nabla_\alpha X(\alpha, t)$ can be an arbitrary 3×3 (real) matrix with determinant 1 and three (complex) eigenvalues λ, λ^{-1}, and 1, with $|\lambda| \geq 1$. When $|\lambda| > 1$, formula (1.51) shows that the vorticity increases when ω_0 aligns roughly with the complex eigenvector associated with λ.

In particular, for ideal fluid flows in two dimensions,

$$\nabla_\alpha X(\alpha, t) = \begin{bmatrix} X^1_{\alpha_1} & X^1_{\alpha_2} & 0 \\ X^2_{\alpha_1} & X^2_{\alpha_2} & 0 \\ 0 & 0 & 1 \end{bmatrix}, \tag{1.52}$$

and $\omega_0(\alpha) = [0, 0, \omega_0(\alpha)]^t$.

This gives us the following corollary.

Corollary 1.2. *Let $X(\alpha, t)$ be the smooth particle trajectories corresponding to a divergence-free velocity field. Then the vorticity $\omega(x, t)$ satisfies*

$$\omega(X(\alpha, t), t) = \omega_0(\alpha), \qquad \alpha \in \mathbb{R}^2, \tag{1.53}$$

and the vorticity $\omega_0(\alpha)$ is conserved along particle trajectories for two-dimensional (2D) inviscid fluid flows.

Next, as a simple corollary to vorticity-transport formula (1.51) we verify the well-known fact that *vortex lines move with an ideal fluid*. Consider a smooth (not necessarily closed) curve $C = \{y(s) \in \mathbb{R}^N \colon 0 < s < 1\}$. We say that C is a *vortex line* at a fixed time t if it is tangent to the vorticity ω at each of its points, i.e., provided that

$$\frac{dy}{ds}(s) = \lambda(s)\omega(y(s), t) \quad \text{for some} \quad \lambda(s) \neq 0. \tag{1.54}$$

Flows having readily observed vortex lines often occur in physical experiments and nature, e.g., centerlines for tornadoes are a prominent example.

An initial vortex line $C = \{y(s) \in \mathbb{R}^N \colon 0 < s < 1\}$ evolves with the fluid to the curve $C(t) = \{X(y(s), t) \in \mathbb{R}^N \colon 0 < s < 1\}$. We want to check that $C(t)$ is also a vortex line, i.e., that

$$\frac{dX}{ds}(y(s), t) = \lambda(s)\omega(X(y(s), t), t).$$

The definition of $C(t)$ implies that

$$\begin{aligned}\frac{dX}{ds}(y(s), t) &= \nabla_\alpha X(y(s), t)\frac{dy}{ds}(s)\\ &= \nabla_\alpha X(y(s), t)\lambda(s)\omega_0(y(s)).\end{aligned}$$

Proposition 1.8, $\omega(X(y(s), t), t) = \nabla_\alpha X(y(s), t)\omega_0(y(s))$, gives

$$\frac{dX}{ds}(y(s), t) = \lambda(s)\omega(X(y(s), t), t),$$

so we have proved Proposition 1.9.

Proposition 1.9. *In inviscid, smooth fluid flows, vortex lines move with the fluid.*

Next we verify formula (1.51). This formula is an immediate application of a much more general geometric formula that does not require the velocity field v to be divergence free. This more general formula is useful in further developments.

Lemma 1.4. *Let $v(s, t)$ be any smooth velocity field (not necessarily divergence free) with associated particle-trajectory mapping $X(\alpha, t)$ satisfying*

$$\frac{dX}{dt} = v(X(\alpha, t), t), \qquad X(\alpha, 0) = \alpha. \tag{1.55}$$

Let $h(x, t)$ be a smooth vector field. Then h satisfies the equation

$$\frac{Dh}{Dt} = h \cdot \nabla v \tag{1.56}$$

if and only if

$$h(X(\alpha, t), t) = \nabla_\alpha X(\alpha, t)h_0(\alpha). \tag{1.57}$$

Formula (1.51) from Proposition 1.8 is a special case of this result with $h = \omega$. Finally we give the proof of Lemma 1.4.

Proof of Lemma 1.4. First we differentiate Eq. (1.55) with respect to α to obtain the general matrix equation

$$\frac{d}{dt}\nabla_\alpha X(\alpha, t) = (\nabla v)|_{(X(\alpha,t),t)} \nabla_\alpha X(\alpha, t).$$

We multiply both sides of this equation by the vector $h_0(\alpha)$ to get

$$\frac{d}{dt}\nabla_\alpha X(\alpha, t)h_0(\alpha) = (\nabla v)|_{(X(\alpha,t),t)} \nabla_\alpha X(\alpha, t)h_0(\alpha).$$

If h satisfies Eq. (1.56), $(Dh/Dt) = h \cdot \nabla v$, then we easily compute that $h(X(\alpha, t), t)$ satisfies

$$\frac{d}{dt}h(X(\alpha, t), t) = (\nabla v)|_{(X(\alpha,t),t)} h(X(\alpha, t), t).$$

Thus both $\nabla_\alpha X(\alpha, t)h_0(\alpha)$ and $h(X(\alpha, t), t)$ satisfy the same linear ODE with the same initial date $h_0(\alpha)$ – so by the uniqueness of solutions to ODEs these two functions coincide. □

Now we discuss some other important properties of the vorticity in ideal fluid flows. Let S be a bounded, open, smooth surface with smooth, oriented boundary C. Then by following the (smooth) particle trajectories $X(\alpha, t)$, S and C evolve with the fluid to $S(t) = \{X(\alpha, t)\colon \alpha \in S\}$ and $C(t) = \{X(\alpha, t)\colon \alpha \in C\}$, respectively (see Fig. 1.10).

We recall the following analog of transport formula (1.15).

Proposition 1.10. *Let C be a smooth, oriented, closed curve and let $X(\alpha, t)$ be the smooth particle trajectories corresponding to a divergence-free velocity field $v(x, t)$.*

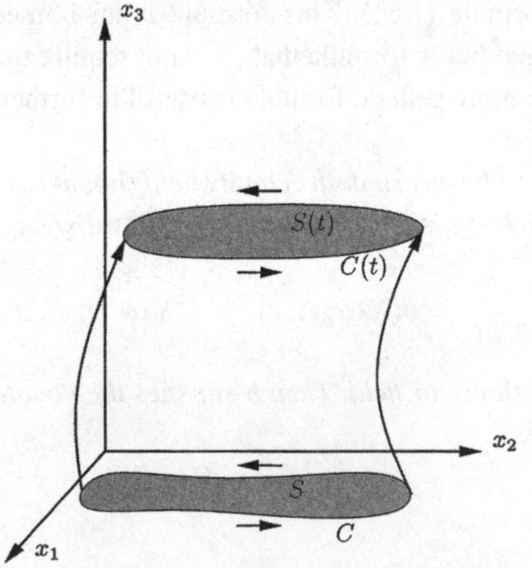

Figure 1.10. The evolution of a bounded, open, smooth surface S with smooth boundary C moving with the fluid velocity.

Then

$$\frac{d}{dt}\oint_{C(t)} v \cdot d\ell = \oint_{C(t)} \frac{Dv}{Dt} \cdot d\ell. \tag{1.58}$$

We leave the proof of Eq. (1.58) as an exercise for the reader.

Because for inviscid flows there are no tangent forces acting on the fluid, intuitively we may expect the rotation to be conserved. Actually, the Euler equation and transport formula (1.52) give

$$\frac{d}{dt}\oint_{C(t)} v \cdot d\ell = -\oint_{C(t)} \nabla p \cdot d\ell = 0$$

because the line integral of the gradient is zero for closed curves.

Thus we have proved the celebrated Proposition 1.11.

Proposition 1.11. *Kelvin's Conservation of Circulation. For a smooth solution v to the Euler equation, the circulation $\Gamma_{C(t)}$ around a curve $C(t)$ moving with the fluid,*

$$\Gamma_{C(t)} = \oint_{C(t)} v \cdot d\ell, \tag{1.59}$$

is constant in time.

As a consequence of this result, Stokes' formula immediately gives Corollary 1.3.

Corollary 1.3. *Helmholtz's Conservation of Vorticity Flux. For a smooth solution v to the Euler equation, the vorticity flux $F_{S(t)}$ through a surface $S(t)$ moving with the fluid,*

$$F_{S(t)} = \int_{S(t)} \omega \cdot ds, \tag{1.60}$$

is constant in time.

Thus we see that if vol $S(t)$ decreases, the vorticity $\omega|_{S(t)}$ must roughly increase (see, e.g., the exact solution for rotating jets – Example 1.1).

The conservation of circulation can be easily generated to viscous flows as well. A simple computation yields

$$\frac{d}{dt}\Gamma_{C(t)} = \nu \oint_{C(t)} \Delta v \cdot d\ell = -\nu \oint_{C(t)} \operatorname{curl} \omega \cdot d\ell, \tag{1.61}$$

so in general $\Gamma_{C(t)}$ is not conserved for $\nu > 0$.

1.7. Conserved Quantities in Ideal and Viscous Fluid Flows

Euler equation (1.1) for incompressible flows of homogeneous, ideal fluids,

$$\frac{Dv}{Dt} = -\nabla p, \tag{1.62}$$

and incompressibility equation (1.2),

$$\operatorname{div} v = 0, \tag{1.63}$$

follow from physical laws of the conservation of momentum and mass, respectively. There are also, however, a number of other important quantities for solutions to Eqs. (1.62) and (1.63) conserved globally in time. Some of them play an important role in later developments in these chapters. We have the following proposition.

Proposition 1.12. *Let v (and $\omega = \operatorname{curl} v$) be a smooth solution to the Euler equation in Eqs. (1.62) and (1.63) in $\mathbb{R}^3$, vanishing sufficiently rapidly as $|x| \nearrow \infty$. Then the following quantities are conserved for all time:*

(i) the total flux V_3 of velocity,

$$V_3 = \int_{\mathbb{R}^3} v\,dx, \tag{1.64}$$

(ii) the total flux Ω_3 of vorticity,

$$\Omega_3 = \int_{\mathbb{R}^3} \omega\,dx, \tag{1.65}$$

(iii) the kinetic energy E_3,

$$E_3 = \frac{1}{2}\int_{\mathbb{R}^3} |v|^2 dx \left[\equiv \int_{\mathbb{R}^3} v \cdot (x \times \omega) dx\right], \tag{1.66}$$

(iv) the helicity H_3,

$$H_3 = \int_{\mathbb{R}^3} v \cdot \omega\,dx, \tag{1.67}$$

(v) the fluid impulse I_3,

$$I_3 = \frac{1}{2}\int_{\mathbb{R}^3} x \times \omega\,dx, \tag{1.68}$$

(vi) the moment M_3 of fluid impulse,

$$M_3 = \frac{1}{3}\int_{\mathbb{R}^3} x \times (x \times \omega) dx. \tag{1.69}$$

Some of these conserved quantities, e.g., the conservation of energy in Eq. (1.66), have important implications in studying the mathematical properties of solutions to the Euler equation – see, for example, Chap. 5.

The fluid impulse I_3 (moment M_3 of fluid impulse) has a dynamical significance as the impulse (moment of impulse) required for a generation of a flow from rest (see, e.g., Batchelor, 1967, p. 518). We do not know, however, any direct applications of H_3, I_3, and M_3 in studying mathematical properties of solutions.

For 2D flows the quantities in Eqs. (1.64)–(1.68) (with obvious changes of definitions) are also conserved. Now, however, the velocity $v = (v^1, v^2)^t$ is orthogonal to the vorticity vector (regarded as a scalar) $\omega = v^2_{x_1} - v^1_{x_2}$, so the reader can easily verify Corollary 1.4.

Corollary 1.4. *Let v (and $\omega = v^2_{x_1} - v^1_{x_2}$) be a smooth solution to the Euler equation in Eqs. (1.62) and (1.63) in $\mathbb{R}^2$, vanishing sufficiently rapidly as $|x| \nearrow \infty$. Then the following quantities are conserved for all time:*

(i) the total 2D flux V_2 of velocity,

$$V_2 = \int_{\mathbb{R}^2} v\, dx, \tag{1.70}$$

(ii) the total flux Ω_2 of vorticity,

$$\Omega_2 = \int_{\mathbb{R}^2} \omega\, dx, \tag{1.71}$$

(iii) the kinetic energy E_2,

$$E_2 = \frac{1}{2}\int_{\mathbb{R}^2} |v|^2 dx \left[\equiv \int_{\mathbb{R}^2} \omega(x_2 v^1 - x_1 v^2)\, dx \right], \tag{1.72}$$

(iv) the L^p, $1 \le |p| \le \infty$, norm of vorticity,

$$\|\omega\|_p = \left(\int_{\mathbb{R}^2} |\omega|^p dx \right)^{1/p}, \tag{1.73}$$

(v) the helicity $H_2 \equiv 0$,

(vi) the fluid impulse I_2,

$$I_2 = \frac{1}{2}\int_{\mathbb{R}^2} (x_2 - x_1)^t \omega\, dx, \tag{1.74}$$

(vii) the moment M_2 of fluid impulse,

$$M_2 = -\frac{1}{3}\int_{\mathbb{R}^2} |x|^2 \omega\, dx. \tag{1.75}$$

For the proof of Proposition 1.12 we need the following well-known lemma.

Lemma 1.5. *Let w be a smooth, divergence-free vector field in $\mathbb{R}^N$ and let q be a smooth scalar such that*

$$|w(x)|\,|q(x)| = \mathcal{O}\left[(|x|)^{1-N}\right] \quad \text{as} \quad |x| \nearrow \infty. \tag{1.76}$$

Then w and ∇q are orthogonal:

$$\int_{\mathbb{R}^N} w \cdot \nabla q \, dx = 0. \tag{1.77}$$

With this orthogonality result, now we are ready to give the proof of Proposition 1.12.

Proof of Proposition 1.12. Integrating Euler equation (1.62) over $\mathbb{R}^3$, we have componentwise

$$\frac{d}{dt}\int_{\mathbb{R}^3} v^i \, dx = -\int_{\mathbb{R}^3} v \cdot \nabla v^i \, dx - \int_{\mathbb{R}^3} p_{x_i} \, dx, \qquad i = 1, 2, 3,$$

so by the Green's formula and Lemma 1.5 we conclude the conservation of velocity flux in Eq. (1.64).

Now we integrate componentwise the vorticity equation $(D\omega/Dt) = \omega \cdot \nabla v$ over $\mathbb{R}^3$:

$$\frac{d}{dt}\int_{\mathbb{R}^3} \omega^i \, dx = -\int_{\mathbb{R}^3} v \cdot \nabla \omega^i \, dx, \qquad i = 1, 2, 3.$$

Because $\operatorname{div} v = 0$ and $\omega = \operatorname{curl} v$, we have $\operatorname{div} \omega \equiv 0$, and by using Lemma 1.5 we conclude the conservation of vorticity flux in Eq. (1.65).

For the proof of energy conservation we apply transport formula (1.15),

$$\frac{d}{dt}\int_{X(\Omega,t)} f \, dx = \int_{X(\Omega,t)} [f_t + \operatorname{div}(fv)] dx,$$

to $\Omega = \mathbb{R}^3$ and $f = \frac{1}{2} v \cdot v$. Using the Euler equation, we get

$$\frac{d}{dt}\int_{\mathbb{R}^3} \frac{1}{2}|v|^2 dx = -\int_{\mathbb{R}^3} v \cdot \nabla p \, dx,$$

so by Lemma 1.5 we conclude the conservation of energy in Eq. (1.66).

Now we prove the conservation of helicity in Eq. (1.67). Because $\operatorname{div} v = 0$, multiplying the Euler equation by ω and using vector identities we get

$$v_t \cdot \omega + \operatorname{div}(v \cdot \omega) - (v \cdot \nabla \omega) v = -\operatorname{div}(p\omega) + p \operatorname{div} \omega.$$

In the same way, multiplying the vorticity equation $(D\omega/Dt) = \omega \cdot \nabla v$ by v, we get

$$\omega_t \cdot v + (v \cdot \nabla \omega) v = \frac{1}{2}\operatorname{div}(\omega v^2) - \frac{1}{2} v^2 \operatorname{div} \omega.$$

Because $\operatorname{div} v = 0$ and $\omega = \operatorname{curl} v$, we have the compatibility condition $\operatorname{div} \omega = 0$, so from the above identities we arrive at

$$(v \cdot \omega)_t + \operatorname{div}\left[v(v \cdot \omega) + \omega \left(p - \frac{1}{2} v^2 \right) \right] = 0.$$

Thus integrating this equation over $\mathbb{R}^3$, by the Green's formula we conclude the conservation of helicity in Eq. (1.67).

Now we apply the transport formula to $\Omega = \mathbb{R}^3$ and $f = x \times \omega$ to get

$$\frac{d}{dt}\int_{\mathbb{R}^3} x \times \omega\, dx = \int_{\mathbb{R}^3} \left(\frac{Dx}{Dt} \times \omega + x \times \frac{D\omega}{Dt} \right) dx.$$

Because $(Dx/Dt) = v \cdot \nabla x \equiv v$ and $(D\omega/Dt) = \omega \cdot \nabla v$, using vector identities we get

$$\frac{d}{dt}\int_{\mathbb{R}^3} x \times \omega\, dx = \int_{\mathbb{R}^3} [v \times \omega + x \times (\omega \cdot \nabla v)] dx = \int_{\mathbb{R}^3} \omega \cdot \nabla (x \times v) dx$$
$$= \sum_{j=1}^{3} \int_{\mathbb{R}^3} [\omega^j (x \times v)]_{x_j} dx.$$

In the last step we have used the condition div $\omega = 0$. The last expression in the above formula is a perfect divergence, so by the Green's formula we conclude the conservation of fluid impulse in Eq. (1.68).

Finally, we prove the conservation of moment of fluid impulse. For simplicity we do it for only 2D flows [see Corollary 1.4, point (vii)], so that

$$M_2 = -\frac{1}{3}\int_{\mathbb{R}^2} x^2 \omega\, dx,$$

where $\omega = v^2_{x_1} - v^1_{x_2}$ is the scalar vorticity. We apply the transport formula to $\Omega = \mathbb{R}^2$ and $f = |x|^2 \omega$ to get

$$\frac{d}{dt}\int_{\mathbb{R}^2} |x|^2 \omega\, dx = \int_{\mathbb{R}^2} \left(\frac{D|x|^2}{Dt}\omega + |x|^2 \frac{D\omega}{Dt} \right) dx.$$

Now, $[(D|x|^2)/Dt] = v \cdot \nabla |x|^2 \equiv 2x \cdot v$ and $(D\omega/Dt) = 0$, so by substituting $\omega = v^2_{x_1} - v^1_{x_2}$ into the preceding equation, we get

$$\frac{d}{dt}\int_{\mathbb{R}^2} |x|^2 \omega\, dx = 2\int_{\mathbb{R}^2} \left\{ \frac{1}{2} x_2 [(v^2)^2]_{x_1} - \frac{1}{2} x_1 [(v^1)^2]_{x_2} + x_1 v^1 v^2_{x_1} - x_2 v^2 v^1_{x_2} \right\} dx.$$

By the Green's formula the first two terms vanish, and the other two terms give

$$\frac{d}{dt}\int_{\mathbb{R}^2} |x|^2 \omega\, dx = -2\int_{\mathbb{R}^2} \left(v^1 v^2 + x_1 v^1_{x_1} v^2 - v^1 v^2 - x_2 v^2_{x_2} v^1 \right) dx.$$

Thus, by using the incompressibility equation $v^1_{x_1} + v^2_{x_2} = 0$ and the Green's formula, we get

$$\frac{d}{dt}\int_{\mathbb{R}^2} |x|^2 \omega\, dx = \int_{\mathbb{R}^2} \{ x_1 [(v^2)^2]_{x_2} - x_2 [(v^1)^2]_{x_1} \} dx = 0,$$

so the moment of fluid impulse M_2 is conserved. □

It remains only to give the proof of Lemma 1.5.

Proof of Lemma 1.5. Because div $w = 0$, by the Green's formula we have

$$\int_{|x|\le R} w \cdot \nabla q \, dx = \int_{|x|=R} q w \cdot n ds.$$

The flux on the right-hand side tends to zero as $R \nearrow \infty$, provided that q and w vanish sufficiently fast as $|x| \nearrow \infty$. Because the surface area of the sphere of radius R is $C_N R^{N-1}$, it is sufficient that q and w satisfy Eq. (1.76), i.e.,

$$|q(x)|\,|w(x)| = \mathcal{O}(|x|^{1-N}),$$

and this concludes the proof. □

Now we study how the above conserved quantities in Eqs. (1.64)–(1.75) change when we add diffusion and consider solutions to the Navier–Stokes equation

$$\frac{Dv}{Dt} = -\nabla p + \nu \Delta v, \tag{1.78}$$

$$\text{div } v = 0. \tag{1.79}$$

Here we compute some of these identities and we leave the rest as an exercise for the interested reader.

The most important identity of this sort gives the effect that viscosity has on the conservation of energy. We apply transport formula (1.15) to $\Omega = \mathbb{R}^N$ and $f = \frac{1}{2} v \cdot v$ from the Navier–Stokes equation to get

$$\frac{d}{dt} \int_{\mathbb{R}^N} \frac{1}{2} |v|^2 dx = - \int_{\mathbb{R}^N} v \cdot \nabla p \, dx + \nu \int_{\mathbb{R}^N} v \cdot \Delta v \, dx,$$

so by using Lemma 1.5 and the Green's formula we compute that the rate of change of energy is given by the following proposition.

Proposition 1.13. *Let v be a smooth solution to the Navier–Stokes equation, vanishing sufficiently rapidly as $|x| \nearrow \infty$. Then the kinetic energy $E(t)$ satisfies the ODE*

$$\frac{d}{dt} E(t) = -\nu \int_{\mathbb{R}^N} |\nabla v|^2 dx. \tag{1.80}$$

This formula explains why the viscosity coefficient ν must be positive if we want to have a dissipation of energy in the Navier–Stokes equation. It also has theoretical applications in the study of the mathematical properties of general solutions to the Navier–Stokes equation in subsequent chapters.

Finally, for 2D fluid flows we calculate the effects of diffusion on the fluid impulse I_2,

$$I_2 = \frac{1}{2} \int_{\mathbb{R}^2} (x_2, -x_1)^t \omega \, dx,$$

and on the moment M_2 of fluid impulse,

$$M_2 = \frac{1}{3}\int_{\mathbb{R}^2} x^2 \omega \, dx.$$

Recall that now the scalar vorticity $\omega = v^2_{x_1} - v^1_{x_2}$ satisfies the vorticity equation

$$\frac{D\omega}{Dt} = \nu \Delta \omega.$$

Because div $v = 0$, the total flux Ω_2 of vorticity is conserved for any $\nu > 0$. For the fluid impulse I_2, by the Green's formula we compute that

$$\frac{d}{dt}\int_{\mathbb{R}^2} x_j \omega \, dx = \nu \int_{\mathbb{R}^2} x_j \Delta\omega \, dx \equiv 0, \qquad j = 1, 2,$$

so the impulse I_2 is conserved in a viscous fluid. For the moment M_2 of fluid impulse we obtain

$$\frac{d}{dt}\int_{\mathbb{R}^2} x^2 \omega \, dx = \nu \int_{\mathbb{R}^2} x^2 \Delta\omega \, dx = 4\nu \int_{\mathbb{R}^2} \omega \, dx.$$

We have just proved the following proposition.

Proposition 1.14. *Let v (and $\omega = v^2_{x_1} - v^1_{x_2}$) be a smooth solution to the Navier–Stokes equation in $\mathbb{R}^2$, vanishing sufficiently rapidly as $|x| \nearrow \infty$. Then for any viscosity $\nu > 0$,*

(i) the vorticity flux Ω_2,

$$\Omega_2 = \int_{\mathbb{R}^2} \omega \, dx,$$

and the fluid impulse I_2,

$$I_2 = \frac{1}{2}\int_{\mathbb{R}^2} (x_2, -x_1)^t \omega \, dx,$$

are conserved for all time, and

(ii) the moment M_2 of fluid impulse,

$$M_2 = -\frac{1}{3}\int_{\mathbb{R}^2} x^2 \omega \, dx,$$

satisfies

$$M_2(t) = M_2(0) - \frac{4}{3}\nu t \Omega_2(0). \tag{1.81}$$

Equation (1.81) determines the second moment of vorticity, $\int_{\mathbb{R}^2} x^2 \omega \, dx$, at time t from the initial values of M_2 and Ω_2. If the initial vorticity ω_0 is such that $\Omega_2(0) \neq 0$, the absolute value of the second moment of vorticity increases in time at the rate proportional to νt. This gives more quantitative support for our intuition that a *diffusion in the Navier–Stokes equation causes solutions to spread and also decay in time. Because the fluid impulse I_2 is conserved, the same applies to the second moment of vorticity about any other fixed point x_0.*

1.8. Leray's Formulation of Incompressible Flows and Hodge's Decomposition of Vector Fields

As we have mentioned in Section 1.1, the Navier–Stokes equation of incompressible flows,

$$\frac{Dv}{Dt} = -\nabla p + \nu\Delta v, \tag{1.82}$$

$$\text{div } v = 0, \tag{1.83}$$

contains time derivatives of only three out of the four unknown functions. Because there is no time derivative of the pressure in Eq. (1.82), we might try to find an equation $p = p(v)$ to eliminate the pressure from Eq. (1.82). Recall that in the proof of Proposition 1.5 we have derived Eq. (1.29):

$$\frac{DV}{Dt} + V^2 = -P + \nu\Delta V,$$

where $V \equiv (v^i_{x_j})$, and $P \equiv (p_{x_i x_j})$ is the Hessian matrix of pressure. Because tr $V =$ div v, by taking the trace of the above equation, we get

$$\frac{D}{Dt}(\text{div } v) + \text{tr } V^2 = -\Delta p + \nu\Delta(\text{div } v).$$

Because div $v = 0$ for incompressible flows, the pressure p and the velocity v are necessarily related by the Poisson equation

$$-\Delta p = \text{tr}(\nabla v)^2 \left(\equiv \sum_{i,j} v^i_{x_j} v^j_{x_i} \right). \tag{1.84}$$

We review some elementary properties for elliptic equations (see Subsection 1.9.2).

Lemma 1.6. *Let f be a smooth function in $\mathbb{R}^N$, vanishing sufficiently rapidly as $|x| \nearrow \infty$. Then the solution u to the Poisson equation*

$$\Delta u = f, \tag{1.85}$$

with ∇u vanishing as $|x| \nearrow \infty$, is given by

$$u(x) = \int_{\mathbb{R}^N} N(x-y) f(y) dy, \tag{1.86}$$

where the fundamental solution (Newtonian potential) N is

$$N(x) = \begin{cases} \frac{1}{2\pi} \ln|x|, & N = 2 \\ \frac{1}{(2-N)\omega_N} |x|^{2-N}, & N \geq 3 \end{cases}, \tag{1.87}$$

and ω_N is the surface area of a unit sphere in $\mathbb{R}^N$.

By applying this result to Eq. (1.84) to obtain p and differentiating under the integral to compute ∇p, we obtain the explicit formula

$$\nabla p(x,t) = C_N \int_{\mathbb{R}^N} \frac{x-y}{|x-y|^N} \mathrm{tr}(\nabla v(y,t))^2 dy. \tag{1.88}$$

Substituting this equation back into the Navier–Stokes equation, we get a closed evolution equation for the unknown v alone:

$$\frac{Dv}{Dt} = -C_N \int_{\mathbb{R}^N} \frac{x-y}{|x-y|^N} \mathrm{tr}(\nabla v(y,t))^2 dy + \nu \Delta v. \tag{1.89}$$

Now we show that the solution v to this equation is automatically divergence free, so incompressibility equation (1.83) can be dropped.

By applying the div operator to Eq. (1.89) and using Eq. (1.84) we get

$$\frac{D}{Dt}(\mathrm{div}\ v) = \nu \Delta(\mathrm{div}\ v). \tag{1.90}$$

We want to deduce that if initially $\mathrm{div}\ v_0 = 0$, then for all time $\mathrm{div}\ v = 0$ also. We prove it by the uniqueness of solutions to Eq. (1.90); namely, if Eq. (1.90) has two scalar solutions u_1 and u_2, each satisfying $\partial_t u_j + v \cdot \nabla u_j = \nu \Delta u_j$, then by multiplying the equation for $u_1 - u_2$ by $u_1 - u_2$ and integrating the product over $\mathbb{R}^N$, we find that the Green's formula gives

$$\begin{aligned}
&\frac{1}{2}\frac{d}{dt}\int_{\mathbb{R}^N} (u_1 - u_2)^2 dx \\
&= -\frac{1}{2}\int_{\mathbb{R}^N} v \cdot \nabla (u_1 - u_2)^2 dx + \nu \int_{\mathbb{R}^N} (u_1 - u_2)\Delta(u_1 - u_2) dx \\
&= \frac{1}{2}\int_{\mathbb{R}^N} (\mathrm{div}\ v)(u_1 - u_2)^2 dx - \nu \int_{\mathbb{R}^N} |\nabla (u_1 - u_2)|^2 dx \\
&\le C \int_{\mathbb{R}^N} (u_1 - u_2)^2 dx,
\end{aligned}$$

where we assume in the last step that $\mathrm{div}\ v$ can at least be bounded in magnitude by a constant C (which is certainly true for smooth solutions on a short time interval). Then Grönwall's lemma gives

$$\int_{\mathbb{R}^N} (u_1 - u_2)^2 dx(T) \le e^{2CT} \int_{\mathbb{R}^N} (u_1 - u_2)^2 dx(0).$$

In particular, if $u_1 = u_2$ at $t = 0$ then $u_1 = u_2$ at any later time T. We have shown that for smooth solutions, $u_1 - u_2 \equiv 0$ for all time and hence the solution to Eq. (1.90) is unique (see Lemma 1.13 in Subsection 1.9.3). The above calculation is an example of energy methods used extensively in subsequent chapters.

By the above uniqueness result, if $\mathrm{div}\ v_0 = 0$, then the solution $\mathrm{div}\ v$ to Eq. (1.90) remains zero for all time, so the incompressibility constraint $\mathrm{div}\ v = 0$ can be dropped when we are solving Eq. (1.89). Moreover, when ∇p is defined by Eq. (1.88), Eq. (1.89) becomes the Navier–Stokes equation and its solution satisfies the incompressibility condition $\mathrm{div}\ v = 0$, provided that $\mathrm{div}\ v_0 = 0$.

Thus we have proved Proposition 1.15.

Proposition 1.15. *Leray's Formulation of the Navier–Stokes Equation. Solving the Navier–Stokes equation in Eqs. (1.82) and (1.83) with smooth initial velocity v_0, $\operatorname{div} v_0 = 0$, is equivalent to solving evolution equation (1.89):*

$$\frac{Dv}{Dt} = C_N \int_{\mathbb{R}^N} \frac{x-y}{|x-y|^N} \operatorname{tr}(\nabla v(y,t))^2 dy + \nu \Delta v,$$
$$v|_{t=0} = v_0.$$

The pressure $p(x,t)$ can be recovered from the velocity $v(x,t)$ by the solution of the Poisson equation as described in Eq. (1.84).

The left-hand side of the above equation is quadratically nonlinear and the right-hand side contains a nonlocal, quadratically nonlinear operator. These facts make the Navier–Stokes equation hard to study analytically.

Although Eq. (1.89) constitutes a closed system for v, this formulation turns out to be not very useful for further analysis except by rather crude methods based on the energy principle (see Chap. 3). The main reason for this is that this formulation hides all the properties of vorticity stretching and interaction between rotation and deformation. Thus later we develop formulations by using the vorticity equation (1.33), in which these interactions are easy to identify.

The reader can note that what we actually did in the derivation of Proposition 1.15 was to project the Navier–Stokes equation on the space of divergence-free vector fields to eliminate ∇p. Then we recovered the pressure p from the gradient part of $v \cdot \nabla v$. Recall that in Lemma 1.5 we proved that a divergence-free field v and the gradient of a scalar p, with both v and p vanishing sufficiently rapidly as $|x| \nearrow \infty$, are necessarily orthogonal in the L^2 inner product $(\cdot,\cdot)_0$:

$$(v, \nabla p)_0 \equiv \int_{\mathbb{R}^N} v \cdot \nabla p \, dx = 0.$$

Thus, by denoting the projection operator on divergence-free vector fields by $\mathcal{P}$ (i.e., $\mathcal{P}v = w$ and $\operatorname{div} w = 0$, $\mathcal{P}v = v$ if and only if $\operatorname{div} v = 0$, $\mathcal{P}\nabla p = 0$), we can denote symbolically Eq. (1.89) as

$$v_t = \mathcal{P}(-v \cdot \nabla v) + \nu \Delta v.$$

This result suggests a more general concept of decomposing vector fields into the divergence-free part and the gradient part. The following fact is very useful for studying incompressible flows.

Proposition 1.16. *Hodge's Decomposition in $\mathbb{R}^N$. Every vector field $v \in L^2(\mathbb{R}^N) \cap C^\infty(\mathbb{R}^N)$ has a unique orthogonal decomposition:*

$$v = w + \nabla q, \qquad \operatorname{div} w = 0, \tag{1.91}$$

with the following properties:

(i) $w, \nabla q \in L^2(\mathbb{R}^N) \cap C^\infty(\mathbb{R}^N)$,

(ii) $w \perp \nabla q$ *in* L^2*, i.e.,* $(w, \nabla q)_0 = 0$*,*
(iii) for any multi-index β *of the derivates* D^β*,* $|\beta| \geq 0$*,*

$$\|D^\beta v\|_0^2 = \|D^\beta w\|_0^2 + \|\nabla D^\beta q\|_0^2.$$

Proof of Proposition 1.16. Given the decomposition in Eq. (1.91), apply the divergence operator to Eq. (1.91) to get $\Delta q = \text{div } v$ for div $w = 0$. Solving this Poisson equation for q, we determine w as $w = v - \nabla q$.

Now we prove that $\nabla q \in L^2(\mathbb{R}^N)$. First we assume that $v \in C_0^\infty(\mathbb{R}^N)$ with supp $v \subset \{|x| \leq R\}$. By using the fundamental solution in Eq. (1.87) (see Lemma 1.12 in Subsection 1.9.2), we have

$$\nabla q(x) = C_N \int_{|y| \leq R} \frac{x - y}{|x - y|^N} \text{div } v(y) dy.$$

Because for $|x| \geq 2R$, $|y| \leq R$ we have the identity

$$|x - y|^{-N} = |x|^{-N} \left| 1 - 2\frac{x \cdot y}{|x|^2} + \frac{y \cdot y}{|x|^2} \right|^{-N/2};$$

by the Taylor expansion for this formula for large $|x|$ we get

$$\nabla q(x) = C_N \frac{x}{|x|^N} \int_{|y| \leq R} \text{div } v(y) dy + \mathcal{O}(|x|^{-N}).$$

By the Green's formula the first term is zero because supp $v \subset \{|x| \leq R\}$ so that $v|_{|y|=R} = 0$. Thus $\nabla q(x) \sim \mathcal{O}(|x|^{-N})$ and $|x| \geq 2R$, and using the polar coordinates we get

$$\int_{|x| \geq 2R} |\nabla q|^2 dx \leq C \int_R^\infty r^{-2N} r^{N-1} dr < \infty.$$

This implies both that $\nabla q \in L^2(\mathbb{R}^N)$ and also that $w = v - \nabla q \in L^2(\mathbb{R}^N)$. Now we show that $w \perp \nabla q$ in $L^2(\mathbb{R}^N)$. Because div $w = 0$, by Proposition 1.13 it is enough to check that the assumption in Eq. (1.76), namely, that

$$|w(x)|\,|q(x)| = \mathcal{O}\big(|x|^{1-N}\big) \qquad \text{as} \quad |x| \nearrow \infty.$$

This immediately follows from properties of the fundamental solution to the Laplace equation (see Lemma 1.11 in Subsection 1.9.2),

$$q(x) \sim \begin{cases} \mathcal{O}(\ln|x|), & N = 2 \\ \mathcal{O}(|x|^{2-N}), & N > 2 \end{cases},$$

and from the previous estimate $w = v - \nabla q \sim \mathcal{O}(|x|^{-N})$. Moreover, the orthogonality of the decomposition in Eq. (1.91) implies its uniqueness and also the formula $\|v\|_0^2 = \|\omega\|_0^2 + \|\nabla q\|_0^2$.

The fact that $q \in C^\infty(\mathbb{R}^N)$ follows from well-known properties of elliptic equations. Here, however, we give an independent argument that is interesting itself. Differentiating Eq. (1.91), for any multi-index β we have

$$\Delta D^\beta q = \operatorname{div}(D^\beta v),$$

so by the above arguments we get

$$\|D^\beta v\|_0^2 = \|D^\beta w\|_0^2 + \|\nabla D^\beta q\|_0^2. \tag{1.92}$$

Now we recall the Sobolev lemma (see, e.g., Folland, 1995) that states that

$$\sum_{|\beta|\le k} \sup_{x\in\mathbb{R}^N} |D^\beta u(x)| \le C \left(\sum_{|\beta|\le s+k} \|D^\beta u\|_0^2 \right)^{\frac{1}{2}} \qquad \text{for} \quad s > \frac{N}{2}. \tag{1.93}$$

Because for any β we have $\|\nabla D^\beta q\|_0 \le \|D^\beta v\|_0$, this implies that $\nabla q \in C^\infty(\mathbb{R}^N)$ and also that $w = v - \nabla q \in C^\infty(\mathbb{R}^N)$.

So far we have concluded the proof for $v \in C_0^\infty(\mathbb{R}^N)$. To do it for a general $v \in L^2(\mathbb{R}^N) \cap C^\infty(\mathbb{R}^N)$, we take $\rho \in C_0^\infty(\mathbb{R})$, $\rho \equiv 1$ for $|x| \le 1$, $\rho \equiv 0$ for $|x| \ge 2$, and define $v_n(x) \equiv v(x)\rho\big(|x|/n\big)$. Then $v_n \in C_0^\infty(\mathbb{R}^N)$, and $v_n \to v$ in L^2, so passing to the limit as $n \nearrow \infty$ we conclude the proof. □

As we have already mentioned, we also study the spatially periodic flows. Although such flows do not have boundary layers caused by viscosity effects in a neighborhood of boundaries, they are significantly different from flows in all of space. We study some interesting properties of these flows in Subsections 2.2.2 and 2.3.2.

Below we formulate the Hodge's decomposition for one-periodic vector fields $v(x) = v(x + e_i)$ that we identify with functions on the torus T^N. The natural tool in studying periodic problems is the Fourier series, which we review briefly in Subsection 1.9.1.

From the proof of Proposition 1.16 we know that if we have a decomposition $v = w + \nabla q$, $\operatorname{div} w = 0$, then q is determined by the Poisson equation $\Delta q = \operatorname{div} v$. The Poisson equation on the torus T^N does not necessarily have a solution, for we have the following proposition.

Proposition 1.17. *Let $f \in C^\infty(T^N)$, and consider the Poisson equation on the torus T^N,*

$$-\Delta u = f, \qquad x \in T^N. \tag{1.94}$$

There exists a solution $u \in C^\infty(T^N)$ if and only if

$$\int_{T^N} f(x)dx = 0. \tag{1.95}$$

Moreover, Fourier transform methods give the explicit solution as

$$u(x) = \sum_{k \neq 0} e^{2\pi i k \cdot x} (4\pi |k|^2)^{-1} \hat{f}(k). \tag{1.96}$$

In our case $f = \operatorname{div} v$, so the compatibility condition in Eq. (1.95) is automatically satisfied, and we can formulate the following proposition.

Proposition 1.18. *The Hodge Decomposition of T^N. Every vector field $v \in C^\infty(T^N)$ on the torus T^N has the unique orthogonal decomposition*

$$v = \bar{v} + w + \nabla q, \qquad \bar{v} = \int_{T^N} v\, dx, \qquad \operatorname{div} w = 0, \tag{1.97}$$

where

$$\nabla q(x) = \sum_{|k| \neq 0} e^{2\pi i x \cdot k} \frac{k \otimes k}{|k|^2} \hat{v}(k) \tag{1.98}$$

and $k \otimes k \equiv (k_i k_j)$. This decomposition has the following properties

(i) $w, \nabla q \in C^\infty(T^N)$,
(ii) $w \perp \nabla q$ *in* $L^2(T^N)$, *i.e.,* $\int_{T^N} w \cdot \nabla q\, dx = 0$,
(iii) $\|v - \bar{v}\|_0^2 = \|w\|_0^2 + \|\nabla q\|_0^2$, *and in general for any multi-index of the derivative* D^β,

$$\|D^\beta v\|_0^2 = \|D^\beta \omega\|_0^2 + \|\nabla D^\beta q\|_0^2.$$

We leave the proof of this proposition as an exercise.

1.9. Appendix

We use the following notation: The space $L^p(\mathbb{R}^N)$, $1 \leq p \leq \infty$, consists of measurable functions satisfying

$$\|v\|_{L^p} \equiv \left[\int_{\mathbb{R}^N} |v(x)|^p\, dx \right]^{1/p}, \qquad \text{for} < \infty,$$

$$\|v\|_{L^\infty} = \operatorname*{ess\,sup}_{x \in \mathbb{R}^N} |v(x)|, \qquad \text{for} \quad p = \infty.$$

We often use the simplified notation $\|v\|_0$ and $|v|_0$ for $p = 2$ and ∞, respectively.

1.9.1. Elementary Properties of the Fourier Series and the Fourier Transform

We review some basic properties of Fourier series and the Fourier transform. For a more extensive discussion we refer the reader to Folland, 1995, and Stein and Weiss, 1971.

Fourier Series

Every smooth one-periodic function f in $\mathbb{R}^N$ has a unique Fourier series expansion

$$f(x) = \sum_{k \in \mathbb{Z}^N} e^{2\pi i x \cdot k} \hat{f}(k), \tag{1.99}$$

where $\hat{f}(k)$ are the Fourier coefficients determined by

$$\hat{f}(k) = \int_{T^N} e^{-2\pi i k \cdot x} f(x) dx. \tag{1.100}$$

Two important properties of the Fourier series are given in the following lemma.

Lemma 1.7. *If f is a smooth one-periodic function, then*

$$\widehat{\partial_x^\beta f(x)} = (2\pi i k)^\beta \hat{f}(k), \tag{1.101}$$

$$\|f\|_{L^2}^2 = \sum_{k \in \mathbb{Z}^N} |\hat{f}(k)|^2. \tag{1.102}$$

Fourier Transform

The Fourier transform is a "continuous" version of the Fourier series. For a function $f \in L^1(\mathbb{R}^N)$, the Fourier transform $\hat{f}$ is

$$\hat{f}(\xi) = \int_{\mathbb{R}^N} e^{-2\pi i x \cdot \xi} f(x) dx. \tag{1.103}$$

Clearly, $\hat{f}(\xi)$ is a well-defined and bounded function on $\mathbb{R}^n$ and $\|\hat{f}\|_{L^\infty} \le \|f\|_{L^1}$. Moreover, if $f, g \in L^1(\mathbb{R}^N)$, Fubini's theorem gives for their convolution $f * g$

$$\begin{aligned} \widehat{f * g}(\xi) &= \int_{\mathbb{R}^N} \int_{\mathbb{R}^N} e^{-2\pi i x \cdot \xi} f(x-y) g(y) dy dx \\ &= \int_{\mathbb{R}^N} \int_{\mathbb{R}^N} e^{-2\pi i (x-y) \cdot \xi} f(x-y) e^{-2\pi i y \cdot \xi} g(y) dy dx \\ &= \hat{f}(\xi) \int_{\mathbb{R}^N} e^{-2\pi i y \cdot \xi} g(y) dy = \hat{f}(\xi) \hat{g}(\xi), \end{aligned}$$

so that we have proved the following lemma.

Lemma 1.8. *If $f, g \in L^1$, then the Fourier transform of their convolution is the product of their Fourier transforms:*

$$\widehat{f * g} = \hat{f} \hat{g}. \tag{1.104}$$

The easiest way to derive other basic properties of the Fourier transform is to consider it first on the Schwartz class $\mathcal{S}(\mathbb{R}^N)$ of smooth functions, which, together with all their derivatives, die out faster than any power of x at infinity. That is, $f \in \mathcal{S}$ if and only if $f \in C^\infty$ and for all multi-indices β_1 and β_2

$$\sup_{x\in\mathbb{R}^N} |x^{\beta_1}\partial_x^{\beta_2} f(x)| < \infty.$$

We have the following lemma.

Lemma 1.9. *If $f \in \mathcal{S}$, then $\hat{f} \in \mathcal{S}$ and*

$$\partial_\xi^\beta \hat{f}(\xi) = [(-2\pi i x)^\beta f(x)]^\wedge, \tag{1.105}$$

$$\widehat{\partial^\beta f}(\xi) = (2\pi i \xi)^\beta \hat{f}(\xi), \tag{1.106}$$

$$\check{(\hat{f})} = f, \tag{1.107}$$

where $\check{}$ *is the inverse Fourier transform:*

$$\check{g}(x) = \int_{\mathbb{R}^N} e^{2\pi i x\cdot\xi} g(\xi)d\xi = \hat{g}(-x). \tag{1.108}$$

From the above results we see that the Fourier transform is an isomorphism of $\mathcal{S}$ onto itself. We see that this holds also for the Hilbert space L^2. Take $f \in \mathcal{S}$ and define $g(x) = \bar{f}(-x)$, where the superscript $^-$ stands for the complex conjugate. We compute

$$\begin{aligned} \|f_{L^2}^2\| &= \int_{\mathbb{R}^N} f(x)g(x-0)dx = f * g(0) = \int_{\mathbb{R}^N} \widehat{f*g}(\xi)d\xi \\ &= \int_{\mathbb{R}^N} \hat{f}(\xi)\bar{\hat{f}}(\xi)d\xi = \|\hat{f}\|_{L^2}^2. \end{aligned}$$

Because the Schwartz space $\mathcal{S}$ is dense in L^2, we have the following lemma.

Lemma 1.10. *Plancherel. The Fourier transform extends uniquely to a unitary isomorphism of L^2 onto itself, so that for all $f \in L^2$*

$$\|f\|_{L^2} = \|\hat{f}\|_{L^2}. \tag{1.109}$$

1.9.2. Elementary Properties of the Poisson Equation

In this subsection we state some elementary properties of the Laplace and the Poisson equations. For an extensive discussion of these problems we refer the reader to Folland, 1995, Gilbarg and Trudinger, 1998, and John, 1982.

Lemma 1.11. *Consider the Laplacian Δ in $\mathbb{R}^N$:*

$$\Delta \equiv \sum_{j=1}^N \frac{\partial^2}{\partial x_j^2}. \tag{1.110}$$

The fundamental solution (or Newtonian potential) N for Δ, i.e., $-\Delta N = \delta$, where δ is the Dirac distribution, is

$$N(x) = \begin{cases} -\frac{1}{2\pi}\ln|x|, & N = 2 \\ -\frac{|x|^{2-N}}{(2-N)\omega_N}, & N > 2 \end{cases}, \tag{1.111}$$

where ω_N the surface area of a unit sphere in $\mathbb{R}^N$:

$$\omega_N = 2\pi^{\frac{N}{2}}/\Gamma\left(\frac{N}{2}\right). \tag{1.112}$$

Lemma 1.12. *Suppose that $f \in L^1(\mathbb{R}^N) \cap C^1(\mathbb{R}^N)$ [and $\int_{|x|\geq 1} |f(x)| \ln|x| dx < \infty$ for $N = 2$]. Then there exists a solution $u \in C^2(\mathbb{R}^N)$ to the Poisson equation*

$$-\Delta u = f, \qquad x \in \mathbb{R}^N, \tag{1.113}$$

given by the convolution

$$u(x) = \int_{\mathbb{R}^N} N(x-y) f(y) dy. \tag{1.114}$$

This solution can be differentiated under the integral to yield

$$\nabla u(x) = \frac{1}{\omega_N} \int_{\mathbb{R}^N} \frac{x-y}{|x-y|^N} f(y) dy, \tag{1.115}$$

where ω_N is the area of a unit sphere in $\mathbb{R}^N$ given in Eq. (1.112).

1.9.3. Elementary Properties of the Heat Equation

In this subsection we state some elementary properties of the heat equation and related convection–diffusion equations. For an extensive discussion of this classical problem we refer the reader to Folland, 1995, and John, 1982.

Lemma 1.13. *Consider the initial value problem for the heat equation*

$$\begin{aligned} u_t^\nu &= \nu \Delta u^\nu, \qquad (x,t) \in \mathbb{R}^N \times (0,\infty), \\ u^\nu|_{t=0} &= u_0, \qquad x \in \mathbb{R}^N. \end{aligned} \tag{1.116}$$

If $u_0 \in L^p(\mathbb{R}^N)$, $1 \leq p \leq \infty$, then for $t > 0$ the unique solution (in the class of bounded functions) is given by

$$u^\nu(x,t) = \int_{\mathbb{R}^N} H(x-y, \nu t) u_0(y) dy, \tag{1.117}$$

where H is the N-dimensional Gaussian kernel:

$$H(x,t) = (4\pi t)^{-\frac{N}{2}} e^{-|x|^2/4t}. \tag{1.118}$$

Lemma 1.14. *Let $u^\nu(x,t)$ be a solution of heat equation (1.116).*

(i) If the initial data u_0 satisfy

$$|u_0|_0 + |\nabla u_0|_0 \leq M, \tag{1.119}$$

then there exists a constant c such that

$$|u^\nu(\cdot,t) - u_0(x)|_0 \leq cM(\nu t)^{\frac{1}{2}}. \tag{1.120}$$

(ii) If the initial data u_0 are smooth functions satisfying

$$\|u_0\|_0 + \|\Delta u_0\|_0 \leq M, \tag{1.121}$$

then there exists a constant c such that

$$\|u^\nu(\cdot,t) - u_0\|_0 \leq cM\nu t. \tag{1.122}$$

First we prove estimate (1.120). Using the explicit solution in Eq. (1.117) we have

$$\begin{aligned} u^\nu(x,t) &= \int_{\mathbb{R}^N} H(x-y,\nu t)u_0(y)dy = \int_{\mathbb{R}^N} H(y,\nu t)u_0(x+y)dy \\ &= (4\pi)^{-\frac{N}{2}} \int_{\mathbb{R}^N} e^{-\frac{|\zeta|^2}{4}} u_0\big(x + (\nu t)^{\frac{1}{2}}\zeta\big)d\zeta \quad \big[y = (\nu t)^{\frac{1}{2}}\zeta\big]. \end{aligned}$$

The mean-value theorem applied to this expression gives

$$\begin{aligned} |v^\nu(x,t) - u_0(x)| &\leq (4\pi)^{-\frac{N}{2}} \int_{\mathbb{R}^N} e^{-\frac{|\zeta|^2}{4}} u_0\big[x + (\nu t)^{\frac{1}{2}}\zeta\big] - u_0(x)|d\zeta \\ &\leq (4\pi)^{-\frac{N}{2}} \int_{\mathbb{R}^N} e^{-\frac{|\zeta|^2}{4}} \sup_{x\in\mathbb{R}^N} |\nabla u_0(x)|(\nu t)^{\frac{1}{2}}|\zeta|d\zeta \\ &\leq \left\{(4\pi)^{-\frac{N}{2}} \int_{\mathbb{R}^N} e^{-\frac{|\zeta|^2}{4}}|\zeta|d\zeta\right\} M(\nu t)^{\frac{1}{2}}, \end{aligned}$$

and this concludes the proof of estimate (1.120).

Now we prove estimate (1.122). We use the elementary properties of the Fourier transform stated in Subsection 1.9.1, namely, for the Fourier transform $\hat{f}$ of a smooth, rapidly decreasing function f, the Plancherel identity holds, $\|f\|_0 = \|\hat{f}\|_0$. Also, for the convolution of H with f we have $\widehat{H * f} = \widehat{H} \cdot \hat{f}$.

Because for the Gaussian kernel H in Eq. (1.118) we have

$$\widehat{H}(\zeta,t) = e^{-\pi^2 4t|\zeta|^2},$$

(see Folland, 1995, for these results and an additional discussion), we compute that

$$\|u^\nu(\cdot,t) - u_0\|_0 = \big\|\big(1 - e^{-\pi^2 4\nu t|\zeta|^2}\big)\hat{u}_0\big\|_0.$$

By the mean-value theorem,

$$\sup_{0\le|\zeta|\le\infty}\left|1-e^{-\pi^2 4\nu t|\zeta|^2}\right|\le c\nu t|\zeta|^2,$$

so that

$$\|u^\nu(\cdot,t)-u_0\|_0\le c\nu t\|\,|\zeta|^2\hat{u}_0\|_0=c\nu t\|\Delta u_0\|_0.$$

Lemma 1.15. *Let $u(x,t)$ be a solution to heat equation (1.116). If u_0 is radial symmetric, i.e., $u_0(x)=u_0(|x|)$, then $u(x,t)$ is also radial symmetric.*

Observe that u_0 is radial symmetric if and only if $u_0(Ux)=u_0(x)$ for any proper linear orthogonal transformation U (i.e., $U^*=U^{-1}$, $\det U>0$). Because U is an isometry, we have

$$\begin{aligned}u(Ux,t)&=\int_{\mathbb{R}^N}H(|Ux-y|,\nu t)u_0(y)dy=\int_{\mathbb{R}^N}H(|U(x-\xi)|,\nu t)u_0(U\xi)d(U\xi)\\&=\int_{\mathbb{R}^N}H(|x-\xi|,\nu t)u_0(\xi)\det U d\xi=u(x,t).\end{aligned}$$

Lemma 1.16. *Let $u(x,t)$ be a solution to heat equation (1.116). Then for a fixed x and the viscosity ν,*

$$u(x,t)\sim(4\pi\nu t)^{-\frac{N}{2}}\int_{\mathbb{R}^N}u_0(y)dy+0\left[(\nu t)^{-\frac{N}{2}}\right],\qquad t\gg 1. \tag{1.123}$$

Because $\lim_{t\nearrow\infty}e^{-|x|^2/4\nu t}=1$, by using the dominated convergence theorem in the exact solution in Eq. (1.117) we have

$$(\nu t)^{\frac{N}{2}}u(x,t)=(4\pi)^{-\frac{N}{2}}\int_{\mathbb{R}^N}e^{-|x-y|^2/4\nu t}u_0(y)dy\underset{t\nearrow\infty}{\longrightarrow}(4\pi)^{-\frac{N}{2}}\int_{\mathbb{R}^N}u_0(y)dy.$$

Finally, we state a result for a scalar convection–diffusion equation that generalizes the one used in Section 1.4 when we discussed the exact solution to the strained shear layer. This result is used in subsequent sections.

Lemma 1.17. *Consider the solution $u(x,t)$ to the convection–diffusion equation given by*

$$\begin{aligned}u_t-\gamma(t)x\cdot\nabla u&=\nu\Delta u,\qquad (x,t)\in\mathbb{R}^N\times(0,\infty),\\u|_{t=0}&=u_0,\end{aligned} \tag{1.124}$$

where $u_0(x)$ is a rapidly decreasing and smooth function and $\gamma(t)$ is a smooth time-varying function. Then the unique bounded solution to Eq. (1.124) is given explicitly by

$$u(x,t)=\int_{\mathbb{R}^N}H[x\Gamma_1(t)-y,\Gamma_2(t)]u_0(y)dy, \tag{1.125}$$

where H is the N-dimensional Gaussian kernel in Eq. (1.118), and

$$\Gamma_1(t) = \exp \int_0^t \gamma(\xi) d\xi, \qquad \Gamma_2(t) = \nu \int_0^t {\Gamma_1}^2(\xi) d\xi. \tag{1.126}$$

Proof of Lemma 1.17. We use the change of variables

$$x \longmapsto \tilde{x} = x \exp \int_0^t \gamma(\xi) d\xi,$$

$$t \longmapsto \tilde{t} = \nu \int_0^t \exp \left[2 \int_0^\xi \gamma(\xi') d\xi' \right] d\xi,$$

which generalizes the one presented earlier in our discussion of the strained shear layer in Section 1.4. Through this change of variables, Eq. (1.124) reduces to the heat equation,

$$u_{\tilde{t}} = \Delta_{\tilde{x}} u,$$

so we conclude the proof by Lemma 1.13. □

Notes for Chapter 1

Although a number of classical fluid dynamics texts (see, e.g., Acheson, 1990, Batchelor, 1967, Lamb, 1945, and Landau and Lifshitz, 1987) address some of the material in this chapter, we wish to present it in a form that provides the intuition for the mathematics in later chapters and illustrates some of the important elementary properties of the flows that we use later.

Table 1.1 lists the notation used in this chapter.

Table 1.1. *Notation used in this chapter*

Notation	Name	Type
v	Fluid velocity	Vector field
p	Fluid pressure	Scalar function
x	Eulerian coordinate	Point in $\mathbb{R}^N$
D/Dt	Material derivative	Scalar differential operator
Δ	Laplacian	Scalar differential operator
R_e	Reynold's number	Positive scalar
ν	Viscosity	Nonnegative scalar
T^N	Torus in N dimensions	N−dimensional manifold
$\mathcal{Q}$	General rotation matrix	$N \times N$ unitary matrix
λ	Change of spatial scale	Real number
τ	Change of time scale	Real number
α	Lagrangian spatial coordinate	Point in $\mathbb{R}^N$
$X(\alpha, t)$	Particle-trajectory map	Bijection on $\mathbb{R}^N$
$\mathcal{D}$	Rate-of-strain matrix	$N \times N$ matrix
Ω	Vorticity rotation matrix	$N \times N$ matrix
ω	Vorticity	Vector field
P	Hessian of pressure	$N \times N$ matrix

References for Chapter 1

Acheson, D. J., *Elementary Fluid Dynamics*, Oxford Univ. Press, Oxford, U.K., 1990.

Batchelor, G. K., *An Introduction to Fluid Dynamics*, 2nd paperback ed., Cambridge Univ. Press, Cambridge, U.K., 1999.

Chorin, A. J. and Marsden, J. E., *A Mathematical Introduction to Fluid Mechanics*, 3rd ed., Springer-Verlag, New York, 1993.

Folland, G. B., *Introduction to Partial Differential Equations*, Princeton Univ. Press, Princeton, NJ, 1995.

Gilbarg, D. and Trudinger, N., *Elliptic Partial Differential Equations*, revised 3rd printing, Springer-Verlag, Berlin, 1998.

John, F., *Partial Differential Equations*, 4th ed., Springer-Verlag, New York, 1991.

Lamb, H., *Hydrodynamics*, Dover, New York, 1945.

Landau, L. D. and Lifshitz, E. M., *Fluid Mechanics*, Vol. 6 of Course in Theoretical Physics Series, 2nd ed., Butterworth-Heinemann, Oxford, U.K., 1987.

Stein, E. M. and Weiss, G., *Introduction to Fourier Analysis on Euclidean Spaces*, Princeton Univ. Press, Princeton, NJ, 1971.

2

The Vorticity-Stream Formulation of the Euler and the Navier-Stokes Equations

In Chap. 1 we introduced the Euler ($\nu = 0$) and Navier–Stokes equations ($\nu > 0$) for incompressible fluids:

$$\frac{Dv}{Dt} = -\nabla p + \nu \Delta v, \tag{2.1}$$

$$\operatorname{div} v = 0, \qquad (x, t) \in \mathbb{R}^N \times [0, \infty), \tag{2.2}$$

$$v|_{t=0} = v_0, \qquad x \in \mathbb{R}^N. \tag{2.3}$$

Here $v(x, t) \equiv (v^1, v^2, \dots, v^N)^t$ is the fluid velocity, $p(x, t)$ is the scalar pressure, D/Dt is the convective derivative (i.e., the derivative along particle trajectories) defined by

$$\frac{D}{Dt} = \frac{\partial}{\partial t} + \sum_{j=1}^{N} v^j \frac{\partial}{\partial x_j}, \tag{2.4}$$

and ν is the kinematic viscosity. We derived some elementary properties of the equations, including scale invariances, energy conservation/dissipation, and velocity and vorticity-flux conservation. We also derived vorticity equation (1.33) and showed that, for inviscid fluids, the vorticity is transported and stretched along particle trajectories for 3D flows and is conserved along particle paths for 2D flows. In this chapter we show that this leads to a natural reformulation of the equations as an evolution equation for the vorticity alone (discussed in Section 2.1). In Section 2.2 we present a general method for constructing exact steady solutions to the 2D Euler equation. We use this method to construct some flows with radial symmetry and with periodicity. In Section 2.3 we construct more exact solutions exhibiting vorticity amplification through 3D effects. We show that every 2D flow always generates a large family of 3D flows with vorticity stretching. Then we introduce a class of flows for which the vorticity is aligned with the velocity field, called Beltrami flows. The final topic of this section is axisymmetric flows, in which the velocity field is independent of the angular coordinate. In Section 2.4 we derive the vorticity-stream formulation for 3D flows. In this case the vorticity equation involves both the velocity field and its gradient. The latter requires a calculation involving singular integral operators (SIOs) that we review briefly. Section 2.5 shows that, for inviscid flows, the transport of vorticity along particle paths leads to yet another reformulation of the equations as an integrodifferential equation for the particle trajectories. It is this reformulation that is used extensively in vortex methods for computing inviscid and high Reynolds number flows (see Chap. 6) and also in Chap. 4 of this book to prove the existence and the uniqueness of solutions to the Euler equations by use of the particle trajectories.

2.1. The Vorticity-Stream Formulation for 2D Flows

Recall that taking the curl of the Navier–Stokes equations leads to the following evolution equation for $\omega = \text{curl } v$:

$$\frac{D\omega}{Dt} = \omega \cdot \nabla v + \nu \Delta \omega. \tag{2.5}$$

For 2D flows the velocity field is $v = (v^1, v^2, 0)^t$, the vorticity $\omega = (0, 0, v^2_{x_1} - v^1_{x_2})^t$ is orthogonal to v, and the vorticity-stretching term vanishes, $\omega \cdot \nabla v \equiv 0$. In this section to simplify notation we denote the velocity field by $v = (v^1, v^2)^t$ and the scalar vorticity by $\omega = v^2_{x_1} - v^1_{x_2}$. Thus vorticity equation (2.5) reduces to the *scalar vorticity equation*

$$\frac{D\omega}{Dt} = \nu \Delta \omega. \tag{2.6}$$

Recall that for inviscid flows ($\nu = 0$) this equation implies conservation of vorticity along particle trajectories $X(\alpha, t)$ (see Section 1.6):

$$\omega(X(\alpha, t), t) = \omega_0(\alpha), \qquad \alpha \in \mathbb{R}^2. \tag{2.7}$$

This special property distinguishes 2D from 3D inviscid flows; general 3D flows have $\omega \cdot \nabla v \neq 0$, and the vorticity stretches according to the formula

$$\omega(X(\alpha, t), t) = \nabla_\alpha X(\alpha, t)\omega_0(\alpha), \qquad \alpha \in \mathbb{R}^3,$$

[see Eq. (1.51) from Proposition 1.8].

We also recall that for shear-layer flows (Examples 1.5–1.8) we could recover the velocity v from the vorticity ω by a nonlocal operator [Eq. (1.36)]. Now we show that for arbitrary 2D flows we can eliminate the velocity from vorticity equation (2.6) to yield a self-contained equation for ω.

Because div $v = 0$, there exists a (unique up to an additive constant) *stream function* $\psi(x, t)$ such that

$$v = (-\psi_{x_2}, \psi_{x_1})^t \equiv \nabla^\perp \psi. \tag{2.8}$$

Computing the curl of Eq. (2.8), we get the Poisson equation for ψ:

$$\Delta \psi = \omega. \tag{2.9}$$

A solution to this equation is given by a convolution with the Newtonian potential with ω (see Lemma 1.12 in the appendix of Chapter 1):

$$\psi(x,t) = \frac{1}{2\pi}\int_{\mathbb{R}^2} \ln|x-y|\omega(y,t)dy, \qquad x \in \mathbb{R}^2.$$

Because we can compute the gradient of ψ by differentiating under the integral, the velocity v can be recovered from ψ by the operator $\nabla^{\perp}$ in Eq. (2.8) as

$$v(x,t) = \int_{\mathbb{R}^2} K_2(x-y)\omega(y,t)dy, \tag{2.10}$$

where the kernel $K_2(\cdot)$ is defined by

$$K_2(x) = \frac{1}{2\pi}\left(-\frac{x_2}{|x|^2}, \frac{x_1}{|x|^2}\right)^t. \tag{2.11}$$

Equation (2.10) is the analog of the well-known *Biot–Savart law* for the magnetic field induced by a current on a wire (Lorrain and Corson, 1970, p. 295). We will use this terminology for Eq. (2.10) later in the book. That the velocity v can be recovered from the vorticity ω by a nonlocal operator is not surprising; recall from Chap. 1 in the Leray formulation of Section 1.8, Eq. (1.89), that we were able to reformulate the pressure in the primitive-variable form of the equation as a nonlocal operator involving the velocity field.

Substituting the velocity in Eq. (2.10) into vorticity equation (2.6) we get a self-contained equation for ω. We have just proved Proposition 2.1.

Proposition 2.1. *The Vorticity-Stream Formulation in All of* $\mathbb{R}^2$. *For 2D flows vanishing sufficiently rapidly as* $|x| \nearrow \infty$, *the Navier–Stokes equations (2.1)–(2.3) are equivalent to the vorticity-stream formulation,*

$$\frac{D\omega}{Dt} = \nu\Delta\omega, \qquad (x,t) \in \mathbb{R}^2 \times [0,\infty),$$

$$\omega|_{t=0} = \omega_0,$$

where the velocity v *is determined from the vorticity* ω *by the Biot–Savart law*

$$v(x,t) = \int_{\mathbb{R}^2} K_2(x-y)\omega(y,t)dy, \qquad x \in \mathbb{R}^2,$$

involving the kernel $K_2 = (2\pi|x|^2)^{-1}(-x_2, x_1)$, *homogeneous of degree* -1. *Similarly, the pressure* p *can be obtained from Poisson equation* (1.84):

$$-\Delta p = \sum_{i,j} v^i_{x_j} v^j_{x_i}.$$

2.2. A General Method for Constructing Exact Steady Solutions to the 2D Euler Equations

In this section we show that the relationship between the stream function and the vorticity leads to a collection of exact steady solutions to the Euler equations. These special solutions serve many purposes. For example, they provide intuition about how well the Euler equation approximates the Navier–Stokes equation in the limit as the viscosity $\nu \searrow 0$. Furthermore, in later chapters of this book we use them as building blocks for vortex methods and for understanding the subtleties in approximations to very weak solutions of the Euler equations such as vortex sheets.

Observe that, when the stream function ψ is used, the scalar vorticity equation $(D\omega/Dt) = \nu\Delta\omega$ is (setting $\nu = 0$)

$$\omega_t + \mathcal{J}(\psi, \Delta\psi) = 0, \tag{2.12}$$

where $\mathcal{J}$ is the Jacobian

$$\mathcal{J}(\psi, \Delta\psi) = \det \begin{bmatrix} \psi_{x_1} & \psi_{x_2} \\ \Delta\psi_{x_1} & \Delta\psi_{x_2} \end{bmatrix}.$$

To have a steady inviscid solution we need $\mathcal{J}(\psi, \Delta\psi) = 0$ that implies that $\nabla^\perp\psi$ and $\nabla^\perp\Delta\psi$ are parallel so that they have the same integral curves, the level curves of ψ and $\Delta\psi$; hence ψ and $\Delta\psi$ must be functionally dependent. We summarize this result in the following proposition.

Proposition 2.2. *A stream function ψ on a domain $\Omega \subset \mathbb{R}^2$ defines a steady solution to the 2D Euler equation on Ω if and only if*

$$\Delta\psi = F(\psi) \tag{2.13}$$

for some function F.

The function F in Eq. (2.13) is, in general nonlinear, and Proposition 2.2 yields many exact steady solutions to the 2D Euler equation. Note that for any 2D steady flow the particle-trajectory equation becomes the 2×2 autonomous Hamiltonian system,

$$\begin{aligned} \frac{dX_1}{dt} &= -\psi_{x_2}(X(\alpha, t)), \\ \frac{dX_2}{dt} &= \psi_{x_1}(X(\alpha, t)), \\ X(\alpha, t) &= (X_1, X_2), \end{aligned}$$

with the stream function ψ as Hamiltonian. Hence for steady flows ψ is constant along the particle trajectories that coincide with the streamlines (level curves of ψ). Stagnation (fixed) points in the steady flow are points α_s where $v(\alpha_s) = 0$ and hence $X(\alpha_s, t) = \alpha_s$. The stagnation points coincide with the critical points (minima, maxima, and saddle points) of the stream function ψ. At minima and maxima of

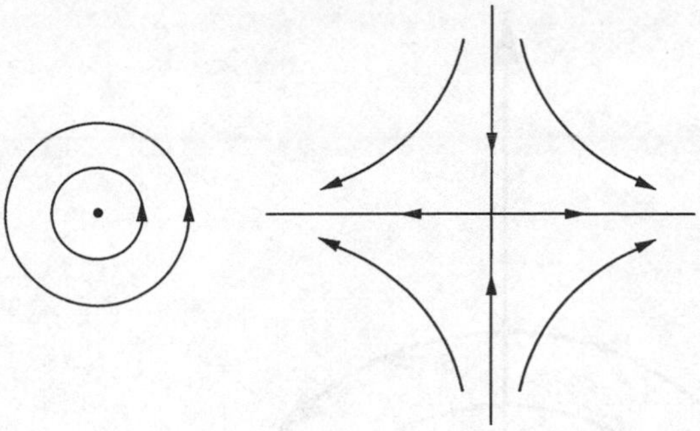

Figure 2.1. Two kinds of stagnation points of the flow: (a) an elliptic fixed point corresponding to a local minimum or maximum of ψ, and (b) a hyperbolic fixed point at a saddle point of ψ.

ψ, locally the flow looks like a pure rotation (an elliptic fixed point), whereas at the saddle points of ψ it looks like a 2D strain flow (a hyperbolic fixed point); see Fig. 2.1.

Next we look at an important special class of inviscid, steady solutions with radial vorticity distribution.

2.2.1. Radial Eddies for the Euler and the Navier–Stokes Equations

Example 2.1. Steady, Inviscid Eddies. Consider a radially symmetric smooth vorticity $\omega_0 = \omega_0(r)$, $r = |x|$. Because the Laplacian Δ is rotationally invariant, the solution ψ_0 to Eq. (2.9), $\Delta\psi_0 = \omega_0$, is also radially symmetric, so $\mathcal{J}(\psi_0, \Delta\psi_0) \equiv 0$. Thus any radially symmetric vorticity $\omega_0(r)$ defines a steady, radially symmetric solution to the 2D Euler equation. By Eq. (2.8) the velocity v is

$$v_0(x) = \left(-\frac{x_2}{r}, \frac{x_1}{r}\right)^t \psi_0'(r)$$

where the prime denotes the differentiation with respect to r. Using polar coordinates, we have

$$\psi_0''(r) + \frac{1}{r}\psi_0'(r) = \omega_0(r),$$

so

$$\psi_0'(r) = \frac{1}{r}\int_0^r s\omega_0(s)ds,$$

and finally

$$v_0(x) = \left(-\frac{x_2}{r^2}, \frac{x_1}{r^2}\right)^t \int_0^r s\omega_0(s)ds. \tag{2.14}$$

Thus all streamlines of the flow are circles, and the fluid rotates depending on the sign of ω_0; see Fig. 2.2.

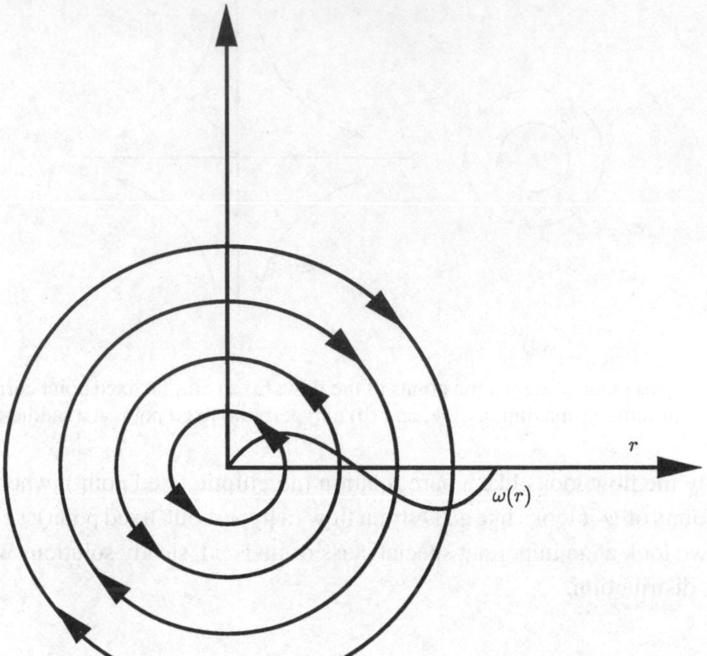

Figure 2.2. An inviscid steady radial eddy. The direction of rotation of the eddy depends on the sign of the vorticity.

Next we incorporate viscous effects into the steady eddies defined in Eq. (2.14) to yield a solution of the Navier–Stokes equations.

Example 2.2. Time-Dependent Viscous Eddies. Consider a smooth radially symmetric initial vorticity $\omega_0 = \omega_0(r)$, $r = |x|$. From Example 2.1 we know that if the vorticity $\omega(x, t)$ is radially symmetric, so is the stream function $\psi(x, t)$, and then $\mathcal{J}(\psi, \Delta\psi) \equiv 0$. Because the heat equation preserves radial symmetry of solutions (see Lemma 1.15 in the appendix to Chap. 1), vorticity equation (2.12) reduces to the 1D heat equation

$$\omega_t = \nu\Delta\omega,$$

$$\omega|_{t=0} = \omega_0,$$

with the solution

$$\omega(x, t) = \frac{1}{4\pi\nu t}\int_{\mathbb{R}^2} e^{-\frac{|x-y|^2}{4\nu t}}\,\omega_0(|y|)dy. \tag{2.15}$$

The velocity field v is given by the same formula as that in Eq. (2.14),

$$v(x, t) = \left(-\frac{x_2}{r^2}, \frac{x_1}{r^2}\right)^t \int_0^r s\omega(s, t)ds. \tag{2.16}$$

Now we use the above exact solutions to get more insight into the fundamental question of how well the Euler equation approximates the Navier–Stokes equation as the viscosity $\nu \searrow 0$. Because the asymptotic behavior of the viscous eddy solution in Eq. (2.15) as time $t \nearrow \infty$ is (see Lemma 1.16 in the appendix of Chap. 1)

$$\omega^\nu(r,t) \sim \frac{1}{4\pi\nu t}\int_{\mathbb{R}^2} \omega_0(|y|)dy + 0\left(\frac{1}{\nu t}\right), \qquad t \gg 1,$$

we see that the inviscid solution $\omega_0(r)$ and the viscous $\omega^\nu(r,t)$ eddy solutions with the same initial data cannot agree as $t \nearrow \infty$. For finite times, however, we can use Lemma 1.14 from the appendix of Chap. 1, giving

$$|\omega^\nu(r,t) - \omega_0(r)| \le c(\nu t)^{\frac{1}{2}}.$$

Inserting this estimate into Eq. (2.14), we get

$$\begin{aligned} |v^\nu(r,t) - v_0(r)| &= \left| \left(-\frac{x_2}{r^2}, \frac{x_1}{r^2}\right)^t \int_0^r s[\omega^\nu(s,t) - \omega_0(s)]ds \right| \\ &\le \frac{1}{r}\int_0^r s|\omega^\nu(s,t) - \omega_0(s)|ds \le cr(\nu t)^{\frac{1}{2}}. \end{aligned}$$

We have just proved Proposition 2.3.

Proposition 2.3. *Let the radially symmetric vorticity $\omega_0(r)$ satisfy $|\omega_0| + |\nabla\omega_0| \le M$, with velocity $v_0(r)$ determined from the vorticity by the (steady) inviscid radial eddy formula in Eq. (2.14). Let $\omega^\nu(r,t)$, $v^\nu(r,t)$ be the viscous radial eddy solution in Eqs. (2.15) and (2.16) with the same initial data $\omega_0(r)$. Then*

$$|\omega^\nu(r,t) - \omega_0(r)| \le c(\nu t)^{\frac{1}{2}}, \tag{2.17}$$

$$|v^\nu(r,t) - v_0(r)| \le cr(\nu t)^{\frac{1}{2}}. \tag{2.18}$$

The use of a more accurate far-field estimate than that in Lemma 1.14 yields Eq. (2.18) without the factor r.

We see that for a fixed T and r, $\omega^\nu(r,t) \to \omega_0(r)$ and $v^\nu(r,t) \to v_0(r)$ in the sup norm uniformly for $0 \le t \le T$ as the viscosity $\nu \searrow 0$. In all examples of exact solutions (see Propositions 1.6 and 1.7) we had square-root convergence. Later in Chap. 3 we prove this "optimal" convergence for arbitrary smooth solutions to the Euler and the Navier–Stokes equations in all of space.

Another family of exact steady solutions to the 2D Euler equation arises from a periodic geometry.

2.2.2. The Vorticity-Stream Formulation for Periodic Flows and More Exact Solutions

From Proposition 1.17 and a Fourier series expansion, we know that, given a periodic vorticity ω, Eq. (2.9), $\Delta\psi = \omega$, has a periodic solution ψ on the torus T^2,

$$\psi(x,t) = -\sum_{k\neq 0} \left(4\pi|k|^2\right)^{-1} e^{2\pi i x\cdot k}\widehat{\omega}(k,t),$$

provided that compatibility condition (1.95) is satisfied:

$$\int_{T^2} \omega\, dx = 0.$$

Because $\omega = v^2_{x_1} - v^1_{x_2}$ and is periodic, this condition is guaranteed. Observe, however, that if we demand that $v = \nabla^\perp \psi$, we necessarily have

$$\int_{T^2} v\, dx = 0.$$

In general we want to consider flows on the torus with a velocity that has a nonzero mean. Hence we split the velocity as $v = \bar{v} + \tilde{v}$, where $\bar{v} = \int_{T^2} v\, dx$ is a constant-velocity field and $\int_{T^2} \tilde{v}\, dx = 0$. Computing the velocity $\tilde{v}$ from the solution ψ, $\tilde{v} = \nabla^\perp \psi$, we get a periodic version of the Biot–Savart law:

$$\tilde{v}(x,t) = \sum_{k\neq 0} (-k_2, k_1)^t \left(2\pi i |k|^2\right)^{-1} e^{2\pi i x\cdot k} \widehat{\omega}(k,t). \tag{2.19}$$

Proposition 2.4. *The Vorticity-Stream Formulation for Period Flows. Let* $v_0 = \bar{v}_0 + \tilde{v}_0$, $\bar{v}_0 = \int_{T^2} v_0\, dx$, div $v_0 = 0$, *be a smooth one-periodic velocity field on the torus* T^2. *Then Navier–Stokes equations (2.1)–(2.3) are equivalent to the vorticity-stream formulation*

$$\begin{aligned} &\omega_t + (\bar{v}_0 + \tilde{v}) \cdot \nabla \omega = \nu \Delta \omega, \qquad (x,t) \in T^2 \times [0,\infty), \\ &\omega|_{t=0} = \omega_0, \end{aligned} \tag{2.20}$$

where $\omega =$ curl $\tilde{v}$ *and the velocity field* $\tilde{v}$ *is computed from the vorticity by*

$$\tilde{v}(x,t) = \sum_{k\neq 0} (-k_2, k_1)^t \left(2\pi i |k|^2\right)^{-1} e^{2\pi i x\cdot k} \widehat{\omega}(k,t).$$

We can recover the pressure p *by solving Poisson equation (1.84) on* T^2,

$$-\Delta p = \sum_{i,j} v^i_{x_j} v^j_{x_i}.$$

We leave the proof to the reader as an exercise.

As an immediate application of the vorticity-stream formulation for periodic flows, we build some explicit examples of such flows. As before in Proposition 2.1, for steady inviscid flows we look for a stream function ψ such that

$$\Delta \psi = F(\psi), \qquad x \in T^2 \tag{2.21}$$

for some function F. Any eigenfunction ψ_k of the Laplacian Δ on the torus T^2,

$$\psi_k(x) = a_k \cos(2\pi k \cdot x) + b_k \sin(2\pi k \cdot x), \qquad k \in \mathbb{Z}^2,$$

satisfies Eq. (2.21) with the linear function $F(\psi) = 4\pi^2|k|^2\psi$,

$$\Delta\psi_k = -4\pi^2|k|^2\psi_k,$$

and hence defines a periodic solution of the 2D Euler equation. Taking a combination of eigenfunctions ψ_k with the same eigenvalue, i.e., $|k|^2 = \ell$ (to have the same linear function $F = -4\pi^2\ell$ for all ψ_k), we have the following proposition.

Proposition 2.5. *Steady Periodic Solutions to the 2D Euler Equation. For a fixed $\ell \in \mathbb{Z}^+$, define the one-periodic stream function on the torus T^2 by*

$$\psi(x) = \frac{1}{2\pi}\sum_{|k|^2=\ell}[a_k\cos(2\pi k\cdot x) + b_k\sin(2\pi k\cdot x)], \tag{2.22}$$

where a_k and b_k are arbitrary real coefficients and $k = (k_1, k_2)$ is an arbitrary integer lattice vector satisfying $|k|^2 = \ell$. Then ψ defines a steady periodic solution to the 2D Euler equation by $v = \nabla^\perp\psi$, i.e.,

$$v(x) = \sum_{|k|^2=\ell}\begin{bmatrix} a_k k_2\sin(2\pi k\cdot x) + b_k k_2\cos(2\pi k\cdot x) \\ -a_k k_1\sin(2\pi k\cdot x) + b_k k_1\cos(2\pi k\cdot x)\end{bmatrix}. \tag{2.23}$$

The examples for $\ell = 1$ yield simple periodic shear flows. However, already for $\ell = 2$ we get a very interesting steady flow with a more complex structure.

Example 2.3. A Steady Inviscid Periodic Solution. Consider the stream function in Eq. (2.22) for $\ell = 2$:

$$\psi(x) = \frac{1}{2\pi}\sin(2\pi x_1)\cos(2\pi x_2), \qquad x \in T^2, \tag{2.24}$$

with the corresponding velocity field given by Eq. (2.23),

$$v(x) = \begin{bmatrix}\sin(2\pi x_1)\sin(2\pi x_2) \\ \cos(2\pi x_1)\cos(2\pi x_2)\end{bmatrix}. \tag{2.25}$$

Recall (see Proposition 2.2) that the particle trajectories coincide with the streamlines $\psi = \text{const}$ and that the stagnation points α_s where $v(\alpha_s) = 0$ are critical points of ψ. The stream function ψ in Eq. (2.24) has six local extrema (maxima or minima),

$$S_1 = \left\{(i, j)\colon\ i = \frac{1}{4}, \frac{3}{4};\ j = 0, \frac{1}{2}, 1\right\},$$

around which the flow looks locally like a pure rotation, and six saddle points

$$S_2 = \left\{(i, j)\colon\ i = 0, \frac{1}{2}, 1;\ j = 1, 3\right\},$$

around which it looks like the 2D strain flow. This flow thus forms six cells with a rotation around their centers S_1 and pure strain flows at their vertices S_2; see Fig. 2.3.

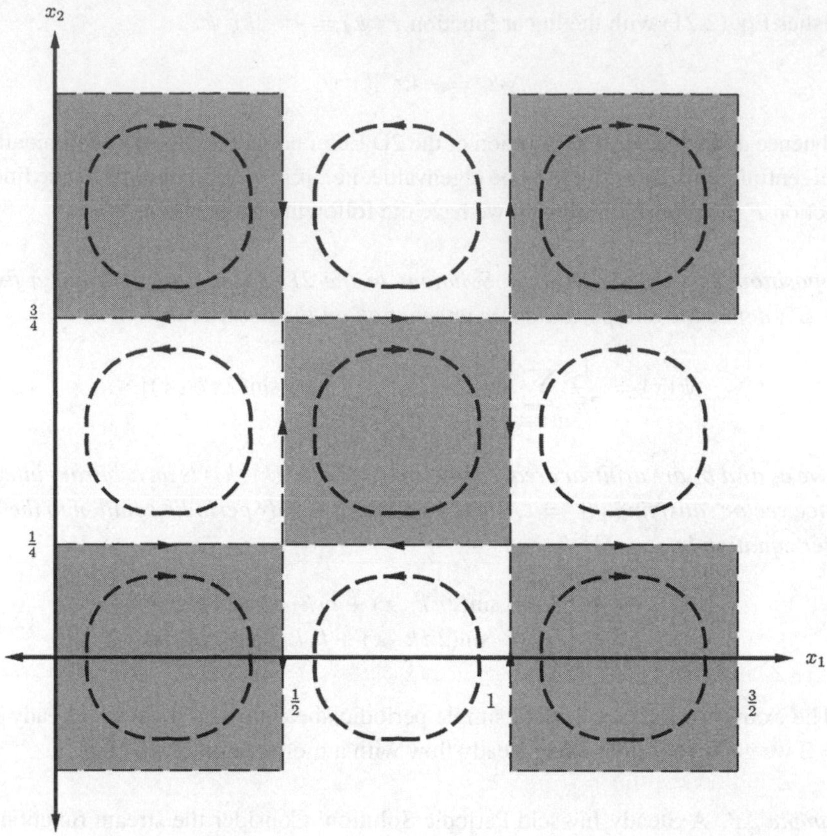

Figure 2.3. A steady inviscid periodic solution with stream function $\psi = [1/(2\pi)]\sin(2\pi x_1)\cos(2\pi x_2)$. The shaded regions indicate negative vorticity, and the unshaded regions have positive vorticity.

This flow is dynamically unstable so that small perturbations result in a very chaotic motion (Dombre et al., 1986; Bertozzi, 1988).

As before for the eddy solutions, now we incorporate the viscous effects into the above periodic solutions. In Proposition 2.5 we have constructed a periodic stream function so that $\mathcal{J}(\psi, \Delta\psi) \equiv 0$. Because the heat equation transforms eigenfunctions of the Laplacian Δ into themselves, vorticity equation (2.12), $\omega_t + \mathcal{J}(\psi, \Delta\psi) = \nu\Delta\omega$, reduces to the periodic heat equation on the torus T^2. Thus by the separation of variables and the Fourier series we get the following proposition.

Proposition 2.6. *The Viscous Periodic Solution. Let $v_0(x)$ be the steady, inviscid, one-periodic solution in Eq. (2.23) to the 2D Euler equation*

$$v_0(x) = \sum_{|k|^2=\ell} \begin{bmatrix} a_k k_2 \sin(2\pi k\cdot x) - b_k k_2 \cos(2\pi k \cdot x) \\ -a_k k_1 \sin(2\pi k \cdot x) + b_k k_1 \cos(2\pi k \cdot x) \end{bmatrix}.$$

Then the viscous one-periodic solution $v^\nu(x,t)$ to the Navier–Stokes equation with *the initial data* $v^\nu|_{t=0} = v_0$ *is*

$$v^\nu(x,t) = e^{-4\pi^2 \ell \nu t} v_0(x), \qquad x \in T^2. \tag{2.26}$$

For these solutions it is extremely easy to compare the Navier–Stokes and the Euler solutions as the viscosity $\nu \searrow 0$. We have

$$\begin{aligned} |v_0(x) - v^\nu(x,t)| &\le |v_0(x)| \left|1 - e^{-4\pi^2 \ell \nu t}\right| \\ &\le c\ell\nu t. \end{aligned} \tag{2.27}$$

For any fixed time t and wave number ℓ, this order of convergence with respect to ν is higher than the square-root convergence observed for solutions in all of space. However, this convergence rate deteriorates rapidly with the increasing wave number ℓ.

These viscous periodic solutions give a nice illustration that the effect of viscosity is frictional, i.e., that fluid particles moving along particle trajectories are slowed down through viscous effects. Indeed, for particle trajectories with the velocity in (2.26),

$$\frac{dX}{dt}(\alpha,t) = e^{-4\pi^2 \ell \nu t} v_0(X(\alpha,t)),$$

thus the velocity along particles trajectories decreases rapidly in time.

Finally we present another important example of the steady solutions to the 2D Euler equation periodic in a single direction. This example, called the cat's-eye flow, arises when flows with strong shear layers are being studied, and it models mixing layers of fluids.

Example 2.4. The Kelvin–Stuart Cat's-Eye Flow. Consider a stream function ψ, 2π periodic in the x_1 variable, that satisfies

$$\Delta\psi = F(\psi) \equiv -e^{2\psi}. \tag{2.28}$$

A one-parameter family of solutions is (see Stuart, 1971)

$$\psi(x) = -\ln(c\cosh x_2 + \sqrt{c^2-1}\cos x_1), \qquad x \in [0, 2\pi] \times \mathbb{R}, \tag{2.29}$$

with the parameter $c \ge 1$. The corresponding velocity v is

$$v(x) = \left(\frac{c\sinh x_2}{c\cosh x_2 + \sqrt{c^2-1}\cos x_1}, \frac{\sqrt{c^2-1}\sin x_1}{c\cosh x_2 + \sqrt{c^2-1}\cos x_1} \right)^t. \tag{2.30}$$

Observe that $c = 1$ gives a shear flow, and for $c > 1$, $v^1 \to \pm 1$, $v^2 \to 0$ as $x_2 \to \pm\infty$. For $c > 1$ the stream function ψ has one local extremum at $(\pi, 0)$ around which the flow rotates locally and two saddle points $(0,0)$, $(2\pi, 0)$ at which it looks like the 2D strain flow. Figure 2.4 shows a plot of the cat's-eye streamlines; in particular, the streamline that outlines the cat's-eye satisfies

$$\cosh x_2 + \sqrt{1 + \frac{1}{c^2}}\cos x_1 = 1 + \sqrt{1 + \frac{1}{c^2}}.$$

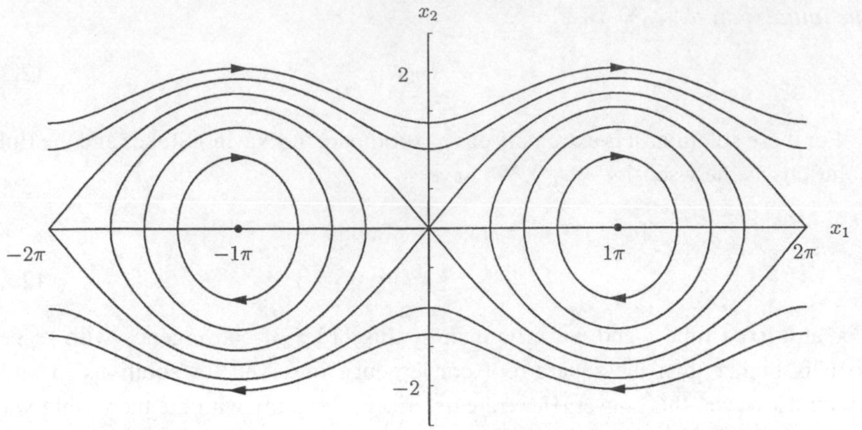

Figure 2.4. The Kelvin–Stuart cat's-eye flow, with stream function $\psi(x) = -\ln(c \cosh x_2 + \sqrt{c^2 - 1} \cos x_1)$.

2.3. Some Special 3D Flows with Nontrivial Vortex Dynamics

In Section 2.2 we showed that for 2D flows the vorticity stretching term $\omega \cdot \nabla v$ in the vorticity equation is identically zero, so the vorticity ω cannot increase (and actually it is conserved along particle trajectories for inviscid flows). For 3D flows, however, $\omega \cdot \nabla v (\equiv \mathcal{D}\omega)$ is, in general, nonzero, so the vorticity can increase and accumulate. The simple examples of exact solutions presented in Chap. 1, e.g., the rotation jets, Example 1.7, and the inviscid and the viscous strained shear-layer solutions, Examples 1.7–1.8, exhibited an increase in vorticity when they roughly align with the eigenvectors corresponding to the positive eigenvalues of the deformation matrix $\mathcal{D} = [(\nabla v) + (\nabla v)]^t$. To build some geometric intuition we construct more exact solutions that exhibit vorticity amplification through 3D effects.

First, we show that every 2D flow always generates a large family of 3D flows with vorticity stretching. Then we introduce Beltrami flows that by definition have the vorticity aligned with the velocity. Finally we introduce 3D axisymmetric flows and show that the case of axisymmetric flow without swirl is similar to that of 2D flow, whereas the case of swirl has a nontrivial 3D behavior.

2.3.1. Two-and-a-Half-Dimensional Flows

In this subsection we show that every 2D flow always generates a large family of 3D flows with vorticity stretching. We refer to these as two-and-a-half-dimensional ($2\frac{1}{2}$D flows) because the flow in the x_3 direction is predetermined by the underlying 2D flow.

Consider a solution $(\tilde{v}, \tilde{p})$ to the 2D Navier–Stokes equations [in the $\tilde{x} = (x_1, x_2) \in \mathbb{R}^2$ variables],

$$\begin{aligned} \frac{\widetilde{D}\tilde{v}}{Dt} &= -\nabla \tilde{p} + \nu \Delta \tilde{v}, \\ \operatorname{div} \tilde{v} &= 0, \qquad (\tilde{x}, t) \in \mathbb{R}^2 \times (0, \infty), \\ \tilde{v}|_{t=0} &= \tilde{v}_0, \qquad x \in \mathbb{R}^2, \end{aligned} \tag{2.31}$$

where $(\widetilde{D}/Dt) = (\partial/\partial t) + \tilde{v} \cdot \nabla_{\tilde{x}}$. Then for any third component $v_0^3(\tilde{x})$ of the velocity (depending on only the $\tilde{x}$ variable) we solve the 3D Navier–Stokes equation with the initial condition $v|_{t=0} = (\tilde{v}_0, v_0^3)$. It is easy to see that $v = (\tilde{v}, v^3)$ is the solution to this problem, provided that the velocity $v^3(\tilde{x}, t)$ does not depend on the x_3 variable and that it solves the linear scalar diffusion equation

$$\begin{aligned} &\frac{\widetilde{D}}{Dt} v^3 = \nu \Delta v^3, \qquad (\tilde{x}, t) \in R^2 \times (0, \infty), \\ &v^3|_{t=0} = v_0^3, \qquad x \in \mathbb{R}^2. \end{aligned} \tag{2.32}$$

Thus we have proved Proposition 2.7.

Proposition 2.7. *3D Flows Depending on Two Space Variables. Let $\tilde{v}(\tilde{x}, t)$, $\tilde{p}(\tilde{x}, t)$, $\tilde{x} \in \mathbb{R}^2$, be a solution to 2D Navier–Stokes equation (2.31), and let $v^3(\tilde{x}, t)$ be a solution to linear scalar diffusion equation (2.32). Then $v = (\tilde{v}, v^3)^t, \tilde{p}$ is a solution to the 3D Navier–Stokes equation.*

Any 2D exact solution, such as those from Section 2.2 or in Chap.1, generates new explicit examples of 3D flows. Now we determine the manner in which the flow amplifies the vorticity. Because the velocity $v = (\tilde{v}, v^3)^t$ does not depend on the x_3 variable, the 3D vorticity equation $(D/Dt)\omega = \nabla v \omega + \nu \Delta \omega$ immediately gives Proposition 2.8.

Proposition 2.8. *Let $v = (\tilde{v}, v^3)^t$ be the solution defined in Proposition 2.7, and let $\omega = \text{curl } v \equiv (\bar{\omega}, \tilde{\omega})^t$, where $\bar{\omega} = (v^3_{x_2}, -v^3_{x_1})^t$ and $\tilde{\omega} = \text{curl } \tilde{v}$. Then $\bar{\omega}$ satisfies the vorticity equation*

$$\frac{\widetilde{D}}{Dt} \bar{\omega} = (\tilde{\nabla} \tilde{v}) \bar{\omega} + \nu \Delta \bar{\omega}, \tag{2.33}$$

where $\tilde{\nabla}\tilde{v}$ is the 2×2 matrix $\frac{1}{2}(\tilde{v}^i_{x_j})$.

As an independent check, Eq. (2.33) follows directly from Eq. (2.32) by differentiation.

From Eq. (2.33) we thus see that the 2D component $\bar{\omega}$ of the vorticity ω increases when it roughly aligns with the eigenvectors corresponding to the positive eigenvalues of the explicit 2×2 matrix $\tilde{\nabla}\tilde{v}$ computed from the known 2D flow $\tilde{v}$. Thus vortex stretching that is characteristic of general 3D flows occurs in $2\frac{1}{2}$D flows; however, the nonlocal interaction between the vorticity and the deformation reduces to the local effect of the known function $\tilde{\nabla}\tilde{v}$ on $\bar{\omega}$.

As an illustration of these results, we give a simple example with stretching of vorticity $\bar{\omega}$.

Example 2.5. Set $\nu = 0$ and take the inviscid shear flow (see Example 1.5) as the 2D flow $\tilde{v}$,

$$\tilde{v}(\tilde{x}, t) = (v^1(x_2), 0)^t. \tag{2.34}$$

Vorticity equation (2.33) for $\bar{\omega}$ is

$$\bar{\omega}_t^1 + v^1 \bar{\omega}_{x_1}^1 = v_{x_2}^1 \bar{\omega}^2,$$
$$\bar{\omega}_t^2 + v^1 \bar{\omega}_{x_1}^2 = 0.$$

By the method of characteristics,

$$\bar{\omega}(\widetilde{X}(\tilde{\alpha}, t), t) = \begin{bmatrix} 1 & v_{x_2}^1(\alpha_2)t \\ 0 & 1 \end{bmatrix} \bar{\omega}_0(\tilde{\alpha}), \tag{2.35}$$

where the particle trajectories $\widetilde{X} = (X_1, X_2)^t$ are

$$X_1(\tilde{\alpha}, t) = \alpha_1 + v^1(\alpha_2)t, \qquad X_2(\tilde{\alpha}, t) = \alpha_2. \tag{2.36}$$

Equation (2.35) is in fact a special case of the identity in Eq. (1.51):

$$\bar{\omega}(\widetilde{X}(\tilde{\alpha}, t), t) = \nabla_{\tilde{\alpha}} \widetilde{X}(\tilde{\alpha}, t) \bar{\omega}_0(\tilde{\alpha}).$$

Observe that by Proposition 1.12 the above solution $v = (\tilde{v}, v^3)^t$ has conserved energy, $\int v^2\, dx = \text{const}$. Nevertheless, the vorticity $\bar{\omega}$ increases in time if $\bar{\omega}_0^2 \neq 0$. In the next chapter we discuss the conjecture that actually the vorticity may blow up in finite time for inviscid 3D flows even though the energy is conserved. In Chap. 5 we discuss this issue in greater detail.

Next we determine when we can use the above principle to generate steady solutions to the 3D Euler equation from a given 2D steady solution. Let $\tilde{\psi}(\tilde{x})$ be a stream function of a 2D steady, inviscid flow $\tilde{v}$. To construct a *steady* $2\frac{1}{2}$D flow from this 2D solution, we must satisfy the steady form of scalar equation (2.32) for $v^3(\tilde{x})$:

$$-\psi_{x_2} v_{x_1}^3 + \psi_{x_1} v_{x_2}^3 = 0.$$

This implies that the velocity $v^3(\tilde{x})$ is constant on level lines of $\tilde{\psi}$, i.e., $v^3(\tilde{x}) = W(\tilde{\psi}(\tilde{x}))$ for some smooth function W. We have just proved the following proposition,

Proposition 2.9. *Let $\tilde{\psi}(\tilde{x})$, $\tilde{x} = (x_1, x_2)^t$ be the stream function of a steady solution $\tilde{v}(\tilde{x})$ to the 2D Euler equation. Then, for a smooth, arbitrary function $W(\tilde{\psi})$,*

$$v = \left[-\tilde{\psi}_{x_2}, \tilde{\psi}_{x_1}, W(\tilde{\psi})\right]^t \tag{2.37}$$

is a steady solution to the 3D Euler equation.

In Subsection 2.3.2 we use this result to generate a class of 3D Beltrami flows.

2.3.2. 3D Beltrami Flows

Experiments show that flows in which the vorticity ω is locally roughly parallel to the velocity generate interesting 3D instabilities. This can result from the self-induced velocity of vortex lines deforming the vortex lines into very unstable horseshoes; see Fig. 2.5.

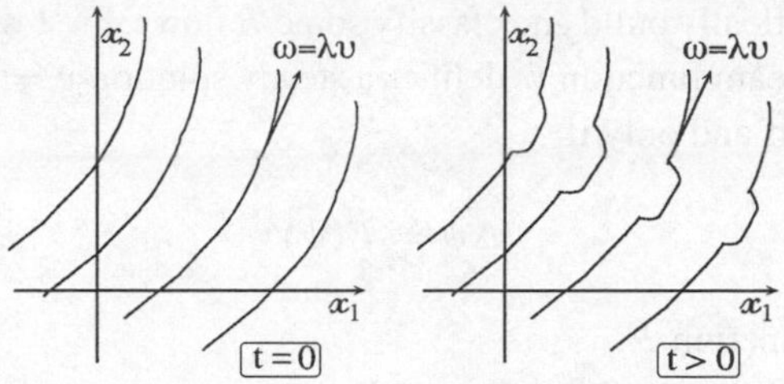

Figure 2.5. The figure on the left shows vortex lines at time $t = 0$. The figure on the right shows what can happen to these lines if the vorticity is roughly aligned with the velocity field as in the case of a Beltrami flow.

In this section we build steady solutions for which the vorticity and the velocity are everywhere collinear.

Definition 2.1. *A steady 3D fluid flow is called a* Beltrami flow *if the vorticity* $\omega =$ curl v *satisfies the Beltrami condition*

$$\omega(x) = \lambda(x)v(x) \qquad \text{for some} \quad \lambda(x) \neq 0 \tag{2.38}$$

for all x.

Suppose that the divergence-free velocity field $v(x)$ satisfies Beltrami condition (2.38). Because $\omega =$ curl v, the vorticity satisfies the compatibility condition div $\omega = 0$, and from Eq. (2.38) we get $0 = \text{div}\, \omega = v \cdot \nabla\lambda + \lambda\, \text{div}\, v$. Thus the first-order differential equation for λ,

$$0 = v \cdot \nabla\lambda, \tag{2.39}$$

is the necessary condition for the velocity field to satisfy Beltrami condition (2.38).

When are such velocity fields solutions to the steady 3D Euler equation? Because the 3D steady, inviscid vorticity equation is

$$v \cdot \nabla\omega = \omega \cdot \nabla v,$$

by substituting $\omega = \lambda v$ we get

$$(v \cdot \nabla\lambda)\omega + \lambda v \cdot \nabla v = \lambda v \cdot \nabla v,$$

so the vorticity equation is satisfied automatically; we have just proved Proposition 2.10.

Proposition 2.10. *Any steady, divergence-free velocity field* $v(x)$ *in* $\mathbb{R}^3$ *that satisfies Beltrami condition (2.38),*

$$\text{curl}\, v(x) = \lambda(x)v(x) \qquad \text{for some} \quad \lambda(x) \neq 0,$$

is a solution to the 3D Euler equation.

Now we systematically build and classify some B flows. First we recall (see Proposition 2.2) that a stream function ψ defines a steady solution $v = (-\psi_{x_2}, \psi_{x_1})^t$ to the 2D Euler equation if and only if

$$\Delta\psi = F(\psi) \tag{2.40}$$

for some smooth function F.

Moreover, by Proposition 2.9 we know that any such stream function ψ generates a solution

$$v = [-\psi_{x_2}, \psi_{x_1}, W(\psi)]^t$$

to the 3D Euler equation for any smooth function W.

Now we use these results to generate the B flows. Because the vorticity of the above velocity field is

$$\omega = [W'(\psi)\psi_{x_2}, -W'(\psi)\psi_{x_1}, F(\psi)]^t,$$

the Beltrami condition $\omega = \lambda v$ implies that the Beltrami coefficient is $\lambda(x) = -W'(\psi)$, and the function W solves the ODE $W'(\psi)W(\psi) = -F(\psi)$.

We have just proved Proposition 2.11.

Proposition 2.11. *Let ψ be a stream function of a steady solution to the 2D Euler equation (so that $\Delta\psi = F(\psi)$ for some smooth function F). Then the velocity field*

$$v = [-\psi_{x_2}, \psi_{x_1}, W(\psi)]^t \tag{2.41}$$

is a 3D inviscid B flow if and only if the function W solves the ODE

$$W'(\psi)W(\psi) = -F(\psi), \tag{2.42}$$

and the Beltrami coefficient $\lambda(x)$ is

$$\lambda(x) = -W'(\psi). \tag{2.43}$$

Now we illustrate this proposition by building several B flows with interesting underlying structures.

Example 2.6. Periodic 3D B Flows. We recall that in Example 2.3 we built periodic solutions to the 2D Euler equation by using the eigenfunctions on the torus T^2:

$$\psi(x) = \sum_{|k|=\ell} [a_k \cos(2\pi x \cdot k) + b_k \sin(2\pi x \cdot k)].$$

These stream functions satisfy Eq. (2.40) with linear functions F:

$$F(\psi) = -4\pi^2|k|^2\psi.$$

Solving ODE (2.41) for this F gives

$$W(\psi) = 2\pi|k|\psi,$$

so the Beltrami coefficient $\lambda(x)$ is

$$\lambda = -2\pi|k|. \tag{2.44}$$

The 3D B flow is thus

$$v(x) = (-\psi_{x_2}, \psi_{x_1}, 2\pi|k|\psi)^t. \tag{2.45}$$

Observe that in the above example the Beltrami coefficient $\lambda(x)$ is constant. These flows are called *strong Beltrami flows*.

Now we build an example of the B flow for which the coefficient $\lambda(x)$ is not constant.

Example 2.7. The Cat's-Eye 3D B Flow. We recall that in Example 2.4 we built cat's-eye solutions to the 2D Euler equation that are 2π periodic in the x_1 variable. The stream function of these solutions satisfies

$$\Delta\psi = F(\psi) \equiv -e^{2\psi}.$$

Now solving ODE (2.42) for this function F we get

$$W(\psi) = W'(\psi) = e^{\psi},$$

so the Beltrami coefficient $\lambda(x)$ is

$$\lambda(x) = -e^{\psi(x)}. \tag{2.46}$$

The 3D B flow is

$$v = (-\psi_{x_2}, \psi_{x_1}, e^{\psi})^t. \tag{2.47}$$

Now we observe that, because of the linearity of the Beltrami condition $\omega = \lambda v$, B flows with a fixed coefficient $\lambda(x)$ constitute a linear space of solutions, i.e., multiplication of a B flow by a constant and any linear combination of B flows with the same $\lambda(x)$ is also a B flow. This observation yields Proposition 2.12.

Proposition 2.12. *Let* $\psi_k(x_i, x_j)$, $k \neq i \neq j$, $k = 1, 2, 3$, *be three eigenfunctions on the torus* T^2 *such that*

$$\Delta\psi_k = \lambda_0\psi_k,$$

where Δ is the 2D Laplacian in the (x_i, x_j) variables. Then, because λ_0 satisfies $\lambda_0 < 0$,

$$v = \begin{bmatrix} -(\psi_3)_{x_2} \\ (\psi_3)_{x_1} \\ (-\lambda_0)^{\frac{1}{2}}\psi_3 \end{bmatrix} + \begin{bmatrix} (-\lambda)_0^{\frac{1}{2}}\psi, \\ -(\psi_1)_{x_3} \\ (\psi_1)_{x_2} \end{bmatrix} + \begin{bmatrix} (\psi_2)_{x_3} \\ (-\lambda)_0^{\frac{1}{2}}\psi_2 \\ -(\psi_2)_{x_1} \end{bmatrix} \tag{2.48}$$

is a 3D B flow.

Even for simple eigenfunctions ψ_k this formula leads to B flows with extremely complicated underlying structures.

Example 2.8. The Arnold–Beltrami–Childress Periodic Flows. We take three periodic shear flows determined by the stream functions

$$\begin{aligned} \psi_1(x) &= a \sin x_3, \\ \psi_2(x) &= b \sin x_1, \\ \psi_3(x) &= c \sin x_2. \end{aligned}$$

Because $-\Delta\psi_k = \psi_k$, we have $\lambda_0 = -1$, and by Proposition 2.12 the 3D B flow is given by

$$v(x) = \begin{pmatrix} a \sin x_3 - c \cos x_2 \\ -a \cos x_3 + b \sin x_1 \\ c \sin x_2 - b \cos x_1 \end{pmatrix}.$$

It is easy to observe that $-v(-x)$ is also the B flow, so we can specialize the above procedure to obtain the famous Arnold–Beltrami–Childress (ABC) flow

$$v(x) = \begin{pmatrix} A \sin x_3 + C \cos x_2 \\ B \sin x_1 + A \cos x_3 \\ C \sin x_2 + B \cos x_1 \end{pmatrix}. \tag{2.49}$$

This ABC flow was analyzed by Dombre et al. (1986) who gave it the name A-B-C because this example was independently introduced by Arnold (1965) and Childress (1970) as an interesting class of Beltrami flows. For some values of the parameters A, B, and C, e.g., $A = B = 0$, this flow is very simple because particle trajectories are helical screw lines. For some other values of the parameters A, B, and C, however, these flows are ergodic and particle trajectories are everywhere dense. The last result is a counterexample to some statements in traditional textbooks on fluid mechanics that vortex lines are either closed or they can not end in the fluid. That is, because for the ABC flows we have $\omega = \lambda v$, vortex lines coincide with the particle trajectories and they are also everywhere dense for some values of the parameters A, B, and C.

Finally we classify strong B flows [with the Beltrami coefficient $\lambda(x) = \text{const}$]. Because for such flows the velocity field satisfies

$$\begin{aligned} \operatorname{curl} v &= \lambda v, \\ \operatorname{div} v &= 0, \end{aligned}$$

we look for the complex eigenvectors

$$v_e(x) = e^{i\lambda x\cdot k}\omega_e(k), \qquad |k| = 1.$$

The first equation, curl $v_e = \lambda v_e$, yields

$$ik \times \omega_e(k) = \omega_e(k),$$

and the second equation, div $v_e = 0$, gives

$$k \cdot \omega_e(k) = 0.$$

Given any unit vector $N(k)$ tangent to the unit sphere, $|N(k)| = 1$ and $N(k)\cdot k = 0$, define $\omega_e(k)$ as

$$\omega_e(k) = A(k)[N(k) + ik \times N(k)],$$

where $A(k)$ is an arbitrary function. Then this choice of $\omega_e(k)$ satisfies the above two equations.

Integrating $v_e(x)$ over the unit sphere gives Proposition 2.13.

Proposition 2.13. *General Strong B Flows. Let μ be any complex measure on the unit sphere S^2, and let $N(k)$ be the μ-measurable unit vector tangent to S^2. Then, for any function $A(k)$, the velocity field*

$$v(x) = \overline{\int_{S^2} e^{i\bar{\lambda}x\cdot k}A(k)[N(k) + ik \times N(k)]d\mu} + \int_{S^2} e^{i\bar{\lambda}x\cdot k}A(k)[N(k) + ik \times N(k)]d\mu \tag{2.50}$$

is a strong B flow.

Given a strong B-flow solution v^0 to the 3D Euler equation with constant $\bar{\lambda}$, we can also solve the Navier–Stokes equation with this as initial data. The explicit solution is

$$v^\nu(x,t) = e^{-\bar{\lambda}^2\nu t}v^0(x).$$

2.3.3. 3D Axisymmetric Flows

Let e_r, e_θ, and e_3 be the standard orthonormal unit vectors defining the cylindrical coordinate system,

$$e_r = \left(\frac{x_1}{r}, \frac{x_2}{r}, 0\right)^t, \qquad e_\theta = \left(\frac{x_2}{r}, -\frac{x_1}{r}, 0\right)^t, \qquad e_3 = (0,0,1)^t, \tag{2.51}$$

where $r = (x_1^2 + x_2^2)^{\frac{1}{2}}$. The velocity field v is *axisymmetric* if

$$v = v^r(r, x_3, t)e_r + v^\theta(r, x_3, t)e_\theta + v^3(r, x_3, t)e_3,$$

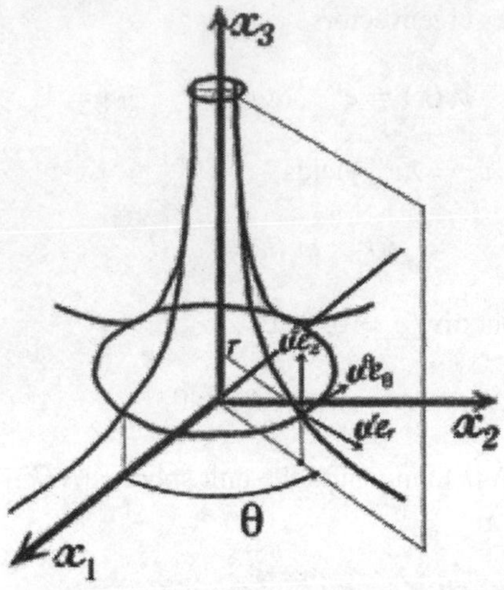

Figure 2.6. Axisymmetric flows without swirl. The velocity field depends on only r and z and moreover has no component in the θ direction.

i.e., if v does not depend on the θ coordinate. These flows are similar to the $2\frac{1}{2}$ D flow from Subsection 2.3.1 because they are the 3D flows that depend on only two coordinates. The velocity v^θ in the e_θ direction is called the *swirl velocity*; see Fig. 2.6. Axisymmetric flows with swirl usually involve interesting interactions of the flow velocities v^r and v^3 in the axial planes with the swirl velocity v^θ.

Axisymmetric Flows without Swirl

Typical examples of such flows are the draining jet in Eq. (1.25) or an idealized smoke ring. If at the time $t = 0$ we start with the velocity

$$v|_{t=0} = v_0^r(r, x_3)e_r + v_0^3(v, x_3)e_3, \tag{2.52}$$

then the solution to the Navier–Stokes equation will remain axisymmetric without a swirl:

$$v = v^r(r, x_3, t)e_r + v^3(r, x_3, t)e_3. \tag{2.53}$$

Because $v^\theta \equiv 0$, the vorticity ω is tangent to e_θ:

$$\omega(r, x_3, t) = \left(v_{x_3}^r - v_r^3\right)e_\theta \equiv \omega^\theta e_\theta. \tag{2.54}$$

Now we derive an equation for the scalar vorticity ω^θ for the inviscid case. We use the following general principle that will also be useful in our discussion in subsequent chapters.

Lemma 2.1. *Given a (not necessarily divergence-free) velocity field $v(x, t)$ in $\mathbb{R}^N$, let $h(x, t)$ be a vector field satisfying*

$$\frac{Dh}{Dt} = h \cdot \nabla v, \tag{2.55}$$

and let $f(x, t)$ be a scalar function solving

$$\frac{Df}{Dt} = 0. \tag{2.56}$$

Then

$$\frac{D}{Dt}(\nabla f \cdot h) = 0. \tag{2.57}$$

The proof follows immediately by the chain rule, and we leave it to the reader as an exercise.

Because ω solves the vorticity equation $(D\omega/Dt) = \omega \cdot \nabla v$, and $(D\theta/Dt) = (v^\theta/r) \equiv 0$, the choices $h = \omega$ and $f = \theta$ satisfy the assumptions of Lemma 2.1. Computing $\nabla\theta \cdot \omega = (\omega^\theta/r) \equiv \xi$ gives the *vortex-dynamics equation*

$$\frac{\widetilde{D}\xi}{Dt} = 0, \tag{2.58}$$

where the material derivative $\widetilde{D}/Dt$ acts in only the axial planes:

$$\frac{\widetilde{D}}{Dt} = \frac{\partial}{\partial t} + v^r\frac{\partial}{\partial r} + v^3\frac{\partial}{\partial x_3}. \tag{2.59}$$

Thus for axisymmetric, inviscid flows without swirl, the quantity ξ is conserved along particle trajectories. This fact, corresponding to the conservation of vorticity in 2D inviscid flows, is used again in Chap. 4, Section 4.3, to show the existence of such flows globally in time.

Now we formulate the vorticity-stream formulation for axisymmetric, inviscid flows without swirl. For velocity fields satisfying the special form of Eq. (2.53), the incompressibility condition $\operatorname{div} v = 0$ in cylindrical coordinates is

$$(rv^r)_r + (rv^3)_{x_3} = 0.$$

Using the axisymmetric stream function $\psi(r, x_3, t)$, we find that the velocities v^r and v^3,

$$v^r = \frac{\psi_{x_3}}{r}, \qquad v^3 = -\frac{\psi_r}{r}, \tag{2.60}$$

satisfy the incompressibility condition. Because $\omega^\theta = v^r_{x_3} - v^3_r$, a short calculation yields the equation

$$L\psi = \xi, \tag{2.61}$$

where L is the linear elliptic operator:

$$L = \frac{1}{r}\frac{\partial}{\partial r}\left(\frac{1}{r}\frac{\partial}{\partial r}\right) + \frac{1}{r^2}\frac{\partial^2}{\partial x_3^2}. \tag{2.62}$$

Solving this equation for ψ in terms of ξ, computing the velocities v^r and v^3 by Eq. (2.61), and substituting them into Eq. (2.58) gives the vorticity equation for ξ alone. We summarize this in Proposition 2.14.

Proposition 2.14. *Let $v(v, x_3, t)$ be an axisymmetric, inviscid flow in Eq. (2.53) without a swirl,*

$$v = v^r e_r + v^3 e_3,$$

with the corresponding vorticity curl $v = \omega^\theta e_\theta$. *Denote $\xi = (\omega/r)$. For such flows the 3D Euler equation is equivalent to the* simplified vorticity-stream formulation,

$$\frac{\widetilde{D}}{Dt}\xi = 0,$$

$$L\psi = \xi,$$

$$v^r = \frac{\psi_{x_3}}{r}, \qquad v^3 = -\frac{\psi_r}{r},$$

where the material derivative $\widetilde{D}/Dt$ acts in only the axial planes,

$$\frac{\widetilde{D}}{Dt} = \frac{\partial}{\partial t} + v^r\frac{\partial}{\partial r} + v^3\frac{\partial}{\partial x_3},$$

and L is the linear elliptic operator:

$$L = \frac{1}{r}\frac{\partial}{\partial r}\left(\frac{1}{r}\frac{\partial}{\partial r}\right) + \frac{1}{r^2}\frac{\partial^2}{\partial x_3^2}.$$

A trivial example of a solution to the above vorticity-stream formulation is furnished by the draining jet in Eq. (1.25),

$$v = \left(-\frac{\gamma}{2}x_1, -\frac{\gamma}{2}x_2, \gamma x_3\right)^t \equiv -\frac{\gamma}{2}re_r + \gamma x_3 e_3,$$

yielding $\xi = 0$ and $\psi = -\frac{\gamma}{2}r^2 x_3$.

In vorticity equation (2.58) there is no interaction between the velocities v^r, v^3 in the axial planes and the swirl velocity $v^\theta \equiv 0$, so that $\xi = (\omega^\theta/r)$ is conserved along particle trajectories.

Much more complicated behavior of the vorticity ω^θ, which is due to the interactions of v^r, v^3, and $v^\theta \neq 0$, occurs for the case of axisymmetric flow with swirl.

Axisymmetric Flows with a General Swirl

We show that the equations for vortex dynamics for such flows are nontrivial but somewhat simpler than the vorticity equations in the general 3D case.

Consider an axisymmetric flow of the form

$$v = v^r(r, x_3, t)e_r + v^\theta(r, x_3, t)e_\theta + v^3(r, x_3, t)e_3. \tag{2.63}$$

The corresponding vorticity $\omega = \text{curl}\, v$ of such flows is

$$\omega = -v^{\theta}_{x_3} e_r + \omega^{\theta} e_{\theta} + \frac{1}{r}(rv^{\theta})_r e_3, \tag{2.64}$$

where $\omega^{\theta} = v^r_{x_3} - v^3_r$. Using the fact that $\partial e_r/\partial\theta = e_{\theta}$ and $\partial e_{\theta}/\partial r = 0$, we find that for axisymmetric flows the e_{θ} component of the Euler equation and the vorticity equations in cylindrical coordinates are

$$\frac{\widetilde{D}}{Dt} v^{\theta} = -\frac{v^r v^{\theta}}{r}, \tag{2.65}$$

$$\frac{\widetilde{D}}{Dt} \omega^{\theta} = \frac{1}{r}(\omega^{\theta} v^r - v^{\theta}\omega^r) + \omega^r v^{\theta}_r + \omega^3 v^{\theta}_{x_3}, \tag{2.66}$$

respectively. The material derivative $\widetilde{D}/Dt$ in Eq. (2.59) acts in only the axial planes:

$$\frac{\widetilde{D}}{Dt} = \frac{\partial}{\partial t} + v^r \frac{\partial}{\partial r} + v^3 \frac{\partial}{\partial x_3}.$$

First observe that Eq. (2.65) yields

$$\frac{\widetilde{D}}{Dt}(rv^{\theta}) = 0. \tag{2.67}$$

Recalling definition (1.59) of the circulation of a velocity field v around the material curve $C(t)$ moving with the fluid,

$$\Gamma_{C(t)} = \oint_{C(t)} v \cdot d\ell,$$

we see that Eq. (2.67) implies conservation of circulation on material circles centered on the axis of symmetry.

With the definition of vorticity in Eq. (2.64), Eq. (2.66) implies that

$$\begin{aligned}
\frac{\widetilde{D}}{Dt}\left(\frac{\omega^{\theta}}{r}\right) &= -\frac{1}{r^2}\omega^r v^{\theta} + \frac{1}{r}\omega^r v^{\theta}_r + \frac{1}{r}\omega^3 v^{\theta}_{x_3} \\
&= \frac{1}{r^2} v^{\theta}_{x_3} v^{\theta} - \frac{1}{r} v^{\theta}_{x_3} v^{\theta}_r + \frac{1}{r}\left(\frac{1}{r} v^{\theta} + v^{\theta}_r\right) v^{\theta}_{x_3},
\end{aligned}$$

so

$$\frac{\widetilde{D}}{Dt}\left(\frac{\omega^{\theta}}{r}\right) = -\frac{1}{r^4}[(rv^{\theta})^2]_{x_3}. \tag{2.68}$$

Thus for axisymmetric, inviscid flows with swirl $v^{\theta} \neq 0$, the quantity ω^{θ}/r is not conserved along particle trajectories, and it changes in response to the swirl velocity v^{θ}. Because the material derivative $\widetilde{D}/Dt$ acts in only the axial planes, this stretching of vorticity ω^{θ} is due to the interaction of the axial-plane velocities v^r and v^3 with the swirl velocity v^{θ}.

Equations (2.67) and (2.68) are simplified equations for general inviscid vortex dynamics. In fact, through the introduction of the axisymmetric stream function

$\psi(r, x_3, t)$ that defines the velocities v^r and v^3 by Eqs. (2.60),

$$v^r = \frac{\psi_{x_3}}{r}, \qquad v^3 = -\frac{\psi_r}{r},$$

these equations can be used to yield a reformulation of the 3D Euler equation for such flows. Namely, we recall that the velocity field in Eq. (2.63) with the velocities v^r and v^3 defined as above is automatically divergence free. The stream function ψ is defined from the vorticity ω^θ by Eq. (2.61),

$$L\psi = \frac{\omega^\theta}{r},$$

where L is the linear elliptic operator:

$$L = \frac{1}{r}\frac{\partial}{\partial r}\left(\frac{1}{r}\frac{\partial}{\partial r}\right) + \frac{1}{r^2}\frac{\partial^2}{\partial x_3^2}.$$

Thus by solving this equation for ψ in terms of the vorticity ω^θ, then computing the velocities by Eq. (2.61) and substituting them into Eqs. (2.67) and (2.68), we get the vorticity equations for ω^θ and v^θ alone – this is the simplified vorticity-stream formulation for axisymmetric, inviscid flows with swirl.

We easily generalize the above arguments to the case of viscous flows by including the viscosity term in Eqs. (2.65) and (2.66). We leave the details to the reader and summarize the results in the following proposition.

Proposition 2.15. *Let $v(r, x_3, t)$ be an axisymmetric, viscous flow in Eq. (2.63) with a swirl*

$$v = v^r e_r + v^\theta e_\theta + v^3 e_3$$

and with the corresponding vorticity

$$\omega = -v^\theta_{x_3} e_r + \omega^\theta e_\theta + \frac{1}{r}(r v^\theta)_r e_3.$$

Defining

$$\xi = \frac{\omega^\theta}{r}, \qquad \Gamma = r v^\theta$$

for such flows, we find that the 3D Navier–Stokes equation is equivalent to the simplified vorticity-stream formulation,

$$\begin{aligned}
\frac{\widetilde{D}}{Dt}\xi &= -\frac{1}{r^4}(\Gamma^2)_{x_3} + \nu\left(\xi_{rr} + \frac{3}{r}\xi_r + \xi_{x_3x_3}\right),\\
\frac{\widetilde{D}}{Dt}\Gamma &= \nu\left(\Gamma_{rr} - \frac{\Gamma_r}{r} + \Gamma_{x_3x_3}\right),\\
L\psi &= \xi,\\
v^r &= \frac{\psi_{x_3}}{r}, \qquad v^3 = -\frac{\psi_r}{r},
\end{aligned}$$

where the material derivative $\widetilde{D}/Dt$ acts in only the axial plane,

$$\frac{\widetilde{D}}{Dt} = \frac{\partial}{\partial r} + v^r \frac{\partial}{\partial r} + v^3 \frac{\partial}{\partial x_3},$$

and L is the elliptic linear operator:

$$L = \frac{1}{r}\frac{\partial}{\partial r}\left(\frac{1}{r}\frac{\partial}{\partial r}\right) + \frac{1}{r^2}\frac{\partial^2}{\partial x_3^2}.$$

For axisymmetric inviscid flows without swirl ($\nu = 0, v^\theta = 0$), the above results reduce to the case in Proposition 2.14.

A simple example of an axisymmetric inviscid flow with swirl is the rotating jet – see Example 1.4; namely, by superimposing the draining jet in Eq. (1.25),

$$v = \left(-\frac{\gamma}{2}x_1, -\frac{\gamma}{2}x_2, \gamma x_3\right)^t \equiv -\frac{\gamma}{2} r e_r + \gamma x_3 e_3,$$

with the vortex solution in Eq. (1.27),

$$v = \left[-\frac{1}{2}\omega(t)x_2, \frac{1}{2}\omega(t)x_1, 0\right]^t \equiv \frac{1}{2}\omega(t) r e_\theta,$$

where $\omega(t)|_{t=0} = \omega_0 =$ const, we produced an axisymmetric flow with swirl velocity $v^\theta = \frac{1}{2}\omega(t)r$ and vorticity given by

$$\omega(t) = \omega_0 e^{\gamma t} e_3.$$

Below we produce a more realistic model of the draining jet flow through a more general exact solution to the vorticity-stream formulation in Proposition 2.15. We see that viscous effects can balance the growth of vorticity and lead to steady solutions, as occurred earlier in the strained shear layers in Chap. 1.

Example 2.9. The Viscous, Rotating Eddy. We superimpose the draining jet solution in Eq. (1.25),

$$v = \left(-\frac{\gamma}{2}x_1, -\frac{\gamma}{2}x_2, \gamma x_3\right)^t \equiv -\frac{\gamma}{2} r e_r = \gamma x_3 e_3,$$

with the radial eddy solution in Eq. (2.16),

$$v = \left(-\frac{x_2}{r^2}, \frac{x_1}{r^2}, 0\right)^t \int_0^r s\omega(s,t)ds \equiv \frac{1}{r}\int_0^r s\omega(s,t)ds\, e_\theta, \tag{2.69}$$

and look for an axisymmetric flow with swirl:

$$v = -\frac{\gamma}{2} r e_r + \frac{1}{r}\int_0^r s\omega(s,t)ds\, e_\theta + \gamma x_3 e_3.$$

The corresponding vorticity $\omega = \text{curl}\, v$ is

$$\omega = \omega(r,t)e_3.$$

Such flows satisfy $\xi = (\omega^\theta/r) \equiv 0$. Denoting $\Gamma = rv^\theta \equiv \int_0^r s\omega(s,t)ds$, we find that the vorticity-stream formulation in Proposition 2.15 reduces to a scalar equation for Γ,

$$\Gamma_t - \frac{\gamma}{2} r\Gamma_r = \nu\left(\Gamma_{rr} - \frac{\Gamma_r}{r}\right),$$

and the stream function is $\psi = -\frac{\gamma}{2} r^2 x_3$.

To solve this equation it is convenient to rewrite it in terms of the vorticity $\omega(r,t)$; because $\Gamma_r = r\omega$, after simple calculations we get

$$\omega_t - \frac{\gamma}{2} r\omega_r = \gamma\omega + \nu\left(\omega_{rr} + \frac{\omega_r}{r}\right).$$

Moreover, because ω is radially symmetric, we have $\Delta_2\omega = \omega_{rr} + (\omega_r/r)$, $(\gamma/2)r\omega_r = (\gamma/2)x \cdot \nabla\omega$, so we can rewrite this equation as a linear diffusion equation in $\mathbb{R}^2$;

$$\omega_t - \frac{\gamma}{2} x \cdot \nabla\omega = \gamma\omega + \nu\Delta_2\omega. \tag{2.70}$$

The solution ω is given explicitly by Lemma 1.17 from the appendix of Chap. 1 as

$$\omega(r,t) = e^{\gamma t}\int_{\mathbb{R}^2} H\left[xe^{\frac{\gamma}{2}t} - y, \frac{\nu}{\gamma}(e^{\gamma t} - 1)\right]\omega_0(|y|)dy, \tag{2.71}$$

where H is the 2D Gaussian kernel.

Equation (2.70) has competing effects of compression from the convection term $-\frac{\gamma}{2} x \cdot \nabla\omega$, the vorticity-stretching term $\gamma\omega$, and the diffusion term $\nu\Delta_2\omega$; we expect that they can balance, giving a steady solution. By applying the dominated convergence theorem to solution (2.71), we can show that solution (2.71) converges in the limit as $t \nearrow \infty$ to the steady solution

$$\lim_{t\to\infty} \omega(r,t) = \omega_\infty(r) = \left(\frac{\gamma}{2\nu}\right) e^{-\frac{\gamma r^2}{4\nu}} \int_0^\infty s\omega_0(s)ds. \tag{2.72}$$

This solution, called the *Burger vortex*, has a swirl velocity v^θ given by formula (2.69), and

$$\lim_{t\to\infty} v^\theta(r,t) = \frac{1 - e^{-\frac{\gamma r^2}{4\nu}}}{r} \int_0^\infty s\omega_0(s)ds. \tag{2.73}$$

Thus the swirl velocity v^θ increases until diffusion causes it to saturate.

Next we build more general explicit solutions with nontrivial swirl velocities by combining the vorticity-stream formulation for swirling flows with a simple principle used in Subsection 2.3.1.

Example 2.10. First we seek axisymmetric viscous flows with the special form

$$v(r, x_3, t) = -\frac{\gamma(t)}{2} r e_r + V^\theta(r, t) e_\theta + \gamma(t) x_3 e_3, \tag{2.74}$$

where $\gamma(t)$ is a given time-dependent strain rate. The corresponding vorticity $\omega =$ curl v is

$$\omega(r, t) = \left(\frac{1}{r} V^\theta + V_r^\theta\right) e_3. \tag{2.75}$$

These flows generalize Example 2.9: The equation for $\Gamma = r V^\theta$ is exactly that from Example 2.9 but has a time-dependent $\gamma(t)$. Using Lemma 1.17 from the appendix to Chap. 1, we find that the swirl velocity V^θ is computed from Eq. (2.70) as

$$V^\theta(r, t) = \frac{\lambda(t)}{r} \int_0^r s \int_{\mathbb{R}^2} H\left[x\lambda^{\frac{1}{2}}(t) - y, \nu \int_0^t \lambda(\tau) d\tau\right] \omega(|y|) dy|_{|x|=s} ds, \tag{2.76}$$

where $\lambda(t) = \exp \int_0^t \gamma(\tau) d\tau$ and $\omega_0(r)$ is any smooth function.

Next we generalize the flows in Eq. (2.74) by utilizing a variant of the idea in Subsection 2.3.1. The exact solution in Eq. (2.74) has a 2D velocity component $\hat{v}$ that solves the 2D Navier–Stokes equation:

$$\hat{v} = (v_1, v_2)^t \equiv -\frac{\gamma(t)}{2} r e_r + V^\theta(r, t) e_\theta. \tag{2.77}$$

The e_3 component of this solution is trivial, $\gamma(t) x_3 e_3$. We seek solutions to the 3D Navier–Stokes equation with the form

$$v = (\hat{v}, 0)^t + (0, 0, \gamma(t) x_3)^t + V^3(x_1, x_2, t)^t. \tag{2.78}$$

We can easily show that the velocity field in Eq. (2.78) is an exact solution to the 3D Navier–Stokes equation, provided that V^3 satisfies the linear convection–diffusion equation

$$\begin{aligned} \frac{\widehat{D}}{Dt} V^3 &= -\gamma(t) V^3 + \nu \Delta_2 V^3, \\ V^3|_{t=0} &= V_0^3. \end{aligned} \tag{2.79}$$

In particular, if we choose V_0^3 to be a function of r alone, Eq. (2.79) has a radially symmetric solution $V^3(r, t)$ that satisfies

$$\begin{aligned} V_t^3 - \frac{\gamma(t)}{2} x \cdot \nabla V^3 &= -\gamma(t) V^3 + \nu \Delta_2 V^3, \\ V^3|_{t=0} &= V_0^3(r). \end{aligned} \tag{2.80}$$

Lemma 1.17 from the appendix to Chap. 1 gives the solution

$$V^3(r, t) = \frac{1}{\lambda(t)} \int_{\mathbb{R}^2} H\left[x\lambda^{\frac{1}{2}}(t) - y, \nu \int_0^t \lambda(\tau) d\tau\right] V^3(|y|)|_{|x|=s} ds, \tag{2.81}$$

where, as before, $\lambda(t) = \exp \int_0^t \gamma(\tau) d\tau$.

To summarize, we have constructed general explicit swirling flows that satisfy the 3D Navier–Stokes equations of the form

$$v(r,t) = -\frac{\gamma(t)}{2} r e_r + V^\theta(r,t) e_\theta + [V^3(r,t) + \gamma(t)x_3] e_3,$$

where $V^\theta(r,t)$ is determined from the initial data by Eq. (2.76) and V^3 is determined from the initial data by Eq. (2.81). The vorticity corresponding to these solutions is

$$\omega(r,t) = -V_r^3 e_r + \left(\frac{1}{r} V^\theta + V_r^\theta\right) e_3. \tag{2.82}$$

We leave it as an exercise for the reader to interpret these solutions by using the vorticity-stream formulation in Proposition 2.15.

Finally, we present the inviscid case of these exact solutions. For $\nu = 0$ the equations for Γ and V^3 are

$$\begin{aligned} \Gamma_t - \frac{\gamma(t)}{2} r \Gamma_r &= 0, \\ V_t^3 - \frac{\gamma(t)}{2} r V_r^3 &= -\gamma(t) V^3. \end{aligned} \tag{2.83}$$

The method of characteristics gives the swirl and the axial velocities as

$$\begin{aligned} V^\theta(r,t) &= \lambda^{\frac{1}{2}}(t) V_0^\theta\left[r\lambda^{\frac{1}{2}}(t)\right], \\ V^3(r,t) &= \lambda^{-1}(t) V_0^3\left[r\lambda^{\frac{1}{2}}(t)\right]. \end{aligned} \tag{2.84}$$

The corresponding vorticities are $\omega^\theta = -V_r^3$ and $\omega^3 = (1/r)V^\theta + V_r^\theta$.

2.4. The Vorticity-Stream Formulation for 3D Flows

We recall that in Section 2.1 we determined the 2D divergence-free velocity field $v = (v^1, v^2)^t$ in terms of its scalar vorticity $\omega = v_{x_1}^2 - v_{x_2}^1$ by the Biot–Savart law [Eq. (2.10)]:

$$v(x,t) = \int_{\mathbb{R}^2} K_2(x-y)\omega(y,t)dy, \qquad x \in \mathbb{R}^2. \tag{2.85}$$

The kernel K_2,

$$K_2(x) = \frac{1}{2\pi}\left(-\frac{x_2}{|x|^2}, \frac{x_1}{|x|^2}\right)^t, \qquad x \in \mathbb{R}^2, \tag{2.86}$$

is homogeneous of degree -1 (in $\mathbb{R}^2$). This derivation was based on the fact that, to solve the equations

$$\begin{cases} v_{x_1}^2 - v_{x_2}^1 = \omega, \\ \operatorname{div} v = 0 \end{cases}, \qquad x \in \mathbb{R}^2 \tag{2.87}$$

for the velocity v, we reduced them to the Poisson equation for the stream function ψ,

$$\Delta\psi = \omega, \qquad x \in \mathbb{R}^2, \tag{2.88}$$

and then we recovered the velocity by $v = (-\psi_{x_2}, \psi_{x_1})^t$. Substituting this velocity into the scalar vorticity equation,

$$\frac{D\omega}{Dt} = \nu\Delta\omega, \tag{2.89}$$

yields an evolution equation for the velocity ω alone.

Now we derive an analogous vorticity-stream formulation for 3D Navier–Stokes flows by determining the 3D divergence-free velocity field v in terms of its vorticity $\omega = \text{curl } v$ that then can be used to solve the 3D vorticity equation:

$$\frac{D\omega}{Dt} = \omega \cdot \nabla v + \nu\Delta\omega.$$

We show below that the evolution equation for the vorticity ω alone obtained by this procedure is equivalent to the Navier–Stokes equation. Moreover, this reformulation shows explicitly how velocity strain stretches vorticity (by means of the term $\omega \cdot \nabla v$). For 3D flows we need to compute both the velocity and the velocity gradients in terms of the vorticity by means of the potential theory. This is much more technical than for the 2D case, and we begin with a useful improvement of the Hodge decomposition discussed at the end of Chap. 1.

2.4.1. Hodge Decomposition, Potential Theory, and the Biot–Savart Law

To determine the velocity v in terms of the vorticity ω, we solve the overdetermined (four equations for the three unknown functions v^1, v^2, v^3) elliptic system

$$\begin{cases} \text{curl } v = \omega, \\ \text{div } v = 0, \end{cases} \qquad x \in \mathbb{R}^3, \tag{2.90}$$

The solution to this problem is summarized in the following proposition.

Proposition 2.16. *Hodge Decomposition. Let $\omega \in L^2$ be a smooth vector field in $\mathbb{R}^3$, vanishing sufficiently rapidly as $|x| \nearrow \infty$. Then*

(i) Eqs. (2.90) have a smooth solution v that vanishes as $|x| \nearrow \infty$ if and only if

$$\text{div } \omega = 0; \tag{2.91}$$

(ii) if $\text{div } \omega = 0$, *then the solution v is determined constructively by*

$$v = -\text{curl}\psi, \tag{2.92}$$

where the vector-stream function ψ solves the Poisson equation

$$\Delta\psi = \omega. \tag{2.93}$$

The explicit formula for v is

$$v(x) = \int_{\mathbb{R}^3} K_3(x-y)\omega(y)dy, \tag{2.94}$$

where the 3×3 matrix kernel K_3 is

$$K_3(x)h = \frac{1}{4\pi}\frac{x \times h}{|x|^3}, \qquad h \in \mathbb{R}^3. \tag{2.95}$$

Again, borrowing notation from electromagnetic theory, we refer to Eqs. (2.94) and (2.95) as the Biot–Savart law.

Proof of Proposition 2.16. First, recall the following two vector identities: for smooth vector fields ψ in $\mathbb{R}^3$,

$$\text{div curl } \psi = 0, \tag{2.96}$$

$$-\text{curl curl } \psi + \nabla \text{ div } \psi = \Delta\psi. \tag{2.97}$$

If v solves Eq. (2.90), then curl $v = \omega$ and, by Eq. (2.96), necessarily div $\omega = 0$. Now we prove the opposite; if div $\omega = 0$, then there exists a vector field v that solves Eqs. (2.90), curl $v = \omega$ and div $v = 0$. We consider Poisson equation (2.93):

$$\Delta\psi = \omega.$$

Recall from Lemma 1.12 that a solution results from convolution with the Newtonian potential

$$\psi(x) = \frac{-1}{4\pi}\int_{\mathbb{R}^3} \frac{\omega(y)}{|x-y|}dy.$$

Define $h = -\text{curl curl } \psi$ and $k = \nabla \text{ div } \psi$. Because $\omega \in L^2$ is smooth and vanishes rapidly as $|x| \nearrow \infty$, we have $h(x) = \mathcal{O}(|x|^{-3})$ and $k(x) = \mathcal{O}(|x|^{-3})$ for $|x| \gg 1$, so $h, k \in L^2$. Taking the L^2 inner product $(\cdot, \cdot)_0$ of identity (2.97) with k yields

$$(k, k)_0 = (\omega, k)_0 - (h, k)_0.$$

By the assumption div $\omega = 0$, div $h = 0$ and k is the gradient of div ψ. Thus by using the orthogonality condition in Lemma 1.5 we get $(\omega, k)_0 = (h, k)_0 = 0$. Thus $(k, k)_0 = 0$, implying that $k \equiv 0$, so identity (2.97) gives

$$\omega = \text{curl}(-\text{curl } \psi).$$

Now define the vector field v by Eq. (2.92), $v = -\text{curl } \psi$: By identity (2.96) this vector is divergence free. It is easy to see that the explicit form of v is given when the curl of the convolution is computed with the Newtonian potential, i.e.,

$$v(x) = \frac{1}{4\pi}\int_{\mathbb{R}^3} \frac{x-y}{|x-y|^3} \times \omega(y)dy;$$

this is the formula in Eqs. (2.94) and (2.95). This vector field v is not in L^2, in general, but it still vanishes as $|x| \nearrow \infty$, provided that ω vanishes sufficiently rapidly as $|x| \nearrow \infty$. □

Observe that in both the 2D and the 3D cases in Eqs. (2.85) and (2.94), the velocity v is recovered from the vorticity ω by nonlocal operators given by the convolution with the kernel K_N:

$$v(x) = \int_{\mathbb{R}^N} K_N(x-y)\omega(y)dy,$$

where K_N is homogeneous of degree $1-N$, i.e., $K_N(\lambda x) = \lambda^{1-N} K_N(x)$, $x \neq 0$. Because we want to eliminate v from the 3D vorticity equation, next we need to compute the gradient of v. The reader may wonder if we can do it by a straightforward differentiation under the integral,

$$\nabla v(x) = \int_{\mathbb{R}^N} \nabla K_N(x-y)\omega(y)dy.$$

Observe that ∇K_N is homogeneous of degree $-N$, so its singularity is not integrable in $\mathbb{R}^N$, and therefore the computation of ∇v requires the use of the distribution derivatives. In fact, as we shall see below, the identity stated above is not always correct.

Next we derive some auxiliary results concerning the differentiation of singular functions and then apply these results to compute ∇v.

2.4.2. *Distribution Derivatives of Functions Homogeneous of Degree* $1-N$ *and Singular Integrals*

We derive some general results for the distribution derivatives of functions K that are smooth away from $x = 0$ and homogeneous of degree $1-N$:

$$K(\lambda x) = \lambda^{1-N} K(x) \qquad \text{for all} \quad \lambda > 0,\ x \in \mathbb{R}^N. \tag{2.98}$$

In Sec. 2.4.3 we apply these results to the specific kernels K_N given in Eqs. (2.87) and (2.95).

Definition 2.2. *The distribution derivative* ∂_{x_j} *of* f *is the linear functional* $\partial_{x_j} f$ *defined operationally by the formula*

$$(\partial_{x_j} f, \varphi)_0 = -(f, \partial_{x_j}\varphi)_0 \qquad \text{for all} \quad \varphi \in C_0^\infty(\mathbb{R}^N). \tag{2.99}$$

For C^1 functions f, this distribution derivative coincides with the classical derivative (see Folland, 1995, pp. 25–33, or John, 1982, pp. 89–92, for a brief discussion of distributions).

Now we compute the distribution derivative of K. Because $K \in L^1_{\mathrm{loc}}(\mathbb{R}^N)$, for any $\varphi \in C_0^\infty$ by the dominated convergence theorem and the Green's theorem:

$$\begin{aligned}(K, \partial_{x_j}\varphi)_0 &= \lim_{\epsilon \searrow 0} \int_{|x| \geq \epsilon} K \partial_{x_j}\varphi \, dx \\ &= \lim_{\epsilon \searrow 0} \left\{ -\int_{|x|\geq\epsilon} \partial_{x_j} K\varphi \, dx + \int_{|x|=\epsilon} K\varphi \frac{x_j}{|x|} ds \right\}.\end{aligned}$$

The first term on the right-hand side of this equation is called the *Cauchy principal-value integral*,

$$\mathrm{PV} \int_{\mathbb{R}^N} f \, dx = \lim_{\epsilon \searrow 0} \int_{|x|\geq\epsilon} f \, dx. \tag{2.100}$$

Changing variables by $x \to \epsilon\ x$ and using the homogeneity of Eq. (2.98) for K turns the second integral into

$$\begin{aligned}\lim_{\epsilon \searrow 0} \int_{|x|=\epsilon} K\varphi \frac{x_j}{|x|} ds &= \lim_{\epsilon \searrow 0} \int_{|x|=1} \epsilon^{1-N} K(x)\varphi(\epsilon x) x_j \epsilon^{N-1} ds \\ &= \varphi(0) \int_{|x|=1} K(x) x_j \, ds.\end{aligned}$$

We have just proved Proposition 2.17.

Proposition 2.17. *Let K be a function homogeneous of degree $1{-}N$ and smooth outside $x = 0$. Then the distribution derivative ∂_{x_j} of K is the linear functional $\partial_{x_j} K$ defined by*

$$(\partial_{x_j} K, \varphi)_0 = -(K, \partial_{x_j}\varphi)_0 = \mathrm{PV} \int_{\mathbb{R}^N} \partial_{x_j} K\varphi \, dx - c_j(\delta, \varphi)_0, \quad \text{for all} \quad \varphi \in C_0^\infty, \tag{2.101}$$

where δ is the Dirac distribution, $(\delta, \varphi)_0 = \varphi(0)$, and

$$c_j = \int_{|x|=1} K(x) x_j \, ds. \tag{2.102}$$

The functions $\partial_{x_j} K$ are homogeneous of degree $-N$. In the next proposition we list some of their important properties that play a crucial role in further mathematical developments in this book (see Chaps. 3 and 4).

Proposition 2.18. *Let K be a function smooth outside $x = 0$ and homogeneous of degree $1 - N$. Then $\partial_{x_j} K$ (homogeneous of degree $-N$) always has mean-value zero on the unit sphere:*

$$\int_{|x|=1} \partial_{x_j} K \, ds = 0. \tag{2.103}$$

In fact, Eq. (2.103) has the following generalization. For any multi-index α and β such that $|\alpha| = |\beta| + 1$, the function $x^\beta D^\alpha K$ is homogeneous of degree $-N$ and has mean-value zero on the unit sphere:

$$\int_{|x|=1} x^\beta D^\alpha K\, ds = 0. \tag{2.104}$$

Proof of Proposition 2.18. Pick a cutoff function $\rho \in C_0^\infty(\mathbb{R})$; $\rho(r) \geq 0$ and $\rho(r) \equiv 0$ for $r \leq A$ and $r \geq B > A > 0$. Then

$$\int_0^\infty \rho'(r)dr = 0, \qquad \int_0^\infty \rho(r)\frac{dr}{r} = c > 0.$$

By the definition of ρ and the Green's theorem,

$$0 = \int_{\mathbb{R}^N} \partial_{x_j}[\rho(|x|)K(x)]dx = \int_{\mathbb{R}^N} \rho'(|x|)\frac{x_j}{|x|}K(x)dx + \int_{\mathbb{R}^N} \rho(|x|)\partial_{x_j}K(x)dx.$$

The first integral on the right-hand side always vanishes because the conversion to polar coordinates and the homogeneity of K gives

$$\int_{\mathbb{R}^N} \rho'(|x|)\frac{x_j}{|x|}K(x)dx = \int_0^\infty \rho'(r)dr \int_{|x|=1} x_j K(x)ds = 0.$$

For the second integral, use polar coordinates to rewrite it as

$$\int_{\mathbb{R}^N} \rho(|x|)\partial_{x_j}K(x)dx = \int_0^\infty \frac{\rho(r)}{r}dr \int_{|x|=1} \partial_{x_j}K(x)ds = c\int_{|x|=1} \partial_{x_j}K(x)ds,$$

so finally, because $c > 0$, Eq. (2.103) holds:

$$\int_{|x|=1} \partial_{x_j}K(x)ds = 0.$$

The proof of Eq. (2.104) follows by the same arguments, and we leave it as a simple exercise for the reader. □

Finally, we observe that the gradient of K is the Kernel $P = \partial_{x_j}K$, which is smooth outside $x = 0$ and homogeneous of degree $-N$,

$$P(\lambda x) = \lambda^{-N}P(x) \qquad \text{for} \quad \lambda > 0,\ x \in \mathbb{R}^N,$$

with mean-value zero on the unit sphere:

$$\int_{|x|=1} P(x)ds = 0.$$

Any function P that is homogeneous of degree $-N$ and has mean-value zero defines a *singular integral operator* (SIO) through the convolution

$$Pf(x) = \mathrm{PV}\int_{\mathbb{R}^N} P(x-y)f(y)dy \equiv \lim_{\epsilon \searrow 0}\int_{|x-y|\geq\epsilon} P(x-y)f(y)dy.$$

Below we show that the computation of the velocity gradients ∇v from the vorticity leads to SIOs. In Chaps. 3 and 4 we derive estimates for these SIOs in various function spaces – the zero mean-value property of P will play a crucial role in the mathematical theory presented there.

Next we give applications of the above formulas to the 3D vorticity-stream formulation.

2.4.3. Computation of the Velocity Gradients from the Vorticity by Means of Singular Integral Operators

As the first application of formula (2.101) we give a direct proof of the following proposition.

Proposition 2.19. *The 2D and the 3D velocity fields in Eqs. (2.85) and (2.94), respectively,*

$$v(x,t) = \int_{\mathbb{R}^N} K_N(x-y)\omega(y,t)dy, \qquad x \in \mathbb{R}^N,$$

are divergence free, $\operatorname{div} v = 0$.

Proof of Proposition 2.19. For $N = 2$ the kernel K_2 is

$$K_2(x) = \frac{1}{2\pi}\left(-\frac{x_2}{|x|}, \frac{x_1}{|x|}\right)^t,$$

so by using Proposition 2.17 we compute

$$v^1_{x_1}(x,t) = \int_{\mathbb{R}^2} \partial_{x_1} K^1_2(x-y)\omega(y,t)dy - \omega(x)\int_{|x|=1} x_1x_2\, ds,$$

$$v^2_{x_2}(x,t) = \mathrm{PV}\int_{\mathbb{R}^2} \partial_{x_2} K^2_2(x-y)\omega(y,t)dy + \omega(x)\int_{|x|=1} x_1x_2\, ds.$$

Now observe that $\int_{|x|=1} x_1x_2\,dx \equiv 0$, and for $x \neq y$, $(\partial_{x_1} K^1_2 + \partial_{x_2} K^2_2)(x-y) \equiv 0$. Thus by the definition of a principal-value integral we have

$$\operatorname{div} v(x) = \lim_{\epsilon \searrow 0} \int_{|x-y|\geq\epsilon} (\partial_{x_1} K^1_2 + \partial_{x_2} K^2_2)(x-y)\omega(y,t)dy \equiv 0. \qquad \square$$

The proof of the 3D case we leave to the reader as an exercise.

As the second application of Proposition 2.17 we compute the explicit form of the velocity gradient ∇v in $\mathbb{R}^3$.

Proposition 2.20. *Let $v(x,t)$ be the 3D velocity field in Eq. (2.94):*

$$v(x,t) = \int_{\mathbb{R}^3} K_3(x-y)\omega(y,t)dy.$$

Then the gradient of v is

$$[\nabla v(x)]h = -\mathrm{PV}\int_{\mathbb{R}^3}\left(\frac{1}{4\pi}\frac{\omega(y)\times h}{|x-y|^3} + \frac{3}{4\pi}\frac{\{[(x-y)\times\omega(y)]\otimes(x-y)\}h}{|x-y|^5}\right)dy$$
$$+\frac{1}{3}\omega(x)\times h, \tag{2.105}$$

where the tensor product $\otimes$ is defined by $z\otimes w = (z_i w_j)$. In the case of a 2D velocity field, ∇v has the simpler form

$$\nabla v(x) = \frac{1}{2\pi}\mathrm{PV}\int\frac{\sigma(x-y)}{|x-y|^2}w(y)dy + \frac{\omega(x)}{2}\begin{bmatrix}0 & -1\\ 1 & 0\end{bmatrix},$$

where

$$\sigma(z) = \frac{1}{|z|^2}\begin{bmatrix}2z_1z_2 & z_2^2 - z_1^2\\ z_2^2 - z_1^2 & -2z_1z_2\end{bmatrix}. \tag{2.106}$$

Proof of Proposition 2.20. For $N = 3$ the kernel K_3 is

$$K_3(x)h = \frac{1}{4\pi}\frac{x\times h}{|x|^3},$$

so, using Proposition 2.17, we compute

$$[\nabla v(x)]h = \mathrm{PV}\int_{\mathbb{R}^3}\nabla_x\left[\frac{1}{4\pi}\frac{(x-y)\times\omega(y)}{|x-y|^3}\right]h\,dy - \frac{1}{4\pi}\int_{|y|=1}[y\times\omega(x)]y\cdot h\,ds.$$

Computing the gradient ∇_x under the integral and using vector identities, we get the first two principal-value integrals in Eq. (2.105). Moreover, in $\mathbb{R}^3$,

$$\int_{|y|=1} y_i y_j\,ds = \begin{cases}\frac{4\pi}{3}, & i = k\\ 0, & i \neq j\end{cases},$$

so we get

$$\frac{1}{4\pi}\int_{|y|=1}[y\times\omega(y)]y\cdot h\,ds = -\frac{1}{3}\omega(x)\times h. \qquad \square$$

From Chap. 1, Eq. (1.32), we know that if $\omega = \mathrm{curl}\ v$ solves the 3D vorticity equation, then the vorticity-stretching term $\omega\cdot\nabla v$ is equal to $\mathcal{D}\omega$, where $\mathcal{D}$ is the deformation matrix of v, $\mathcal{D} = \frac{1}{2}(\nabla v + \nabla v^t)$. We observe that the contribution from $\omega\times h$ to $\mathcal{D}h$ is zero, so by taking the symmetric part of matrix formula (2.105) we get the following corollary.

Corollary 2.1. *Let ∇v be the gradient of the 3D velocity v, as defined in Eq. (2.105), and $\omega = \mathrm{curl}\ v$. Then the deformation matrix of v, $\mathcal{D} = \frac{1}{2}(\nabla v + \nabla v^t)$, is the SIO*

$$\mathcal{D}(x) = \mathrm{PV}\int_{\mathbb{R}^3} P(x-y)\omega(y)dy, \tag{2.107}$$

where the matrix kernel P acts on a vector h by

$$P(x)h = -\frac{3}{8\pi}\frac{(x \times h) \otimes x + x \otimes (x \times h)}{|x|^5}. \tag{2.108}$$

2.4.4. The Vorticity-Stream Formulation in All of $\mathbb{R}^3$

Substituting the velocity v from Eq. (2.94) and the velocity gradient ∇v from Eq. (2.105) into the 3D vorticity equation $(D\omega/Dt) = \omega \cdot \nabla v + \nu\Delta\omega$, we get a self-contained evolution equation for the vorticity ω alone. We prove Proposition 2.21.

Proposition 2.21. *The Vorticity-Stream Formulation with ∇v in All of $\mathbb{R}^3$. For the 3D smooth flows that vanish sufficiently rapidly as $|x| \nearrow \infty$, Navier–Stokes equations (2.1)–(2.3) are equivalent to the following self-contained evolution equation for the vorticity alone*

$$\begin{aligned}
&\frac{D\omega}{Dt} = \omega \cdot \nabla v + \nu\Delta\omega,\\
&(x, t) \in \mathbb{R}^3 \times (0, \infty),\\
&\omega|_{t=0} = \omega_0 \equiv \text{curl } v_0,
\end{aligned} \tag{2.109}$$

where the velocity v is determined from the vorticity ω by the Biot–Savart law in Eq. (2.94),

$$v(x, t) = \int_{\mathbb{R}^3} K_3(x - y)\omega(y, t)dy, \qquad x \in \mathbb{R}^3,$$

with the kernel

$$K_3(x)h = \frac{1}{4\pi}\frac{x \times h}{|x|^3},$$

and the velocity gradient ∇v is computed from ω by Eq. (2.105):

$$\begin{aligned}
\nabla v(x)h = -\text{PV}\int_{\mathbb{R}^3}\left\{\frac{1}{4\pi}\frac{\omega(y) \times h}{|x - y|^3} + \frac{3}{4\pi}\frac{\{[(x - y) \times \omega(y)] \otimes (x - y)\}}{|x - y|^5}h\right\}dy\\
+\frac{1}{3}\omega(x) \times h.
\end{aligned}$$

We give the proof of this proposition at the end of this section.

Because $\omega = \text{curl } v$, the stretching term $\omega \cdot \nabla v$ in the vorticity equation is equal to $\mathcal{D}\omega$, where $\mathcal{D} = \frac{1}{2}(\nabla v + \nabla v^t)$ is the deformation matrix. Thus we can substitute formula (2.94) for the velocity v and the simpler formula for the deformation matrix $\mathcal{D}$ as determined by the vorticity in Eq. (2.107) into the 3D vorticity equation $(D\omega/Dt) = \mathcal{D}\omega + \nu\Delta\omega$ to get a simpler self-contained equation for the vorticity ω alone. We have the following proposition.

Proposition 2.22. *The Simplified Vorticity-Stream Formulation with $\mathcal{D}$ in All of $\mathbb{R}^3$. Let $v(x,t)$ be a smooth velocity field vanishing sufficiently rapidly as $|x| \nearrow \infty$. Assume that $v(x,t)$ satisfies the 3D Navier–Stokes equations (2.1)–(2.3). Then the vorticity $\omega = \text{curl } v$ is a solution to the integrodifferential equation*

$$\frac{D\omega}{Dt} = \mathcal{D}\omega + \nu\Delta\omega,$$
$$(x,t) \in \mathbb{R}^3 \times (0,\infty),$$
$$\omega|_{t=0} = \omega_0 \equiv \text{curl } v_0, \tag{2.110}$$

where the velocity v is determined from the vorticity ω by Eq. (2.94),

$$v(x,t) = \int_{\mathbb{R}^3} K_3(x-y)\omega(y,t)dy, \qquad x \in \mathbb{R}^3,$$

and the deformation matrix $\mathcal{D}$ is determined by Eqs. (2.107) and (2.108):

$$\mathcal{D}(x,t) = \text{PV} \int_{\mathbb{R}^3} P(x-y)\omega(y,t)dy$$

with

$$P(x)h = -\frac{3}{8\pi}[(x \times h) \otimes x + x \times (x \times h)]/|x|^5.$$

We emphasize that the above proposition *does not* state that the 3D Navier–Stokes equation is equivalent to the vorticity-stream formulation in Eq. (2.110). The reason is that the formula in Eqs. (2.107) and (2.108) for the deformation matrix is valid provided $\omega = \text{curl } v$, a fact we must prove to show that Eq. (2.110) is equivalent to the Navier–Stokes equation – see the proof of Proposition 2.21 below.

Many of the exact solutions studied in Chap. 1 illustrate the fact that the vorticity is amplified if it roughly aligns with the eigenvectors corresponding to the positive eigenvalues of the deformation matrix $\mathcal{D}$. Thus in order to explain vorticity amplification for arbitrary flows, we must answer the following difficult question: When does the deformation matrix $\mathcal{D}$, defined from the vorticity ω by the SIO in Eqs. (2.107) and (2.108), have eigenvectors with the positive eigenvalues roughly aligned in the direction ω? This is a new type of question regarding matrix singular integrals – a simple model problem of this type is discussed in Chap. 5.

Finally we give the proof of Proposition 2.21. In this proof we use the following general fact.

Lemma 2.2. *Let v be a smooth, divergence-free vector field in $\mathbb{R}^N$, vanishing sufficiently rapidly as $|x| \nearrow \infty$. Let the vector field $h(x,t)$ solve the convection–diffusion equation*

$$\frac{Dh}{Dt} = h \cdot \nabla v + \nu\Delta h. \tag{2.111}$$

Then div h *satisfies the scalar equation*

$$\frac{D}{Dt}\text{div}\, h = \nu\Delta\ \text{div}\, h. \tag{2.112}$$

We give the proof of this lemma at the end of this section. Now we give the proof of Proposition 2.21.

Proof of Proposition 2.21. If v solves the Navier–Stokes equation, then by the derivation of v in Eq. (2.94) and ∇v in Eq. (2.105) the vorticity $\omega = \text{curl}\ v$ solves vorticity equation (2.109).

Now we prove the opposite. Let ω solve vorticity equation (2.109), $(D\omega/Dt) = \omega \cdot \nabla v + \nu\Delta\omega$, with v defined by Eq. (2.94) and ∇v by Eq. (2.105)

By Proposition 2.19 we know that div $v = 0$ (regardless of whether div $\omega = 0$), so applying Lemma 2.2 yields that the divergence of vorticity, div ω, solves the scalar convection–diffusion equation

$$\frac{D}{Dt}\ \text{div}\ \omega = \nu\Delta\ \text{div}\ \omega,$$
$$\text{div}\ \omega|_{t=0} = 0.$$

Thus by the uniqueness of smooth solutions to this scalar parabolic equation (see Lemma 1.13 in the appendix to Chap. 1), we get div $\omega \equiv 0$ for all $t \geq 0$.

Because div $\omega = 0$, by the Hodge decomposition formula in Proposition 2.16 there exists the velocity field v given by Eq. (2.94),

$$v(x, t) = \int_{\mathbb{R}^3} K_3(x - y)\omega(y, t)dy,$$

such that curl $v = \omega$ (and div $v = 0$). It remains only to prove that this velocity field v solves the Navier–Stokes equation. Because $\omega = \text{curl}\ v$, it is easy to see that the vorticity equation

$$\frac{D}{Dt}\omega = \omega \cdot \nabla v + \nu\Delta\omega$$

is identical to the equation

$$\text{curl}\left(\frac{Dv}{Dt} - \nu\Delta v\right) = 0.$$

The last equation implies that $(Dv/Dt) - \nu\Delta v$ is the gradient of some function, say $-p$, i.e., that v solves the Navier–Stokes equation

$$\frac{Dv}{Dt} = -\nabla p + \nu\Delta v. \qquad \square$$

Finally we give the proof of Lemma 2.2.

Proof of Lemma 2.2. Differentiating Eq. (2.111) component-wise gives

$$h^i_{x_i t} + \sum_j v^j_{x_i} h^i_{x_j} + \sum_j v^j h^i_{x_i x_j} = \sum_j h^j_{x_i} v^i_{x_j} + \sum_j h^j v^i_{x_i x_j} + \nu \Delta h^i_{x_i}.$$

Because div $v = 0$, summing these equations over i we get

$$\operatorname{div} h_t + v \cdot \nabla \operatorname{div} h = \nu \Delta \operatorname{div} h + \sum_{i,j} h^j_{x_i} v^i_{x_j} - \sum_{i,j} v^j_{x_i} h^i_{x_j},$$

yielding

$$\frac{D}{Dt} \operatorname{div} h = \nu \Delta \operatorname{div} h. \qquad \square$$

2.5. Formulation of the Euler Equation as an Integrodifferential Equation for the Particle Trajectories

Proposition 2.21 showed that the Euler equation,

$$\frac{\partial v}{\partial t} + v \cdot \nabla v = -\nabla p,$$
$$\operatorname{div} v = 0,$$
$$v|_{t=0} = v_0(x).$$

is equivalent to the *vorticity-stream form* of Eq. (2.109), an evolution equation for the vorticity, $\omega = \operatorname{curl} v$:

$$\frac{D\omega}{Dt} = \omega \cdot \nabla v,$$
$$\omega|_{t=0} = \omega_0. \tag{2.113}$$

The divergence-free velocity v in Eq. (2.113) is recovered from ω by the Biot–Savart law:

$$v(x, t) = \int_{\mathbb{R}^N} K_N(x - y)\omega(y, t)dy, \qquad x \in \mathbb{R}^N. \tag{2.114}$$

The kernels K_N are homogeneous of degree $1 - N$:

$$K_2(x) = \frac{1}{2\pi} {}^t\left(-\frac{x_2}{|x|^2}, \frac{x_1}{|x|^2}\right), \qquad x \in \mathbb{R}^2,$$

$$K_3(x)h = \frac{1}{4\pi} \frac{x \times h}{|x|^3}, \qquad x, h \in \mathbb{R}^3.$$

We now derive a third reformulation of the equations as an integrodifferential equation for the particle trajectories. Recall from Eq. (1.13) that for a given smooth velocity field $v(x, t)$ the particle trajectories $X(\alpha, t)$ satisfy

$$\frac{dX}{dt}(\alpha, t) = v(X(\alpha, t), t), \qquad X(\alpha, t)|_{t=0} = \alpha. \tag{2.115}$$

The particle-trajectory mapping $X(\cdot,t):\mathbb{R}^N\to\mathbb{R}^N;\alpha\mapsto X(\alpha,t)=x\in\mathbb{R}^N$ is $1-1$ and onto, with the inverse $X^{-1}(\cdot,t)$. Incompressibility, div $v=0$, implies $\det(\nabla_\alpha X(\alpha,t))=1$. The time-dependent map $X(\cdot,t)$ connects the Lagrangian reference frame (with variable α) to the Eulerian reference frame (with variable x).

Recall from Chap. 1 that Eq. (2.113) implies the vorticity-transport formula (from Proposition 1.8)

$$\omega(X(\alpha,t),t)=\nabla_\alpha X(\alpha,t)\omega_0(\alpha), \tag{2.116}$$

which says that vorticity is stretched by $\nabla_\alpha X(\alpha,t)$ along particle trajectories. Changing variables $x'=X(\alpha',t)$ and using Eq. (2.116), we rewrite Eq. (2.114) as an integral in the Lagrangian frame:

$$v(x,t)=\int_{\mathbb{R}^3}K_3[x-X(\alpha',t)]\nabla_\alpha X(\alpha',t)\omega_0(\alpha')d\alpha'. \tag{2.117}$$

Substituting this velocity into Eq. (2.115), we get the 3D integrodifferential equation for the particle trajectories $X(\alpha,t)$:

$$\begin{aligned}\frac{dX}{dt}(\alpha,t)&=\int_{\mathbb{R}^3}K_3[X(\alpha,t)-X(\alpha',t)]\nabla_\alpha X(\alpha',t)\omega_0(\alpha')d\alpha',\\ X(\alpha,t)|_{t=0}&=\alpha.x\end{aligned} \tag{2.118}$$

In the case of 2D inviscid flows, vorticity cannot amplify because it is conserved on particle trajectories. The solution in Eq. (2.116) reduces to

$$\omega(X(\alpha,t),t)=\omega_0(\alpha),$$

and hence yields the simplified velocity

$$v(x,t)=\int_{\mathbb{R}^2}K_2[x-X(\alpha',t)]\omega_0(\alpha')d\alpha'.$$

The 2D integrodifferential equation for the particle trajectories $X(\alpha,t)$ is

$$\begin{aligned}\frac{dX}{dt}(\alpha,t)&=\int_{\mathbb{R}^2}K_2[X(\alpha,t)-X(\alpha',t)]\omega_0(\alpha')d\alpha',\\ X(\alpha,t)|_{t=0}&=\alpha.\end{aligned} \tag{2.119}$$

In Chap. 4 we show rigorously that the right-hand side of Eq. (2.118) is a bounded nonlinear operator on $X(\alpha,t)$. The smoothing properties of the kernel K_3 compensate for the loss of regularity in differentiating $X(\alpha,t)$ in the term $\nabla_\alpha X$. Also, Eq. (2.118) involves the explicit term $\nabla_\alpha X(\alpha,t)\omega_0(\alpha)$ that represents the stretching of vorticity. In Section 4.2 we show a direct link between the accumulation of vorticity and the existence of smooth solutions globally in time. As a consequence we obtain global existence of solutions in two dimensions because vorticity is conserved along particle trajectories. In Chap. 6 we show that a discretization of the particle-trajectory formulation leads to *computational vortex methods* that have been very successful in accurately computing inviscid Euler solutions. Part of their success comes from the fact that the particle-trajectory reformulation contains only bounded operators.

A solution $X(\alpha, t)$ to Eq. (2.118) or Eq. (2.119) defines a velocity v either from Eq. (2.115) or Eq. (2.117) and a vorticity ω from Eq. (2.116). To complete this section, we show that for sufficiently smooth solutions, the particle-trajectory formulation is equivalent to the Euler equation. We state the 3D version. The 2D version follows in a similar way.

Proposition 2.23. *Let $v_0(x)$ be a smooth velocity field satisfying* div $v_0 = 0$, *and* $\omega_0 =$ curl v_0. *Let $X(\alpha, t)$ solve*

$$\frac{dX}{dt}(\alpha, t) = \int_{\mathbb{R}^3} K_3[X(\alpha, t) - X(\alpha', t)]\nabla_\alpha X(\alpha', t)\omega_0(\alpha')d\alpha',$$
$$X(\alpha, t)|_{t=0} = \alpha. \tag{2.120}$$

Define the velocity field v by

$$v(x, t) = \int_{\mathbb{R}^3} K_3[x - X(\alpha', t)]\nabla_\alpha X(\alpha', t)\omega_0(\alpha')d\alpha'. \tag{2.121}$$

Then integrodifferential equation (2.120) for the particle trajectories is equivalent to the 3D Euler equation for sufficiently smooth solutions with a rapidly decreasing vorticity.

Proof of Proposition 2.23. The derivation of particle-trajectory equation (2.120) shows that if a smooth velocity v solves the Euler equation, then the particle trajectories X defined by Eq. (2.115), $(dX/dt)(\alpha, t) = v(X(\alpha, t), t)$, solve integrodifferential equation (2.120).

Now we prove the converse. In Proposition 2.21 we show that the 3D Euler equation is equivalent to the vorticity stream formulation; hence it suffices to show that a solution to Eq. (2.120) yields a $v(x, t)$ and an $\omega(x, t)$ such that

$$\text{curl } v = \omega, \qquad \text{div } v = 0,$$
$$\frac{D\omega}{Dt} = \omega \cdot \nabla v, \qquad \omega|_{t=0} = \omega_0.$$

Let $X(\alpha, t)$ be a smooth solution to integrodifferential equation (2.120). $X(\cdot, t)$ is $1 - 1$ and onto and has an inverse $X^{-1}(\cdot, t)$. Define the function $\omega(x, t)$ by the formula

$$\omega(x, t) = \nabla_\alpha X(\alpha, t)\omega_0(\alpha)|_{\alpha = X^{-1}(x,t)}.$$

By Lemma 1.4 this function solves the equation

$$\frac{D\omega}{Dt} = \omega \cdot \nabla v,$$

where $v = K_3 * \omega$. Proposition 2.21 implies that the velocity $v(x, t)$ in Eq. (2.117) is divergence free, div $v = 0$. To complete the proof we show that $\omega =$ curl v. We

Table 2.1. *New notation*

Notation	Name	Type
D/Dt	Material derivative	Differential operator
$\nabla^{\perp}$	Perpendicular grad	2D differential operator
$\mathrm{PV}\int$	Principle-value singular integral	Integral operator

apply Lemma 2.2; if a vector field h solves $(Dh/Dt) = h \cdot \nabla v$ and $\operatorname{div} v = 0$, then $(D/Dt)\operatorname{div} h = 0$. Setting $h = \omega$ gives

$$\frac{D}{Dt}\operatorname{div} \omega = 0.$$

Moreover, the initial condition satisfies $\omega_0 = \operatorname{curl} v_0$ and $\operatorname{div} v_0 = 0$, so we necessarily have $\operatorname{div} \omega_0 = 0$. By the uniqueness of solutions for ODEs, we conclude that $\operatorname{div} \omega = 0$ for all $t \geq 0$.

We show that $\operatorname{div} \omega = 0$ implies $\omega = \operatorname{curl} v$. Because $\operatorname{div} v = 0$, $\det \nabla_\alpha X = 1$ and we rewrite v as in Eq. (2.114):

$$v(x, t) = \int_{\mathbb{R}^3} K_3(x - x')\omega(x', t)dx'.$$

Applying the Hodge decomposition (Proposition 2.16) to this velocity field yields $\operatorname{div} \omega = 0$ if and only if $\omega = \operatorname{curl} v$. □

Remark: The arguments of the proof require merely that v, X, and ω be differentiable in space and time. In particular, the arguments work for the class of solutions we consider in Chap. 4. This equivalence allows us to prove local-in-time existence and uniqueness of solutions to the Euler equations by proving that the result is true for the particle trajectory equations. One might wonder if the Proposition remains true for *weak* solutions with less regularity. In Chaps. 8 and 9 we introduce two notions of weak solutions that explicitly show that in fact the two formulations are not equivalent without sufficient regularity of the solutions.

Notes for Chapter 2

Table 2.1 lists the notation used in this chapter.

References for Chapter 2

Arnold, V. I., "Sur 1a topologic des écoulements stationnaires des fluides parfaits," *C. R. Acad. Sci. Paris* **261**, 17–20, 1965.

Bertozzi, A. L., "Heteroclinic orbits and chaotic dynamics in planar fluid flows," *SIAM J. Math. Anal.* **19**, 1271–1294, 1970.

Dombre, T., Frisch, U., Greene, J., Henon, M., Mehr, A., and Soward, A. J., "Chaotic streamlines in the ABC flows," *J. Fluid Mech.* **167**, 353–391, 1986.

Folland, G. B., *Introduction to Partial Differential Equations*, Princeton Univ. Press, Princeton, NJ, 1995.

John, F., *Partial Differential Equations*, Springer-Verlag, New York, 1982.

Lorrain, P. and Corson, D. R., *Electromagnetic Fields and Waves*, Freeman, San Francisco, 1970.

Stuart, J. T., "Stability problems in fluids," in *Mathematical Problems in the Geophysical Sciences*, Vol. 13 of Lectures in Applied Mathematics Series, American Mathematical Society, Providence, RI, 1971, pp. 139–155.

3

Energy Methods for the Euler and the Navier–Stokes Equations

In the first two chapters of this book we described many properties of the Euler and the Navier–Stokes equations, including some exact solutions. A natural question to ask is the following: Given a general smooth initial velocity field $v(x, 0)$, does there exist a solution to either the Euler or the Navier–Stokes equation on some time interval $[0, T)$? Can the solution be continued for all time? Is it unique? If the solution has a finite-time singularity, so that it cannot be continued smoothly past some critical time, in what way does the solution becomes singular? This chapter and Chap. 4 introduce two different methods for proving existence and uniqueness theory for smooth solutions to the Euler and the Navier–Stokes equations. In this chapter we introduce classical *energy methods* to study both the Euler *and* the Navier–Stokes equations. The starting point for these methods is the physical fact that the kinetic energy of a solution of the homogeneous Navier–Stokes equations decreases in time in the absence of external forcing. The next chapter introduces a particle method for proving existence and uniqueness of solutions to the inviscid Euler equation. As is true for all partial differential evolution equations, the challenge in proving that the evolution is well posed lies in understanding the effect of the unbounded spatial differential operators. The particle method exploits the fact that, without viscosity, the vorticity is transported (and stretched in three dimensions) along particle paths. By reformulating the Euler equation as an integrodifferential equation, we obtain a nonlinear but bounded operator on a Banach space. Chapter 4 then uses the abstract Picard theorem (Theorem 4.1) on a Banach space to prove local existence and uniqueness of smooth solutions. However, the machinery of Chapter 4 does not directly handle the diffusion in the Navier–Stokes equation.

A class of techniques applicable to both the Euler and the Navier–Stokes equations is that of *energy methods*. In lieu of reformulating the equations to remove unbounded operators, energy methods use an approximation scheme combined with a priori estimates for various "energies" associated with the equations. Other techniques for proving existence of solutions to partial differential equations (PDEs) include the method of finite differences, Galerkin methods, and fixed-point methods. For example, John (1982, pp. 172–181) discusses in detail the classical method of finite differences. For a concise proof of existence of solutions to the Euler and the Navier–Stokes equations by use of a fixed-point method, see Kato (1972).

We discuss some elementary features of energy methods in Section 3.1, including the fact that solutions to the Euler equation with initial velocity v_0 are good approximations to solutions of the Navier–Stokes equation with the same initial velocity v_0 and small viscosity ν. In Section 3.2 we use the energy method to prove local existence and uniqueness of solutions to both the Euler and the Navier–Stokes equation. In Section 3.3 we show that continuation of the solution depends on the boundedness of the magnitude of the vorticity. Last, in Section 3.4 we discuss *viscous-splitting algorithms* used to numerically compute solutions of the Navier–Stokes equation. We use energy methods to show that such algorithms converge and to estimate their accuracy.

Energy methods derive their name from the fact that they involve a priori estimates on Sobolev norms of quantities like the velocity v. For example, the *kinetic energy* of a fluid is simply $\int_{\mathbb{R}^N} |v|^2 dx$ or the square of the L^2 norm of the velocity. We discuss Sobolev spaces in detail in the beginning of Section 3.2. In that section we also introduce the technique of regularization by mollifiers, in which bounded operators replace the unbounded operators ∇ and $\nu\Delta$ of the Navier–Stokes equation. In Section 3.2 we show that specific a priori bounds resulting from a special choice of regularization allow us to prove the existence of a solution of the original equation. The proof involves passing to a limit in a subsequence and hence does not readily give uniqueness of the solution. To prove uniqueness we need a separate a priori bound on the difference of two solutions with the same initial data. Fortunately, such a result is simple for the Navier–Stokes and the Euler equations, and appears as a corollary to the basic energy estimate in Section 3.1.

3.1. Energy Methods: Elementary Concepts

The Euler and the Navier–Stokes equations possess a natural physical energy, $E = \frac{1}{2}\int_{\mathbb{R}^N} |v|^2 dx$, the kinetic energy of the fluid. In Section 1.7 of Chap. 1, we showed that, for inviscid flows, this energy is conserved† whereas, for viscous flows, this energy dissipates:‡

$$\frac{dE}{dt}(t) = -\nu \int_{\mathbb{R}^N} |\nabla v|^2 dx < 0. \tag{3.1}$$

Note that the case $\nu = 0$ reduces to energy conservation for inviscid flows. Equation (3.1) yields an a priori bound for the L^2 norm of the velocity. In Section 3.2 we derive analogous a priori bounds for higher derivatives of v to study existence, uniqueness, and properties of solutions to the Navier–Stokes and the Euler equations.

3.1.1. A Basic Energy Estimate

Our focus in this chapter is to use energy methods to study existence, uniqueness, and continuation of smooth solutions of the Euler and the Navier–Stokes equations.

† See Eq. (1.66) in Proposition 1.12.
‡ Proposition 1.13.

Energy methods are important tools in nonlinear PDEs, and this chapter serves as a good example of its use in studying complex nonlinear problems.

All energy methods involve multiplying the differential equations by a functional of the solution and integrating by parts. To familiarize the reader with the technique and to lay some groundwork, we begin by deriving a basic energy estimate for smooth solutions that has a number of consequences, including uniqueness of solutions and approximation of Navier–Stokes solutions by Euler solutions. Let v_1 and v_2 be two smooth solutions to the Euler or the Navier–Stokes equation with respective external forces F_1 and F_2:

$$\begin{aligned} \frac{Dv_i}{Dt} &= -\nabla p_i + \nu \Delta v_i + F_i, \\ \operatorname{div} v_i &= 0 \\ v_i|_{t=0} &= v_{0_i}. \end{aligned} \tag{3.2}$$

Assume that these solutions exist on a common time interval $[0, T]$ and that they vanish sufficiently rapidly as $|x| \nearrow \infty$ so that, in particular, $v_i \in L^2(\mathbb{R}^N)$. Denote $\tilde{v} = v_1 - v_2$, $\tilde{p} = p_1 - p_2$, and $\tilde{F} = F_1 - F_2$, and take the difference of Eqs. (3.2) to obtain

$$\tilde{v}_t + v_1 \cdot \nabla \tilde{v} + \tilde{v} \cdot \nabla v_2 = -\nabla \tilde{p} + \nu \Delta \tilde{v} + \tilde{F}.$$

We now proceed in a similar fashion to the arguments in Section 1.8 on the Leray formulation of the Navier–Stokes equation and the Hodge decomposition theorem. Take the L^2 inner product $(\cdot, \cdot)$ of this equation with $\tilde{v}$ to get

$$(\tilde{v}_t, \tilde{v}) + (v_1 \cdot \nabla \tilde{v}, \tilde{v}) + (\tilde{v} \cdot \nabla v_2, \tilde{v}) = -(\nabla \tilde{p}, \tilde{v}) + \nu(\Delta \tilde{v}, \tilde{v}) + (\tilde{F}, \tilde{v}).$$

Integration by parts, by use of the divergence theorem,§ gives

$$\int \Delta \tilde{v}_i \tilde{v}_i = \int_{\mathbb{R}^N} (\nabla \cdot \nabla v_i) \tilde{v}_i = -\int_{\mathbb{R}^N} |\nabla \tilde{v}_i|^2,$$

where the boundary terms vanish because we assume that v decays sufficiently rapidly at infinity. Likewise

$$\int_{\mathbb{R}^N} (v_1 \cdot \nabla \tilde{v}_i) \tilde{v}_i = \int_{\mathbb{R}^N} v_1 \cdot \nabla \left(\frac{1}{2} \tilde{v}_i^2 \right) = \int_{\mathbb{R}^N} \operatorname{div} \left(\frac{1}{2} v_1 \tilde{v}_i^2 \right) = 0.$$

We can also integrate by parts on the pressure term to obtain $-(\nabla \tilde{p}, \tilde{v}) = (\tilde{p}, \operatorname{div} \tilde{v}) = 0$ because $\tilde{v}$ is divergence free. The result is the *basic energy identity*:

$$(\tilde{v}_t, \tilde{v}) + \nu(\nabla \tilde{v}, \nabla \tilde{v}) = -(\tilde{v} \cdot \nabla v_2, \tilde{v}) + (\tilde{F}, \tilde{v}).$$

§ From Folland (1995): If $\Omega \in \mathbb{R}^N$ is a bounded domain with C^1 boundary $S = \partial\Omega$, $\eta(y)$ is the unit normal to S, and v is a C^1 vector field on $\bar{\Omega}$, then

$$\int_S v(y) \cdot \eta(y) d\sigma(y) = \int_\Omega \operatorname{div} v(x) dx.$$

In particular, if v is $C^1(\mathbb{R}^N)$ and decays fast enough at infinity, then $\int_{\mathbb{R}^N} \operatorname{div} v(x) dx = 0$.

Denote by $\|\cdot\|_0$ the L^2 norm on $\mathbb{R}^N$,

$$\|f\|_0 = \left(\int_{\mathbb{R}^N} |f|^2 dx\right)^{1/2}. \tag{3.3}$$

This norm satisfies a *Schwarz inequality* (see, e.g., Rudin, 1987, p. 63),

$$\left|\int_{\mathbb{R}^N} fg\right| \le \|f\|_0 \|g\|_0, \tag{3.4}$$

which gives

$$\frac{1}{2}\frac{d}{dt}\|\tilde{v}(\cdot,t)\|_0^2 + \nu\|\nabla\tilde{v}(\cdot,t)\|_0^2 \le |\nabla v_2(\cdot,t)|_{L^\infty}\|\tilde{v}(\cdot,t)\|_0^2 + \|\tilde{F}(\cdot,t)\|_0\|\tilde{v}(\cdot,t)\|_0. \tag{3.5}$$

Here $|\cdot|_{L^\infty}$ denotes the L^∞ norm (see Rudin, 1987, pp. 65–66) that for continuous functions is equivalent to the supremum norm $|f|_{C^0} = \sup_{x\in\mathbb{R}^N}|f(x)|$. Because $\nu \ge 0$, we can drop the term $\nu\|\nabla\tilde{v}\|_0^2$ so that

$$\frac{d}{dt}\|\tilde{v}\|_0 \le |\nabla v_2|_{L^\infty}\|\tilde{v}\|_0 + \|\tilde{F}\|_0. \tag{3.6}$$

We now apply the well-known Grönwall's lemma (Coddington and Levinson, 1955, p. 37).

Lemma 3.1. *Grönwall's Lemma. If u, q, and $c \ge 0$ are continuous on $[0, t]$, c is differentiable, and*

$$q(t) \le c(t) + \int_0^t u(s)q(s)ds,$$

then

$$q(t) \le c(0)\exp\int_0^t u(s)ds + \int_0^t c'(s)\left[\exp\int_s^t u(\tau)d\tau\right]ds.$$

Applying this lemma to relation (3.6) gives Proposition 3.1.

Proposition 3.1. *Basic Energy Estimate. Let v_1 and v_2 be two smooth solutions to the Navier–Stokes equation with external forces F_i and the same viscosity $\nu \ge 0$. Suppose that these solutions exist on a common time interval $[0, T]$, and, for fixed time, decay fast enough at infinity to belong to $L^2(\mathbb{R}^N)$. Then*

$$\sup_{0\le t\le T}\|v_1 - v_2\|_0 \le \left[\|(v_1 - v_2)|_{t=0}\|_0 + \int_0^T \|F_1 - F_2\|_0 dt\right]\exp\left(\int_0^T |\nabla v_2|_{L^\infty}dt\right). \tag{3.7}$$

As an immediate consequence we obtain uniqueness of smooth solutions. Note that the right-hand side of this inequality does not depend explicitly on the viscosity.

Corollary 3.1. *Uniqueness of Solutions. Let v_1 and v_2 be two smooth L^2 solutions to Eqs. (3.2) on $[0, T]$ with the same initial data and forcing F. Then $v_1 = v_2$.*

We observe that energy estimate (3.7) does not depend explicitly on the viscosity ν.

The basic energy estimate also implies a control of gradients of viscous solutions. Namely, integrating relation (3.5) in time gives

$$\begin{aligned}\frac{1}{2}\|\tilde{v}(\cdot, T)\|_0^2 - \frac{1}{2}\|\tilde{v}(\cdot, 0)\|_0^2 + \nu \int_0^T \|\nabla\tilde{v}(\cdot, t)\|_0^2 dt \\ \leq \sup_{0\leq t\leq T} \|\tilde{v}(\cdot, t)\|_0^2 \int_0^T |\nabla v_2(\cdot, t)|_{L^\infty} dt + \sup_{0\leq t\leq T} \|\tilde{v}(\cdot, t)\|_0 \int_0^T \|\tilde{F}(\cdot, t)\|_0 dt.\end{aligned} \tag{3.8}$$

Applying estimate (3.7) to the right-hand side of relation (3.8) implies an estimate of the form

$$\begin{aligned}\nu \int_0^T \|\nabla(v_1 - v_2)(\cdot, t)\|_0^2 dt \\ \leq c(v_2, T)\left(\|(v_1 - v_2)(\cdot, 0)\|_0^2 + \left(\int_0^T \|(F_1 - F_2)(\cdot, t)\|_0 dt\right)^2\right)\end{aligned} \tag{3.9}$$

where the constant $c(v_2, T)$ depends on $\int_0^T |\nabla v_2|_{L^\infty} dt$.

3.1.2. Approximation of Inviscid Flows by High Reynold's Number Viscous Flows

Chapters 1 and 2 contain several examples in which a zero-viscosity solution v^0 to the Euler equation approximates well a solution v^ν to the Navier–Stokes equation with small viscosity ν. Such closeness of solutions is in fact true in general for smooth solutions. However, this is a rather subtle issue when the Euler equation does not have a smooth solution. See, for example, Chaps. 9 and 11 for a discussion of vortex sheets. Provided that all of the solutions exist, the basic energy estimate shows that for fixed initial data, smooth Navier–Stokes solutions v^ν converge to a solution v^0 (also with this same initial data) of the Euler equation as the viscosity $\nu \searrow 0$. That is, we can regard v^0 as a solution to the Navier–Stokes equation with the external force $-\nu\Delta v^0$. Note that when ν is small and v^0 is smooth, this is a small force. We can bound its contribution to the evolution by applying basic energy estimate (3.7) to $v_1 = v^\nu$, $F_1 = 0$, $v_2 = v^0$, and $F_2 = -\nu\Delta v^0$ to obtain

$$\begin{aligned}\sup_{0\leq t\leq T} \|v^\nu - v^0\|_0 \leq \int_0^T \|\nu\Delta v^0\|_0 dt \exp\left(\int_0^T |\nabla v^0|_{L^\infty} dt\right) \\ \leq \nu T c(v^0, T).\end{aligned}$$

Estimate (3.9) gives

$$\begin{aligned}\int_0^T \|\nabla(v^\nu - v^0)\|_0 dt &\le \left(T\int_0^T \|\nabla(v^\nu - v^0)\|_0^2 dt\right)^{1/2} \\ &\le c(v^0, T)\left(\frac{T}{\nu}\right)^{1/2}\int_0^T \|\nu\Delta v^0\|_0 dt \\ &\le \nu^{1/2}T^{3/2}c(v^0, T).\end{aligned}$$

We have just proved the following proposition.

Proposition 3.2. *Approximation of Smooth Solutions to the Navier–Stokes Equation by Smooth Solutions to the Euler Equation. Given fixed initial data, let* v^ν, $\nu \ge 0$, *be a smooth solution to the Euler* ($\nu = 0$) *or the Navier–Stokes* ($\nu > 0$) *equation. Suppose that for* $0 \le \nu \le \nu_0$ *these solutions exist on a common time interval* $[0, T]$ *and vanish sufficiently rapidly as* $|x| \nearrow \infty$. *Then*

$$\sup_{0\le t\le T} \|v^\nu - v^0\|_0 \le c(v^0, T)\nu T, \tag{3.10}$$

$$\int_0^T \|\nabla(v^\nu - v^0)\|_0\, dt \le c(v^0, T)\nu^{1/2}T^{3/2}, \tag{3.11}$$

where $c(v^0, T) = c[\sup_{0\le t\le T} \|\Delta v^0(\cdot, t)\|_0, \int_0^T |\nabla v^0|_{L^\infty} dt]$.

This proposition proves (first order in ν) convergence of $v^\nu \to v^0$ in the $L^\infty\{[0, T]; L^2(\mathbb{R}^N)\}$ norm:

$$\sup_{0\le t\le T} \|v^\nu - v^0\|_0 = \mathcal{O}(\nu).$$

It also proves (square root in ν) convergence for the gradients $\nabla v^\nu \to \nabla v^0$ in the $L^1\{[0, T]; L^2(\mathbb{R}^N)\}$ norm:

$$\int_0^T \|\nabla(v^\nu - v^0)\|_0 dt = \mathcal{O}(\nu^{1/2}).$$

Recall that, for the family of exact solutions in Proposition 1.6, v^ν converges to v^0 in the supremum norm as $(\nu T)^{1/2}$ (square-root convergence). The higher order of convergence in relation (3.10) is due to the fact that the L^2 norm is less sensitive to local gradients.

We reiterate that the above estimates assume that the inviscid solution v^0 and the viscous solution v^ν exist and are smooth on a common time interval. In Sections 3.2 and 3.3 we show that for smooth initial data there exists such a common time interval and that the continuation of solutions is connected to the control on the magnitude of the vorticity. Constantin (1986) proves that for sufficiently small viscosity ν, the time interval of existence of the viscous Navier–Stokes solution is at least as large as the time interval of existence of the Euler ($\nu = 0$) solution with the same initial data. We refer the reader to this paper for the details.

3.1.3. Kinetic Energy for 2D Fluids

The above techniques require that the velocity field decay fast enough at infinity so that $v \in L^2(\mathbb{R}^N)$. This is a reasonable assumption in the case of three dimensions but is not so reasonable for the case of two dimensions. In three dimensions typical velocity fields decay fast enough in the far field so that the integral $\int |v|^2 dx$ is finite. Such a typical velocity field is one with smooth vorticity of compact support. In two dimensions, however, this same class of velocity fields does not decay fast enough at infinity to be globally in L^2 and hence does not have globally finite kinetic energy. This is a consequence of the decay properties of the kernel K_N in the Biot–Savart law. In two dimensions K_2 decays at infinity like $1/r$, which is not square integrable, whereas in three dimensions K_3 decays like $1/r^2$.

However, 2D solutions typically have *locally* finite kinetic energy, that is, for any radius R, $\int_{|x|<R} |v|^2 dx \leq C(R)$. For example, consider any 2D smooth vorticity ω satisfying

$$\operatorname{supp} \omega \subset \{|x| \leq R\}, \qquad R > 0. \tag{3.12}$$

Recall from the Biot–Savart law that the velocity field of the flow is

$$v(x,t) = \int_{\mathbb{R}^2} K_2(x-y)\omega(y,t)dy, \tag{3.13}$$

$$K_2(x) = \frac{1}{2\pi}\frac{1}{|x|^2}\begin{pmatrix} -x_2 \\ x_1 \end{pmatrix}. \tag{3.14}$$

Note that for $|x| \neq 0$

$$|x-y|^{-2} = |x|^{-2}\left(1 - \frac{2x \cdot y}{|x|^2} + \frac{|y|^2}{|x|^2}\right)^{-1}.$$

If $|y| \leq R$ and $|x| \geq 2R$, then for large $|x|$,

$$|x-y|^{-2} = |x|^{-2} + \mathcal{O}\big(|x|^{-3}\big).$$

Because $\omega(y)$ has support inside $|y| < R$ we see that, for large $|x|$,

$$v(x,t) = K_2(x)\int_{\mathbb{R}^2} \omega(y,t)dy + \mathcal{O}\big(|x|^{-2}\big). \tag{3.15}$$

Combining this with the fact that

$$\int_{\mathbb{R}^N} (1+|x|^2)^{-l/2} dx < \infty, \Longleftrightarrow l > N \tag{3.16}$$

we obtain the following proposition.

Proposition 3.3. *A 2D incompressible velocity field with vorticity of compact support in $\mathbb{R}^2$ has finite kinetic energy, $\int_{\mathbb{R}^2} v^2 dx < \infty$, if and only if*

$$\int_{\mathbb{R}^2} \omega(y,t)dy = 0. \tag{3.17}$$

There are many 2D solutions to the Navier–Stokes equations for which ω does not satisfy Eq. (3.17); for instance, the case in which ω is of fixed sign. The basic energy estimate does not apply to such 2D examples because it demands globally finite kinetic energy. We can, however, derive an analogous estimate by writing v as the sum of two parts, one with globally finite kinetic energy and one that is a radially symmetric exact solution, either the inviscid eddy from Example 2.1 or the viscous eddy from Example 2.2 in Chap. 2.

Note that any smooth solution to either the Euler equation or the Navier–Stokes equation with vorticity $\omega(\cdot, t) \in L^1(\mathbb{R}^2)$ satisfies

$$\int \omega(x, t)dx = \int \omega(x, 0)dx. \tag{3.18}$$

Hence we make the following definition.

Definition 3.1. *Radial-Energy Decomposition. A smooth incompressible velocity field $v(x)$ in $\mathbb{R}^2$ has a* radial-energy decomposition *if there exists a smooth radially symmetric vorticity $\bar{\omega}(|x|)$ such that*

$$v(x) = u(x) + \bar{v}(x),$$
$$\int |u(x)|^2 dx < \infty, \qquad \text{div } u = 0,$$

where $\bar{v}$ is defined from $\bar{\omega}$ by means of the Biot–Savart law:

$$\bar{v}(x) = \begin{pmatrix} -x_2 \\ x_1 \end{pmatrix} |x|^{-2} \int_0^{|x|} s\bar{\omega}(s)ds.$$

The radial-energy decomposition is not unique. It depends on the specific choice of $\bar{\omega}$. For time-dependent solutions to the Euler or the Navier–Stokes equations, Eq. (3.18) implies that the radial part $\bar{\omega}$ of the energy decomposition can be determined by the initial data alone. Note also that the radial-energy decomposition is *not* an orthogonal decomposition. In particular we cannot define it by means of a projection operator onto a subspace as in the case of the Hodge decomposition discussed in Chap. 1.

Formula (3.15) shows that the total integral of the vorticity determines the part of v that has infinite energy. In particular, v has globally finite kinetic energy if and only if the integral of the vorticity is zero (the vorticity has global mean zero). Hence we have the following lemma.

Lemma 3.2. *Any smooth incompressible velocity field with vorticity*

$$\omega = \text{curl } v \in L^1(\mathbb{R}^2)$$

has a radial-energy decomposition.

We now introduce special radial-energy decompositions for 2D time-dependent solutions to the Euler and the Navier–Stokes equations.

For the case of the Euler equation, recall from Chap. 2 the inviscid eddies:

$$v_0(x) = \left(\frac{-x_2}{r^2} \frac{x_1}{r^2}\right)^t \int_0^r s\omega_0(s)ds. \tag{3.19}$$

These are time-independent exact solutions to the Euler equation with radially symmetric vorticity $\omega_0(r)$.

For the case of the Navier–Stokes equation, recall that the viscous eddies are exact radial solutions with initial vorticity $\omega_0(r)$. The radial symmetry implies that $\omega(x, t)$ satisfies the heat equation:

$$\omega_t = \nu\Delta\omega, \tag{3.20}$$

$$\omega|_{t=0} = \omega_0, \tag{3.21}$$

The solution to Eq. (3.20) has an explicit form that uses the heat kernel (see Folland, 1995, Chap. 4, Section A):

$$\omega(x, t) = \frac{1}{4\pi\nu t} \int_{\mathbb{R}^2} e^{\frac{|x-y|^2}{4\nu t}} \omega_0(|y|)dy.$$

The velocity field v is given by the same formula as that in Eq. (3.19):

$$v(x, t) = \left(\frac{-x_2}{r^2}, \frac{x_1}{r^2}\right)^t \int_0^r s\omega(s, t)ds. \tag{3.22}$$

We now explicitly construct time-dependent decompositions for solutions to both the Euler and the Navier–Stokes equation. Given an initial smooth vorticity $\omega_0(x) \in L^1(\mathbb{R}^2)$ and respective initial velocity

$$v_0(x) = \int_{\mathbb{R}^2} K_2(x - y)\omega_0(y)dy,$$

let $\bar{\omega}_0(r)$ be the radial part of an energy decomposition of $\omega_0(x)$, $v_0(x)$ defined in Definition 3.1.

Let $\bar{v}(r, t), \bar{\omega}(r, t)$ be the exact radial Euler [Eq. (3.19)] or Navier–Stokes [Eq. (3.22)] solution with initial vorticity $\bar{\omega}_0(r)$. Then $\bar{v}(r, t)$ is the radial part of an energy decomposition of a solution to the respective Euler or Navier–Stokes equation with initial vorticity $\omega_0(x)$.

Note that $u(x, t) = v(x, t) - \bar{v}(x, t)$ satisfies

$$\frac{\partial u}{\partial t} + u \cdot \nabla u + \bar{v} \cdot \nabla u + u \cdot \nabla \bar{v} = -\nabla p + \nu\Delta u + F. \tag{3.23}$$

Because $\bar{v}$ is a known smooth velocity field obtained from the initial data, classical energy estimates on u give estimates for the full solution $v(x, t)$. We begin by deriving a 2D basic energy estimate by using the radial-energy decomposition. Let $v_1 = u_1 + \bar{v}_1$ and $v_2 = u_2 + \bar{v}_2$ be respective finite-energy decompositions for two solutions to the

2D Navier–Stokes equation. Denote $\tilde{u} = u_1 - u_2$, $\tilde{\bar{v}} = \bar{v}_1 - \bar{v}_2$, $\tilde{F} = F_1 - F_2$, etc. The difference $\tilde{u}$ satisfies

$$\frac{\partial \tilde{u}}{\partial t} + u_1 \cdot \nabla \tilde{u} + \tilde{u} \cdot \nabla u_2 + \bar{v}_1 \cdot \nabla \tilde{u} + \tilde{\bar{v}} \cdot \nabla u_2 + \tilde{u} \cdot \nabla \bar{v}_1 + u_2 \cdot \nabla \tilde{\bar{v}}$$
$$= -\nabla \tilde{p} + \nu \Delta \tilde{u} + \tilde{F}. \tag{3.24}$$

Then using the same integration-by-parts procedure used to derive basic energy estimate (3.7), we see that the difference between two solutions satisfies the bound

$$\frac{d}{dt}\frac{1}{2}\|\tilde{u}\|_0^2 + \nu\|\tilde{u}\|_0^2 \le \|\tilde{u}\|_0[\|\tilde{u}\|_0(|\nabla u_2|_{L^\infty} + |\nabla \bar{v}_1|_{L^\infty})$$
$$+ |\nabla(\bar{v}_1 - \bar{v}_2)|_{L^\infty}\|u_2\|_0 + \|\tilde{F}\|_0 + |\bar{v}_1 - \bar{v}_2|_{L^\infty}\|\nabla u_2\|_0].$$

As in the case of Proposition 3.1, we apply Grönwall's lemma, given here as Lemma 3.1, to obtain the following proposition.

Proposition 3.4. *2D Basic Energy Estimate and Gradient Control. Let v_1 and v_2 be two smooth divergence-free solutions to the Navier–Stokes equation with radial-energy decomposition $v_i(x, t) = u_i(x, t) + \bar{v}_i(x)$ and with external force F_1, F_2 and presures p_1, p_2. Then we have the following 2D basic energy estimate,*

$$\sup_{0\le t\le T} \|u_1 - u_2\|_0 \le \|(u_1 - u_2)(\cdot, 0)\|_0$$
$$+ e^{\left[\int_0^T (|\nabla u_2|_{L^\infty} + |\nabla \bar{v}_1|_{L^\infty})dt\right]} \int_0^T \big[\|(F_1 - F_2)(\cdot, t)\|_0 + |\bar{v}_1 - \bar{v}_2|_{L^\infty}\|\nabla u_2(\cdot, t)\|_0$$
$$+ |\nabla \bar{v}_1 - \nabla \bar{v}_2|_{L^\infty}\|u_2(\cdot, t)\|_0\big]dt, \tag{3.25}$$

and the gradient control,

$$\nu \int_0^T \|\nabla(u_1 - u_2)\|_0^2$$
$$\le C(v_2, \bar{v}_0, T)\Bigg\{\|(v_1 - v_2)(\cdot, 0)\|_0^2 + \Bigg[\int_0^T (\|F_1(\cdot, t) - F_2(\cdot, t)\|_0$$
$$+ |\bar{v}_1 - \bar{v}_2|_{L^\infty}\|\nabla u_2(\cdot, t)\|_0 + |\nabla \bar{v}_1 - \nabla \bar{v}_2|_{L^\infty}\|u_1(\cdot, t)\|_0)dt\Bigg]^2\Bigg\}.$$

Remark: As in Corollary 3.1, relation (3.25) implies uniqueness of 2D solutions to the Euler and the Navier–Stokes equations with local finite-energy decomposition. Note here that the energy of $u(x, t)$, $\|u(\cdot, t)\|_0^2$ is not bounded by the initial energy $\|u(\cdot, 0)\|_0$ as in the case of the original basic energy estimate. This is because the radial part of the radial-energy decomposition can feed energy into the evolution of the $\tilde{u}$. In Chap. 8 we discuss a related pseudoenergy for 2-D solutions to the Euler equation that is in fact conserved for all time. This pseudoenergy is important for proving the existence of solutions to the 2D Euler equation with vortex-sheet initial data.

We also can approximate viscous flows by inviscid flows. As in the 3D case, let $F_1=0$ and $F_2=-\nu\Delta v^0$. Viscous and inviscid solutions with the same initial data always have local finite-energy decompositions with the same time-independent part, $\bar{v}$. Hence we get the same estimate for $v^0(x,t)-v^\nu(x,t)$ as that in the 3D case.

Proposition 3.5. *Approximation of 2D Navier–Stokes Solutions by Euler Solutions. Given $v^0(x,t)$ and $v^\nu(x,t)$, two solutions to the Euler and the Navier–Stokes equations, respectively, with the same initial data, then $v^0(x,t)-v^\nu(x,t)$ has finite kinetic energy and*

$$\sup_{0\le t\le T}\|v^0(x,t)-v^\nu(x,t)\|_0\le C(T)\nu.$$

3.2. Local-in-Time Existence of Solutions by means of Energy Methods

In this section we prove the local-in-time existence of solutions to the full Navier–Stokes equation:

$$\frac{\partial v}{\partial t}+v\cdot\nabla v=-\nabla p+\nu\Delta v, \tag{3.26}$$

$$\operatorname{div} v=0. \tag{3.27}$$

Although this is a now classical result, we present a somewhat different proof than can be found in previous references. Our proof treats the Euler and the Navier–Stokes equations simultaneously, using estimates that are independent of the viscosity ν in the limit as $\nu\to 0$. For simplicity we consider the equation without forcing; however, the results of this section can be easily extended to the case of the Navier–Stokes equation with a sufficiently smooth L^2 forcing. We leave this as an exercise for the reader. The case of two dimensions is complicated by the fact that a large class of solutions does not have globally finite kinetic energy. We handle this by considering a radial-energy decomposition as described in the preceding section. We present the details of the 3D proof and, at the end of this section, discuss the 2D proof. In the next section, we use the 2D result to prove global-in-time existence of 2D solutions to both the Euler and the Navier–Stokes equations.

Basic energy estimate (3.7), derived for smooth solutions, yields many useful results, including uniqueness of solutions and approximation of inviscid solutions by high Reynolds number solutions. Note that the basic energy estimate is an a priori estimate that assumes the existence of smooth solutions and then proves results about them. In this section we show that for general smooth initial data such solutions exist on a finite-time interval. In Chap. 4 we re-prove this result for the Euler equation by using a different method, based on the integrodifferential equation for the particle trajectories, derived in Section 2.5 of Chap. 2.

The *energy method* presented in this chapter uses Sobolev norms that involve L^2 bounds for higher derivatives of $v(x, t)$. The strategy of this section has two parts. First we find an approximate equation and approximate solutions that have two properties: (1) the existence theory for all time for the approximating solutions is easy, (2) the solutions satisfy an analogous energy estimate. The second part is the passage to a limit in the approximation scheme to obtain a solution to the Navier–Stokes equation. The "art" of designing this method lies in choosing the correct approximation procedure that succeeds in achieving the above two steps. Here we construct the approximate equations by using a smoothing procedure called mollification to convert unbounded differential operators into bounded operators.

We begin with some basic properties of the Sobolev spaces. The proofs of these lemmas are in the appendix to this chapter.

3.2.1. Calculus Inequalities for Sobolev Spaces and Properties of Mollifiers

The Sobolev space $H^m(\mathbb{R}^N)$, $m \in \mathbb{Z}^+ \cup \{0\}$, consists of functions $v \in L^2(\mathbb{R}^N)$ such that $D^\alpha v \in L^2(\mathbb{R}^N)$, $0 \le |\alpha| \le m$, where D^α is the distribution derivative $D^\alpha = \partial_1^{\alpha_1} \cdots \partial_n^{\alpha_n}$. The H^m norm, denoted as $\|\cdot\|_m$, is

$$\|v\|_m = \left(\sum_{0 \le |\alpha| \le m} \|D^\alpha v\|_0^2 \right)^{1/2}. \tag{3.28}$$

The Sobolev space H^m generalizes to the case $m = s \in \mathbb{R}$. Consider the functional

$$\|\ \|_s : \mathcal{S}(\mathbb{R}^N) \to \mathbb{R}^+ \cup \{0\}$$

defined by

$$\|v\|_s = \left[\int_{\mathbb{R}^N} (1 + |\xi|^2)^s |\hat{v}(\xi)|^2 d\xi \right]^{1/2}, \tag{3.29}$$

acting on the Schwarz space of rapidly decreasing smooth functions $\mathcal{S}(\mathbb{R}^N)$. Here $\hat{v}$ denotes the Fourier transform of v. The Sobolev space $H^s(\mathbb{R}^N)$, $s \in \mathbb{R}$, is the completion of $\mathcal{S}(\mathbb{R}^N)$ with respect to the norm $\|\cdot\|_s$. For $s = m$ the two norms are equivalent.

We review some properties of Sobolev spaces. First we state a well-known and important fact, that an L^2 bound on a higher derivative implies a pointwise bound on a lower derivative (see Folland, 1995).

Lemma 3.3. *Sobolev Inequality. The space $H^{s+k}(\mathbb{R}^N)$, $s > N/2$, $k \in \mathbb{Z}^+ \cup \{0\}$, is continuously embedded in the space $C^k(\mathbb{R}^N)$. That is, there exists $c > 0$ such that*

$$|v|_{C^k} \le c\|v\|_{s+k} \qquad \forall\, v \in H^{s+k}(\mathbb{R}^N). \tag{3.30}$$

Also, Sobolev spaces satisfy the following calculus inequalities proved in the appendix.

Lemma 3.4. *Calculus Inequalities in the Sobolev Spaces.*

(i) For all $m \in \mathbb{Z}^+ \cup \{0\}$, there exists $c > 0$ such that, for all $u, v \in L^\infty \cap H^m(\mathbb{R}^N)$,

$$\|uv\|_m \le c\{|u|_{L^\infty}\|D^m v\|_0 + \|D^m u\|_0 |v|_{L^\infty}\}, \tag{3.31}$$

$$\sum_{0\le|\alpha|\le m} \|D^\alpha(uv) - uD^\alpha v\|_0 \le c\{|\nabla u|_{L^\infty}\|D^{m-1}v\|_0 + \|D^m u\|_0 |v|_{L^\infty}\}. \tag{3.32}$$

(ii) For all $s > (N/2)$, $H^s(\mathbb{R}^N)$ is a Banach algebra. That is, there exists $c > 0$ such that, for all $u, v \in H^s(\mathbb{R}^N)$,

$$\|uv\|_s \le c\|u\|_s\|v\|_s. \tag{3.33}$$

Our first step is to modify the equation in order to easily produce a family of global smooth solutions. The technique of regularization by mollifiers involves convolution of a less smooth function with a smooth kernel to produce a smooth function. It has the added feature that when composed with unbounded differential operators, mollification produces bounded operators with, albeit, a bad dependence on the regularization parameter. The difficulty in designing an approximation for the Navier–Stokes/Euler equation lies in choosing a regularization that, despite containing operators with bad dependence on the regularization parameter, yields energy estimates for quantities like the kinetic energy that are independent of the regularization parameter.

We begin by defining a regularizing operator called a *mollifier*. We then show how to use this operator to design an appropriate regularization of the Navier–Stokes equations that turns the unbounded differential operators into bounded operators on the Sobolev spaces. Given any radial function

$$\rho(|x|) \in C_0^\infty(\mathbb{R}^N), \qquad \rho \ge 0, \qquad \int_{\mathbb{R}^N} \rho\, dx = 1, \tag{3.34}$$

define the mollification $\mathcal{J}_\epsilon v$ of functions $v \in L^p(\mathbb{R}^N)$, $1 \le p \le \infty$, by

$$(\mathcal{J}_\epsilon v)(x) = \epsilon^{-N} \int_{\mathbb{R}^N} \rho\left(\frac{x-y}{\epsilon}\right) v(y) dy, \qquad \epsilon > 0. \tag{3.35}$$

Mollifiers have several well-known properties (see, for example, Taylor, 1991).

Lemma 3.5. *Properties of Mollifiers. Let $\mathcal{J}_\epsilon$ be the mollifier defined in Eq. (3.35). Then $\mathcal{J}_\epsilon v$ is a C^∞ function and*

(i) for all $v \in C^0(\mathbb{R}^N)$, $\mathcal{J}_\epsilon v \to v$ uniformly on any compact set Ω in $\mathbb{R}^N$ and

$$|\mathcal{J}_\epsilon v|_{L^\infty} \le |v|_{L^\infty}. \tag{3.36}$$

(ii) Mollifiers commute with distribution derivatives,

$$D^\alpha \mathcal{J}_\epsilon v = \mathcal{J}_\epsilon D^\alpha v, \qquad \forall\, |\alpha| \le m, \quad v \in H^m. \tag{3.37}$$

(iii) *For all* $u \in L^p(\mathbb{R}^N)$, $v \in L^q(\mathbb{R}^N)$, $(1/p) + (1/q) = 1$,

$$\int_{\mathbb{R}^N} (\mathcal{J}_\epsilon u) v \, dx = \int_{\mathbb{R}^N} u(\mathcal{J}_\epsilon v) dx. \tag{3.38}$$

(iv) *For all* $v \in H^s(\mathbb{R}^N)$, $\mathcal{J}_\epsilon v$ *converges to* v *in* H^s *and the rate of convergence in the* H^{s-1} *norm is linear in* ϵ:

$$\lim_{\epsilon \searrow 0} \|\mathcal{J}_\epsilon v - v\|_s = 0, \tag{3.39}$$

$$\|\mathcal{J}_\epsilon v - v\|_{s-1} \leq C\epsilon \|v\|_s. \tag{3.40}$$

(v) *For all* $v \in H^m(\mathbb{R}^N)$, $k \in \mathbb{Z}^+ \cup \{0\}$, *and* $\epsilon > 0$,

$$\|\mathcal{J}_\epsilon v\|_{m+k} \leq \frac{c_{mk}}{\epsilon^k} \|v\|_m, \tag{3.41}$$

$$|\mathcal{J}_\epsilon D^k v|_{L^\infty} \leq \frac{c_k}{\epsilon^{N/2+k}} \|v\|_0. \tag{3.42}$$

The proof of Lemma 3.5 is also in the appendix. Finally, we recall some properties of the Leray projection operator P on the space of divergence-free functions in $H^m(\mathbb{R}^N)$. Proposition 1.16 proves the Hodge decomposition in L^2. A more general version is Lemma 3.6

Lemma 3.6. *The Hodge Decomposition in* H^m.[‖] *Every vector field* $v \in H^m(\mathbb{R}^N)$, $m \in \mathbb{Z}^+ \cup \{0\}$, *has the unique orthogonal decomposition*

$$v = w + \nabla\varphi$$

such that the Leray's projection operator $Pv = \omega$ *on the divergence-free functions satisfies*

(i) $Pv, \nabla\varphi \in H^m$, $\int_{\mathbb{R}^N} Pv \cdot \nabla\varphi \, dx = 0$, $\operatorname{div} Pv = 0$, *and*

$$\|Pv\|_m^2 + \|\nabla\varphi\|_m^2 = \|v\|_m^2, \tag{3.43}$$

(ii) P *commutes with the distribution derivatives,*

$$PD^\alpha v = D^\alpha Pv, \qquad \forall v \in H^m, \quad |\alpha| \leq m, \tag{3.44}$$

(iii) P *commutes with mollifiers* $\mathcal{J}_\epsilon$,

$$P(\mathcal{J}_\epsilon v) = \mathcal{J}_\epsilon(Pv), \qquad \forall\, v \in H^m, \quad \epsilon > 0. \tag{3.45}$$

(iv) P *is symmetric,*

$$(Pu, v)_m = (u, Pv)_m.$$

‖ We leave the proof of this result as an exercise for the reader.

3.2.2. Global Existence of Solutions to a Regularization of the Euler and the Navier–Stokes Equations

Recall that our strategy for this section begins with designing an approximation of the equations for which we can easily prove a global existence of solutions and for which we can also show an analogous energy estimate that is independent of the regularization parameter. The need for such a regularization comes from the fact that the equations

$$
\begin{aligned}
v_t + v \cdot \nabla v &= -\nabla p + \nu \Delta v, \\
\operatorname{div} v &= 0, \\
v|_{t=0} &= v_0,
\end{aligned}
\tag{3.46}
$$

contain unbounded operators so that we cannot directly apply the Picard theorem for ODEs in a Banach space. We use the Picard theorem, which plays a central role in Chap. 4.

Theorem 3.1. *Picard Theorem on a Banach Space . Let $O \subseteq \mathbf{B}$ be an open subset of a Banach space $\mathbf{B}$ and let $F : O \to B$ be a mapping that satisfies the following parameters:*

(i) $F(X)$ maps O to $\mathbf{B}$.
(ii) F is locally Lipschitz continuous, i.e., for any $X \in O$ there exists $L > 0$ and an open neighborhood $U_X \subset O$ of X such that

$$\|F(\tilde{X}) - F(\hat{X})\|_{\mathbf{B}} \le L\|\tilde{X} - \hat{X}\|_{\mathbf{B}} \qquad \text{for all} \quad \tilde{X}, \hat{X} \in U_X.$$

Then for any $X_0 \in O$, there exists a time T such that the ODE

$$\frac{dX}{dt} = F(X), \qquad X|_{t=0} = X_0 \in O, \tag{3.47}$$

has a unique (local) solution $X \in C^1[(-T, T); O]$.

In the preceding theorem, $\|\cdot\|_{\mathbf{B}}$ denotes a norm in the Banach space $\mathbf{B}$, and $C^1[(-T, T); \mathbf{B}]$ denotes the space of C^1 functions $X(t)$ on the open interval $(-T, T)$ with values in $\mathbf{B}$.[¶] Theorem 3.1 is a generalization of the classical Picard theorem for ODEs in $\mathbb{R}^N$ to the case of infinite-dimensional Banach spaces (see, for example, Hartman, 1982). In fact, the proof is identical to the finite-dimensional version, which uses a contraction mapping. The main difference between finite- and infinite-dimensional ODEs on Banach spaces lies in the continuation properties of the solutions. Unlike ODEs in $\mathbb{R}^N$, nonautonomous[#] infinite-dimensional ODEs cannot, in general, be continued in time as in the finite-dimensional case. However, if F does

[¶] Note that $C^1[(-T, T); \mathbf{B}]$ is also a Banach space equipped with the norm $||X|| = \sup_{t\in(-T,T)} ||X(t)||_{\mathbf{B}} + \sup_{t\in(-T,T)} ||\frac{\partial}{\partial t} X(t)||_{\mathbf{B}}$.

[#] That is, explicitly time dependent.

not depend explicitly on time, solutions to Eq. (3.47) can be continued until they leave the set O. We use this result (see Theorem 3.3) to prove that the approximate equation has a global-in-time solution.

We create an approximate equation to the Navier–Stokes equation that satisfies the condition of the Picard theorem by regularizing the equations by using $\mathcal{J}_\epsilon$ as defined in Eq. (3.35):

$$v_t^\epsilon + \underline{\mathcal{J}_\epsilon}[(\mathcal{J}_\epsilon v^\epsilon) \cdot \nabla(\mathcal{J}_\epsilon v^\epsilon)] = -\nabla p^\epsilon + \nu \underline{\mathcal{J}_\epsilon}(\mathcal{J}_\epsilon \Delta v^\epsilon),$$
$$\operatorname{div} v^\epsilon = 0, \tag{3.48}$$
$$v^\epsilon|_{t=0} = v_0.$$

The need for the mollifiers underlined in Eq. (3.48) is not obvious – they provide a necessary balance of terms in the integration by parts. In particular, the balance of terms yields a simple energy estimate for the regularized problem.

Equations (3.48) explicitly contain the pressure p^ϵ. Following Leray, we eliminate p^ϵ and the incompressibility condition $\operatorname{div} v^\epsilon = 0$ by projecting these equations onto the space of divergence-free functions:

$$V^s = \{v \in H^s(\mathbb{R}^N) : \operatorname{div} v = 0\}. \tag{3.49}$$

Because the Leray projection operator P commutes with derivatives and mollifiers (Lemma 3.6) and $Pv^\epsilon = v^\epsilon$, we have

$$v_t^\epsilon + P\mathcal{J}_\epsilon[(\mathcal{J}_\epsilon v^\epsilon) \cdot \nabla(\mathcal{J}_\epsilon v^\epsilon)] = \nu \mathcal{J}_\epsilon^2 \Delta v^\epsilon. \tag{3.50}$$

The regularized Euler or Navier–Stokes equation in Eqs. (3.48) reduces to an ODE in the Banach space V^s:

$$\frac{dv^\epsilon}{dt} = F_\epsilon(v^\epsilon),$$
$$v^\epsilon|_{t=0} = v_0, \tag{3.51}$$

where

$$F_\epsilon(v^\epsilon) = \nu \mathcal{J}_\epsilon^2 \Delta v^\epsilon - P\mathcal{J}_\epsilon[(\mathcal{J}_\epsilon v^\epsilon) \cdot \nabla(\mathcal{J}_\epsilon v^\epsilon)] = F_\epsilon^1(v^\epsilon) - F_\epsilon^2(v^\epsilon). \tag{3.52}$$

Our goal in this subsection is to prove the following theorem.

Theorem 3.2. *Global Existence of Regularized Solutions. Given an initial condition $v_0 \in V^m$, $m \in \mathbb{Z}^+ \cup \{0\}$, for any $\epsilon > 0$ there exists for all time a unique solution $v^\epsilon \in C^1([0, \infty); V^m)$ to regularized equation (3.51).*

The strategy is first to show that the existence and the uniqueness of solutions v^ϵ locally in time follow from the Picard theorem (Theorem 3.1). Then we use the continuation property of ODEs on a Banach space combined with an a priori estimate to show that the solution exists globally in time. First we prove the following proposition.

Proposition 3.6. *Consider an initial condition $v_0 \in V^m$, $m \in \mathbb{Z}^+ \cup \{0\}$. Then*

(i) for any $\epsilon > 0$ there exists the unique solution $v^\epsilon \in C^1([0, T_\epsilon); V^m)$ to the ODE in Eq. (3.51), where $T_\epsilon = T(\|v_0\|_m, \epsilon)$;

(ii) on any time interval $[0, T]$ on which this solution belongs to $C^1([0, T]; V^0)$,

$$\sup_{0\le t\le T} \|v^\epsilon\|_0 \le \|v_0\|_0. \tag{3.53}$$

Note that energy estimate (3.53) is independent of ϵ and ν. This is in fact an optimal estimate as it agrees with the basic energy estimate (Proposition 3.4) for the unregularized Euler and Navier–Stokes equations. To obtain this optimality we needed the special choice of regularization in Eq. (3.52).

Proof of Proposition 3.6. First we prove the existence of regularized solutions v^ϵ locally in time. We show that the function F_ϵ in Eq. (3.52) maps V^m into V^m and is locally Lipschitz continuous. First note that $F_\epsilon : V^m \to V^m$ because $\operatorname{div} v^\epsilon = 0$, P maps into divergence-free vector fields, and $\mathcal{J}_\epsilon$ commutes with derivatives.

The definition of Sobolev spaces and estimate (3.41) for mollifiers implies that

$$\begin{aligned}\left\|F_\epsilon^1(v^1) - F_\epsilon^1(v^2)\right\|_m &= \nu\left\|\mathcal{J}_\epsilon^2\Delta(v^1 - v^2)\right\|_m\\ &\le \nu\left\|\mathcal{J}_\epsilon^2(v^1 - v^2)\right\|_{m+2}\\ &\le \frac{c\nu}{\epsilon^2}\|v^1 - v^2\|_m.\end{aligned}$$

Calculus inequality (3.31) and commutation property (3.45) of P and $\mathcal{J}_\epsilon$ imply that

$$\begin{aligned}&\left\|F_\epsilon^2(v^1) - F_\epsilon^2(v^2)\right\|_m\\ &\quad\le \|P\mathcal{J}_\epsilon\{(\mathcal{J}_\epsilon v^1)\cdot\nabla[\mathcal{J}_\epsilon(v^1 - v^2)]\}\|_m + \|P\mathcal{J}_\epsilon\{[\mathcal{J}_\epsilon(v^1 - v^2)]\cdot\nabla(\mathcal{J}_\epsilon v^2)\}\|_m\\ &\quad\le c\{|\mathcal{J}_\epsilon v^1|_{L^\infty}\|D^m\mathcal{J}_\epsilon\nabla(v^1 - v^2)\|_0 + \|D^m\mathcal{J}_\epsilon v^1\|_0|\mathcal{J}_\epsilon\nabla(v^1 - v^2)|_{L^\infty}\\ &\qquad + |\mathcal{J}_\epsilon(v^1 - v^2)|_{L^\infty}\|D^m\mathcal{J}_\epsilon\nabla v^2\|_0 + \|D^m\mathcal{J}_\epsilon(v^1 - v^2)\|_0|\mathcal{J}_\epsilon\nabla v^2|_{L^\infty}\}.\end{aligned}$$

Mollifier properties (3.41) and (3.42) then give

$$\left\|F_\epsilon^2(v^1) - F_\epsilon^2(v^2)\right\|_m \le \frac{c}{\epsilon^{N/2+1+m}}(\|v^1\|_0 + \|v^2\|_0)\|v^1 - v^2\|_m.$$

The final result is

$$\|F_\epsilon(v^1) - F_\epsilon(v^2)\|_m \le c(\|v^j\|_0, \epsilon, N)\|v^1 - v^2\|_m, \tag{3.54}$$

so that F_ϵ is locally Lipschitz continuous on any open set:

$$O^M = \{v \in V^m \mid \|v\|_m < M\}.$$

Thus the Picard theorem (Theorem 3.1) implies that, given any initial condition $v_0 \in H^m$, there exists a unique solution $v^\epsilon \in C^1([0, T_\epsilon); V^m \cap O^M)$, $m \in \mathbb{Z}^+ \cup \{0\}$, for some $T_\epsilon > 0$.

Remark: The Picard theorem actually gives local existence of solutions to the regularized equation both forward and backward in time. However, we make use of only the forward-in-time existence of regularized solutions when proving existence of solutions to the original Navier–Stokes equation. This is because the Navier–Stokes equations are ill posed if time is reversed; as in the case of the heat equation, the diffusion term necessitates a direction for time to proceed in order to obtain certain a priori Sobolev bounds. In the next subsection we discuss these bounds in great detail.

Now we prove energy bound (3.53); if $v^\epsilon \in C^1([0,T); V^0)$, then

$$\sup_{0\le t\le T} \|v^\epsilon\|_0 \le \|v_0\|_0.$$

Take the L^2 inner product of Eq. (3.51) with v^ϵ to obtain

$$\frac{1}{2}\frac{d}{dt}\int_{\mathbb{R}^N} |v^\epsilon|^2 dx = \nu \int_{\mathbb{R}^N} v^\epsilon \mathcal{J}_\epsilon^2 \Delta v^\epsilon dx - \int_{\mathbb{R}^N} v^\epsilon P\mathcal{J}_\epsilon[(\mathcal{J}_\epsilon v^\epsilon)\cdot\nabla(\mathcal{J}_\epsilon v^\epsilon)]dx.$$

The properties of mollifiers and the operator P from Lemmas 3.5 and 3.6 imply, after integration by parts, that

$$\begin{aligned}\frac{1}{2}\frac{d}{dt}\int_{\mathbb{R}^N} (v^\epsilon)^2 dx &= \nu \int_{\mathbb{R}^N} (\mathcal{J}_\epsilon v^\epsilon)\Delta(\mathcal{J}_\epsilon v^\epsilon)dx + \frac{1}{2}\int_{\mathbb{R}^N} (\mathcal{J}_\epsilon v^\epsilon)\cdot\nabla(\mathcal{J}_\epsilon v^\epsilon)^2 dx \\ &= -\nu \int_{\mathbb{R}^N} (\mathcal{J}_\epsilon \nabla v^\epsilon)^2 dx - \frac{1}{2}\int_{\mathbb{R}^N} (\operatorname{div} \mathcal{J}_\epsilon v^\epsilon)(\mathcal{J}_\epsilon v^\epsilon)^2 dx,\end{aligned}$$

so that

$$\frac{d}{dt}\|v^\epsilon\|_0^2 + 2\nu\|\nabla \mathcal{J}_\epsilon v^\epsilon\|_0^2 = 0.$$

The specific choice of the regularization of Eq. (3.51) provides a balance of terms for the integration by parts. Because $\nu \ge 0$, we obtain energy bound (3.53). □

To end this subsection, we complete the proof of Theorem 3.2, that the regularized equations have global-in-time solutions.

It remains to show that the solution derived in Proposition 3.6 can be continued for all time. We invoke the continuation property of ODEs on a Banach space, that solutions can be continued in time provided that the functional F is time independent. That is, we have Theorem 3.3.

Theorem 3.3. *Continuation of an Autonomous ODE on a Banach Space. Let $O \subset \mathbf{B}$ be an open subset of a Banach space $\mathbf{B}$, and let $F : O \to \mathbf{B}$ be a locally Lipschitz continuous operator. Then the unique solution $X \in C^1([0,T); O)$ to the autonomous ODE,*

$$\frac{dX}{dt} = F(X), \qquad X|_{t=0} = X_0 \in O,$$

either exists globally in time, or $T < \infty$ and $X(t)$ leaves the open set O as $t \nearrow T$.

For a proof, the reader is referred to Hartman (1982). The solution can be continued for all time, provided that we can show an a priori bound on $\|v^\epsilon(\cdot,t)\|_m$. To do this we note that relation (3.54) with $v^2(x,t) \equiv 0$ gives the bound

$$\frac{d}{dt}\|v^\epsilon(\cdot,t)\|_m \leq C(\|v^\epsilon\|_0, \epsilon, N)\|v^\epsilon\|_m.$$

Energy bound (3.53) gives

$$\frac{d}{dt}\|v^\epsilon(\cdot,t)\|_m \leq C(\|v_0\|_0, \epsilon, N)\|v^\epsilon\|_m,$$

which, by Grönwall's lemma, Lemma 3.1, implies that the a priori bound $\|v^\epsilon(\cdot,T)\|_m \leq e^{cT}$. □

3.2.3. Local-In-Time Existence of Solutions to the Euler and the Navier–Stokes Equations

In the previous Subsection we proved global existence and uniqueness of solutions to the regularized equation. That is, for any fixed viscosity $\nu \geq 0$ and any regularization parameter $\epsilon > 0$ there exist solutions $v^\epsilon \in C^1([0,\infty); V^m)$, $m \in \mathbb{Z}^+ \cup \{0\}$, to regularized equation (3.51):

$$v_t^\epsilon = \nu\mathcal{J}_\epsilon^2 \Delta v^\epsilon - P\mathcal{J}_\epsilon[(\mathcal{J}_\epsilon v^\epsilon)\cdot\nabla(\mathcal{J}_\epsilon v^\epsilon)],$$
$$v^\epsilon|_{t=0} = v_0.$$

We now show that, provided that $m > N/2 + 2$, there exists a time interval of existence $[0,T]$ and a subsequence (v^ϵ) convergent to a limit function $v \in C([0,T]; C^2) \cap C^1([0,T]; C)$ that solves the Euler or the Navier–Stokes equation. More precisely, we have Theorem 3.4.

Theorem 3.4. *Local-in-Time Existence of Solutions to the Euler and the Navier–Stokes equations. Given an initial condition $v_0 \in V^m$, $m \geq [\frac{N}{2}] + 2$, then the following results*

(i) there exists a time T with the rough upper bound

$$T \leq \frac{1}{c_m\|v_0\|_m}, \tag{3.55}$$

such that for any viscosity $0 \leq \nu < \infty$ there exists the unique solution $v^\nu \in C\{[0,T]; C^2(\mathbb{R}^3)\} \cap C^1\{[0,T]; C(\mathbb{R}^3)\}$ to the Euler or the Navier–Stokes equation. The solution v^ν is the limit of a subsequence of approximate solutions, v^ϵ, of Eqs. (3.51) and (3.52).

(ii) The approximate solutions v^ϵ and the limit v^ν satisfy the higher-order energy estimates

$$\sup_{0\leq t\leq T}\|v^\epsilon\|_m \leq \frac{\|v_0\|_m}{1 - c_m T\|v_0\|_m}, \tag{3.56}$$

$$\sup_{0\leq t\leq T}\|v^\nu\|_m \leq \frac{\|v_0\|_m}{1 - c_m T\|v_0\|_m}.$$

(iii) The approximate solutions and the limit v^ν are uniformly bounded in the spaces $L^\infty\{[0,T], H^m(\mathbb{R}^3)\}$, $\mathrm{Lip}\{[0,T]; H^{m-2}(\mathbb{R}^3)\}$, and $C_W\{[0,T]; H^m(\mathbb{R}^3)\}$.

Definition 3.2. *The space $C_W\{[0,T]; H^s(\mathbb{R}^3)\}$ denotes continuity on the interval $[0,T]$ with values in the weak topology of H^s, that is, for any fixed $\varphi \in H^s$, $[\varphi, u(t)]_s$ is a continuous scalar function on $[0,T]$, where*

$$(u, v)_s = \sum_{\alpha \le s} \int_{\mathbb{R}^3} D^\alpha u \cdot D^\alpha v \, dx.$$

The strategy for the local-existence proof, which we implement below, is to first prove the bounds (3.56) in the high norm, then show that we actually have a contraction in the $H^0 = L^2$ norm. We then apply an interpolation inequality to prove convergence as $\epsilon \to 0$.

We establish, following the local-existence proof, Theorem 3.5, that the solution v^ν is actually continuous in time in the highest norm H^m and can be continued in time provided that its H^m norm remains bounded. We need this fact in order to discuss, in Section 3.3, the link between the existence of these solutions globally in time and the accumulation of vorticity. Our idea in that section is to show that a quantity, related to the accumulation of vorticity, controls the H^m norm of the solution.

Theorem 3.5. *Continuity in the High Norm. Let v^ν be the solution described in Theorem 3.4. Then*

$$v^\nu \in C([0,T); V^m) \cap C^1([0,T); V^{m-2}).$$

Recall from Subsection 3.2.2 that solutions to the regularized equation satisfy energy bound (3.53):

$$\sup_{0 \le t \le T} \|v^\epsilon\|_0 \le \|v_0\|_0. \tag{3.57}$$

This is due to the special form of regularization of Eqs. (3.48). Note, however, that the a priori estimates

$$\frac{d}{dt}\|v^\epsilon(\cdot,t)\|_m \le C(\|v_0\|_0, \epsilon, N)\|v^\epsilon\|_m$$

have coefficients $C(\|v_0\|_0, \epsilon, N)$ with a bad dependence on ϵ. To prove Theorem 3.4 we need a priori higher derivative estimates that are also independent of ϵ. We achieve this with the following *higher-order energy estimate*.

Proposition 3.7. *The H^m Energy Estimate. Let $v_0 \in V^m$. Then the unique regularized solution $v^\epsilon \in C^1([0,\infty); V^m)$ to Eq. (3.51) satisfies*

$$\frac{d}{dt}\frac{1}{2}\|v^\epsilon\|_m^2 + \nu\|\mathcal{J}_\epsilon \nabla v^\epsilon\|_m^2 \le c_m|\nabla \mathcal{J}_\epsilon v^\epsilon|_{L^\infty}\|v^\epsilon\|_m^2. \tag{3.58}$$

Proof of Proposition 3.7. Let v^ϵ be a smooth solution to Eq. (3.51):

$$v_t^\epsilon = \nu \mathcal{J}_\epsilon^2 \Delta v^\epsilon - P\mathcal{J}_\epsilon[(\mathcal{J}_\epsilon v^\epsilon)\cdot\nabla(\mathcal{J}_\epsilon v^\epsilon)].$$

Recall the techniques used to prove basic energy estimate (3.7) from the first section. In particular, we used integration by parts by means of the divergence theorem. Following a similar procedure, we take the derivative D^α, $|\alpha| \le m$ of this equation and then the L^2 inner product with $D^\alpha v^\epsilon$:

$$\begin{aligned}\left(D^\alpha v_t^\epsilon, D^\alpha v^\epsilon\right) &= \left(D^\alpha \mathcal{J}_\epsilon^2 \Delta v^\epsilon, D^\alpha v^\epsilon\right) - \{D^\alpha P\mathcal{J}_\epsilon[(\mathcal{J}_\epsilon v^\epsilon)\cdot\nabla(\mathcal{J}_\epsilon v^\epsilon)], D^\alpha v^\epsilon\}\\ &= -\nu\|\mathcal{J}_\epsilon D^\alpha \nabla v^\epsilon\|_0^2 - \{P\mathcal{J}_\epsilon[(\mathcal{J}_\epsilon v^\epsilon)\cdot\nabla(D^\alpha\mathcal{J}_\epsilon v^\epsilon)], D^\alpha v^\epsilon\}\\ &\quad - (\{D^\alpha P\mathcal{J}_\epsilon[(\mathcal{J}_\epsilon v^\epsilon)\cdot\nabla(\mathcal{J}_\epsilon v^\epsilon)] - P\mathcal{J}_\epsilon\{(\mathcal{J}_\epsilon v^\epsilon)\cdot\nabla(D^\alpha\mathcal{J}_\epsilon v^\epsilon)]\}, D^\alpha v^\epsilon).\end{aligned}$$

Lemmas 3.5 and 3.6 and the divergence theorem imply that

$$\begin{aligned}\{P\mathcal{J}_\epsilon[(\mathcal{J}_\epsilon v^\epsilon)\cdot\nabla(D^\alpha\mathcal{J}_\epsilon v^\epsilon)], D^\alpha v^\epsilon\} &= \frac{1}{2}\left[\mathcal{J}_\epsilon v^\epsilon, \nabla(\mathcal{J}_\epsilon D^\alpha v^\epsilon)^2\right]\\ &= -\frac{1}{2}\left(\operatorname{div}\mathcal{J}_\epsilon v^\epsilon, |\mathcal{J}_\epsilon D^\alpha v^\epsilon|^2\right) = 0.\end{aligned}$$

Again, the specific choice of the regularization gives a balance of terms during the integration by parts. Summing over $|\alpha| \le m$, we find that calculus inequality (3.32) implies that

$$\begin{aligned}&\frac{1}{2}\frac{d}{dt}\|v^\epsilon\|_m^2 + \nu\|\mathcal{J}_\epsilon\nabla v^\epsilon\|_m^2\\ &\quad\le \|v^\epsilon\|_m \sum_{|\alpha|\le m}\|D^\alpha[(\mathcal{J}_\epsilon v^\epsilon)\cdot\nabla(\mathcal{J}_\epsilon v^\epsilon)] - [(\mathcal{J}_\epsilon v^\epsilon)\cdot\nabla(D^\alpha\mathcal{J}_\epsilon v^\epsilon)]\|_0\\ &\quad\le c_m\|v^\epsilon\|_m(|\nabla\mathcal{J}_\epsilon v^\epsilon|_{L^\infty}\|D^{m-1}\nabla\mathcal{J}_\epsilon v^\epsilon\|_0 + \|D^m\mathcal{J}_\epsilon v^\epsilon\|_0|\nabla\mathcal{J}_\epsilon v^\epsilon|_{L^\infty})\\ &\quad\le c_m|\mathcal{J}_\epsilon\nabla v^\epsilon|_{L^\infty}\|\mathcal{J}_\epsilon v^\epsilon\|_m^2,\end{aligned}$$

so that

$$\frac{d}{dt}\frac{1}{2}\|v^\epsilon\|_m^2 + \nu\|\mathcal{J}_\epsilon\nabla v^\epsilon\|_m^2 \le c_m|\mathcal{J}_\epsilon\nabla v^\epsilon|_{L^\infty}\|v^\epsilon\|_m^2. \qquad \square$$

We can now complete the proof of Theorem 3.4.

First we show that the family (v^ϵ) of regularized solutions is uniformly bounded in H^m. Energy estimate (3.58) and Sobolev inequality (3.30) imply that the time derivative of $\|v^\epsilon\|_m$ can be bounded by a quadratic function of $\|v^\epsilon\|_m$ independent of ϵ, provided that $m > N/2 + 1$:

$$\frac{d}{dt}\|v^\epsilon\|_m \le c_m|\mathcal{J}_\epsilon\nabla v^\epsilon|_{L^\infty}\|v^\epsilon\|_m \le c_m\|v^\epsilon\|_m^2, \tag{3.59}$$

and hence, for all ϵ,

$$\sup_{0\le t\le T}\|v^\epsilon\|_m \le \frac{\|v_0\|_m}{1 - c_m T\|v_0\|_m} = \|v_0\|_m + \frac{\|v_0\|_m^2 c_m T}{1 - c_m T\|v_0\|_m}. \tag{3.60}$$

Thus the family (v^ϵ) *is uniformly bounded in* $C([0,T];H^m)$, $m > N/2$, provided that $T < (c_m\|v_0\|_m)^{-1}$.

Furthermore, the family of time derivatives (dv^ϵ/dt) is uniformly bounded in H^{m-2}. Equation (3.51) implies that, for $m > (N/2)+2$,

$$\left\|\frac{dv^\epsilon}{dt}\right\|_{m-2} \le \nu\left\|\mathcal{J}_\epsilon^2\Delta v^\epsilon\right\|_{m-2} + \|P\mathcal{J}_\epsilon[(\mathcal{J}_\epsilon v^\epsilon)\cdot\nabla(\mathcal{J}_\epsilon v^\epsilon)]\|_{m-2}$$
$$\le c\nu\|v^\epsilon\|_m + c\|v^\epsilon\|_m^2.$$

Hence the previous estimates yield that for a given $0 \le \nu < \infty$ the family (dv^ϵ/dt) *is uniformly bounded in* H^{m-2}.

We now show that the solutions v^ϵ to regularized equation (3.51) form a contraction in the low norm $C\{[0,T]; L^2(\mathbb{R}^3)\}$. Specifically we prove Lemma 3.7.

Lemma 3.7. *The family* v^ϵ *forms a Cauchy sequence in* $C\{[0,T]; L^2(\mathbb{R}^3)\}$. *In particular, there exists a constant* C *that depends on only* $\|v_0\|_m$ *and the time* T *so that, for all* ϵ *and* ϵ',

$$\sup_{0<t<T}\|v^\epsilon - v^{\epsilon'}\|_0 \le C\max(\epsilon,\epsilon').$$

Proof of Lemma 3.7. Using Eq. (3.51), we have that

$$\frac{d}{dt}\frac{1}{2}\|v^\epsilon - v^{\epsilon'}\|_0^2 = \nu\left(\mathcal{J}_\epsilon^2\Delta v^\epsilon - \mathcal{J}_{\epsilon'}^2\Delta v^{\epsilon'}, v^\epsilon - v^{\epsilon'}\right) - \{P\mathcal{J}_\epsilon[(\mathcal{J}_\epsilon v^\epsilon)\cdot\nabla(\mathcal{J}_\epsilon v^\epsilon)]$$
$$- P\mathcal{J}_{\epsilon'}[(\mathcal{J}_{\epsilon'}v^{\epsilon'})\cdot\nabla(\mathcal{J}_{\epsilon'}v^{\epsilon'})], v^\epsilon - v^{\epsilon'}\}$$
$$= T1 + T2.$$

We can estimate the first term, $T1$, by means of integration by parts and mollifier property (3.40):

$$\left(\mathcal{J}_\epsilon^2\Delta v^\epsilon - \mathcal{J}_{\epsilon'}^2\Delta v^{\epsilon'}, v^\epsilon - v^{\epsilon'}\right) = \left[(J_\epsilon^2 - \mathcal{J}_{\epsilon'}^2)\Delta v^\epsilon, v^\epsilon - v^{\epsilon'}\right] - \|\mathcal{J}_{\epsilon'}\nabla(v^\epsilon - v^{\epsilon'})\|_0^2.$$
$$\le C\max(\epsilon,\epsilon')\|v^\epsilon\|_3\|v^\epsilon - v^{\epsilon'}\|_0.$$

We can estimate the second term, $T2$, by also using the same tools and the fact that v^ϵ is divergence free:

$$\{P\mathcal{J}_\epsilon[(\mathcal{J}_\epsilon v^\epsilon)\cdot\nabla(\mathcal{J}_\epsilon v^\epsilon)] - P\mathcal{J}_{\epsilon'}[(\mathcal{J}_{\epsilon'}v^{\epsilon'})\cdot\nabla(\mathcal{J}_{\epsilon'}v^{\epsilon'})], v^\epsilon - v^{\epsilon'}\}$$
$$= \{(\mathcal{J}_\epsilon - \mathcal{J}_{\epsilon'})[\mathcal{J}_\epsilon v^\epsilon\cdot\nabla(\mathcal{J}_\epsilon v^\epsilon)], v^\epsilon - v^{\epsilon'}\}$$
$$+ \{\mathcal{J}_{\epsilon'}[(\mathcal{J}_\epsilon - \mathcal{J}_{\epsilon'})v^\epsilon\cdot\nabla(\mathcal{J}_\epsilon v^\epsilon)], v^\epsilon - v^{\epsilon'}\}$$
$$+ \{\mathcal{J}_{\epsilon'}[\mathcal{J}_{\epsilon'}(v^\epsilon - v^{\epsilon'})\cdot\nabla(\mathcal{J}_\epsilon v^\epsilon)], v^\epsilon - v^{\epsilon'}\}$$
$$+ [\mathcal{J}_{\epsilon'}(\mathcal{J}_{\epsilon'}\{v^{\epsilon'}\cdot\nabla[(\mathcal{J}_\epsilon - \mathcal{J}_{\epsilon'})v^\epsilon]\}), v^\epsilon - v^{\epsilon'}]$$
$$+ [\mathcal{J}_{\epsilon'}(\mathcal{J}_{\epsilon'}\{v^{\epsilon'}\cdot\nabla[\mathcal{J}_{\epsilon'}(v^\epsilon - v^{\epsilon'})]\}), v^\epsilon - v^{\epsilon'}]$$
$$= R1 + R2 + R3 + R4 + R5.$$

Using calculus inequality (3.108) and Sobolev inequality (3.30), we get

$$\begin{aligned}|R1| &\le C\max(\epsilon,\epsilon')\|[v^\epsilon\cdot\nabla(v^\epsilon)]\|_1\|v^\epsilon - v^{\epsilon'}\|_0\\ &\le C\max(\epsilon,\epsilon')(|v^\epsilon|_{L^\infty} + |\nabla v^\epsilon|_{L^\infty})\|\nabla(v^\epsilon)\|_1\|v^\epsilon - v^{\epsilon'}\|_0\\ &\le C\max(\epsilon,\epsilon')\|v^\epsilon\|_m^2\|v^\epsilon - v^{\epsilon'}\|_0.\end{aligned}$$

A similar estimate holds for $R2$ and $R4$,

$$|R3| \le C\|v^\epsilon - v^{\epsilon'}\|_0^2\|v^\epsilon\|_m,$$

and, finally, integration by parts and the fact that $v^{\epsilon'}$ is divergence free shows that

$$\begin{aligned}R5 &= \{\mathcal{J}_{\epsilon'}v^{\epsilon'}\cdot\nabla[\mathcal{J}_{\epsilon'}(v^\epsilon - v^{\epsilon'})], \mathcal{J}_{\epsilon'}(v^\epsilon - v^{\epsilon'})\}\\ &= \frac{1}{2}\int_{\mathbb{R}^3}\mathcal{J}_{\epsilon'}v^{\epsilon'}\cdot\nabla\left[|\mathcal{J}_{\epsilon'}(v^\epsilon - v^{\epsilon'})|^2\right]dx = 0.\end{aligned}$$

Putting this all together gives

$$\frac{d}{dt}\|v^\epsilon - v^{\epsilon'}\|_0 \le C(M)[\max(\epsilon,\epsilon') + \|v^\epsilon - v^{\epsilon'}\|_0],$$

where M is an upper bound, from relation (3.60) for the $\|v^\epsilon\|_m$ on $[0, T]$. Integrating this yields

$$\begin{aligned}\sup_{0<t<T}\|v^\epsilon - v^{\epsilon'}\|_0 &\le e^{C(M)T}\left[\max(\epsilon,\epsilon') + \left\|v_0^\epsilon - v_0^{\epsilon'}\right\|_0\right] - \max(\epsilon,\epsilon')\\ &\le C(M,T)\max(\epsilon,\epsilon'), \qquad (3.61)\end{aligned}$$

where we establish the final inequality by recalling that $v_0^\epsilon = v_0^{\epsilon'}$.

Thus v^ϵ is a Cauchy sequence in $C\{[0, T]; L^2(\mathbb{R}^3)\}$ so that it converges strongly to a value $v^\nu \in C\{[0, T]; L^2(\mathbb{R}^3)\}$. □

We have just proved the existence of a v such that

$$\sup_{0\le t\le T}\|v^\epsilon - v\|_0 \le C\epsilon. \qquad (3.62)$$

We now use the fact that the v^ϵ are uniformly bounded in a high norm to show that we have strong convergence in all the intermediate norms. To do this, we need the following interpolation lemma for the Sobolev spaces.

Lemma 3.8. *Interpolation in Sobolev Spaces. (Adams, 1995). Given $s > 0$, there exists a constant C_s so that for all $v \in H^s(\mathbb{R}^N)$ and $0 < s' < s$,*

$$\|v\|_{s'} \le C_s\|v\|_0^{1-s'/s}\|v\|_s^{s'/s}. \qquad (3.63)$$

We now apply the interpolation lemma to the difference $v^\epsilon - v$. Taking $s = m$ and using relations (3.60) and (3.62) gives

$$\sup_{0\le t\le T} \|v^\epsilon - v\|_{m'} \le C(\|v_0\|_m, T)\epsilon^{1-m'/m}.$$

Hence for all $m' < m$ we have strong convergence in $C\{[0, T]; H^{m'}(\mathbb{R}^3)\}$. With $0 < 7/2 < m' < m$, this implies strong convergence in $C\{[0, T]; C^2(\mathbb{R}^3)\}$. Also, from the equation

$$v_t^\epsilon = \nu \mathcal{J}_\epsilon^2 \Delta v^\epsilon - P\mathcal{J}_\epsilon[(\mathcal{J}_\epsilon v^\epsilon)\cdot\nabla(\mathcal{J}_\epsilon v^\epsilon)],$$

so that v_t^ϵ converges in $\{C[0, T], C(\mathbb{R}^3)\}$ to $\nu\Delta v - P(v\cdot\nabla v)$. Because $v^\epsilon \to v$, the distribution limit of v_t^ϵ must be v_t so, in particular, v is a classical solution of the Navier–Stokes (Euler) equations.

To show part (iii) of Theorem 3.4 we introduce the notion of weak convergence.

Definition 3.3. *Given a Banach space* **B** *and a sequence* f_ϵ *in* **B**, f_ϵ *converges* strongly *in* **B** *if there exists* $f \in \mathbf{B}$ *so that* $\|f_\epsilon - f\|_{\mathbf{B}} \to 0$ *as* $\epsilon \to 0$. *We use the notation* $f_\epsilon \to f$ *to denote strong convergence. Given a Hilbert space* **H** *with inner product* $(u, v)_{\mathbf{H}}$, *the sequence* u_ϵ *is said to converge* weakly *to* u *in* **H** ($u_\epsilon \rightharpoonup u$) *if for all* $v \in \mathbf{H}$, $(u_\epsilon, v) \to (u, v)$ *as* $\epsilon \to 0$.[††]

One consequence of the Banach–Alaoglu theorem (see Royden, 1968, Chap. 10) is that a bounded sequence $\|u_\epsilon\|_m \le C$ in $H^m(\mathbb{R}^N)$ has a subsequence that converges *weakly* to some limit in H^m, $u_\epsilon \rightharpoonup u$.

The preceding arguments show that

$$\sup_{0\le t\le T} \|v^\epsilon\|_m \le M, \tag{3.64}$$

$$\sup_{0\le t\le T} \left\|\frac{\partial v^\epsilon}{\partial t}\right\|_{m-2} \le M_1. \tag{3.65}$$

Hence v^ϵ is uniformly bounded in the Hilbert space $L^2\{[0, T]; H^m(\mathbb{R}^3)\}$ so there exists a subsequence that converges weakly to

$$v \in L^2\{[0, T]; H^m(\mathbb{R}^3)\}. \tag{3.66}$$

Moreover, if we fix $t \in [0, T]$, the sequence $v^\epsilon(\cdot, t)$ is uniformly bounded in H^m, so that it also has a subsequence that converges weakly to $v(t) \in H^m$. Thus we see that for each t, $\|v\|_m$ is bounded. This, combined with relation (3.66), implies that $v \in L^\infty([0, T]; H^m)$. A similar argument, applied to estimate (3.65), shows that $v \in \mathrm{Lip}\{[0, T]; H^{m-2}(\mathbb{R}^3)\}$.

In addition, v is continuous in the weak topology of H^m. To prove that $v \in C_W\{[0, T]; H^m(\mathbb{R}^3)\}$, let $[\varphi, u]$, $\varphi \in H^{-m}$ denote the dual pairing of H^{-m} [defined

[††] For an elementary discussion of Banach and Hilbert spaces and strong and weak convergence, see Riesz and Nagy (1990, Chap. V).

by means of Eq. (3.29)] and H^m through the L^2 inner product. Because $v^\epsilon \to v$ in $C([0, T]; H^{m'})$, it follows that $[\varphi, v^\epsilon(\cdot, t)] \to [\varphi, v(\cdot, t)]$ uniformly on $[0, T]$ for any $\varphi \in H^{-m'}$. Using relation (3.64) and the fact that $H^{-m'}$ is dense in H^{-m} for $m' < m$, by means of an $\epsilon/2$ argument with relation (3.64), we have $[\varphi, v^\epsilon(\cdot, t)] \to [\varphi, v(\cdot, t)]$ uniformly on $[0, T]$ for any $\varphi \in H^{-m}$. This fact implies that $v \in C_W([0, T]; H^m)$.

This concludes the proof of Theorem 3.4. □

Proof of Theorem 3.5. We now prove Theorem 3.5, i.e., that the limiting solution is in $C([0, T]; H^m) \cap C^1([0, T]; H^{m-2})$. By virtue of the Navier–Stokes equations it is sufficient to show that $v \in C([0, T]; H^m)$. It is important to obtain this sharper result of continuity in time with values in H^m in order to prove the global-in-time results of the following section.

By virtue of the fact that $v \in C_W\{[0, T]; H^m(\mathbb{R}^3)\}$, it suffices to show that the norm $\|v(t)\|_m$ is a continuous function of time. This proof uses one strategy for the case $\nu = 0$, involving the time reversibility of the Euler equations, and a different strategy, involving the diffusive smoothing from the viscosity in the Navier–Stokes equations, for the case $\nu > 0$.

Case 1: $\nu = 0$

Recall relation (3.60) for the uniform $\|\cdot\|_m$ norm bound for the approximations. Passing to the limit in this equation and using the fact that for fixed t, $\limsup_{\epsilon\to 0} \|v^\epsilon\|_m \geq \|v\|_m$, we obtain

$$\sup_{0\leq t\leq T} \|v\|_m - \|v_0\|_m \leq \frac{\|v_0\|_m^2 c_m T}{1 - c_m T\|v_0\|_m}. \tag{3.67}$$

From the fact that $v \in C_W\{[0, T]; H^m(\mathbb{R}^3)\}$, we have $\liminf_{t\to 0+} \|v(\cdot, t)\|_m \geq \|v_0\|_m$. Estimate (3.67) gives $\limsup_{t\to 0+} \|v(\cdot, t)\|_m \leq \|v_0\|_m$. In particular, $\lim_{t\to 0^+} \|v(\cdot, t)\|_m = \|v_0\|_m$. This gives us strong right continuity at $t = 0$. Becaus the analysis that we performed for the Euler equations is time reversible, we could likewise show strong left continuity at $t = 0$.

It remains to prove continuity of the $\|\cdot\|_m$ norm of the solution at times other than the initial time.

Consider a time $T_0 \in [0, T]$ and the solution $v(\cdot, T_0)$. At this fixed time, $v(\cdot, T_0) \equiv v_0^{T_0} \in H^m(\mathbb{R}^3)$ and, from relation (3.60),

$$\left\|v_0^{T_0}\right\|_m \leq \|v_0\|_m + \frac{\|v_0\|_m^2 c_m T_0}{1 - c_m T_0\|v_0\|_m}. \tag{3.68}$$

So we can take $v_0^{T_0}$ as initial data and construct a forward- and backward-in-time-solution as above by solving regularized equation (3.51). We obtain approximate solutions $v_{T_0}^\epsilon(\cdot, t)$ that satisfy relation (3.59):

$$\frac{d}{dt}\left\|v_{T_0}^\epsilon\right\|_m \leq c_m\left|\mathcal{J}_\epsilon \nabla v_{T_0}^\epsilon\right|_{L^\infty}\left\|v_{T_0}^\epsilon\right\|_m \leq c_m\left\|v_{T_0}^\epsilon\right\|_m^2;$$

we can pass to a limit in $v^{\epsilon}_{T_0}$ as before and find a solution $\tilde{v}$ to the Euler equation on a time interval $[T_0 - T', T_0 + T']$ with initial data v_{T_0}. Following the same estimates as above, we obtain that the time T' satisfies the constraint

$$0 < T' < \frac{1}{c_m \|v_0\|_m} - T_0.$$

Furthermore, this solution $\tilde{v}$ must agree with v on $[T_0 - T', T_0 + T'] \cap [0, T]$ by virtue of Corollary 3.1 and the fact that v and $\tilde{v}$ agree at $t = T_0 \in [0, T]$. Following the argument above used to show that $\|v\|_m$ is continuous at $t = 0$, we conclude that $\|\tilde{v}\|_m$ is continuous at T_0; hence $\|v\|_m$ itself must be continuous at T_0. Because $T_0 \in [0, T]$ is arbitrary, we have just showed that $\|v\|_m$ is a continuous function on $[0, T]$ and hence by the fact that $v \in C_W\{[0, T]; H^m(\mathbb{R}^3)\}$, we obtain $v \in C\{[0, T]; H^m(\mathbb{R}^3)\}$.

Case 2: $\nu > 0$

Following the beginning of the argument in Case 1, we obtain that v has strong right continuity at $t = 0$. For the Navier–Stokes equation, with $\nu > 0$, the equation is time irreversible so we cannot use this to prove left continuity. Instead, however, the forward diffusion introduces smoothing into the evolution. This introduces a gain of derivatives after the initial time $t = 0$ that allows us to prove continuity in time for all $t > 0$. To see this, note that H^m energy estimate (3.58) implies an additional piece of information for the case $\nu > 0$. That is, $\nu \int_0^T \|\mathcal{J}_\epsilon \nabla v^\epsilon\|_m^2$ is bounded independent of ϵ. This is sufficient to guarantee that the limit v is in $L^2([0, T]; H^{m+1})$ (the bound will depend badly on ν and is not true for the Euler equation with $\nu = 0$). Thus, for almost every $T_0 \in [0, T]$, $v(\cdot, T_0) \in H^{m+1}$. In particular, for any $\delta > 0$, there exists $T_0 < \delta$ with $v(\cdot, T_0) \equiv v_0^{T_0} \in H^{m+1}$. Taking $v_0^{T_0}$ as initial data and repeating the above existence construction with $m+1$ instead of m, we have a solution in $C([T_0, T'], H^{\tilde{m}})$ for all $\tilde{m} < m + 1$. The time interval $[T_0, T']$ is any time interval on which we can get bounds for the H^{m+1} norm of the approximating solutions independent of ϵ. From the H^m energy estimate, we see that we merely need to control $\|\nabla \mathcal{J} v^{\epsilon}_{T_0}\|_{L^\infty}$ independent of ϵ. In particular, it suffices to control $\|v^{\epsilon}_{T_0}\|_m$ independent of ϵ. By following the estimates for the time interval T' in terms of the initial data, as before, we can show that $T \leq T' < 1/c_m\|v_0\|_m$. Again, by uniqueness of solutions (Corollary 3.1) this solution is identical to v on its interval of existence. Because $\delta > 0$ was arbitrary this implies in fact that $v \in C\{(0, T]; H^m\}$. Combining this with the strong right continuity at $t = 0$ gives the desired result, that $v \in C([0, T]; H^m)$. □

Note that in the above construction, the initial data alone control the time of existence of the solution:

$$T < \frac{1}{c_m \|v_0\|_m}. \tag{3.69}$$

The fact that the solution $v \in C([0, T]; H^m)$ implies that it can be continued in time provided that $\|v(\cdot, t)\|_m$ remains bounded. That is, construct a solution on a time interval $[0, T]$ for which T satisfies inequality (3.69). At time T, choose $v(x, T)$ as initial data for a new solution and repeat the process by continuing the solution on a time interval $[T, T_1]$ for which

$$T_1 < \frac{c}{\|v(\cdot, T)\|_m}.$$

Clearly the process can be continued either for all time or until $\|v(\cdot, t)\|_m$ becomes infinite. We state this as a corollary.

Corollary 3.2. *Given an initial condition $v_0 \in V^m$, $m \geq [\frac{N}{2}]+2$, then for any viscosity $\nu \geq 0$, there exists a maximal time of existence T^* (possibly infinite) and a unique solution $v^\nu \in C([0, T^*); V^m) \cap C^1([0, T^*); V^{m-2})$ to the Euler or the Navier–Stokes equation. Moreover, if $T^* < \infty$ then necessarily $\lim_{t \to T^*} \|v^\nu(\cdot, t)\|_m = \infty$.*

In the next Subsection we show that the norm $\sup_{0 \leq t \leq T} \|v(\cdot, T)\|_m$ is controlled a priori by the $L^1\{[0, T]; L^\infty(\mathbb{R}^N)\}$ norm of the vorticity. This allows us to link the accumulation of vorticity with the global existence of solutions to the Navier–Stokes and the Euler equations.

3.2.4. Existence of 2D Solutions with Locally Finite-Energy Decomposition

The existence proof featured in this Subsection specifically requires the velocity to have globally finite kinetic energy, $v \in L^2(\mathbb{R}^N)$. As we saw at the end of the previous subsection, there exists a large class of 2D flows that do not have globally finite kinetic energy. Recall that any 2D flow with vorticity in $L^1(\mathbb{R}^2)$ has a *radial-energy decomposition* of the form

$$v(x, t) = u(x, t) + \bar{v}(x, t), \tag{3.70}$$

$$\int |u(x)|^2 dx < \infty, \qquad \operatorname{div} u = 0, \tag{3.71}$$

$$\bar{v}(x) = \begin{pmatrix} -x_2 \\ x_1 \end{pmatrix} |x|^{-2} \int_0^{|x|} s\bar{\omega}(s, t) ds, \tag{3.72}$$

where $\bar{\omega}(r, t)$ is a smooth radially symmetric vorticity. When $\nu = 0$, $\bar{\omega}(r, t) = \bar{\omega}_0(r)$ is an exact inviscid eddy solution to the Euler equation. When $\nu > 0$, $\bar{\omega}(r, t)$ is an exact viscous eddy solution of the Navier–Stokes equation.

With minor adaptations to the preceding arguments, the energy method proves the local existence of 2D solutions. We apply the preceding analysis to the evolution equation for the finite-energy field u satisfying

$$\frac{\partial u}{\partial t} + u \cdot \nabla u + \bar{v} \cdot \nabla u + u \cdot \nabla \bar{v} = -\nabla p + \nu \Delta u. \tag{3.73}$$

We present an outline for the local-existence proof. Because $\bar{v}(x, t)$ is a known smooth velocity field, the estimates here immediately imply estimates for the smoothness of the full solution v that is due to the radial-energy decomposition. First we regularize Eq. (3.73) and project onto the space of divergence-free functions as in

Eq. (3.50) to obtain

$$\frac{\partial u^\epsilon}{\partial t} + P\{\mathcal{J}_\epsilon[(\mathcal{J}_\epsilon u^\epsilon)\cdot\nabla\mathcal{J}u^\epsilon] + \mathcal{J}_\epsilon(\bar{v}\cdot\nabla\mathcal{J}_\epsilon u^\epsilon) + \mathcal{J}_\epsilon[(\mathcal{J}_\epsilon u^\epsilon)\cdot\nabla\bar{v}]\} = \nu\mathcal{J}_\epsilon^2\Delta u^\epsilon. \tag{3.74}$$

We now straightforwardly apply the arguments of Subsections 3.2.2 and 3.2.3 to Eq. (3.73) by using regularization (3.74).

As in Subsection 3.2.2, regularization (3.74) reduces to an ODE on the Banach space:

$$V^s = \{u \in H^s(\mathbb{R}^2) : \operatorname{div} u = 0\}.$$

Following the arguments used to prove Theorem 3.2 we have Theorem 3.2A.

Theorem 3.2A. *Given an initial 2D velocity $v_0(x)$ and a corresponding local finite-energy decomposition $u_0(x) + \bar{v}_0(x)$, $u_0 \in V^m(\mathbb{R}^2)$, $m \in \mathbb{Z}^+ \cup \{0\}$, and $\bar{v}_0(x)$ is a smooth function of the form of Eq. (3.72). Then for any $\epsilon > 0$ there exists for all time a unique solution $u^\epsilon \in C^1([0,\infty); V^m)$ to regularized equation (3.74).*

As in Subsection 3.2.2 we prove Theorem 3.2A by using the following proposition.

Proposition 3.6A. *Consider an initial 2D velocity that satisfies the conditions of Theorem 3.2 A. Then*

(i) for any $\epsilon > 0$ there exists the unique solution $u^\epsilon \in C^1([0, T_\epsilon); V^m)$ to ODE. (3.51), where $T_\epsilon = T(\|u_0\|_m, \epsilon)$;
(ii) on any time interval $[0, T]$ on which this solution belongs to $C^1([0, T]; V^0)$,

$$\sup_{0\le t\le T} \|u^\epsilon(\cdot,t)\|_0 \le \|u^\epsilon(\cdot,0)\|_0\left(\exp\int_0^T \|\nabla\bar{v}\|_{L^\infty} dt\right). \tag{3.75}$$

The proof of Proposition 3.6A is so similar to the proof of Proposition 3.6 that we leave it as an exercise for the reader. The time dependence of relation (3.75) follows from Grönwall's lemma combined with integration by parts as in the proof of energy bound (3.53). Note, however, that $\|u^\epsilon(\cdot,t)\|_0$ does not have an upper bound independent of T because of energy contributions to $u(\cdot,t)$ from the radial term $\bar{v}$. This growth rate is also apparent in the 2D basic energy estimate from Proposition 3.4 in Subsection 3.1.3.

To prove local existence of solutions to the Euler or the Navier–Stokes equations, we follow the same procedure detailed in Subsection 3.2.3 for the 3D case. In particular, we require that the 2D regularized solutions also satisfy the following proposition.

Proposition 3.7A. *2D H^m Energy Estimate. Given $u_0 \in V^m$, $m \in \mathbb{Z}^+ \cup \{0\}$, the regularized solution $u^\epsilon \in C^1([0, T_\epsilon); V^m)$ to Eq. (3.74) satisfies*

$$\frac{d}{dt}\|u^\epsilon\|_m \le c_m[|\nabla\mathcal{J}_\epsilon u^\epsilon|_{L^\infty} + |\nabla\bar{v}(x)|_{L^\infty}]\|u^\epsilon\|_m. \tag{3.76}$$

As in Subsection 3.2.3, choosing $m > 2$ and applying the Sobolev inequality gives

$$\frac{d}{dt}\|u^\epsilon\|_m \le c_m\|u^\epsilon\|_m^2 + C(\bar{v})\|u^\epsilon\|_m,$$

where $C(\bar{v}) = |\nabla\bar{v}|_{L^\infty}$.

This estimate gives the the following bound independent of ϵ:

$$\|u^\epsilon\|_m \le \frac{C(\bar{v})C_1(u_0)e^{C(\bar{v})t}}{1 - c_m C_1(u_0)e^{C(\bar{v})t}}.$$

Here

$$C_1(u_0) = \left[\frac{\|u_0\|_m}{C(\bar{v}) + c_m\|u_0\|_m}\right].$$

In particular we have the existence of a time interval $[0, T]$ on which we can show that there exists M and M_1 independent of ϵ such that

$$\|u^\epsilon(\cdot, t)\|_m \le M,$$

$$\left\|\frac{\partial u^\epsilon}{\partial t}(\cdot, t)\right\|_m \le M_1.$$

The passage to the limit in the regularization parameter ϵ and the time continuity of the m norm then follow exactly as in the end of Subsection 3.2.3.

In the next section we show that an a priori pointwise bound on the vorticity in two dimensions gives the global existence of solutions to the Euler and the Navier–Stokes equation.

3.3. Accumulation of Vorticity and the Existence of Smooth Solutions Globally in Time

In the previous section we used energy methods to show that the Euler and the Navier–Stokes equations have a unique classical solution $v \in C^1([0, T]; C^2 \cap V^m)$, where $V^m = \{v \in H^m(\mathbb{R}^N) : \text{div } v = 0\}$, $m > N/2 + 2$. We also showed that the solution can be continued in time provided that $\|v(\cdot, t)\|_m$ remains bounded. It is not known in general if smooth solutions exist globally in time. In this section we show that global existence is linked to accumulation of vorticity. We prove the same result in Chap. 4, Section 4.2, by using the particle method for the case of the Euler equation. Specifically we show that the supremum of the H^m norm of v on any time interval $[0, T]$ is controlled a priori by the $L^1\{[0, T]; L^\infty(\mathbb{R}^N)\}$ norm of the vorticity $\omega(x, t)$. This result gives global-in-time existence of solutions for both the 2D Euler and Navier–Stokes equations.

Recall from Chap. 1 that we showed the existence of smooth exact solutions to the 3D Euler and Navier–Stokes equations in which the vorticity grows without bound. Specifically, we saw in Example 1.4 (rotating jet) and in Example 1.7 (inviscid strained shear layer) that the vorticity can grow exponentially without bound. Despite the unboundedness of the vorticity, these were both global solutions in time. An important point in this chapter is that control of the maximum of the vorticity is sufficient to control the smoothness of a solution to either the Euler or the Navier–Stokes equation. Specifically, in this section we show that, independent of the size of the viscosity, if for any $T > 0$ there exists $M_1 > 0$ such that the time integral of the supremum norm of the vorticity $\omega = \text{curl } v$ is bounded,

$$\int_0^t |\omega(\cdot, s)|_{L^\infty} ds \le M_1, \qquad \forall\, 0 \le t \le T,$$

then the classical solution v exists globally in time. Conversely, if the maximal time T of the existence of classical solutions is finite, then necessarily

$$\lim_{t \nearrow T} \int_0^t |\omega(\cdot, s)|_{L^\infty} ds = \infty.$$

For 2D inviscid flows, the vorticity is bounded by its initial maximum; hence such flows exist globally in time. These results have very bad dependence on ν and do not hold in the limit as $\nu \to 0$. The point of proving the result here is that the proof works regardless of the size of ν and in particular works in the limit as $\nu \to 0$. In the inviscid limit this result is the sharpest result known (which is due to Beale et al, 1984). It is not known whether or not solutions to the Euler or Navier–Stokes equations can be smoothly continued for all time. In Chap. 5 we discuss various issues involving finite-time singularities in solutions to the Euler and the Navier–Stokes equations.

In this section we use energy methods to prove this result for both the Euler and the Navier–Stokes equations. We use Eulerian variables combined with some potential theory estimates on Sobolev spaces. The method here is in contrast to the techniques in Chap. 4 that combine potential theory estimates on Hölder spaces, Lagrangian variables, and particle trajectories.

By the construction of a local-in-time solution to the Euler or the Navier–Stokes equation from Section 3.2, the solution can be continued in time provided that $\|v\|_m$ remains bounded. That is, T is the maximal time of the existence of smooth solutions $v \in C^1([0, T); V^m)$ if and only if $\lim_{t \nearrow T} \|v(\cdot, t)\|_m = \infty$. We present a necessary condition for such a maximal time.

Theorem 3.6. *L^∞ Vorticity Control and Global Existence. Let the initial velocity $v_0 \in V^m$, $m > N/2 + 2$, so that there exists a classical solution $v \in C^1([0, T); C^2 \cap V^m)$ to the 3D Euler or the Navier–Stokes equation. Then:*

(i) If for any $T > 0$ there exists $M_1 > 0$ such that the vorticity $\omega = \text{curl } v$ satisfies

$$\int_0^T |\omega(\cdot, \tau)|_{L^\infty} d\tau \le M_1, \tag{3.77}$$

then the solution v exists globally in time, $v \in C^1([0, \infty); C^2 \cap V^m)$.

(ii) If the maximal time T of the existence of solutions $v \in C^1([0,T); C^2 \cap V^m)$ is finite, then necessarily the vorticity accumulates so rapidly that

$$\lim_{t \nearrow T} \int_0^t |\omega(\cdot,\tau)|_{L^\infty} d\tau = \infty. \tag{3.78}$$

In particular this theorem holds in the limit as $\nu \to 0$ and for the case $\nu = 0$. In fact, classical theory of the Navier–Stokes equations shows that sharper continuation results are true for fixed viscosity $\nu > 0$ (see, for example, Constantin and Foias, 1988, Chap. 10). In Chap. 4 we derive Theorem 4.3 for the Euler equation ($\nu = 0$) by using the Hölder spaces and the particle-trajectory formulation. Theorem 3.6 still does not answer the outstanding question of whether smooth solutions v develop singularities in a finite time.

Before proving this theorem, we state its consequences for the 2D case. With slight modifications to the upcoming arguments, we can show that Theorem 3.6 is true for 2D solutions with a radial-energy decomposition. Specifically we use the modified H^m energy estimate from Proposition 3.7A instead of the 3D H^m energy estimate in Proposition 3.7. However, for 2D flows vorticity equation (3.81) reduces to the scalar equation

$$\omega_t + v \cdot \nabla\omega = \nu\Delta\omega.$$

For inviscid flows ($\nu = 0$) ω is conserved along particle trajectories $X(\alpha, t)$, $\omega(X(\alpha,t)) = \omega_0(\alpha)$ so that $|\omega(\cdot,t)|_{L^\infty} = |\omega_0|_{L^\infty}$. For viscous flows ($\nu > 0$) the maximum principle for parabolic equations[‡‡] implies that

$$|\omega(\cdot,t)|_{L^\infty} \leq |\omega_0|_{L^\infty}, \qquad \forall\, 0 \leq t \leq T.$$

Theorem 3.6 immediately gives the following well-known Corollary.

Corollary 3.3. *Global Existence of 2D Solutions. Given an initial 2D velocity field v_0 with locally finite-energy decomposition $v_0 = u_0 + \bar{v}$ with $u_0 \in H^m(\mathbb{R}^2)$, $m > 3$, and* curl $\bar{v} = \omega_0(r) \in C^\infty(\mathbb{R}^2) \cap L^2(\mathbb{R}^2)$, *then there exists for all time a unique smooth solution*

$$v(x,t) = u(x,t) + \bar{v}(x,t)$$

to the 2D Euler (respectively, Navier–Stokes) equation, with $u(x,t) \in L^\infty\{[0,T]; H^m(\mathbb{R}^2)\}$ on any time interval $[0,T]$ and $\bar{v}(x,t)$ an exact inviscid (respectively, viscous) eddy solution.

Proof of Theorem 3.6. We present the 3D case; the 2D case follows in an analogous fashion by use of radial-energy decomposition.

H^m energy estimate (3.58) from Section 3.2 that we proved for regularized solutions is also true for solutions to the Navier–Stokes and the Euler equations. This is a special

‡‡ See, for example, John, 1982, Chap. 7.

case of general estimate with mollifiers. It states that for all $m \in \mathbb{Z}^+ \cap \{0\}$ an H^m solution to Euler or Navier–Stokes equations satisfies

$$\frac{d}{dt}\|v\|_m \leq c_m |\nabla v|_{L^\infty} \|v\|_m.$$

Grönwall's lemma, Lemma 3.1, then gives that any solution $v \in C^1([0,T); C^2 \cap V^m)$ satisfies

$$\|v(\cdot,T)\|_m \leq \|v_0\|_m e^{\int_0^T c_m |\nabla v(\cdot,t)|_{L^\infty} dt}. \tag{3.79}$$

Hence the solution exists in V^m provided we have an a priori bound on $\int_0^T |\nabla v(\cdot,t)|_{L^\infty} dt$. A similar bound holds for the L^2 norm of the vorticity:

$$\|\omega(\cdot,t)\|_0 \leq \|\omega_0\|_0 \exp\left(c \int_0^T |\nabla v(\cdot,t)|_{L^\infty} dt \right), \tag{3.80}$$

We now show that bound (3.80) results from the vorticity equation and a simple integration by parts. Recall from Section 2.1 that the Navier–Stokes equation has an equivalent vorticity-stream formulation:

$$\begin{aligned} \omega_t + v \cdot \nabla \omega &= \omega \cdot \nabla v + \nu \Delta \omega, \qquad (x,t) \in \mathbb{R}^3 \times [0,\infty), \\ \omega|_{t=0} &= \omega_0, \end{aligned} \tag{3.81}$$

Take the L^2 inner product of vorticity equation (3.81) with ω to get

$$\begin{aligned} \frac{1}{2}\frac{d}{dt}\|\omega\|_0^2 + \nu\|\nabla\omega\|_0^2 &= \int_{\mathbb{R}^3} (\omega \cdot \nabla v)\omega \, dx - \int_{\mathbb{R}^3} (v \cdot \nabla \omega)\omega \, dx \\ &= \int_{\mathbb{R}^3} (\omega \cdot \nabla v)\omega \, dx, \end{aligned}$$

because $\operatorname{div} v = 0$. For $\nu \geq 0$ we have

$$\frac{d}{dt}\|\omega\|_0 \leq |\nabla v(\cdot,t)|_{L^\infty} \|\omega\|_0. \tag{3.82}$$

Grönwall's lemma, Lemma 3.1, gives bound (3.80).

We now introduce a potential theory estimate for Sobolev spaces. The proof, which uses ideas from the theory of SIOs, introduced in Subsection 2.4.2 of Chap. 2 and later developed in Chap. 4, is presented at the end of this section.

Proposition 3.8. *Potential Theory Estimate. Let v be a smooth, $L^2 \cap L^\infty$, divergence-free velocity field and let $\omega = \operatorname{curl} v$. Then*

$$|\nabla v|_{L^\infty} \leq c(1 + \ln^+ \|v\|_3 + \ln^+ \|\omega\|_0)(1 + |\omega|_{L^\infty}). \tag{3.83}$$

Here $\ln^+(x)$ denotes $\ln(x)$ for $x > 1$, and 0 otherwise. Note that this estimate is "kinematic" in that it does not depend explicitly on time.

Plugging relation (3.79) for $m = 3$ and bound (3.80) into relation (3.83), we have

$$|\nabla v(\cdot,t)|_{L^\infty} \le C\left[1 + \int_0^t |\nabla v(\cdot,s)|_{L^\infty} ds)(1 + |\omega(\cdot,t)|_{L^\infty}\right].$$

Grönwall's lemma, Lemma 3.1, implies that

$$|\nabla v(\cdot,t)|_{L^\infty} \le |\nabla v_0|_0 e^{\int_0^t |\omega(\cdot,s)|_{L^\infty} ds},$$

which by relation (3.79) gives an a priori bound for $\|v(\cdot,t)\|_m$ for all $m \ge 3$ provided that $\int_0^t |\omega(\cdot,s)|_{L^\infty} ds$ is bounded. □

Remark: Potential theory estimate (3.83) and the H^m energy estimate provide a maximal growth rate for 2D solutions to the Euler or the Navier–Stokes equations. That is, for $m > 3$, we have

$$\frac{d}{dt}\|v\|_m \le C\|v\|_m(1 + \ln^+\|v\|_m). \tag{3.84}$$

Here we use the fact that in two dimensions both $\|\omega\|_0$ and $\|\omega\|_{L^\infty}$ are a priori bounded by the initial data. Note that relation (3.84) implies that whenever $\|v\|_m > 1$, $d/dt(1 + \ln^+\|v\|_m) \le C(1 + \ln^+\|v\|_m)$ or $1 + \ln^+\|v\|_m \le (1 + \ln^+\|v_0\|_m)e^{Ct}$. Hence we have the a priori bound

$$\|v(\cdot,t)\|_m \le e^{\exp(Ct)-1}\|v_0\|_m^{\exp(Ct)}$$

that depends on only the initial data and t. We directly use relation (3.84) in the next section to show the stability of viscous-splitting algorithms in two dimensions.

Proof of Proposition 3.8. Potential Theory Estimate. Recall that the vorticity ω determines the velocity v in Eq. (3.81) by means of the nonlocal operator

$$v(x,t) = \int_{\mathbb{R}^N} K_N(x-y)\omega(y,t)dy, \qquad x \in \mathbb{R}^N. \tag{3.85}$$

The kernel K_N is homogeneous of degree $1 - N$:

$$K_N(\lambda x) = \lambda^{1-N} K_N(x), \qquad \forall\lambda > 0, \quad 0 \ne x \in \mathbb{R}^N.$$

In Proposition 2.20 we showed that ∇v can also be computed from ω,

$$\nabla v(x) = c\omega(x) + P_3\omega(x), \tag{3.86}$$

where $P_3\omega$ is a SIO defined by the Cauchy principal-value integral:

$$P_3\omega(x) = \text{PV}\int_{\mathbb{R}^3} \nabla K_3(x-y)\omega(y)dy.$$

Consider a cutoff function $\rho(r)$ such that $\rho(r) = 0$ for $r > 2R_0$, $\rho(r) = 1$ for $r < R_0$, and $\rho \geq 0$. Decompose $P_3\omega(x)$ into two parts:

$$\begin{aligned} P_3\omega(x) &= \mathrm{PV}\int_{\mathbb{R}^3} \nabla K_3(x-y)\rho(|x-y|)\omega(y)dy \\ &\quad + \mathrm{PV}\int_{\mathbb{R}^3} \nabla K_3(x-y)[1-\rho(|x-y|)]\omega(y)dy \\ &= (\nabla v)_1(x) + (\nabla v)_2(x). \end{aligned}$$

We now use a potential theory estimate (see Lemma 4.6). Because this result and many similar results are used in Chap. 4 we refer the reader to those pages for the proof. If ω has support inside a ball of radius R, then

$$\|\nabla v\|_{L^\infty} \leq c\left\{|\omega|_{C^\gamma}\epsilon^\gamma + \max\left(1, \ln\frac{R}{\epsilon}\right)\|\omega\|_{L^\infty}\right\}, \qquad \forall \epsilon > 0. \tag{3.87}$$

Applying relation (3.87) to $(\nabla v)_1$ implies that for any $\epsilon > 0$

$$|(\nabla v)_1|_{L^\infty} \leq c\left\{|\omega|_{C^\gamma}\epsilon^\gamma + \max\left(1, \ln\frac{R_0}{\epsilon}\right)|\omega|_{L^\infty}\right\},$$

where $|\cdot|_{C^\gamma}$, $0 < \gamma < 1$, denotes the Hölder norm. Sobolev inequality (3.30) implies that $\|\omega\|_\gamma \leq c\|\omega\|_2$ for all $\omega \in H^2(\mathbf{R}^3)$ and thus

$$\begin{aligned} |(\nabla v)_1|_{L^\infty} &\leq c\left\{\|\omega\|_2\epsilon^\gamma + \max\left(1, \ln\frac{R_0}{\epsilon}\right)|\omega|_{L^\infty}\right\} \\ &\leq c\left\{\|v\|_3\epsilon^\gamma + \max\left(1, \ln\frac{R_0}{\epsilon}\right)|\omega|_{L^\infty}\right\}. \end{aligned}$$

Furthermore, Schwarz inequality (3.4) gives

$$|(\nabla v)_2|_{L^\infty} \leq C R_0^{-N/2}\|\omega\|_0.$$

Taking $0 < \epsilon < R_0$ as $\epsilon = 1$ if $\|v\|_3 \leq 1$ and $\|v\|_3^{-\gamma}$ otherwise, and $R_0^{N/2} = \|\omega\|_0$ finally gives relation (3.83), we have

$$|\nabla v|_{L^\infty} \leq c\{(1 + \ln^+\|v\|_3 + \ln^+\|\omega\|_0)(1 + |\omega|_{L^\infty})\}. \qquad \square$$

3.4. Viscous-Splitting Algorithms for the Navier–Stokes Equation

One key point of this chapter is that, for domains without boundaries, smooth solutions to the Euler equation are good approximations to smooth solutions to the Navier–Stokes equation with small viscosity. Specifically, in Proposition 3.2 the difference between v^ν and v^0 is

$$\sup_{0\leq t\leq T} \|v^\nu(\cdot, t) - v^0(\cdot, t)\|_0 \leq c\nu T.$$

We use this idea to show that we can well approximate a solution to the Navier–Stokes equation by means of a method of alternatively solving, over small time steps, the inviscid Euler equation and a simple diffusion process. We call such approximations *viscous-splitting algorithms* because they are a form of operator splitting in which the viscous term $\nu \Delta v$ is "split" from the inviscid part of the equation. Operator splitting is a very general technique with many applications beyond hydrodynamics and the Navier–Stokes equations.

In practical terms, for hydrodynamic applications, we use viscous splitting to numerically compute solutions of the Navier–Stokes equation when in possession of a good method for solving the Euler equations and a separate heat equation. The inviscid transport and stretching of vorticity [as described by vorticity equation (3.81)] leads to a simple yet powerful class of vortex methods for computing inviscid solutions (discussed in Chap. 6).

We do not address, in this chapter, specific spatial discretizations for numerically computing the inviscid or the viscous dynamics. Chapter 6 introduces the random-vortex method, a form of viscous splitting, for computing solutions of the high Reynolds number Navier–Stokes equation. This fractional-step algorithm splits the dynamics on each time step into an inviscid evolution computed by the vortex-blob method and a viscous evolution simulated by a random walk of the blobs and vorticity generation at the boundaries. We present the details of the random-vortex method in Chap. 6.

In this chapter we discuss the error associated with two general viscous-splitting methods. We introduce one algorithm, based on the Trotter product formula (Taylor, 1996a, Chap. 11, Appendix A) that we show converges at the rate $C\nu(\Delta t)$, where Δt is the time-step size. We then introduce a second splitting algorithm, based on Strang splitting, that has the advantage of converging as $C\nu(\Delta t)^2$ with no added computational expense.

3.4.1. Viscous-Splitting Algorithm

We consider a general algorithm that decomposes the evolution into two steps, an inviscid step followed by a purely diffusive step. In this Subsection we compute the error associated with splitting up the evolution this way.

The two basic splitting techniques were considered for linear hyperbolic problems by Strang (1968). He deduced the order of convergence by considering a Taylor expansion in time of the true solution compared with the approximation. To illustrate the methods and their order of convergence, we first consider a simpler problem. Consider the vector evolution problem

$$u_t = (A + B)u,$$
$$u_{t=0} = u_0,$$

where u is an N vector and A and B are noncommuting $N \times N$ matrices. The solution is

$$u(t) = e^{(A+B)t} u_0 \tag{3.88}$$

and can be approximated by the two splitting algorithms:

$$\tilde{u}(t) = (e^{A\Delta t} e^{B\Delta t})^n u_0, \tag{3.89}$$

$$\hat{u}(t) = (e^{A\Delta t/2} e^{B\Delta t} e^{A\Delta t/2})^n u_0. \tag{3.90}$$

We see that the Trotter formula in Eq. (3.89) converges to Eq. (3.88) as Δt and that the Strang splitting formulated in Eq. (3.90) converges as $(\Delta t)^2$ by computation of the three Taylor expansions:

$$\begin{aligned}
e^{(A+B)\Delta t} &= I + (A+B)\Delta t + (A^2 + AB + BA + B^2)\Delta t^2/2! + \ldots, \\
e^{A\Delta t} e^{B\Delta t} &= (I + A\Delta t + A^2 \Delta t^2/2! + \cdots)(I + B\Delta t + B^2 \Delta t^2/2! + \cdots) \\
&= I + (A+B)\Delta t + (A^2 + 2AB + B^2)\Delta t^2/2! + \ldots, \\
e^{A\Delta t/2} e^{B\Delta t} e^{A\Delta t/2} &= (I + A\Delta t/2 + A^2 \Delta t^2/8 + \cdots)(I + B\Delta t + B^2 \Delta t^2/2! + \cdots) \\
&\quad \times (I + A\Delta t/2 + A^2 \Delta t^2/8 + \cdots) \\
&= I + (A+B)\Delta t + (A^2 + AB + BA + B^2)\Delta t^2/2! + \cdots.
\end{aligned}$$

If the matrices A and B do not commute then on each time step Δt, the error between Eqs. (3.89) and (3.88) is of the order of $(\Delta t)^2$ whereas the error between Eqs. (3.90) and (3.88) is of the order of $(\Delta t)^3$. We can show that on a time interval $T = n\Delta t$ the errors are, respectively, $\mathcal{O}(T\Delta t)$ and $\mathcal{O}[T(\Delta t)^2]$.

For the Navier–Stokes equation the splitting is done as follows. Discretize time as $t_n = n\Delta t$, and on each time step Δt first solve the Euler equation, then the heat equation to simulate effects of the diffusion term $\nu \Delta v^\nu$. Below we define this algorithm formally.

Denote the solution operator to the Euler equation by $E(t)$, so that $u(t) = E(t)u_0$ solves

$$\begin{aligned}
u_t + P(u \cdot \nabla u) &= 0, \\
\operatorname{div} u &= 0, \\
u|_{t=0} &= u_0.
\end{aligned} \tag{3.91}$$

Also denote the solution operator to the heat equation by $H(t)$, so that $w(t) = H(t)w_0$ solves

$$w_t = \nu \Delta w, \qquad w|_{t=0} = w_0. \tag{3.92}$$

Because $\operatorname{div} w_0 = 0$ implies $\operatorname{div} w = 0$ for all time, we define the first viscous-splitting algorithm by means of a Trotter formula:

$$\tilde{v}_n = [H(\Delta t)E(\Delta t)]^n v_0, \tag{3.93}$$

where $\tilde{v}_n$ is the approximate value of the exact solution v^ν at time $t_n = n\Delta t$. The second viscous-splitting algorithm follows Strang's method:

$$\hat{v}_n = \left[H\left(\frac{\Delta t}{2}\right) E(\Delta t) H\left(\frac{\Delta t}{2}\right) \right]^n v_0. \tag{3.94}$$

Note that algorithms (3.93) and (3.94) are different because in general the operators E and H do not commute. The fact that they do not commute is crucial in proving the respective convergence results of the two algorithms. We show that (3.93) converges as $\nu(\Delta t)$ and (3.94) converges as $\nu(\Delta t)^2$. Hence, algorithm (3.94) is in fact a more accurate method. We see that the accuracy is due to the symmetry of terms in approximating the operator. Before proceeding, we make the observation that the second method, algorithm (3.94), is no more computationally expensive than the first method. This is because of the simple fact that the operator $H(\Delta t/2)$ applied twice is simply the single operator $H(\Delta t)$ so that we can rewrite algorithm (3.94) as

$$\hat{v}_n = H\left(\frac{\Delta t}{2}\right)E(\Delta t)\{H(\Delta t)E(\Delta t)\}^{n-1}H\left(\frac{\Delta t}{2}\right)v_0. \tag{3.95}$$

Before discussing the convergence of the respective algorithms, we present an example that illustrates that viscous splitting is more accurate for approximating solutions to the Navier–Stokes equation than a purely inviscid approximation.

Example 3.1. Recall the viscous eddy solutions (see Example 2.2 and the radial-energy decomposition in Definition 3.1):

$$v^\nu(x,t) = {}^t\left(\frac{-x_2}{|x|^2}, \frac{x_1}{|x|^2}\right)\int_0^{|x|} r\omega^\nu(r,t)dr,$$

where

$$\omega^\nu(|x|,t) = \begin{cases} \omega_0(|x|), & \nu = 0 \\ \dfrac{1}{4\pi\nu t}\displaystyle\int_{\mathbb{R}^2}\exp\left(-\frac{|x-y|}{4\nu t}\right)\omega_0(|y|)dy, & \nu > 0 \end{cases}. \tag{3.96}$$

These solutions have radial symmetry. Hence $\omega^\nu(r,t)$ is an exact solution to the heat equation

$$\frac{d\omega}{\partial t} = \nu\Delta\omega.$$

Furthermore, recall that for any radially symmetric vorticity $\omega(r)$,

$$v(x) = {}^t\left(\frac{-x_2}{|x|^2}, \frac{x_1}{|x|^2}\right)\int_0^{|x|} r\omega(r)dr,$$

is an exact steady solution to the Euler equation. Thus, either viscous-splitting algorithm (3.93) or (3.94) gives a solution with zero error:

$$\|v^\nu(\cdot,n\Delta t) - \tilde{v}_n\|_{L^\infty} = \|v^\nu(\cdot,n\Delta t) - \hat{v}_n\|_{L^\infty} = 0.$$

A purely inviscid approximation $v^{\text{inv}}(\cdot,t)$ gives $v^{\text{inv}}(x,t) = v^0(x)$ and the error is given by Proposition 2.3:

$$\left|v^\nu(x,t) - v_n^{\text{inv}}(x)\right| \le c|x|(\nu t)^{1/2}.$$

This example illustrates the fact that a careful simulation of viscosity effects through splitting algorithms can improve the accuracy of approximate solutions for small viscosity $\nu \ll 1$. We prove this general result below.

The actual implementation of a viscous-splitting algorithm requires a choice of a numerical method for solving Euler equation (3.91) and heat equation (3.92). We save the discussion of specific choices to Chap. 6, in which we introduce the random-vortex method.

3.4.2. Convergence of the Viscous-Splitting Algorithm

We discuss in detail only 2D solutions. The convergence in two dimensions is made simpler by the fact that by Corollary 3.3 the solutions to the Navier–Stokes equation exist globally in time. We discuss briefly the ideas of the 3D proof at the end of this Subsection. Furthermore, for simplicity we consider solutions that have a globally finite kinetic energy. However, the methods here can easily be applied to the radial-energy decomposition. The details are left to the reader.

Theorem 3.7. *Convergence of $\tilde{v}_n \to v^\nu$ in $L^2(\mathbb{R}^2)$. Let $v_0 \in H^\sigma(\mathbb{R}^2)$, σ sufficiently large,* div $v_0 = 0$, $\nu > 0$, $T < \infty$, *and let v^ν be the solution to the 2D Navier–Stokes equation with initial data v_0. Then the solution $\tilde{v}_n$ to viscous-splitting algorithm (3.93),*

$$\tilde{v}_n = [H(\Delta t)E(\Delta t)]^n v_0,$$

is L^2 convergent to v^ν, and

$$\max_{0 \le n\Delta t \le T} \|\tilde{v}_n - v^\nu(\cdot, n\Delta t)\|_0 \le c(T, \|v_0\|_\sigma)\nu\Delta t. \tag{3.97}$$

Furthermore the solution $\hat{v}_n$ to viscous-splitting algorithm (3.94),

$$\hat{v}_n = [H(\Delta t/2)E(\Delta t)H(\Delta/2)]^n v_0,$$

is L^2 convergent to v^ν, and

$$\max_{0 \le n\Delta t \le T} \|\hat{v}_n - v^\nu(\cdot, n\Delta t)\|_0 \le c(T, \|v_0\|_\sigma)\nu(\Delta t)^2. \tag{3.98}$$

To prove Theorem 3.7 we follow the general philosophy that *stability* and *consistency* of a numerical algorithm imply its convergence. This is exactly true for linear PDEs but must be made rigorous in the case of nonlinear PDEs. We therefore break the proof up into two steps.

The first step (stability) ensures that the algorithm produces an approximate solution that is a priori controlled in an appropriate norm. Here we show that either splitting method (3.93) or (3.93) is $H^m(\mathbb{R}^2)$ stable.

Definition 3.4. *An algorithm is said to be H^m* stable *if there exists a constant C that depends on only T and $\|v_0\|_m$ so that the corresponding solution v_n satisfies*

$$\|v_n\|_m, \|\hat{v}_n\|_m \le c(T, \|v_0\|_m), \qquad \forall 0 \le n\Delta t \le T.$$

Proposition 3.9. *Stability. Let* $v_0 \in H^m(\mathbb{R}^2)$, $m \geq 3$, $\operatorname{div} v_0 = 0$, $\nu > 0$, *and* $T < \infty$. *Then the solutions* $\tilde{v}_n$ *and* $\hat{v}_n$ *to the respective viscous-splitting algorithms (3.93) and (3.94),*

$$\tilde{v}_n = [H(\Delta t)E(\Delta t)]^n v_0,$$
$$\hat{v}_n = \left[H\left(\frac{\Delta t}{2}\right)E(\Delta t)H\left(\frac{\Delta t}{2}\right)\right]^n v_0,$$

are H^m *stable.*

By using this H^m stability we can prove that the algorithms are *consistent* with the PDE by deriving an estimate for the local L^2 error between $\tilde{v}_n$ (respectively, $\hat{v}_n$) and v^ν.

Proposition 3.10. *Consistency. [The Local Error between* $\tilde{v}_n$ *(respectively,* $\hat{v}_n$*) and* v^ν *in* $L^2(\mathbb{R}^2)$*]. Let* $v_0 \in H^\sigma(\mathbb{R}^2)$, $\operatorname{div} v_0 = 0$, $v_0 > \nu > 0$, $T < \infty$, *with* σ *sufficiently large. Let* v^ν *be the viscous solution to the 2D Navier–Stokes equation with initial data* v_0 *and let* $\tilde{v}_n$ *be an* H^σ*-stable solution to viscous-splitting algorithm (3.93). Then the local error*

$$\tilde{r}_n(s) = H(s)E(s)\tilde{v}_n - v^\nu(n\Delta t + s), \qquad 0 \leq s \leq \Delta t,\ (n+1)\Delta t \leq T, \tag{3.99}$$

satisfies

$$\|\tilde{r}_n(s)\|_0 \leq \|\tilde{r}_n(0)\|_0 \exp(cs) + c\nu\Delta t \cdot s, \tag{3.100}$$

where $c = c(T, \|v_0\|_\sigma, \nu_0)$. *If we alternatively consider an* H^σ*-stable solution to viscous-splitting algorithm (3.94), then the local error,*

$$\tilde{r}_n(s) = H(s)E(s)\hat{v}_n - v^\nu(n\Delta t + s), \qquad 0 \leq s \leq \Delta t,\ (n+1)\Delta t \leq T, \tag{3.101}$$

satisfies

$$\|\tilde{r}_n(s)\|_0 \leq \|\tilde{r}_n(0)\|_0 \exp(cs) + c(\Delta t) \cdot s^2, \tag{3.102}$$

where $c = c(T, \|v_0\|_\sigma)$.

Taking in particular $\tilde{r}_n(0) \equiv 0$, from relation (3.100) and (3.102) we get the *consistency* of viscous-splitting algorithms (3.93) and (3.94).

Using the above stability and consistency results we have the proof of Theorem 3.7.

Proof of Theorem 3.7. Convergence. By Proposition 3.10 we estimate the error $\tilde{r}_n(\Delta t) = \tilde{v}_{n+1} - v^\nu[(n+1)\Delta t]$ as

$$\|\tilde{v}_{n+1} - v^\nu[(n+1)\Delta t]\|_0 \leq \|\tilde{v}_n - v^\nu(n\Delta t)\|_0 \exp(c\Delta t) + c\nu(\Delta t)^2.$$

Standard induction implies that

$$\|\tilde{v}_n - v^\nu(n\Delta t)\|_0 \le c\nu(\Delta t)^2 \sum_{j=1}^{n-1} \exp(jc\Delta t) = c\nu(\Delta t)^2 \frac{\exp(cn\Delta t) - 1}{\exp(c\Delta t) - 1}$$

$$\le (\exp(cT) - 1)\nu\Delta t.$$

A similar argument holds for algorithm (3.94). □

It remains to prove Propositions 3.9 and 3.10.

Proof of Proposition 3.9. Stability. Define on $0 < s < \Delta t$,

$$u(s + t_{n-1}) = E(s)\tilde{v}_{n-1}, \quad \tilde{v}(s + t_{n-1}) = H(s)u(s + t_{n-1}) = H(s)E(s)\tilde{v}_{n-1}.$$

Note that when $s = \Delta t$, $\tilde{v}(t_n) = \tilde{v}_n$. The energy principle for parabolic equations gives, for $m \ge 0$,

$$\|\tilde{v}(s + t_{n-1})\|_m = \|H(t)u(t)\|_m \le \|u(t)\|_m, \qquad 0 \le t \le \Delta t.$$

For 2D inviscid flows the supremum norm of vorticity, $|\omega|_{L^\infty}$, is preserved; hence the evolution operator $E(t)$ satisfies differential inequality (3.84) derived in the remark at the end of the proof of Theorem 3.6:

$$\frac{d}{dt}\|E(t)v_{n-1}\|_m \le C[1 + \ln^+\|E(t)v_{n-1}\|_m]\|E(t)v_{n-1}\|_m.$$

Furthermore, the vorticity in the heat equation step satisfies a maximum principle. Hence we have the following bound:

$$\|E(\Delta t)\tilde{v}_{n-1}\|_m \le (e\|\tilde{v}_{n-1}\|)^{\exp(c\Delta t)}.$$

Combining the above results gives

$$\|\tilde{v}_n\|_m \le \|v_0\|^{\exp(cn\Delta t)}.$$

□

To prove the consistency result in Proposition 3.10, we recall the example from Subsection 3.4.1 in which we derived the different convergence rates of the two algorithms, (3.93) and (3.94), for a simpler model problem describing vector evolution. This approach does not directly give the ν dependence of the convergence. To obtain this we need some energy estimates of the difference between the exact solution and the approximation. Using the same approach, we can show that the Taylor series of $\tilde{v}_n$ and $\hat{v}_n$ indicate their respective convergence rates *with respect to* Δt.

For the operator splitting of the Navier–Stokes equation, we motivate the proof of consistency by first considering the Taylor series

$$v^\nu(t_n + s) = v^\nu(t_n) + s v^\nu_t(t_n) + \frac{s^2}{2} v^\nu_{tt}(t_n) + \mathcal{O}(s^3), \tag{3.103}$$

$$\tilde{v}(t_n + s) = \tilde{v}(t_n) + s \tilde{v}_t(t_n) + \frac{s^2}{2} \tilde{v}_{tt}(t_n) + \mathcal{O}(s^3), \tag{3.104}$$

$$\hat{v}(t_n + s) = \hat{v}(t_n) + s \hat{v}_t(t_n) + \frac{s^2}{2} \hat{v}_{tt}(t_n) + \mathcal{O}(s^3). \tag{3.105}$$

We use shorthand notation for the bilinear form:

$$B\{u, v\} = -P(u \cdot \nabla v).$$

The first and the second time derivatives for v^ν are

$$\begin{aligned}
v^\nu_t(t) &= \nu \Delta v(t) + B\{v(t), v(t)\},\\
v^\nu_{tt}(t) &= \nu \Delta(\nu \Delta v + B\{v, v\}) + B\{\nu \Delta v + B\{v, v\}, v(t)\} + B\{v, \nu \Delta v + B\{v, v\}\}\\
&= \nu^2 \Delta^2 v + \nu \Delta B\{v, v\} + B\{\nu \Delta v, v\} + B\{v, \nu \Delta v\}\\
&\quad + B\{v, B\{v, v\}\} + B\{B\{v, v\}, v\}.
\end{aligned}$$

The reader can check that the first and the second§§ time derivatives for $\tilde{v}$ are

$$\begin{aligned}
\tilde{v}_t(t_n + s)|_{s=0} &= \nu \Delta \tilde{v}(t_n) + B\{\tilde{v}(t_n), \tilde{v}(t_n)\},\\
\tilde{v}_{tt}(t_n + s)|_{s=0} &= \nu^2 \Delta^2 \tilde{v} + 2\nu \Delta B\{\tilde{v}, \tilde{v}\} + B\{B\{\tilde{v}, \tilde{v}\}, \tilde{v}\} + B\{\tilde{v}, B\{\tilde{v}, \tilde{v}\}\}.
\end{aligned}$$

Note that the first derivatives of $\tilde{v}$ and v^ν would agree at $t_n = n\Delta t$ if $v^\nu(t_n) = \tilde{v}(t_n)$ but their second derivatives would not. Thus the best we can expect to prove is Δt convergence for algorithm (3.93).

Now we show how to prove consistency. To compute the first derivative of $\hat{v}$ we note that $E(t)$ is not a linear operator; hence we need the following lemma.

Lemma 3.9. *The function $F(t, v_0) = E(t)v_0$ from H^s to H^{s-2} has a derivative with respect to v_0 that is the linear operator $dF(t, v_0)$ taking w_0 to the solution at time t of*

$$w_t + B\{v, w\} + B\{w, v\} = 0, \qquad w(0) = w_0, \quad v(t) = E(t)v_0.$$

Using this, we leave it as an exercise to the reader to show that

$$\hat{v}_t(t_n + s)|_{s=0} = \nu \Delta \hat{v}(t_n) + B\{\hat{v}(t_n), \hat{v}(t_n)\}.$$

§§ Right-hand derivatives, i.e., at the beginning of the time interval.

Moreover, using the definition of dF to evaluate dF_t, we have

$$\begin{aligned}\hat{v}_{tt}(t_n+s)|_{s=0} &= \nu^2\Delta^2\hat{v} + \nu\Delta B\{\hat{v},\hat{v}\}\\ &+B\{B\{\hat{v},\hat{v}\},\hat{v}\}\\ &+B\{\hat{v},B\{\hat{v},\hat{v}\}\}\\ &+B\{\hat{v},\nu\Delta\hat{v}\}+B\{\nu\Delta\hat{v},\hat{v}\}.\end{aligned}$$

Hence series (3.103) and (3.105) agree up to the order of $(\Delta t)^2$ but series (3.103) and (3.104) agree only up to the order of Δt because of the lack of commutativity of the operators E and H and the lack of symmetry in algorithm (3.93). This fact necessarily implies the slower convergence rate of algorithm (3.93) with respect to (Δt).

Proof of Proposition 3.10. Consistency. We now use the stability of algorithms (3.93) and (3.94) to show the consistency estimates, including their dependence on small ν. First consider initial data $v_0 \in H^\sigma(\mathbb{R}^2)$ with σ sufficiently large (to be chosen later in the proof). Note that the stability from Proposition 3.9 guarantees that $\tilde{v}_n$ and $\hat{v}_n$ are always bounded in $H^\sigma(\mathbb{R}^2)$. We use this fact in the following way. We need to estimate the L^2 norm of the approximation error, that is the difference between $v^\nu(x,t)$ and $\tilde{v}(x,t)$ or $\hat{v}(x,t)$. We begin with the an estimate for $\tilde{v}$. Consider the time interval (t_n, t_{n+1}). On this time interval $\tilde{v}$ satisfies $\tilde{v}(t_n+s) = H(s)E(s)\tilde{v}_n$ and hence

$$\begin{aligned}\frac{\partial}{\partial t}\tilde{v} &= \nu\Delta H(s)E(s)v_n + H(s)B(E(s)v_n, E(s)v_n)\\ &= \nu\Delta\tilde{v} + [H(s)B - BH(s)]E(s)v_n + B(\tilde{v},\tilde{v})\\ &= \nu\Delta\tilde{v} + B(\tilde{v},\tilde{v}) + \tilde{f}_n(s)\\ \tilde{f}_n(s) &= [H(s)B - BH(s)]\tilde{v}_n.\end{aligned}$$

Thus the difference $\tilde{r}_n(s)$ between and $\tilde{v}$ and v^ν satisfies

$$\frac{\partial}{\partial t}\tilde{r}_n(s) = \nu\Delta\tilde{r}_n(s) + + B(v^\nu,\tilde{r}_n) + B(\tilde{r}_n,\tilde{v}) + \tilde{f}_n(s), \qquad 0 \le s \le \Delta t. \tag{3.106}$$

Take the L^2 inner product of Eq. (3.106) with $\tilde{r}_n$. Because $(Pu, v) = (u, v)$ if $\operatorname{div} v = 0$,

$$\frac{1}{2}\frac{d}{dt}\|\tilde{r}_n(s)\|_0^2 + \nu\|\nabla\tilde{r}_n(s)\|_0^2 = -(v^\nu\cdot\nabla\tilde{r}_n,\tilde{r}_n) - (\tilde{r}_n\cdot\nabla\tilde{v}_n,\tilde{r}_n) + (\tilde{f}_n,\tilde{r}_n).$$

The first term on the right-hand side is zero because $\operatorname{div} v^\nu = 0$. Moreover, Schwarz inequality (3.4) gives

$$\begin{aligned}|(\tilde{r}_n\cdot\nabla\tilde{v}_n,\tilde{r}_n)| &\le |\nabla\tilde{v}_n|_0\|\tilde{r}_n\|_0^2 \le c\|\tilde{v}\|_3\,\|\tilde{r}_n\|_0^2,\\ |(\tilde{f},\tilde{r}_n)| &\le \|\tilde{f}\|_0\|\tilde{r}_n\|_0,\end{aligned}$$

so that by Sobolev inequality (3.30),

$$\frac{d}{dt}\|\tilde{r}_n(t)\|_0 \le c\|\tilde{v}(\cdot,t)\|_3\|\tilde{r}_n(t)\|_0 + \|\tilde{f}(t)\|_0. \tag{3.107}$$

The last step is to estimate the function f_n:

$$\tilde{f}_n(s) = [H(s)B - BH(s)]E(s)v_n.$$

By definition, $H(t)$ satisfies

$$H(t)w_0 = H(0)w_0 + \nu\int_0^t \Delta H(\tau)w_0 d\tau,$$

so that

$$H(t) = I + \nu\int_0^t \Delta H(\tau)d\tau \equiv I + \nu\bar{H}(t).$$

Rewrite $\tilde{f}_n$ as [denoting $\omega(s) = E(s)v_n$]

$$\begin{aligned}\tilde{f}_n &= \{[I+\nu\bar{H}(s)]B - B[I+\nu\bar{H}(s)]\}\omega(s)\\ &= [I+\nu\bar{H}(s)]B[\omega(s),\omega(s)] - B\{[I+\nu\bar{H}(s)]\omega(s),[I+\nu\bar{H}(s)]\omega(s)\}\\ &= \nu\{(\bar{H}B - \nu B\bar{H})\omega(s) - B(\bar{H}\omega(s),\omega(s)) - B(\omega(s),\bar{H}\omega(s))\}\end{aligned}$$

Hence, for σ large enough, there is a bound independent of ν for the $C^1\{[0,T];L^2(\mathbb{R}^2)\}$ norm of $\tilde{f}_n/\nu$. Also $\tilde{f}(n\Delta t) = 0$ so that there exists a constant that depends on only σ and v_0 such that

$$\|\tilde{f}_n(s)\|_0 \le C\nu s, \qquad \forall 0 \le s \le \Delta t.$$

Hence relation (3.107) combined with the H^m stability of algorithm (3.93) gives

$$\frac{d}{dt}\|\tilde{r}_n(s)\|_0 \le c\|\tilde{r}_n(s)\|_0 + sc\nu.$$

Grönwall's lemma, Lemma 3.1, gives

$$\|\tilde{r}_n(t)\|_0 \le \|\tilde{r}_n(0)\|_0\exp(ct) + c\nu\Delta t\cdot t, \qquad 0 \le t \le \Delta t.$$

This proves the result for first-order splitting.

To prove the analogous result for $\hat{v}(x,t)$, we note that by using the same procedure in the above arguments, we can write

$$\frac{\partial}{\partial t}\hat{v}(n\Delta t + s) = \nu\Delta\hat{v} - B(\hat{v},\hat{v}) + \hat{f}(s),$$

where

$$\begin{aligned}\hat{f}_n(s) &= [H(s/2)B - BH(s/2)]E(s)H(s/2)v_n\\ &\quad + (\nu/2)H(dE\Delta - \Delta E)H(s/2)v_n\\ &= f_1 + f_2.\end{aligned}$$

Using an argument identical to the one we used to estimate $\tilde{f}_n$ above, we can show that, for σ sufficiently large, $\hat{f}_n/\nu$ is bounded in $C^2\{[0,\Delta t];L^2(\mathbb{R}^2)\}$. By definition,

$\hat{f}_n$ is equal to $\hat{v}_t - v^\nu_{nt}$, where v^ν_{nt} is the time derivative of the solution to the Navier–Stokes equation with initial data $\hat{v}_n$. From the Taylor series calculations for $\hat{v}$ and v^ν we see that in fact both $\hat{f}_n(0) = 0$ and $\hat{f}'_n(0) = 0$. This gives

$$\|\hat{f}_n(s)\|_0 \leq C\nu s^2, \qquad \forall 0 \leq s \leq \Delta t,$$
$$\frac{d}{dt}\|\hat{r}_n(s)\|_0 \leq c\|\hat{r}_n(s)\|_0 + s^2 c\nu\Delta t.$$

Grönwall's lemma, Lemma 3.1, gives

$$\|\hat{r}_n(t)\|_0 \leq \|\hat{r}_n(0)\|_0 \exp(ct) + c\nu\Delta t \cdot t^2, \qquad 0 \leq t \leq \Delta t,$$

which proves the result for second-order splitting. □

Remark on the Proof in Three Dimensions: As in the 2D case, we can prove the respective convergence rates for algorithms (3.93) and (3.94), provided that we are on a time interval $[0, T]$ for which H^m stability is guaranteed. One possible choice, which is discussed in Beale and Majda (1981), is to prove the result for a time interval $[0, T]$ bounded by the blowup time [see relation (3.55)]:

$$T \leq 1/c_m\|v_0\|_m.$$

Another possibility is to consider a time interval $[0, T]$ for which we know a priori that the solution to the inviscid Euler equation stays smooth. We can show if a given initial condition produces a smooth solution to the inviscid Euler equation on a time interval $[0, T]$, then the solution to the Navier–Stokes equation with the same initial data and sufficiently small viscosity necessarily is smooth on the same time interval $[0, T]$. See, for example, Constantin and Foias (1988, Chap. 11) for a proof. This result can be used to prove H^m stability of either splitting algorithm on the time interval $[0, T]$ provided that the viscosity ν is sufficiently small.

3.5. Appendix for Chapter 3

In this section we state some calculus inequalities in the Sobolev spaces and properties of mollifiers that are used for the energy method.

3.5.1. Calculus Inequalities in the Sobolev Spaces

We present the proof of Lemma 3.4.

Lemma 3.4. *Calculus Inequalities in the Sobolev Spaces.*

(i) $\forall m \in \mathbb{Z}$ there exists $c > 0$ such that, $\forall u, v \in L^\infty \cap H^m(\mathbb{R}^N)$,

$$\|uv\|_m \leq c\{|u|_{L^\infty}\|D^m v\|_0 + \|D^m u\|_0 |v|_{L^\infty}\}, \tag{3.108}$$

$$\sum_{0\leq|\alpha|\leq m} \|D^\alpha(uv) - uD^\alpha v\|_0 \leq c\{|\nabla u|_{L^\infty}\|D^{m-1}v\|_0 + \|D^m u\|_0 |v|_{L^\infty}\}. \tag{3.109}$$

(ii) $\forall s > N/2\ \exists c > 0$ *such that,* $\forall u, v \in H^s(\mathbb{R}^N)$,

$$\|uv\|_s \le c\|u\|_s\|v\|_s. \tag{3.110}$$

Proof of Lemma 3.4. (from Klainerman and Majda, 1981). The Hölder inequality and the Leibnitz differentiation formula give

$$\begin{aligned}\|D^\alpha(uv)\|_0 &\le c_\alpha \sum_{\beta\le\alpha} \|D^\beta u D^{\alpha-\beta} v\|_0 \\ &\le c_\alpha \sum_{\beta\le\alpha} \|D^\beta u\|_{L^{2m/|\beta|}} \|D^{\alpha-\beta} v\|_{L^{2m/|\alpha-\beta|}}.\end{aligned}$$

The Gagliardo–Nirenberg inequality

$$\|D^i u\|_{L^{2r/i}} \le c_r \|u\|_{L^\infty}^{1-i/r} \|D^r u\|_0^{i/r}, \qquad 0 \le i \le r,$$

implies that

$$\begin{aligned}\|D^\alpha(uv)\|_0 &\le c_\alpha \sum_{\beta\le\alpha} c_{\beta\alpha} |u|_{L^\infty}^{1-|\beta|/m} \|D^m u\|_0^{|\beta|/m} |v|_{L^\infty}^{1-|\alpha-\beta|/m} \|D^m v\|_0^{|\alpha-\beta|/m} \\ &\le c_\alpha \sum_{\beta\le\alpha} c_{\beta\alpha} (|u|_{L^\infty}\|D^m v\|_0)^{|\alpha-\beta|/m} (|v|_{L^\infty}\|D^m u\|_0)^{|\beta|/m} \\ &\le c_m \{|u|_{L^\infty}\|D^m v\|_0 + \|D^m u\|_0 |v|_{L^\infty}\}.\end{aligned}$$

Summing over all $|\alpha| \le m$ gives calculus inequality (3.108):

$$\|uv\|_m \le c_m\{|u|_{L^\infty}\|D^m v\|_0 + \|D^m u\|_0|v|_{L^\infty}\},$$

Now we prove the sharp calculus inequality (3.109). For $|\alpha| \le m$, the Leibnitz differentiation formula yields

$$\begin{aligned}\|D^\alpha(uv) - uD^\alpha v\|_0 &\le c_\alpha \sum_{\substack{\beta\le\alpha \\ |\beta|>0}} \|D^\beta u D^{\alpha-\beta} v\|_0 \\ &= c_\alpha \sum_{|\beta'|+|\alpha-\beta'|\le m-1} \|D^{\beta'}(Du)D^{\alpha-\beta'} v\|_0.\end{aligned}$$

Applying calculus inequality (3.108) and summing over $|\alpha| \le m$ gives calculus inequality (3.109):

$$\sum_{0\le|\alpha|\le m} \|D^\alpha(uv) - uD^\alpha v\|_0 \le c_m\{|\nabla u|_{L^\infty}\|D^{m-1} v\|_0 + \|D^m u\|_0 |v|_{L^\infty}\}.$$

Finally observe that calculus inequality (3.110), $\|uv\|_m \le c\|u\|_m\|v\|_m$ for $s = m > N/2$, follows from Sobolev inequality (3.30) and calculus inequality (3.108). □

3.5.2. Properties of Mollifiers

A standard technique of energy methods involves a regularization of equations to produce a family of smooth solutions. We recall some properties of regularization by mollifiers. Given a function $\rho \in C_0^\infty(\mathbb{R}^N)$, $\rho \geq 0$, $\int_{\mathbb{R}^N} \rho\, dx = 1$, define the mollification operator $\mathcal{J}_\epsilon$ acting on a function $v \in L^p(\mathbb{R}^N)$, $1 \leq p \leq \infty$, by

$$(\mathcal{J}_\epsilon v)(x) = \epsilon^{-N} \int_{\mathbb{R}^N} \rho\Big(\frac{x-y}{\epsilon}\Big) v(y) dy. \tag{3.111}$$

Lemma 3.5. *Properties of Mollifiers. Let $\mathcal{J}_\epsilon$ be the mollifier defined in Eq. (3.111). Then*

(i) $\forall v \in C^0(\mathbb{R}^N)$, $\mathcal{J}_\epsilon v \to v$ uniformly on any compact set $\Omega \in \mathbb{R}^N$ and

$$|\mathcal{J}_\epsilon v|_{L^\infty} \leq |v|_{L^\infty}. \tag{3.112}$$

(ii) Mollifiers commute with distribution derivatives, i.e.,

$$D^\alpha \mathcal{J}_\epsilon v = \mathcal{J}_\epsilon D^\alpha v, \qquad \forall\, |\alpha| \leq m, \;\; v \in H^m. \tag{3.113}$$

(iii) $\forall u \in L^p(\mathbb{R}^N)$, $v \in L^q(\mathbb{R})$, $1/p + 1/q = 1$,

$$\int_{\mathbb{R}^N} (\mathcal{J}_\epsilon u) v\, dx = \int_{\mathbb{R}^N} u(\mathcal{J}_\epsilon v) dx. \tag{3.114}$$

(iv) $\forall v \in H^s(\mathbb{R}^N)$,

$$\lim_{\epsilon \searrow 0} \|\mathcal{J}_\epsilon v - v\|_s = 0, \tag{3.115}$$

$$\|\mathcal{J}_\epsilon v - v\|_{s-1} \leq C\epsilon \|v\|_s. \tag{3.116}$$

(v) $\forall v \in H^m(\mathbb{R}^N)$, $k \in \mathbb{Z}^+ \cup \{0\}$, and $\epsilon > 0$,

$$\|\mathcal{J}_\epsilon v\|_{m+k} \leq \frac{c_{mk}}{\epsilon^k} \|v\|_m, \tag{3.117}$$

$$|\mathcal{J}_\epsilon D^k v|_{L^\infty} \leq \frac{c_k}{\epsilon^{N/2+k}} \|v\|_0. \tag{3.118}$$

Proof of Lemma 3.5. Without loss of generality assume that $\rho(x/\epsilon) = 0$ for $|x| \geq \epsilon$. For a given compact set $\Omega \subset \mathbb{R}^N$ define the compact set $\Omega_\epsilon = \{x \in \mathbb{R}^N : \text{dist}(x, \Omega) \leq \epsilon\}$. Because $v|_{\Omega_\epsilon}$ is uniformly continuous, given ϵ' there exists $\delta > 0$ such that for all $x, y \in \Omega_\epsilon$, $|x - y| \leq \delta$, then $|v(x) - v(y)| \leq \epsilon'$. If $x \in \Omega$, $\epsilon < \delta$, then

$$\begin{aligned} |\mathcal{J}_\epsilon v(x) - v(x)| &= \left| \epsilon^{-N} \int_{\mathbb{R}^N} \rho\Big(\frac{x-y}{\epsilon}\Big)[v(y) - v(x)] dy \right| \\ &\leq \epsilon' \int_{\mathbb{R}^N} \epsilon^{-N} \rho\Big(\frac{y}{\epsilon}\Big) dy = \epsilon', \end{aligned}$$

so that $\mathcal{J}_\epsilon v \to v$ uniformly on Ω. Calculus inequality (3.112) is obvious.

Now we prove that mollifiers commute with the distribution derivatives. Integrating by parts gives

$$\begin{aligned}(D^\alpha \mathcal{J}_\epsilon v)(x) &= \epsilon^{-N} \int_{\mathbb{R}^N} D_x^\alpha \rho\Big(\frac{x-y}{\epsilon}\Big) v(y)dy \\ &= (-1)^{|\alpha|}\epsilon^{-N} \int_{\mathbb{R}^N} D_y^\alpha \rho\Big(\frac{x-y}{\epsilon}\Big) v(y)dy \\ &= \epsilon^{-N} \int_{\mathbb{R}^N} \rho\Big(\frac{x-y}{\epsilon}\Big) D_y^\alpha v(y)dy = (\mathcal{J}_\epsilon D^\alpha v)(x),\end{aligned}$$

so that Eq. (3.113) follows.

Formula (3.114) is obvious, and the proof of Eq. (3.115), $\lim_{\epsilon\searrow 0} \|\mathcal{J}_\epsilon v - v\|_s = 0$, follows by the Lebesgue dominated convergence theorem and density arguments. We find that the proof of relation (3.116) follows from first noting that from the definition of the mollifier in function (3.34), $\int \rho = 1$, and the fact that ρ is radial implies that $\hat{\rho}(0) = 1$ and moreover $|\hat{\rho}(\xi) - 1| \le C\xi^2$. Also, $\hat{\rho}(\xi) \le C$, so that

$$\left|\frac{(\hat{\rho}(\epsilon\xi)-1)^2}{(1+|\xi|^2)}\right| < C^2\epsilon^4\delta^2, \qquad |\xi| < \delta,$$

$$\left|\frac{(\hat{\rho}(\epsilon\xi)-1)^2}{(1+|\xi|^2)}\right| \le \frac{C}{\delta^2}, \qquad |\xi| \ge \delta.$$

Taking $\delta = 1/\epsilon$ in the above, we can estimate

$$\begin{aligned}\|\mathcal{J}_\epsilon v - v\|_{s-1}^2 &= \int_{\mathbb{R}^3} [\hat{\rho}(\epsilon\xi) - 1]^2 \hat{v}^2 (1+|\xi|^2)^{s-1} d\xi \\ &\le \left|\frac{[\hat{\rho}(\epsilon\xi)-1]^2}{(1+|\xi|^2)}\right|_{L^\infty} \|v\|_s^2 \le C\epsilon^2 \|v\|_s^2.\end{aligned}$$

Now we prove calculus inequality (3.117):

$$\|\mathcal{J}_\epsilon v\|_{m+k} \le \frac{c_{mk}}{\epsilon^k}\|v\|_m.$$

Let $|\alpha| \le m$, $|\beta| \le k$, and $\rho_\beta(x) = D^\beta\rho(x)$. We compute

$$\begin{aligned}D^\beta D^\alpha \mathcal{J}_\epsilon v(x) &= \epsilon^{-N} D^\alpha \int_{\mathbb{R}^N} D_x^\beta \rho\Big(\frac{x-y}{\epsilon}\Big) v(y)dy \\ &= \epsilon^{-|\beta|-N} D^\alpha \int_{\mathbb{R}^N} \rho_\beta\Big(\frac{x-y}{\epsilon}\Big) v(y)dy \\ &= (-1)^{|\alpha|}\epsilon^{-|\beta|-N} \int_{\mathbb{R}^N} D_y^\alpha \rho_\beta\Big(\frac{x-y}{\epsilon}\Big) v(y)dy \\ &= \epsilon^{-|\beta|-N} \int_{\mathbb{R}^N} \rho_\beta\Big(\frac{x-y}{\epsilon}\Big) D_y^\alpha v(y)dy\end{aligned}$$

Because $\rho_\beta \in C_0^\infty$, the Schwarz inequality implies that

$$|D^\beta D^\alpha \mathcal{J}^\epsilon v(x)|^2 \le \epsilon^{-2|\beta|}\left[\epsilon^{-N}\int_{\mathbb{R}^N}\left|\rho_\beta\left(\frac{x-y}{\epsilon}\right)\right|dy\right]$$
$$\times\left[\epsilon^{-N}\int_{\mathbb{R}^N}\left|\rho_\beta\left(\frac{x-y}{\epsilon}\right)\right|\left|D_y^\alpha v(y)\right|^2 dy\right]$$
$$\le \frac{c_\beta}{\epsilon^{2|\beta|}}\,\epsilon^{-N}\int_{\mathbb{R}^N}\left|\rho_\beta\left(\frac{x-y}{\epsilon}\right)\right|\left|D_y^\alpha v(y)\right|^2 dy.$$

Fubini's theorem gives

$$\sum_{\substack{|\beta|\le k\\|\alpha|\le m}}\|D^\beta D^\alpha \mathcal{J}_\epsilon v\|_0^2 \le \sum_{\substack{|\beta|\le k\\|\alpha|\le m}}\frac{c_\beta}{\epsilon^{2|\beta|}}\int_{\mathbb{R}^N}\epsilon^{-N}\int_{\mathbb{R}^N}\left|\rho_\beta\left(\frac{x-y}{\epsilon}\right)\right|\left|D_y^\alpha v(y)\right|^2 dy\,dx$$
$$=\sum_{\substack{|\beta|\le k\\|\alpha|\le m}}\frac{c_\beta}{\epsilon^{2|\beta|}}\int_{\mathbb{R}^N}\left|D_y^\alpha v(y)\right|^2\left(\epsilon^{-N}\int_{\mathbb{R}^N}\left|\rho_\beta\left(\frac{x-y}{\epsilon}\right)\right|dx\right)dy$$
$$\le \frac{c_k}{\epsilon^{2k}}\sum_{|\alpha|\le m}\|D^\alpha v\|_0^2,$$

so that calculus inequality (3.117) follows.

Finally, we prove calculus inequality (3.118):

$$|\mathcal{J}_\epsilon D^k v|_{L^\infty} \le \frac{c_k}{\epsilon^{N/2+k}}\|v\|_0.$$

The Schwarz inequality, denoting $\rho_k(x) = D^k\rho(x)$, implies that

$$|\mathcal{J}_\epsilon D^k v(x)| = \epsilon^{-N-k}\left|\int_{\mathbb{R}^N}\rho_k\left(\frac{x-y}{\epsilon}\right)v(y)dy\right|$$
$$\le \epsilon^{-N/2-k}\left[\epsilon^{-N}\int_{\mathbb{R}^N}\left|\rho_k\left(\frac{x-y}{\epsilon}\right)\right|^2 dy\right]^{1/2}\left[\int_{\mathbb{R}^N}|v(y)|^2 dy\right]^{1/2}$$
$$\le \frac{c_k}{\epsilon^{N/2+k}}\|v\|_0,$$

which gives calculus inequality (3.118). □

Notes for Chapter 3

Whether or not solutions to the 3D Euler and Navier–Stokes equations blow up in finite time is still an open problem. We discuss some details in Chap. 5.

The Existence theory for smooth solutions to the incompressible Euler and Navier–Stokes equations can be found in numerous references including Constantin and Foias (1988), Ebin and Marsden (1970), Kato (1972, 1975), Klainerman and Majda (1981), Taylor (1996b), and Temam (1986). However, the inviscid case and the viscous case are usually treated separately. Logarithmic estimate (3.83) is crucial for proving that the L^∞ norm of the vorticity controls global regularity of solutions to the Euler and the Navier–Stokes equations. This estimate is highlighted in the well-known proof that is due to Beale et al. (1984). The paper by Kato (1986) gives a simple existence theorem for Euler and Navier–Stokes in $\mathbb{R}^2$, uniform in ν, using the log estimate. Another example that we discuss in detail in Chap. 8 is an analogous estimate for the boundary of a vortex patch that we use to prove global regularity. See Appendix B of Taylor (1991) for a discussion of such inequalities. Some recent results concerning the connection between direction of vorticity and global existence for Navier–Stokes are discussed in Constantin and Fefferman (1993).

Beale and Majda (1981) first proved the convergence rate $\mathcal{O}(\nu \Delta t^2)$ for viscous splitting algorithm (3.94). Our proof of the convergence rates of algorithm (3.93) and (3.94) is based on the arguments from their paper.

There exist some results for viscous splitting in the presence of bounded domains. All of these results concern an Euler-Stokes splitting, involving alternating fractional steps by means of an Euler solution with $\mathbf{v} \cdot \mathbf{n} = 0$ at the boundary and a linear Stokes solution with $\boldsymbol{v} = 0$ on the boundary. The estimates deteriorate in the limit of vanishing viscosity. Zhang (1993) proves first-order-in-time L^2 convergence of the Euler-Stokes splitting. Beale and Greengard (1994) prove first-order-in-time convergence in all L^p, $p < \infty$. In addition, Ying (1992) proves convergence in $H^{s,2}$, $s < 1/2$.

Chorin's random-vortex method is described in Chorin (1973).

References for Chapter 3

Adams, R. A., *Sobolev Spaces*, Academic, New York, 1975.
Beale, J. T. and Greengard, C., "Convergence of Euler–Stokes splitting of the Navier–Stokes equations," *Commun. Pure Appl. Math.* **47**, 1083–1115, 1994.
Beale, J. T., Kato, T., and Majda, A., "Remarks on the breakdown of smooth solutions for the 3-D Euler equations," *Commun. Math. Phys.* **94**, 61–66, 1984.
Beale, J. T. and Majda, A., "Rates of convergence for viscous splitting of the Navier–Stokes equations," *Math. Comput.* **37**(156), 243–259, 1981.
Chorin, A. J., "Numerical study of slightly viscous flow," *J. Fluid Mech.* **57**, 785–796, 1973.
Coddington, E. A. and Levinson N., *Theory of Ordinary Differential Equations*, McGraw-Hill, New York, 1955.
Constantin, P., "Note on loss of regularity for solutions of the 3-D incompressible Euler and related equations," *Common. Math. Phys.* **104**, 311–329, 1986.
Constantin, P. and Fefferman, C., "Direction of vorticity and the problem of global regularity for the Navier–Stokes equations," *Indiana Univ. Math. J.* **42**, 775–789, 1993.
Constantin, P. and Foias, C., *Navier–Stokes Equations*, Univ. of Chicago Press, Chicago, 1988.
Ebin, D. G. and Marsden, J. E., "Groups of diffeomorphisms and the motion of an incompressible fluid," *Ann. Math.* **92**, 102–163, 1970.
Folland, G. B., *Introduction to Partial Differential Equations*, Princeton Univ. Press, Princeton, NJ, 1995.
Hartman, P., *Ordinary Differential Equations*. Birkhäuser, Boston, 1982.
John, F., *Partial Differential Equations*. Springer-Verlag, New York, 1982.
Kato, T., "Nonstationary flows of viscous and ideal fluids in $\mathbb{R}^3$," *J. Funct. Anal.* **9**, 296–305, 1972.
Kato, T., "The Cauchy problem for quasi-linear symmetric hyperbolic systems," *Arch. Ration. Mech. Anal.* **58**(3),181–205, 1975.
Kato, T., "Remarks on the Euler and Navier–Stokes equations in $\mathbb{R}^2$, nonlinear functional analysis and its applications," in *Proceedings of the Symposium on Pure Mathematics*, Vol. 45 of the AMS Proceedings Series, American Mathematical Society, Providence, RI, 1986, pp. 1–7.
Klainerman, S. and Majda, A., "Singular limits of quasilinear hyperbolic systems with large

parameters and the incompressible limit of compressible fluids," *Commun. Pure Appl. Math.* **34**, 481–524, 1981.

Riesz, F. and Nagy, B. S., *Functional Analysis*, Dover, Mineola, NY, 1990. Dover edition, republication of 1955 Ungar edition with additional appendix published separately by Ungar, New York, 1960.

Royden, H. L., *Real Analysis*, Macmillan, New York, 1968.

Rudin, W., *Real and Complex Analysis*, 3rd ed., McGraw-Hill, New York, 1987.

Strang, G., "On the construction and comparison of difference schemes," *SIAM J. Numer. Anal.* **5**, 506–517, 1968.

Taylor, M. E., *Pseudodifferential Operators and Nonlinear PDE*, Birkhäuser, Boston, 1991.

Taylor, M. E., *Partial Differential Equations, II, Qualitative Studies of Linear Equations*, Vol. 116 of Applied Mathematical Sciences Series, Springer-Verlag, New York, 1996a.

Taylor, M. E., *Partial Differential Equations III, Nonlinear Equations*, Vol. 117 of Applied Mathematical Sciences Series, Springer-Verlag, New York, 1996b.

Temam, R., *Navier–Stokes Equations*, 2nd ed., North-Holland, Amsterdam, 1986.

Ying, L. A., "Optimal error estimates for a viscosity splitting formula," in *Proceedings of the Second Conference on Numerical Methods for Partial Differential Equations*, World Scientific, River Edge, NJ 1992, pp. 139–147.

Zhang, P., "A sharp estimate of simplified viscosity splitting scheme," *J. Comput. Math.* **11**, 205–210, 1993.

4

The Particle-Trajectory Method for Existence and Uniqueness of Solutions to the Euler Equation

In Chaps. 3 and 4 we address the questions of existence, uniqueness, and continuation of solutions to the Euler and the Navier–Stokes equations. In Chap. 3 we introduced the technique of *energy methods* to address these issues. In this chapter we introduce the particle-trajectory method to study the inviscid Euler equation ($\nu = 0$):

$$\frac{\partial v}{\partial t} + v \cdot \nabla v = -\nabla p,$$

$$\operatorname{div} v = 0.$$

The Euler and the Navier–Stokes equations are both evolution equations for the velocity field $v(x, t)$. At the end of Chap. 2 we derived integrodifferential equations (2.118) and (2.119) for the particle trajectories $X(\alpha, t)$ in two and three dimensions. Proposition 2.23 showed that the integrodifferential equations are equivalent to the Euler equation when the flow is sufficiently smooth. In this chapter we make use of the fact that the particle-trajectory equations define an ODE on an infinite-dimensional Banach space. The Euler equation in primitive-variable form contains the unbounded operator $v \cdot \nabla v$. A key feature of the particle-trajectory equations is that they reformulate the problem in terms of a bounded nonlinear operator on a Banach space. In Section 4.1 we show that standard existence and uniqueness theories for such equations yield local-in-time existence and uniqueness of solutions to the particle-trajectory equation and hence to the Euler equation. Recall from Chap. 3 that we presented for both the Euler and the Navier–Stokes equation a sufficient condition, involving accumulation of vorticity, for global-in-time existence of solutions in three dimensions. We prove the same result here in Section 4.2 by using Hölder norms and potential theory. Because vorticity does not accumulate in two dimensions, we obtain global-in-time existence of smooth solutions for 2D Euler equations. The result from Section 4.2 has a special implication for 3D flows with certain geometry constraints. For example, in Section 4.3 we show that finite-time singularities are impossible for smooth solutions to 3D Euler equations that are axisymmetric without swirl. In Section 4.4 we discuss higher regularity of solutions and in particular show that the condition in Section 4.2 controls blowup of all higher derivatives of the solution. Such a direct ODE method serves only to study time-reversible systems.† In particular, a

† A simple example of a time-irreversible equation is the heat equation $T_t = \Delta T$. See, for example, Folland (1995).

particle-trajectory method applied to the Navier–Stokes equation in which viscosity introduces diffusion requires the machinery of stochastic differential equations. We introduce this idea in Chap. 6 in conjunction with the *random-vortex method*, a numerical scheme for computing high Reynolds number solutions of the Navier–Stokes equation.

Taking the curl of the Euler equation,

$$\frac{\partial v}{\partial t} + v \cdot \nabla v = -\nabla p,$$
$$\operatorname{div} v = 0,$$
$$v|_{t=0} = v_0(x).$$

gives evolution equation (2.109) for the vorticity, $\omega = \operatorname{curl} v$:

$$\frac{D\omega}{Dt} = \omega \cdot \nabla v,$$
$$\omega|_{t=0} = \omega_0. \tag{4.1}$$

The divergence-free velocity v in Eq. (4.1) is recovered from ω by the Biot–Savart law:

$$v(x,t) = \int_{\mathbb{R}^N} K_N(x-y)\omega(y,t)dy, \qquad x \in \mathbb{R}^N. \tag{4.2}$$

The kernels K_N are homogeneous of degree $1 - N$:

$$K_2(x) = \frac{1}{2\pi}{}^t\!\left(-\frac{x_2}{|x|^2}, \frac{x_1}{|x|^2}\right), \qquad x \in \mathbb{R}^2,$$
$$K_3(x)h = \frac{1}{4\pi}\frac{x \times h}{|x|^3}, \qquad x, h \in \mathbb{R}^3.$$

In Section 2.5 of Chap. 2, we used Eqs. (4.1) and (4.2) to derive an integrodifferential equation for the particle trajectories by combining their evolution equation,

$$\frac{dX}{dt}(\alpha,t) = v(X(\alpha,t),t), \qquad X(\alpha,t)|_{t=0} = \alpha, \tag{4.3}$$

with the Biot–Savart law,

$$v(x,t) = \int_{\mathbb{R}^3} K_3[x - X(\alpha',t)]\nabla_\alpha X(\alpha',t)\omega_0(\alpha')d\alpha', \tag{4.4}$$

which we rewrite here in Lagrangian form by means of the change of variables $x' = X(\alpha', t)$ and the vorticity-amplification identity

$$\omega(X(\alpha,t),t) = \nabla_\alpha X(\alpha,t)\omega_0(\alpha). \tag{4.5}$$

Combining Eqs. (4.4) and (4.3) gives the *3D integrodifferential equation for the particle trajectories* $X(\alpha, t)$:

$$\frac{dX}{dt}(\alpha,t) = \int_{\mathbb{R}^3} K_3[X(\alpha,t) - X(\alpha',t)]\nabla_\alpha X(\alpha',t)\omega_0(\alpha')d\alpha',$$
$$X(\alpha,t)|_{t=0} = \alpha. \tag{4.6}$$

In the case of 2D inviscid flows, vorticity cannot amplify because it is conserved on particle trajectories. The solution in Eq. (4.5) reduces to

$$\omega\left(X(\alpha, t), t\right) = \omega_0(\alpha),$$

and hence yields the simplified velocity

$$v(x, t) = \int_{\mathbb{R}^2} K_2[x - X(\alpha', t)]\omega_0(\alpha')d\alpha'.$$

The *2D integrodifferential equation for the particle trajectories* $X(\alpha, t)$ is

$$\begin{aligned} \frac{dX}{dt}(\alpha, t) &= \int_{\mathbb{R}^2} K_2[X(\alpha, t) - X_0(\alpha', t)]\omega_0(\alpha')d\alpha', \\ X(\alpha, t)|_{t=0} &= \alpha. \end{aligned} \tag{4.7}$$

In the next section we show rigorously that the right-hand side of Eq. (4.6) is a bounded nonlinear operator on $X(\alpha, t)$. The smoothing properties of the kernel K_3 compensate for the loss of regularity in differentiating $X(\alpha, t)$ in the term $\nabla_\alpha X$. Also, Eq. (4.6) involves the explicit term $\nabla_\alpha X(\alpha, t)\omega_0(\alpha)$ that represents the stretching of vorticity. In Section 4.2 we show a direct link between the accumulation of vorticity and the existence of smooth solutions globally in time. In Chap. 6 we show that a discretization of the particle-trajectory formulation leads to *computational vortex methods* that have been very successful in accurately computing inviscid Euler solutions. Part of their success comes from the fact the particle-trajectory reformulation contains only bounded operators. Recall from Section 2.5, Proposition 2.23, that for sufficiently smooth solutions the particle-trajectory formulation is equivalent to the Euler equation.

4.1. The Local-in-Time Existence of Inviscid Solutions

Our goal in this section is to prove the local-in-time existence and uniqueness of solutions to the Euler equation. We first show that the particle-trajectory equations formulate the problem as an ODE on a Banach space. Then we invoke the standard Picard theorem for ODEs on a Banach space to prove local existence and uniqueness. In this section we consider only 3D solutions. The 2D case is simpler, and the method is identical.

The 3D integrodifferential equation for the particle trajectories $X(\alpha, t)$ is,

$$\begin{aligned} \frac{d}{dt}X(\alpha, t) &= F(X(\alpha, t)) \\ X(\alpha, t)|_{t=0} &= \alpha, \end{aligned} \tag{4.8}$$

where the nonlinear mapping $F(X)$ is

$$F(X(\alpha, t)) = \int_{\mathbb{R}^3} K_3[X(\alpha, t) - X(\alpha', t)]\nabla_\alpha X(\alpha', t)\omega_0(\alpha')d\alpha'. \tag{4.9}$$

We view Eqs. (4.8) and (4.9) as an ODE on an infinite-dimensional Banach space. Our goal in this section is to show that there is an appropriate Banach space **B** on which the operator F is locally Lipschitz continuous. We then evoke a standard local-existence theorem for ODEs.

Theorem 4.1. *Picard Theorem on a Banach Space. Let $O \subseteq \mathbf{B}$ be an open subset of a Banach space $\mathbf{B}$, and let $F(X)$ be a nonlinear operator satisfying the following criteria:*

(i) $F(X)$ maps O to$\mathbf{B}$.
(ii) $F(X)$ is locally Lipschitz continuous, i.e., for any $X \in O$ there exists $L > 0$ and an open neighborhood $U_X \subset O$ of X such that

$$\|F(\tilde{X}) - F(\hat{X})\|_{\mathbf{B}} \leq L\|\tilde{X} - \hat{X}\|_{\mathbf{B}} \quad \textit{for all} \quad \tilde{X}, \hat{X} \in U_X.$$

Then for any $X_0 \in O$, there exists a time T such that the ODE

$$\frac{dX}{dt} = F(X), \qquad X|_{t=0} = X_0 \in O, \tag{4.10}$$

has a unique (local) solution $X \in C^1[(-T, T); O]$.

In the above $|| \cdot ||_{\mathbf{B}}$ denotes a norm in the Banach space $\mathbf{B}$, and $C^1[(-T, T); \mathbf{B}]$ denotes the space of C^1 functions $X(t)$ on the interval $(-T, T)$ with values in $\mathbf{B}$.‡

Recall that this abstract theorem appeared in Chap. 3 in the proof of global existence of solutions to an approximating problem for the Euler and the Navier–Stokes equations. As we saw in Chap. 3, if F does not depend explicitly on time, as in the case of Eq. (4.9), solutions to Eq. (4.10) can be continued until they leave the set O. We explicitly use the continuation property to study higher regularity and global existence of solutions in the following sections. In Section 4.2 we find a sufficient condition involving the accumulation of vorticity for global existence of solutions in three dimensions. As a consequence we obtain global existence of solutions in two dimensions because vorticity is conserved along particle trajectories. In Section 4.3 we present a special class of 3D flows that also have global existence of solutions that is due to their inherent two dimensionality. These are the 3D axisymmetric flows without swirl from Subsection 2.3.3. In Section 4.4 we show that higher spatial derivatives are controlled by the lower ones yielding so that the result from Section 4.2 is true for the C^∞ case. The reformulation of the Euler equation in terms of the particle-trajectory equations is crucial here. A direct treatment of the inviscid 3D Euler equation could never use this approach because the convective derivative,

$$\frac{\partial v}{\partial t} + v \cdot \nabla v,$$

involves the term $v \cdot \nabla$, an unbounded operator on standard Banach spaces.

In the following three subsections, we (1) choose the appropriate Banach space $\mathbf{B}$, (2) choose an appropriate family of open sets O_M, and (3) show that F satisfies the assumption of the Picard theorem for the choices made in (1) and (2).

‡ Note that $C^1[(-T, T); \mathbf{B}]$ is also a Banach space equipped with the norm $||X|| = \sup_{t\in(-T,T)} ||X(t)||_{\mathbf{B}} + \sup_{t\in(-T,T)} ||\frac{\partial}{\partial t} X(t)||_{\mathbf{B}}$.

4.1.1. The Choice of the Banach Space B and Elementary Calculus Inequalities for Hölder Continuous Functions

First we carefully choose the Banach space **B** and the subset $O \subseteq \mathbf{B}$ on which the nonlinear mapping,

$$F(X(\alpha,t)) = \int_{\mathbb{R}^3} K_3[X(\alpha,t) - X(\alpha',t)]\nabla_\alpha X(\alpha',t)\omega_0(\alpha')d\alpha', \tag{4.11}$$

has certain desired properties. The open set O must include maps $X(\cdot,t) : \mathbb{R}^3 \to \mathbb{R}^3$ that are $1-1$ and onto, including the identity map. **B** must allow us to prove the Lipschitz continuity in the Picard theorem.

Because Eqs. (4.8) and (4.9) explicitly contain the gradient $\nabla_\alpha X$, **B** must contain information about

$$|\nabla_\alpha X|_0 \equiv \sup_{\alpha\in\mathbb{R}^3} |\nabla_\alpha X(\alpha)| < \infty.$$

However, we cannot take $X \in C^1$ – the space of bounded, continuously differentiable functions. There are two reasons for this. The first is simply that if $X : \mathbb{R}^3 \to \mathbb{R}^3$ is $1-1$ and onto then X cannot be bounded as $|\alpha| \nearrow \infty$. A second, more technical reason is that the equation for $\nabla_\alpha X$, computed from the gradient of Eq. (4.9), contains a SIO (as discussed in Chap. 2). Such operators do not map bounded functions to bounded functions but map them instead to the larger class of functions of bounded mean oscillation (BMO). This fact becomes important in the formulation of the weak L^∞ solution to the vorticity-stream equation, discussed in detail in Chap. 8. In this chapter we use the fact that SIOs are bounded operators on the Hölder spaces (see below).

Define the Banach space **B** as

$$\mathbf{B} = \{X : \mathbb{R}^3 \to \mathbb{R}^3 \text{ such that } |X|_{1,\gamma} < \infty\}, \tag{4.12}$$

where $|\cdot|_{1,\gamma}$ is the norm defined by

$$|X|_{1,\gamma} = |X(0)| + |\nabla_\alpha X|_0 + |\nabla_\alpha X|_\gamma. \tag{4.13}$$

In the above $|\cdot|_0$ is the supremum or C^0 norm and $|\cdot|_\gamma$ is the Hölder seminorm defined by

$$|X|_\gamma = \sup_{\substack{\alpha,\alpha'\in\mathbb{R}^3\\ \alpha\neq\alpha'}} \frac{|X(\alpha) - X(\alpha')|}{|\alpha-\alpha'|^\gamma}, \qquad 0<\gamma\le 1. \tag{4.14}$$

To simplify the notation we define the Hölder$-\gamma$ norm $\|\cdot\|_\gamma$ by

$$\|X\|_\gamma = |X|_0 + |X|_\gamma, \qquad 0<\gamma\le 1. \tag{4.15}$$

We denote by C^γ the space of functions with bounded $\|\cdot\|_\gamma$ norm. The space **B** in Eq. (4.12) is a Banach space; it is linear, normed, and complete. It consists of all functions for which $\nabla_\alpha X$ is Hölder continuous. The term $|X(0)| < \infty$ in the norm allows X to be unbounded as $|\alpha| \nearrow \infty$. The Hölder seminorm $|\nabla_\alpha X|_\gamma < \infty$ allows us to show in Subsection 4.1.3 that F is bounded and Lipschitz continuous on a subset of $\mathbf{B}$.

Next we state some elementary calculus inequalities for functions in $\mathbf{B}$. The proofs are straightforward and hence are presented in the appendix.

Lemma 4.1. *Let $X, Y : \mathbb{R}^N \to \mathbb{R}^N$ be smooth, bounded functions. Then for $\gamma \in (0, 1]$ there exists $c > 0$ such that*

$$|XY|_\gamma \leq |X|_0 |Y|_\gamma + |X|_\gamma\, |Y|_0, \tag{4.16}$$

$$\|XY\|_\gamma \leq c\|X\|_\gamma\, \|Y\|_\gamma. \tag{4.17}$$

Lemma 4.2. *Let $X : \mathbb{R}^N \to \mathbb{R}^N$ be a smooth, invertible transformation with*

$$|\det \nabla_\alpha X(\alpha)| \geq c_1 > 0.$$

Then for $0 < \gamma \leq 1$ there exists $c > 0$ such that

$$\|(\nabla_\alpha X)^{-1}\|_\gamma \leq c\|\nabla_\alpha X\|_\gamma^{2N-1}, \tag{4.18}$$

$$|X^{-1}|_{1,\gamma} \leq c|X|_{1,\gamma}^{2N-1}. \tag{4.19}$$

Lemma 4.3. *Let $X : \mathbb{R}^N \to \mathbb{R}^N$ be an invertible transformation with*

$$|\det \nabla_\alpha X(\alpha)| \geq c_1 > 0,$$

and let $f : \mathbb{R}^N \to \mathbb{R}^M$ be a smooth function. Then for $0 < \gamma \leq 1$ the composition $f \circ X$ and $f \circ X^{-1}$ satisfies

$$|f \circ X|_\gamma \leq |f|_\gamma |\nabla_\alpha X|_0^\gamma, \tag{4.20}$$

$$\|f \circ X\|_\gamma \leq \|f\|_\gamma \left(1 + |X|_{1,\gamma}^\gamma\right), \tag{4.21}$$

$$\|f \circ X^{-1}\|_\gamma \leq \|f\|_\gamma \left[1 + c\,|X|_{1,\gamma}^{\gamma(2N-1)}\right]. \tag{4.22}$$

4.1.2. The Choice of the Open Subset $\mathbf{O} \subseteq \mathbf{B}$

To apply the Picard theorem we must choose an open subset of $\mathbf{B}$ that contains the identity map (our initial condition) on which we can show that

$$F(X(\alpha, t)) = \int_{\mathbb{R}^3} K_3[X(\alpha, t) - X(\alpha', t)]\nabla_\alpha X(\alpha', t)\omega_0(\alpha')d\alpha' \tag{4.23}$$

is locally Lipschitz continuous. The incompressibility condition $\nabla \cdot v = 0$ forces $\det \nabla_\alpha X = 1$. However, this condition is too restrictive because it defines a hypersurface of functions in $\mathbf{B}$. The Picard theorem requires an *open* set on which we show the Lipschitz continuity of F. Define

$$O_M = \left\{ X : X \in \mathbf{B} \middle| \inf_{\alpha \in \mathbb{R}^3} \det \nabla_\alpha X(\alpha) > \frac{1}{2} \text{ and } |X|_{1,\gamma} < M \right\}. \tag{4.24}$$

The identity map belongs to O_M for $M > 1$. If $X \in O_M$, then by the inverse function theorem X is locally 1-1 (local homeomorphism). However, particle-trajectory maps are *global* homeomorphisms; thus we require all elements of O_M to also have this property. A lemma that is due to Hadamard (Berger, 1977, p. 222) ensures that this is the case.

Lemma 4.4. *Hadamard. Suppose that $X \in \mathbf{B}$ is a local homeomorphism, and there exists $c > 0$ such that*

$$\sup_{\alpha \in \mathbb{R}^3} \left|(\nabla_\alpha X)^{-1}(\alpha)\right| \leq c.$$

Then X is a homeomorphism of $\mathbb{R}^3$ onto $\mathbb{R}^3$.

We now prove the following proposition.

Proposition 4.1. *For any $M > 0, 0 < \gamma \leq 1$, the set O_M in Eq. (4.24), is nonempty, open, and it consists of $1 - 1$ mappings of $\mathbb{R}^3$ onto $\mathbb{R}^3$.*

Proof of Proposition 4.1. The set O_M is nonempty because it contains multiples of the identity map $cI, c < M$.

The mapping $\inf_{\alpha \in \mathbb{R}^3} \det \nabla_\alpha : \mathbf{B} \to \mathbb{R}$ and the norm $|\cdot|_{1,\gamma} : \mathbf{B} \to [0, \infty)$ are continuous; thus inverse images of the open subsets $(\frac{1}{2}, \infty)$ and $[0, M)$ are open subsets in $\mathbf{B}$, and the set O_M is open in $\mathbf{B}$.

It remains only to show that O_M consists of $1 - 1$ and onto mappings. By the definition of O_M $\det \nabla_\alpha X > \frac{1}{2}$ and $|X|_{1,\gamma} < M$. Calculus inequality (4.19) implies that

$$\left|(\nabla_\alpha X)^{-1}(\alpha)\right| \leq |X^{-1}|_{1,\gamma} \leq c|X|_{1,\gamma}^{2N-1} \leq cM^{2N-1}.$$

Direct application of the Hadamard lemma, Lemma 4.4, completes the proof. □

We now have all the necessary ingredients to apply the Picard theorem (Theorem 4.1).

4.1.3. Local Existence and Uniqueness of Solutions $X(\alpha, t)$ and Potential Theory Estimates

Recall that the integrodifferential equations for the particle trajectories form an ODE on a Banach space:

$$\begin{aligned} \frac{dX}{dt}(\alpha, t) &= F(X(\alpha, t)), \\ X(\alpha, t)|_{t=0} &= \alpha, \end{aligned} \tag{4.25}$$

where the nonlinear mapping $F(X)$ is defined by

$$F(X(\alpha, t)) = \int_{\mathbb{R}^3} K_3[X(\alpha, t) - X(\alpha', t)] \nabla_\alpha X(\alpha', t) \omega_0(\alpha') d\alpha'. \tag{4.26}$$

In this section we show that F is an bounded, nonlinear, Lipschitz continuous operator on the Banach space:

$$\mathbf{B} = \{X : \mathbb{R}^3 \to \mathbb{R}^3 \text{ such that } |X|_{1,\gamma} < \infty\}, \tag{4.27}$$

where $|\cdot|_{1,\gamma}$ is the norm defined by

$$|X|_{1,\gamma} = |X(0)| + |\nabla_\alpha X|_0 + |\nabla_\alpha X|_\gamma. \tag{4.28}$$

Hence we prove Theorem 4.2.

Theorem 4.2. *Consider a compactly supported initial vorticity $\omega_0 \in C^\gamma$, $\gamma \in (0, 1)$, $\omega_0 = \operatorname{curl} v_0$, $\operatorname{div} v_0 = 0$. Then for any $M > 0$ there exists $T(M) > 0$ and a unique volume-preserving solution*

$$X \in C^1((-T(M), T(M)); O_M)$$

to particle-trajectory equations (4.8) and (4.9).

The assumption about the support of ω_0 is not necessary to prove the theorem. We make it here to simplify the technical arguments in the potential theory estimates. Theorem 4.2 is not valid for the Hölder exponent $\gamma = 1$. This is a result of a standard property of SIOs – see Lemma 4.6.

The proof of Theorem 4.2 follows from the Picard theorem (Theorem 4.2) provided that we can prove the following proposition.

Proposition 4.2. *Let $|\omega_0|_\gamma < \infty$ and $0 < \gamma < 1$. Let $F : O_M \to \mathbf{B}$ be defined by*

$$F(X(\alpha, t)) = \int_{\mathbb{R}^3} K_3[X(\alpha, t) - X(\alpha', t)]\nabla_\alpha X(\alpha', t)\omega_0(\alpha')d\alpha'.$$

Then F satisfies the assumptions of the Picard theorem, i.e., F is bounded and locally Lipschitz continuous on O_M.

To prove this proposition we make use of the fact that the kernel $P_N(x) = \nabla K_N(x)$ defines a standard SIO. Recall from Chap. 2, Subsection 2.4.2, that

$$K_N f(x) = \int_{\mathbb{R}^N} K_N(x - y) f(y) dy, \tag{4.29}$$

where K_N is homogeneous of degree $1 - N$,

$$K_N(\lambda x) = \lambda^{1-N} K_N(x), \qquad \forall \lambda > 0, \quad x \neq 0, \tag{4.30}$$

and $P_N = \nabla K_N$ is homogeneous of degree $-N$,

$$P_N(\lambda x) = \lambda^N P_N(x), \qquad \forall \lambda > 0, \quad x \neq 0. \tag{4.31}$$

Furthermore, P_N has mean-value zero on the unit sphere,

$$\int_{|x|=1} P_N \, ds = 0, \tag{4.32}$$

and it defines the principal-value SIO

$$P_N f(x) = \mathrm{PV} \int_{\mathbb{R}^N} P_N(x-y) f(y) dy. \tag{4.33}$$

Any operator K_N satisfying the above conditions yields the following potential theory estimates.

Lemma 4.5. *Let K_N satisfy Eqs. (4.30)–(4.33). Consider $f \in C^\gamma(\mathbb{R}^N; \mathbb{R}^N)$, $0 < \gamma < 1$ with compact support and define $m_f = m(\operatorname{supp} f) < \infty$. Define the length scale R by $R^N = m_f$. Then there exists a constant c, independent of f and R, so that*

$$|K_N f|_0 \leq cR|f|_0. \tag{4.34}$$

Lemma 4.6. *Let f, γ and R satisfy the assumptions of Lemma 4.5. Let the SIO P_N satisfy Eqs. (4.31)–(4.33). Then there exists a constant c, independent of R and f, so that*

$$|P_N f|_0 \leq c\left\{ |f|_\gamma \epsilon^\gamma + \max\left(1, \ln \frac{R}{\epsilon}\right) |f|_0 \right\}, \qquad \forall \epsilon > 0, \tag{4.35}$$

$$|P_N f|_\gamma \leq c\|f\|_\gamma. \tag{4.36}$$

We give the proofs of Lemmas 4.5 and 4.6 in the appendix. The assumption on the support of f is not necessary but simplifies the analysis. In particular, the proofs contain a concise argument with estimates that depend on the quantity m_f.

Proof of Proposition 4.2. First we show that the operator $F : O_M \to \mathbf{B}$ is bounded, i.e., $|F(X)|_{1,\gamma} < \infty \; \forall X \in O_M$. Because the set O_M consists of 1 – 1 and onto functions, we use the change of variables $X^{-1}(x) = \alpha$ and rewrite $F(X)$ as

$$\begin{aligned} F(X) &= K_3 f \circ X^{-1}, \\ K_3 f(x) &= \int_{\mathbb{R}^3} K_3(x - x') f(x') dx', \qquad (4.37) \\ f(x') &= \nabla_\alpha X(\alpha) \omega_0(\alpha)|_{\alpha = X^{-1}(x')} \det \nabla_x X^{-1}(x'). \end{aligned}$$

By calculus inequality (4.19) we have

$$|F(X)|_{1,\gamma} \leq |K_3 f|_{1,\gamma} \, |X^{-1}|_{1,\gamma} \leq c|K_3 f|_{1,\gamma} |X|_{1,\gamma}^2,$$

so it suffices to estimate $|K_3 f|_{1,\gamma}$.

Potential theory estimate (4.34) implies that

$$|K_3 f(0)| \leq |K_3 f|_0 \leq c|f|_0. \tag{4.38}$$

Moreover, recall from Section 2.4.3 that the gradient of $K_N f$ is

$$\nabla[K_N f(x)] = \mathrm{PV} \int_{\mathbb{R}^N} P_N(x - x') f(x') dx' + c f(x),$$

where $P_N(x) = \nabla K_N(x)$ is homogeneous of degree $-N$. Potential theory estimate (4.36) gives

$$\|\nabla K_3 f\|_\gamma \leq c\|f\|_\gamma + \|P_3 f\|_\gamma \leq c\|f\|_\gamma$$

and from relation (4.38) implies that

$$|K_3 f|_{1,\gamma} \leq c\|f\|_\gamma. \tag{4.39}$$

It remains to estimate the norm $\|f\|_\gamma$. By the definition of f in Eq. (4.38) and calculus inequalities (4.17), (4.19), and (4.22) we estimate

$$\begin{aligned}
\|f\|_\gamma &\leq \|(\nabla_\alpha X \omega_0) \circ X^{-1}\|_\gamma \|\det \nabla_x X^{-1}\|_\gamma \\
&\leq \|\nabla_\alpha X \omega_0\|_\gamma \left(1 + c|X|_{1,\gamma}^{\gamma(2N-1)}\right)|X|_{1,\gamma}^{2N-1} \\
&\leq c(M)\|\omega_0\|_\gamma
\end{aligned}$$

because $X \in O_M$ implies that $|X|_{1,\gamma} < M$. Hence

$$|K_3 f|_{1,\gamma} \leq c(M)\|\omega_0\|_\gamma,$$

so the operator $F : O_M \to \mathbf{B}$ is bounded.

We prove that F is locally Lipschitz continuous on O_M by showing a sufficient condition. If the derivative $F'(X)$ is bounded as a linear operator from O_M to $\mathbf{B}$, $\|F'(X)\| < \infty \; \forall X \in O_M$, then the mean-value theorem implies that

$$\begin{aligned}
|F(\tilde{X}) - F(\hat{X})|_{1,\gamma} &= \left| \int_0^1 \frac{d}{d\epsilon} F[\tilde{X} + \epsilon(\hat{X} - \tilde{X})] d\epsilon \right|_{1,\gamma} \\
&\leq \int_0^1 \|F'[\tilde{X} + \epsilon(\hat{X} - \tilde{X})]\| ds |\tilde{X} - \hat{X}|_{1,\gamma},
\end{aligned}$$

so F is locally Lipschitz continuous on O_M. The derivative $F'(X)$ is

$$\begin{aligned}
F'(X)Y &= \frac{d}{d\epsilon} F(X + \epsilon Y)|_{\epsilon=0} \\
&- \frac{d}{d\epsilon} \int_{\mathbb{R}^3} K_3\{X(\alpha) - X(\alpha') + \epsilon[Y(\alpha) - Y(\alpha')]\} \\
&\cdot \nabla_\alpha [X(\alpha') + \epsilon Y(\alpha')] \omega_0(\alpha') d\alpha'|_{\epsilon=0} \\
&= \int_{\mathbb{R}^3} K_3[X(\alpha) - X(\alpha')] \nabla_\alpha Y(\alpha') \omega_0(\alpha') d\alpha' \\
&+ \int_{\mathbb{R}^3} \nabla K_3[X(\alpha) - X(\alpha')][Y(\alpha) - Y(\alpha')] \nabla_\alpha X(\alpha') \omega_0(\alpha') d\alpha' \\
&\equiv G_1(X)Y + G_2(X)Y.
\end{aligned}$$

In the above derivation the term $Y(\alpha) - Y(\alpha')$ kills the singularity of ∇K_3, eliminating the need for a principal-value integral in $G_2(X)Y$.

We estimate $G_1(X)Y$ exactly in the same way as in relation (4.39), which gives

$$|G_1(X)Y|_{1,\gamma} \leq c\|(\nabla_\alpha Y\omega_0) \circ X^{-1}\|_\gamma.$$

Calculus inequalities (4.17), (4.19) and (4.22) imply that

$$|G_1(X)Y|_{1,\gamma} \leq c\|\nabla_\alpha Y\|_\gamma \|\omega_0\|_\gamma \left(1 + c|X|_{1,\gamma}^{\gamma(2N-1)}\right) \leq c(M)\|\omega_0\|_\gamma |Y|_{1,\gamma}. \tag{4.40}$$

An estimate of $G_2(X)Y$ requires some new calculations that are straightforward but somewhat technical. The details are in the appendix. The result is

$$|G_2(X)Y|_{1,\gamma} \leq c(M)\|\omega_0\|_\gamma \ |Y|_{1,\gamma}. \tag{4.41}$$

Combining relations (4.40) and (4.41) gives

$$|F'(X)Y|_{1,\gamma} \leq c(M)\|\omega_0\|_\gamma \ |Y|_{1,\gamma}.$$

The derivative $F'(X) : O_M \to \mathbf{B}$ is a linear bounded operator, and hence the operator $F : O_M \to \mathbf{B}$ is locally Lipschitz continuous. □

4.2. Link Between Global-in-Time Existence of Smooth Solutions and the Accumulation of Vorticity through Stretching

Using the integrodifferential equations for the particle trajectories, we showed in the previous section that the Euler equation must have a solution locally in time. In this section, we give a sufficient condition for the solution to exist for all time. Recall from Chap. 2 that the 3D Euler equation allows for vorticity amplification in time by means of stretching that is due to the local velocity gradients. In particular we showed some exact solutions in which the vorticity increased without bound (these specific examples had amplification in infinite time). In this section, as in Section 3.3., we show that if the magnitude of the vorticity remains bounded, then the solution exists globally in time. In fact, because vorticity does not grow in two dimensions, solutions to the Euler equation exist globally in time.

Our main goal in this section is to prove the following theorem, which is due to Beale et al. (1984), for continuation of solutions forward in time. An analogous statement holds for continuation backward in time.

Theorem 4.3. *Beale–Kato–Majda. Consider a compactly supported initial vorticity* $\omega_0 = \operatorname{curl} v_0$, $\operatorname{div} v_0 = 0$, *with Hölder norm* $\|\omega_0\|_\gamma < \infty$ *for some* $\gamma \in (0, 1)$. *Let* $|\omega(\cdot, s)|_0$ *be the supremum norm at fixed time* s *of a solution to the Euler equation, with initial data* ω_0.

(i) Suppose that for any $T > 0$ *there exists* $M_1 > 0$ *such that the vorticity* $\omega(x, t)$ *satisfies*

$$\int_0^T |\omega(\cdot, s)|_0 ds \leq M_1. \tag{4.42}$$

Then, for any T there exists $M > 0$ such that $X \in C^1([0, T); O_M)$ (the solution X exists globally in time).

(ii) Suppose that for any $M > 0$ there is a finite maximal time $T(M)$ of existence of solutions $X \in C^1\{[0, T(M)); O_M\}$ and that $\lim_{M\to\infty} T(M) = T^* < \infty$; then necessarily the vorticity accumulates so rapidly that

$$\lim_{t\nearrow T^*} \int_0^t |\omega(\cdot, s)|_0 ds = \infty.$$

The assumption on the size of support of ω_0 serves to simplify the calculations below. In the next section we show that there exists an a priori bound (4.42) for the special case of 3D axisymmetric flows without swirl. Hence such solutions exist globally in time. In Section 3.4 we showed that bound (4.42) controls all higher derivatives of the particle trajectories to derive a C^∞ version of Theorem 4.3. In this chapter we show that this result is true for the Navier–Stokes equation (with viscosity) as well. In Chap. 5 we discuss the question of a finite-time breakdown of solutions to the Euler equation and study some relevant model problems. To date, it is not known if there actually are solutions to the 3D Euler equation that blow up in finite time.

One consequence of Theorem 4.3 is that the 2D Euler equation has global-in-time existence of solutions. This is due to the simple fact that vorticity is conserved along particle trajectories in two dimensions and hence cannot become unbounded in magnitude. We state this as a corollary.

Corollary 4.1. *Global Existence of Solutions to the Euler Equation in Two Dimensions. Consider the 2D Euler equation with compactly supported initial vorticity $\omega_0 = \text{curl } v_0$, $\text{div } v_0 = 0$ satisfying $\|\omega_0\|_\gamma < \infty$ for some $\gamma \in (0, 1)$. Then there exists a unique solution $X(\cdot, t) \in C^1[(-\infty, \infty); \mathbf{B}]$ to the particle-trajectory equation in the time interval $(-\infty, \infty)$.*

We prove Theorem 4.3 by using both the particle trajectories in Lagrangian variables and the vorticity-stream form in Eulerian variables. We also use the Hölder norm potential theory estimates. The proof in this chapter for both the Euler and the Navier–Stokes equations uses energy methods and the H^s Sobolev spaces.

We first guarantee the existence of smooth solutions on any given time interval provided that $\int_0^t |\nabla v(\cdot, s)|_0 ds$ remains bounded on that time interval. Then we show that $\int_0^t |\nabla v(\cdot, s)|_0 ds$ is a priori controlled by $\int_0^t |\omega(\cdot, s)|_0 ds$.

First recall that the particle-trajectory equation

$$\frac{dX}{dt}(\alpha, t) = \int_{\mathbb{R}^3} K_3[X(\alpha, t) - X(\alpha', t)]\nabla_\alpha X(\alpha', t)\omega_0(\alpha')d\alpha',$$
$$X(\alpha, t)|_{t=0} = \alpha, \tag{4.43}$$

has a unique solution $X \in C^1\{[0, T(M)); O_M\}$ locally in time for some $T(M) > 0$:

$$O_M = \left\{X \in \mathbf{B} : \inf_{\alpha\in\mathbb{R}^3} \det \nabla_\alpha X(\alpha) > \frac{1}{2}, |X|_{1,\gamma} < M\right\},$$
$$|X|_{1,\gamma} = |X(0)| + |\nabla_\alpha X|_0 + |\nabla_\alpha X|_\gamma, \qquad 0 < \gamma < 1.$$

We now determine when it is possible to continue the solution $X(\alpha, t)$ further in time. In fact, for particle trajectory equation (4.43) we can link the ability to continue the solution directly with the absence of blowup in the norm $|\cdot|_{1,\gamma}$. Such a continuation property is not in general true for nonlinear PDEs (Ball, 1977). Even the Picard theorem on an infinite-dimensional Banach space does not in general have a continuation result as in the finite-dimensional case. However, when the Lipschitz function F is time independent, the continuation result does hold. The following theorem is a special case of Theorem 5.6.1 from p. 161 of Ladas and Lakshmikamtham (1972).

Theorem 4.4. *Continuation of an Autonomous ODE on a Banach Space. Let $O \subset \mathbf{B}$ be an open subset of a Banach space $\mathbf{B}$, and let $F : O \to \mathbf{B}$ be a locally Lipschitz continuous operator. Then the unique solution $X \in C^1([0, T); O)$ to the autonomous ODE,*

$$\frac{dX}{dt} = F(X), \qquad X|_{t=0} = X_0 \in O,$$

either exists globally in time or $T < \infty$ and $X(t)$ leaves the open set O as $t \nearrow T$.

This fact is the key to proving a sufficient condition for global existence. Because the particle trajectories are volume preserving, the only way that a solution can leave the set O_M is if $|X|_{1,\gamma}$ grows larger than M. We cease to have a solution at a finite time T^* only if $|X|_{1,\gamma}$ becomes unbounded as $t \to T^*$. Hence we obtain a sufficient condition for global-in-time existence by finding a sufficient condition for $|X|_{1,\gamma}$ to be a priori bounded.

We now prove the following proposition.

Proposition 4.3. *A sufficient condition for $|X(\cdot, t)|_{1,\gamma}$ to be a priori bounded is an a priori bound on $\int_0^t |\nabla v(\cdot, s)|_0 ds$.*

Proof of Proposition 4.3. Recall that $X(\alpha, t)$ satisfies

$$\frac{d}{dt} X(\alpha, t) = v(X(\alpha, t), t)$$

$$v(x, t) = \int_{\mathbb{R}^3} K_3[x - X(\alpha', t)] \nabla_\alpha X(\alpha', t) \omega_0(\alpha') d\alpha', \tag{4.44}$$

and hence

$$\frac{d}{dt} \nabla_\alpha X(\alpha, t) = \nabla v(X(\alpha, t), t) \nabla_\alpha X(\alpha, t). \tag{4.45}$$

Recall from Lemma 4.5 that because v can be obtained from w by the Biot–Savart law, estimate (4.34) implies, at a fixed time s, that $|v(\cdot, s)|_0 \leq C R(s) |\omega(\cdot, s)|_0$, where $R(s)^3$ is the measure of the support of the vorticity $\omega(\cdot, s)$ at time s. Because $X(\cdot, t)$ is a volume-preserving map, the size of the support of $\omega(\cdot, t)$ is time independent and we have $|v(\cdot, s)|_0 \leq C |\omega(\cdot, s)|_0$. Equation (4.44) implies that

$$|X(0, t)| \leq \int_0^t |v(\cdot, s)|_0 ds \leq C \int_0^t |\omega(\cdot, s)|_0 ds. \tag{4.46}$$

Also, Eq. (4.45) implies that

$$\frac{d}{dt}|\nabla_\alpha X(\cdot,t)|_0 \le |\nabla v(\cdot,t)|_0 |\nabla_\alpha X(\cdot,t)|_0,$$

and hence

$$|\nabla_\alpha X(\cdot,t)|_0 \le e^{\int_0^t |\nabla v(\cdot,s)|_0 ds}. \tag{4.47}$$

Bound (4.47) is the simplest example of a Grönwall inequality. We use this idea in more generality throughout this book. The usual form for Grönwall's lemma is the following lemma.

Lemma 4.7. *Grönwall's Lemma. If u, q, and $c \ge 0$ are continuous on $[0,t]$, c is differentiable, and*

$$q(t) \le c(t) + \int_0^t u(s)q(s)ds,$$

then

$$q(t) \le c(0)\exp\int_0^t u(s)ds + \int_0^t c'(s)\left[\exp\int_s^t u(\tau)d\tau\right]ds.$$

To bound $|\nabla_\alpha X(\cdot,t)|_\gamma$ note that Eq. (4.45) implies that

$$\begin{aligned}\frac{d}{dt}|\nabla_\alpha X(\cdot,t)|_\gamma &\le |\nabla v(X(\cdot,t),t)|_\gamma |\nabla_\alpha X(\cdot,t)|_0 + |\nabla v(\cdot,t)|_0 |\nabla_\alpha X(\cdot,t)|_\gamma \\ &\le |\nabla v(\cdot,t)|_\gamma |\nabla_\alpha X(\cdot,t)|_0^{1+\gamma} + |\nabla v(\cdot,t)|_0 |\nabla_\alpha X(\cdot,t)|_\gamma \\ &\le C|\omega(\cdot,t)|_\gamma e^{(1+\gamma)\int_0^t |\nabla v(\cdot,s)|_0 ds} + |\nabla v(\cdot,t)|_0 |\nabla_\alpha X(\cdot,t)|_\gamma.\end{aligned}$$

In the above, steps one and two use calculus inequalities (4.16) and (4.20), respectively. The last step uses bound (4.47). To finish the proof of Proposition 4.3, we need to estimate the term $|\omega(\cdot,t)|_\gamma$ in terms of $|\nabla v(\cdot,t)|_0$. We use the following lemma

Lemma 4.8. *Assume that $\omega_0(x) = \omega_0 \in C^\gamma(\mathbb{R}^3)$ and for $|t| \le T$ and $\omega(x,t)$ satisfies the vorticity-stream form*

$$\frac{\partial \omega}{\partial t} + v\cdot\nabla w = \omega\cdot\nabla v$$

and that $X(\alpha,t)$ are the particle trajectories associated with the velocity v that satisfy Eqs. (4.44) and (4.45). Then

$$|\omega(\cdot,t)|_\gamma \le |\omega_0|_\gamma \exp\left[(C_0+\gamma)\int_0^t |\nabla v(\cdot,s)|_0 ds\right], \tag{4.48}$$

The proof uses the vorticity-stream form of the Euler equation and is somewhat technical. We therefore save the proof for the end of this section. Bound (4.48) implies

$$\frac{d}{dt}|\nabla_\alpha X(\cdot,t)|_\gamma \le C|\omega_0|_\gamma e^{C_1\int_0^t |\nabla v(\cdot,s)|_0 ds} + |\nabla v(\cdot,t)|_0 |\nabla_\alpha X(\cdot,t)|_\gamma.$$

The time integral is [by use of $X(\alpha, 0) = \alpha$]

$$|\nabla_\alpha X(\cdot, t)|_\gamma \leq C|\omega_0|_\gamma \int_0^t e^{C_1 \int_0^s |\nabla v(\cdot,\tau)|_0 d\tau} ds + \int_0^t |\nabla v(\cdot, s)|_0 |\nabla_\alpha X(\cdot, s)|_\gamma ds.$$

Grönwall's lemma then gives

$$\begin{aligned} |\nabla_\alpha X(\cdot, t)|_\gamma &\leq C|\omega_0|_0 \int_0^t e^{C_1 \int_0^s |\nabla v(\cdot,\tau)|_0 d\tau} e^{\int_s^t |\nabla v(\cdot,\tau)|_0 d\tau} ds \\ &\leq C|\omega_0|_0 t\, e^{C_1 \int_0^t |\nabla v(\cdot,s)|_0 ds} \int_0^t e^{\int_0^s |\nabla v(\cdot,\tau)|_0 d\tau} ds. \end{aligned}$$

Hence $|\nabla_\alpha X(t)|_{1,\gamma}$ is a priori bounded provided that $\int_0^t |\nabla v(\cdot, s)|_0 ds$ is a priori bounded. □

To complete the proof of Theorem 4.3 we need to show that $\int_0^t |\omega(\cdot, s)|_0 ds$ controls $\int_0^t |\nabla v(\cdot, s)|_0 ds$. Recalling that $\omega(\cdot, t)$ has support of constant measure because $X(\alpha, t)$ is volume preserving, we note that the gradient of the Biot–Savart law,

$$\nabla v(x, t) = \mathrm{PV} \int_{\mathbb{R}^N} P_N(x - x')\omega(x', t)dx' + c\omega(x, t). \tag{4.49}$$

can be bounded if we use relation (4.35) from Lemma 4.6:

$$|\nabla v(\cdot, t)|_0 \leq c|\omega(\cdot, t)|_\gamma \epsilon^\gamma + |\omega(\cdot, t)|_0[1 + \log(R/\epsilon)], \qquad \forall \epsilon > 0,$$

where R is a fixed constant independent of time [from Lemma 4.6 R^N is the time independent measure of the support of $\omega(\cdot, t)$].

At each fixed time t, set $\epsilon = |\omega(\cdot, t)|_\gamma^{-1/\gamma}$ to obtain

$$|\nabla v(\cdot, t)|_0 \leq |\omega(\cdot, t)|_0[|\log(R|\omega(\cdot, t)|_\gamma)| + C]. \tag{4.50}$$

Recall from Lemma 4.8 that $|\omega(\cdot, t)|_\gamma$ satisfies bound (4.48):

$$|\omega(\cdot, t)|_\gamma \leq |\omega_0|_\gamma \exp\left[(C_0 + \gamma) \int_0^t |\nabla v(\cdot, s)|_0 ds\right].$$

Plugging this into bound (4.50) gives

$$\begin{aligned} |\nabla v(\cdot, t)|_0 &\leq C(\omega_0)|\omega(\cdot, t)|_0 \left[1 + \int_0^t |\nabla v(\cdot, s)|_0 ds\right], \\ \frac{|\nabla v(\cdot, t)|_0}{\left[1 + \int_0^t |\nabla v(\cdot, s)|_0 ds\right]} &\leq C(\omega_0)|\omega(\cdot, t)|_0, \\ \frac{d}{dt} \ln \left[1 + \int_0^t |\nabla v(\cdot, s)|_0 ds\right] &\leq C(\omega_0)|\omega(\cdot, t)|_0, \\ \ln \left[1 + \int_0^t |\nabla v(\cdot, s)|_0 ds\right] &\leq (\omega_0) \int_0^t |\omega(\cdot, s)|_0 ds, \\ 1 + \int_0^t |\nabla v(\cdot, s)|_0 ds &\leq e^{C(\omega_0) \int_0^t |\omega(\cdot,s)|_0 ds}. \end{aligned}$$

This last inequality combined with Proposition 4.3 yields the proof of Theorem 4.3. We conclude this section with a proof of Lemma 4.8.

Proof of Lemma 4.8. We want to show that

$$|\omega(\cdot,t)|_\gamma \leq |\omega_0|_\gamma \exp\left[(C_0+\gamma)\int_0^t |\nabla v(\cdot,s)|_0 ds\right]. \tag{4.51}$$

The first step is to prove the inequality

$$|\omega\cdot\nabla v(\cdot,t)|_\gamma \leq C_0|\nabla v(\cdot,t)|_0|\omega(\cdot,t)|_\gamma. \tag{4.52}$$

Calculus inequality (4.16) from Lemma 4.1 implies that

$$|\omega\cdot\nabla v(\cdot,t)|_\gamma \leq |\omega(\cdot,t)|_\gamma|\nabla v|_0 + |\omega(\cdot,t)|_0|\nabla v(\cdot,t)|_\gamma.$$

Equation (4.49) coupled with potential theory estimate (4.36) from Lemma 4.6 imply that $|\nabla v(\cdot,t)|_\gamma \leq C|\omega(\cdot,t)|_\gamma$. Because ω is the curl of v we trivially have $|\omega(\cdot,t)|_0 \leq 2|\nabla v(\cdot,t)|_0$. Combining all of these results gives inequality (4.52), which we use below.

Recall that $\omega(x,t)$ satisfies the transport equation

$$\frac{\partial\omega}{\partial t} + v\cdot\nabla\omega = \omega\cdot\nabla v. \tag{4.53}$$

Using the inverse of the particle-trajectory map in Eulerian variables, we rewrite transport equation (4.53) as

$$\omega(x,t) = \omega_0(X^{-1}(x,t)) + \int_0^t \omega\cdot\nabla v[X^{-1}(x,t-s),s]ds.$$

We use this to estimate the Hölder seminorm of ω. Taking the seminorm $|\cdot|_\gamma$ of both sides gives

$$\begin{aligned}
&|\omega(x,t)-\omega(x',t)| \\
&\quad\leq |\omega_0(X^{-1}(x,t)) - \omega_0(X^{-1}(x',t))| \\
&\qquad + \left|\int_0^t \omega\cdot\nabla v[X^{-1}(x,t-s),s] - \omega\cdot\nabla v[X^{-1}(x',t-s),s]ds\right| \\
&\quad\leq |\omega_0|_\gamma|\nabla X^{-1}(\cdot,t)|_0^\gamma|x-x'|^\gamma \\
&\qquad + \int_0^t |\omega\cdot\nabla v(\cdot,s)|_\gamma|\nabla X^{-1}(\cdot,t-s)|_0^\gamma|x-x'|^\gamma ds \\
&\quad\leq |\omega_0|_\gamma \exp\left[\gamma\int_0^t |\nabla v(\cdot,s)|_0 ds\right]|x-x'|^\gamma \\
&\qquad + \int_0^t |\omega\cdot\nabla v(\cdot,s)|_\gamma \exp\left[\gamma\int_s^t |\nabla v(s',\cdot)|_0 ds'\right]|x-x'|^\gamma ds.
\end{aligned}$$

Here we use the fact that the backward particle trajectories satisfy

$$\frac{\partial X^{-1}(x,t)}{dt} = -v(X^{-1}(x,t),t)$$

and hence, as in the derivation of bound (4.47),

$$|\nabla X^{-1}(x, t-s)|_0 \leq \exp\left[\int_s^t |\nabla v(s', \cdot)|_0 ds'\right].$$

Dividing by $|x - x'|^\gamma$ gives

$$|\omega(\cdot, t)|_\gamma \leq |\omega_0|_\gamma \exp\left[\gamma \int_0^t |\nabla v(\cdot, s)|_0 ds\right] + \int_0^t |\omega \cdot \nabla v(\cdot, s)|_\gamma \exp\left[\gamma \int_s^t |\nabla v(\cdot, s')|_0 ds'\right] ds.$$

Here we use the fact that $\nabla_x X^{-1}$ satisfies

$$|\nabla X^{-1}(x, t-s)|_0 \leq \exp\left[\int_s^t |\nabla v(s', \cdot)|_0 ds'\right].$$

Writing $Q(s) = |\nabla v(\cdot, s)|_0$, we find that relation (4.52) implies that

$$|\omega(\cdot, t)|_\gamma \leq |\omega_0|_\gamma \exp\left[\gamma \int_0^t Q(s) ds\right] + C \int_0^t Q(s)|\omega(\cdot, s)|_\gamma \exp\left[\gamma \int_s^t Q(s') ds'\right] ds.$$

Multiplying both sides by $\exp[-\gamma \int_0^t Q(s')ds']$ yields

$$|\omega(\cdot, t)|_\gamma \exp\left[-\gamma \int_0^t Q(s') ds'\right] \leq |\omega_0|_\gamma + C_0 \int_0^t Q(s)|\omega(\cdot, s)|_\gamma \exp\left[-\gamma \int_0^s Q(s') ds'\right] ds,$$

so that $|\omega(\cdot, t)|_\gamma \exp[-\gamma \int_0^t Q(s')ds'] = G(t)$ satisfies

$$G(t) \leq |\omega_0|_\gamma + C_0 \int_0^t Q(s) G(s) ds,$$

and thus, by Grönwall's lemma, satisfies

$$G(t) \leq |\omega_0|_\gamma \exp\left[C_0 \int_0^t Q(s) ds\right],$$

which gives

$$|\omega(\cdot, t)|_\gamma \leq |\omega_0|_\gamma \exp\left[(C_0 + \gamma) \int_0^t |\nabla v(\cdot, s)|_0 ds\right]. \qquad \square$$

4.3. Global Existence of 3D Axisymmetric Flows without Swirl

In Section 4.2 we proved that if for any $T > 0$ there exists $M_1 > 0$ such that

$$\int_0^T |\omega(\cdot, t)|_0 dt \leq M_1,$$

then the solution v to the 3D Euler equation exists globally in time, $v \in C^1([0, \infty); C^2)$. In particular we concluded that the 2D flows exist globally in time because $|\omega(\cdot, t)|_0 \le |\omega_0|_0$. Global existence of 2D solutions suggests that a 2D geometry can inhibit blowup of solutions to the 3D Euler equations. An example of this is the class of 3D axisymmetric flows without swirl. Recall from Subsection 2.3.3 that v is an axisymmetric flow if

$$v = v^r(r, x_3, t)e_r + v^\theta(r, x_3, t)e_\theta + v^3(r, x_3, t)e_3, \tag{4.54}$$

where e_r, e_θ, and e_3 are the standard orthonormal unit vectors defining the cylindrical coordinate system

$$e_r = {}^t\left(\frac{x_1}{r}, \frac{x_2}{r}, 0\right), \qquad e_\theta = {}^t\left(\frac{x_2}{r}, -\frac{x_1}{r}, 0\right), \qquad e_3 = {}^t(0, 0, 1),$$

where $r = (x_1^2 + x_2^2)^{1/2}$. These 3D flows depend on only two space coordinates, $-r$ and x_3. The velocity v^θ in the e_θ direction is called the *swirl velocity*. The corresponding vorticity $\omega = \operatorname{curl} v$ is

$$\omega = -v^\theta_{x_3} e_r + \omega^\theta e_\theta + \frac{1}{r}(r v^\theta)_r e_3, \tag{4.55}$$

where $\omega^\theta = v^r_{x_3} - v^3_r$. The general question of whether or not smooth axisymmetric solutions in system (4.54) exist globally in time is still an open problem. Below we identify a nontrivial class of flows without swirl ($v^\theta \equiv 0$) that exist globally in time.

Proposition 4.4. *Let $v \in C^1\{[0, T); C^{1,\gamma}(\mathbb{R}^3)\}$, $0 < \gamma < 1$, be a 3D inviscid, axisymmetric flow without swirl,*

$$v = v^r(r, x_3, t)e_r + v^3(r, x_3, t)e_3, \tag{4.56}$$

with the corresponding vorticity $\omega = \operatorname{curl} v$ given by

$$\omega = \left(v^r_{x_3} - v^3_4\right)e_\theta \equiv \omega^\theta e_\theta. \tag{4.57}$$

Assume that the initial vorticity ω_0^θ satisfies

$$|\omega_0^\theta(r, x_3)| \le cr, \tag{4.58}$$

$$\omega_0^\theta(r, x_3) \text{ has compact support.} \tag{4.59}$$

Then the solution v exists globally in time, $v \in C^1([0, \infty); C^{1,\gamma}(\mathbb{R}^3))$.

Condition (4.58) actually follows from the smoothness of ω_0^θ. Assumption (4.59) is for simplicity of exposition.

Proof of Proposition 4.4. First we recall some properties of 3D inviscid, axisymmetric flows without swirl, discussed in Subsection 2.3.3. For such flows the vorticity equation reduces to Eq. (2.58),

$$\frac{\tilde{D}}{Dt}\xi = 0, \tag{4.60}$$

where $\xi = (\omega^\theta / r)$ and $\tilde{D}/Dt$ denotes the material derivative:

$$\frac{\tilde{D}}{Dt} = \frac{\partial}{\partial t} + v^r \frac{\partial}{\partial r} + v^3 \frac{\partial}{\partial x_3}.$$

Thus ξ is conserved along particle trajectories.

We now show that for any $T > 0$ there exists $M_1 > 0$ such that

$$\int_0^T |\omega^\theta(\cdot, t)|_0 dt \leq M_1,$$

so that by Theorem 4.3 we have the existence of a smooth solution $X(\alpha, t) \in C^1([0, \infty); \mathbf{B})$ globally in time.

First note that Eq. (4.60) implies that $\xi = (\omega^\theta / r)$ is conserved along particle trajectories $X(\alpha, t) = (X', X_3)$. Hence

$$\frac{\omega^\theta(X(\alpha, t), t)}{|X'(\alpha, t)|} = \frac{\omega_0^\theta(\alpha', \alpha_3)}{|\alpha'|}.$$

Assumption (4.58) implies that

$$|\omega^\theta(X(\alpha, t), t)| \leq c|X'(\alpha, t)|. \tag{4.61}$$

It suffices to show that $|X'(\alpha, t)|$ remains bounded on the support of ω_0^θ. This, however, is a direct result of Lemma 4.5. Define $R(t)$ to be the radius of the support of v. That is, $R(t) = \sup_{\alpha \in \text{supp}\, \omega_0} |X(\alpha, t)|$. Lemma 4.5 states that if $\tilde{R}(t)^3$ is the 3D Lebesgue measure of the support of $\omega(t)$, then

$$|v(X(\alpha, t, t))| \leq \tilde{R}(t)|\omega(\cdot, t)|_0.$$

The incompressibility of the velocity field and the fact that the support of the vorticity is convected along particle paths implies that $\tilde{R}(t)$ is a constant, independent of time. Combining this with relation (4.61) yields

$$|v(X(\alpha, t, t))| \leq \tilde{R}(0) \sup_{\alpha \in \text{supp}\, \omega_0} |X'(\alpha, t)|.$$

The growth rate of $R(t)$ is thus bounded by

$$\begin{aligned}\frac{d}{dt}|R(t)| &\leq \sup_{\alpha \in \text{supp}\, \omega_0} |v(X(\alpha, t), t)| \\ &\leq cR(0)|\omega(\cdot, t)|_0 \\ &\leq c_1 R(0) R(t).\end{aligned}$$

Grönwall's lemma implies that $R(t)$ is a priori bounded by

$$R(t) \leq R(0)e^{R(0)c_1 t}. \tag{4.62}$$

Hence the vorticity remains a priori pointwise bounded on any time interval and hence by Theorem 4.3 the solution exists globally in time. □

Relation (4.62) coupled with relation (4.61) gives an exponential-in-time bound for the growth of $|\omega(\cdot, t)|_0$. Even if the vorticity accumulates, it must do so in infinite time. In Chap. 8 we obtain a superexponential growth-rate bound for the Lagrangian tangent vector to a patch of constant vorticity that allows us to show, by using a similar philosophy, that the vortex-patch boundary remains smooth for all time.

4.4. Higher Regularity

So far we have considered the existence and the uniqueness of solutions in the Banach space:

$$\mathbf{B} = \{X : \mathbb{R}^3 \to \mathbb{R}^3 \text{ such that } |X|_{1,\gamma} < \infty\}, \tag{4.63}$$

where $|\cdot|_{1,\gamma}$ is the norm defined by

$$|X|_{1,\gamma} = |X(0)| + |\nabla_\alpha X|_0 + |\nabla_\alpha X|_\gamma. \tag{4.64}$$

In the above $|\cdot|_0$ is the supremum or C^0 norm and $|\cdot|_\gamma$ is the Hölder seminorm defined by

$$|X|_\gamma = \sup_{\substack{\alpha,\alpha' \in \mathbb{R}^3 \\ \alpha \neq \alpha'}} \frac{|X(\alpha) - X(\alpha')|}{|\alpha - \alpha'|^\gamma}, \qquad 0 < \gamma \leq 1. \tag{4.65}$$

Recall the simplified notation

$$\|X\|_\gamma = |X|_0 + |X|_\gamma, \qquad 0 < \gamma \leq 1. \tag{4.66}$$

In this section we address the question of higher regularity, that is, given an initial vorticity in the space $\omega_0(x) \in C^{m,\gamma}$, what kind of regularity will the solution sustain on its interval of existence? We show here that the same condition that controls the $|X|_{1,\gamma}$ norm of the particle trajectories controls higher derivatives as well.

Recall that the integrodifferential equation for the particle trajectories,

$$\begin{aligned} \frac{dX}{dt}(\alpha, t) &= F(X(\alpha, t)), \\ X(\alpha, t)|_{t=0} &= \alpha, \end{aligned} \tag{4.67}$$

where the nonlinear mapping $F[X]$, defined by

$$F(X(\alpha, t)) = \int_{\mathbb{R}^3} K_3[X(\alpha, t) - X(\alpha', t)]\nabla_\alpha X(\alpha', t)\omega_0(\alpha')d\alpha', \tag{4.68}$$

is actually an ODE on a Banach space. F is an bounded, nonlinear, Lipschitz continuous operator on this Banach space. We now consider initial data in the class $\omega_0 \in C^{m,\gamma}$. We begin by defining a Banach space $\mathbf{B}_m$ that contains information about higher derivatives.

As in Section 4.1, we define a norm for the particle-trajectory maps:

$$|X|_{m,\gamma} = |X(0)| + \sup_{1 \le |\mathbf{k}| \le m} \left(\left| \frac{\partial^{\mathbf{k}} X}{\partial x^{\mathbf{k}}} \right|_0 + \left| \frac{\partial^{\mathbf{k}} X}{\partial x^{\mathbf{k}}} \right|_\gamma \right),$$

$$\frac{\partial^{\mathbf{k}}}{\partial x^{\mathbf{k}}} = \frac{\partial^{k_1}}{\partial x^{k_1}} \cdots \frac{\partial^{k_N}}{\partial x^{k_N}}. \tag{4.69}$$

To prove higher regularity of solutions to the Euler equation, we show that $\omega_0 \in C^{m,\gamma}$ implies that F is locally Lipschitz continuous on the Banach space $\mathbf{B}_m$ defined with the (4.69). Define the open set

$$O_M^m = \left\{ X : X \in \mathbf{B}_m \middle| \inf_{\alpha \in \mathbb{R}^3} \det \nabla_\alpha X(\alpha) > \frac{1}{2} \text{ and } |X|_{m,\gamma} < M \right\}. \tag{4.70}$$

We have the following proposition.

Proposition 4.5. *Let $|\omega_0|_{m-1,\gamma} < \infty$ and $0 < \gamma < 1$, $m \ge 1$. Let $F : O_M^m \to \mathbf{B}_m$ be defined by Eq. (4.68):*

$$F(X(\alpha, t)) = \int_{\mathbb{R}^3} K_3[X(\alpha, t) - X(\alpha', t)] \nabla_\alpha X(\alpha', t) \omega_0(\alpha') d\alpha'.$$

Then F satisfies the assumptions of the Picard theorem, i.e., F is bounded and locally Lipschitz continuous on O_M^m.

The proof of this proposition is very similar to the proof of Proposition 4.2 from Section 4.1. We present the main idea. First, note that the sets O_M^m are open by the same argument as that in Proposition 4.1. To show that F is a bounded operator, we need estimates for the higher derivatives of $F(X(\alpha, t))$. We discuss the case $m = 2$. Estimates of successive derivatives follow in a similar fashion. In order to do this, we need to make sense of the derivative of a singular integral of the form

$$\mathrm{PV} \int P_N(x - y) f(y) dy. \tag{4.71}$$

Differentiating the kernel P_N produces an even more singular kernel. However, if f is differentiable then

$$g(x) = \int K_N(x - y) f(y) dy = \int K_N(-y) f(x + y) dy,$$

$$\nabla g(x) = \int K_N(-y) \nabla f(x + y) dy = \int K_N(x - y) \nabla f(y) dy,$$

providing a nonsingular integral for ∇g. We compute $\partial_{x_i} \nabla g$ as in Proposition Section 2.4.3

$$\partial_{x_i} \nabla g(x) = \mathrm{PV} \int \nabla K_N(x - y) \partial_{x_i} \nabla f(y) dy + c \nabla f(x)$$

where the principal-value integral converges, provided that ∇f is Hölder continuous with exponent $\gamma \in (0, 1]$. We apply this idea to $F(X(\alpha, t))$ and use the calculus inequalities and potential theory estimates from Section 4.2. A straightforward repetition of the arguments from Section 4.2 gives the following estimate.

Lemma 4.9. *Let F be given by Eq. (4.68) with $\omega_0 \in C^{m-1,\gamma}$. Then for all $X \in O_M^m$,*

$$|F(X(\cdot, t))|_{m,\gamma} \leq C[|X(\cdot, t)|_{m-1,\gamma}, |\omega_0|_{C^{1,\gamma}}]|X(\alpha, t)|_{m,\gamma}. \tag{4.72}$$

Lemma 4.9 provides two essential ingredients: first, that F is a bounded operator on $O_m^m \to \mathbf{B}^m$; the second is that the bound on F depends *linearly* on the highest norm. This is the key ingredient to show that higher derivatives are controlled by lower ones.

For local existence, we need that F is Lipschitz continuous on O_M^m by showing that, for $X \in O_M^m$,

$$G_1(X)Y = \int_{\mathbb{R}^3} K_3[X(\alpha) - X(\alpha')]\nabla_\alpha Y(\alpha')\omega_0(\alpha')d\alpha', \tag{4.73}$$

$$\begin{aligned} G_2(X)Y = \int_{\mathbb{R}^3} & \nabla K_3[X(\alpha) - X(\alpha')][Y(\alpha) - Y(\alpha')] \\ & \times \nabla_\alpha X(\alpha')\omega_0(\alpha')d\alpha' \end{aligned} \tag{4.74}$$

satisfy

$$|G_1(X)Y|_{m,\gamma}, |G_2(X)Y|_{m,\gamma} \leq C(M)|\omega_0|_{C^{m-1,\gamma}}|Y|_{m,\gamma}.$$

The proof is so similar to that of the first-order case that we leave it as an exercise for the reader.

We now prove the higher-derivative version of Theorem 4.3 on the continuation of solutions. We show that all higher derivatives are controlled by $|\omega(\cdot, t)|_0$.

Theorem 4.5. *Beale–Kato–Majda for $C^{m,\gamma}$ Initial Data. Consider a compactly supported initial vorticity $\omega_0 = \text{curl } v_0$, $\text{div } v_0 = 0$, with Hölder norm $\|\omega_0\|_{m,\gamma} < \infty$ for some $\gamma \in (0, 1)$. Let $|\omega(\cdot, s)|_0$ be the supremum norm at fixed time s of a solution to the Euler equation, with initial data ω_0.*

(i) Suppose that for any $T > 0$ there exists $M_1 > 0$ such that the vorticity $\omega(x, t)$ satisfies

$$\int_0^T |\omega(\cdot, s)|_0 ds \leq M_1. \tag{4.75}$$

Then, for any T there exists $M > 0$ such that $X \in C^1([0, T); O_M)$ (the solution X exists globally in time).

(ii) Suppose that for any $M > 0$ there is a finite maximal time $T(M)$ of existence of solutions $X \in C^1\{[0, T(M)); O_M\}$ and that $\lim_{M\to\infty} T(M) = T^* < \infty$; then

necessarily the vorticity accumulates so rapidly that

$$\lim_{t \nearrow T^*} \int_0^t |\omega(\cdot, s)|_0 ds = \infty.$$

As in the case of Theorem 4.3 this result also holds for solutions on the backward time interval $(-T_-^*, 0]$.

Proof of Theorem 4.5. The proof of this theorem is straightforward. We use an induction argument on the number of Lagrangian derivatives, m, of the particle trajectories $X(\alpha, t)$. Recall from Theorem 4.4 that we can continue the solution to an *autonomous* ODE on a Banach space as long as it remains in the open set O_M^m. Hence we have global existence of a solution in B_M^m provided that the $|\cdot|_{m,\gamma}$ norm of the particle-trajectory map $X(\alpha, t)$ remains bounded. We now use the fact that estimate (4.72) depends linearly on the highest norm. Grönwall's lemma combined with this bound gives

$$\begin{aligned}
\frac{d}{dt}|X(\cdot, t)|_{m,\gamma} &\leq |F[X(\cdot, t)]|_{m,\gamma} \\
&\leq C[|X(\cdot, t)|_{m-1,\gamma}, |\omega_0|_{m-1,\gamma}]|X(\alpha, t)|_{m,\gamma} \\
|X(\cdot, t)|_{m,\gamma} &\leq |X(\cdot, 0)|_{m,\gamma} \exp\left\{\int_0^t C[|X(\cdot, s)_{m-1,\gamma}, |\omega_0|_{m-1,\gamma}]ds\right\}.
\end{aligned}$$

Hence, Grönwall's lemma, Lemma 4.7, shows us that $|X(\cdot, t)|_{m,\gamma}$ is a priori controlled given a bound on $|X(\cdot, t)|_{m-1,\gamma}$. Because this fact holds for all $m \geq 2$, an induction argument combined with Theorem 4.3 finishes the proof. □

4.5. Appendixes for Chapter 4

4.5.1. Calculus Inequalities in the Hölder Spaces

We review the definition of Hölder continuity and prove the calculus inequalities stated in Section 4.1.

Let Ω be an open, bounded domain in $\mathbb{R}^N$. The Hölder space $C^{0,\gamma}(\bar{\Omega})$, $0 < \gamma \leq 1$, is a subspace of $C^0(\bar{\Omega})$ (continuous functions with the supremum norm, $|\cdot|_0$) of functions X that satisfy the additional condition

$$|X|_\gamma = \sup_{\substack{\alpha,\alpha' \in \Omega \\ \alpha \neq \alpha'}} \frac{|X(\alpha) - X(\alpha')|}{|\alpha - \alpha'|^\gamma} < \infty.$$

$|\cdot|_\gamma$ is often referred to as the Hölder seminorm. The full $C^{0,\gamma}(\bar{\Omega})$ norm is

$$\|X\|_\gamma = |X|_0 + |X|_\gamma.$$

The spaces $C^0(\bar{\Omega})$ and $C^{0,\gamma}(\bar{\Omega})$ are Banach spaces; hence all Cauchy sequences in these spaces converge to an element of the space. The space $C^{m,\gamma}(\bar{\Omega})$ of m-times differentiable functions with Hölder continuous derivatives has a norm

$$|f|_{m,\gamma} = |f|_0 + \sup_{1\le|\mathbf{k}|\le m} \left|\frac{\partial^{\mathbf{k}} f}{\partial x^{\mathbf{k}}}\right|_\gamma .$$

In the case in which the spatial domain Ω is unbounded, e.g., $\Omega = \mathbb{R}^N$, the spaces $C^{m,\gamma}(\Omega)$ contain bounded functions that cannot describe homeomorphisms like the particle-trajectory maps, $X(\alpha, t)$. In Section 4.1 we introduce the Banach space $\mathbf{B}$ equipped with the norm

$$|X|_{1,\gamma} = |X(0)| + |\nabla_\alpha X|_0 + |\nabla_\alpha X|_\gamma, \qquad 0 < \gamma \le 1,$$

allowing functions $X \in \mathbf{B}$ to map from $\mathbb{R}^N$ onto $\mathbb{R}^N$.

We restate and prove some calculus inequalities of these Hölder spaces. We begin with calculus inequalities for products of functions.

Lemma 4.1. *Let $X, Y : \mathbb{R}^N \to \mathbb{R}^N$ be smooth, bounded functions. Then for $\gamma \in (0, 1)$ there exists $c > 0$ such that*

$$|XY|_\gamma \le \|X\|_0 |Y|_\gamma + |X|_\gamma \, \|Y\|_0, \tag{4.76}$$

$$\|XY\|_\gamma \le c\|X\|_\gamma \, \|Y\|_\gamma. \tag{4.77}$$

Proof of Lemma 4.1. If $X, Y \in C^{0,\gamma}$, the definition of the Hölder seminorm $|\cdot|_\gamma$ implies that

$$\begin{aligned} |X(\alpha)Y(\alpha) - X(\alpha')Y(\alpha')| &\le |Y(\alpha)|\, |X(\alpha) - X(\alpha')| + |X(\alpha')|\, |Y(\alpha) - Y(\alpha')| \\ &\le |\alpha - \alpha'|^\gamma \{|Y|_0 |X|_\gamma + |X|_0 \, |Y|_\gamma\} \end{aligned}$$

so that

$$|XY|_\gamma \le |X|_0 \, |Y|_\gamma + |X|_\gamma \, |Y|_0.$$

The fact that $|XY|_0 \le |X|_0 \, |Y|_0$ is obvious. □

Lemma 4.2. *Let $X : \mathbb{R}^N \to \mathbb{R}^N$ be a smooth, invertible transformation with*

$$|\det \nabla_\alpha X(\alpha)| \ge c_1 > 0.$$

Then for $0 < \gamma \le 1$ there exists $c > 0$ such that

$$\|(\nabla_\alpha X)^{-1}\|_\gamma \le c\|\nabla_\alpha X\|_\gamma^{2N-1}, \tag{4.78}$$

$$|X^{-1}|_{1,\gamma} \le c|X|_{1,\gamma}^{2N-1}. \tag{4.79}$$

Proof of Lemma 4.2. Because the function X is globally invertible, by differentiating the identity $X^{-1}[X(\alpha)] = \alpha$ we get $1 = \nabla_x X^{-1}(x) \nabla_\alpha X(\alpha)$, so

$$\nabla_x X^{-1}(x) = [\nabla_\alpha X(\alpha)]^{-1} = \frac{\mathrm{Co}[\nabla_\alpha X(\alpha)]}{\det \nabla_\alpha X(\alpha)},$$

where $\mathrm{Co}(\nabla_\alpha X)$ denotes the cofactor matrix of $\nabla_\alpha X$. Estimate (4.77) implies that

$$\begin{aligned}\|(\nabla_\alpha X)^{-1}\|_\gamma &\le \|\mathrm{Co}(\nabla_\alpha X)\|_\gamma \|\det[\nabla_\alpha X(\alpha)]^{-1}\|_\gamma \\ &\le c\|\nabla_\alpha X\|_\gamma^{2N-1},\end{aligned}$$

and likewise

$$|X^{-1}|_{1,\gamma} \le c|X|_{1,\gamma}^{2N-1}. \qquad \square$$

Lemma 4.3. *Let $X : \mathbb{R}^N \to \mathbb{R}^N$ be an invertible transformation with*

$$|\det \nabla_\alpha X(\alpha)| \ge c_1 > 0,$$

and let $f : \mathbb{R}^N \to \mathbb{R}^M$ be a smooth function. Then for $0 < \gamma \le 1$ the composition $f \circ X$ and $f \circ X^{-1}$ satisfies

$$|f \circ X|_\gamma \le |f|_\gamma \|\nabla_\alpha X\|_0^\gamma, \tag{4.80}$$

$$\|f \circ X\|_\gamma \le \|f\|_\gamma (1 + |X|_{1,\gamma}^\gamma), \tag{4.81}$$

$$\|f \circ X^{-1}\|_\gamma \le \|f\|_\gamma \left[1 + c\,|X|_{1,\gamma}^{\gamma(2N-1)}\right]. \tag{4.82}$$

Proof of Lemma 4.3. For all $\alpha \ne \alpha'$, the mean-value theorem implies that

$$\begin{aligned}|f \circ X(\alpha) - f \circ X(\alpha')| &= \frac{|f \circ X(\alpha) - f \circ X(\alpha')|}{|X(\alpha) - X(\alpha')|^\gamma} |X(\alpha) - X(\alpha')|^\gamma \\ &\le \sup_{\substack{X,X' \in \mathbb{R}^3 \\ X \ne X'}} \frac{|f(X) - f(X')|}{|X - X'|^\gamma} \left\{ \sup_{s \in [0,1]} |\nabla_\alpha X[\alpha + s(\alpha' - \alpha)]| \right\}^\gamma |\alpha - \alpha|^\gamma \\ &\le |f|_\gamma |\nabla_\alpha X|_0^\gamma |\alpha - \alpha'|^\gamma,\end{aligned}$$

so that

$$|f \circ X|_\gamma \le |f|_\gamma \|\nabla_\alpha X\|_\gamma^\gamma.$$

Moreover, $|f \circ X|_0 = |f|_0$; hence

$$\|f \circ X\|_\gamma \le \|f\|_\gamma (1 + |X|_{1,\gamma}^\gamma).$$

Finally, estimate (4.79) implies that

$$\|f \circ X^{-1}\|_\gamma \le \|f\|_\gamma (1 + |X^{-1}|_{1,\gamma}^\gamma) \le \|f\|_\gamma \left[1 + c|X|_{1,\gamma}^{\gamma(2N-1)}\right]. \qquad \square$$

4.5.2. Potential Theory Estimates in the Hölder Spaces

We prove the potential theory estimates from Section 4.2.

Let K_N be the integral operator defined by

$$K_N f(x) = \int_{\mathbb{R}^N} K_N(x - y) f(y) dy, \tag{4.83}$$

where the kernel K_N is smooth outside $x = 0$ and homogeneous of degree $1 - N$:

$$K_N(\lambda x) = \lambda^{1-N} K_N(x), \qquad \forall \lambda > 0,\ x \neq 0. \tag{4.84}$$

The kernel $P_N = \nabla K_N$ is homogeneous of degree $-N$:

$$P_N(\lambda x) = \lambda^{-N} P_N(x), \qquad \forall \lambda > 0, x \neq 0. \tag{4.85}$$

P_N has mean-value zero on the unit sphere,

$$\int_{|x|=1} P_N\, ds = 0, \tag{4.86}$$

and defines the principal-value SIO P_N:

$$P_N f(x) = \text{PV} \int_{\mathbb{R}^N} P_N(x - y) f(y) dy. \tag{4.87}$$

First we show that the operator K_N is linear and bounded in the space C^0 of continuous functions. That is, we have Lemma 4.5.

Lemma 4.5. *Let K_N satisfy Eqs. (4.83) and (4.84). Consider $f \in C^\gamma(\mathbb{R}^N; \mathbb{R}^N)$, $0 < \gamma < 1$ with compact support, where $m(\text{supp } f) = m_f < \infty$. Define the length scale R by $R^N = m_f$. Then there exists a constant c independent of f and R so that*

$$|K_N f|_0 \leq cR|f|_0. \tag{4.88}$$

Proof of Lemma 4.5. We split the intergal up into two parts:

$$\begin{aligned}
|K_n f(x)| &\leq \left| \int_{|x-y|<R} K_N(x - y) f(y) dy \right| + \left| \int_{|x-y|\geq R} K_N(x - y) f(y) dy \right| \\
&\leq c_1 |f|_0 \int_{|x'|<R} \frac{1}{|x'|^{N-1}} dx' + c_2 |f|_0 R^{-N+1} m_f \\
&\leq cR|f|_0
\end{aligned}$$

The last step uses the fact that $m(\text{supp } f) = m_f = R^N$. □

The above estimates were made simple by the fact that K_N is not a singular kernel. That is, it is contained in $L^1_{\text{loc}}(\mathbb{R}^N)$. When the kernel is singular, as in the case of $P_N(x) = \nabla K_N(x)$, we need to use the special cancellation properties of the kernel to derive useful potential theory estimates. If f is in $C_0^\infty(\mathbb{R}^N)$, the Calderon–Zygmund inequality gives

$$\|P_N f\|_{L^p} \leq c_p \|f\|_{L^p}, \qquad 1 < p < \infty.$$

Hence P_N is a linear, bounded (thus continuous) operator on the spaces of L^p, $1 < p < \infty$, integrable functions. For $p = \infty$ this estimate is not valid. We discuss this case in greater detail in Chap. 6 when we introduce weak L^∞ solution to the vorticity-stream form. In this chapter we make use of the fact that, as in the case of the L^p spaces for $1 < p < \infty$, SIOs are bounded on the Hölder spaces $C^{0,\gamma}$ with the exponent $0 < \gamma < 1$.

Lemma 4.6. *Let f, γ, and R satisfy the assumptions of Lemma 4.5. Let the SIO P_N satisfy Eqs. (4.85)–(4.87). Then there exists a constant c, independent of R and f so that*

$$\|P_N f\|_0 \leq c\left\{|f|_\gamma \epsilon^\gamma + \max\left(1, \ln\frac{R}{\epsilon}\right)\|f\|_0\right\}, \qquad \forall\epsilon > 0, \tag{4.89}$$

$$|P_N f|_\gamma \leq c|f|_\gamma. \tag{4.90}$$

The above potential theory estimates for singular kernels are not valid on the Lipschitz spaces, i.e., for $\gamma = 1$.

Proof of Lemma 4.6. First split the integration over $\mathbb{R}^N$ as

$$P_N f(x) = \left(\mathrm{PV}\int_{|x-y|\leq\epsilon} + \int_{|x-y|\geq\epsilon}\right) P_N(x-y) f(y) dy = I_1(x) + I_2(x).$$

Using the cancellation property for the kernel P_N,

$$\int_{|x|=1} P_N\, ds = 0,$$

rewrite the first integral, I_1, as

$$I_1(x) = \mathrm{PV}\int_{|y|\leq\epsilon} P_N(y)\{f(x-y) - f(x)\}dy.$$

Hence

$$\begin{aligned}|I_1(x)| &\leq \int_{|y|\leq\epsilon} |P_N(x)|\frac{|f(x-y) - f(x)|}{|y|^\gamma}|y|^\gamma dy \\ &\leq c|f|_\gamma \int_{|y|\leq 2\epsilon} |y|^{-N+\gamma} dy \leq c|f|_\gamma \epsilon^\gamma, \qquad 0 < \gamma < 1. \end{aligned}\tag{4.91}$$

The Hölder continuity of f gives the factor $|y|^\gamma$ that compensates for the singularity $|y|^{-N}$ of the kernel P_N.

The second integral, I_2, satisfies

$$\begin{aligned}|I_2(x)| &\leq \int_{R>|x-y|\geq 2\epsilon} |P_N(x-y)f(y)|dy + \int_{R\leq|x-y|} |P_N(x-y)f(y)|dy \\ &\leq c|f|_0 \ln(R/\epsilon) + cR^{-N}|f|_0 m_f \\ &\leq c|f|_0(\ln(R/\epsilon) + 1).\end{aligned}$$

The above estimates imply the sharp potential theory estimate (4.89):

$$|P_N f|_0 \le c\left\{|f|_\gamma \epsilon^\gamma + \max\left(1, \ln\frac{R}{\epsilon}\right)|f|_0\right\}, \qquad \epsilon > 0.$$

We prove potential theory estimate (4.90). The Hölder continuity of f with the exponent $\gamma \in (0, 1)$ is essential for the proof. Write $P_N f(x) - P_N f(x+h)$ as

$$\begin{aligned}
&\mathrm{PV}\int P_N(x-y)[f(x)-f(y)]dy - \mathrm{PV}\int P_N(x+h-y)[f(x+h)-f(y)]dy \\
&= \mathrm{PV}\int_{|x-y|<2h} P_N(x-y)[f(x)-f(y)]dy \\
&\quad - \mathrm{PV}\int_{|x-y|<2h} P_N(x+h-y)[f(x+h)-f(y)]dy \\
&\quad + \mathrm{PV}\int_{|x-y|\ge 2h} P_N(x-y)[f(x)-f(x+h)]dy \\
&\quad + \mathrm{PV}\int_{|x-y|\ge 2h} [P_N(x-y)-P_N(x+h-y)][f(x+h)-f(y)]dy \\
&= (1)+(2)+(3)+(4).
\end{aligned}$$

Clearly $|(1)|, |(2)| \le C_\gamma |f|_\gamma h^\gamma$, as in the derivation of relation (4.91) above. (3) is zero because of the cancellation property of P_N. Also, we have

$$|(4)| \le \int_{|x-y|\ge 2h} h\frac{C}{|x-y|^{n+1-\gamma}}|f|_\gamma dy \le C_\gamma h^\gamma |f|_\gamma.$$

In the last estimate, we use the mean-value theorem. That is, for $|x-y| \ge 2h$, $|x+h-y| \ge h$, and thus

$$|P_N(x-y) - P_N(x+h-y)| \le \sup_{0\le\xi\le 1} |\nabla P_N(x-y+\xi h)||h| \le ch|x-y|^{-N-1},$$

because

$$\frac{1}{2}|x-y| \le -h + |x-y| \le |x-y+\xi(h)| \le h + |x-y| \le \frac{3}{q2}|x-y|,$$

and by the homogeneity of P_N, $|\nabla P_N(x)| \le c|x|^{-N-1}$. The above estimates imply that, for all x and h,

$$|P_N f(x) - P_N f(x+h)| \le c|f|_\gamma |h|^\gamma,$$

and hence, by the definition of $|\cdot|_\gamma$,

$$|P_N f|_\gamma \le c|f|_\gamma, \qquad 0 < \gamma < 1. \qquad \square$$

Finally we derive the estimate used in the proof of Proposition 4.2. Recall that

$$O_M = \left\{X \in \mathbf{B} : \inf_{\alpha\in\mathbb{R}^3} \det \nabla_\alpha X(\alpha) > \frac{1}{2} \text{ and } |X|_{1,\gamma} < M\right\},$$

where

$$|X|_{1,\gamma} = |X(0)| + \|\nabla_\alpha X\|_\gamma, \qquad 0 < \gamma < 1.$$

We must show the following lemma.

Lemma 4.10. *Let the smooth vorticity ω_0 have the support of a bounded measure and let $\|\omega_0\|_\gamma < \infty$ for some $0 < \gamma < 1$. Given $X, Y \in O_M$, define*

$$G_2(X)Y = \int_{\mathbf{R}^3} \nabla K_3[X(\alpha) - X(\alpha')][Y(\alpha) - Y(\alpha')]\nabla_\alpha X(\alpha)\omega_0(\alpha')d\alpha' \tag{4.92}$$

Then there exists $c > 0$, independent of the particular choice of X and Y, such that

$$|G_2(X)Y|_{1,\gamma} \leq c(M)\|\omega_0\|_\gamma |Y|_{1,\gamma}. \tag{4.93}$$

Proof of Lemma 4.10. Although the kernel ∇K_3 is homogeneous of degree -3, Eq. (4.92) is not a singular integral because the term $Y(\alpha) - Y(\alpha')$ kills the singularity of ∇K_3.

The estimate of $|G_2(X)Y|_0$ follows in a similar fashion as that of $|K_3 f|_0$ in Lemma 4.5. The set O_M consists of 1-1 and onto functions; the change of variables $X^{-1}(x) = \alpha$ gives

$$G_2(X)Y \circ X^{-1}(x) = \int_{\mathbf{R}^3} \nabla K_3(x - x')(Y(X^{-1}(x)) - Y(X^{-1}(x')))f(x')dx', \tag{4.94}$$

where

$$f(x') = \nabla_\alpha X(\alpha')\omega_0(\alpha')|_{\alpha'=X^{-1}(x')} \det \nabla_x X^{-1}(x'). \tag{4.95}$$

Splitting the integration over $\mathbb{R}^N$ into two parts yields

$$\begin{aligned}
&G_2(X)Y \circ X^{-1}(x) \\
&= \left(\int_{|x-x'|\leq 1} + \int_{|x-x'|\geq 1}\right) \nabla K_3(x - x')\{Y[X^{-1}(x)] - Y[X^{-1}(x')]\}f(x')dx' \\
&\equiv I_1 + I_2.
\end{aligned}$$

The mean-value theorem implies that

$$|Y[X^{-1}(x)] - Y[X^{-1}(x')]| \leq |(\nabla_\alpha Y \circ X^{-1})\nabla_x X^{-1}|_0 |x - x'|.$$

Because $|\nabla K_3(x)| \leq c|x|^{-3}$,

$$\begin{aligned}
|I_1| &\leq c|(\nabla_\alpha Y \circ X^{-1})\nabla_x X^{-1}|_0 |f|_0 \int_{|x-x'|\leq 1} |x - x'|^{-3+1}dx' \\
&\leq c|(\nabla_\alpha Y \circ X^{-1})\nabla_x X^{-1}|_0 |f|_0.
\end{aligned}$$

The second term, I_2, satisfies

$$
\begin{aligned}
|I_2| &\le c|(\nabla_\alpha Y \circ X^{-1})\nabla_x X^{-1}|_0 |f|_0 \int_{\substack{|x-x'|\ge 1 \\ x' \in \operatorname{supp} f}} |x - x'|^{-3+1} dx' \\
&\le c|(\nabla_\alpha Y \circ X^{-1})\nabla_x X^{-1}|_0 \|f\|_\gamma,
\end{aligned}
$$

because the vorticity has the support of a bounded measure. The proof of Proposition 4.2 implies that

$$
\|f\|_\gamma \le c(M)\|\omega_0\|_\gamma. \tag{4.96}
$$

Inequalities (4.76)–(4.82) yield

$$
\begin{aligned}
|G_2(X)Y|_0 &\le c(M)\|\omega_0\|_\gamma \|\nabla_\alpha Y\|_\gamma \left(1 + |X|_{1,\gamma}^{5\gamma}\right) \|\nabla_\alpha X\|_\gamma^5 \\
&\le c(M)\|\omega_0\|_\gamma |Y|_{1,\gamma}.
\end{aligned} \tag{4.97}
$$

An estimate of the Hölder norm of $\nabla G_2(X)Y$ first requires a calculation of the distribution derivative of $H(x) = G_2(X)Y \circ X^{-1}(x)$. Recall from Chap. 2 that the distribution derivative ∂_{x_k} of f is the linear functional $\partial_{x_k} f$ defined operationally by the formula

$$
(\partial_{x_k} f, \varphi)_0 = -(f, \partial_{x_k}\varphi)_0 \qquad \text{for all } \varphi \in C_0^\infty(\mathbf{R}^N).
$$

Rewrite $H(x)$ as

$$
\begin{aligned}
H(x) &= \int_{\mathbb{R}^3} R(x, x') f(x') dx' \\
R(x, x') &= \nabla K_3(x - x')\{Y[X^{-1}(x)] - Y[X^{-1}(x')]\}.
\end{aligned}
$$

As in the proof of proposition 2.[ ], we compute the distribution derivative of the kernel $R(x + x', x')$. Note that R has the property

$$
\lim_{h\to 0} R(x, x + h\mathbf{a})h^{N-1} = \nabla_\alpha Y[X^{-1}(x)]\nabla_x X^{-1}(x)\mathbf{c}(\mathbf{a}).
$$

Thus

$$
\begin{aligned}
(-\partial_{x_i}\varphi, R(x + x', x')) &= -\int \partial_{x_i}\varphi(x) R(x + x', x') dx \\
&= -\lim_{\epsilon\searrow 0} \int_{|x|\ge\epsilon} \partial_{x_i}\varphi(x) R(x + x', x') dx \\
&= \lim_{\epsilon\searrow 0} \int_{|x|\ge\epsilon} \varphi(x)\partial_{x_i} R(x + x', x') dx \\
&\quad + \lim_{\epsilon\searrow 0} \int_{|x|=\epsilon} \varphi(x) R(x + x', x') \frac{(x_i)}{|x|} ds \\
&= \mathrm{PV} \int \varphi(x)\partial_{x_i} R(x + x', x') dx \\
&\quad + \nabla_\alpha Y(X^{-1}(x))\nabla_x X^{-1}(x) c_i \varphi(0).
\end{aligned}
$$

This formula implies that the distribution derivative of $H(x)$ is

$$\begin{aligned}\partial_{x_k} H(x) &= \mathrm{PV}\int_{\mathbf{R}^3} \partial_{x_k} \nabla K_3(x-x')[Y(X^{-1}(x)) - Y(X^{-1}(x'))]f(x')dx' \\ &\quad + \mathrm{PV}\int_{\mathbf{R}^3} \nabla K_3(x-x')\nabla_\alpha Y[X^{-1}(x)]\partial_{x_k} X^{-1}(x) f(x')dx' \\ &\quad + \nabla_\alpha Y[X^{-1}(x)]\nabla_x X^{-1}(x) c_k f(x) \\ &\equiv \mathcal{J}_1 + \mathcal{J}_2 + \mathcal{J}_3.\end{aligned}$$

To estimate the Hölder norm of $H(x)$, repeat the arguments from the proof of Lemma 4.6 for a kernel of the type $\nabla_x R(x, x')$.

The result is

$$\|\partial_{x_k} G_2(X)Y\|_\gamma \le c\|(\nabla_\alpha Y \circ X)\nabla_x X^{-1}\|_\gamma \|f\|_\gamma .$$

Combining this with estimates (4.96) and (4.97) gives

$$|G_2(X)Y|_{1,\gamma} \le c(M)\|\omega_0\|_\gamma |Y|_{1,\gamma}. \qquad \square$$

Notes for Chapter 4

This chapter is designed to be self-contained so that a student unfamiliar with SIOs and ODEs on abstract spaces will be able to follow the arguments. We recommend several good texts for those who wish to augment their background.

The theory of ODEs in $\mathbb{R}^N$ has several classical texts, including Hartman (1982) and Coddington and Levinson (1955). Both references contain a discussion of Grönwall's lemma. Hartman has a discussion of the Picard theorem on a Banach space in Chap. XX. Ladas and Lakshmikamtham (1972) have a comprehensive treatment of evolution equations on a Banach spaces including continuation properties of such solutions. The subject of singular integrals on Hölder spaces is classical. References include Mikhlin and Prössdorf (1986), Stein (1970), and Torchinsky (1986). The Calderon–Zygmund (1952) theorem is a fundamental result in harmonic analysis. Detailed discussion of the space BMO can be found in Chap. 6 and in John and Nirenberg (1961), Stein (1967), and Torchinsky (1986).

The continuation result in Theorem 4.3, which is true for initial data satisfying $\int_{R^3} |v(\cdot, 0)|^2 dx < \infty$, is due to Beale et al. (1984). The proof we present here is similar to arguments in Bertozzi and Constantin (1993) for the related problem of the regularity of the boundary of a vortex patch. We discuss vortex patches in detail in Chap. 8. A simple proof of Proposition 4.4 for axisymmetric flows without swirl was given by Majda (1986). Without potential theory estimate (4.34) from Lemma 4.5 he proved global existence by assuming fixed sign of the vorticity and by using the fact that the fluid impulse of Eq. (1.68) in Proposition 1.12, $I_3 = \frac{1}{2}\int_{\mathbb{R}^3} x \times \omega\, dx$, is conserved for all time. For axisymmetric flows without a swirl, this yields

$$\int_{\mathbb{R}^3} |x'|\omega^\theta(x,t)dx = \int_{\mathbb{R}^3} |x'|\omega_0^\theta(x)dx, \tag{4.98}$$

where $x' = {}^t(x_1, x_2)$.

It is still not known whether solutions to 3D Euler can develop singularities in finite time. Constantin et al. (1985) showed that such singularities are possible for a simple 1D model.

Recent results show that Euler singularities, if they occur, must be accompanied by the generation of an arbitrarily small-scale geometric structure (work by P. Constantin, C. Fefferman, and A. Majda). There is much numerical work on this subject with conflicting results. Some references are Bell and Marcus (1992), Kerr (1993), and Pumir and Siggia (1992). An excellent review article discussing singularities in Euler is Majda (1986).

References for Chapter 4

Ball, J. M., "Remarks on blow-up and nonexistence theorems for nonlinear evolution equations," *Q. J. Math. Oxford* **28**, 473–486, 1977.

Beale, J. T., Kato, T., and Majda, A., "Remarks on the breakdown of smooth solutions for the 3-D Euler equations," *Commun. Math. Phys.* **94**, 61–66, 1984.

Bell, J. B. and Marcus, D. L., "Vorticity intensification and transition to turbulence in three dimensional Euler equations," *Commun. Math. Phys.* **147**, 371–394, 1992.

Berger, M. S., "Nonlinearity and functional analysis," *Lectures on Nonlinear Problems in Mathematical Analysis*, Academic, New York, 1977.

Bertozzi, A. L. and Constantin, P., "Global regularity for vortex patches," *Commun. Math. Phys.* **152**, 19–28, 1993.

Calderon, A. P. and Zygmund, A., "On the existence of certain singular integrals," *Acta Math.* **88**, 85–139, 1952.

Coddington, E. A. and Levinson, N., *Theory of Ordinary Differential Equations*, McGraw-Hill, New York, 1955.

Constantin, P., Lax, P. D., and Majda, A., "A simple one dimensional model for the three dimensional vorticity equation," *Commun. Pure Appl. Math.* **38**, 715–724, 1985.

Folland, G. B., *Introduction to Partial Differential Equations*, Princeton Univ. Press, Princeton, NJ, 1995.

Hartman, P., *Ordinary Differential Equations*, Birkhäuser, Boston, 1982.

John, F. and Nirenberg, L., "On functions of bounded mean oscillation," *Commun. Pure Appl. Math.* **14**, 415–426, 1961.

Kerr, R. M., "Evidence for a singularity of the three-dimensional incompressible Euler equations," *Phys. Fluids A* **5**, 1725–1746, 1993.

Ladas, G. E. and Lakshmikamtham, V., *Differential Equations in Abstract Spaces*, Academic, New York, 1972.

Majda, A., "Vorticity and the mathematical theory of incompressible fluid flow," *Commun. Pure Appl. Math.* **39**, 5187–5220, 1986.

Mikhlin, S. G. and Prössdorf, S., *Singular Integral Operators.* Springer-Verlag, Berlin, 1986. Extended English version of *Singuläre Integraloperatoren.*

Pumir, A. and Siggia, E. D., "Development of singular solutions to the axisymmetric Euler equations," *Phys. Fluids A* **4**, 1472–1491, 1992.

Stein, E. M., "Singular integrals, harmonic functions, and differentiability properties of functions of several variables," in *Proceedings of the Symposium on Pure Mathematics*, Vol. 10 of the AMS Proceedings Series, American Mathematical Society, Providence, RI, 1967, pp. 316–355.

Stein, E. M., *Singular Integrals and Differentiability Properties of Functions*, Princeton Univ. Press, Princeton, NJ, 1970.

Torchinsky, A., *Real Variable Methods in Harmonic Analysis*, Academic, New York, 1986.

5
The Search for Singular Solutions to the 3-D Euler Equations

An important unsolved research problem for incompressible flow and the current attempts to make progress on this problem are the main focus of this chapter. These attempts involve the interaction of ideas from mathematical analysis, large-scale computation, and qualitative modeling, and we demonstrate this multifaceted approach here. Next we precisely state this outstanding problem.

Consider smooth solutions of the 3D Euler equations with finite energy,

$$\frac{Dv}{Dt} = -\nabla p,$$

$$\operatorname{div} v = 0,$$

$$v|_{t=0} = v_0, \tag{5.1}$$

where $(D/Dt) = [(\partial/\partial t) + v \cdot \nabla]$. If v_0 is smooth enough, i.e., v_0 belongs to the Sobolev space H^s, with $s \geq 3$ in either all of space or the periodic setting, then the mathematical theory from Chap. 3 guarantees the existence of smooth solutions on some maximal interval $[0, T_*)$ as well as several equivalent characterizations of this maximal interval of existence. The main topic here is the following *outstanding unsolved problem:*

> Are there smooth solutions with finite energy of the 3D Euler equations that develop singularities in a finite time? More precisely, are there smooth initial data in the Sobolev space H^s, $s \geq 3$, in all of space or in the periodic setting so that the maximal interval of smooth existence, $[0, T_*)$, is finite, i.e., $T_* < \infty$? (5.2)

One of the basic results in Chap. 3 is a precise characterization of the maximal interval of existence of a smooth solution in terms of the accumulation of vorticity. We recall the following fact: $[0, T_*)$ with $T_* < \infty$ is a maximal interval of smooth existence for the 3D Euler equations for smooth initial data with finite energy if and only if the vorticity $\omega = \operatorname{curl} v$ accumulates so rapidly that

$$\int_0^T \max_{\mathbf{x}} |\omega(x, s)| ds \to \infty \tag{5.3}$$

as T approaches T_*. We also recall that the vorticity satisfies the equation

$$\frac{D\omega}{Dt} = \mathcal{D}(\omega)\omega \tag{5.4}$$

where $\mathcal{D}(\omega)$ is the symmetric, traceless, 3×3 deformation or rate-of-strain matrix given by

$$\mathcal{D} = (\mathcal{D}_{ij}) = \left[\frac{1}{2}\left(\frac{\partial v_i}{\partial x_j} + \frac{\partial v_j}{\partial x_i}\right)\right]. \tag{5.5}$$

Thus, from criterion (5.3) and Eqs. (5.4) and (5.5), we see that the search for singular solutions to the 3D Euler equations involves the quantitative understanding of the creation of extremely large values of vorticity by means of the nonlinear vortex-amplification process. As we indicate through the numerous references discussed below, the search for singular solutions for the 3D Euler equations is a much studied and important topic in the physics and engineering communities, precisely because it involves the fundamental mechanism of nonlinear intensification of vorticity as well as the generation of small scales in turbulent fluid flows at very high Reynolds numbers. On the other hand, such problems are extremely difficult because the numerous examples from Chap. 2 involving exact solutions indicate that the growth of vorticity in time is a ubiquitous phenomena in smooth nonsingular solutions of the 3D Euler equations. Therefore, in an evolving smooth solution with a sea of actively growing interacting regimes of vorticity, we must decide whether there is merely a huge growth of vorticity or actually that criterion (5.3) is satisfied and a finite-time singularity develops. This is a grand challenge for both numerical methods and mathematical theory that remains unsolved at the present time. Nevertheless, the alliance of mathematical theory and careful numerical computations in the attempt to solve this outstanding problem has substantially increased the current knowledge regarding nonlinear accumulation of vorticity and generation of small scales. Next we outline the remaining contents of this chapter.

In Section 5.1, we discuss both various intuitive mathematical criteria that are equivalent to criterion (5.3) and the use of these criteria in recent numerical simulations of the Euler equations, regarding problem (5.2) as an important diagnostic in monitoring the numerical simulations in a self-consistent fashion. With the subtle and novel issues regarding potential singular solutions for the 3D Euler equations, it is very natural to develop simpler analog models that retain some of the features of the 3D Euler equations in a simpler and more transparent context.

In Section 5.2 we introduce and study such a model in a single-space dimension, and we discuss a simpler analog model in two space dimensions in Section 5.3 involving a surface quasi-geostrophic flow. In this context, there is an important recent connection between specific geometric configurations and nonlinear intensification that we develop briefly in the analog problem; we also indicate the current generalizations for the 3D Euler equations. We believe that this interplay between geometric configurations and nonlinearity is an important direction for further research.

In Section 5.4, we discuss the search for singular solutions within the special context of 3D axisymmetric flows with the swirl described earlier in Chap. 2 as well as another related analog problem involving the 2D Boussinesq equations for a variable-density flow. Finally, in Section 5.5 we briefly discuss further possible research directions.

5.1. The Interplay between Mathematical Theory and Numerical Computations in the Search for Singular Solutions

It is obviously interesting to provide alternative equivalent conditions to the accumulation of vorticity in criterion (5.3) through other physical quantities as all of these conditions provide self-consistent checks for any numerical computation that is searching for potentially singular solutions. We discuss these conditions next.

Besides the vorticity ω, we might naturally conjecture that the symmetric part of the deformation matrix $\mathcal{D}$ from Eq. (5.5) also controls the potential formation of singular solutions. In fact, we will see below that a more subtle and natural physical quantity related to the deformation matrix controls potential singularity formation.

To introduce this quantity, we calculate how the length of a vortex line changes in time. Recall from Section 1.6 that, by definition, the vorticity is tangent to a vortex line and vortex lines move with the fluid for inviscid flow. Thus the evolving length of a piece of a vortex line is given by

$$\mathcal{L}(t) = \int_{s_0}^{s_1} |\omega|[\mathbf{X}(\alpha(s), t), t]ds, \tag{5.6}$$

where $\alpha(s)$, $s_0 \leq s \leq s_1$, parameterizes the vortex line at $t = 0$. We have

$$\frac{d\mathcal{L}(t)}{dt} = \int_{s_0}^{s_1} \frac{d}{dt}|\omega|[\mathbf{X}(\alpha(s), t), t]ds. \tag{5.7}$$

Using Eq. (5.4), we easily calculate the identity

$$\frac{d}{dt}|\omega|(\mathbf{X}(\alpha, t), t) = \mathcal{S}(\mathbf{X}(\alpha, t), t)|\omega|(\mathbf{X}(\alpha, t), t), \tag{5.8}$$

where the vortex-line-stretching factor $\mathcal{S}(\mathbf{x}, t)$ is defined through the deformation matrix $\mathcal{D}$ and the direction of vorticity, $\xi = (\omega/|\omega|)$, by

$$\mathcal{S}(\mathbf{x}, t) = \begin{cases} \mathcal{D}\xi \cdot \xi, & \xi \neq 0 \\ 0, & \xi = 0 \end{cases}. \tag{5.9}$$

From Eqs. (5.6)–(5.8), we see that the infinitesimal length of all vortex lines will remain bounded as $t \nearrow T$ with $T < \infty$, provided that

$$\int_0^T \max_{\mathbf{x} \in R^3} \mathcal{S}(\mathbf{x}, t)dt < \infty \tag{5.10}$$

and furthermore, from identity (5.8),

$$|\omega(\mathbf{X}(\alpha, t), t)| \leq \exp\left[\int_0^T \max_{\mathbf{x} \in R^3} \mathcal{S}(\mathbf{x}, t)dt\right] |\omega_0(\alpha)| \tag{5.11}$$

for any $\alpha \in R^3$.

Thus we anticipate that the infinitesimal vortex-line-stretching factor $\mathcal{S}(\mathbf{x}, t)$ from Eq. (5.9) controls potential singularity formation. In fact, we have the following theorem.

Theorem 5.1. *Equivalent Physical Conditions for Potential Singular Solutions. The following conditions are equivalent:*

(1) The time interval, $[0, T_*)$ with $T_* < \infty$ is a maximal interval of smooth H^s existence for the 3D Euler equations.

(2) The vorticity ω accumulates so rapidly in time that

$$\int_0^t |\omega|_{L^\infty}(s)ds \to \infty \quad as \quad t \nearrow T_*.$$

(3) The deformation matrix $\mathcal{D}$ accumulates so rapidly in time that

$$\int_0^t |\mathcal{D}|_{L^\infty}(s)ds \to \infty \quad as \quad t \nearrow T_*.$$

(4) The stretching factor $\mathcal{S}(\mathbf{x}, t)$ defined in Eq. (5.9) accumulates so rapidly in time that

$$\int_0^t \left[\max_{\mathbf{x}\in R^3} \mathcal{S}(\mathbf{x}, s)\right] ds \nearrow \infty \quad as \quad t \nearrow T_*.$$

Proof of Theorem 5.1. Recall from Chap. 3 that $[0, T_*)$ with $T_* < \infty$ is a maximal interval of H^s existence for $s > (N/2) + 1$ if and only if $\|v(t)\|_s \to \infty$ as $t \nearrow T_*$. From Chap. 3 we already know that condition (1) is equivalent to condition (2). The logic of our proof is as follows: First, we show that the negation of condition (1) implies that negation of condition (3) that implies the negation of condition (4). This step is extremely easy because

$$\begin{aligned} \max_{\mathbf{x}\in R^3} \mathcal{S}(\mathbf{x}, t) &\le |\mathcal{D}(t)|_{L^\infty} \le C|\nabla v(t)|_{L^\infty} \\ &\le C_3\|\nabla v\|_{s-1}(t) \le C_s\|v\|_s(t). \end{aligned} \tag{5.12}$$

We conclude the proof by establishing that the negation of condition (4) guarantees the negation of condition (2), which is equivalent to the negation of condition (1). This is an immediate consequence of relation (5.11), which yields

$$\max_{0\le t\le T} |\omega(t)|_{L^\infty} \le \exp\left[\int_0^T \max_{\mathbf{x}\in R^3} \mathcal{S}(\mathbf{x}, t)dt\right] |\omega_0|_{L^\infty}, \tag{5.13}$$

and this concludes the proof. □

Direct numerical simulation is an obvious strategy to gain insight into the issues of potential singularity formation for the 3D Euler equations. However, this is a very subtle undertaking because an artificial blowup might occur as a purely numerical artifact or, conversely, numerical viscosity or smoothing might prevent the detection of a physical singularity. These are especially subtle issues for the 3D Euler equations because typical solutions always have amplifying vorticity, and sufficiently rapid accumulation of vorticity controls the potential singularity formation. Theorem 5.1 provides several physically interesting and easily implemented diagnostic quantities to monitor in direct numerical simulations. In this fashion, mathematical theory provides several self-consistent checks on potential candidate singular solutions from numerical simulation. The first author (Majda, 1991) has advocated such an approach for many years since the first results of Beale et al. (1984).

5.1.1. The Search for Potential Singularities Through Direct Numerical Simulation

Here we briefly mention some of the numerical computations for the full 3D Euler equations without special symmetries that yield insight into the nonlinear amplification of vorticity and potential singular solutions.

Historically, the first candidate proposed for a singular solution for the 3D Euler equations was the Taylor–Green vortex, which arises in periodic geometry from the single-mode velocity field, $v = (v_1, v_2, v_3)$, with

$$\begin{aligned} v_1 &= \sin x \cos y \cos z, \\ v_2 &= -\cos x \sin y \cos z, \\ v_3 &= 0. \end{aligned} \tag{5.14}$$

Green and Taylor (1937) proposed a finite-time singularity for this initial data on the basis of calculating several successive terms of the Taylor series in time of the solution that seemed to diverge. Brachet et al. (1983) utilized pseudospectral methods in a direct numerical solution of 3D Euler equations with the Taylor–Green initial data and compared their results with a Padé approximation procedure borrowed from condensed-matter physics and predicted a singularity at early times. Brachet (1991) subsequently studied the Taylor–Green vortex for much longer times and systematically monitored the accumulation of vorticity condition (2) from Theorem 5.1. In fact, he found no evidence for singularities at the early times claimed by Brachet et al. (1983), and in fact the maximum vorticity amplified by only a factor of 4 compared with the initial vorticity at the previously conjectured singularity time (Brachet, private communication)!! This weak amplification of vorticity occurs despite the fact that there is a complex structure developing in the evolving solution. The above discussion provides a simple way in which mathematical criteria such as those in Theorem 5.1 can constrain the interpretation of potential singular solutions through numerical procedures. Chorin (1982) developed another interesting early study of the amplification of vorticity by means of numerical vortex methods (see Chap. 6).

In the authors' opinion, the most significant direct simulation with insight into potential singularities for 3D Euler equations is due to Kerr (1993). There have been many other attempts at direct numerical simulation of singularities for 3D Euler equations over the past many years that we do not list here [the interested reader can consult the bibliographies of Majda (1991) and Kerr (1993), as well as more recent issues of *Physics of Fluids* and *Journal of Fluid Mechanics*]. Kerr considers suitable perturbations of the antiparallel vortex pair in periodic geometry as the initial data. Kerr systematically monitors all the mathematical conditions in Theorem 5.1 for a potential singularity and uses pseudospectral methods at a resolution that utilizes the full capacity of the largest available supercomputers. Numerical evolution of the mathematical diagnostics leads to a potential numerical singularity in the 3D Euler equations with a complex geometric structure involving highly anisotropic layers of intense vorticity peaking in a symmetric conical fashion (see Kerr, 1993). Nevertheless, numerical resolution necessarily limits the amplification of vorticity to a factor of only 20 times the initial data and the potential singularity is conjectured by extrapolating both the physical structure and the trends in the solution utilizing the mathematical diagnostics. Deciding whether the interaction of this highly anisotropic geometric structure of vorticity eventually leads to the depletion of nonlinearity and prevents singularity formation, as we discuss later in Sections 5.3 and 5.5 in simpler analytic-geometric contexts, is a fundamental open problem.

5.2. A Simple 1D Model for the 3D Vorticity Equation

In Section 2.3 of Chap. 2, we have seen that 3D Euler equations (5.1) can be reformulated in terms of the quadratic nonlinear equation for vorticity in Eq. (5.4) with nontrivial vortex stretching that involves the product of the deformation matrix and the vorticity. This interaction is nonlocal and quadratically nonlinear in the vorticity because the deformation matrix is determined from the vorticity by convolution through a matrix-valued strongly singular integral. This formulation suggests an instructive qualitative 1D model (Constantin et al., 1985), which we present and analyze here.

The operator relating ω to $\mathcal{D}\omega$ is a linear SIO that commutes with translation, i.e., it is given by the convolution of ω with a kernel homogeneous of degree -3 and with a mean value on the unit sphere equal to zero. In the one-space dimension, there is only one such operator, the Hilbert transform,

$$H(\omega) = \frac{1}{\pi}\,\mathrm{PV}\int \frac{\omega(y)}{(x-y)}dy. \tag{5.15}$$

The quadratic term $H(\omega)\omega$ is a scalar 1D analog of the vortex-stretching term $\mathcal{D}(\omega)\omega$. We replace the convective derivative D/Dt with $\partial/\partial t$ in order to have a 1D incompressible flow and arrive at the *model vorticity equation*:

$$\frac{\partial\omega}{\partial t} = H(\omega)\omega,$$
$$\omega(x,0) = \omega_0(x). \tag{5.16}$$

For the Euler equations, the velocity is determined from the vorticity by convolution with a mildly singular kernel, homogeneous of degree $1-N$, and the analog of the velocity for the model is defined within a constant by such a convolution, i.e.,

$$v = \int_{-\infty}^{x} \omega(y, t) dy. \tag{5.17}$$

Because the Hilbert transform is a skew-symmetric operator,

$$\int_{-\infty}^{\infty} H(\omega)\omega \, dy = (H\omega, \omega) = 0.$$

Integrating Eqs. (5.16) with respect to y on $\mathbb{R}$ shows then that all smooth solutions of Eqs. (5.16) that decay sufficiently rapidly as $|y| \to \infty$ satisfy for all t

$$\frac{d}{dt} \int_{-\infty}^{\infty} \omega(y, t) dy = 0. \tag{5.18}$$

Thus, if $\omega_0(x)$ is the derivative of a function vanishing for $|x| \to \infty$, smooth solutions of Eqs. (5.16) also retain this property for $t > 0$. Many studies for the Euler equations concentrate on periodic fluid flow; there is an obvious analog of the Hilbert transform in Eq. (5.15) on the circle and the periodic model vorticity equation can be defined as in Eqs. (5.16). In this case, the mean of ω per period is conserved and v is defined unambiguously by

$$v = \int_{x_0}^{x} \omega(y, t) dy$$

provided that the initial data $\omega_0(x)$ satisfies

$$\int_{x_0}^{x_0+p} \omega_0(y) dy = 0$$

where p is the period.

5.2.1. Integration of the Model Vorticity Equation and Explicit Breakdown of Solutions

The nonlinear equation in Eqs. (5.16) is well posed in many standard function spaces, for example, $H^1(\mathbb{R})$, the Sobolev space of functions that are square integrable with a square-integrable first derivative. The local existence and uniqueness follow from the fact that $H^1(\mathbb{R})$ is a Banach algebra of continuous functions and the Hilbert transform maps $H^1(\mathbb{R})$ continuously onto itself so that, as in Chap. 4, standard existence and uniqueness results for Lipschitz nonlinear ODEs in Banach space apply. We have the following explicit solution formula for the model equation in Eqs. (5.16).

Theorem 5.2. *Suppose that $\omega_0(x)$ is a smooth function decaying sufficiently rapidly as $|x| \to \infty$ [$\omega_0 \in H^1(\mathbb{R})$ suffices]. Then the solution to the model vorticity equation in Eqs. (5.16) is given explicitly by*

$$\omega(x, t) = \frac{4\omega_0(x)}{(2 - tH\omega_0(x))^2 + t^2\omega_0^2(x)}. \tag{5.19}$$

Remark: The proof also provides an explicit expression for $H(\omega)$:

$$(H\omega)(x,t) = \frac{2H\omega_0(x)(2 - tH\omega_0(x)) - 2t\omega_0^2(x)}{(2 - tH\omega_0(x))^2 + t^2\omega_0^2(x)}. \tag{5.20}$$

Formula (5.19) immediately yields the following corollary.

Corollary 5.1. *Breakdown of Smooth Solutions for the Model Vorticity Equation. The smooth solution to the differential equation in Eqs. (5.16) blows up in finite time if and only if the set Z, defined by*

$$Z = \{x|\omega_0(x) = 0 \text{ and } H\omega_0(x) > 0\}, \tag{5.21}$$

is not empty. In this case, $\omega(x,t)$ becomes infinite as $t \uparrow T$, where the blowup time is given explicitly by $T = 2/M$ with $M = \sup\{H\omega_0)_+(x)|\omega_0(x) = 0\}$.

Proof of Theorem 5.2. To display the generality of the proof, we use the following identities for the Hilbert transform on the line and discuss their proof later:

$$H(Hf) = f, \tag{5.22a}$$

$$H(fg) = fHg + gHf + H(Hf \cdot Hg). \tag{5.22b}$$

From these identities it follows that

$$H(fHf) = \frac{1}{2}\left[(Hf)^2 - f^2\right]. \tag{5.23}$$

By applying H to the model vorticity equation and using Eq. (5.23), we obtain an equation satisfied by $(H\omega)(x,t)$:

$$\frac{\partial}{\partial t}H\omega = \frac{1}{2}\left[(H\omega)^2 - \omega^2\right]. \tag{5.24}$$

We introduce the quantity

$$z(x,t) = H\omega(x,t) + i\omega(x,t), \tag{5.25}$$

and by combining Eqs. (5.16) and (5.24), we see that $z(x,t)$ satisfies the local equation

$$\frac{\partial z}{\partial t}(x,t) = \frac{1}{2}z^2(x,t) \tag{5.26}$$

with the explicit solution

$$z(x,t) = \frac{z_0(x)}{1 - \frac{1}{2}tz_0(x)}. \tag{5.27}$$

Formulas (5.19) and (5.20) are the real and the imaginary parts of Eqs. (5.27), respectively.

Formula (5.27) defines $z(x,t)$ as an analytic function in Im $x < 0$. It is well known that the Hilbert transform on the line can be interpreted in terms of complex-valued functions z on the real axis, which are boundary values of analytic functions in the lower half-plane that tend to zero sufficiently fast at infinity. The Hilbert transform relates the imaginary part of such a function to its real part on the real axis. Thus a function of the form

$$z = H\omega + i\omega$$

is always the boundary value of a function analytic in the lower half-plane. Identity (5.22a) is then merely the observation that if z is analytic in the lower half-plane, so is iz. Identity (5.22b) is the observation that if z and w are analytic in the lower half-plane, so is their product, $z \cdot w$. Equation (5.16) is then the imaginary part of Eq. (5.26); by analyticity Eq. (5.26) holds. □

Next, we give an instructive explicit example in the 2π periodic case.

Example 5.1. We choose $\omega_0(x) = \cos(x)$ so that $H[\omega_0(x)] = \sin(x)$ and compute that

$$\begin{aligned}\omega(x, 2t) &= \frac{\cos(x)}{1 + t^2 - 2t\sin(x)},\\ v(x, 2t) &= \int_0^x \omega(x', 2t)dx' = (2t)^{-1}\log[1 + t^2 - 2t\sin(x)].\end{aligned} \tag{5.28}$$

In this specific example, the breakdown time is $T = 2$ and, as $t \nearrow T$, $\omega(x, t)$ develops a nonintegrable local singularity like $1/x$ near $x = 0$. There are two interesting facets to this breakdown process. First,

$$\int_{-\pi}^{\pi} |\omega(x, t)|^p dx \nearrow \infty \qquad \text{as} \quad t \nearrow T \tag{5.29}$$

for any fixed p with $1 \le p < +\infty$. Also, there are finite constants M_p such that

$$\int_{-\pi}^{\pi} |v(x, t)|^p dx \le M_p \qquad \text{as} \quad t \nearrow T \tag{5.30}$$

for any p with $1 \le p < \infty$. In particular, the kinetic energy of v remains uniformly bounded as t approaches the breakdown time T. The behavior in the above example is typical for solutions of the model vorticity equation as the following corollary of Theorem 5.2 indicates.

Corollary 5.2. *Given the initial data $\omega_0(x)$ for the model vorticity equation, assume that the points x_0 with $\omega_0(x_0) = 0$ and defining the breakdown time T are simple zeros of $\omega_0(x)$. Then as $t \nearrow T$, where T is the breakdown time, $\omega(x, t)$ and $v(x, t)$ have the same properties as given in expressions (5.29) and (5.30)*

We omit the proof because it is a straightforward but tedious calculation that uses the explicit solution formulas. It is easy to show that, if the initial data ω_0 satisfy the assumption of Corollary 5.2, then, for $T < t < T + \tau$, where T is the breakdown time and τ is small enough, the analytic function

$$1 - \frac{1}{2} t z_0(x)$$

has a zero in the lower half-plane $\mathrm{Im}\, x < 0$. Thus for such t the function $z(x, t)$ defined by Eq. (5.27) has a pole in the lower half-plane, and therefore its imaginary part, given by formula (5.19), is *not* related to its real part by the Hilbert transform. In particular, for such t, formula (5.20) does *not* hold and therefore $\omega(x, t)$ as given in formula (5.19) does *not* continue the solution of (5.16) for $T < t < T + \tau$.

5.3. A 2D Model for Potential Singularity Formation in 3D Euler Equations

There is a set of nonlocal quadratically nonlinear equations in two-space dimensions, the 2D quasi-geostrophic (QG) active scalar, with a strikingly powerful analogy to 3D Euler equations (5.1) (Constantin et al., 1994). These equations are given by

$$\frac{D\theta}{Dt} = \frac{\partial \theta}{\partial t} + v \cdot \nabla \theta = 0 \tag{5.31}$$

where the 2D velocity, $v = (v_1, v_2)$, is determined from θ by a stream function

$$(v_1, v_2) = \left(-\frac{\partial \psi}{\partial x_2}, \frac{\partial \psi}{\partial x_1} \right) = \nabla^{\perp} \psi, \tag{5.32}$$

and the stream function ψ satisfies

$$(-\Delta)^{\frac{1}{2}} \psi = -\theta. \tag{5.33}$$

Thus the stream function ψ is recovered from the scalar θ by the formula

$$\psi = -\int_{R^2} \frac{1}{|y|} \theta(x + y, t) dy. \tag{5.34}$$

5.3.1. Analogies between the 2D QG Active Scalar and the 3D Euler Equations

We begin our list of physical, geometric, and analytic analogies between the 2D QG active scalars in Eqs. (5.31)–(5.33) and the 3D Euler equations in vorticity-stream form from Eq. (5.4) by introducing

$$\nabla^{\perp} \theta = {}^{t}(-\theta_{x_2}, \theta_{x_1}). \tag{5.35}$$

We claim that the vector field $\nabla^{\perp}\theta$ has a role for the 2D QG active scalar equations that is completely analogous to the vorticity in 3D incompressible fluid flow, i.e.,

$$\nabla^{\perp} \theta \longleftrightarrow \omega. \tag{5.36}$$

By differentiating Eq. (5.31), we obtain the evolution equation for $\nabla^{\perp}\theta$ given by

$$\frac{D \nabla^{\perp} \theta}{Dt} = (\nabla v) \nabla^{\perp} \theta \tag{5.37}$$

with $(D/Dt) = [(\partial/\partial t) + v \cdot \nabla]$ and $v = \nabla^{\perp}\psi$ so that div $v = 0$. With identification (5.36), evolution equation (5.37) for $\nabla^{\perp}\theta$ clearly has a similar superficial structure resembling the equation for vorticity in the 3D Euler equation from Eq. (5.4). Next we show that this analogy extends considerably beyond this superficial level to detailed geometric and analytic properties of solutions.

The Geometric Analogy of Level Sets with Vortex Lines

From Eq. (5.31) it follows that the level sets, $\theta =$ constant, move with the fluid flow and that $\nabla^{\perp}\theta$ is tangent to these level sets; these facts are analogs for the 2D QG active scalar of the well-known facts for 3D incompressible fluid flow established in Chap. 1 that the vorticity, by definition, is tangent to vortex lines and vortex lines move with the fluid. Thus

for the 2D QG active scalar, the level sets of θ are analogous to vortex lines for 3D Euler equations. (5.38)

Recall from Eqs. (5.7)–(5.9) above that the infinitesimal length of a vortex line is given by the magnitude of ω, $|\omega|$ and that the evolution of this infinitesimal length is given by

$$\frac{D|\omega|}{Dt} = \mathcal{S}|\omega| \tag{5.39}$$

where $\mathcal{S}(\mathbf{x}, t)$ is defined in Eq. (5.9) by means of the deformation matrix. Similarly, for the 2D QG active scalar the infinitesimal length of a level set for θ is given by $|\nabla^{\perp}\theta|$ and, from Eq. (5.37), the evolution equation for the infinitesimal arc length of a level set is given by

$$\frac{D|\nabla^{\perp}\theta|}{Dt} = \mathcal{S}|\nabla^{\perp}\theta|, \tag{5.40}$$

where

$$\mathcal{S}(x, t) = \mathcal{D}\xi \cdot \xi \tag{5.41}$$

where $\xi(x, t)$ is the unit direction of $\nabla^{\perp}\theta$, i.e., $\xi = [(\nabla^{\perp}\theta)/(|\nabla^{\perp}\theta|)]$, and

$$\mathcal{D} = \frac{1}{2}(\nabla v + {}^{T}\nabla v) \tag{5.42}$$

is the symmetric part of the deformation matrix. With the analogy between level sets for the 2D QG active scalar and vortex lines and similar equations (5.39) and (5.41), it should be evident to the reader that there is a powerful geometric analogy between these two problems in two and three dimensions, respectively.

The Analytic Analogy with Vortex Stretching

We demonstrate that Eqs. (5.4) and (5.37) are remarkably similar in their analytic structure. Recall from Section 2.3 in Chap. 2 that the velocity v in Eq. (5.1) is determined from the vorticity ω by the familiar Biot–Savart law,

$$v(x) = -\frac{1}{4\pi}\int_{R^3}\left(\nabla^{\perp}\frac{1}{|y|}\right)\times\omega(x+y)dy, \tag{5.43}$$

and the strain matrix $\mathcal{D}$, which is the symmetric part of the velocity gradient,

$$\mathcal{D} = \frac{1}{2}\left[(\nabla v) + {}^{T}(\nabla v)\right]$$

is given in terms of the vorticity by the strongly singular integral,

$$\mathcal{D}(x) = \frac{3}{4\pi}\,\mathrm{PV}\int_{R^3} M[\hat{y},\omega(x+y)]\frac{dy}{|y|^3}. \tag{5.44}$$

In Eq. (5.44) the matrix M is a function of two variables, the first a unit vector, the second a vector, and is given by the formula

$$M(\hat{y},\omega) = \frac{1}{2}[\hat{y}\otimes(\hat{y}\times\omega) + (\hat{y}\times\omega)\otimes\hat{y}] \tag{5.45}$$

where $a\otimes b = (a_i b_i)$ is the matrix formed by the tensor product of two vectors. For the 2D QG active scalar, by differentiating Eq. (5.34) with $v = \nabla^{\perp}\psi$, we obtain

$$v = -\int_{R^2}\frac{1}{|y|}\nabla^{\perp}\theta(x+y)dy. \tag{5.46}$$

Next, we compute the symmetric part of the matrix, ∇v, from Eq. (5.42). With Eq. (5.46) we calculate that the matrix $\mathcal{D}(x)$ has the singular integral representation

$$\mathcal{D}(x) = \mathrm{PV}\int_{R^2} N[\hat{y},(\nabla^{\perp}\theta)(x+y)]\frac{dy}{|y|^2} \tag{5.47}$$

where $\hat{y} = (y/|y|)$ and N is a function of two variables, the first a unit vector in R^2 and the second a vector with

$$N(\hat{y},\omega) = \frac{1}{2}(\hat{y}^{\perp}\otimes\omega^{\perp} + \omega^{\perp}\otimes\hat{y}^{\perp}). \tag{5.48}$$

The function N has mean zero on the unit circle for fixed w and thus, as discussed in Chap. 2, the operator in Eq. (5.47) is a legitimate strongly singular integral. As in the situation for 3D fluid flow, we also see from Eqs. (5.40)–(5.42) above that only the symmetric part $\mathcal{D}(x)$ contributes to strong and potentially singular front formation. We have displayed only the formula for the symmetric part of ∇v for simplicity in the exposition.

With formulas (5.43)–(5.48), we develop the analytic analogy between Eqs. (5.4) and (5.37). From Eqs. (5.43) and (5.46), the velocity is given in terms of w, either ω or $\nabla^{\perp}\theta$ in three or two dimensions, respectively, by

$$v = \int_{R^d} K_d(y)w(x+y)dy$$

where $K_d(y)$ is homogeneous of degree $1-d$ in R^d for $d=2,3$, i.e., $K_d(\lambda y)=\lambda^{1-d}K_d(y)$ for $\lambda>0$. Furthermore, from Eqs. (5.44) and (5.47),

$$\frac{\nabla v+{}^T(\nabla v)}{2}=\mathcal{D}$$

has a representation formula in terms of $w(x+y)$ by means of a strongly SIO defined through a kernel homogeneous of degree $-d$, in R^d, for $d=2,3$ with specific cancellation properties; also the geometric formulas for these strongly singular operators given in Eqs. (5.45) and (5.48) are very similar in structure. Thus, with the identification of $\nabla^\perp\theta$ and vorticity, the evolution equation for $\nabla^\perp\theta$ from Eq. (5.37) has a completely parallel analytic structure in two dimensions as the equation for the evolution of vorticity, ω, in Eq. (5.4) for 3D incompressible flow.

The Search for Potential Singularities through Direct Numerical Simulation

With all of the analogous geometric/analytic structures for the surface QG active scalar in two dimensions established in this subsection, it is evident that there is a version of Theorem 5.1 for this equation that yields equivalent physical conditions for potential singular solutions. We need only to replace ω in Theorem 5.1 with $\nabla^\perp\theta$. The interested reader can check this as an exercise (see Constantin et al., 1994).

Constantin et al. introduced the analog of Taylor–Green initial data in Eq. (5.14) for the 2D equations (5.31)–(5.33) given by

$$\theta(x,0)=\sin x_1\sin x_2+\cos(x_2). \tag{5.49}$$

All of the diagnostics from Theorem 5.1 were utilized to monitor the numerical solution. A strongly nonlinear intensification of $|\nabla^\perp\theta|$ was observed throughout the history of the numerical simulation, and if this behavior was extrapolated to somewhat longer times in a self-consistent fashion, there was the possibility of a finite-time singularity. Ohkitani and Yamada (1997) utilized higher resolution with a longer time integration as well as the same self-consistent diagnostics and showed that the growth of $|\nabla^\perp\theta|$ in the numerical simulation had a better fit with a double exponential rather than an algebraic singularity; thus, the numerical evidence is that there is tremendous nonlinear amplification of $|\nabla^\perp\theta|$, the analog of vorticity, in the 2D analog problem but no singularity at finite time, according to recent computations.

The simulations of Constantin et al. (1994) also emphasized the role of geometry of the level sets of θ, the analog of vortex lines, in producing depletion of nonlinearity. Next we describe the simplest analytic/geometric criteria for the depletion of nonlinearity (Constantin et al., 1994) motivated by these numerical simulations.

5.3.2. Rigorous Analytic and Geometric Constraints on Singular Solutions

The structure of Eqs. (5.31)–(5.33) automatically guarantees that the geometric structure of the level sets of the initial data is preserved by the solution at later times. Here we illustrate how the simplest geometric structure of level sets can be combined with the specific nonlinear structure from Eqs. (5.40)–(5.42) to establish, under mild a priori hypotheses, that no singularity is possible. More refined results of this type are developed elsewhere (Constantin et al., 1994; Cordoba, 1998). In particular, in an important paper, Cordoba (1998) has ruled out the geometric structures of a hyperbolic saddle as a potential singular structure in finite time under some mild a priori hypotheses similar to those given below; this result is significant because the refined diagnostic analysis of numerical simulations (Constantin et al., 1994; Ohkitani and Yamada, 1997) gave the vicinity of hyperbolic saddles as the most nonlinear and potentially singular event. All of the other elementary geometric structures for level sets, for example, for the initial data in Eq. (5.49), are ruled out by earlier work (Constantin et al., 1994). The theorem that we state below has analog for the 3D Euler equations (Constantin et al., 1996) with a more sophisticated proof following the same strategy.

The idea of the theorem that we state below is that if the direction field, $\xi = [(\nabla^{\perp}\theta)/(|\nabla^{\perp}\theta|)]$, remains smooth in a suitably weak sense on an open set moving with the 2D QG thermal fluid flow, then, regardless of the amplitude of $|\nabla^{\perp}\theta|$ on this evolving set, no singularity is possible. Next, we state this result precisely. Recall that $\xi = [(\nabla^{\perp}\theta)/(|\nabla^{\perp}\theta|)]$ is the direction field tangent to the level sets. We say that a set Ω_0 is *smoothly directed* if there exists $\rho > 0$ such that

$$\sup_{q\in\Omega_0} \int_0^T |v(X(q,t),t)|^2\, dt < \infty, \tag{5.50a}$$

$$\sup_{q\in\Omega_0^*} \int_0^T \|\nabla\xi(\cdot,t)\|^2_{L^\infty\{B_\rho(X(q,t))\}} dt < \infty, \tag{5.50b}$$

where $B_\rho(X)$ is the ball of radius ρ centered at X and

$$\Omega_0^* = \{q \in \Omega_0;\ |\nabla\theta_0(q)| \neq 0\}.$$

If Ω_0 is a set, we denote by Ω_t its image at time t under the particle-trajectory map

$$\Omega_t = X(\Omega_0, t)$$

and by $\mathcal{O}_T(\Omega_0)$ the semiorbit,

$$\mathcal{O}_T(\Omega_0) = \{(x,t) | x \in \Omega_t, 0 \le t \le T\}.$$

Theorem 5.3. *Assume that Ω_0 is smoothly directed. Then*

$$\sup_{\mathcal{O}_T(\Omega_0)} |\nabla\theta(x,t)| < \infty,$$

i.e., if the direction field is smooth locally on a set moving with the fluid in the precise sense of inequality (5.50b), then no singularity is possible in that set.

We consider a QG active scalar $\theta(x,t)$ with smooth initial data and suppose the scalar is defined and smooth for $(x,t) \in \mathbb{R}^2 \times [0,T)$. Recall the particle trajectories $X(q,t)$ for the 2D QG thermal fluid flow, which are solutions of

$$\frac{dX}{dt} = v(X,t).$$

Proof of Theorem 5.3. This proof relies on a special formula for the level-set stretching factor $\mathcal{S}$ from Eqs. (5.40) and (5.41) together with a local version of the proof that we utilized in Theorem 5.1 to demonstrate that condition (4) implies condition (2) with the identification $\nabla^{\perp}\theta \sim \omega$. We start by computing the full gradient of the velocity field v from formula (5.46):

$$(\nabla v)(x) = -\nabla_x \int \frac{1}{|y|}(\nabla^{\perp}\theta)(x+y)dy.$$

Differentiating under the integral sign, we get

$$(\nabla v)(x) = -\int \frac{1}{|y|}\left(\nabla_y \nabla_y^{\perp}\theta\right)(x+y)dy.$$

We write the integral as a limit as $\epsilon \to 0$ of integrals on $|y| > \epsilon$. Because the two gradients applied to θ commute, we can choose any one of them and integrate by parts. The limit of the contributions from $|y| = \epsilon$ vanishes. In this fashion, we obtain the formula

$$(\nabla v)(x) = -\mathrm{PV}\int \{\hat{y}^{\perp} \otimes [\nabla\theta(x+y)]\}\frac{dy}{|y|^2}. \tag{5.51}$$

Writing

$$\nabla^{\perp}\theta = A\xi, \qquad A = |\nabla^{\perp}\theta|,$$

and using the definition from Eqs. (5.37) and (5.40) that

$$\mathcal{S}(x) = \{[(\nabla v)(x)]\xi(x)\} \cdot \xi(x),$$

we deduce the representation for the stretching factor $\mathcal{S}$:

$$\mathcal{S}(x) = \mathrm{PV}\int \{[\hat{y} \cdot \xi^{\perp}(x)][\xi(x+y) \cdot \xi^{\perp}(x)]\}A(x+y)\frac{dy}{|y|^2}. \tag{5.52}$$

Next we consider a number $\rho > 0$ and decompose

$$\mathcal{S}(x) = s_{\mathrm{in}}(x) + s_{\mathrm{out}}(x),$$

where

$$s_{\mathrm{in}}(x) = \mathrm{PV}\int \chi\left(\frac{|y|}{\rho}\right)\cdots,$$

$$s_{\mathrm{out}}(x) = \mathrm{PV}\int \left[1 - \chi\left(\frac{|y|}{\rho}\right)\right]\cdots,$$

where $\chi(r)$ is a smooth nonnegative function of one positive variable satisfying $\chi(r) = 1$ for $0 \le r \le \frac{1}{2}$, $\chi(r) = 0$ for $r \ge 1$.

It is easy to prove the estimate

$$|\alpha_{\text{out}}(x)| \leq C\rho^{-2}\|\theta\|_{L^2}. \tag{5.53}$$

Indeed by using the representation in Eq. (5.52) and the fact that $A\xi = \nabla^{\perp}\theta$ and by integrating by parts we obtain

$$|\alpha_{\text{out}}(x)| \leq C\rho^{-1}\int_{|y|\geq\frac{1}{2}\rho} |\theta(x+y)|\frac{dy}{|y|^2}.$$

We consider the situation, which corresponds to the hypothesis of Theorem 5.3, in which the direction field ξ is smooth in the ball of center x and radius ρ and use the representation in Eq. (5.52). We denote by G the maximum of the gradient of ξ there:

$$G = \sup_{|y|\leq\rho} |\nabla\xi(x+y)|.$$

Clearly

$$|\xi(x+y)\cdot\xi^{\perp}(x)| \leq G|y|$$

for $|y| \leq \rho$. So we deduce from Eq. (5.52) that

$$|s_{\text{in}}(x)| \leq G\int \chi\left(\frac{|y|}{\rho}\right) A(x+y)\frac{dy}{|y|}.$$

Now we use the fact that

$$A = \xi\cdot(\nabla^{\perp}\theta)$$

and integrate by parts:

$$\int \chi\left(\frac{|y|}{\rho}\right) A(x+y)\frac{dy}{|y|} = -\int \theta(x+y)\nabla_y^{\perp}\cdot\left[\xi(x+y)\chi\left(\frac{|y|}{\rho}\right)\frac{1}{|y|}\right]dy.$$

We carry out the differentiation and obtain three terms, which we denote as I, II, and III:

$$\text{I} = -\int [\nabla_y^{\perp}\cdot\xi(x+y)]\theta(x+y)\chi\left(\frac{|y|}{\rho}\right)\frac{dy}{|y|},$$

$$\text{II} = -\int \theta(x+y)\xi(x+y)\cdot\nabla_y^{\perp}\left[\chi\left(\frac{|y|}{\rho}\right)\right]\frac{dy}{|y|},$$

$$\text{III} = \text{PV}\int [\xi(x+y)\cdot\hat{y}^{\perp}]\theta(x+y)\chi\left(\frac{|y|}{\rho}\right)\frac{dy}{|y|^2}.$$

The first two can be estimated in a straightforward manner:

$$|\text{I}| \leq C\rho G\|\theta\|_{L^\infty},$$

$$|\text{II}| \leq C\|\theta\|_{L^\infty}.$$

We write in the third term $\xi(x+y) = \xi(x) + [\xi(x+y) - \xi(x)]$, and therefore

$$\mathrm{III} = \xi(x) \cdot \mathrm{PV} \int \hat{y}^{\perp} \theta(x+y) \frac{dy}{|y|^2} + \mathrm{III}',$$

with

$$|\mathrm{III}'| \le C(\rho G \|\theta\|_{L^\infty} + \rho^{-1} \|\theta\|_{L^2}).$$

We observe that

$$\mathrm{PV} \int \hat{y}^{\perp} \theta(x+y) \frac{dy}{|y|^2} = -v(x);$$

thus

$$|s_{\mathrm{in}}(x)| \le CG[|v(x)| + (\rho G + 1)\|\theta\|_{L^\infty} + \rho^{-1}\|\theta\|_{L^2}]. \tag{5.54}$$

Combining this with estimate (5.53) we proved Lemma 5.1.

Lemma 5.1. *Assume that x is such that*

$$G := \sup_{|y| \le \rho} |\nabla \xi(x+y)|$$

and $|v(x)|$ are finite. Then $|\mathcal{S}(x)|$ is bounded by

$$|\mathcal{S}(x)| \le C[G|v(x)| + (\rho G + 1)(G\|\theta\|_{L^\infty} + \rho^{-2}\|\theta\|_{L^2})].$$

Next we finish the proof by applying the kinematic estimate for the stretching factor $\mathcal{S}$ in Lemma 5.1 together with Eq. (5.40) to obtain a bound on $|\nabla\theta|$ moving with the fluid flow.

If Ω_0 is smoothly directed then we can apply Lemma 5.1 with $x = X(q,t)$, for any $q \in \Omega_0$ and any $t \in [0,T)$. Using ODE (5.40),

$$\frac{d}{dt}|\nabla^{\perp}\theta(X(q,t),t)| = \mathcal{S}(X(q,t),t)|\nabla^{\perp}\theta(X(q,t),t)|$$

and the bound in Lemma 5.1 we obtain

$$\sup_{\mathcal{O}_T(\Omega_0)} |\nabla\theta(x,t)| \le e^{Q} \sup_{\Omega_0} |\nabla\theta_0|$$

where

$$\begin{aligned} Q &= C \sup_{q\in\Omega_0} \int_0^T E(t)dt, \\ E(t) &= \{G(x)|v(x)| + [\rho G(x) + 1][G(x)\|\theta_0\|_{L^\infty} + \rho^{-2}\|\theta_0\|_{L^2}]\}, \\ x &= X(q,t), \\ G(x) &= \sup_{B_\rho(x)} |\nabla\xi|. \end{aligned}$$

□

This completes the proof of Theorem 5.3. In the above, we used the fact that the L^2 norm of θ is conserved, $\|\theta\|^2_{L^2}(t) = \|\theta_0\|^2_{L^2}$, for all times t. Also note that $\|\theta\|_{L^p}(t)$ is conserved for all $1 \leq p \leq \infty$.

5.4. Potential Singularities in 3D Axisymmetric Flows with Swirl

Besides searching for potential singular solutions with general initial data or in 1D or 2D analog problems, as we have described in Sections 5.1, 5.2, and 5.3, respectively, it is a natural mathematical question to search for potential singular solutions in 3D Euler flows with geometric symmetry. The 3D axisymmetric flows with swirl discussed earlier in Chap. 2, Subsection 2.2.2, are a natural candidate for such an investigation (Grauer and Sideris, 1991). However, the reader should be warned that nonlinear behavior in such solutions with high symmetry may not be the generic behavior for nonlinear intensification of vorticity in 3D Euler flows; such axisymmetric flows with swirl are typically highly unstable to symmetry-breaking perturbations. Of course, the difficult resolution issues for numerical computations for general 3D flows are significantly reduced for axisymmetric flows with swirl to essentially a 2D computational problem.

Recall from Eqs. (2.67) and (2.68) the basic vorticity equations for axisymmetric swirling flow:

$$\begin{aligned} \frac{\tilde{D}}{Dt}(rv^\theta) &= 0, \\ \frac{\tilde{D}}{Dt}\left(\frac{\omega^\theta}{r}\right) &= -\frac{1}{r^4}\left[(rv^\theta)^2\right]_{x_3}, \end{aligned} \tag{5.55}$$

where

$$\begin{aligned} \frac{\tilde{D}}{Dt} &= \frac{\partial}{\partial t} + v^r\frac{\partial}{\partial r} + v^3\frac{\partial}{\partial x_3}, \\ \omega^\theta &= \frac{\partial v^r}{\partial x_3} - \frac{\partial v^3}{\partial r}. \end{aligned} \tag{5.56}$$

Thus ω^θ, the axial vorticity, intensifies through gradients of the quantity $(rv^\theta)^2$, which moves with the fluid:

$$\frac{\tilde{D}}{Dt}(rv^\theta)^2 = 0. \tag{5.57}$$

As described in Subsection 2.2.2, we compute the velocity components v^r and v^3 from ω^θ by investing a variable coefficient elliptic equation for a stream function. Equations (5.55)–(5.57) represent highly simplified equations for vortex stretching for axisymmetric flows with swirl. A recent theorem (Chae and Kim, 1996) refining the Beale–Kato–Majda criterion yields the expected result that ω^θ controls the breakdown of these special solutions.

Grauer and Sideris (1991) were the first to study the potential formation of singular solutions in such special swirling flows through a standard symmetric finite-difference scheme applied to the vorticity-stream form of the equations for swirling flows described in Eqs. (5.55)–(5.57) above. These authors systematically monitored criterion (2) of Theorem 5.1 regarding accumulation of vorticity: The vorticity maximum in these flows increases from an initial size of 10^{-1} to values of approximately 400 as time evolves; there is corresponding generation of small scales in the solution far away from the axis singularity at $r = 0$. At roughly the time at which the vorticity amplification reaches 400, strong perturbations of the order of the mesh size are generated, so with this lack of numerical resolution, the "blowup" reported by Grauer and Sideris is almost certainly a numerical artifact.

The calculations of Pumir and Siggia (1992) also claim to yield a blowup in finite time, but these results are more suspect for serious numerical reasons. These authors applied a highly nonuniform adaptive stretched mesh that focused on a detailed region of potential singularity and brutally violated energy conservation in the far field; none of the self-consistent mathematical criteria from Theorem 5.1 were utilized to monitor the potential singularity formation. Nevertheless, the blowup reported by Pumir and Siggia (1992) is far from the axis of singularity at $r = 0$.

Next, we discuss the reasons that potential finite-time singularity formation for the initial data in either the study of Grauer and Sideris (1991) or Pumir and Siggia (1992) is rather unlikely by describing an important contribution of E and Shu (1994). Away from the singularity region, $r = 0$, the 3D axisymmetric flows with swirl have transparently analogous structure to the 2D Boussinesq equations for a stratified flow.

5.4.1. The 2D Boussinesq Equations as a Model for 3D Axisymmetric Swirling Flows

The 2D Boussinesq equations in vorticity-stream form are given by

$$\begin{aligned} \frac{D\rho}{Dt} &= 0, \\ \frac{D\omega}{Dt} &= -\rho_x, \\ -\Delta\psi &= \omega, \end{aligned} \tag{5.58}$$

where ρ is the density (perturbation) and $(D/Dt) = [(\partial/\partial t) + u(\partial/\partial x) + v(\partial/\partial y)]$, ω is the 2D vorticity, and ψ is the stream function:

$$\omega = \frac{\partial u}{\partial y} - \frac{\partial v}{\partial x}, \qquad u = -\psi_y, \quad v = \psi_x. \tag{5.59}$$

If the y axis is identified with the vertical axis, Eqs. (5.58) and (5.59) describe the motion of lighter or denser fluid under the influence of gravitational forces. Although these equations describe interesting physics in their own right, here we utilize them as a simpler analog problem for the axisymmetric swirling flows described earlier in Section 2.3.3. We have the following correspondence between the two sets of equations in Eqs. (5.55)–(5.57) and (5.58) and (5.59):

$$\begin{aligned} x_3 &\longleftrightarrow x, \\ r &\longleftrightarrow y, \\ \omega^\theta &\longleftrightarrow \omega, \\ (rv^\theta)^2 &\longleftrightarrow \rho, \\ v^r &\longleftrightarrow v, \\ v^3 &\longleftrightarrow u. \end{aligned} \tag{5.60}$$

With this correspondence, we see that the 2D Boussinesq equations are formally identical to the equations for 3D axisymmetric swirling flows provided that we evaluate all external variable coefficients in Eqs. (5.55)–(5.57) at $r = 1$. Thus, away from the axis of singularity $r = 0$ for swirling flows, we expect the qualitative behavior of solutions for the two systems of equations to be identical.

E and Shu (1994) exploited this analogy and studied the nonlinear development of potential singularities in the 2D Boussinesq equations with initial data completely analogous to those of Grauer and Sideris (1991) or Pumir and Siggia (1992). Recall that those studies concluded that there is potential singular behavior far away from the singularity axis $r = 0$. E and Shu monitored the mathematical diagnostics from Theorem 5.1 systematically in their numerical solutions; they found no evidence for singular solutions and found that the regions of largest nonlinear intensification according to the rigorous mathematical criteria were at completely different spatial locations in the flow compared with those reported by Pumir and Siggia (1992). This work of E and Shu provides convincing evidence that 3D swirling flows, at least for the initial data studied, do not become singular in finite time.

5.5. Do The 3D Euler Solutions Become Singular in Finite Times?

The evidence on the simpler model problems described in Sections 5.3 and 5.4 suggests the strong possibility that there might always be a combination of geometric and nonlinear structures that eventually depletes the strongly nonlinear growth and leads to the absence of singular behavior in general 3D Euler solutions. The only rigorous mathematical evidence of this type for the 3D Euler equations is the work of Constantin et al. (1996) with a very special geometric structure that we discussed briefly in Subsection 5.3.2. The numerical solutions of Kerr (1993) discussed earlier in Section 5.1 are the most attractive current candidates for potential singular behavior. Thus it is an important mathematical problem to see if it is possible to generalize the work on geometric–analytic depletion of nonlinearity (Constantin et al., 1994, 1996; Cordoba, 1998) to the highly anisotropic configurations of vorticity in Kerr's numerical solutions. Similar results for 3D axisymmetric swirling flows and the 2D Boussinesq equations are also very interesting.

Notes for Chapter 5

Theorem 5.1 subsumes, with a much simpler proof, the remarks of Ponce (1985) on characterizing potential blowup through the deformation matrix. Our proof of Theorem 5.1 is similar to the one in Constantin (1994) for the surface QG equations. The 1D model for vorticity described in Section 5.2 actually arises as an asymptotic model in a very different physical context involving the Hall effect in thin films (Chukbar, 1990). The 2D model in Section 5.3 for the 3D Euler equations is actually a set of physical equations describing the temperature on the boundary of rapidly rotating stably stratified flow (Pedlosky, 1987; Garner et al., 1995). The strong nonlinear events from the numerical simulations described in Section 5.3 occur at times far beyond the regime of validity of the equations in describing thermal fronts. Majda and Tabak realized the strong geometric and analytic analogy of these simpler 2D equations and the 3D Euler equations after conversations with Isaac Held of GFDL at Princeton; at approximately the same time, Constantin realized that these 2D equations were the one 2D active scalar with a structure resembling that of 3D Euler equations (Constantin, 1994).

References for Chapter 5

Beale, J. T., Kato, T., and Majda, A., "Remarks on the breakdown of smooth solutions of the 3-D Euler equations," *Commun. Math. Phys.* **94**, 61–66, 1984.

Brachet, M. E., "Direct simulation of three-dimensional turbulence in the Taylor–Green vortex," *Fluid Dyn. Res.* **8**, 1–8, 1991.

Brachet, M. E., Frisch, U., Meiron, D. I., Morf, R. H., Nickel, B. G., and Orszag, S. A., "Small-scale structure of the Taylor–Green vortex," *J. Fluid Mech.* **130**, 411–000, 1983.

Chae, D. and Kim, N., "On the breakdown of axisymmetric smooth solutions for the 3-D Euler equations," *Commun. Math. Phys.* **178**, 391–398, 1996.

Chorin, A., "The evolution of a turbulent vortex," *Commun. Math. Phys.* **83**, 517, 1982.

Chukbar, K. V., "Hall effect in thin films," *Sov. Phys. JETP* **70**, 769–773, 1990.

Constantin, P., "Geometric statistics in turbulence," *SIAM Rev.* **36**, 73–98, 1994.

Constantin, P., Fefferman, C., and Majda, A., "Geometric constraints on potential singularity formulation in the 3-D Euler equations," *Commun. Partial Different Eqns.* **21** (3–4), 559–571, 1996.

Constantin, P., Lax, P. D., and Majda, A., "A simple one-dimensional model for the three-dimensional vorticity equations," *Commun. Pure Appl. Math.* **38**, 715–724, 1985.

Constantin, P., Majda, A., and Tabak, E., "Formation of strong fronts in the 2-D quasi-geostrophic thermal active scalar," *Nonlinearity* **7**, 1495–1533, 1994.

Cordoba, D., "Nonexistence of simple hyperbolic blow-up for the quasi-geostrophic equation," *Ann. Math.* **148**, 1135–1152, 1998.

E, W. and Shu, C., "Small-scale structures in Boussinesq convection," *Phys. Fluids* **6**, 49–58, 1994.

Garner, S., Held, I. M., Pierrehumbert, R. T., and Swanson, K., "Surface quasigeostrophic dynamics," *J. Fluid Mech.* **282**, 1–20, 1995.

Grauer, R. and Sideris, T., "Numerical computation of three dimensional incompressible ideal fluids with swirl," *Phys. Rev. Lett.* **67**, 3511–3514, 1991.

Green, A. E. and Taylor, G. I., "Mechanism of the production of small eddies from large ones," *Proc. R. Soc. London A* **158**, 499, 1937.

Kerr, R., "Evidence for a singularity of the three-dimensional incompressible Euler equations," *Phys. Fluids A* **5**, 1725–1746, 1993.

Majda, A. J., "Vorticity, turbulence and acoustics in fluid flow," *SIAM Rev.* **33**, 349–388, 1991.

Ohkitani, K. and Yamada, M., "Inviscid and inviscid-limit behavior of a surface quasi-geostrophic flow," *Phys. Fluids* **9**, 876–882, 1997.

Pedlosky, J., *Geophysical Fluid Dynamics*, Springer-Verlag, New York, 1987.

Ponce, G., "Remark on a paper by J. T. Beale, T. Kato and A. Majda," *Commun. Math. Phys.* **98**, 349–353, 1985.

Pumir, A., and Siggia, E. D., "Development of singular solutions to the axisymmetric Euler equations," *Phys. Fluids A* **4**, 1472–1491, 1992.

6

Computational Vortex Methods

The topic of this chapter is a class of computational methods for solutions of the Euler equation and the Navier-Stokes equation with high Reynold's numbers. These methods are particularly useful for simulations dominated by vortex dynamics. Applications range from accurate predictions of hurricane paths to control of huge vortices shed by large aircraft to the design of efficient internal combustion engines.

The efficiency and the accuracy of vortex methods lie in the fact that they directly simulate the particle trajectories of the flow. We discuss two types of vortex methods, the inviscid-vortex method for simulating Euler solutions and the random-vortex method, an extension of the inviscid-vortex method that incorporates the diffusion of the Navier–Stokes equation through a Brownian motion of the particles. *Inviscid-vortex methods* are based on the integrodifferential equation for particle trajectories:

$$\frac{dX}{dt}(\alpha, t) = \int_{\mathbb{R}^2} K_3[X(\alpha, t) - X(\alpha', t)]\nabla_\alpha X(\alpha', t)\omega_0(\alpha')d\alpha',$$
$$X(\alpha, t)|_{t=0} = \alpha. \tag{6.1}$$

In Section 4.1, we showed that the smoothing properties of the integral operator with the kernel K_3 compensate for the loss of regularity that is due to the differentiation in the term $\nabla_\alpha X$; this formulation contains a bounded space operator that is advantageous for numerical approximations. It also contains the explicit term $\omega(X(\alpha, t), t) = \nabla_\alpha X(\alpha, t)\omega_0(\alpha)$ that describes the stretching of vorticity that increases the accuracy of a 3D simulation in which accumulation of vorticity is present. Methods based on direct approximation of the Euler equation or the vorticity-stream formulation tend to have a significant amount of numerical viscosity and also do not accurately compute local accumulation of vorticity. We discuss 2D inviscid-vortex methods in Section 6.2 and 3D inviscid-vortex methods in Section 6.3. The convergence of 2D vortex methods is proved in Section 6.4. In Section 6.5 we discuss the computational performance of inviscid-vortex methods.

Vortex methods are also powerful tools for simulating Navier–Stokes solutions with small viscosity $\nu \ll 1$. Recall from Chap. 3 that such solutions v^ν are well approximated by inviscid solutions v^0, with a typical estimate of $|v^\nu - v^0| \leq c\nu^{1/2}$. Moreover, viscous-splitting algorithms that solve the inviscid Euler equation and the heat equation alternately in each time step provide an accurate way to compute these slightly viscous flows. The use of such an algorithm requires an effective numerical method for inviscid flows. The random-vortex method is an unusual algorithm for introducing diffusion into the vortex method. In particular it uses a Brownian motion of the fluid particles to simulate diffusion. In Section 6.1 we introduce the ideas of the random-vortex method through the simple model problem of *viscous strained shear layers*. In Section 6.6 we discuss the full random-vortex method in two dimensions, including a proof of the optimal rate convergence for this algorithm.

The ancestor of inviscid-vortex methods is the *point-vortex method* first used by Rosenhead (1932) to simulate vortex sheets in 2D inviscid fluid flows. This algorithm is based on the idea of approximating a solution of the 2D inviscid Euler equations by a finite superposition of interacting elementary exact solutions. For many years, this algorithm was considered suspect (see, e.g., Birkhoff and Fisher, 1959; Birkhoff, 1962). A point vortex induces a singular velocity field

$$v_\Gamma = \frac{\Gamma}{2\pi|x|^2}\begin{pmatrix} -x_2 \\ x_1 \end{pmatrix} \tag{6.2}$$

with $|x|^2 = x_1^2 + x_2^2$. Recall that a given velocity field v has locally finite kinetic energy if

$$\int_{|x|\le R} |v|^2 dx \le C(R) \qquad \text{for any} \quad R > 0.$$

However, point vortices have locally infinite energy:

$$\int_{|x|\le R} |v_\Gamma|^2 dx = \infty \qquad \text{for any} \quad R > 0.$$

This locally infinite energy might lead to spurious nonphysical behavior when solutions, with locally finite energy; are approximated by a superposition of a finite number of point vortices with infinite energy. Careful numerical experiments for both vortex sheets and smooth flows show this to be true (Beale and Majda, 1985; Chorin and Bernard, 1973; Krasny, 1986). However, recent results (Goodman et al., 1990; Hou and Lowengrub, 1990; Cottet et al., 1991) show that point-vortex methods actually do converge to a solution of the Euler equation. On the other hand, the discretization threshold below which we achieve stability and convergence of the method may be prohibitively small (Hou, 1991). The development of computationally effective vortex methods began with the seminal paper of Chorin (1973). This paper contains the first large-scale simulations of flow past a 2D cylinder at high Reynolds numbers. The three novel ideas he used are (1) finite vortex cores to cut off the energy singularity from the point-vortex method, (2) simulating diffusion by a random walk of the cores, and (3) mimicking no-slip boundary conditions by introducing new vortices at the boundary. In this chapter we discuss in detail ideas (1) and (2).

The inviscid-vortex method uses a straightforward strategy for discretizing the integrodifferential equation for the particle trajectories. It has the following four steps:

(i) smoothing the kernel K_3,
(ii) approximating the integral by a numerical quadrature,

(iii) approximating the vortex-stretching term, and
(iv) a closing of the equations to obtain a finite-dimensional system of ODEs.

In Section 6.6 we discuss an analog of this strategy for solving stochastic ODEs. We end this introduction by listing some of the computational advantages of vortex algorithms:

(1) Computational elements are needed only in regions with nonzero vorticity; the algorithms automatically follow the concentration of vorticity as time evolves.
(2) Computation of the pressure is unnecessary, and the approximation automatically satisfies incompressibility.
(3) In two dimensions a superposition of interacting elementary exact solutions approximates a general solution.
(4) The method minimizes numerical viscosity by using a Lagrangian method in which the incompressible fluid velocity is the only characteristic. As a measure of the lack of viscosity, 2D vortex methods conserve a discrete approximation of the energy.
(5) The algorithm is a discretization of an evolution equation with a bounded space operator. Hence there are no stability constraints on the time step Δt. Such constraints are always present in direct simulations of the primitive-variable form of the Navier–Stokes equation.

One important disadvantage of vortex methods is the fact that each particle interacts with every other particle, creating $\mathcal{O}(N^2)$ operations per time step if N is the number of particles. Recent progress has been made to overcome this problem. See Majda (1988) and Puckett (1993) for a discussion.

6.1. The Random-Vortex Method for Viscous Strained Shear Layers

In this section we illustrate the random-vortex method for a special class of solutions to the Navier–Stokes equations. These are explicit solutions to the Navier–Stokes equations that we can directly compare with the discrete approximation. Also, this example reduces the nonlinear Navier–Stokes equations to a linear PDE, which also simplifies the analysis. This special application of the random-vortex method was discussed in Long (1987).

Recall the *viscous strained shear layers*, introduced in Example 1.8 in Chap. 1, with a velocity field of the form

$$v(x, y, z, t) = \gamma(-x, 0, z) + [0, u(x, t), 0],$$

where $\gamma > 0$. The term $\gamma(-x, 0, z)$ denotes the straining, and the term $u(x, t)$ denotes the shear layer that is evolving under the competition between external straining flows and viscous diffusion. If $v(x, y, z, t)$ solves the Navier–Stokes equation, then the equation for $u(x, t)$ is a 1D *linear* convection–diffusion equation:

$$\frac{\partial u}{\partial t} - \gamma x \frac{\partial u}{\partial x} = \nu \frac{\partial^2 u}{\partial x^2}. \tag{6.3}$$

The vorticity vector is aligned with the z axis with strength $\omega = (\partial u/\partial x)$ satisfying

$$\frac{\partial \omega}{\partial t} - \frac{\partial}{\partial x}(\gamma x \omega) = \nu \frac{\partial^2 \omega}{\partial x^2}. \tag{6.4}$$

Note that, as in the case of the vorticity-stream form for the Navier–Stokes equations, the velocity field can be recovered from the vorticity in Eq. (6.4) by the solution of the 1D equivalent of the Biot–Savart law (2.10), namely,

$$u(x) = \int_{-\infty}^{x} \omega(y)dy = \int_{\mathbb{R}} H(x-y)\omega(y)dy. \tag{6.5}$$

Here H denotes the Heaviside function $H(x) = 1,\ x > 0,\ H(x) = 0,\ x \le 0$.

If $\nu = 0$, then Eq. (6.4) reduces to the first-order PDE,

$$\frac{\partial \tilde{\omega}}{\partial t} - \frac{\partial}{\partial x}(\gamma x \tilde{\omega}) = 0, \tag{6.6}$$

with characteristic curves given by the solutions $x(t) = \alpha e^{-\gamma t}$ of the ODE,

$$\frac{dx}{dt} = -\gamma x, \qquad x(0) = \alpha. \tag{6.7}$$

Hence the solution of Eq. (6.6) is

$$\tilde{\omega}(x,t) = e^{\gamma t}\tilde{\omega}(xe^{\gamma t}, 0) \tag{6.8}$$

and we see that the straining flow causes vorticity to be pushed toward the origin along the x axis and amplified. The effects of diffusion combat this by diffusing vorticity away from the origin and lowering the vorticity concentration.

Solutions of diffusive equation (6.4) can also be computed explicitly. As in Chap. 1, the change of variables $\xi = xe^{\gamma t}$, $\tau = [(\nu/2\gamma)(e^{2\gamma t} - 1)]$ reduces Eq. (6.4) to the heat equation:

$$\tilde{\omega}_\tau = \tilde{\omega}_{\xi\xi}, \qquad \tilde{\omega}|_{\tau=0} = \omega_0, \quad \xi \in \mathbb{R}.$$

Hence the solution to equation (6.4) is

$$\omega(x,t) = e^{\gamma t}\left[2\pi \frac{\nu}{\gamma}(e^{2\gamma t} - 1)\right]^{-1} \int_{\mathbb{R}} e^{-|xe^{\gamma t} - \xi|^2 / \frac{\nu}{\gamma}(e^{2\gamma t}-1)} \omega_0(\xi) d\xi. \tag{6.9}$$

As an elementary model, we now illustrate the use of a random-vortex method in simulating solution (6.9) to Eq. (6.4). In the inviscid case, particle position can be determined by the solution of the simple ODE (6.7). Equation (6.7) is a simplified version of particle-trajectory equation (6.1) for inviscid Euler dynamics. When viscosity is present, vorticity is no longer carried along particle paths and Eq. (6.1) no longer holds. Instead, vorticity is diffused at the same time it is convected. We can capture this dynamics by introducing a random walk of the particles; that is, we let $W(t)$ be a *Wiener process* or Brownian motion. A little later in this section we discuss how to directly simulate such a process. We introduce diffusion into the model by replacing deterministic differential equation (6.7) with the *stochastic differential equation* (SDE)

$$dX(t) = -\gamma X(t)dt + \sqrt{2\nu}\, dW(t), \qquad X(0) = \alpha, \tag{6.10}$$

which reduces to Eq. (6.7) when $\nu = 0$. The probability $p(x, t; \alpha, s)$ that a particle at point α at time s reaches point x at time $t > s$ is called the *transition probability density* of the diffusion process of SDE (6.10). The theory of SDEs (Kloeden and Platen, 1992) says that p satisfies a "forward" or Fokker–Planck equation (Gard, 1988; Kloeden and Platen, 1992) of the form of Eq. (6.4). In particular, simulating SDE (6.10) provides a solution to Eq. (6.4) by means of the formula

$$\omega(x, t) = \int_{\mathbb{R}} p(x, t; \alpha, 0)\omega_0(\alpha)d\alpha.$$

Solution (6.9) tells us that the transition probability density of SDE (6.10) is

$$p(x, t; \alpha, s) = \frac{1}{\sqrt{2\pi\sigma^2}} \exp\left[-(x - q)^2/2\sigma^2\right] \tag{6.11}$$

where $q = e^{-\gamma(t-s)}\alpha$, and

$$\sigma^2 = (\nu/\gamma)\left[1 - e^{-2\gamma(t-s)}\right]. \tag{6.12}$$

Note that, as $t \longrightarrow \infty$,

$$p(x, t; \alpha, 0) \longrightarrow \frac{1}{\sqrt{2\pi\sigma_\infty^2}} \exp\left(-x^2/w\sigma_\infty^2\right),$$

where $\sigma_\infty^2 = \nu/\gamma$.

Because $p(x, t; \alpha, 0)$ is uniformly bounded for $t \geq t_0 > 0$, the dominated convergence theorem implies that

$$\begin{aligned}
\lim_{t\longrightarrow\infty} \omega(x, t) &= \lim_{t\longrightarrow\infty} \int_{\mathbb{R}} G(x, t; \alpha, 0)\omega(\alpha, 0)d\alpha \\
&= \int_{\mathbb{R}} \frac{1}{\sqrt{2\pi\sigma_\infty^2}} \exp\left(\frac{-x^2}{2\sigma_\infty^2}\right)\omega(\alpha, 0)d\alpha \\
&= \frac{\nu_0}{\sqrt{2\pi\sigma_\infty^2}} \exp\left(\frac{-x^2}{2\sigma_\infty^2}\right),
\end{aligned} \tag{6.13}$$

where $\nu_0 = \int_{\mathbb{R}} \omega(\alpha, 0)d\alpha$ is the total vorticity, a conserved quantity. Limit (6.13) is a steady solution of Eq. (6.6).

Before discussing the numerical approximation of the viscous strained shear layers, we motivate the approximation scheme by considering separately the quadrature error in the inviscid problem and the sampling error in the purely diffusive case without advection.

6.1.1. The Discrete Problem with Only Diffusion and Advection

We motivate the analysis in the next subsection by considering separately the case with only diffusion and the case with only convection. In the convection-only case, we first consider the equivalent of the point-vortex approximation and then show how this can be improved by considering smoothing of the point vorticies into blobs. In the purely diffusive case, we use point vorticies and see that the discretization error is not quite as good as in the case with no diffusion because of the randomness of particles. Moreover, there is an additional sampling error.

Case 1. No Diffusion

In this case, we do not need to simulate a stochastic equation. The vortex method for this case consists of discretizing the equation for the particle paths with a finite number of particles,

$$\frac{dX_i}{dt} = \gamma X_i, \qquad X_i(0) = \alpha_i,$$

where we use the particle positions to discretize Eq. (6.5) for the velocity field in terms of the vorticity. First we write Eq. (6.5) by using the Lagrangian variable α. Note that, on making the change of variables $y = X(\alpha, t)$ and using Eq. (6.8), we obtain that

$$u(x,t) = \int_{\mathbb{R}} H[x - X(\alpha,t)]\omega_0(\alpha)d\alpha.$$

One possibility is to consider the simple discretization

$$\tilde{u}(x,t) = \sum_i H[x - X_i(t)]\omega_i h, \tag{6.14}$$

in which the Lagrangian variable α is divided into equally spaced intervals of size h. We can think of Eq. (6.14) as approximating the vorticity by a finite number of Dirac delta functions, i.e., Eq. (6.14) is equivalent to

$$\tilde{u}(x,t) = \sum_i \int_{-\infty}^{x} \delta[y - X_i(t)]\omega_i h\, dy. \tag{6.15}$$

Using the fact that Eq. (6.14) is equivalent to

$$\tilde{u}(x,t) = \sum_{i, X(t)_i \le x} \omega_i h,$$

we see that

$$|\tilde{u}(\cdot,t) - u(\cdot,t)|_{L^\infty} \le Ch. \tag{6.16}$$

The vortex method exploits the fact that a higher-order rate of convergence than that of relation (6.16) can be obtained when the Dirac delta functions in Eq. (6.15) are replaced with smooth approximations, which we refer to here as blobs.

Recall from Chap. 3, Subsection 3.2.1, the definition of a mollifier, which we define by considering a smooth function $\psi(x)$ with $\int_{\mathbb{R}} \psi = 1$. Denote by ψ_δ the rescaled function $\frac{1}{\delta}\psi(x/\delta)$, also of integral 1. Then $H_\delta(x) = \psi_\delta * H$ defines a smoothed Heaviside function. We now define u^{blob} to be the velocity field:

$$\int_{\mathbb{R}} H_\delta[x - X(\alpha, t)]\omega_0(\alpha)d\alpha. \tag{6.17}$$

Later in Section 6.1.2, we show that if ψ has the property that $\int \psi(x)x^n\, dx$ vanishes for all integers $n < m$ then $|u - u^{\text{blob}}| \le C\delta^m$. In general, if we use a quadrature rule of the order of r, the convergence depends on a bound of the rth derivatives of the integrand. That is,

$$\left| \int g(x)dx - \sum g_i w_i \right| \le C\|g\|_{W^{r,1}} h^r,$$

where w_i represent weight functions for a quadrature rule of the order of r and h is the mesh size for the discretization of the integral. $W^{r,1}$ denotes the Sobolev norm associated with functions that have r derivatives in L^1. Because $\partial_x^L H_\delta$ scales as δ^{-L} and has characteristic width of the order of δ, the integrand in expression (6.17) has a $W^{L,1}$ norm that behaves as δ^{1-L}. Thus for a trapezoidal approximation of the integral in expression (6.17), we expect a convergence rate of the form

$$|u^{\text{comp}} - u^{\text{blob}}|_{L^\infty} \le C\delta(h/\delta)^L.$$

In particular, we cannot take $\delta \to 0$ without simultaneously taking $h \to 0$. The net result is that

$$|u(x,t) - u^{\text{comp}}| \le C\left[\delta^m + \delta(h/\delta)^L\right].$$

Note that we actually get a better approximation of the velocity field by representing the vorticity by means of a sum of blob functions instead of the velocity field directly. If u itself is approximated by a sum of blobs, then

$$|u^{\text{blob}} - u^{\text{exact}}|_{L^\infty} \sim |u^{\text{blob}}|_{L^\infty} \sim \frac{1}{\delta}.$$

On the other hand, the above blob approximation for the *vorticity* has an the error introduced in the *velocity* field that is quite small. That is, if $\omega^{\text{blob}} = \sum_i \varphi^\delta(x - x_i)$ $w_i\omega_i$, then

$$u^{\text{blob}} = \int_{-\infty}^{x} \omega^{\text{blob}}(y)dy, \tag{6.18}$$

and

$$|u^{\text{blob}} - u^{\text{exact}}|_{L^\infty} \le C\delta$$

if δ is of the order of the mesh size. Because accurate pointwise representation of the velocity field is essential for a vortex-method calculation, we in general want to approximate the vorticity by blobs instead of the velocity field.

Case 2. No Advection

We now consider solutions to the heat equation $\omega_t = \nu\Delta\omega$. We wish to approximate solutions to this equation by a Brownian motion.

To do this, we introduce the standard Wiener process $W(t)$, an example of a continuous-time stochastic process with independent increments. The Wiener process is a Gaussian random variable for which $W(0) = 0$ with probability one, $E(W(t)) = 0$, and $\mathrm{Var}[W(t) - W(s)] = t - s$ for all $0 \le s \le t$. The characteristic function for the Wiener process is $E[e^{ikW(t)}] = e^{-k^2t/2}$. For more information about the Wiener process and the theory of SDEs, the reader is referred to Friedman (1975), Gard (1988), Kloeden and Platen (1992), and Pollard (1984).

We now show that the random variable

$$\omega(x) = \omega_0[x + \sqrt{2\nu}W(t)] \tag{6.19}$$

has an expected value that is the solution of the initial value problem:

$$\frac{\partial\omega}{\partial t} = \nu\Delta\omega,$$
$$\omega(\cdot, 0) = \omega_0(x).$$

Consider the inverse Fourier transform

$$\omega(x) = \int e^{ikx}\hat{\omega}(k)dk = \int e^{ik[x+\sqrt{2\nu}W(t)]}\hat{\omega}_0(k)dk.$$

Then we have

$$E[\omega(x)] = \int e^{ikx}E\left[e^{ik\sqrt{2\nu}W(t)}\right]\hat{\omega}_0(k)dk$$
$$= \int e^{ikx}e^{-k^2\nu t}\hat{\omega}_0(k)dk.$$

The only fact from stochastic theory that we use here is that $E[e^{ik\sqrt{2\nu}W(t)}] = e^{-k^2\nu t}$.

We now consider the discretization and sampling error associated with approximating such a Brownian motion by the motion of a finite number of particles. Recall that the velocity field u is obtained from the vorticity ω by the formula

$$u(x,t) = \int_{-\infty}^{x}\omega(y,t)dy = \int_{\mathbb{R}} EH[x - W(t;\alpha)]\omega_0(\alpha)d\alpha. \tag{6.20}$$

Consider the following discretization of this formula: Let $W_i(t)$ be the Wiener process with initial condition $W_i(0) = \alpha_i \in \Lambda^h$, where $\Lambda^h = \{\alpha_i : \alpha_i = h \cdot i,\ i \in \mathbb{Z},\ \omega_i = \omega(\alpha_i, 0) \neq 0\}$. Denote by ω_i the vorticity $\omega_0(\alpha_i)$. We approximate the vorticity field u at time t in Eq. (6.20) by

$$\sum_i \omega_i h H[x - W_i(t)]. \tag{6.21}$$

The error between the the actual u in Eq. (6.20) and the discretized u in expression (6.21) can be separated into two parts, the sampling error, given by

$$\sum_i H[x - W_i(t)]\omega_i h - \sum_i EH[x - W_i(t)]\omega_i h,$$

and the discretization error, given by

$$\sum_i EH[x - W_i(t)]\omega_i h - \int_{-\infty}^{x} \omega(y,t)dy = \int_{\mathbb{R}} EH[x - W(t;\alpha)]\omega_0(\alpha)d\alpha.$$

After the arguments that are detailed in the following subsection, we show that the discretization error is $\mathcal{O}(h|\ln h|^{1/2})$. Note that this is logarithmically in h worse than error (6.16) for the point-vortex approximation with only advection. The log term results from the randomness of the system. The sampling error is $\mathcal{O}[(\nu^{1/4}h^{1/2} + h)|\ln h|]$, where the unusual ν dependence is due to the fact that the kernel H is discontinuous.

6.1.2. Analysis of the Schemes with Both Diffusion and Advection

We now consider the viscous strained shear layers with both diffusion and convection. In lieu of simulating deterministic PDE (6.4), we simulate SDE (6.10) with the following discretization of SDE (6.10). Let $X_i(t)$ be the solution of the SDE

$$\begin{aligned} dX_i(t) &= -\gamma X_i(t)dt + \sqrt{2\nu}\, dW(t), \\ X_i(0) &= \alpha_i \in \Lambda^h, \end{aligned} \tag{6.22}$$

where $\Lambda^h = \{\alpha_i : \alpha_i = h \cdot i,\ i \in \mathbb{Z},\ \omega_i = \omega(\alpha_i, 0) \neq 0\}$ and $h > 0$ is the grid size. We assume that the initial vorticity $\omega(\alpha, 0)$ has compact support. Once we have a solution to SDE (6.22) we can directly obtain an approximate velocity at time t by either the point-vortex formula

$$u_{\text{pv}}(x,t) = \sum_i H[x - X_i(t)]\omega_i h, \tag{6.23}$$

where H is the unsmoothed velocity kernel (Heaviside function)

$$H(x) = \begin{cases} 1 & x > 0 \\ 0 & x \leq 0 \end{cases},$$

or from the vortex-blob formula

$$u_{\text{blob}}(x,t) = \sum_i H_\delta[x - X_i(t)]\omega_i h, \tag{6.24}$$

where $H_\delta = H * \psi_\delta$ is a smoothed kernel with $\psi_\delta(x) = \delta^{-1}\psi(\delta^{-1}x)$, $\delta > 0$, $\int_{\mathbb{R}} \psi(x)dx = 1$ and $\psi \in M^{L,m}$. Here $M^{L,m}$ is defined by the following definition.

Definition 6.1. *A mollifier ψ belongs to the class $M^{L,m}$, $L \geq 0$, $m \geq 2$, provided that*

(i) $\int_{\mathbb{R}^2} \psi \, dx = 1$,
(ii) $\int_{\mathbb{R}^2} x^\alpha \psi(x) dx = 0$, $1 \leq |\alpha| \leq p-1$, *and* $\int_{\mathbb{R}^2} |x|^m |\psi(x)| dx < \infty$,
(iii) $\psi \in C^L(\mathbb{R}^2)$, *and* $|x|^{2+|\beta|} |\partial^\beta \psi(x)| \leq c \; \forall \, |\beta| \leq L$

Property (ii) of vanishing certain moments of the mollifier ψ is essential to prove the higher-order convergence of the inviscid method. The convergence results of Eqs. (6.23) and (6.24) are summarized in Theorem 6.1 and Theorem 6.2, respectively. In Theorem 6.1 we allow the initial vorticity to be rough because the velocity u_{pv} is computed from the discontinuous kernel H. The rate $(h + \sigma^{1/2} h^{1/2}) |\ln h|$ of convergence does not improve when restricted to smooth initial vorticities. On the other hand, in Theorem 6.2 we assume that the initial vorticities $\omega(\alpha, 0)$ are $C^2(\mathbb{R})$ because we want to show theoretically that the smoothing does improve the accuracy.

Theorem 6.1. *Assume that the initial vorticity $\omega(\alpha, 0)$ has compact support and is Lipschitz continuous with the Lipschitz constant M; then, for any $x \in \mathbb{R}$, $0 \leq t \leq T$,*

$$|u_{\text{pv}}(x, t) - u(x, t)| \leq C(h + \sigma^{1/2} h^{1/2}) |\ln h| \tag{6.25}$$

with high probability, where the constant C depends on only γ, T, M, the diameter of supp $\omega(\alpha, 0)$, *and* $\|\omega(\alpha, 0)\|_{L^\infty}$. *Here σ^2 is the explicit variance defined in Eq. (6.12):*

$$\sigma^2 = (\nu/\gamma)\left[1 - e^{-2\gamma(t-s)}\right].$$

The error estimate consists of two parts: the discretization error and the sampling error. The discretization error is of the order of $h |\ln h|^{1/2}$ and the sampling error is of the order of $(\sigma^{1/2} h^{1/2} + h) |\ln h|$. Contrary to what we might expect, the ν dependence of the lowest-order $h^{1/2}$ in the sampling error is a factor of $\nu^{1/4}$ (not $\nu^{1/2}$) because $\sigma^2 = (\nu/\gamma)(1 - e^{-2\gamma(t-s)})$ is proportional to ν. This unusual ν dependence is due to the jump discontinuity of the kernel H, as we shall see in the proof.

Theorem 6.2. *Assume that the initial vorticity $\omega(\alpha, 0)$ has compact support and $\omega(\alpha, 0) \in C^L(\mathbb{R})$, $L \geq 2$. If $h \leq C_0 \sigma$ for some fixed constant C_0, then for any $x \in \mathbb{R}$, $0 \leq t \leq T$,*

$$|u_{\text{blob}}(x, t) - u(x, t)| \leq C\left(\delta^m + \left\{\left[\left(\frac{h}{\delta}\right)^L \delta\right]^{1/2} + \sigma^{1/2}\right\} h^{1/2} |\ln h|\right) \tag{6.26}$$

with high probability, where the constant C depends on only C_0, γ, T, L, m, the diameter of supp $\omega(\alpha, 0)$, *and* $\max_{0 \leq |\beta| \leq L} \|\partial^\beta \omega(\alpha, 0)\|_{L^\infty}$.

The sampling error has two components: $(h/\delta)^{L/2}(\delta h)^{1/2}|\ln h|$ and $(\sigma h)^{1/2}|\ln h|$. The discretization error does not appear in relation (6.26) because its order $(h/\delta)^L\delta$ is smaller than the first component of the sampling error. By comparing relations (6.25) and (6.26), we find that the smoothing improves the accuracy for $h \ll 1$ by reducing the first component of the sampling error from the order of $h|\ln h|$ to the order of $(h/\delta)^{L/2}(\delta h)^{1/2}|\ln h|$. From the assumption in Theorem 6.2, this effect can be seen only when the ratio h/σ is not too large. If the grid size h is too large compared with σ, then we can hardly expect that the computational results from the vortex method offer good resolution on the much smaller scale of viscous effects. We shall see in the proof that the right-hand side of relation (6.26) approaches that of relation (6.25) as the ratio $h/\sigma \longrightarrow \infty$.

Actually, a stronger version of Theorem 6.1 and Theorem 6.2 with $|u_{\text{comp}}(x,t) - u(x,t)|$ replaced with $\|u_{\text{comp}}(x,t) - u(x,t)\|_{L^\infty(\mathbb{R}\times[0,T])}$ is true. The proof of the stronger version is found in Section 6.4. To show the basic idea, we prove the weaker version here.

Proof of Theorem 6.1. Write the velocity u in terms of expectations

$$\begin{aligned} u(x,t) &= \int_{-\infty}^{x} \omega(y,t)dy \\ &= \int_{-\infty}^{x}\int_{\mathbb{R}} G(y,t;\alpha,0)\omega(\alpha,0)d\alpha dy \\ &= \int_{\mathbb{R}} \omega(\alpha,0)\int_{-\infty}^{x} G(y,t;\alpha,0)dyd\alpha \\ &= \int_{\mathbb{R}} \omega(\alpha,0)\int_{\mathbb{R}} H(x-y)G(y,t;\alpha,0)dyd\alpha \\ &= \int_{\mathbb{R}} EH[x-X(t;\alpha)]\omega(\alpha,0)d\alpha \end{aligned}$$

and decompose the error as

$$\begin{aligned} |u_{\text{pv}}(x,t) - u(x,t)| &\le \left|\sum_i H[x-X_i(t)]\omega_i h - \sum_i EH[x-X_i(t)]\omega_i h\right| \\ &\quad + \left|\sum_i EH[x-X_i(t)]\omega_i h - \int_{\mathbb{R}} EH[x-X(t;\alpha)]\omega(\alpha,0)d\alpha\right| \\ &= \text{sampling error} + \text{discretization error}. \end{aligned}$$

To estimate the discretization error, we need to estimate on the local Lipschitz constants of $EH[x-X(t;\alpha)]$ by calculating its derivative with respect to α. From Eq. (6.11),

$$EH[x-X(t;\alpha)] = \int_{-\infty}^{x} \frac{1}{\sqrt{2\pi\sigma^2}} \exp\left[-(y-q)^2/2\sigma^2\right]dy = \Phi(x-q),$$

with $0 \le \Phi \le 1$, $\Phi(-\infty) = 0$, $\Phi(\infty) = 1$, $\Phi(0) = 1/2$. It follows that

$$\frac{d}{d\alpha}\Phi(x - q) = -e^{\gamma t}\Phi'(x - e^{-\gamma t}\alpha) = -e^{-\gamma t} \cdot G(x, t; \alpha, 0).$$

Therefore the local Lipschitz constant at t, α is of the order of $G(x, t; \alpha, 0)$. To estimate the discretization error, consider the following two cases.

(1) $\sigma^2 \ge e^{-1}$: The Lipschitz constant of $EH[x - X(t; \alpha)]\omega(\alpha, 0)$ is bounded and $\sum_i EH[x - X_i(t)]\omega_i h$ is just the Riemann sum approximation of the integral $\int_{\mathbb{R}} EH[x - X(t; \alpha)]\omega(\alpha, 0)d\alpha$. Therefore the error is bounded by Ch.

(2) $\sigma^2 < e^{-1}$: For $|x - e^{-\gamma t}\alpha| \ge \sigma(-2\ln\sigma)^{1/2}$, we have the estimate

$$0 < G(x, t; \alpha, 0) \le \frac{1}{\sqrt{2\pi\sigma^2}} \exp\{-[\sigma(-2\ln\sigma)^{1/2}]^2/2\sigma^2\} = \frac{1}{\sqrt{2\pi}}.$$

There are two subcases.

(a) $\sigma(-2\ln\sigma)^{1/2} \le h$: The local Lipschitz constant of $EH[x - X(t; \alpha)]\omega(\alpha, 0)$ is bounded except on an interval of length $2h$ where the integrand is bounded. Therefore we can still conclude that the discretization error is bounded by Ch.

(b) $\sigma > h$: Because $\sigma^2 < e^{-1}$, $\sigma > h$ implies that $\sigma(-2\ln\sigma)^{1/2} < h$. The local Lipschitz constant of the integrand on the interval $|x - e^{-\gamma t}\alpha| < \sigma(-2\ln\sigma)^{1/2}$ is bounded by $C_1 + C_2\sigma^{-1}$. Hence the discretization error is bounded by

$$\begin{aligned} C'h + (C_1 + C_2\sigma^{-1})h^2 \cdot \frac{2\sigma(-2\ln\sigma)^{1/2}}{h} & \\ \le C'h + 2\sqrt{2}(C_1\sigma + C_2)h(-\ln h)^{1/2} & \\ \le C'h + C''h|\ln h|^{1/2} & \\ \le Ch|\ln h|^{1/2}. & \end{aligned}$$

In summary, the discretization error is bounded by $Ch|\ln h|^{1/2}$.

We estimate the sampling error by applying Bennett's inequality.

Lemma 6.1. *Bennett's Inequality (Pollard, 1984, Appendix B). Let Y_i be independent bounded random variables (not necessarily identically distributed) with mean zero, variances σ_i^2, and $|Y_i| \le M$. Let $S = \sum_i Y_i$, $V \ge \sum_i \sigma_i^2$; then, for all $\eta > 0$,*

$$P\{|S| \ge \eta\} \le 2\exp\left[-\frac{1}{2}\eta^2 V^{-1} B(M\eta V^{-1})\right], \tag{6.27}$$

where $B(\lambda) = 2\lambda^{-2}[(1 + \lambda)\ln(1 + \lambda) - \lambda]$, $\lambda > 0$, $\lim_{\lambda\to 0^+} B(\lambda) = 1$, and $B(\lambda) \sim 2\lambda^{-1}\ln\lambda$ as $\lambda \to \infty$. If Y_i are random vectors in $\mathbb{R}^2$, relation (6.27) applied to each component gives

$$P\{|S| \ge \eta\} \le 4\exp\left[-\frac{1}{4}\eta^2 V^{-1} B(M\eta V^{-1})\right].$$

The proof of this inequality can be found in the appendix.

To apply Bennett's inequality, let

$$Y_i = \omega_i h\{H[x - X_i(t)] - EH[x - X_i(t)]\}.$$

We have $EY_i = 0$, $|Y_i| \le C_1 h$, and $\text{Var } Y_i = p_i(1 - p_i)\omega_i^2 h^2$, where

$$p_i = EH[x - X_i(t)] = \Phi[x - \exp(-\gamma t)\alpha].$$

It follows that

$$\sum_i \text{Var } Y_i = h \sum_i p_i(1 - p_i)\omega_i^2 h,$$

and the summation on the right-hand side is just a Riemann sum of the integral

$$\int_{\mathbb{R}} \Phi(x - e^{-\gamma t}\alpha)[1 - \Phi(x - e^{-\gamma t}\alpha)]\omega^2(\alpha, 0)d\alpha. \tag{6.28}$$

Integral (6.28) is bounded by $C\sigma$. By following the same argument as in the estimate of the discretization error, we conclude that

$$\sum_i \text{Var } Y_i \le h(C\sigma + C'h|\ln h|^{1/2}) \le C_2 h(\sigma + h)|\ln h|^{1/2} := V.$$

For any fixed constant $\kappa > 0$, let

$$\eta = \kappa C_2^{1/2} h^{1/2}(\sigma + h)^{1/2}|\ln h|.$$

Because $(\sigma + h)^{1/2} \le \sigma^{1/2} + h^{1/2}$, it follows from Bennett's inequality that

$$\begin{aligned}
P&\left\{\left|\sum_i Y_i\right| \ge \kappa C_2^{1/2} h^{1/2}(\sigma^{1/2} + h^{1/2})|\ln h|\right\} \\
&\le P\left\{\left|\sum_i Y_i\right| \ge \eta\right\} \\
&\le 2\exp\left[-\frac{1}{2}\eta^2 V^{-1} \cdot B(C_1 h \eta V^{-1})\right] \\
&\le 2\exp\left[-\frac{1}{2}\kappa^2 |\ln h|^{3/2} \cdot B\left(C_1 C_2^{-1/2} \kappa h^{1/2}(\sigma + h)^{-1/2}|\ln h|^{1/2}\right)\right] \\
&\le 2\exp\left[-\frac{1}{2}\kappa^2 |\ln h|^{3/2} \cdot B\left(C'\kappa |\ln h|^{1/2}\right)\right] \\
&\le 2\exp\left(-C''\kappa^2 |\ln h|^{3/2} \cdot \kappa^{-1}|\ln h|^{-1/2}\right) \\
&\le 2h^{C''\kappa} \\
&\le 2N^{-C\kappa}
\end{aligned}$$

as $h \longrightarrow 0$. The probability approaches zero more rapidly than any polynomial rate if κ is chosen sufficiently large. This finishes the proof of Theorem 6.1. □

Proof of Theorem 6.2. Decompose the error into

$$
\begin{aligned}
|u_{\text{blob}}(x,t) - u(x,t)| &= \left| \sum_i H_\delta[x - X_i(t)]\omega_i h - \int_{\mathbb{R}} H(x-y)\omega(y,t)dy \right| \\
&\le \left| \sum_i H_\delta[x - X_i(t)]\omega_i h - \sum_i E H_\delta[x - X_i(t)]\omega_i h \right| \\
&\quad + \left| \sum_i E H_\delta(x - X_i(t))\omega_i h - \int_{\mathbb{R}} H_\delta(x-y)\omega(y,t)dy \right| \\
&\quad + \left| \int_{\mathbb{R}} H_\delta(x-y)\omega(y,t)dy - \int_{\mathbb{R}} H(x-y)\omega(y,t)dy \right| \\
&= \text{sampling error} + \text{discretization error} + \text{moment error},
\end{aligned}
$$

where we have used the identities

$$
u(x,t) = \int_{-\infty}^{x} \omega(y,t)dy = \int_{\mathbb{R}} H(x-y)\omega(y,t)dy,
$$

$$
\begin{aligned}
\int_{\mathbb{R}} H_\delta(x-y)\omega(y,t)dy &= \int_{\mathbb{R}}\int_{\mathbb{R}} H_\delta(x-y)G(y,t;\alpha,0)\omega(\alpha,0)d\alpha dy \\
&= \int_{\mathbb{R}} \omega(\alpha,0)\left[\int_{\mathbb{R}} H_\delta(x-y)G(y,t;\alpha,0)dy\right]d\alpha \\
&= \int_{\mathbb{R}} E H_\delta[x - X(t;\alpha)]\omega(\alpha,0)d\alpha.
\end{aligned}
$$

The moment error is bounded by

$$
\left| \int_{\mathbb{R}} H_\delta(x-y)\omega(y,t)dy - \int_{\mathbb{R}} H(x-y)\omega(y,t)dy \right| \le C\delta^p.
$$

This is a special case of Lemma 6.5, which we prove in Section 6.4. To estimate the discretization error we require the following lemma on the error associated with trapezoidal quadrature. We prove the result in two-space dimensions as we will use it later in this chapter. The result is the same in one-space dimension. □

Lemma 6.2. *Trapezoidal Quadrature Error. Let $g \in C_0^r(\mathbb{R}^2)$, $r \ge 3$. Then*

$$
\left| \int_{\mathbb{R}^3} g(x)dx - \sum_{j\in\Lambda^h} g(jh)h^2 \right| \le c_r \|g\|_{W^{r,1}(\mathbb{R}^2)} h^r, \tag{6.29}
$$

where $\|\cdot\|_{W^{r,1}}$ is the Sobolev norm.

Proof of Lemma 6.2. By the Poisson summation formula (Stein and Weiss, 1971, p. 252),

$$
h^2 \sum_{j\in\Lambda^h} g(jh) = \sum_{j\in\Lambda^h} \hat{g}\left(\frac{j}{h}\right),
$$

we have

$$\left|\int_{\mathbb{R}^2} g(x)dx - \sum_{j\in\Lambda^h} g(jh)h^2\right| = \left|\hat{g}(0) - \sum_{j\in\Lambda^h} \hat{g}\left(\frac{j}{h}\right)\right| = \left|\sum_{\substack{j\in\Lambda^h\\ j\neq 0}} \hat{g}\left(\frac{j}{h}\right)\right|.$$

For $g \in C_0^r(\mathbb{R}^2)$ we have

$$|\hat{g}(\xi)| \leq \frac{c\|g\|_{W^{r,1}}}{(1+|\xi|)^r} \leq \frac{c\|g\|_{W^{r,1}}}{|\xi|^r}, \qquad \xi \neq 0,$$

$$\sum_{\substack{j\in\Lambda^h\\ j\neq 0}} \frac{1}{|j|^r} = \sum_{\substack{k\geq 1\\ |j|=k}} \frac{1}{|j|^r} = \sum_{k\geq 1} \frac{8k}{k^r} < \infty \qquad \text{for} \quad r \geq 3.$$

Hence

$$\left|\sum_{\substack{j\in\Lambda^h\\ j\neq 0}} \hat{g}\left(\frac{j}{h}\right)\right| \leq c\|g\|_{W^{r,1}} \sum_{\substack{j\in\Lambda^h\\ j\neq 0}} \left|\frac{h}{j}\right|^r \leq c\|g\|_{W^{r,1}} h^r, \qquad r \geq 3. \qquad \square$$

Using the trapezoidal quadrature lemma and the fact that $dH_\delta/dx = \psi_\delta$, we obtain

$$\left|\sum_i EH_\delta[x - X_i(t)]\omega_i h - \int_{\mathbb{R}} EH_\delta[x - X(t;\alpha)]\omega(\alpha,0)d\alpha\right|$$

$$\leq C'h^L\left\{\sum_{0\leq\gamma\leq L}\int_{|x-y|\leq R}\left|\frac{d^\gamma H_\delta}{dx^\gamma}(x-y)\right|dy + \sum_{0\leq\gamma\leq L}\sup_{|x-y|>R}\left|\frac{d^\gamma H_\delta}{dx^\gamma}(x-y)\right|\right\}$$

$$\leq C'h^L\{C''\delta^{1-L} + C'''\}$$

$$\leq C(h/\delta)^L\delta,$$

where we choose $R = \max[1, \text{diameter}(\Omega)]$ and the constant C depends on only γ, T, L, m, the diameter of $\operatorname{supp}\omega(\alpha, 0)$, and $\max_{0\leq|\beta|\leq L}\|\partial^\beta\omega(\alpha,0)\|_{L^\infty}$.

We begin to estimate the sampling error. Let

$$Z_i = \omega_i h[H_\delta(x - X_i(t)) - EH_\delta(x - X_i(t))].$$

We have $EZ_i = 0$, $|Z_i| \leq C_1 h$, and the sum of variances

$$\sum_i \operatorname{Var} Z_i = h\sum_i \omega_i^2 h \cdot \operatorname{Var} H_\delta(x - X_i(t))$$

$$= h\sum_i \omega_i^2 h \cdot \left\{E[H_\delta(x - X_i(t))]^2 - [EH_\delta(x - X_i(t))]^2\right\}. \qquad (6.30)$$

Likewise, we can apply Lemma 6.2 to approximate Eq. 6.30 by the integral

$$h\int_{\mathbb{R}} \operatorname{Var} H_\delta[x - X(t;\alpha)]\omega^2(\alpha,0)d\alpha \qquad (6.31)$$

within an error $C'(h/\delta)^L\delta$. Integral (6.31) is bounded by $C\sigma$ for $0 \le \sigma \le \sigma_0$, $0 \le \delta \le \delta_0$ because Var $H_\delta[x - X(t;\alpha)]$ depends smoothly on σ and δ and it is identically zero when $\sigma = 0$. Therefore

$$\sum_i \mathrm{Var} Z_i \le h\big[C\sigma + C'(h/\delta)^L\delta\big] \le C_2 h\big[\sigma + (h/\delta)^L\delta\big] := V.$$

For any fixed constant $\kappa > 0$, let

$$\eta = \kappa C_2^{1/2} h^{1/2}\big[\sigma + (h/\delta)^L\delta\big]^{1/2}|\ln h|.$$

Because $[\sigma + (h/\delta)^L\delta]^{1/2} \le \sigma^{1/2} + (h/\delta)^{L/2}\delta^{1/2}$, it follows from Bennett's inequality that

$$\begin{aligned}
P\left\{\left|\sum_i Z_i\right| \ge \kappa C_2^{1/2} h^{1/2}(\sigma^{1/2} + (h/\delta)^{L/2}\delta^{1/2})\right\}
&\le P\left\{\left|\sum_i Z_i\right| \ge \eta\right\} \\
&\le 2\exp\left[-\frac{1}{2}\eta^2 V^{-1}\cdot B(C_1 h\eta V^{-1})\right] \\
&\le 2\exp\left(-\frac{1}{2}\kappa^2|\ln h|^2\cdot B\big\{C_1 C_2^{-1/2}\kappa h^{1/2}\big[\sigma + (h/\delta)^L\delta\big]^{-1/2}|\ln h|\big\}\right) \\
&\le 2\exp\left[-\frac{1}{2}\kappa^2|\ln h|^2\cdot B(C'\kappa|\ln h|)\right] \\
&\le 2\exp\left(-C''\kappa^2|\ln h|^2\cdot \kappa^{-1}|\ln h|^{-1}\right) \\
&\le 2h^{C''\kappa} \\
&\le 2N^{-C\kappa}
\end{aligned}$$

as $h \longrightarrow 0$. The probability approaches zero more rapidly than any polynomial rate if κ is chosen sufficiently large.

In the case $h/\sigma \longrightarrow \infty$, except for a factor of $|\ln h|$,

$$B(M\eta V^{-1}) \sim h^{-1/2}\big[\sigma + (h/\delta)^L\delta\big]^{1/2}, \tag{6.32}$$

which goes to zero. To ensure that the probability goes to one, we divide the sampling error in relation (6.26) by approximation (6.32).

6.1.3. Computational Performance of the 2D Random-Vortex Method on Viscous Strained Shear Layers

We now review some computations presented in Long (1987). The first computation takes initial vorticity as the Dirac measure at 0. Because the initial data are not smooth, it makes sense to use u_{pv} to simulate the dynamics. Long considers variances $\sigma^2 = 10^{-2}, 10^{-3}, 10^{-4}, 10^{-5}$, and 10^{-6}, and computes the L^1 and the L^∞ norms. The

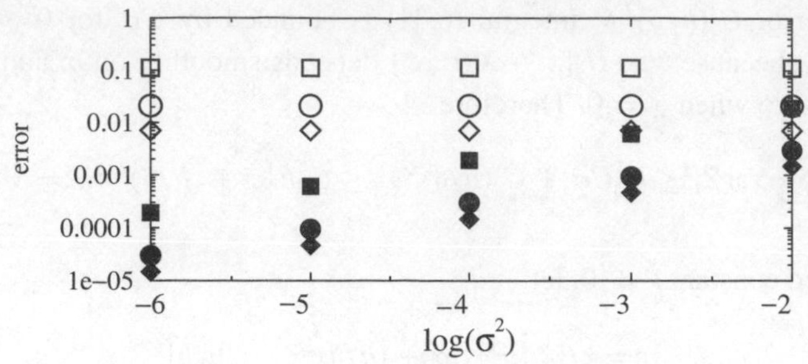

Figure 6.1. No smoothing, Dirac initial data. Shown are computations of the numerical error for $N = 100$ (squares), $N = 1000$ (circles) and $N = 10{,}000$ (diamonds). The L^∞ error is indicated by the open symbols and is independent of σ, and the L^1 error, denoted by the filled symbols, scales as σ.

advantage of using the Dirac measure as the initial data is that there are no moment and discretization errors. The only error is the sampling error. In this case, the exact solutions with different σ are similar to one another by a simple scaling. The same is true for the computed solutions. Note from Fig. 6.1 that the L^∞ error is the same for different σ whereas the L^1 error scales as σ. This is because the essential support of the vorticity scales like 8σ. The constancy of the L^∞ norm is due to the fact that the largest variance of u_{pv} occurs at $x = 0$ where its value of $1/4N$ is independent of σ. Figure 6.1 also shows that the sampling error scales as $N^{-1/2} \sim h^{1/2}$.

The second computation takes step-function initial vorticity

$$\omega(x, 0) = \begin{cases} 1/2 & \text{if } -1 \le x \le 1 \\ 0 & \text{otherwise} \end{cases}. \tag{6.33}$$

In this case the initial data are piecewise Lipschitz continuous so we expect the L^∞ error to show dependence on σ. This is clearly shown in Fig. 6.2, top graph, where we see that the error scales as $\sigma^{1/2}$, as predicted. Note, however, that the degree of σ dependence decreases as the straining rate γT increases. The variation in the L^∞ errors for $\gamma T = 4.0$ (down triangle) is much smaller than that at $\gamma T = 0.1$ (circle). The reason for this is that, in all cases, the solutions converge at long times to the same equilibrium. Note from Fig. 6.2 that the L^1 error is not affected by γT as we might expect. For fixed σ, the L^1 error decreases as γT increases because the essential support of the vorticity decreases. The data also support the theoretical result that the error is of the order of $h^{1/2}$.

We now discuss the effects of smoothing. If we take given piecewise smooth initial data, for example, Eq. (6.33), and use a higher-order smoothing, then as the smoothing parameter $\delta = h^q$ increases, the moment error increases while the discretization error and the sampling error decrease. The optimal result is reached when the three effects balance. As an example, Long showed that for initial data (6.33), $N = 10{,}000$, $\gamma T = 1.0$, there was an optimal q at which the L^∞ and the L^1 errors reached a minimum. For the L^∞ norm it was $q = 0.82$ whereas for the L^1 norm it was $q = 0.76$. He states that similar effects were found for other cases as well.

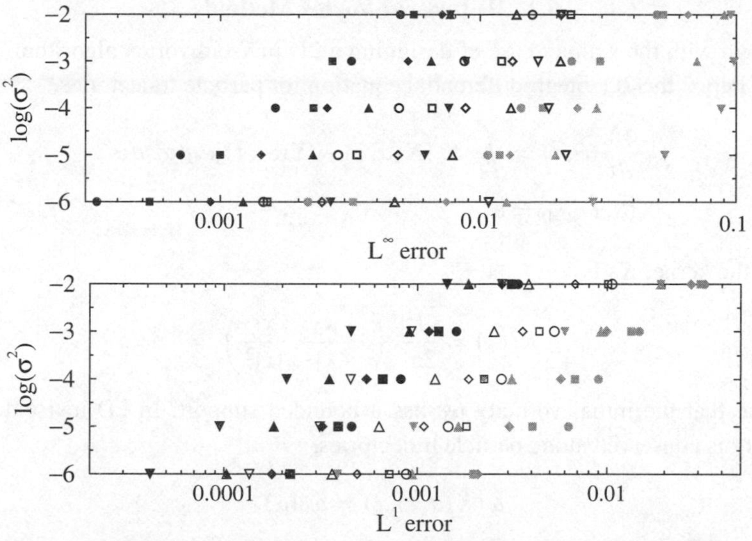

Figure 6.2. No smoothing, step initial data (6.33). Shown are computations of the L^∞ error (top) and the L^1 error (bottom) for $N = 100$ (shaded symbols), $N = 1000$ (open symbols) and $N = 10{,}000$ (filled symbols). The errors at $\gamma T = 0.1$, 0.5, 1.0, 2.0, and 4.0 are shown as circles, squares, diamonds, up triangles, and down triangles, respectively.

Long compares his computations with no smoothing to those with smoothing. He finds that for the cases $N = 100$ and $N = 1000$ smoothing does not give improved errors. However, the smoothing does show an improvement over no smoothing for $N = 10{,}000$. This supports the theoretical result in Theorem 6.2 that the grid size h has to be sufficiently small for the effect of smoothing to be observable. Figure 6.3 shows $N = 10{,}000$ with second-order smoothing and step initial data (6.33). Note that the most significant improvement of the accuracy occurs when $\sigma^2 = 10^{-2}$ and the improvement diminishes as σ decreases. This illustrates the fact that the condition $h \le C_0\sigma$ is significant. At $\sigma^2 = 10^{-6}$, the ratio $h/\sigma = 0.1$. This suggests that the practical value of C_0 is very small. Small values of h/σ suggest that the sampling error becomes the dominant component.

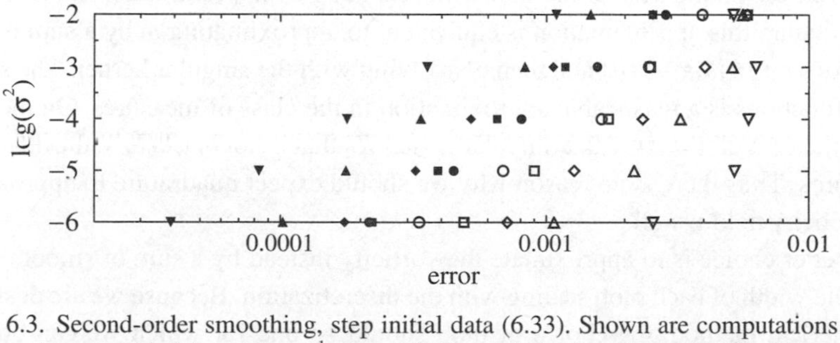

Figure 6.3. Second-order smoothing, step initial data (6.33). Shown are computations of the L^∞ error (open symbols) and the L^1 error (filled symbols) for $N = 10{,}000$. The errors at $\gamma T = 0.1$, 0.5, 1.0, 2.0, and 4.0 are shown as circles, squares, diamonds, up triangles, and down triangles, respectively.

6.2. 2D Inviscid Vortex Methods

We begin with the simpler case of designing a 2D inviscid vortex algorithm. Recall from Chap. 2 the 2D integrodifferential equation for particle trajectories,

$$\frac{dX}{dt}(\alpha, t) = \int_{\mathbb{R}^2} K_2[X(\alpha, t) - X(\alpha', t)]\omega_0(\alpha')d\alpha', \tag{6.34}$$

$$X(\alpha, t)|_{t=0} = \alpha, \tag{6.35}$$

where the kernel K_2 is

$$K_2(x) = \frac{1}{2\pi}^t\left(-\frac{x_2}{|x|^2}, \frac{x_1}{|x|^2}\right). \tag{6.36}$$

Assume that the initial vorticity ω_0 has a bounded support. In 2D inviscid flows, vorticity is conserved along particle trajectories,

$$\omega(X(\alpha, t), t) = \omega_0(\alpha), \tag{6.37}$$

and hence the design of a 2D vortex method for Eq. (6.35) does not involve the stretching of vorticity as in the 3D case.

The following subsections address the three main steps in the 2D vortex method: Subsection 6.2.1, smoothing of the kernel K_2; Subsection 6.2.2, approximation of the integral; and Subsection 6.2.3, closing of a finite system of ODEs.

6.2.1. Smoothing of the Kernel K_2

First we motivate the need for the smoothing of the kernel K_2. In constructing a discretization of the equations, we need to approximate the velocity field

$$v(x, t) = \int_{\mathbb{R}^2} K_2(x - x')\omega(x')dx'. \tag{6.38}$$

An approximation by trapezoidal quadrature would be

$$v(x, t) \sim \sum_{j\in\mathbb{Z}^n} K_2(x - jh)\omega(jh)h^2. \tag{6.39}$$

However, the singularity in the kernel makes such an approximation a bad choice. In particular, this approximation is equivalent to approximating ω by a sum of delta functions $\omega(jh)\delta(x - jh)$ and then convolving with the singular kernel. The sum of delta functions is a reasonable approximation in the class of measures. On the other hand, the convolution is with a singular kernel K_2 that is not in $C^0(\mathbb{R}^2)$, the dual to the measures. Thus there is no reason why we should expect quadrature to approximate the velocity field v well.

A better choice is to approximate the vorticity instead by a sum of smooth blobs, with the width of each blob scaling with the discretization. Because we are designing a numerical method, the choice of blob should be one for which we can obtain a simple closed-form expression for the approximate velocity field.

We now show that an approximation of the vorticity by smooth blobs is equivalent to a smoothing of the kernel K_2.

Consider a radially symmetric function $\psi \in C_0^\infty$ such that $\int_{\mathbb{R}^2} \psi \, dx = 1$ and define the mollifier $\psi_\delta(x) = \delta^{-2}\psi(x/\delta)$, $\delta > 0$. Regularize the kernel K_2 by convolving it componentwise with ψ_δ:

$$K_2^\delta(x) = K_2 * \psi_\delta(x). \tag{6.40}$$

This regularized kernel K_2^δ is smooth and bounded at $x = 0$. Using it instead of the singular kernel K_2 gives the *regularized integrodifferential equation* for the particle trajectories $X^\delta(\alpha, t)$:

$$\begin{aligned} \frac{dX^\delta}{dt}(\alpha, t) &= \int_{\mathbb{R}^2} K_2^\delta[X^\delta(\alpha, t) - X^\delta(\alpha', t)]\omega_0(\alpha')d\alpha', \\ X^\delta(\alpha, t)|_{t=0} &= \alpha. \end{aligned} \tag{6.41}$$

We show an equivalent interpretation of regularization (6.40) that gives a simpler formula for computing K_2^δ. Recall that the velocity v and its vorticity ω satisfy the equations

$$\operatorname{curl} v = \omega, \qquad \operatorname{div} v = 0. \tag{6.42}$$

For a radially symmetric vorticity ω, the solution v to this equation is the inviscid steady eddy from Example 2.1:

$$v(x) = {}^t\left(-\frac{x_2}{|x|^2}, \frac{x_1}{|x|^2}\right) \int_0^r s\omega(s)ds.$$

Observe that regularization (6.40) implies that the regularized kernel K_2^δ satisfies Eqs. (6.42),

$$\operatorname{curl} K_2^\delta = \psi_\delta, \qquad \operatorname{div} K_2^\delta = 0,$$

so that

$$\begin{aligned} K_2^\delta(x) &= {}^t\left(-\frac{x_2}{|x|^2}, \frac{x_1}{|x|^2}\right) \int_0^r s\psi_\delta(s)ds \\ &= \frac{1}{2\pi}\,{}^t\left(-\frac{x_2}{|x|^2}, \frac{x_1}{|x|^2}\right) 2\pi \int_0^{r/\delta} s\psi(s)ds. \end{aligned}$$

Thus we have the first equivalent interpretation: *Smoothing (6.40) of the kernel K_2 is equivalent to cutting off the singularity of K_2,*

$$K_2^\delta(x) = K_2(x) f\left(\frac{r}{\delta}\right), \tag{6.43}$$

where the cutoff function f is related to the mollifier ψ by

$$f(r) = 2\pi \int_0^r s\psi(s)ds. \tag{6.44}$$

The definition of ψ implies that the cutoff function f satisfies

(i) $f(r) = \mathcal{O}(r)$ for $r \ll 1$ to cut off the singularity r^{-1} of K_2, and
(ii) $f(r) \nearrow 1$ as $r \nearrow \infty$ to satisfy the condition $\int_{\mathbb{R}^2} \psi \, dx = 1$.

Two examples of cutoffs are

$$\psi(r) = \frac{1}{2\pi}\frac{1}{(1+r^2)^2 r}, \qquad f(r) = \frac{r^2}{1+r^2}, \tag{6.45}$$

discussed in Chap. 9 for the desingularization of vortex-sheet motion, and

$$\psi(r) = \frac{1}{\pi}e^{-r^2}, \qquad f(r) = 1 - e^{-r^2}, \tag{6.46}$$

which gives

$$K_2^\delta(x) = \frac{1 - e^{-(\frac{r}{\delta})^2}}{2\pi r^2}(-x_2, x_1)^T. \tag{6.47}$$

There is a second equivalent interpretation of smoothing (6.40) of the kernel K_2. Because convolution is associative, $(F * G) * H = F * (G * H)$, the velocity v^δ corresponding to K_2^δ is

$$v^\delta(x,t) = \int_{\mathbb{R}^2} K_2^\delta(x-y)\omega(y,t)dy \tag{6.48}$$

$$= \int_{\mathbb{R}^2} K_2(x-y)\omega^\delta(y,t)dy, \tag{6.49}$$

where ω^δ is the regularized vorticity:

$$\omega^\delta(x) = \omega * \psi_\delta(x). \tag{6.50}$$

The second equivalent interpretation is that *Smoothing (6.40) of the kernel K_2 is equivalent to smoothing (6.50) of the vorticity ω.*

6.2.2. Approximation of the Integral

To approximate the integral in Eq. (6.41), we can use any quadrature technique. For simplicity we consider *trapezoidal quadrature*, defined by

$$\int_{\mathbb{R}^N} g(\alpha)d\alpha \cong \sum_{j\in\mathbb{Z}^N} g(jh)h^N, \tag{6.51}$$

where the multi-index $j = (j_1, j_2, \dots, j_N)$ defines a uniform grid of points jh, $h > 0$. If the vorticity has a compact support, we need a finite number of grid points, $jh \in \operatorname{supp}\omega_0$, $j \in \Lambda^h$. Use of quadrature (6.51) to discretize regularized equation (6.41) gives an approximate velocity field at the point $X^{\delta h}(\alpha, t)$:

$$v^{\delta h}(X^{\delta h}(\alpha,t),t) = (\alpha,t) \cong \sum_{j\in\Lambda^h} K_2^\delta\big[X^{\delta h}(\alpha,t) - X_j^{\delta h}(t)\big]\omega_{0_j}h^2. \tag{6.52}$$

6.2.3. Closing of the Equation by means of a Finite-Dimensional System of ODEs

To completely discretize Eq. (6.52) we use the collocation method of evaluating this equation at the points $\{X_i^{\delta h}(t)\}_{i\in\Lambda^h}$. In such a way we approximate the particle-trajectory mapping $X(\cdot, t)$ by a finite system of particle trajectories $\{X_j^{\delta h}(t)\}_{j\in\Lambda^h}$ that satisfies the system of ODEs that defines the 2D inviscid-vortex method:

$$\frac{dX_i^{\delta h}}{dt}(t) = \sum_{j\in\Lambda^h} K_2^\delta\left[X_i^{\delta h}(t) - X_j^{\delta h}(t)\right]\omega_{0_j} h^2,$$

$$X_i^{\delta h}(t)|_{t=0} = ih, \qquad i \in \Lambda^h. \tag{6.53}$$

The approximate velocity field corresponding to the particle trajectories $X_i^{\delta h}$ is

$$v^{\delta h}(x,t) = \sum_{j\in\Lambda^h} K_2^\delta\left[x - X_j^{\delta h}(t)\right]\omega_{0_j} h^2, \tag{6.54}$$

and the approximate vorticity is

$$\omega^{\delta h}(x,t) = \sum_{j\in\Lambda^h} \psi_\delta\left[x - X_j^{\delta h}(t)\right]\omega_{0_j} h^2. \tag{6.55}$$

Note that the regularized kernel K_2^δ produces a divergence-free velocity $v^{\delta h}$ in Eq. (6.54). Also, $\omega^{\delta h}$ satisfies the compatibility condition div $\omega^{\delta h} = 0$.

Formula (6.55) for the approximate vorticity $\omega^{\delta h}$ illustrates the dual interpretations of the regularization of K_2. Namely, if we approximate the vorticity $\omega(x,t)$ by smooth *vortex blobs* $\psi_\delta[x - X_j^{\delta h}(t)]\omega_{0_j} h^2$ as in Eq. (6.55) and interpret formula (6.54) for the velocity $v^{\delta h}$ as the Riemann sum for Eq. (6.49) over the partitions formed by current images of initial squares, we get

$$\int_{\mathbb{R}^2} K_2[x - X(\alpha',t)]\omega_0(\alpha')d\alpha' \cong \sum_{j\in\Lambda^h} K_2^\delta\left[x - X_j^{\delta h}(t)\right]\omega_{0_j} h^2.$$

For these reasons algorithm (6.53) is often called the *vortex-blob formulation*.

Inviscid-vortex algorithm (6.53) depends on two parameters, the regularization parameter $\delta > 0$ and the discretization parameter $h > 0$, and on the function ψ (or f). Setting $\delta = 0$ gives the point-vortex method that has been shown to have poor convergence as $h \searrow 0$ (Beale and Majda, 1985). The convergence of approximate solutions to the vortex method in algorithm (6.53) to exact solutions depends on the relation between δ and h and properties of the function ψ (or f). We discuss this rigorously in Section 6.4.

6.3. 3D Inviscid-Vortex Methods

3D inviscid-vortex methods are based on 3D integrodifferential equation (4.6) for the particle trajectories:

$$\frac{dX}{dt}(\alpha,t) = \int_{\mathbb{R}^3} K_3[X(\alpha,t) - X(\alpha',t)]\nabla_\alpha X(\alpha',t)\omega_0(\alpha')d\alpha',$$

$$X(\alpha,t)|_{t=0} = \alpha, \tag{6.56}$$

$$K_3(x) = \frac{1}{4\pi}\frac{x}{|x|^3}\times. \tag{6.57}$$

We assume that the initial vorticity ω_0 has a compact support. The additional challenge in designing a 3D vortex method lies in the treatment of the stretching of vorticity given by the term

$$\omega(X(\alpha,t),t) = \nabla_\alpha X(\alpha,t)\omega_0(\alpha).$$

In Subsection 6.3.3 we consider both a Lagrangian and an Eulerian method for computing the stretching of vorticity. We begin with a discussion of the smoothing of the kernel K_3 in Subsection 6.3.1, and then in Subsection 6.3.2 we briefly discuss the discretization of the integral. In Subsection 6.3.4 we finish with a discussion of the collocation method used to create a finite system of ODEs.

6.3.1. Smoothing of the Kernel K_3

As in the 2D case we consider a radially symmetric function $\psi \in C_0^\infty$, $\int_{\mathbb{R}} \psi\, dx = 1$, which defines the mollifier $\psi_\delta(x) = \delta^{-3}(x/\delta)$, $\delta > 0$. Regularize the kernel K_3 by convolving it componentwise with ψ_δ:

$$K_3^\delta(x) = K_3 * \psi_\delta(x). \tag{6.58}$$

This regularized kernel K_3^δ is smooth and bounded at $x = 0$ and defines the *regularized 3D integrodifferential equation* for the particle trajectories $X^\delta(\alpha, t)$:

$$\frac{dX^\delta}{dt}(\alpha,t) = \int_{\mathbb{R}} K_3^\delta[X^\delta(\alpha,t) - X^\delta(\alpha',t)]\omega(X^\delta(\alpha',t),t)d\alpha',$$
$$X^\delta(\alpha,t)|_{t=0} = \alpha, \tag{6.59}$$

where ω is also an unknown function to be determined.

As in the 2D case, there is a dual interpretation of regularization (6.58) as a cutoff of the singularity r^{-2} of K_3. Later we use cutoff functions rather than mollifiers ψ_δ to derive explicit forms of regularized kernels K_3^δ.

Recall that the Hodge decomposition (Proposition 2.16) gives

$$v = K_3 * \omega \equiv \operatorname{curl}(G_3 * \omega),$$

where $G_3(r) = [1/(4\pi r)]$ is the Newtonian potential in $\mathbb{R}^3$. A simple calculation shows that

$$K_3^\delta * \omega(x) = \int_{\mathbb{R}^3} \frac{\partial}{\partial r} G_3^\delta(|x-y|)\frac{x-y}{|x-y|} \times \omega(y)dy,$$

where $G_3^\delta = G_3 * \psi_\delta$. Thus

$$K_3^\delta(x) = \frac{\partial}{\partial r} G_3^\delta(|x|)\frac{x}{|x|}\times,$$

and K_3^δ will have a simple expression whenever $(\partial/\partial r)G_3^\delta$ does. In particular, if

$$\frac{\partial}{\partial r} G_3^\delta(r) = -\frac{f\left(\frac{r}{\delta}\right)}{4\pi r^2},$$

where f is a cutoff function, then the kernel K_3^δ is

$$K_3^\delta(x) = -\frac{f\left(\frac{r}{\delta}\right)}{4\pi r^3} \times x. \tag{6.60}$$

Because $\Delta G_3^\delta = \psi_\delta$, in spherical coordinates we have

$$\Delta G_3^\delta(x) = \frac{1}{r^2}\frac{\partial}{\partial r}\left[r^2\frac{\partial}{\partial r}G_3^\delta(r)\right] = -\frac{f'\left(\frac{r}{\delta}\right)}{4\pi r^2}\delta^{-1}.$$

Hence the relation between the mollifier ψ and the cutoff function f is

$$\psi(r) = -\frac{f'(r)}{4\pi r^2}. \tag{6.61}$$

The above discussion shows that the smooth cutoff function f must satisfy (i) $f(r) = \mathcal{O}(r^2)$ for $r \ll 1$ to cut off the singularity r^{-2} of K_3, and (ii) $f(r) \nearrow 1$ as $r \nearrow \infty$ to satisfy the condition $\int_{\mathbb{R}^3} \psi\, dx = 1$.

For example, a simple choice $f(r) = 1 - e^{-r^3}$ leads to $\psi(r) = [3/(4\pi)]e^{-r^3}$, and

$$K_3^\delta(x) = -\frac{1 - e^{-\left(\frac{r}{\delta}\right)^3}}{4\pi r^3} \times x. \tag{6.62}$$

6.3.2. Approximation of the Integral

As in the 2D case, we use trapezoidal quadrature (6.51),

$$\int_{\mathbb{R}^N} g(\alpha)d\alpha \cong \sum_{j\in\Lambda^h} g(jh)h^N,$$

for computing the integrals in the regularized integrodifferential equation for the particle trajectories $X^\delta(\alpha, t)$ in Eq. (6.59):

$$\frac{dX^{\delta h}}{dt}(\alpha, t) = \sum_{j\in\Lambda^h} K_3^\delta\left[X^{\delta h}(\alpha, t) - X_j^{\delta h}(t)\right]\omega\left[X_j^{\delta h}(t), t\right]h^3,$$

$$X^{\delta h}(\alpha, t)|_{t=0} = \alpha, \tag{6.63}$$

where the finite set Λ^h is defined by $jh \in \operatorname{supp}\omega_0$ for $j \in \Lambda^h$.

Unlike in the 2D case, the vorticity $\omega(X_j^{\delta h}(t), t)$ is as yet unknown. The next subsection gives two methods for approximating the vorticity.

6.3.3. Two Approximations of the Vorticity

Approximation 1: Lagrangian Stretching

Recall from Chap. 2 that in the 3D integrodifferential equation for the particle trajectories, the vorticity ω is determined from the initial condition ω_0 and the particle trajectories $X(\alpha, t)$ by a Lagrangian stretching:

$$\omega(X(\alpha, t), t) = \nabla_\alpha X(\alpha, t)\omega_0(\alpha). \tag{6.64}$$

This stretching term can lead to amplification of the vorticity field that is impossible in two dimensions. In Chap. 3 we showed, by using the particle-trajectory equations, a direct link between vorticity amplification and the smoothness of a solution to the Euler equation (Theorem 4.3). The vorticity stretching introduces another term that must be discretized when a 3D computational vortex method is being constructed.

The first approximation of the vorticity that we consider comes from simply approximating the gradient ∇_α in Eq. (6.64) by a finite-difference operator ∇_α^h based on the grid points jh, $j \in \Lambda^h$:

$$\frac{\partial X_i}{\partial \alpha_j}(\alpha, t) \cong [X_i(\alpha + he_j, t) - X_i(\alpha - he_j, t)]/2h. \tag{6.65}$$

This leads to the approximation of the vorticity $\omega(X_j(t), t)$ by $\omega_j^h(t)$ given by

$$\omega_j^h(t) = \nabla_\alpha^h X_j(t)\omega_{0_j}, \tag{6.66}$$

so that we determine ω_j^h in terms of its initial values ω_{0_j} and the particle trajectories $X_j(t)$, $j \in \Lambda^h$.

Approximation 2: Eulerian Stretching

Recall that vorticity-stretching formula (6.64) is actually derived from 3D vorticity equation (1.32), $(D\omega/Dt) = \omega \cdot \nabla v$. Evaluating this equation on the particle trajectories $X(\alpha, t)$ gives

$$\frac{d}{dt}\omega(X(\alpha, t), t) = \omega(X(\alpha, t), t) \cdot \nabla_x v|_{(X(\alpha,t),t)}. \tag{6.67}$$

We can compute the vorticity $\omega(X(\alpha, t), t)$ by a direct approximation of this ODE, given the velocity field v corresponding to the particle trajectories $X_j(t)$. That is, we compute $\omega_i^h \cong \omega(X_j(t), t)$ as the solutions to

$$\frac{d\omega_i^h}{dt}(t) = \omega_i^h(t) \cdot \nabla_x v|_{(X_i(t),t)},$$
$$\omega_i^h(t)|_{t=0} = \omega_{0_i}. \tag{6.68}$$

6.3.4. Closing of the Equations by means of a Finite-Dimensional System of ODEs

We formulate the 3D inviscid-vortex methods corresponding to the two different approximations of the vorticity ω discussed in Subsection 6.2.3.

To get a discrete version of Eq. (6.68), we use the collocation method, evaluating this equation at the points $\{X_i^{\delta h}(t)\}_{i \in \Lambda^h}$:

$$\frac{dX_i^{\delta h}}{dt}(t) = v^{\delta h}\left(X_i^{\delta h}(t), t\right), \tag{6.69}$$

where

$$v^{\delta h}(x,t) = \sum_{j\in\Lambda^h} K_3^\delta\big[x - X_j^{\delta h}(t)\big]\omega\big[X_j^{\delta h}(t),t\big]h^3. \tag{6.70}$$

The regularized kernel K_3^δ always produces a divergence-free velocity field $v^{\delta h}$ [Eq. (6.70)]. Equations (6.69) and (6.70) are not closed because the vorticity $\omega(X_j^{\delta h}(t),t)$ is unknown – here we use the two different approximations, introduced in Subsection 6.3.3, to close these equations.

Vorticity approximation (6.66) gives the 3D inviscid-vortex method with Lagrangian stretching:

$$\frac{dX_i^{\delta h}}{dt}(t) = \sum_{j\in\Lambda^h} K_3^\delta\big[X_i^{\delta h}(t) - X_j^{\delta h}(t)\big]\nabla_\alpha^h X_j^{\delta h}(t)\omega_{0_j}h^3,$$

$$X_i^{\delta h}(t)|_{t=0} = ih, \qquad i\in\Lambda^h. \tag{6.71}$$

Because the finite-difference operator ∇_α^h acts on only the points $\{X_j^{\delta h}(t)\}_{j\in\Lambda^h}$, the above system of ODEs is closed. The corresponding approximate velocity $v^{\delta h}$ and vorticity $\omega^{\delta h}$ are

$$v^{\delta h}(x,t) = \sum_{j\in\Lambda^h} K_3^\delta\big[x - X_j^{\delta h}(t)\big]\nabla_\alpha^h X_j^{\delta h}(t)\omega_{0_j}h^3, \tag{6.72}$$

$$\omega^{\delta h}(x,t) = \sum_{j\in\Lambda^h} \nabla_\alpha^h X_j^{\delta h}(t)\omega_{0_j}h^3. \tag{6.73}$$

To obtain a second 3D vortex method we approximate the vorticity ω by solving Eq. (6.68). Given an explicit formula for K_3^δ, the gradient of the approximate velocity in Eq. (6.70) is precisely

$$\nabla v^{\delta h}(x,t) = \sum_{j\in\Lambda^h} \nabla K_3^\delta\big[x - X_j^{\delta h}(t)\big]\omega\big(X_j^{\delta h}(t),t\big)h^3.$$

Substituting this approximate velocity gradient into Eq. (6.68) gives a 3D vortex method with Eulerian stretching:

$$\frac{dX_i^{\delta h}}{dt}(t) = \sum_{j\in\Lambda^h} K_3^\delta\big[X_i^{\delta h}(t) - X_j^{\delta h}(t)\big]\omega_j^{\delta h}(t)h^3,$$

$$\frac{d\omega_i^{\delta h}}{dt}(t) = \omega_i^{\delta h}(t)\cdot\sum_{j\in\Lambda^h} \nabla K_3^\delta\big[X_i^{\delta h}(t) - X_j^{\delta h}(t)\big]\omega_j^{\delta h}(t)h^3,$$

$$X_i^{\delta h}(t)|_{t=0} = ih, \qquad \omega_i^{\delta h}(t)|_{t=0} = \omega_{0_i}, \qquad i\in\Lambda^h. \tag{6.74}$$

This is a closed system of equations with six unknowns $X_i^{\delta h}(t)$, $\omega_i^{\delta h}(t)$, for each $i\in\Lambda^h$.

Recall that in the 2D inviscid-vortex method approximation (6.55) of the vorticity by smooth vorticity blobs was equivalent to representing the velocity field by a smoothing and discretizing the kernel K_2 as in Eq. (6.54). Because of the stretching of vorticity in three dimensions, the most natural approximation of vorticity is different. We observe that the approximate vorticity $\omega_i^{\delta h}(t)$ does not define vortex lines in the flow, but still it is divergence free, as required from the exact vorticity ω.

As in the 2D case, the 3D inviscid-vortex methods depend on two parameters, the regularization parameter $\delta > 0$ and the discretization parameter $h > 0$, and on the function ψ(or f). The convergence of approximate solutions to the vortex methods in Eqs. (6.71) and (6.74) to exact solutions depends on the relation between δ and h and properties of the function ψ (or f).

6.4. Convergence of Inviscid-Vortex Methods

In this section we study the convergence of the inviscid-vortex method. Because the method involves both the discretization parameter h and the regularization parameter δ, we expect that δ depends on h. We also expect that the convergence depends on the mollifiers ψ used to desingularize the kernel K in the velocity formula $v = K * \omega$. We state both 2D and 3D convergence results. We present the proof of the 2D convergence. The 3D proof is extremely technical. The interested reader is referred to the many sources mentioned in the notes at the end of this chapter, in particular, to Beale and Majda (1982a, 1982b).

Recall that the design of inviscid-vortex methods is based on the integrodifferential equation on particle trajectories $X(\alpha, t)$, which for 2D flows is

$$\frac{dX}{dt}(\alpha, t) = \int_{\mathbb{R}^2} K[X(\alpha, t) - X(\alpha', t)]\omega_0(\alpha')d\alpha',$$
$$X(\alpha, t)|_{t=0} = \alpha \in \mathbb{R}^2, \tag{6.75}$$

with the kernel K given by

$$K(x) = \frac{1}{2\pi}{}^t\left(-\frac{x_2}{|x|^2}, \frac{x_1}{|x|^2}\right), \qquad x \in \mathbb{R}^2. \tag{6.76}$$

The velocity v is

$$v(x, t) = \int_{\mathbb{R}^2} K[x - X(\alpha', t)]\omega_0(\alpha')d\alpha'. \tag{6.77}$$

The 2D inviscid vortex methods consisted of (see Section 6.2)

(i) a desingularization of the kernel K by convolving it with a mollifier ψ_δ, $K^\delta = K * \psi_\delta$,
(ii) an approximation of the integral in Eq. (6.75) by trapezoidal quadrature,
(iii) closing of a finite-dimensional system of ODEs by the collocation method.

These steps lead to the 2D inviscid-vortex methods given in Eq. (6.53):

$$\frac{d\tilde{X}_i}{dt}(t) = \sum_{j\in\Lambda^h} K^\delta[\tilde{X}_i(t) - \tilde{X}_j(t)]\omega_{0_j} h^2,$$
$$\tilde{X}_i(t)|_{t=0} = ih, \qquad i \in \Lambda. \tag{6.78}$$

Equation (6.78) is a discretization of the particle-trajectory equation. The corresponding approximate velocity $\tilde{v}^{\delta h}$ is

$$\tilde{v}^{\delta h}(x,t) = \sum_{j\in\Lambda^h} K^\delta[x - \tilde{X}_j(t)]\omega_{0_j} h^2, \tag{6.79}$$

and the approximate vorticity $\tilde{\omega}^{\delta h}$ is

$$\tilde{\omega}^{\delta h}(x,t) = \sum_{j\in\Lambda^h} \psi_\delta[x - \tilde{X}_j(t)]\omega_{0_j} h^2. \tag{6.80}$$

Now we formulate the convergence results for the 2D inviscid-vortex method in Eq. (6.78). First we recall from Section 6.1 the class of mollifiers ψ in $M^{L,m}$ used to desingularized the kernel K.

Definition 6.1. *A mollifier ψ belongs to the class $M^{L,m}$, $L \geq 0$, $m \geq 2$, provided that*

(i) $\int_{\mathbb{R}^2} \psi\, dx = 1$,
(ii) $\int_{\mathbb{R}^2} x^\alpha \psi(x) dx = 0$, $1 \leq |\alpha| \leq m-1$, *and* $\int_{\mathbb{R}^2} |x|^m\, |\psi(x)| dx < \infty$,
(iii) $\psi \in C^L(\mathbb{R}^2)$, *and* $|x|^{2+|\beta|}|\partial^\beta \psi(x)| \leq c$, $\forall\, |\beta| \leq L$

Property (ii) of vanishing certain moments of the mollifier ψ is essential to prove the error of an approximation of K by K^δ. The convergence result is the following theorem.

Theorem 6.3. *The Convergence of 2D Inviscid-Vortex Methods. Let $v \in C^1\{[0,T];$ $C^{L+1}(\mathbb{R}^2)\}$, $L \geq 3$, be a solution to the 2D Euler equation for the initial vorticity ω_0 with a compact support. Let $\psi \in M^{L,m}$, $m \geq 2$, and let $\tilde{x}$ and $\tilde{v}^{\delta h}$ be solutions to the inviscid-vortex method in Eqs. (6.78) and (6.79). Then for all $0 \leq t \leq T$ and $1 < p < \infty$, there exists a constant C such that*

$$\|x(\cdot,t) - \tilde{x}(\cdot,t)\|_{L_h^p} \leq C(\delta^m + h^L\delta^{1-L}), \tag{6.81}$$
$$\|v(\cdot,t) - \tilde{v}^{\delta h}(\cdot,t)\|_{L_h^p} \leq C(\delta^m + h^L\delta^{1-L}), \tag{6.82}$$

provided that

$$C(\delta^m + h^L\delta^{1-L}) \leq \frac{1}{2}h\delta. \tag{6.83}$$

The constant c depends on T and v.

We prove this theorem later in this section. In relations (6.81) and (6.82), $\|\cdot\|_{L_h^p}$ denotes the discrete L^p norm:

$$\|f(\cdot,t)\|_{L_h^p} = \left\{ \sum_{j\in\Lambda^h} |f(x(\alpha_j,t),t)|^p h^2 \right\}^{1/p}. \tag{6.84}$$

If we set $\delta = h^q$, $0 < q < 1$, then $\delta^m + h^L \delta^{1-L} = h^{mq} + h^{L+q(1-L)}$. If we choose L large enough so that $mq < L(1-q)+q$, then $c[h^{mq} + h^{L+q(1-L)}] \le ch^{mq}$, $\forall\, 0 < h \le h_0$, so the order of convergence is $h^{m-\epsilon}$, $0 < \epsilon \ll 1$. We state this as a corollary.

Corollary 6.1. *The Order of Convergence. Let the assumptions of Theorem 6.3 be satisfied. If $L = \infty$, i.e., $\psi \in M^{\infty,m}$, then for all $0 \le t \le T$*

$$\|X(\cdot,t) - \tilde{X}(\cdot,t)\|_{L_h^p} \le ch^{m-\epsilon}, \qquad 0 < \epsilon \ll 1, \tag{6.85}$$

$$\|v(\cdot,t) - \tilde{v}^{\delta h}(\cdot,t)\|_{L_h^p} \le ch^{m-\epsilon}. \tag{6.86}$$

Thus by a proper choice of the mollifiers ψ we can obtain as high an order of convergence as we wish.

The 3D inviscid-vortex methods are more complicated because the vorticity is stretched along particle trajectories:

$$\omega(X(\alpha,t),t) = \nabla_\alpha X(\alpha,t)\omega_0(\alpha).$$

In this case the integrodifferential equation on particle trajectories is

$$\frac{dX}{dt}(\alpha,t) = \int_{\mathbb{R}^3} K[X(\alpha,t) - X(\alpha',t)]\nabla_\alpha X(\alpha',t)\omega_0(\alpha')d\alpha',$$
$$X(\alpha,t)|_{t=0} = \alpha \in \mathbb{R}^3, \tag{6.87}$$

with the kernel K given by

$$K(x)h = \frac{1}{4\pi}\frac{x\times h}{|x|^3}, \qquad x,h \in \mathbb{R}^3. \tag{6.88}$$

The simplest 3D inviscid-vortex method, which also involves an approximation of the gradient ∇_α by a finite-difference operator ∇_α^h, is in Eq. (6.71):

$$\frac{d\tilde{X}_i}{dt}(t) = \sum_{j\in\Lambda^h} K^\delta[\tilde{X}_i(t) - \tilde{X}_j)]\nabla_\alpha^h \tilde{X}_j(t)\omega_{0_j}h^3,$$
$$\tilde{X}_i(t) = ih, \qquad i \in \Lambda. \tag{6.89}$$

The convergence of these approximate solutions depends also on the finite-difference operator ∇_α^h used.

Definition 6.2. *Let ∇_α^h be a finite-difference operator.*

(i) We say that ∇_α^h is stable if there exists a fixed constant c so that, for all $h \le h_0$,

$$\left\|\nabla_\alpha^h f\right\|_{-1,h} \le c\|f\|_{L_h^p}, \tag{6.90}$$

where $\|\cdot\|_{-1,h}$ is a norm on the dual space to the discrete Sobolev space H^1.

(ii) We say that ∇_α^h is rth-order accurate if there exists a fixed constant c so that

$$\left\|\nabla_\alpha^h f - \nabla_\alpha f\right\|_{L_h^p} \le ch^r. \tag{6.91}$$

The convergence result is the following theorem.

Theorem 6.4. *The Convergence of the 3D Inviscid-Vortex Methods. Let v be a sufficiently smooth solution to the 3D Euler equation for the initial vorticity ω_0 with a compact support. Let the mollifier $\psi \in M^{L,m}$, $L > 5$. Let $m \ge 4$ if $L = \infty$ and $m(L-1)/(m+L) > 3$ if $L < \infty$. Let the finite-difference operator ∇_α^h be stable and rth-order accurate, with $r > 3$. Take $\delta = h^q$, where $q = 1 - \epsilon_0$ if $L = \infty$ and $q = (L - 1 - \epsilon_0)/(m+L)$ with ϵ_0 small enough so that $mq > 3$.*

Then, for all $0 \le t \le T$, the solution $\tilde{X}$ and $\tilde{v}^{\delta h}$ to inviscid-vortex method (6.89) satisfies

$$\|X(\cdot,t) - \tilde{X}(\cdot,t)\|_{L_h^2} + \|\omega(\cdot,t) - \tilde{\omega}^{\delta h}(\cdot,t)\|_{H^{-1},h} \le c(h^{mq} + h^r), \tag{6.92}$$

$$\|v(\cdot,t) - \tilde{v}^{\delta h}(\cdot,t)\|_{L_h^2} \le c(h^{mq} + h^r). \tag{6.93}$$

The constant c depends on T, v, L, m, ϵ_0, and r. Here H_h^{-1} denotes the dual of H_h^1.

6.4.1. Proof of Convergence of the 2D Vortex Method

Now we prove the convergence of the 2D inviscid-vortex methods. We need to compute the discrete L^2 errors between the exact solution $X(\alpha, t)$ with velocity

$$v(x(\alpha,t),t) = \int_{\mathbb{R}^2} K[x(\alpha,t) - X(\alpha',t)]\omega_0(\alpha')d\alpha' \tag{6.94}$$

and the approximate numerical solution $\tilde{X}(\alpha, t)$ with velocity

$$\tilde{v}^{\delta h}(\tilde{X}(\alpha,t),t) = \sum_{j\in\Lambda^h} K^\delta[\tilde{X}(\alpha,t) - \tilde{X}_j(t)]\omega_{0_j}h^2. \tag{6.95}$$

Introducing the approximate velocity evaluated on the exact particle trajectories,

$$v^{\delta h}(X(\alpha,t),t) = \sum_{j\in\Lambda^h} K^\delta[X(\alpha,t) - X_j(t)]\omega_{0_j}h^2, \tag{6.96}$$

we split the velocity error $\|v(\cdot,t) - \tilde{v}^{\delta h}(\cdot,t)\|_{L_h^p}$ into the *consistency error* $\|v(\cdot,t) - v^{\delta h}(\cdot,t)\|_{L_h^p}$ and the *stability error* $\|v^{\delta h}(\cdot,t) - \tilde{v}^{\delta h}(\cdot,t)\|_{L_h^p}$:

$$\|v(\cdot,t) - \tilde{v}^{\delta h}(\cdot,t)\|_{L_h^p} \le \|v(\cdot,t) - v^{\delta h}(\cdot,t)\|_{L_h^p} + \|v^{\delta h}(\cdot,t) - \tilde{v}^{\delta h}(\cdot,t)\|_{L_h^p}. \tag{6.97}$$

Recall from Chapter 3, Section 3.4, on viscous-splitting algorithms that a proof of convergence for a general computational algorithm has two parts, consistency and stability. Hence the proof of Theorem 6.3 relies on the following two lemmas proved in the next two subsections.

Lemma 6.3. *The Consistency Error. Let* $v \in C^1\{[0,T]; C^{L+1}(\mathbb{R}^2)\}$, $L \geq 3$, *be a solution to the 2D Euler equation with compactly supported initial vorticity* ω_0. *Let the mollifier* $\psi \in M^{L,m}$, $m \geq 2$. *Then, for all* $0 \leq t \leq T$,

$$\|v(\cdot,t) - v^{\delta h}(\cdot,t)\|_{L^\infty} \leq c(\delta^m + h^L \delta^{1-L}).$$

Lemma 6.4. *The Stability Error. Let* $v \in C^1\{[0,T]; C^{L+1}(\mathbb{R}^2)\}$, $L \geq 3$, *be a solution to the 2D Euler equation for the initial vorticity* ω_0 *with a compact support and let the mollifier* $\psi \in M^{L,m}$, $m \geq 2$. *Then, for all* $0 \leq t \leq T$ *and* $1 < p < \infty$, *there exists a constant C such that*

$$\|v^{\delta h}(\cdot,t) - \tilde{v}^{\delta h}(\cdot,t)\|_{L^p_h} \leq C\|X(\cdot,t) - \tilde{X}(\cdot,t)\|_{L^p_h},$$

provided that

$$\max_j |X_j(t) - \tilde{X}_j(t)| \leq \delta.$$

We estimate the consistency error in a fashion analogous to the approximation of the discretization error in the convergence proof of the random-vortex method for the viscous strained shear-layer model of Section 6.1.

Using the above consistency and stability errors, we are now ready to give the following proof.

Proof of Theorem 6.3. Denote $e(\alpha,t) = X(\alpha,t) - \tilde{X}(\alpha,t)$. Integrodifferential equations (6.75) and (6.78) imply that

$$\frac{d}{dt}\|e(\cdot,t)\|_{L^p_h} \leq \|v(\cdot,t) - v^{\delta h}(\cdot,t)\|_{L^p_h} + \|v^{\delta h}(\cdot,t) - \tilde{v}^{\delta h}(\cdot,t)\|_{L^p_h}.$$

The consistency error in Lemma 6.3 and the stability error in Lemma 6.4 give

$$\frac{d}{dt}\|e(\cdot,t)\|_{L^p_h} \leq c\|e(\cdot,t)\|_{L^p_h} + c(\delta^m + h^L\delta^{1-L}),$$

provided that $\max_j |e_j(t)| \leq \delta$. Let $T^* = \min\{T, \inf\{t : \|e(\cdot,t)\|_{L^p_h} \geq \delta\}\}$. Grönwall's inequality implies that

$$\|e(\cdot,t)\|_{L^p_h} \leq cT^* e^{cT^*}(\delta^m + h^L\delta^{1-L}), \qquad \forall\, 0 \leq t \leq T^*.$$

Suppose that $T^* < T$. Because $\max_j |e_j(t)|^2 h^2 \leq \|e(\cdot,t)\|^2_{L^p_h}$, by imposing the condition $cT^* e^{cT^*}(\delta^m + h^L\delta^{1-L}) \leq \delta h/2$, for all $0 \leq t \leq T^*$, we have $\max_j |e_j(t)| \leq h^{-1}\|e(\cdot,t)\|_{L^p_h} \leq \delta/2$, so that we can continue the solution $\|e(\cdot,t)\|_{L^p_h}$ up to $T^{**} > T^*$. Repeating these arguments, we conclude the proof. □

6.4.2. The Consistency Error

To prove Lemma 6.3, first observe that

$$
\begin{aligned}
|v(\cdot,t) - v^{\delta h}(\cdot,t)| &\le \left| \int_{\mathbb{R}^2} (K - K^\delta)(x - x')\omega(x',t)dx' \right| \\
&\quad + \left| \int_{\mathbb{R}^2} K^\delta(x - x')\omega(x',t)dx' - \sum_{j\in\Lambda^h} K^\delta[x - X_j(t)]\omega_{0_j} h^2 \right| \\
&= e_m(x,t) + e_d(x,t).
\end{aligned}
$$

Thus the consistency error $\|v(\cdot,t) - v^{\delta h}(\cdot,t)\|_{L^\infty}$ can be split into the *moment error* $\|e_m(\cdot,t)\|_{L^\infty}$ of a desingularization of the kernel K and the *discretization error* $\|e_d(\cdot,t)\|_{L^\infty}$ of the trapezoidal quadrature. We estimate these errors below.

Lemma 6.5. *The Moment Error. Let the mollifier $\psi \in M^{L,m}$, $m \ge 2$. Then, for all $0 \le t \le T$,*

$$
\|e_m(\cdot,t)\|_{L^\infty} \le c\delta^m. \tag{6.98}
$$

Proof of Lemma 6.5. Denote $g(x,t) = [(K - K^\delta) * \omega](x,t)$. Because $g(\cdot,t) \in L^1(\mathbb{R}^2)$, we can take its Fourier transform $\widehat{g}$:

$$
\begin{aligned}
\widehat{g}(\xi,t) &= (\widehat{K - K^\delta})(\xi)\widehat{\omega}(\xi,t) = [\widehat{K}(\xi) - \widehat{K}(\xi)\widehat{\psi}_\delta(\xi)]\widehat{\omega}(\xi,t) \\
&= \widehat{K}(\xi)\widehat{\omega}(\xi)[1 - \widehat{\psi}_\delta(\xi)].
\end{aligned}
$$

The definition $\psi_\delta(x) = \delta^{-2}\psi(x/\delta)$ gives $\widehat{\psi}_\delta(x) = \widehat{\psi}(\delta\xi)$. Because $\widehat{\psi}(0) = \int_{\mathbb{R}^2}\psi(x)\,dx = 1$, we get

$$
\widehat{g}(\xi,t) = \widehat{K}(\xi)\widehat{\omega}(\xi,t)[\widehat{\psi}(0) - \widehat{\psi}(\delta\xi)]. \tag{6.99}
$$

Now we expand $\widehat{\psi}(\xi)$ as a Taylor series,

$$
\widehat{\psi}(\xi) = \sum_{|\alpha|\le m-1} \partial_\xi^\alpha \widehat{\psi}(0)\xi^\alpha + \sum_{|\alpha|=m} \partial_\xi^\alpha \widehat{\psi}(h_\alpha\xi)\xi^\alpha,
$$

and compute

$$
\begin{aligned}
\partial_\xi^\alpha \widehat{\psi}(0) &= \partial_\xi^\alpha \int_{\mathbb{R}^2} e^{-2\pi i x\cdot\xi}\psi(x)dx|_{\xi=0} = \int_{\mathbb{R}^2} (-2\pi i x)^\alpha e^{-2\pi i x\cdot\xi}\psi(x)dx|_{\xi=0} \\
&= c\int_{\mathbb{R}^2} x^\alpha \psi(x)dx.
\end{aligned}
$$

From Definition 6.1 of the class $M^{L,m}$ we have $\int_{\mathbb{R}^2} x^\alpha\psi(x)dx = 0$ for $1 \le |\alpha| \le m-1$ and $\int_{\mathbb{R}^2} |x|^m|\psi(x)|dx < \infty$ from the preceding equation we get

$$
|\widehat{\psi}(0) - \widehat{\psi}(\xi)| \le c|\xi|^m.
$$

Now observe that, if $\omega(\cdot,t) \in C_0^{m+2}(\mathbb{R}^2)$, then

$$|\widehat{\omega}(\xi,t)| \le \frac{M_{m+2}}{(1+|\xi|)^{m+2}}.$$

To estimate $|\widehat{K}(\xi)|$, recall that $-\Delta\Psi = \omega$ and $v = K * \omega = \nabla^T\Psi$, so that

$$\widehat{K}(\xi)\widehat{\omega}(\xi) = (-2\pi i\xi)^\beta\widehat{\psi}(\xi) \text{ with some } |\beta| = 1.$$

Because $\widehat{\omega}(\xi) = -(-2\pi i\xi)^\alpha\widehat{\psi}(\xi)$ with $|\alpha| = 2$, we get

$$|\widehat{K}(\xi)| \le \frac{c}{|\xi|}.$$

Combining the above estimates, from Eq. (6.99) we get

$$|\widehat{g}(\xi,t)| \le \frac{c}{|\xi|}\frac{M_{m+2}}{(1+|\xi|)^{m+2}}c|\delta\xi|^m \le c\delta^m\frac{|\xi|^{m-1}}{(1+|\xi|)^{m+2}},$$

so that $\widehat{g}(\cdot,t) \in L^1(\mathbb{R}^2)$. Thus

$$\|e_m(\cdot,t)\|_{L^\infty} = \|g(\cdot,t)\|_{L^\infty} \le \|\widehat{g}(\cdot,t)\|_{L^1} \le c\delta^m,$$

which concludes the proof. □

To estimate the discretization error $\|e_d(\cdot,t)\|_{L^\infty}$, first we show that the approximation of an integral by the trapezoidal quadrature is of an arbitrary high accuracy. We recall the following lemma from Section 6.1.

Lemma 6.2. *Trapezoidal Quadrature Error. Let $g \in C_0^r(\mathbb{R}^2)$, $r \ge 3$. Then*

$$\left|\int_{\mathbb{R}^2} g(x)dx - \sum_{j\in\Lambda^h} g(jh)h^2\right| \le c_r\|g\|_{W^{r,1}(\mathbb{R}^2)}h^r, \tag{6.100}$$

where $\|\cdot\|_{W^{r,1}}$ is the Sobolev norm.

We also need the following estimates for the regularized kernel K^δ:

Proposition 6.1. *Let $\psi \in M^{L,m}$, $L \ge 0$, $m \ge 2$, be a radially symmetric mollifier, and $K^\delta = K * \psi_\delta$. Then for all $|\beta| \le L$ there exists a constant c_β such that*

$$|\partial^\beta K^\delta(x)| \le \begin{cases} c_\beta|x|^{-1-|\beta|}, & |x| \ge \delta \\ c_\beta\delta^{-1-|\beta|}, & x \in \mathbb{R}^2 \end{cases}, \tag{6.101}$$

$$\|\partial^\beta K^\delta\|_{L^1[B(0,R)]} \le \begin{cases} c, & |\beta| = 0 \\ c\ln\delta, & |\beta| = 1 \\ c\delta^{1-|\beta|}, & |\beta| \ge 2 \end{cases}. \tag{6.102}$$

We prove estimates (6.102) at the end of this section.

Now we are ready to give Lemma 6.6.

Lemma 6.6. *The Discretization Error. Let the assumptions of consistency error Lemma 6.3 be satisfied. Then, for all* $0 \leq t \leq T$,

$$\|e_d(\cdot, t)\|_{L^\infty} \leq ch^L \delta^{1-L}. \tag{6.103}$$

Proof of Lemma 6.6. Because the discretization error is

$$e_d(x, t) = \left| \int_{\mathbb{R}^2} K^\delta[x - X(\alpha', t)]\omega(X(\alpha', t), t)d\alpha' \right.$$
$$\left. - \sum_{j \in \Lambda^h} K^\delta[x - X(\alpha_j, t)]\omega(X(\alpha_j, t), t)h^2 \right|,$$

applying error estimate (6.100) for the trapezoidal quadrature gives

$$|e_d(x, t)| \leq c\|K^\delta[x - X(\cdot, t)]\omega(X(\cdot, t), t)\|_{W^{L,1}} h^L, \qquad L \geq 3.$$

The Leibnitz formula gives

$$\partial_\alpha^\beta\{K^\delta[x - X(\alpha, t)]\omega(X(\alpha, t), t)\} = \sum_{\gamma \leq \beta} \binom{\beta}{\gamma} \partial_\alpha^{\beta-\gamma} K^\delta[x - X(\alpha, t)] \partial_\alpha^\gamma \omega(X(\alpha, t), t).$$

Because $v \in C^1\{[0, T]; C^{L+1}(\mathbb{R}^2)\}$, $\|\omega(\cdot, t)\|_{C^L} \leq c$ and $\text{supp } \omega(\cdot, t) \subset \Omega$, a bounded domain. Thus by using estimates (6.102) for all $x \in \Omega$ we have

$$|e_d(x, t)| \leq c\|K^\delta\|_{W^{L,1}(\Omega)} h^L \leq ch^L \delta^{1-L},$$

so that $\|e_d(\cdot, t)\|_{L^\infty} \leq ch^L \delta^{1-L}$. □

A higher-order convergence for the discretization error was proved by Beale, who used the fact that the kernel K_2 is odd (Beale, 1986).

To finish the analysis of consistency, it remains only to give the proof of Proposition 6.1.

Proof of Proposition 6.1. The mollification $K * \psi_\delta$, for a radially symmetric mollifier, is equivalent to the multiplication of K by the cutoff function f_δ, $K^\delta(x) = K(x) f_\delta(x)$, where

$$f_\delta(x) = 2\pi \int_0^{|x|/\delta} s\psi(s)ds.$$

Let $|x| \geq \delta$. Using the Leibnitz formula and the form of K, we estimate

$$|\partial^\beta K^\delta(x)| \leq \sum_{\alpha \leq \beta} \binom{\beta}{\alpha} |\partial^{\beta-\alpha} K(x)|\, |\partial^\alpha f_\delta(x)| \leq \sum_{\alpha \leq \beta} \frac{c_{\alpha\beta}}{|x|^{1+|\beta|-|\alpha|}} |\partial^\alpha f_\delta(x)|$$

Now we estimate the derivatives of f_δ. We have $|f_\delta(x)| \leq c$, $\forall x \in \mathbb{R}^2$. Moreover,

$$\frac{\partial}{\partial r} f_\delta(r) = \frac{2\pi}{\delta^2} r\psi\left(\frac{r}{\delta}\right),$$

and, in general,

$$\frac{\partial^k}{\partial r^k} f_\delta(r) = (k-1)\frac{2\pi}{\delta^k}\psi^{(k-2)}\left(\frac{r}{\delta}\right) + \frac{2\pi}{\delta^{k+1}} r\psi^{(k-1)}\left(\frac{r}{\delta}\right), \qquad k \geq 2.$$

Because $\psi \in M^{L,m}$, we have $|x|^{2+|\beta|}|\partial^\beta \psi(x)| \leq c \quad \forall\, |\beta| \leq L$, so that, in particular,

$$\frac{1}{\delta^{2+|\beta|}}\left|\psi^{(\beta)}\left(\frac{r}{\delta}\right)\right| \leq \frac{c}{r^{2+|\beta|}}, \qquad |\beta| \leq L.$$

Using this estimate we get

$$\left|\frac{\partial^k}{\partial r^k} f_\delta(r)\right| \leq \frac{c_k}{r^k}, \qquad k \geq 0,$$

so that finally we can estimate

$$|\partial^\beta K^\delta(x)| \leq \sum_{\alpha \leq \beta} \frac{c_{\alpha\beta}}{|x|^{1+|\beta|-|\alpha|}} \frac{c_\alpha}{|x|^{|\alpha|}} \leq \frac{c_\beta}{|x|^{1+|\beta|}}, \qquad |x| \geq \delta.$$

This proves the first estimate in estimates (6.101).

The proof of the second estimate in estimates (6.101) follows easily from a rescaling of K:

$$\begin{aligned}
|\partial^\beta K^\delta(x)| &= \left|\int_{\mathbb{R}^2} K(x')\delta^{-2-|\beta|}\partial^\beta \psi\left(\frac{x-x'}{\delta}\right)dx'\right| \\
&= \left|\int_{\mathbb{R}^2} K(\delta x')\delta^{-|\beta|}\partial^\beta \psi\left(\frac{x}{\delta} - x'\right)dx'\right| \\
&= \delta^{-1-|\beta|}\left|\int_{\mathbb{R}^2} K(x')\partial^\beta \psi\left(\frac{x}{\delta} - x'\right)dx'\right| \\
&\leq \delta^{-1-|\beta|}\left\{\|\partial^\beta \psi\|_{C^0}\int_{|x'|\leq 1} |K(x')|dx' + \sup_{|x'|\geq 1}|K(x')|\, \|\partial^\beta \psi\|_{L^1}\right\} \\
&\leq c\delta^{-1-|\beta|},
\end{aligned}$$

because $|K(x)| \leq (c/|x|)$ and $\psi \in M^{L,m}$. Finally we prove estimates (6.102). For $\delta < R$ we have

$$\|\partial^\beta K^\delta\|_{L^1[B(0,R)]} = \left[\int_{B(0,R)\cap B(x,\delta)} + \int_{B(0,R)-B(x,\delta)}\right]|\partial^\beta K^\delta(x')|dx'.$$

The first term we estimate by using estimates (6.101) as

$$\int_{B(0,R)\cap B(x,\delta)} |\partial^\beta K^\delta(x')|dx' \leq c\delta^{-1-|\beta|}\pi\delta^2 = c\delta^{1-|\beta|}.$$

The second term we estimate as

$$\begin{aligned}
&\int_{B(0,R)-B(x,\delta)} |\partial^\beta K^\delta(x')|dx' \\
&\qquad \leq c\int_\delta^{2R} r^{-1-|\beta|+1}dr = \begin{cases} c\big(R-\delta\big), & |\beta| = 0 \\ c\ln\frac{R}{\delta}, & |\beta| = 1. \\ c\big[R^{1-|\beta|} - \delta^{1-|\beta|}\big], & |\beta| \geq 2 \end{cases}
\end{aligned}$$

Picking the correct values of the constants c concludes the proof of estimates (6.102):

$$\|\partial^\beta K^\delta\|_{L^1[B(0,R)]} \leq \begin{cases} c, & |\beta| = 0 \\ c \ \ln \delta, & |\beta| = 1. \\ c\delta^{1-|\beta|}, & |\beta| \geq 2 \end{cases}$$

□

6.4.3. The Stability Error

We prove Lemma 6.4.

Lemma 6.4. *The Stability Error. Let* $v \in C^1\{[0, T];^{L+1}(\mathbb{R}^2)\}$, $L \geq 3$, *be a solution to the 2D Euler equation with compactly supported initial vorticity* ω_0. *Let the mollifier* $\psi \in M^{L,m}$, $m \geq 2$. *Then for all* $0 \leq t \leq T$, $1 < p < \infty$,

$$\|v^{\delta h}(\cdot, t) - \tilde{v}^{\delta h}(\cdot, t)\|_{L_h^p} \leq C_p \|X(\cdot, t) - \tilde{X}(\cdot, t)\|_{L_h^p} \tag{6.104}$$

provided that

$$\max_j |X_j(t) - \tilde{X}_j(t)| \leq \delta. \tag{6.105}$$

Proof of Lemma 6.4. We split the difference between $v^{\delta h}$ and $\tilde{v}^{\delta h}$ as

$$\begin{aligned} &\tilde{v}^{\delta h}(\tilde{X}_i(t), t) - v^{\delta h}(X_i(t), t) \\ &\quad = \sum_{j \in \Lambda} \{K^\delta[X_i(t) - \tilde{X}_j(t)] - K^\delta[X_i(t) - X_j(t)]\}\omega_j h^2 \\ &\qquad + \sum_{j \in \Lambda} \{K^\delta[\tilde{X}_i(t) - \tilde{X}_j(t)] - K^\delta[X_i(t) - \tilde{X}_j(t)]\}\omega_j h^2 \qquad (6.106) \\ &\quad = v_i^{(1)} + v_i^{(2)}. \qquad (6.107) \end{aligned}$$

Applying the mean-value theorem gives

$$v_i^{(1)} = \sum_{j \in \Lambda} \nabla K^\delta[X_i(t) - X_j(t) + y_{ij}]e_j \omega_j h^2,$$

where $|y_{ij}| = \theta_{ij}|e_j| \leq \delta$ and $e_j = X_j(t) - \tilde{X}_j(t)$. Because the mapping $X(\cdot, t)$ is volume preserving, squares Q_j determined by the lattice Λ^h (which constitute a partition of $\mathbb{R}^2$) are mapped onto curvilinear quadrilaterals B_j, which also constitute a partition of $\mathbb{R}^2$, and meas $B_j =$ meas $Q_j = h^2$. With this interpretation, the preceding function $v_i^{(1)}$ is equivalent to

$$v^{(1)}(x) = \int_S G^\delta(x - x')g(x')dx', \qquad x \in B_i,$$

where $G^\delta(x - x') = \nabla K^\delta[X_i(t) - X_j(t) + y_{ij}]$, $x \in B_i$, $x' \in B_j$; $g(x') = e_j \omega_j$, $x' \in B_j$; and $S = \cup_{j \in \Lambda_0} B_j$. Now we approximate the kernel $G^\delta(x - x')$ by the kernel

$\nabla K^{\delta}(x - x')$, so that

$$\begin{aligned} v^{(1)}(x) &= \int_S \nabla K^{\delta}[X_i(t) - x']g(x')dx' \\ &\quad + \int_S \{G^{\delta}(x - x') - \nabla K^{\delta}(X_i(t) - x')\}g(x')dx' \\ &= \tilde{v}^{(1)}(x) + \hat{v}^{(1)}(x). \end{aligned}$$

Note that $\|v^{(1)}\|_{L^2(S)} = \|v^{(1)}\|_{L_h^p}$ and $\|g\|_{L^2(S)} = \|e\omega\|_{L_h^p} \leq c\|e\|_{L_h^p}$ because ω_j is uniformly bounded. The Calderon–Zygmund inequality implies that

$$\left\|\tilde{v}^{(1)}\right\|_{L^2(S)} \leq c\|g\|_{L^2(S)} \leq c\|e\|_{L_h^p}.$$

For the second term, $\hat{v}^{(1)}$, we apply the mean-value theorem and the well-known result that convolution operators with L^1 kernels are bounded operators on L^p spaces† to get

$$\left\|\hat{v}^{(1)}\right\|_{L^2(S)} \leq \|\partial^2 K^{\delta}\|_{L^1(S)}\|g\|_{L^2(S)} \sup_{i,j}|y_{ij}|,$$

so that, by using estimate (6.102) from Proposition 6.1, we have

$$\left\|\hat{v}^{(1)}\right\|_{L^2(S)} \leq c\delta^{-1}\|e\|_{L_h^p}\delta = c\|e\|_{L_h^p}.$$

This proves that for the first term $v^{(1)}$ in Eq. (6.107) we have

$$\left\|v^{(1)}\right\|_{L_h^p} = \|v^{(1)}\|_{L^2(S)} \leq c\|e\|_{L_h^p},$$

provided that $\max_j |e_j| \leq \delta$.

Now we estimate the second term $v^{(2)}$ in Eq. (6.107). We apply the mean-value theorem to get

$$v_i^{(2)} = \sum_{j\in\Lambda} \nabla K^{\delta}[X_i(t) - \tilde{X}_j(t) + y_{ij}]e_i\omega_j h^2.$$

Because $\tilde{v}^{\delta h}(\cdot, t)$ is divergence free, the mapping $\tilde{x}(\cdot, t)$ is volume conserving, so that we can repeat the same argument as before to get

$$\left\|v^{(2)}\right\|_{L^2(S)} \leq \sup_{x\in S}\int_S |G^{\delta}(x - x')g(x')|dx' \; \|e\|_{L^2(S)},$$

where $G^{\delta}(x - x') = \nabla K^{\delta}[X_i(t) - \tilde{X}_j(t) + y_{ij}],\ x \in \tilde{B}_i,\ x' \in \tilde{B}_j$, and $g(x') = \omega_j$, $x' \in \tilde{B}_j$. Now we approximate the kernel G^{δ} by ∇K^{δ} and g by ω, so that

$$\begin{aligned} \int_S G^{\delta}(x - x')g(x')dx' &= \int_S \nabla K^{\delta}[X_i(t) - x']\omega(x')dx' \\ &\quad + \int_S \nabla K^{\delta}[X_i(t) - x'][g(x') - \omega(x')]dx' \\ &\quad + \sum_{j\in\Lambda}\{\nabla K^{\delta}[X_i(t) - \tilde{X}_j(t) + y_{ij}] - \nabla K^{\delta}[X_i(t) - x']\}\omega_j h^2. \end{aligned}$$

† See, for example, Folland (1995, Section 0.C) on the generalized Young's inequality.

Using the mean-value theorem, estimate (6.102) and the Calderon–Zygmund inequality, we get

$$\sup_{x \in S} \int_S |G^\delta(x - x')g(x')|dx' \le c,$$

so that finally $\|v^{(2)}\|_{L_h^p} = \|v^{(2)}\|_{L^2(S)} \le c\|e\|_{L_h^p}$. □

6.5. Computational Performance of the 2D Inviscid-Vortex Method on a Simple Model Problem

We now illustrate by means of a simple example the computational performance of the 2D inviscid-vortex method.

First we discuss various ways of obtaining simple forms for higher-order kernels. Recall from the preceding sections that we approximate the singular kernel

$$K(x) = \frac{(-x_2, x_1)^t}{|x|^2}$$

in the Biot–Savart law by a nonsingular kernel $K_\delta = K * \psi_\delta$, $\psi_\delta(x) = \delta^{-N}\psi(z/\delta)$. We choose ψ subject to the conditions that

(i) ψ is smooth and rapidly decreasing, i.e.,

$$|D^\beta \psi(z)| \le C_{\beta j}\left(1 + |z|^2\right)^{-j}$$

for every multi-index β and every integer ;

(ii) $\int \psi(z)d^N z = 1$;

(iii) $\int z^\beta \psi(z)d^N z = 0$, $1 \le |\beta| \le m - 1$, where m is an integer.

The results of Section 6.4 show that vortex methods satisfying conditions (i)–(iii) converge provided that δ and h are suitably related. If $\delta = h^q$ the error is of the order of $\delta^m = h^{mq}$, that is, the method is mth order. Our object here is to choose ψ such that K_δ has a simple expression consistent with these requirements. As we now show, choices of ψ satisfying $m = 2$ can produce new ψ satisfying $m \ge 4$. Condition (i) implies that the Fourier transform of ψ, as well as ψ itself, is smooth and rapidly decreasing. For simplicity, we take $\psi(r)$, a function of the radial variable r. Hence condition (iii) is satisfied by symmetry for $|\beta|$ odd so that we can assume that m is even. For any choice of a radial ψ, condition (iii) is always satisfied with $m = 2$. Condition (i) can be relaxed somewhat to allow a ψ that is not very smooth at $z = 0$. In fact, the simplest choice in three dimensions has this property (Beale and Majda, 1985). On the other hand, in two dimensions there is a family of very smooth kernels that produce an arbitrarily high degree of accuracy. We discuss this family here.

In two dimensions, a natural choice of ψ is the Gaussian $\psi^{(2)}(r) = e^{-r^2}/\pi$. Conditions (i)–(iii) are satisfied with $m = 2$. We write $K_\delta = K * \psi_\delta$ as

$$K_\delta = (\partial_{x_2}, -\partial_{x_1})G_\delta, \qquad G_\delta = G * \psi_\delta.$$

Because ψ is radial, G_δ is also, and

$$\nabla^2 G_\delta = \nabla^2(G * \psi_\delta) = -\psi_\delta \quad \text{or} \quad \frac{1}{r}D_r(rD_rG_\delta) = -\psi_\delta(r) = -\frac{1}{\pi\delta^2}e^{-r^2/\delta^2}.$$

After integrating, we have

$$D_r G_\delta = \frac{1}{2\pi r}\left(e^{-r^2/\delta^2} - 1\right).$$

The constant of integration is uniquely determined by the fact that G_δ must be smooth. Hence the corresponding velocity kernel is

$$K_\delta^{(2)}(z) = \frac{(-x_2, x_1)}{2\pi r^2}\left(1 - e^{-r^2/\delta^2}\right),$$

where the superscript (2) denotes the order of the kernel.

Now we show how to use the above kernel, which is second order, to obtain a fourth-order kernel. We choose $\psi = \psi^{(4)}$ as a combination of two Gaussians with different scalings:

$$\psi^{(4)}(r) = c_1\psi^{(2)}(R) + c_2\psi^{(2)}(r/a),$$

where a is arbitrary except that $a \neq 1$. To satisfy condition (ii) we must have

$$c + a_2^c = 1.$$

Because of symmetry, condition (ii) will hold with $m = 4$ provided that

$$\psi^{(4)}(r)r^2 \cdot r\, dr = 0.$$

This in turn holds if

$$c_1 + a^4 c_2 = 0,$$

and the two equations determine $\psi^{(4)}$ in terms of a. We can now find $K_\delta^{(4)}$, just as in the preceding case:

$$K_\delta^{(4)}(z) = \frac{(-x_2, x_1)}{2\pi r^2}\left(1 - c_1 e^{-r^2/\delta^2} - c_2 a^2 e^{-r^2/a^2\delta^2}\right).$$

For example, the choice $a^2 = 2$ leads to

$$\begin{aligned} K_\delta^{(4)}(z) &= \frac{(-x_2, x_1)}{2\pi r^2}\left(1 - 2e^{-r^2/\delta^2} + e^{-r^2/2\delta^2}\right) \\ &= \frac{(-x_2, x_1)}{2\pi r^2}\left(1 - e^{-r^2/\delta^2}\right)\left(1 + 2e^{-r^2/2\delta^2}\right). \end{aligned}$$

We can construct higher-order kernels by including more in ψ terms with different scalings. An example of a sixth-order kernel is

$$K_\delta^{(6)}(z) = \frac{(-x_2, x_1)}{2\pi r^2}\left(1 - \frac{8}{3}e^{-r^2/\delta^2} + 2e^{-r^2/2\delta^2} - 1/3e^{-r^2/4\delta^2}\right).$$

A class of simpler high-order kernels comes from the choice $\psi(r) = P(r)e^{-r^2}$, where P is an even polynomial in r. Such a choice produces a kernel of the form

$$K_\delta(z) = K(z)\left[1 - Q(r/\delta)e^{-r^2/\delta^2}\right],$$

where Q is a related polynomial. The moment condition (iii) requires

$$\int_0^\infty r^{2j} P(r) e^{-r^2} r\, dr = 0, \qquad 1 \le j \le (m-2)/2,$$

or, after integration by parts,

$$\int_0^\infty r^{2j-1} Q(r) e^{-r^2} dr = 0.$$

Hence the moment conditions reduce to a set of linear equations for the coefficients of Q. A kernel of the order of m has $(m-2)/2$ conditions, thus we seek a polynomial of degree $m-2$. The fourth, sixth, and eighth kernels are

$$\begin{aligned} Q^{(4)}(r) &= 1 - r^2, \\ Q^{(6)}(r) &= 1 - 2r^2 + r^4/2, \\ Q^{(8)}(r) &= 1 - 3r^2 + 3r^4 - r^6/6. \end{aligned}$$

These are the Laguerre polynomials of r^2, normalized so that the constant term is 1.

6.5.1. Numerical Results

Here we present some results of numerical experiments that illustrate the higher-order (superquadratic) accuracy of the explicit kernels derived in the first section for 2D flows and moderate integration times. Included in this example is an illustration of the poorer performance of the point-vortex method. These calculations are from the paper by Beale and Majda (1985). Hald and Del Prete (1978) studied the accuracy of a variety of low-order accurate vortex methods for moderate integration times.

We consider a class of test problems with radial vorticity $\omega(r)$. Recall from Chap. 2 that such a vorticity yields a simple exact expression for the velocity field:

$$u(x, y) = (u_1, u_2) = \frac{1}{2\pi r^2}(-y, x) \int_0^r s\omega(s) ds.$$

Particle paths are circles about the origin. Because the choice of vorticity is arbitrary, the angular velocity can vary considerably with the radius, causing substantial shearing.

We now summarize the results of a numerical study presented in Beale and Majda (1985). Consider a simple choice of vorticity,

$$\omega(r) = \begin{cases} (1 - r^2)^3, & r < 1 \\ 0, & r > 1 \end{cases}. \tag{6.108}$$

Because the support of w lies inside $r < 1$, denote by U the normalized mean for $r < 1$:

$$U^2 = \frac{1}{\pi} \int\int_{r<1} |u|^2 dx dy = 2 \int_0^1 |u|^2 r\, dr. \tag{6.109}$$

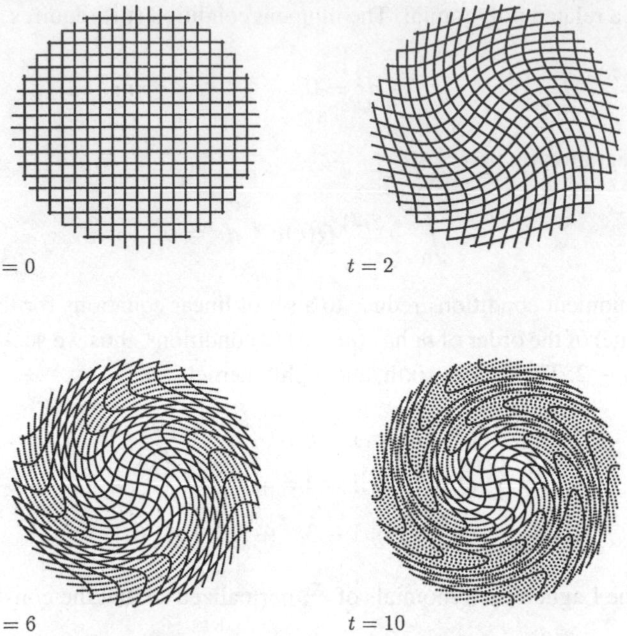

Figure 6.4. The deformation of an initially square grid that is due to the induced velocity field at times $t = 0, 2, 6$, and 10.

This choice of ω introduces significant shearing in the velocity field. Figure 6.4 shows the deformation of an initially square grid that is due to the induced velocity field at times $t = 0, 2, 6$, and 10.

The simulation uses a square grid centered at the origin with each square of size $h \times h$. The initial particles are the centers of the squares. The initial vorticity associated with each particle is the exact value of the vorticity at the center of each square multiplied by h^2. This choice of grid purposely does not take advantage of the special radial geometry in order to test the accuracy of the method.

The particles are advected according to the 2D vortex method described in Section 6.2. At a given time, the advected particles $\tilde{x}_i$ determine the numerical velocity field according to the expression

$$u_{\text{comp}}(z, t) = u^h(z, t) = \sum_k K_\delta(x - \tilde{x}_k)\omega_k h^2$$

where

$$u_{j,\text{comp}} = \sum_k K_\delta(\tilde{x}_j - \tilde{x}_k)\omega_k h^2$$

is the associated discrete velocity field computed at the particles.

Consider the following two different nondimensional ways for measuring the error. First denote the normalized mean velocity U for $r < 1$ by

$$U^2 = \frac{1}{\pi}\int\int_{r<1} |u|^2 dxdy = 2\int_0^1 |u|^2 r\, dr.$$

Particle error, e_{part}: Let N be the number of particles used in the calculation. The mean-square average error in their velocities is E_{part}, with

$$(E_{\text{part}})^2 = \frac{1}{N}\sum_{j=1}^{N} |u_{j,\text{comp}} - u_{j,\text{exact}}|^2.$$

The nondimensional relative mean-square error in the velocity field is then $e_{\text{part}} = E_{\text{part}}/U$.

Ray error, e_{ray}: We measure the error along the ray $\{(x_1, x_2)|0 < x_1 < 1; x_2 = 0\}$. The rotation symmetry of the problem indicates that this error is typical of all rays going out from the origin. The mean-square average error is

$$\frac{1}{\pi}\cdot 2\pi \int_0^1 |u_{\text{comp}}(x,t) - u_{\text{exact}}(x,t)|^2 x\, dx.$$

To compute this integral, sample the computed velocity at the 10 points along the ray, $x = j/10$, $j = 1, \ldots, 10$, $y = 0$, and use the trapezoidal rule given by

$$E_{\text{ray}}^2 = 2\sum_{j=1}^{1} 0|u_{\text{comp}}(j/10,t) - u_{\text{exact}}(j/10,t)|^2 \frac{j}{100} f_j\, dx,$$

where $f_j = 1$, $j = 1, \ldots, 9$, and $f_{10} = \frac{1}{2}$. The nondimensional relative mean-square error in the ray velocity is $e_{\text{ray}} = E_{\text{ray}}/U$.

This second error is useful in determining the accuracy of the field away from the particles themselves. Such accuracy is important, for example, in problems with boundaries that require the computation of an irrotational flow to satisfy the boundary conditions and also for passive tracer fields.

We now summarize the results of simulations from Beale and Majda (1985) by using the higher-order kernels

$$K_\delta^{(m)}(x) = K(x)\left[1 - Q_m\left(\frac{r}{\delta}\right)e^{-r^2/\delta}\right]$$

with orders $m = 2, 4, 6, 8$. Note that $m = 0$ is the point-vortex method in which u_{comp} and $u_{j,\text{comp}}$ are computed by means of

$$u_{\text{comp}}(x,t) = \sum_k K(x - \tilde{x}_k)\omega_k h^2,$$

$$u_{j,\text{comp}} = \sum_k K(\tilde{x}_j - \tilde{x}_k)\omega_k h^2.$$

We compute the time step by using a standard fourth-order Runge–Kutta scheme with dt sufficiently small to produce errors dominated by the spatial discretization. Table 6.1 shows the errors that are due to various orders of smoothing with $h = 0.125$ and 208 particles with nonzero vorticity.

Figure 6.4 indicates that an initially square grid undergoes much distortion by time $T = 6$; hence the numerical simulation of this problem up to time $T = 12$ is a good test for the inviscid 2D vortex method.

Table 6.1. *Errors that are due to various orders of smoothing*

	m:	0–Point Vortices	2	3	6	8
Time (T)	$c_m = \delta/h$:		1	2	2.5	2.5
0	e_{part}	0.009	0.027	0.012	0.0054	0.0015
	e_{ray}	0.021	0.028	0.012	0.0053	0.0015
3	e_{part}	0.013	0.027	0.012	0.0054	0.0016
	e_{ray}	0.128	0.028	0.012	0.0053	0.0015
6	e_{part}	0.033	0.028	0.012	0.0054	0.0017
	e_{ray}	0.159	0.028	0.012	0.0053	0.0016
9	e_{part}	0.050	0.030	0.013	0.0060	0.0046
	e_{ray}	0.111	0.029	0.012	0.0053	0.0040
12	e_{part}	0.051	0.034	0.014	0.0077	0.0086
	e_{ray}	0.366	0.033	0.014	0.0086	0.0111

$\omega(r) = (1 - r^2)^3$, $h = 0.125$ and 208 particles with nonzero vorticity.

A striking feature of Table 6.1 is the comparatively poor performance of the point-vortex method in the ray error. In particular, at $T = 12$, this error has grown to 36% of the average velocity. This indicates that a vortex-blob method is preferable to a point-vortex method for any simulation that requires accurate representations of the velocity field off of the particle paths. Examples include flows with boundaries and smooth inviscid shear layers.

Another feature in Table 6.1 is the reliability of smoothed-core vortex methods in both the particle and the ray errors.

6.6. The Random-Vortex Method in Two Dimensions

So far we have discussed in detail both 2D and 3D inviscid-vortex methods for computing solutions to the Euler equation. In Section 6.1 we introduced, by means of a simple model problem, a way of solving the viscous Navier–Stokes equations by means of a vortex method with particle diffusion. In this section we present the details of the random-vortex method for the general 2D case. The idea is to implement an inviscid-vortex method coupled with a Brownian motion of the vortex blobs to simulate diffusion. 3D random-vortex methods are discussed in the notes at the end of this chapter.

6.6.1. Motivation for the Need for a Stochastic Particle Method: The Failure of Core Spreading

Recall that the inviscid-vortex method approximates the vorticity by a finite number of radially symmetric blobs and then advects the blobs by their own velocity field. For the viscous 2D Navier–Stokes equations, we showed in Example 2.2 of Chap. 2 that, for radial functions, the 2D Navier–Stokes equation reduces to the radially symmetric heat equation in $\mathbb{R}^2$. Thus we might guess that a natural way to incorporate viscosity into a vortex method is to have the blobs "spread" by means of solutions to the radially symmetric heat equation. This idea led to a class of methods known as *core spreading*.

In 1985, Greengard (Greengard, 1985) showed that for general 2D flows the core-spreading method approximates an equation different from the Navier–Stokes equation. The problem is that the actions of advection and diffusion do not commute. We rederive his calculation here because it follows naturally from the machinery we have developed so far.

The original core-spreading idea involves advecting the particles by a velocity field

$$v(x,t) = \sum_{i \in \mathbb{Z}^2} K_2^{\delta(t)}[x - X_i(t)]\omega_i h^2,$$

where

$$K_2^{\delta(t)} = G_t * K_2,$$
$$G_t(x) = \frac{1}{4\pi t \nu} e^{-|x|^2/4t\nu}$$

is the standard heat kernel. Note that as $h \to 0$, we are solving the equation

$$\frac{\partial \tilde{X}(\alpha,t)}{\partial t} = \int_{\mathbb{R}^2} K_2^{\delta(t)}[\tilde{X}(\alpha,t) - \tilde{X}(\alpha',t)]\omega_o(\alpha')d\alpha'.$$

Define

$$\tilde{v}(x) = \int_{\mathbb{R}^2} K_2^{\delta(t)}[x - \tilde{X}(\alpha',t)]\omega_o(\alpha')d\alpha'.$$

Let ξ denote the passive transport of ω_0 by the particle-trajectory map $\tilde{X}$, i.e.,

$$\xi(\tilde{X}(\alpha,t),t) = \omega_0(\alpha),$$

so that in Eulerian variables

$$\frac{\partial \xi}{\partial t} + \tilde{v} \cdot \nabla \xi = 0.$$

Note that $\tilde{\omega} = \tilde{v}$ satisfies

$$\begin{aligned}\tilde{\omega} &= \text{curl} \int K_2^{\delta(t)}[x - \tilde{X}(\alpha',t)]\omega_0(\alpha')d\alpha' \\ &= \text{curl} \int K_2[x - \tilde{X}(\alpha',t)]G_t * \omega_0(\alpha')d\alpha' \\ &= G_t * \xi.\end{aligned}$$

We now compare $\tilde{\omega}$ with a solution of the Navier–Stokes equations (vorticity-stream form)

$$\frac{\partial \omega}{\partial t} + v \cdot \nabla \omega = \nu \Delta \omega$$

that have the same initial data ω_0.

Core spreading gives

$$\begin{aligned}\frac{\partial \tilde{\omega}}{\partial t} &= \Delta(G_t * \xi) + G_t * \frac{\partial \xi}{\partial t} \\ &= \nu \Delta \tilde{w} - G_t * (\tilde{v} \cdot \nabla \xi).\end{aligned}$$

Note that at $t=0$, $(\partial\tilde{\omega}/\partial t)$ and $(\partial\omega/\partial t)$ and hence $(\partial\tilde{v}/\partial t)$ and $(\partial v/\partial t)$ are equivalent.

However,

$$\frac{\partial^2\tilde{\omega}}{\partial t^2} = \nu\Delta\frac{\partial\tilde{\omega}}{\partial t} - \nu\Delta(\tilde{v}\cdot\nabla\xi) - G_t * \left(\frac{\partial\tilde{v}}{\partial t}\cdot\nabla\xi\right) + G_t * [\tilde{v}\cdot\nabla(\tilde{v}\cdot\nabla\xi)],$$

and the Navier–Stokes equation satisfies

$$\frac{\partial^2\omega}{\partial t^2} = \nu\Delta\frac{\partial\omega}{\partial t} - \frac{\partial v}{\partial t}\cdot\nabla\omega - \nu v\cdot\nabla(\Delta\omega) + v\cdot\nabla(v\cdot\nabla\omega),$$

so that, in particular, at $t=0$,

$$\frac{\partial^2\tilde{\omega}}{\partial t^2} - \frac{\partial^2\omega}{\partial t^2} = \nu[v_0\cdot\nabla(\Delta\omega_0) - \Delta(v_0\cdot\nabla\omega_0)],$$

which is in general nonzero.

Note that for radial solutions the two equations are equivalent.

6.6.2. Implementation of the Random-Vortex Method

The random-vortex method views the motion of fluid particles as a system of interacting diffusion processes. The Navier–Stokes equation for incompressible viscous fluids,

$$\frac{\partial v}{dt} + v\cdot\nabla v = -\nabla p + \nu\Delta v,$$
$$\operatorname{div} v = 0,$$

yields in two dimensions the vorticity-stream form

$$\frac{\partial\omega}{dt} + v\cdot\nabla\omega = \nu\Delta\omega, \tag{6.110}$$
$$v(x,t) = K_2 * \omega(x,t). \tag{6.111}$$

The 2D inviscid-vortex method exploited the fact that in two dimensions vorticity is conserved along particle trajectories. This is not the case when diffusion is present; however, we can simulate its effects by means of Brownian motion. We begin by noting that a solution to Eq. (6.110) satisfies a maximum principle,

$$\sup_{x\in\mathbb{R}^2}\omega(x,t) \le \sup_{x\in\mathbb{R}^2}\omega(x,s) \qquad \text{for} \quad t\ge s.$$

Furthermore, the total vorticity

$$V(t) = \int_{\mathbb{R}^2}\omega(x,t)dx$$

is conserved.

The design of a random-vortex method requires a probabilistic interpretation of Eq. (6.110). We have already discussed this interpretation in Section 6.1 on an exactly solvable model problem. The reader might find it useful to review the discussion there. First we use the incompressibility condition $\operatorname{div} v = 0$ to rewrite Eq. (6.110) in divergence form:

$$\frac{\partial \omega}{dt} + \operatorname{div}(vw) = \nu \Delta \omega. \tag{6.112}$$

For a given velocity field v, Eq. (6.112) is the forward or Fokker–Planck equation for a diffusion process $X(\alpha; t)$ satisfying

$$\begin{aligned} dX(\alpha; t) &= v(X(\alpha; t), t)dt + \sqrt{2\nu}\, dW(t), \\ X(\alpha; 0) &= \alpha. \end{aligned} \tag{6.113}$$

$W(t)$ is a standard Wiener process (Brownian motion) in $\mathbb{R}^2$. The 1D Wiener process was introduced in Case 2 in Subsection 6.1.1. Here we take the obvious generalization to two dimensions. In this case, the diffusion process has the physical interpretation of particle paths. PDE (6.112) is then thought of as an evolution equation for the probability density $p(x, t)$ of $X(\alpha; t)$. Because the diffusion coefficient $\sqrt{2\nu}$ is a constant in Eq. (6.113) we can directly integrate Eq. (6.113) in time without introducing an Ito integral:

$$X(\alpha; t) = \alpha + \int_0^t v(X(\alpha; s), s)ds + \sqrt{2\nu} W(t). \tag{6.114}$$

With a probability of 1, $W(t)$ has continuous sample paths and Eq. (6.114) can be solved sample path by sample path. For each continuous sample path $\xi(t)$ of the process $W(t)$, the deterministic integral equation

$$\xi(\alpha; t) = \alpha + \int_0^t v(\xi(\alpha; s), s)ds + \sqrt{2\nu}\xi(t) \tag{6.115}$$

has a unique continuous solution. We use this special property to prove convergence of the random-vortex method. The fundamental solution (the Green's function) $G(x, t; \alpha, s)$ of PDE (6.112) is the transition probability density of the diffusion process X in Eq. (6.114). That is, $G(x, t; \alpha, s)$ is the probability that a particle reaches the position x at time t from position α and time $s < t$.

A probability density $p_0(x)$ in $\mathbb{R}^2$ is well approximated weakly by N particles with equal weights N^{-1} at time $t = 0$ if any small cube centered at x with side length Δx satisfies

$$p_0(x) \cdot (\Delta x)^2 \sim \Delta N / N$$

for any $x \in \mathbb{R}^2$, where ΔN is the number of particles in the cube. If we let the particles move according to the diffusion process in Eq. (6.114) then the ensemble of N particles at time $t > 0$ approximates the probability density

$$p(x, t) = \int G(x, t; \alpha, 0) p_0(\alpha) d\alpha,$$

provided that N is large enough. Because $p(x, t)$ is the unique solution of PDE (6.112) with initial data $p_0(x)$, we can compute $p(x, t)$ or any continuous functional of $p(x,t)$, for example $K_\delta * p(x,t)$, by simulating the diffusion process X. The particles may carry different weights with positive or negative signs, corresponding to the value of the vorticity at that point.

6.6.3. Formulation and Main Theorem

From the above discussion, the numerical method is designed from SDE (6.113). We can write the velocity in the following way:

$$\begin{aligned} v(x,t) &= \int_{\mathbb{R}^2} K(x-y)\omega(y,t)dy \\ &= \int_{\mathbb{R}^2} K(x-y)\left[\int_{\mathbb{R}^2} G(y,t;\alpha,0)\omega(\alpha,0)d\alpha\right]dy \\ &= \int_{\mathbb{R}^2}\left[\int_{\mathbb{R}^2} K(x-y)G(y,t;\alpha,0)dy\right]\omega(\alpha,0)d\alpha \\ &\int_{\mathbb{R}^2} E[K(x-X(\alpha;t)]\omega(\alpha,0)d\alpha, \end{aligned}$$

where EY denotes the expectation of the random vector Y. It follows that SDE (6.113) is equivalent to

$$dX(\alpha;t) = \left(\int_{\mathbb{R}^2} E'\{K[X(\alpha;t)-X(\alpha';t)]\}\omega(\alpha',0)d\alpha'\right)dt + \sqrt{2\nu}\,dW(t), \tag{6.116}$$

where $E'K[X(\alpha;t)-X(\alpha';t)]$ is $EK[x-X(\alpha';t)]$ evaluated at $x = X(\alpha;t)$. Equation (6.116) closely resembles particle-trajectory formulation (6.1) in the inviscid case. When the same discretization as in inviscid-vortex method (6.78) is used, the random-vortex method approximates Eq. (6.116) by a diffusion process $\tilde{X}_i(t)$ satisfying the system of SDEs

$$d\tilde{X}_i(t) = \left\{\sum_j K_\delta[\tilde{X}_i(t)-\tilde{X}_j(t)]\omega_j h^2\right\}dt + \sqrt{2\nu}\,dW_i(t) \tag{6.117}$$

with the initial data

$$\tilde{X}_i(0) = \alpha_i = h\cdot i.$$

Here the $W_i(t)$ are independent standard Wiener processes in $\mathbb{R}^2$. Likewise we denote by $X_i(T)$ the solution of the system of SDEs

$$dX_i(t) = v(X_i(t),t)dt + \sqrt{2\nu}\,dW_i(t) \tag{6.118}$$

with the same initial data $X_i(0) = \alpha_i$ and where $v(\cdot,t)$ is the unique solution of the Navier–Stokes equation. The main difficulty in proving convergence of the random-vortex method lies in controlling the sampling error that is due to the Browning motion of the particles. In this section we prove the following sharp convergence theorem.

Theorem 6.5. *Convergence of 2D Random-Vortex Methods. Given the same assumptions as in Theorem 6.3, in particular $\psi \in M^{L,m}$, $m \geq 2$, for all $0 \leq t \leq T$, we have the following estimates:*

(i) Convergence of particle paths:

$$\max_{0\leq t\leq T} \|\tilde{X}_i(t) - X_i(t)\|_{L_h^p} \leq C\left[\delta^m + (h/\delta)^L\delta + h|\ln h|\right].$$

(ii) Convergence of discrete velocity:

$$\max_{0\leq t\leq T} \left\|\tilde{v}_i^h(t) - v(X_i(t), t)\right\|_{L_h^p} \leq C\left[\delta^m + (h/\delta)^L\delta + h|\ln h|\right].$$

(iii) Convergence of continuous velocity:

$$\max_{0\leq t\leq T} \|\tilde{v}^h(\cdot, t) - v(\cdot, t)\|_{L_h^p} \leq C\left[\delta^m + (h/\delta)^L\delta + h|\ln h|\right],$$

except for an event of probability less than $h^{C'C}$, provided that $C \geq C''$, where the constants $C', C'' > 0$ depend on only the same parameters as those described in Theorem 6.3.

We can choose a larger constant C to make the probability $h^{C'C}$ approach zero faster. There is a trade-off between the accuracy and the probability.

Remark: The sharp convergence bound $h|\ln h|$ is due to the sampling error. The central limit theorem guarantees that this is the best bound attainable when a linear heat equation is approximated by a discretized Brownian motion.

Notation: We use the shorthand $a \preceq b$ in the rest of this section to denote $a \leq b$ except for an event of probability's approaching zero faster than any polynomial rate by choosing the constant C sufficiently large.

6.6.4. Preliminary Estimates

Because the initial positions of the particles are chosen on the lattice points instead of being chosen randomly, the following lemma for the quadrature error is essential. It generalizes the corresponding inviscid Lemma 6.2 for the trapezoidal quadrature to the case with diffusion.

Lemma 6.7. *Key Stochastic Quadrature Lemma. Let $X(\alpha; t)$ be the solution of the SDE*

$$dX(\alpha; t) = v(X(\alpha; t), t)dt + \sqrt{2\nu}\, dW(t)$$

in $\mathbb{R}^2$ with initial data $X(\alpha; 0) = \alpha \in \mathbb{R}^2$. Assume that $v \in C^L(\mathbb{R}^2 \times [0, T])$, $L \geq 3$, that is, all the spatial derivatives of $v(\cdot, t)$ up to the order of L are uniformly bounded. Given $\mathbb{R}^2$ – valued functions $f \in C_0^L(\mathbb{R}^2)$, $g \in C_0^L(\mathbb{R}^2)$, define $\Gamma(\alpha, t) =$

$\{Ef[X(\alpha;t)]\} \cdot g(\alpha)$ *so that the support of* Γ *is contained in a bounded domain* $\text{supp}\,\Gamma \subset \text{supp}\, g$. *Then we have the following estimate for the quadrature error:*

$$\max_{0\le t\le T}\left|\sum_{i\in\mathbb{Z}^2}\Gamma(h\cdot i,t)h^2-\int_{\mathbb{R}^2}\Gamma(\alpha,t)d\alpha\right|$$

$$\le Ch^L \max_{0\le|\beta|\le L}\|\partial^\beta g\|_{L^\infty} \tag{6.119}$$

$$\times\left[\sum_{0\le|\beta|\le L}\int_{|x|\le R}|\partial^\beta f(x)|dx+\sum_{0\le|\beta|\le L}\sup_{|x|>R}|\partial^\beta f(x)|\right],$$

where $R>0$ *is arbitrary and* C *depends on only* T, L, *the diameter* Ω, *and* $\max_{1\le|\gamma|\le L}\|\partial^\gamma v\|_{L^\infty(\mathbb{R}^2\times[0,T])}$.

The proof of Lemma 6.7 is presented in the appendix. We also need Bennett's inequality, an elementary lemma on large deviations, which we recall from Section 6.1.

Lemma 6.1. *Bennett's Inequality (Pollard, 1984, Appendix B). Let* Y_i *be independent bounded random variables (not necessarily identically distributed) with mean zero, variances* σ_i^2, *and* $|Y_i|\le M$. *Let* $S=\sum_i Y_i$, $V\ge\sum_i\sigma_i^2$; *then, for all* $\eta>0$,

$$P\{|S|\ge\eta\}\le 2\exp\left[-\frac{1}{2}\eta^2V^{-1}B(M\eta V^{-1})\right], \tag{6.120}$$

where $B(\lambda)=2\lambda^{-2}[(1+\lambda)\ln(1+\lambda)-\lambda]$, $\lambda>0$, $\lim_{\lambda\to0^+}B(\lambda)=1$, *and* $B(\lambda)\sim 2\lambda^{-1}\ln\lambda$ *as* $\lambda\to\infty$. *If* Y_i *are random vectors in* $\mathbb{R}^2$, *relation (6.120) applied to each component gives*

$$P\{|S|\ge\eta\}\le 4\exp\left[-\frac{1}{4}\eta^2V^{-1}B(M\eta V^{-1})\right].$$

The proof of Lemma 6.1 can also be found in the appendix.

6.6.5. Consistency Error for a Fixed Time

We decompose the consistency error into three components:

$$\begin{aligned}|v^h(x,t)-v(x,t)| &= \left|\sum_i K_\delta[x-X_i(T)]\omega_i h^2 - v(x,t)\right|\\ &\le \left|\sum_i K_\delta[x-X_i(t)]\omega_i h^2-\sum_i EK_\delta[x-X_i(t)]\omega_i h^2\right|\\ &\quad+\left|\sum_i EK_\delta[x-X_i(t)]\omega_i h^2-\int_{\mathbb{R}^2}K_\delta(x-y)\omega(y,t)dy\right|\\ &\quad+\left|\int_{\mathbb{R}^2}K_\delta(x-y)\omega(y,t)dy-\int_{\mathbb{R}^2}K(x-y)\omega(y,t)dy\right|\\ &= \textit{sampling error} + \textit{discretization error} + \textit{moment error}.\end{aligned}$$

The *moment error* is exactly the same as in the inviscid case. From Lemma 6.5, relation (6.98),

$$\left| \int_{\mathbb{R}^2} K_\delta(x-y)\omega(y,t)dy - \int_{\mathbb{R}^2} K(x-y)\omega(y,t)dy \right| \le C\delta^m.$$

Note that in the *discretization error* $\int K_\delta(x-y)\omega(y,t)dy$ is a disguised version of $\int E\{K_\delta[x - X(\alpha;t)]\}\omega(\alpha,0)d\alpha$. Applying Lemma 6.7 with $f(y) = K_\delta(x-y)$, $g(\alpha) = \omega(\alpha, 0)$ gives

$$\left| \sum_i E K_\delta[x - X_i(t)]\omega_i h^2 - \int_{\mathbb{R}^2} E\{K_\delta[x - X(\alpha;t)]\}\omega(\alpha,0)d\alpha \right|$$
$$\le Ch^L \left[\sum_{0\le|\gamma|\le L} \int_{|x-y|\le R} |\partial^\gamma K_\delta(x-y)dy + \sum_{0\le|\gamma|\le r} \sup_{|x-y|>R} |\partial^\gamma K_\delta(x-y)| \right].$$

Choose $R = \max[1, \text{diameter}(\Omega)]$. Proposition 6.1 then gives

$$\left| \sum_i E K_\delta[x - X_i(t)]\omega_i h^2 - \int_{\mathbb{R}^2} K_\delta(x-y)\omega(y,t)dy \right| \le C(h/\delta)^L\delta,$$

where the constant C depends on only L, the L^∞ bounds of the velocity field $v(x,t)$, and its spatial derivatives up to the order of $L+1$ and the diameter of Ω.

The main step in this section is to estimate the *sampling error* by application of Bennett's inequality (Lemma 6.1).

Let

$$Y_i = \omega_i h^2\{K_\delta[x - X_i(t)] - E K_\delta[x - X_i(t)]\}.$$

We have $EY_i = 0$, $|Y_i| \le Ch^2\delta^{-1} \equiv M$, and

$$\sum_i \text{Var}\, Y_i = h^2 \sum_i \{E|K_\delta[x - X_i(t)]|^2 - |E K_\delta[x - X_i(t)]|^2\}\omega_i^2 h^2$$
$$\le h^2 \sum_i \{E|K_\delta[x - X_i(t)]|^2\}\omega_i^2 h^2.$$

For simplicity we drop the term $|E K_\delta[x - X_i(t)]|^2$, thus losing information about the dependence of the sampling errors on the viscosity ν. We apply Lemma 6.7 again with $f(y) = |K_\delta(x-y)|^2$, $g(\alpha) = \omega^2(\alpha,0)$ to approximate $\sum_i E|K_\delta[x - X_i(t)]|^2\omega_i^2 h^2$ by the integral

$$\int_\Omega E|K_\delta[x - X(\alpha;t)]|^2\omega(\alpha,0)d\alpha \tag{6.121}$$

with an error of $C(h/\delta)^L$, which follows from Proposition 6.1 and the fact that

$|K(x)|^2 = \text{constant}\cdot|\nabla K(x)|$. Note that

$$\begin{aligned}
&\int_\Omega E|K_\delta[x - X(\alpha;\,,t)]|^2\omega^2(\alpha,0)d\alpha \\
&\quad = \int_\Omega\int_{\mathbb{R}^2} |K_\delta(x-y)|^2 G(y,t;\alpha,0)\omega^2(\alpha,0)dyd\alpha \\
&\quad = \int_{\mathbb{R}^2} |K_\delta(x-y)|^2\left[\int_\Omega G(y,t;\alpha,0)\omega^2(\alpha,0)d\alpha\right]dy,
\end{aligned}$$

where G is the Green's function associated with PDE (6.112). Because $\int_\Omega G(y,t;\alpha,0)\omega^2(\alpha,0)d\alpha$ is a solution of both Eqs. (6.110) and (6.112), its L^∞ and L^1 norms are bounded by those of the initial data. Choosing $R = \max[1, \text{diameter}(\Omega)]$ bounds integral (6.121) by

$$\begin{aligned}
&C_1\left[\int_{|x-y|\le R} |K_\delta(x-y)|^2 dy + \sup_{|x-y|>R} |K_\delta(x-y)|^2\right] \\
&\quad = C_1\left[\int_{0\le|x|\le\delta} |K_\delta(x)|^2 dx + \int_{\delta\le|x|\le R} |K_\delta(x)|^2 dx + C_2\right] \\
&\quad \le C_3\left(\delta^{-2}\cdot\delta^2 + \ln\frac{R}{\delta} + C_2\right) \\
&\quad \le C_4|\ln\delta|, \\
&\qquad\qquad \sum_i \operatorname{Var} Y_i \le Ch^2|\ln\delta| \equiv V,
\end{aligned}$$

where the constant C depends on only L, the L^∞ bound on the velocity $v(x,t)$ and its spatial derivatives up to the order of $L+1$ and the diameter of Ω. Bennett's inequality (Lemma 6.1) implies that

$$\begin{aligned}
&P\left\{\left|\sum_i Y_i\right| \ge Ch|\ln h|\right\} \\
&\quad \le 4\exp\left\{-\frac{1}{4}C^2(h|\ln h|)^2V^{-1}B[M(Ch|\ln h|)V^{-1}]\right\} \\
&\quad \le \exp[-C_1C^2|\ln h|^2|\ln\delta|^{-1}\cdot B(C_2Ch\delta^{-1}|\ln h||\ln\delta|^{-1})] \\
&\quad \le \exp(-C_3C|\ln h|^2|\ln\delta|^{-1}) \\
&\quad \le \exp(-C_3C|\ln h|)\ (\text{because } h\le\delta) \\
&\quad = h^{C_3C}.
\end{aligned}$$

The above arguments bound the consistency error at fixed x and t:

$$|v^h(x,t) - v(x,t)| \le C[\delta^m + (h/\delta)^L\delta + h|\ln h|], \tag{6.122}$$

except for an event of probability of less than h^{C_3C} with $C_3 > 0$. For the lattice points

$z_k = h \cdot k$ in any ball $B(R_0)$, it follows from the pointwise estimate above that

$$\max_k |v^h(z_k, t) - v(z_k, t)| \leq C\left[\delta^m + (h/\delta)^L \delta + h|\ln h|\right], \tag{6.123}$$

except for an event of probability of less than $C_4 h^{-2} \cdot h^{C_3 C}$ with some constant $C_4 > 0$, provided that C is large enough. Let X_i' be an independent copy of X_i. Consistency bound (6.122) and $K_\delta(0) = 0$ imply that

$$\max_i \left| v_i^h(t) + K_\delta[X_i(t) - X_i'(t)]\omega_i h^2 - v(X_i(t), t)\right| \preceq C\left[\delta^m + (h/\delta)^L \delta + h|\ln h|\right].$$

Because $|K_\delta(x)| \leq C\delta^{-1}$,

$$\max_i \left| v_i^h(t) - v(X_i(t), t)\right| \preceq C\left[\delta^m + (h/\delta)^L \delta + h|\ln h|\right]. \tag{6.124}$$

Relations (6.123) and (6.124) are the consistency estimates for the discrete velocities at any fixed time t.

6.6.6. Stability Estimate for All Time

As in inviscid Lemma 6.4 we have a stability result for the random-vortex method.

Lemma 6.8. *Stability Error for 2D Random-Vortex Methods. Assume that*

$$\max_{0 \leq t \leq T_*} \max_{h \cdot i \in \Lambda^h} |\tilde{X}_i(t) - X_i(t)| \preceq C_0 \delta$$

for some $T_* \leq T$; then

$$\left\| \tilde{v}_i^h(t) - v_i^h(t) \right\|_{L_h^p} \preceq C \|\tilde{X}_i(t) - X_i(t)\|_{L_h^p}$$

uniformly for $t \in [0, T_*]$, where the constant C depends on only T, L, p, q, the diameter of supp ω_0, *and the bounds for a finite number of derivatives of the velocity field v. C is independent of T_*.*

As in the proof of Lemma 6.4, we write

$$\begin{aligned}
\tilde{v}_i^h - v_i^h &= \sum_j [K_\delta(X_i - \tilde{X}_j) - K_\delta(X_i - X_j)]\omega_j h^2 \\
&\quad + \sum_j [K_\delta(\tilde{X}_i - \tilde{X}_j) - K_\delta(X_i - \tilde{X}_j)]\omega_j h^2 \\
&= v_i^{(1)} + v_i^{(2)}.
\end{aligned}$$

In the proof of Lemma 6.4 we estimate $v_i^{(1)}$ by considering a partition of the support of $\omega(\cdot, t)$ into curvilinear quadrilaterals $B_i = \Phi^t(Q_i)$, where Φ^t is the flow map determined by the velocity field v and Q_i is the square determined by the lattice Λ^h. The fact that the flow map is volume preserving allowed for a direct application of the Calderon–Zygmund inequality to estimate $v_i^{(1)}$.

For the random-vortex method this approach requires modification because the positions of the vortices $\tilde{X}_i(t)$, $X_j(t)$ are random. In fact it is not necessary to use the flow map to define a partition of $\mathbb{R}^2$. We merely require a partition satisfying the following conditions:

(1) the sizes of the cells are of the order of δ or less,
(2) the densities of the vortices in the cells are uniformly bounded with high probability in order to apply the Calderón–Zygmund inequality to estimate $v_i^{(1)}$.

An obvious choice for the partition is $\{Q_k\}$, the squares Q_k centered at lattice points $\delta \cdot k$, with the side length δ and where $k \in \mathbb{Z}^2$.

The following lemma concerns the uniform boundedness of the density of the vortices in balls. It is equivalent to the uniform boundedness in squares.

Lemma 6.9. *Let $N(x, r, t) = \#\{X_i(t) : |X_i(T) - x| \le r\}$, with $r > 0$, $0 \le t \le T$, be the number of vortices in the ball $B(x, r)$. If $r \ge h|\ln h|$, then*

$$h^2 \cdot N(x, r, t) \preceq Cr^2.$$

For a proof see the appendix. We apply this lemma with $r \sim \delta$. The condition $r \ge h|\ln h|$ is satisfied if we choose $\delta = h^q, 0 \le q \le 1$. Next we give a generalization of some of the estimates from Proposition 6.1.

Lemma 6.10. *Let $M_{ij}^{(l)} = \max_{|y| \le C_0\delta} \max_{|\beta|=l} |\partial^\beta K_\delta(X_i - X_j + y)|$. Then*

$$\sum_j M_{ij}^{(l)} h^2 \preceq \begin{cases} C|\ln \delta| & \text{if } l = 1 \\ C\delta^{-1} & \text{if } l = 2 \end{cases}. \tag{6.125}$$

For a proof the reader is referred to Long (1988). Finally we are ready to estimate $v^{(1)}$. We write

$$\begin{aligned} v^{(1)} &= \sum_j [K_\delta(X_i - \tilde{X}_j) - K_\delta(X_i - X_j)]\omega_j h^2 \\ &= \sum_j \nabla K_\delta(X_i - X_j + Y_{ij}) \cdot e_j \omega_j h^2, \end{aligned}$$

where $e_j = X_j - \tilde{X}_j$ and we ignore the fact that Y_{ij} may depend on the components. Let $Z_i \in \delta \cdot \mathbb{Z}^2$ be the closest lattice point to X_i. If there is more than one lattice point closest to X_i, then we chose one of them arbitrarily. Then

$$v_i^{(1)} = \sum_j \nabla K_\delta(Z_i - Z_j) \cdot e_j \omega_j h^2 + r_i^{(1)},$$

where

$$r_i^{(1)} = \sum_j [\nabla K_\delta(X_i - X_j + Y_{ij}) - \nabla K_\delta(Z_i - Z_j)] \cdot e_j \omega_j h^2.$$

For each $z_k = \delta \cdot k = \delta \cdot (k_1, k_2)$, we define f_k to be the average of all $e_j \omega_j$, where X_j is in the square Q_k, namely,

$$f_k = \delta^{-2} \sum_{X_j \in Q_k} e_j \omega_j h^2$$

with the convention that $f_k = 0$ if Q_k contains none of the vortices X_j. It follows from Lemma 6.9 that

$$\begin{aligned}
\|f_k\|_{L_h^p}^p &= \sum_k \delta^{-2p+2} \left| \sum_{X_j \in Q_k} e_j \omega_j h^2 \right|^p \\
&\preceq \sum_k \delta^{-2p+2} \cdot (C\delta^2)^{p-1} \sum_{X_j \in Q_k} |e_j \omega_j|^p h^2 \\
&= C^{p-1} \|e_j \omega_j\|_{L_h^p}^p.
\end{aligned}$$

Furthermore

$$\left\| \sum_j \nabla K_\delta(Z_i - Z_j) \cdot e_j \omega_j h^2 \right\|_{L_h^p} \preceq C \left\| \sum_{k'} \nabla K_\delta(z_k - z_{k'}) \cdot f_{k'} \delta^2 \right\|_{L_h^p}$$

because

$$\begin{aligned}
\left\| \sum_j \nabla K_\delta(Z_i - Z_j) \cdot e_j \omega_j h^2 \right\|_{L_h^p}^p &= \left\| \sum_{k' \in \mathbb{Z}^2} \nabla K_\delta(Z_i - z_{k'}) \cdot f_{k'} \delta^2 \right\|_{L_h^p}^p \\
&= \sum_i \left| \sum_{k'} \nabla K_\delta(Z_i - z_{k'}) \cdot f_{k'} \delta^2 \right|^p \cdot h^2 \\
&= \sum_i \left| \sum_{k'} \nabla_\delta(Z_i - z_{k'}) \cdot f_{k'} \delta^2 \right|^p \cdot h^2 \\
&= \sum_k \#(X_i \in Q_k) \cdot \left| \sum_{k'} \nabla_\delta(z_i - z_{k'}) \cdot f_{k'} \delta^2 \right|^p \cdot h^2 \\
&\preceq C \sum_k \left| \sum_{k'} \nabla_\delta(z_i - z_{k'}) \cdot f_{k'} \delta^2 \right|^p \cdot \delta^2 \\
&= C \left\| \sum_{k'} \nabla_\delta(z_i - z_{k'}) \cdot f_{k'} \delta^2 \right\|_{L_\delta^p}^p
\end{aligned}$$

by Lemma 6.9.

Following the same procedure as in the inviscid case we have

$$\left\| \sum_{k \in \mathbb{Z}^2} \nabla K_\delta(z_i - z_k) \cdot f_k \delta^2 \right\|_{L_\delta^p} \leq C \|f_k\|_{L_\delta^p},$$

which implies that

$$\left\| \sum_j \nabla K_\delta(Z_i - Z_k) \cdot e_j \omega_j h^2 \right\|_{L^p_\delta} \preceq C \|e_j \omega_j\|_{L^p_h},$$

provided that $\delta \le h|\ln h|$. By the mean-value theorem,

$$r_i^{(1)} = \sum_j [\Delta K_\delta(X_i - X_j + Y_{ij} + Y''_{ij}) \cdot Y'_{ij}] \cdot e_j \omega_j h^2,$$

where $Y'_{ij} = Y_{ij} + (X_i - Z_i) - (X_i - Z_j)$. We ignore the dependence of Y'_{ij} and Y''_{ij} on the components. Because $|Y_{ij}| \le \delta$, $|Y''_{ij}| \le |Y'_{ij}| \le 4\delta$,

$$|r_i^{(1)}| \le \sum_j M_{ij}^{(2)} \cdot 4\delta \cdot |e_j \omega_j| h^2.$$

By Young's inequality and Lemma 6.10 with $C_0 = 5$ we conclude that

$$\begin{aligned} \|r_i^{(1)}\|_{L^p_h} &\le 4\delta \cdot \max \sum_j M_{ij}^{(2)} h^2 \cdot \sum_i M_{ij}^{(2)} h^2 \cdot \|e_j \omega_j\|_{L^p_h} \\ &\preceq C \|e_j \omega_j\|_{L^p_h}. \end{aligned}$$

This finishes the estimate of $v^{(1)}$.

To estimate $v^{(2)}$ we apply the mean-value theorem:

$$\begin{aligned} v_i^{(2)} &= \sum_j \nabla K_\delta(X_i - X_j + Y_{ij}) \cdot (\tilde{X}_i - X_i) \cdot \omega_j h^2 \\ &= \left[\sum_j \nabla K_\delta(X_i - X_j + Y_{ij}) \cdot \omega_j h^2 \right] \cdot (\tilde{X}_i - X_i). \end{aligned}$$

We want to show that

$$\max_i \left| \sum_j \nabla K_\delta(X_i - X_j + Y_{ij}) \cdot \omega_j h^2 \right| \preceq C.$$

Because $|\nabla K_\delta(X_i - X_j + Y_{ij}) - |\nabla K_\delta(X_i - X_j)| \le \delta M_{ij}^{(2)}$, it suffices to prove the uniform boundedness of $\sum_j \nabla K_\delta(X_i - X_j) \cdot \omega_j h^2$ by Lemma 6.10 and Young's inequality as in the estimate for $r^{(1)}$. The proof is essentially the same as the consistency estimate, with K_δ replaced with ∇K_δ. For any $x \in \mathbb{R}^2$

$$\begin{aligned} \sum_j \nabla K_\delta(x - X_j) \cdot \omega_j h^2 &= \sum_j E \nabla K_\delta(x - X_j) \cdot \omega_j h^2 \\ &\quad + \sum_j [\nabla K_\delta(x - X_j) - E \nabla K_\delta(x - X_j)] \cdot \omega_j h^2 \\ &= (III) + (IV). \end{aligned}$$

By applying Lemma 6.7 with $f(y) = \nabla K_\delta(x-y)$, $g(\alpha) = \omega(\alpha, 0)$, we approximate (III) by the integral

$$\int_{\mathbb{R}^2} E\{\nabla K_\delta(x - X(\alpha;t))]\} \cdot \omega(\alpha,0)d\alpha = \int_{\mathbb{R}^2} \nabla K_\delta(x-y)\cdot w(y,t)dy$$

with an error $C(h/\delta)^L$ that follows from Proposition 6.1. (III) is uniformly bounded because

$$\left|\int_{\mathbb{R}^2} \nabla K_\delta(x-y)\cdot\omega(y,t)dy\right| = \left|\int_{\mathbb{R}^2} K_\delta(x-y)\cdot\nabla\omega(y,t)dy\right|$$
$$\leq \|\nabla\omega\|_{L^\infty}\|K_\delta\|_{L^1(B)} + \|\nabla\omega\|_{L^1}\|K_\delta\|_{L^\infty(\mathbb{R}^2/B)}$$

is uniformly bounded with respect to x, where $B = \{x \in \mathbb{R}^2 : |x| < 1\}$. Estimate (IV) requires more work. Let

$$Y_j = \nabla K_\delta(x - X_j)\cdot\omega_j h^2 - E\nabla K_\delta(x - X_j)\cdot\omega_j h^2.$$

Assuming that $h|\ln h| \leq \delta$ as before,

$$|Y_j| \leq Ch2\delta^{-2} \leq C|\ln h|^{-2} \equiv M.$$

To bound the variance

$$V = \sum_j \operatorname{Var} Y_j \leq h^2 \sum_j E|\nabla K_\delta(X - X_j)|^2 \cdot \omega_j h^2,$$

again apply Lemma 6.7 to approximate the sum above by the integral

$$\int_{\mathbb{R}^2} |\nabla K_\delta(x-y)|^2 \left[\int_\Omega G(y,t;\alpha,0)\omega^2(\alpha,0)d\alpha\right]dy$$

with an error $Ch^L\delta^{-2-L} \leq C\delta^{-2}$. Hence $V \leq Ch^2\delta^{-2} \leq C|\ln h|^{-2}$ and Bennett's inequality (Lemma 6.1) gives

$$\begin{aligned}
P\left(\left|\sum_j Y_j\right| \geq C\right) &\leq 4\exp\left[-\frac{1}{4}C^2V^{-1}B(MCV^{-1})\right]\\
&\leq \exp\left[-C^2C_1|\ln h|^2B(C_2C)\right]\\
&\leq \exp\left(-C_3C^2|\ln h|^2\cdot C^{-1}\right)\\
&\leq \exp\left(-C_3C|\ln h|^2\right)\\
&\leq h^{C_3C|\ln h|}.
\end{aligned}$$

For any fixed time t, $\sum_j \nabla K_\delta(X_i - X_j)\cdot\omega_j h^2$ is uniformly bounded by C except for an event of probability of less than $h^{C_4C|\ln h|}$. We need to extend the previous stability result to all time $t \in [0, T_*]$. Here is the strategy. Let t_n, $n = 0, \ldots, N$ divide $[0, T_*]$ into N subintervals with lengths less than h^l for some $l > 0$ to be determined. Because the stability estimate holds for any fixed time except for an event of probability of less than $h^{C_1 C|\ln h|}$ it holds on $\{t_n\}_{n=1}^N$ except for an event of probability of less than $C_2 h^{C_1 C|\ln h|-l}$ where $C_1 C > l$ when C is chosen large enough. The length of each time interval approaches zero as the number of vortices approaches infinity. Therefore if we can prove that the positions of the vortices $\tilde{X}_i(t)$, $X_i(t)$ for $t \in [t_n, t_{n+1}]$ are close to their own positions at time $t = t_n$ then we can pass the stability estimate on $t_0, \ldots, t_n$ to the stability estimate for all time; that is, we write

$$\begin{aligned}\tilde{X}_i(t) - X_i(t) &= \{\tilde{X}_i(t_n) + [\tilde{X}_i(t) - \tilde{X}_i(t_n) + X_i(t_n) - X_i(t)]\} - X_i(t_n)\\ &= [\tilde{X}_i(t_n) + Y_i(t)] - X_i(t_n)\\ &= \tilde{X}_i(t) - X_i(t_n)\end{aligned}$$

for $t \in [t_n, t_{n+1}]$. Because our stability estimate requires only the statistics of $X_i(t_n)$ and we can treat $\tilde{X}_i(t)$ as a small perturbation from $X_i(t_n)$ and because the stability estimate

$$\left\|\tilde{v}_i^h(t) - v_i^h(t)\right\|_{L_h^p} \preceq C\|\tilde{X}_i(t) - X_i(t_n)\|_{L_h^p} = C\|\tilde{X}_i(t) - X_i(t)\|_{L_h^p}$$

holds for all t, $t_n < t < t_{n+1}$ as long as $Y_i(t) \preceq constant \cdot \delta$ because our estimate does not make use of the statistics of $\tilde{X}_i(t)$.

To prove $\max_n \max_{t_n \le t \le t_{n+1}} |\tilde{X}_i(t) - X_i(t_n)| \preceq \delta$ we need the following elementary property of Wiener processes.

Lemma 6.11. *Let $W(t)$ be a standard Wiener process in $\mathbb{R}^2$. Then*

$$P\left[\max_{t \le s \le t+\Delta t} |W(s) - W(t)| \ge b\right] \le C_1(\sqrt{\Delta t}/b)\exp(-C_2 b^2/\Delta t)$$

where $b > 0$.

For a proof see Freedman (1971, p. 18). Because

$$X_i(t) - X_i(s) = \int_s^t v(X_i(\tau), \tau)d\tau + \sqrt{2\nu}[W(t) - W(s)],$$

it follows that for all $t \in [t_n, t_{n+1}]$

$$\begin{aligned}|X_i(t) - X_i(t_n)| &\le C_1|t - t_n| + \sqrt{2\nu}|W(t) - W(t_n)|\\ &\le Ch^l + \sqrt{2\nu}|W(t) - W(t_n)|.\end{aligned}$$

By Lemma 6.11

$$P\left[\max_{t_n \le t \le t_{n+1}} |W(t) - W(t_n)| \ge h\right] \le C_1 h^{(l/2)-1}\exp(-C_2 h^{-l+2}),$$

which implies that, for $l > 2$,

$$P\left[\max_n \max_{t_n \le t \le t_{n+1}} |W(t) - W(t_n)| \ge h\right] \le C_1 h^{-(l+2)/2}\exp(-C2h^{-l+2}) \to 0$$

faster than any polynomial in h as $h \to 0$. This justifies the inequality

$$\max_n \max_{t_n \le t \le t_{n+1}} |X_i(t) - X_i(t_n)| \le Ch^l + \sqrt{2\nu}|W(t) - W(t_n)| \\ \preceq Ch^l + \nu^{1/2}h.$$

This completes the proof of stability.

6.6.7. *Convergence*

We first prove the convergence of the discrete velocity. To do that we need to extend the consistency estimate for the discrete velocity to all time. We can derive it simply by combining the consistency estimate for a finite number of times $0 = t_0 < t_1 < \cdots < t_n = T$ and the stability estimate for all time, where $\max_n |t_{n+1} - t_n| < h^l$ with $l > 2$. For each $t \in [t_n, t_{n+1})$, we apply the triangle inequality to obtain

$$\begin{aligned} \left\|v_i^h(t) - v(X_i(t), t)\right\|_{L_h^p} &\le \left\|v_i^h(t) - v_i^h(t_n)\right\|_{L_h^p} + \left\|v(t_n) - v_i^h(X_i(t_n), t_n)\right\|_{L_h^p} \\ &\quad + \left\|v_i^h(X_i(t_n), t_n)) - v(X_i(t), t)\right\|_{L_h^p} \\ &\preceq C_1\|X_i(t) - X_i(T_n)\|_{L_h^p} + C_2[\delta^m + (h/\delta)^L\delta + h|\ln h|] \\ &\preceq C[\delta^m + (h/\delta)^L\delta + |h|\ln h|] \end{aligned}$$

by the estimate $|X_i(t) - X_i(t_n)| \preceq C(h^l + \nu^{1/2}h)$ and the assumption that v has bounded derivatives. Therefore

$$\max_{0 \le t \le T} \left\|v_i^h(t) - v(X_i(T), t)\right\|_{L_h^p} \preceq C[\delta^m + (h/\delta)^L\delta + h|\ln h|]. \tag{6.126}$$

Similarly,

$$\max_{0 \le t \le T} \|v^h(z_k, t) - v(z_k, t)\|_{L_h^p} \preceq C[\delta^m + (h/\delta)^L\delta + h|\ln h|] \tag{6.127}$$

for the lattice points $z_k = h \cdot k$ in any ball $B(R_0)$. This finishes the (discrete) consistency estimate for all time. Note that the L^∞ estimate is excluded because we applied the stability lemma.

The convergence of particle paths and the continuous velocity can be proved by the same argument as that in the inviscid case. The details are presented in Long (1988).

6.7. Appendix for Chapter 6

We present a proof of Bennett's inequality from Pollard (1984, Appendix B).

Lemma 6.1. *Bennett's Inequality (Pollard, 1984 Let Y_i be independent bounded random variables (not necessarily identically distributed) with mean zero, variances σ_i^2, and $|Y_i| \le M$. Let $S = \sum_i Y_i$, $V \ge \sum_i \sigma_i^2$; then, for all $\eta > 0$,*

$$P\{|S| \ge \eta\} \le 2\exp\left[-\frac{1}{2}\eta^2 V^{-1} B(M\eta V^{-1})\right], \tag{6.128}$$

where $B(\lambda) = 2\lambda^{-2}[(1+\lambda)\ln(1+\lambda) - \lambda]$, $\lambda > 0$, $\lim_{\lambda\to 0^+} B(\lambda) = 1$, *and* $B(\lambda) \sim 2\lambda^{-1}\ln\lambda$ *as* $\lambda \to \infty$. *If* Y_i *are random vectors in* $\mathbb{R}^2$, *relation (6.128) applied to each component gives*

$$P\{|S| \geq \eta\} \leq 4\exp\left[-\frac{1}{4}\eta^2 V^{-1} B(M\eta V^{-1})\right].$$

Proof of Lemma 6.1. The central limit theorem leads us to expect sums of independent random variables to behave as if they were normally distributed, that is, tail probabilities for standardized sums can be approximated by normal tail probabilities. The tails of normal distributions decay rapidly. For $\eta > 0$,

$$\left(\frac{1}{\eta} - \frac{1}{\eta^3}\right)\frac{\exp(-\frac{1}{2}\eta^2)}{\sqrt{2\pi}} < P\{N(0,1) \geq \eta\} < \frac{1}{\eta}\frac{\exp(-\frac{1}{2}\eta^2)}{\sqrt{2\pi}},$$

where the important factor is $\exp(-\frac{1}{2}\eta^2)$. Bennett's inequality provides a similar upper bound for the tail probabilities of a sum of independent random variables $Y_1, \ldots, Y_n$. Set $S = Y_1 + \cdots + Y_n$. For each $t > 0$,

$$P\{S \geq \eta\} \leq \exp(-\eta t) P\exp(tS) = \exp(-\eta t)\prod_{i=1}^{n} P\exp(tY_i). \tag{6.129}$$

The trick is to find a t that makes the last product small.

To prove relation (6.128) it suffices to establish the corresponding one-sided inequality. The two-sided inequality will then follow when it is combined with the companion inequality for $\{-Y_i\}$.

Bound the moment-generating function of Y_i. Dropping the subscript i temporarily, we have

$$\begin{aligned} Pe^{tY} &= 1 + tPY + \sum_{k=2}^{\infty}(t^k/k!)P(Y^2Y^{k-2}) \\ &\leq 1 + \sum_{k=2}^{\infty}(t^k/k!)\sigma^2 M^{k-2} \\ &= 1 + \sigma^2 g(t) \qquad \text{where} \quad g(t) = (e^{tM} - 1 - tM)/M^2 \\ &\leq e^{\sigma^2 g(t)}. \end{aligned}$$

From relation (6.129) we deduce $P\{S \geq \eta\} \leq \exp[Vg(t) - \eta t]$. We differentiate to find the minimizing value, $t = M^{-1}\log(1 + M\eta V^{-1})$, which is positive. □

We present the proofs of the lemmas used in Section 6.6.

Lemma 6.7. *Key Stochastic Quadrature Lemma. Let* $X(\alpha; t)$ *be the solution of the SDE*

$$dX(\alpha; t) = v(X(\alpha; t), t)dt + \sqrt{2\nu}\, dW(t)$$

in $\mathbb{R}^2$ *with initial data* $X(\alpha; 0) = \alpha \in \mathbb{R}^2$. *Assume that* $v \in C^L(\mathbb{R}^2 \times [0, T])$, $L \geq 3$, *that is, all the spatial derivatives of* $v(\cdot, t)$ *up to the order of* L *are uniformly bounded. Given* $\mathbb{R}^2$*-valued functions* $f \in C_0^L(\mathbb{R}^2)$, $g \in C_0^L(\mathbb{R}^2)$, *define* $\Gamma(\alpha, t) = [Ef(X(\alpha; t))] \cdot g(\alpha)$ *so that the support of* Γ *is contained in a bounded domain* $\operatorname{supp} \Gamma \subset \operatorname{supp} g$. *Then we have the following estimate for the quadrature error:*

$$\max_{0\leq t\leq T} \left| \sum_{i\in\mathbb{Z}^2} \Gamma(h \cdot i, t)h^2 - \int_{\mathbb{R}^2} \Gamma(\alpha, t) d\alpha \right| \leq Ch^L \max_{0\leq|\beta|\leq L} \|\partial^\beta g\|_{L^\infty} \times \left[\sum_{0\leq|\beta|\leq L} \int_{|x|\leq R} |\partial^\beta f(x)| dx + \sum_{0\leq|\beta|\leq L} \sup_{|x|>R} |\partial^\beta f(x)| \right] \tag{6.130}$$

where $R > 0$ *is arbitrary and* C *depends on only* T, L, *the diameter* Ω, *and* $\max_{1\leq|\gamma|\leq L} \|\partial^\gamma v\|_{L^\infty(\mathbb{R}^2\times[0,T])}$.

Proof of Lemma 6.7. We directly apply Lemma 6.2 on the trapezoidal quadrature. Hence we need to estimate $\|\partial_1^L \Gamma\|_{L^1}$ and $\|\partial_2^L \Gamma\|_{L^1}$. By direct differentiation we know that $\partial_1^L \Gamma$ and $\partial_2^L \Gamma$ are sums of finite terms of the form

$$E\left\{ (\partial^\beta f)(X(\alpha, t)) \cdot \prod_{1\leq|\gamma|\leq L} (\partial^\gamma X(t; \alpha))^{\kappa(\gamma)} \right\} \partial^\mu (g(\alpha)), \tag{6.131}$$

where β, γ, κ, and μ are multiple indices with $0 \leq |\beta|, |\gamma|, |\kappa|, |\mu| \leq L$. We need a bound for $\partial^\gamma X(\alpha, t)$. By differentiating

$$X(\alpha, t) = \alpha + \int_0^1 u(X(\alpha, s), s) ds + \sqrt{2\nu} W(t),$$

we obtain integral equations

$$\frac{\partial}{\partial \alpha_1} X(\alpha, t) = \begin{pmatrix} 1 \\ 0 \end{pmatrix} + \int_0^t [\nabla u(X(\alpha; s), s)] \cdot \left[\frac{\partial}{\partial \alpha_1} X(\alpha; s) \right] ds,$$

$$\frac{\partial}{\partial \alpha_2} X(\alpha; t) = \begin{pmatrix} 1 \\ 0 \end{pmatrix} + \int_0^t [\nabla u(X(\alpha; s), s)) \cdot \left[\frac{\partial}{\partial \alpha_2} X(\alpha; s) \right] ds,$$

for $[\partial/(\partial\alpha_2)]\, X(\alpha; t)$ and $[\partial/(\partial\alpha_2)]\, X(\alpha; t)$. It follows that

$$\left| \frac{\partial}{\partial \alpha_1} X(\alpha; t) \right| \leq 1 + \int_0^t \|\nabla u\|_{L^\infty} \cdot \left| \frac{\partial}{\partial \alpha_1} X(\alpha; s) \right| ds,$$

$$\left| \frac{\partial}{\partial \alpha_2} X(\alpha; t) \right| \leq 1 + \int_0^t \|\nabla u\|_{L^\infty} \cdot \left| \frac{\partial}{\partial \alpha_2} X(\alpha; s) \right| ds.$$

Therefore

$$\left| \frac{\partial}{\partial \alpha_1} X(\alpha; t) \right|, \quad \left| \frac{\partial}{\partial \alpha_2} X(\alpha; t) \right| \leq \exp\left(\|\nabla u\|_{L^\infty(\mathbb{R}^2\times[0,T])} T \right)$$

by Grönwall's inequality.

The integral equation for higher-order derivatives $\partial^\gamma X(\alpha;t)$ are

$$\partial^\gamma X(\alpha;t) = \int_0^t Y(s)ds + \int_0^t [\nabla u(X(\alpha;s),s)]\cdot\partial^\gamma X(\alpha;s)ds$$

where $Y(s)$ is some function of $(\partial^\lambda u)(X(\alpha;s))$ and $\partial^\rho X(\alpha;s)$ with $1\le|\lambda|\le|\gamma|$ and $1\le|\rho|\le|\gamma-1|$. By Grönwall's inequality and induction on γ, we conclude that

$$|\partial^\gamma X(\alpha;t)|\le C, \qquad 1\le|\gamma|\le L, \tag{6.132}$$

where C depends on only $\max_{1\le|\gamma|\le L}\|\partial^\gamma u\|_{L^\infty(\mathbb{R}^2\times[0,T])}$, L, and T.

It follows from relations (6.131) and (6.132) that

$$\begin{aligned}
\|\partial_1^L\Gamma\|_{L^1} &= \int_{\mathbb{R}^2}\left|\frac{\partial^L}{\partial\alpha_1^L}\Gamma(\alpha)\right|d\alpha = \int_\Omega\left|\frac{\partial^L}{\partial\alpha_1^L}\Gamma(\alpha)\right|d\alpha\\
&\le C\max_{0\le|\beta|\le L}\|\partial^\beta g\|_{L^\infty}\sum_{0\le|\beta|\le L}\int_\Omega|E(\partial^\beta f)(X(\alpha;t))|d\alpha\\
&\le C\max_{0\le|\beta|\le L}\|\partial^\beta g\|_{L^\infty}\sum_{0\le|\beta|\le L}\int_\Omega E|(\partial^\beta f)(X(\alpha;t))|d\alpha
\end{aligned} \tag{6.133}$$

Similarly, $\|\partial_2^L\Gamma\|_{L^1}$ is bounded by Eq. (6.133). We can write

$$\begin{aligned}
\int_\Omega E|(\partial^\beta f)(X(\alpha;t))|d\alpha &= \int_\Omega\int_{\mathbb{R}^2}|\partial^\beta f(x)|G(x,t;\alpha,0)dx\,d\alpha\\
&= \int_{\mathbb{R}^2}|\partial^\beta f(x)|\left[\int_\Omega G(x,t;\alpha,0)d\alpha\right]dx,
\end{aligned} \tag{6.134}$$

where G is the fundamental solution of vorticity equation (6.112). The function $w(x,t)=\int_\Omega G(x,t;\alpha,0)d\alpha$ is the solution of vorticity equation (6.112) with the initial data $w(\cdot,0)=\mathcal{X}_\Omega$, where

$$\mathcal{X}_\Omega = \begin{cases}1 & \text{if } x\in\Omega\\ 0 & \text{if } x\notin\Omega\end{cases}$$

is the characteristic function of Ω. Because $w(x,t)$ satisfies backward equation (6.110) and forward equation (6.112), we have

$$0<w(x,t)<1 \quad\text{and}\quad \int_{\mathbb{R}^2}w(x,t)dx = \text{area}(\Omega),\ \forall t>0,$$

by the maximum principle and the conservation of vorticity, respectively. Therefore

Eq. (6.134) can be estimated by

$$\begin{aligned}
&\int_{\mathbb{R}^2} |\partial^\beta f(x)| \left[\int_\Omega G(x,t;\alpha,0)d\alpha \right] dx \\
&\quad = \int_{\mathbb{R}^2} |\partial^\beta f(x)| \cdot w(x,t)dx \qquad (6.135)\\
&\quad \leq \int_{|x|\leq R} |\partial^\beta f(x)|dx + \text{area}(\Omega) \cdot \sup_{|x|>R} |\partial^\beta f(x)|.
\end{aligned}$$

We proved the lemma by combining Eqs. (6.133), (6.134), (6.135), and Lemma 6.2. □

Lemma 6.9. *Let $N(x,r,t) = \#\{X_i(t) : |X_i(T) - x| \leq r\}$, with $r > 0, 0 \leq t \leq T$, be the number of vortices in the ball $B(x,r)$. If $r \geq h|\ln h|$, then*

$$h^2 \cdot N(x,r,t) \preceq Cr^2.$$

Proof of Lemma 6.9. Let

$$H(y) = \begin{cases} 1 & \text{if } y \in B(x,r) \\ 0 & \text{if } y \notin B(x,r) \end{cases}$$

be the characteristic function of the ball $B(x,r)$. We can write

$$\begin{aligned}
h^2 \cdot \mathcal{N}(x,r,t) &= h^2 \sum_i H(X_i(t)) \\
&= h^2 \sum_i EH(X_i(t)) + h^2 \sum_i [H(X_i(t)) - EH(X_i(t))] \\
&= (I) + (II)
\end{aligned}$$

where (I) is h^2 times the expected number of vortices inside $B(x,r)$ and (II) is the fluctuation from the expectation. We wish to apply Lemma 6.7 on stochastic quadrature and thus need a smooth function $\tilde{H} \geq H$. Let $\varphi \in C_0^\infty(\mathbb{R}^2)$ be defined by

$$\varphi(x) = \begin{cases} \exp[-x^2/(1-x^2)] & \text{if } |x| < 1 \\ 0 & \text{if } |x| \geq 1 \end{cases}.$$

Define the function $\tilde{H}$ by

$$\tilde{H}(y) = \begin{cases} H(y) = 1 & \text{if } y \in B(x,r) \\ \varphi(|y-x|/r - 1) & \text{if } r \leq |y-x| \leq 2r \\ 0 & \text{otherwise} \end{cases}.$$

It is obvious that $\tilde{H} \in C_0^\infty(\mathbf{R^2})$ and its partial derivatives of the order of L are bounded by Cr^{-L}, where C depends on only L. Because $\tilde{H} \geq H$, we can bound (I) by $h^2 \sum_i E\tilde{H}(X_i(t))$.

By applying Lemma 6.7 with $f(y) = \tilde{H}(y), g(\alpha) = 1$, we can approximate $h^2 \sum_i E\tilde{H}(X_i(t))$ by the integral $\int_\Omega E\tilde{H}(X(\alpha;t))d\alpha$ within an error

$$C_1 h^L \cdot 4\pi r^2 \sum_{l=0}^{L} r^{-1} \leq C_2 r^2 h^L \{L \cdot [\min(1,r)]^{-L}\} \leq Cr^2,$$

provided that $h \leq r$. Moreover

$$\begin{aligned}\int_\Omega E\tilde{H}(X(\alpha;t))d\alpha &\leq \int_\Omega \int_{B(x,2r)} G(y,t;\alpha,0)dyd\alpha \\ &\leq \int_{B(x,2r)} dy = 4\pi r^2.\end{aligned}$$

Thus we arrive at the conclusion that $(I) \leq Cr^2$ if $r \geq h$.

We use Bennett's inequality to estimate (II). Let

$$Y_i = h^2[H(X_i(t)) - EH(X_i(t))].$$

We have $EY_i = 0, |Y_i| \leq h^2$, and

$$\sum_i \operatorname{Var} Y_i \leq h^4 \sum_i E[H(X_i(t))]^2 \leq h^4 \sum_i E[\tilde{H}(X_i(t))]^2.$$

We apply Lemma 6.7 with $f(y) = E[\tilde{H}(X_i(t))]^2, g(\alpha) = 1$, to approximate $h^2 \sum_i E[\tilde{H}(X,(t))]^2$ by integral $\int_\Omega E[\tilde{H}(X(\alpha;t)]^2 d\alpha$ within an error

$$C_1 h^L \cdot 4\pi r^2 \sum_{l=0}^{L} r^{-1} \leq C_2 r^2 h^L \{L \cdot [\min(1,r)]^{-L}\} \leq Cr^2,$$

provided that $h \leq r$. Furthermore

$$\begin{aligned}\int_\Omega E[\tilde{H}(X(\alpha;t))]^2 d\alpha &\leq \int_\Omega \int_{B(x,2r)} G(Y,t;\alpha,0)dyd\alpha \\ &\leq \int_{B(x,2r)} dy = 4\pi r^2.\end{aligned}$$

Therefore $\sum_i \operatorname{Var} Y_i \leq Cr^2h^2$ if $r \geq h$.

According to Bennett's inequality, it follows that

$$\begin{aligned}&P\left\{\left|\sum_i Y_i\right| \geq Crh|\ln h|\right\} \\ &\leq 4\exp\left[-\frac{1}{4}(Crh|\ln h|)^2 \cdot C_1 r^{-2}h^{-2} \cdot B[h^2 Crh|\ln h| C_1 r^{-2} h^{-2}]\right] \\ &\leq \exp\left[-C_2C^2|\ln h|^2 \cdot B(C_2 C r^{-1} h|\ln h|)\right] \\ &\leq \exp\left(-C_3 C|\ln h|^2\right) \\ &= h^{C_3 C|\ln h|},\end{aligned}$$

provided that $r^{-1}h|\ln h|$ stays bounded.

Hence the fluctuation of $h^2 \cdot \mathcal{N}(x, r, t)$ is

$$|(II)| \leq Crh|\ln h| \leq Cr^2$$

if $h|\ln h| \leq r$. □

Notes for Chapter 6

We present a brief historical summary of inviscid-vortex methods, focusing on the principal developments leading to the classical convergence theory. The inviscid-vortex method first introduced by Chorin (1973) was very successful in actual numerical computations. However, for several years some remained skeptical about its validity. The main objection was that when a fluid undergoes large local deformations, the vortex blobs $\psi_\delta[x - \tilde{\psi}_j(t)]\omega_{0_j}h^2$ in the approximation remain unchanged and hence might not approximate the actual vorticity ω. Hald and Del Prete (1978) first addressed these concerns with a convergence proof for 2D vortex methods. However, their proof applied for only short times $T_* \ll T$, with exponential loss of accuracy. Hald's (1979) paper on 2D vortex methods was the first pioneering breakthrough in the numerical analysis of vortex methods. He proved second-order convergence for arbitrarily long time intervals $[0, T]$ for a special class of blob functions. Beale and Majda (1982a, 1982b), by improving Hald's stability argument in an essential way, found that vortex methods could be designed that converge with arbitrarily high-order accuracy. They also gave the first convergence proof for the 3D vortex algorithms with Lagrangian stretching [Eq. (6.71)]. Exploiting the similarity to the 2D Euler equations, Cottet and Raviart (1984) gave simple convergence proofs for particle methods for the 1D Vlasov–Poisson equations. Cottet extended these ideas and developed a simplified approach to the consistency arguments for vortex algorithms. Anderson and Greengard (1985) gave a further simplified version of Cottet's consistency argument for the trapezoidal rule by means of the Poisson summation formula; their paper also contains a simple proof of convergence of multistep time discretization for 2D vortex methods. Anderson and Greengard first suggested the 3D vortex methods with Eulerian stretching, presenting them in Anderson and Greengard (1985). With very different proofs, Beale (1986) and Cottet (1985) independently proved the convergence of the 3D vortex algorithm in (6.74). Cottet (1987) developed a new approach to the numerical analysis of vortex methods that does not use a direct analysis of the stability and consistency of the ODEs for the particle trajectories. His approach leads to the convergence of 2D vortex-in-cell algorithms, etc. Building on preceding work, Hald (1988) gave a number of definitive results for the classical convergence theory for 2D vortex methods, including the convergence of Runge–Kutta time differencing for 2D vortex methods. In addition to the above references we recommend the survey papers of Anderson and Greengard (1985), Beale and Majda (1984, 1985), Hald (1991), Majda (1988), Puckett (1993), and Sethian (1991). In particular, the papers of Beale and Majda (1985) and Perlman (1985) contain careful and detailed numerical studies of 2D inviscid-vortex methods on simple model problems. A discussion of the literature on vortex-filament techniques can be found in Leonard (1985) and Majda (1988).

The random-vortex method dates back to Chorin (1973). More recently, Marchioro and Pulvirenti (1982) considered a continuous-time random-vortex method with Gaussian random walks replaced with independent Wiener processes (Brownian motions). Goodman (1987) first proved a result containing a quantitative rate of convergence. The optimal rate of convergence for the random-vortex method was proved by Long (1988). Our Section 6.4 on random-vortex methods comes from the paper of Long. Details about the ν dependence of the convergence of random-vortex methods can be found in Long (1987). Cottet (1988) and Mas–Gallic (1990) proposed deterministic vortex methods, in contrast to the random-vortex method, based on an equivalent systems of equations to the Navier–Stokes equations. Also, Rossi (1993) has recently introduced a deterministic core-spreading method that corrects the inconsistency of standard core spreading by means of a splitting algorithm.

References for Chapter 6

Anderson, C. and Greengard, C., "On vortex methods," *SIAM J. Numer. Anal.* **22**, 413–440, 1985.

Beale, J. T., "A convergent 3-D vortex method with grid free stretching," *Math. Comput.* **46**, 402–424, S15–S20, 1986.

Beale, J. T. and Majda, A., "Vortex methods I: convergence in three dimensions," *Math. Comput.* **39**(159), 1–27, 1982a.

Beale, J. T. and Majda, A., "Vortex methods II: higher order accuracy in two and three dimensions," *Math. Comput.* **39**(159), 29–52, 1982b.

Beale, J. T. and Majda, A., "Vortex methods for fluid flows in two or three dimensions," *Contemp. Math.* **28**, 221–229, 1984.

Beale, J. T. and Majda, A., "High order accurate vortex methods with explicit velocity kernels," *J. Comput. Phys.* **58**, 188–208, 1985.

Birkhoff, G., *Helmholtz and Taylor Instability*, American Mathematical Society, Providence, RI, 1962, pp. 55–76.

Birkhoff, G. and Fisher, J., "Do vortex sheets roll-up?" *Rend. Circ. Math. Palermo* **8**, 77–90, 1959.

Chorin, A. J., "Numerical study of slightly viscous flow," *J. Fluid Mech.* **57**, 785–796, 1973.

Chorin, A. J. and Bernard, P. J., "Discretization of vortex sheet with an example of roll-up," *J. Comput. Phys.* **13**, 423–428, 1973.

Cottet, G., "A new approach for the analysis of vortex methods in 2 and 3 dimensions," *Ann. Inst. H. Poincaré Anal. Non Linéare* vol.5 no. 3 pp. 227–285, 1988.

Cottet, G. H., "Boundary conditions and deterministic vortex methods for the Navier–Stokes equations," in R. E. Caflisch, ed., *Mathematical Aspects of Vortex Dyanamics*, Society for Industrial and Applied Mathematics, Philadelphia, PA, 1988, pp. 128–143.

Cottet, G. H., Goodman, J., and Hou, T. Y., "Covergence of the grid-free point vortex method for the three dimensional Euler equations," *SIAM J. Numer. Anal.* **28**, 291–307, 1991.

Cottet, G. and Raviart, P., "Particle methods for the 1-D Vlasov–Poisson equations," *SIAM J. Numer. Anal.* **21**, 52–76, 1984.

Folland, G. B., *Introduction to Partial Differential Equations*, Princeton Univ. Press, Princeton, NJ, 1995.

Freedman, D. A., *Brownian Motion and Diffusion*, Holden-Day, San Francisco, 1971.

Friedman, A., *Stochastic Differential Equations and Applications*, Academic, New York, 1975.

Gard, T. C., *Introduction to Stochastic Differential Equations*, Marcel Dekker, New York, 1988.

Goodman, J., "Convergence of the random vortex method," *Commun. Pure Appl. Math.* **40**, 189–220, 1987.

Goodman, J., Hou, T. Y., and Lowengrub, J., "The convergence of the point vortex method for the 2-D Euler equations," *Commun. Pure and Appl. Math.* **43**, 415–430, 1990.

Greengard, C., "The core spreading vortex method approximates the wrong equation," *J. Comput. Phys.* **61**, 345–348, 1985.

Hald, O., "The convergence of vortex methods II," *SIAM J. Numer. Anal.* **16**, 726–755, 1979.

Hald, O., "Convergence of vortex methods for Euler's equations III," *SIAM J. Numer. Anal.* **24**, 538–582, 1987.

Hald, O. H., "Convergence of vortex methods," in K. A. Gustafson and J. A. Sethian, eds., *Vortex Methods and Vortex Motion*, Society for Industrial and Applied Mathematics, Philadelphia, PA, 1991, pp. 33–58.

Hald, O. and Del Prete, V. M., "Convergence of vortex methods for Euler's equations," *Math. Comput.* **32**, 791–809, 1978.

Hou, T. Y., "Vortex dynamics and vortex methods," in C. Anderson and C. Greengard, eds., *A Survey on Convergence Analysis for Point Vortex Methods*, Vol. 28 of AMS Lectures in Applied Mathematics Series, American Mathematical Society, Providence, RI, 1991, pp. 327–339.

Hou, T. Y. and Lowengrub, J., "Convergence of a point vortex method for the 3-d Euler equation," *Commun. Pure Appl. Math.* **43**, 965–981, 1990.

Kloeden, P. E. and Platen, E., *Numerical Solutions of Stochastic Differential Equations*, Springer-Verlag, New York, 1992; 2nd printing, 1995.

Krasny, R., "A study of singularity formation in a vortex sheet by the point vortex approximation," *J. Fluid Mech.* **167**, 292–313, 1986.

Leonard, A., "Computing three dimensional flows with vortex elements," *Ann. Rev. Fluid Mech.* **17**, 523–559, 1985.

Long, D.-G., "Convergence of the random vortex method," Ph.D. dissertation, University of California, Berkeley, CA, 1987.

Long, D.-G., "Convergence of the random vortex method in two dimensions," *J. Am. Math. Soc.* **1**, 779–804, 1988.

Majda, A., "Vortex dynamics: numerical analysis, scientific computing, and mathematical theory," in *ICIAM '87: Proceedings of the First International Conference on Industrial and Applied Mathematics*, Society for Industrial and Applied Mathematics, Philadelphia, PA, 1988, pp. 153–182.

Marchioro, C. and Pulvirenti, M., "Hydrodynamics in two dimensions and vortex theory," *Commun. Math. Phys.* **84**, 483–502, 1982.

Mas-Gallic, S., "Deterministic particle method: diffusion and boundary conditions," Université Pierre et Marie Curie, Centre National de la Recherche Scientifique, Paris, 1990.

Perlman, M., "On the accuracy of vortex methods," *J. Comput. Phys.* **59**, 200–223, 1985.

Pollard, D., *Convergence of Stochastic Processes*, Springer-Verlag, New York, 1984.

Puckett, E. G., "Vortex methods: an introduction and survey of selected research topics," in M. D. Gunzburger and R. A. Nicolaides, eds., *Incompressible Computational Fluid Dynamics Trends and Advances*, Cambridge Univ. Press, New York, Chap. 11, pp. 335–408.

Rosenhead, L., "The point vortex approximation of a vortex sheet or the formation of vortices from a surface of discontinuity-check," *Proc. R. Soc. London A* **134**, 170–192, 1932.

Rossi, L. F., "Resurrecting core spreading vortex methods: a new scheme that is both deterministic and convergent," *SIAMJ. Sci. Comput.* **17**, 370–397, 1996.

Sethian, J. A., "A brief overview of vortex methods," in K. A. Gustafson and J. A. Sethian, eds., *Vortex Methods and Vortex Motion*, Society for Industrial and Applied Mathematics, Philadelphia, PA, 1991, pp. 33–58.

Stein, E. M. and Weiss, G., *Introduction to Fourier Analysis on Euclidean Spaces*, Princeton Univ. Press, Princeton, NJ, 1971.

7

Simplified Asymptotic Equations for Slender Vortex Filaments

Here we discuss a research direction with a very different flavor than that of the preceding six chapters. Recall from Chap. 2, Section 2.1, that we derived the vorticity-stream form of the Euler and the Navier–Stokes equations. This reformulation highlights the important difference between 2D and 3D flows; namely, in two dimensions the vorticity is convected along particle paths in the inviscid case and diffused and convected in the viscous case. On the other hand, in three dimensions the vorticity can stretch, because of the interaction of the strain matrix with the vorticity vector. In Chap. 3 we showed that in the absence of vortex stretching, we could prove the global existence of solutions to the Euler and the Navier–Stokes equations. Moreover, in Chap. 5 we explored various possible routes to blowup in three dimensions and showed that, in addition to vorticity amplification, the direction of vorticity must become singular in the event of a blowup, causing the vortex lines or tubes to form kinks or corners. In this chapter we use formal but concise asymptotic expansions to study the motion of slender tubes of vorticity at high Reynolds numbers. Beautiful simplified asymptotic equations emerge with remarkable properties that give qualitative insight into the folding, wrinkling, and bending of vortex tubes.

The key technical idea is to expand the Biot–Savart law [Eqs. (2.94) and (2.95)], determining the velocity from the vorticity

$$v(x,t) = \frac{1}{4\pi}\int x \times \frac{\omega}{|x|^3}(y,t)dy \tag{7.1}$$

in a suitable asymptotic expansion, provided that the vorticity, ω, is large and confined to a narrow tube around a curve $X(s,t)$ that defines the centerline. Of course, it is crucial to develop such expansions in a fashion consistent with the Navier–Stokes equations at high Reynolds numbers. In this manner, simplified dynamical equations emerge for the motion of the curve $X(s,t)$ in various asymptotic regimes as well as equations for the interaction of many such strong vortex filaments represented by a finite collection of such curves $\{X_j(s,t)\}_{j=1}^N$.

In Section 7.1 we describe the simplest such theory involving the self-induction approximation for the motion of a single vortex filament. We also describe Hasimoto's remarkable transformation that reduces these simplified vortex dynamics to that for the cubic nonlinear Schrödinger equation, a well-known completely integrable PDE with soliton behavior and heteroclinic instabilities (Ablowitz and Segur, 1981; Lamb,

1980). We further demonstrate that despite the great beauty of the self-induction approximation, it fails to allow for any vortex stretching, one of the most prominent features in 3D incompressible fluid flow with small viscosity! Some of these shortcomings are addressed by a recent asymptotic theory (Klein and Majda, 1991a, 1991b) that allows for self-stretch of a single filament. This theory is described in Section 7.2.

In Sections 7.3 and 7.4 we discuss simplified equations for the dynamics of many interacting vortex filaments. Such situations frequently arise in many practical contexts such as the two trailing-wake vortices shed from the wingtips of aircraft (Van Dyke, 1982). In Section 7.3 we briefly discuss the simplified dynamics for the interaction of a finite number of exactly parallel vortex filaments that reduces to the familiar interaction of point vortices in the plane (Chorin and Marsden, 1990; Aref, 1983). In Section 7.4 we describe a recent asymptotic theory (Klein et al., 1995) for the interaction of *nearly parallel vortex filaments* with remarkable properties that incorporate features of both self-induction from Section 7.1 and mutual point-vortex interaction as described in Section 7.3. We end this chapter in Section 7.5 with a discussion as well as a list of interesting open mathematical problems related to the material in this chapter.

7.1. The Self-Induction Approximation, Hasimoto's Transform, and the Nonlinear Schrödinger Equation

First we give some precise geometric assumptions on the nature and strength of the evolving vortex filament that lead to a concise self-consistent asymptotic development of the Biot–Savart law in Eq. (7.1), consistent with the Navier–Stokes equations. Here, as elsewhere in this chapter, we do not supply the lengthy details of the asymptotic derivation but instead refer the interested reader to the published literature.

Consider a concentrated thin tube of vorticity so that the following assumptions are satisfied:

(1) The vorticity is essentially nonzero only inside a tube of cross-sectional radius δ, with $\delta \ll 1$.
(2) The vorticity ω is large inside this tube so that the circulation $\Gamma = \int \omega \cdot \mathbf{n}\, ds$ satisfies $R_e = \Gamma/\nu \to \infty$, where ν is the viscosity and R_e is the Reynolds number.
(3) Despite (1) and (2) the radius of curvature of the filament centerline remains bounded strictly away from zero.
(4) The cross-sectional radius δ from (1) and the Reynolds number R_e from (2) are balanced so that $\delta = (R_e)^{-\frac{1}{2}}$. (7.2)

Assumptions (7.2) indicate that there are primarily two scales in the assumed vortex-filament motion, an outer scale that is of the order of $\mathcal{O}(1)$ defined by the radius of curvature of the filament, and an inner scale, δ, $\delta \ll 1$, describing the cross-sectional radius of the filament. The large value of the vorticity within the filament is expressed concisely in nondimensional terms through the requirement in (4) of assumptions (7.2) in which $\delta = (R_e)^{-\frac{1}{2}}$.

For such a thin filament of vorticity, consider the centerline curve $\mathcal{L}(t)$: $s \to X(s, t)$. To a high degree of approximation for $\delta \ll 1$ this centerline should move with the fluid so that

$$\frac{dX}{dt} = v(X(s,t),t). \tag{7.3}$$

To get the velocity $v(X(s,t),t)$, we need to simplify Biot–Savart law (7.1) near the centerline $X(s,t)$ under geometric assumptions (7.2). Under these circumstances, concise matched asymptotic expansions (Callegari and Ting, 1978; Ting and Klein, 1991) establish that the right-hand side of Eq. (7.3) is given by

$$\frac{\partial X}{\partial t}(s,t) = \kappa \mathbf{b}(s,t) + \left[\ln\left(\frac{1}{\delta}\right)\right]^{-1} [\tilde{C}(t)\mathbf{b} + \mathbf{Q}_f] + \mathcal{O}(1) \tag{7.4}$$

for $\delta \ll 1$. Here $\mathbf{b} = \mathbf{t} \times \mathbf{n}$ is the binormal to the curve, where $\mathbf{t}$ is the tangent vector, $\mathbf{n}$ is the normal vector, and κ is the curvature of the curve. We change the time variable in Eq. (7.4) by $\bar{t} = \ln(1/\delta)t$ and retain only the leading-order asymptotic term in Eq. (7.4) to obtain the *self-induction equation for an isolated vortex filament:*

$$\frac{\partial X}{\partial \bar{t}} = \kappa \mathbf{b}. \tag{$*$}$$

This equation was first derived through a different procedure by Arms and Hama (1965).

7.1.1. Hasimoto's Transform and the Cubic Nonlinear Schrödinger Equation

For further developments in this chapter we consider a generalization of the self-induction equation involving the perturbed binormal law, *the Hasimoto transform for a perturbed binormal law*:

$$\frac{\partial X(\bar{s},\bar{t})}{\partial \bar{t}} = (\kappa \mathbf{b})(\bar{s},\bar{t}) + \tilde{\delta} v(\bar{s},\bar{t}), \tag{7.5}$$

where $\bar{s}$ is the arc length along the curve and $\tilde{\delta} v(\bar{s},\bar{t})$ is a small general perturbation velocity. Closely following the procedure of Hasimoto for the case of $\tilde{\delta} \equiv 0$, we derive a connection between Eq. (7.5) and a perturbed Schrödinger equation for the related filament function.

First we recall some basic notation and identities. The Serret–Frenet formulas (do Carmo, 1976),

$$X_{\bar{s}} = \mathbf{t}, \qquad \mathbf{t}_{\bar{s}} = \kappa \mathbf{n}, \qquad \mathbf{n}_{\bar{s}} = T\mathbf{b} - \kappa \mathbf{t}, \qquad \mathbf{b}_{\bar{s}} = -T\mathbf{n}, \tag{7.6}$$

describe the variation of the intrinsic basis $\mathbf{t}, \mathbf{n}, \mathbf{b}$ along $\mathcal{L}$ in terms of the curvature κ and torsion T of the curve. Hasimoto considers the *filament function*

$$\psi(\bar{s},\bar{t}) = \kappa(\bar{s},\bar{t})e^{i\Phi}, \qquad \text{with} \quad \Phi = \int_0^{\bar{s}} T(s',\bar{t})ds'. \tag{7.7}$$

He replaces the principal and binormal unit vectors $\mathbf{n}$ and $\mathbf{b}$ with the complex vector function

$$\mathbf{N}(\bar{s},\bar{t}) = (\mathbf{n} + i\mathbf{b})(\bar{s},\bar{t}) \exp[i\Phi(\bar{s},\bar{t})], \tag{7.8}$$

where Φ is as in Eq. (7.7), and with its complex conjugate $\overline{\mathbf{N}}$ to span the planes normal to $\mathbf{t}$. The new basis satisfies the orthogonality relations

$$\mathbf{t}\cdot\mathbf{t}=1, \qquad \mathbf{N}\cdot\mathbf{N}=0, \qquad \mathbf{N}\cdot\overline{\mathbf{N}}=2, \tag{7.9}$$

and the Serret–Frenet equations yield its variations along $\mathcal{L}$:

$$\mathbf{N}_{\tilde{s}}=-\psi\mathbf{t}, \qquad \mathbf{t}_{\tilde{s}}=\frac{1}{2}(\overline{\psi}\mathbf{N}+\psi\overline{\mathbf{N}}). \tag{7.10}$$

Differentiating perturbed binormal law (7.5) with respect to $\tilde{s}$, we obtain the dynamic behavior of the tangent vector

$$\mathbf{t}_{\tilde{t}}=\kappa_{\tilde{s}}\mathbf{b}-\kappa T\mathbf{n}+\tilde{\delta}v_{\tilde{s}}, \tag{7.11}$$

which may be expressed in terms of the new basis as

$$\mathbf{t}_{\tilde{t}}=\frac{1}{2}i(\psi_{\tilde{s}}\overline{\mathbf{N}}+\overline{\psi}_{\tilde{s}}\mathbf{N})+\tilde{\delta}v_{\tilde{s}}. \tag{7.12}$$

We next derive an evolution equation for the filament function by computing two independent representations of the cross derivative $\mathbf{N}_{\tilde{s}\tilde{t}}$ of N and by comparing their respective components along $\mathbf{N}$ and $\mathbf{t}$. We first decompose $\mathbf{N}_{\tilde{t}}$ as

$$\mathbf{N}_{\tilde{t}}=\alpha\mathbf{N}+\beta\overline{\mathbf{N}}+\gamma\mathbf{t}. \tag{7.13}$$

Then orthogonality relations (7.9) and (7.12) yield the identities

$$\begin{aligned}
\alpha+\bar{\alpha}&=\frac{1}{2}(\mathbf{N}\cdot\overline{\mathbf{N}}_t+\mathbf{N}_{\tilde{t}}\cdot\overline{\mathbf{N}})=\frac{1}{2}(\mathbf{N}\cdot\overline{\mathbf{N}})_{\tilde{t}}\equiv 0,\\
\beta&=\frac{1}{4}(\mathbf{N}\cdot\mathbf{N})_{\tilde{t}}\equiv 0,\\
\gamma&=\mathbf{t}\cdot\mathbf{N}_{\tilde{t}}=-\mathbf{t}_{\tilde{t}}\cdot\mathbf{N}=-i\psi_{\tilde{s}}-\tilde{\delta}\mathbf{N}\cdot v_{\tilde{s}}.
\end{aligned} \tag{7.14}$$

Thus we find

$$\mathbf{N}_{\tilde{t}}=i[R\mathbf{N}-(\psi_{\tilde{s}}-i\tilde{\delta}\mathbf{N}\cdot v_{\tilde{s}})\mathbf{t}], \tag{7.15}$$

where $R=R(\tilde{s},\tilde{t})$ is an as yet unknown real-valued function. The two representations of $\mathbf{N}_{st}$ now follow from the partial differentiation of Eqs. (7.10) and (7.15) with respect to $\tilde{t}$ and $\tilde{s}$, respectively:

$$\begin{aligned}
N_{\tilde{s}\tilde{t}}&=-\psi_{\tilde{t}}\mathbf{t}-\frac{1}{2}i(\psi\psi_{\tilde{s}}\overline{\mathbf{N}}-\psi\overline{\psi}_{\tilde{s}}\mathbf{N})-\tilde{\delta}\psi v_{\tilde{s}},\\
N_{\tilde{t}\tilde{s}}&=i\{R_{\tilde{s}}\mathbf{N}-R\psi\mathbf{t}-[\psi_{\tilde{s}\tilde{s}}-i\tilde{\delta}(\mathbf{N}\cdot v_{\tilde{s}})_{\tilde{s}}]\mathbf{t}\\
&\quad-\frac{1}{2}(\psi_{\tilde{s}}-i\tilde{\delta}\mathbf{N}\cdot v_{\tilde{s}})(\overline{\psi}\mathbf{N}+\psi\overline{\mathbf{N}})\}.
\end{aligned} \tag{7.16}$$

Equating the coefficients of $\mathbf{t}$ and $\mathbf{N}$ in these relations, we obtain

$$\psi_{\tilde{t}} + \tilde{\delta}\psi v_{\tilde{s}} \cdot \mathbf{t} = i[R\psi + \psi_{\tilde{s}\tilde{s}} - i\tilde{\delta}(\mathbf{N} \cdot v_{\tilde{s}})_{\tilde{s}}], \tag{7.17}$$

$$\mathbf{R}_{\tilde{s}} = \frac{1}{2}(\psi_{\tilde{s}}\overline{\psi} + \psi\overline{\psi}_{\tilde{s}}) + \frac{1}{2}i\tilde{\delta}(\psi\overline{\mathbf{N}} - \overline{\psi}\mathbf{N}) \cdot v_{\tilde{s}}. \tag{7.18}$$

At this stage we recover Hasimoto's result,

$$R = \frac{1}{2}|\psi|^2 \qquad \text{for} \quad \tilde{\delta} = 0, \tag{7.19}$$

by removing the perturbation term from Eq. (7.18). This yields the cubic nonlinear Schrödinger equation when inserted into Eq. (7.17). Hasimoto points out that an arbitrary integration function $A(\tilde{t})$ that might appear on integration of Eq. (7.18) may be eliminated without loss of generality through a shift,

$$\tilde{\Phi}(\tilde{s}, \tilde{t}) = \Phi(\tilde{s}, \tilde{t}) - \int_0^{\tilde{t}} A(t)dt,$$

of the phase function Φ in Eq. (7.7). In fact, all the geometric information contained in the filament function is

$$|\psi| = \kappa, \qquad T = [\arg(\psi)]_{\tilde{s}} = \Phi_{\tilde{s}} = \tilde{\Phi}_{\tilde{s}}, \tag{7.20}$$

so that the phase shift does not influence the geometry of $\mathcal{L}$. In general, we can integrate Eq. (7.18) and insert its result into Eq. (7.17) to obtain the *exact perturbed Schrödinger equation*:

$$\left(\frac{1}{i}\right)\psi_{\tilde{t}} = \psi_{\tilde{s}\tilde{s}} + \frac{1}{2}|\psi|^2\psi - \tilde{\delta}\left\{ i[(\mathbf{N}\cdot v_{\tilde{s}})_{\tilde{s}} - \psi v_{\tilde{s}}\cdot\mathbf{t}] + \psi \int_{\tilde{s}_0}^{\tilde{s}} \operatorname{Im}(\psi\overline{\mathbf{N}})\cdot v_{\tilde{s}}\, d\tilde{s} \right\}. \tag{7.21}$$

In particular, for the special case with $\tilde{\delta} = 0$ in Eq. (7.21), we see that, through the filament function

$$\psi(\tilde{s}, \tilde{t}) = \kappa(\tilde{s}, \tilde{t})e^{i\int_0^{\tilde{s}} T(s', \tilde{t})ds'}, \tag{7.22}$$

the self-induction equation reduces to the cubic nonlinear Schrödinger equation,

$$\frac{1}{i}\psi_t = \psi_{\tilde{s}\tilde{s}} + \frac{1}{2}|\psi|^2\psi. \tag{7.23}$$

Equation (7.23) is a famous, exactly solvable equation by the inverse-scattering method (Ablowitz and Segur, 1981; Lamb, 1980), so that we can utilize exact solutions of Eq. (7.23) to infer properties of vortex-filament motion under the self-induction approximation. Next, we point out that, despite the beauty of the self-induction equation for a vortex filament, the length of the curve $X(s, t)$, which represents the vortex tube, cannot increase in time, i.e., the *vortex filament cannot have any self-stretching within the self-induction approximation*. Thus, if we want to include other effects of bending and folding of a single vortex filament, including local self-stretching, different asymptotic models are needed, with new assumptions beyond those of assumptions (7.2). This is the topic of Section 7.2.

To establish this result we consider the general evolution of a curve $X(s, t)$ according to the law

$$\frac{\partial X}{\partial t} = \beta(s, t)\mathbf{n} + \gamma(s, t)\mathbf{b}, \tag{7.24}$$

where $\mathbf{n}$ is the normal and $\mathbf{b}$ is the binormal. Here the parameter s is not necessarily the arc length. Without loss of generality, we omitted the term $\alpha(s, t)\mathbf{t}$ on the right-hand side of law (7.24) as such terms yield motion along only the curve $X(s, t)$ itself and can always be eliminated by reparameterization. Consider the time rate of change of the infinitesimal arc length of the curve, $(X_s \cdot X_s)^{\frac{1}{2}}$. We compute from law (7.24) that

$$\frac{\partial}{\partial t}(X_s \cdot X_s)^{\frac{1}{2}} = \mathbf{t} \cdot \frac{\partial}{\partial s}[\beta(s, t)\mathbf{n} + \gamma(s, t)\mathbf{b}]. \tag{7.25}$$

Utilizing Serret–Frenet formulas (7.6), we obtain the general identity

$$\frac{\partial}{\partial t}(X_s \cdot X_s)^{\frac{1}{2}} = -\kappa\beta(X_s \cdot X_s)^{\frac{1}{2}}, \tag{7.26}$$

where κ is the curvature. In particular, if the motion of the curve is always along the binormal, we have $\beta \equiv 0$ in Eq. (7.26) and the infinitesimal arc length of a curve cannot increase in time. Thus the self-induction approximation alone does not allow for vortex stretching!

We end this section by computing the linearized self-induction equation about a straight-line filament parallel to the x_3 axis. First we rewrite the self-induction equation in the form

$$\frac{\partial X}{\partial t} = C_0 \mathbf{t} \times \frac{\partial^2 X}{\partial \sigma^2}, \tag{7.27}$$

where C_0 is a constant that depends on the choice of reference time. We consider small perturbations of a straight-line filament parallel to the x_3 axis that, without loss of generality, have the form

$$X = (0, 0, \sigma) + \epsilon[x(\sigma, t), y(\sigma, t), 0], \tag{7.28}$$

and $\varepsilon \ll 1$. Inserting Eq. (7.28) into Eq. (7.27) and calculating the leading-order behavior in ε, we obtain the *linearized self-induction equations along a straight-line filament*,

$$\frac{\partial X}{\partial t} = J\left[C_0 \frac{\partial^2 X}{\partial \sigma^2}\right] \tag{7.29}$$

for $X = [x(\sigma, t), y(\sigma, t)]$, where J is the standard skew-symmetric matrix

$$J = \begin{bmatrix} 0 & -1 \\ 1 & 0 \end{bmatrix}. \tag{7.30}$$

If we introduce the complex notation

$$\psi(\sigma, t) = x(\sigma, t) + iy(\sigma, t), \tag{7.31}$$

then Eq. (7.29) becomes the linear Schrödinger equation

$$\frac{1}{i}\frac{\partial \psi}{\partial t} = C_0 \frac{\partial^2 \psi}{\partial \sigma^2}. \tag{7.32}$$

We hope that our slight abuse of notation in still denoting the perturbation quantity $[x(\sigma, t), y(\sigma, t)]$ by X in Eq. (7.29) has not confused the reader. We utilize Eqs. (7.29) and (7.32) in interpreting the asymptotic equations for nearly parallel interacting vortex filaments described in Section 7.4.

7.2. Simplified Asymptotic Equations with Self-Stretch for a Single Vortex Filament

Here we describe a simplified asymptotic equation that allows for some features of self-stretch of a single vortex filament including incipient formation of kinks, folds, and hairpins (Klein and Majda, 1991a, 1991b, 1993; Klein et al., 1992). This simplified asymptotic equation has attractive features in that it incorporates some of the nonlocal features of vortex stretching represented in Biot–Savart law (7.1) through a filament function ψ defined by means of the same Hasimoto transform introduced in Eq. (7.7). In this fashion a singular perturbation of cubic nonlinear Schrödinger equation (7.23) emerges where the nonlocal terms represented by $I[\psi]$ compete directly with the cubic nonlinear terms in Eq. (7.23) that, as we have seen earlier, arise from local self-induction.

The key idea in the derivation of these equations beyond those conditions needed in the derivation of self-induction equation $(*)$ is to allow the vortex filament to have wavy perturbations on a scale ε that are short wavelength relative to the $\mathcal{O}(1)$ radius of curvature of the vortex filament but are long wavelength relative to the core thickness. Thus the wavy perturbations of the vortex filament satisfy

$$1 \gg \epsilon \gg \delta.$$

The choice of such scalings is motivated by numerical simulations (Chorin, 1982, 1994), demonstrating that such kinds of perturbations often develop folds and hairpins in vortex filaments.

More precisely, the asymptotic filament equations with self-stretch described below arise from small-amplitude short-wavelength perturbations of a straight-line vortex filament aligned along the x_3 axis. These perturbations of the vortex centerline have the form

$$X(s, \bar{t}) = \bar{s}\mathbf{e}_3 + \varepsilon^2 X^{(2)}\left(\frac{\bar{s}}{\varepsilon}, \frac{\bar{t}}{\varepsilon^2}\right) + o(\varepsilon^2), \tag{7.33}$$

where

$$X^{(2)} = \left[x^{(2)}\left(\frac{\bar{s}}{\varepsilon}, \frac{\bar{t}}{\varepsilon^2}\right), y^{(2)}\left(\frac{\bar{s}}{\varepsilon}, \frac{\bar{t}}{\varepsilon^2}\right), 0 \right]. \tag{7.34}$$

Although we do not present the details here, the time scale $\bar{t}$ is the same one used for the local self-induction approximation so that the perturbations in Eq. (7.33) are not only short wavelength but also occur on rapid time scales relative to the local self-induction time scale $\bar{t}$. We also note that although the vortex filaments in Eq. (7.33) vary rapidly in space, nevertheless they have only an $\mathcal{O}(1)$ effect on the radius of curvature because

$$\frac{d^2 X(s,t)}{ds^2} = \left[\frac{d^2 X^{(2)}}{d\sigma^2}\right]\left(\sigma, \frac{\bar{t}}{\varepsilon^2}\right)\Big|_{\sigma = \frac{\bar{s}}{\varepsilon}}.$$

The key assumptions that emerge in the derivation (Klein and Majda, 1991a) require that the vortex-filament thickness δ, the Reynolds number R_e defined in (2) of assumptions (7.2), and the wavelength parameter satisfy the distinguished limits

$$\varepsilon^2 \left[\ln\left(\frac{2\varepsilon}{\delta}\right) + c \right] = 1$$

$$\delta = (R_e)^{-1/2}, \qquad \delta \ll 1. \tag{7.35}$$

The first condition in Eqs. (7.35) guarantees that the requirement $1 \gg \varepsilon \gg \delta$ is satisfied, the second condition in Eqs. (7.35) is already familiar from (4) of assumptions (7.2), and the form of the perturbations in Eq. (7.33) automatically enforces (3) of assumptons (7.2).

7.2.1. Sketch of the Asymptotic Derivation

As in the derivation sketched in Section 7.1 for the self-induction equation, we need to develop a suitable asymptotic expansion of Biot-Savart law (7.1) with the ansatz in Eq. (7.33) for the filament equations under assumptions (7.35) and then insert this expansion into Eq. (7.3) to obtain a dynamical equation for the vortex-filament centerline, $X(s, \bar{t})$. Klein and Majda (1991a, Sections 3 and 4) show that the following *perturbed binormal law* emerges from this procedure under assumptions (7.35):

$$\frac{\partial X}{\partial \bar{t}} = \kappa \mathbf{b} + \varepsilon^2 I\left[X^{(2)}\right] \times \mathbf{e}_3. \tag{7.36}$$

Here $I[w]$ is the linear nonlocal operator:

$$I[w] \stackrel{\text{def}}{=} \int_{-\infty}^{x} \frac{1}{|h^3|}\left[w(\sigma+h) - w(\sigma) - h w_\sigma(\sigma+h) + \frac{1}{2} h^2 \mathcal{H}(1-|h|) w_{\sigma\sigma}(\sigma) \right] dh, \tag{7.37}$$

where we use the notation $w_\sigma = (\partial w/\partial\sigma)$ and we use the Heaviside distribution $\mathcal{H}$. The operator $I[X^{(2)}]$ incorporates the leading-order nonlocal effects of Biot–Savart law (7.1) under assumptions (7.33) and (7.35).

We recall the Hasimoto transform from Eq. (7.7) that maps a curve $X(\bar{s},\bar{t})$ to a *filament function* $\psi(\bar{s},\bar{t})$ by means of

$$\psi(\bar{s},\bar{t}) = \kappa(\bar{s},\bar{t})e^{i\int_0^{\bar{s}} T(s',\bar{t})ds'}. \tag{7.38}$$

We claim that, under the Hasimoto transform, perturbed binormal law (7.36) with the ansatz in Eq. (7.33) becomes the *asymptotic filament equation with self-stretching*,

$$\frac{1}{i}\psi_\tau = \psi_{\sigma\sigma} + \varepsilon^2\left(\frac{1}{2}|\psi|^2\psi - I[\psi]\right), \tag{7.39}$$

in which the expansion parameter ε appears as a coupling constant. In Eq. (7.39) the arguments σ and τ are given by $\sigma = (\bar{s}/\varepsilon^2)$ and $\tau = (\bar{t}/\varepsilon^2)$ up to minor arc-length corrections (see Klein and Majda, 1991a) that we ignore here and in our discussion below. As stated earlier in this section, asymptotic equation (7.39) has a linear nonlocal term $I[\psi]$ that competes directly at the same order with the local self-induction represented by the cubic nonlinear terms $\frac{1}{2}|\psi|^2\psi$.

How can we compute the local rate of filament curve stretching from the solution ψ of Eq. (7.39)? With the ansatz in Eq. (7.33), we have the infinitesimal arc length given by

$$\ell\left(\frac{\bar{s}}{\varepsilon},\frac{\bar{t}}{\varepsilon^2}\right) \underset{\text{Def}}{=} (X_{\bar{s}}\cdot X_{\bar{s}})^{1/2}.$$

A lengthy calculation (Klein and Majda, 1991a, Section 5) utilizing Eqs. (7.25), (7.26), and (7.36) establishes that the time derivative $\dot{\ell}(\sigma,\tau)$ satisfies the *local curve-stretching identity*

$$\frac{\dot{\ell}(\sigma,\tau)}{\ell} = \frac{\varepsilon^2}{4}i\int_{-\infty}^{\infty}\frac{1}{|h|}[\overline{\psi}(\sigma+h)\psi(\sigma) - \psi(\sigma+h)\overline{\psi}(\sigma)]dh. \tag{7.40}$$

We know that vortex stretching is a quadratically nonlinear process, yet the effects of nonlocal self-stretch are incorporated into filament equation (7.39) through the linear operator $I[\psi]$. Formula (7.40) resolves this apparent paradox because the relative changes in arc length are quadratically nonlinear functionals of ψ.

We proceed with the derivation of filament equation (7.39) from perturbed binormal law (7.36) by applying the general Hasimoto transform for a perturbed binormal law derived in Eqs. (7.5)–(7.19) while respecting the rapid space–time scalings implicit in the special ansatz from Eq. (7.33). From binormal law (7.36) we have the correspondence

$$\varepsilon^2 I\left[X^{(2)}\right]\times \mathbf{e}_3 = \tilde{\delta}v(\bar{s},\bar{t}) \tag{7.41}$$

in perturbed binormal law (7.5). For small-amplitude high-frequency perturbations of the centerline with the form $\varepsilon^2 X^{(2)}(\bar{s}/\varepsilon,\bar{t}/\varepsilon^2)$, we rescale the exact perturbed nonlinear Schrödinger equation (7.19) by means of the variables $\sigma = (s/\varepsilon)$ and $\tau = (t/\varepsilon^2)$ so the evolution equation for the filament function,

$$\frac{1}{i}\psi_\tau = \psi_{\sigma\sigma} + \left[\varepsilon^2 \frac{1}{2}|\psi|^2\psi - i\tilde{\delta}(\mathbf{N}\cdot v_\sigma)_\sigma\right], \tag{7.42}$$

emerges at leading order with **N** as defined in vector function (7.8). Next, identifying $\tilde{\delta}v(\bar{s},\bar{t})$ by utilizing Eq. (7.41), we obtain the evolution equation

$$\frac{1}{i}\psi_\tau = \psi_{\sigma\sigma} + \varepsilon^2\frac{1}{2}|\psi|^2\psi - i\varepsilon^2\left(\mathbf{N}\cdot I\left[X_\sigma^{(2)}\right]\times \mathbf{e}_3\right)_\sigma. \tag{7.43}$$

Applying the identity

$$-i(\mathbf{N}\cdot v_\sigma)_\sigma = -I[\psi] \tag{7.44}$$

to evolution equation (7.43) completes the derivation of asymptotic filament equation (7.39) from perturbed binormal law (7.36). The proof of identity (7.44) is lengthy and is given in Section 5 of Klein and Majda (1991a). The perceptive reader will note that our derivation of the perturbed Schrödinger equation from the perturbed binormal law in Section 7.1 utilized arc-length coordinates although s in Eq. (7.33) is not exactly an arc-length coordinate, $\tilde{s}$; in fact, we have $s = \tilde{s}[1 + \mathcal{O}(\varepsilon^2)]$. However, for pedagogical simplicity, we ignore these minor differences in the sketch presented here.

7.2.2. The Mathematical Structure of the Asymptotic Equation

Here we discuss the mathematical properties of the filament equations with self-stretching (7.39). We develop Fourier representations for the nonlocal operator $I[w]$ in Eq. (7.37). These formulas reveal the fact that filament equation (7.39) becomes a novel singular perturbation of the linear Schrödinger equation.

Explicit Fourier Representation of the Nonlocal Operators

We discuss the operator $I[\cdot]$ in Eq. (7.37) that acts on functions defined for all of $\mathbb{R}^1$ with rapid decay. The operator $I[\cdot]$ commutes with translation and therefore is given by convolution with a distribution kernel. For all of space and rapidly decreasing functions, the Fourier inversion formula yields

$$f(\sigma) = (2\pi)^{-1}\int e^{i\sigma\xi}\hat{f}(\xi)d\xi,$$

$$\hat{f}(\xi) = \int e^{-i\sigma\xi}f(\sigma)d\sigma.$$

Thus we have

$$I[w](\sigma) = (2\pi)^{-1}\int e^{i\sigma\xi}\hat{I}(\xi)\widehat{w}(\xi)d\xi \tag{7.45}$$

with the Fourier symbol $\hat{I}(\xi)$ of I defined by

$$\hat{I}(\xi) = e^{-i\sigma\xi} I[e^{i\sigma\xi}]. \tag{7.46}$$

From the definitions in Eqs. (7.37) and (7.46), we compute explicitly that

$$\hat{I}(\xi) = \int_0^\infty \frac{1}{h^3}\left[(e^{i\xi h} + e^{-i\xi h} - 2) - ih\xi(e^{i\xi h} - e^{-i\xi h}) - \theta(h)(\xi h)^2\right] dh, \tag{7.47}$$

with $\theta(h) = \mathcal{H}(1 - |h|)$. Next we determine the scaling properties of $\hat{I}(\xi)$. First we introduce real trigonometric functions in Eq. (7.47) and change variables with $\eta = \xi h$ to obtain

$$\hat{I}(\xi) = \xi^2 \int_0^\infty \frac{1}{\eta^3}\left[2(\cos\eta - 1) + 2\eta\sin\eta - \theta\left(\frac{\eta}{|\xi|}\right)\eta^2\right] d\eta. \tag{7.48}$$

Note that only the function $\theta(\eta/|\xi|)$ has explicit dependence on ξ in the integrand in Eq. (7.48). We split the integral in Eq. (7.48) into a sum of convergent factors, obtaining the decomposition

$$\begin{aligned}\hat{I}(\xi) = \xi^2 &\left[\int_0^1 \frac{2(\cos\eta - 1) + 2\eta\sin\eta - \eta^2}{\eta^3} d\eta + \int_1^\infty \frac{2(\cos\eta - 1) + 2\eta\sin\eta}{\eta^3} d\eta\right] \\ &- \xi^2 \int_0^\infty \left[\theta\left(\frac{\eta}{|\xi|}\right) - \theta(\eta)\right]\eta^{-1} d\eta. \end{aligned} \tag{7.49}$$

The term in brackets in the second line of Eq. (7.49) contributes a finite constant, and, by the definition of the function θ, we compute that

$$\int_0^\infty \left[\theta\left(\frac{\eta}{|\xi|}\right) - \theta(\eta)\right]\eta^{-1} d\eta = \ln|\xi|. \tag{7.50}$$

With Eqs. (7.49) and (7.50), we obtain the formula

$$\hat{I}(\xi) = -\xi^2 \ln|\xi| + C_0\xi^2, \tag{7.51}$$

with the constant C_0 determined by the expression in brackets in the second line of Eq. (7.49). We can evaluate the constant C_0 by a lengthy calculation by utilizing the method of stationary phase. With that information, we derive the *explicit formula for the symbol of* $I[\cdot]$:

$$\hat{I}(\xi) = -\xi^2 \ln|\xi| + C_0\xi^2,$$

where

$$C_0 = \frac{1}{2} - \gamma = -0.0772, \ldots,$$

and γ is given by Euler's constant.

Spectrum of the Linearized Operator and Singular Perturbation

The linearization of asymptotic filament equation (7.39) about the state $\psi \equiv 0$ is the linearized equation

$$\frac{1}{i}\psi_t = \psi_{\sigma\sigma} - \epsilon^2 I[\psi]. \tag{7.52}$$

The operator on the right-hand side of Eq. (7.52) is symmetric, and we calculate that the dispersion relation is

$$\omega = P_\epsilon(\xi),$$
$$P_\epsilon(\xi) = -|\xi|^2 - \epsilon^2 C_0|\xi|^2 + \epsilon^2\xi^2 \ln|\xi|. \tag{7.53}$$

The generator of linearized equation (7.52) is the operator

$$\psi_{\sigma\sigma} - \epsilon^2 I[\psi]. \tag{7.54}$$

The spectrum of the linearized generator $\epsilon = 0$ is the familiar spectrum for the Schrödinger operator given by the interval $(-\infty, 0]$; the spectrum of the linearized generator in operator (7.54) is radically different for any $\epsilon > 0$ and reflects the singular character of this perturbation. We compute that $P_\epsilon(\xi)$ from dispersion relation (7.53) is a generalized eigenvalue of linearized operator (7.54) for any $\epsilon > 0$ and $\xi \in \mathbb{R}^1$. It follows immediately that, in contrast to the case with $\epsilon = 0$, the spectrum of operator (7.54) is $[a(\epsilon), +\infty]$ for any $\epsilon > 0$, where $a(\epsilon) \downarrow -\infty$ rapidly as $\epsilon \downarrow 0$; in fact, $|a(\epsilon)| = \mathcal{O}[\epsilon^2 \exp(2/\epsilon^2)]$. Thus a complete half-interval of a new spectrum in $[0, \infty)$ is created by the effect of $I[\psi]$ for any $\epsilon > 0$. We conclude that $-\epsilon^2 I[\psi]$ is in fact a singular perturbation of the linear Schrödinger operator. We invite the reader to graph the dispersion relation $\omega = P_\epsilon(\xi)$ and to compare this graph with that for $\omega = -|\xi|^2$ from the linear Schrödinger equation.

Remark: The nature of the singular perturbation depends crucially on the sign of the coefficient that precedes $I[\psi]$ in filament equation (7.39). For comparison we consider the equation

$$\frac{1}{i}\psi_t^+ = \psi_{\sigma\sigma}^+ + \epsilon^2\left(\frac{1}{2}|\psi^+|^2\psi^+ + I[\psi^+]\right). \tag{7.55}$$

Then the dispersion relation for the linearized operator derived from the above equation is given by

$$\omega = P_\epsilon^+(\xi),$$
$$P_\epsilon^+(\xi) = -|\xi|^2 - \epsilon^2\xi^2 \ln|\xi| + \epsilon^2 C_0|\xi|^2,$$

and the spectrum of the linearized operator is a $(-\infty, a_+(\epsilon)]$, where $a_+(\epsilon) \downarrow 0$ rapidly as $\epsilon \downarrow 0$; in this case, the spectrum for $\epsilon > 0$ is always well approximated by that of the linearized Schrödinger operator. Thus, at the linearized level, the perturbation in

Eq. (7.55) is much milder than the one actually occurring in the asymptotic filament equation.

Asymptotic filament equation (7.39) has global existence of solutions for all times. The elementary proof, sketched in Klein and Majda (1991b), reflects the character of the nonlocal operator $I[\psi]$ as a singular perturbation. However, solutions of these filament equations can lose their asymptotic validity as approximate solutions of the actual fluid motion at finite times (see Klein and Majda, 1991b).

7.2.3. Asymptotic Equations for the Stretching of Vortex Filaments in a Background Flow Field

The axisymmetric vortices with swirl from Subsection 2.3.3, including the Burgers vortex [Eq. (2.72)], provide simple analytic examples of how a background strain flow can interact with a vortex tube and create vortex stretching. Localized vortex filaments with large strength and narrow cross section are also prominent fluid-mechanical structures in applications such as secondary 3D instability in mixing layers (Corcos and Lin, 1984), in which larger-scale vortices provide a slowly varying background flow field in which strong vortex filaments are embedded. Motivated by these and other considerations, we now develop asymptotic equations that generalize filament equation (7.39) with self-stretching to the case with a background flow field.

In the beginning of this section we considered a straight-line vortex filament with small-amplitude wavy distortions of the form of Eqs. (7.33) and (7.34) that satisfy asymptotic conditions (7.35). We now include, in addition to the effects of local induction and self-stretch, the effect of a known background flow field of the form

$$\delta v_b(x,t) = \frac{1}{\varepsilon^2} A x, \tag{7.56}$$

where the 3×3 matrix A, representing the velocity gradients,

$$A = \begin{bmatrix} s_{11} & -\frac{1}{2}\omega + s_{12} & 0 \\ \frac{1}{2}\omega + s_{12} & s_{22} & 0 \\ 0 & 0 & s_{33} \end{bmatrix}, \tag{7.57}$$

satisfies the incompressibility condition

$$\operatorname{tr} A = s_{11} + s_{22} + s_{33} = 0. \tag{7.58}$$

Recall from Section 1.4 of Chap. 1 that the background flow field described in Eqs. (7.56) and (7.57) represents the local Taylor expansion of a general 3D incompressible flow field that satisfies the additional *crucial requirement that the centerline along the x_3 axis for the unperturbed vortex filament remain invariant under the particle-trajectory flow map associated with the background flow in* Eq. (7.56). This requirement is guaranteed by the block diagonal structure flow in matrix (7.57). The coefficient s_{33} represents the rate of strain along the vortex filament that is due to the background flow, and the 2×2 matrix

$$S_2 = \begin{bmatrix} s_{11} & s_{12} \\ s_{12} & s_{22} \end{bmatrix} \tag{7.59}$$

represents the contribution of the deformation matrix perpendicular to the unperturbed vortex filament.

The same general procedure, as that outlined in Subsection 7.2.1 can be utilized to pass from a perturbed binormal law such as law (7.36), which also includes the background flow field in Eq. (7.56), to a modified filament equation, including the effects of the background flow from Eqs. (7.56) and (7.57), by means of Hasimoto transform (7.7) or (7.38). The reader interested in the details can consult Section 2 of the paper by Klein et al. (1992). The result of this derivation is the *asymptotic filament equation with self-stretch in a background flow field*:

$$\frac{1}{i}\psi_\tau = \psi_{\sigma\sigma} + \epsilon^2\left(\frac{1}{2}|\psi|^2\psi - I[\psi]\right) + \epsilon^2\left[\left(\omega + \frac{5}{2}is_{33}\right)\psi + is_{33}\sigma\psi_\sigma + \left(s_{12} - i\frac{s_{11} - s_{22}}{2}\right)\overline{\psi}\right]. \tag{7.60}$$

The first two terms on the right-hand side of filament equation (7.60) are already familiar to the reader from the filament equations with self-stretch without a background flow field in filament equation (7.39). The new terms that are due to the background flow field in Eq. (7.56) are in brackets on the right-hand side of Eq. (7.60).

We consider each of these three terms separately in order to extract a physical interpretation of these effects. The separate effect of the first term yields the equation

$$\frac{1}{i}\psi_\tau = \varepsilon^2\omega\psi. \tag{7.61}$$

Thus the effect of background rotation around the axis $s\mathbf{t}_0$ on the perturbed filament corresponds to the rotation of ψ in the complex plane at a frequency $\varepsilon^2\omega$. The separate effect of the second background flow term in Eq. (7.60) is the advection operator

$$\psi_\tau = -\varepsilon^2 s_{33}\left(\frac{5}{2}\psi + \sigma\psi_\sigma\right), \tag{7.62}$$

which corresponds to growth or decay of amplitudes that is due to flow convergence or divergence in the normal plane and to advection along the x_3 axis. The simple axisymmetric swirling flows constructed in Subsection 2.3.3 of Chap. 2 have a similar physical structure as exact solutions. Finally, the last term in Eq. (7.60) involving the background flow yields the equation

$$\left(\frac{1}{i}\right)\psi_\tau = \varepsilon^2\left\{s_{12} - i\left[\frac{(s_{11} - s_{22})}{2}\right]\right\}\overline{\psi}, \tag{7.63}$$

which represents the effect of the planar strain normal to the filament axis and given by S_2. We note that the complex conjugate of ψ, $\overline{\psi}$, appears on the right-hand side of Eqs. (7.60) and (7.63). ODE (7.63) has both damping and growing solutions corresponding to the strain axes of the background flow field; this term both damps and drives the solutions of asymptotic filament equation (7.60).

7.2.4. Analytical Properties of the Filament Equation with a Background Flow

Here we develop a criterion for the nonlinear stability of vortex filaments to appropriate short-wavelength perturbations through Eq. (7.60). We also describe several properties of the linearization of Eq. (7.60) in an instructive special case involving planar-strain flows in directions orthogonal to the unperturbed vortex filament.

The Nonlinear Stability of Vortex Filaments in a Background Flow Field

Recent numerical simulations at moderately large Reynolds numbers (Ashurst et al., 1987; Jackson and Orszag, 1990; Majda, 1991) reveal that nearly columnar, strong vortex filaments in a background flow field persist for some time despite the fact that the planar strains for the swirling flow in directions orthogonal to the filament axis are stronger than the strain component aligned with the vortex axis. Here we give some precise nonlinear stability conditions by utilizing Eq. (7.60) that are consistent with this case and provide additional analytic insight.

From Hasimoto transform (7.7), we see that $|\psi(\sigma, \tau)|$ has the physical significance of the curvature of the perturbed filament so that a natural quantity for measuring the stability of the columnar vortex is

$$\int |\psi(\sigma, \tau)|^2 d\sigma.$$

The unperturbed vortex filament is stable if there is a constant $\bar{\alpha} > 0$ such that

$$\frac{\partial}{\partial \tau} \int |\psi(\sigma, \tau)|^2 d\sigma \leq -\bar{\alpha} \int |\psi(\sigma, \tau)|^2 d\sigma \tag{7.64}$$

for all solutions of Eq. (7.60) and for all $\tau \geq 0$. Condition (7.64) guarantees that the mean-square average of the curvature decays at least exponentially with time so that the columnar filament is stable for the specific class of short-wavelength filament perturbations from Eq. (7.33), consistent with the asymptotic derivation.

Using Eq. (7.60), we calculate that

$$\begin{aligned}\frac{\partial}{\partial \tau} \int |\psi|^2 d\sigma &= 2 \operatorname{Re} \int_\tau \psi_\tau \overline{\psi}\, d\sigma \\ &= 2\epsilon^2 \left\{ -s_{33}(\tau) \left[\int \operatorname{Re} \left(\sigma \psi_\sigma \overline{\psi} + \frac{5}{2} \psi \overline{\psi} \right) d\sigma \right] \right\} \\ &\quad + \left\{ \int \operatorname{Re} \left[\left(\frac{s_{11} - s_{22}}{2} + i s_{12} \right) \overline{\psi^2} \right] d\sigma \right\},\end{aligned} \tag{7.65}$$

where Re denotes the real part. We compute further that

$$\int \operatorname{Re} \left(\sigma \psi_\sigma \overline{\psi} + \frac{5}{2} \psi \overline{\psi} \right) d\sigma = 2 \int |\psi|^2 d\sigma, \tag{7.66}$$

$$\operatorname{Re} \left[\left(\frac{s_{11} - s_{22}}{2} + i s_{12} \right) \overline{\psi}^2 \right] = \psi \cdot M \psi, \tag{7.67}$$

where $\psi = {}^t(\psi_1, \psi_2)$, $\psi = \psi_1 + i\psi_2$, and M is the symmetric 2×2 matrix,

$$M = \begin{bmatrix} \frac{s_{11} - s_{22}}{2} & -s_{12} \\ -s_{12} & \frac{s_{22} - s_{11}}{2} \end{bmatrix}. \tag{7.68}$$

The matrix M has the two real eigenvalues

$$\lambda_{\pm} = \pm \left[\frac{(s_{11} - s_{22})^2}{4} + s_{12}^2 \right]^{1/2},$$

so that

$$\psi \cdot M\psi \le \lambda_+ |\psi|^2. \tag{7.69}$$

Combining relations (7.65)–(7.69), we obtain that

$$\frac{\partial}{\partial \tau} \int |\psi|^2 d\sigma \le 2\epsilon^2(-2s_{33} + \lambda_+) \int |\psi|^2 d\sigma. \tag{7.70}$$

Looking back at the definition of stability in condition (7.64), we see that the *columnar vortex filament is nonlinearly stable to the short-wavelength perturbations in* Eq. (7.33), *provided that the strain matrix from the background flow satisfies*

$$\left[\frac{(s_{11} - s_{22})^2}{4} + s_{12}^2 \right]^{1/2} \le 2s_{33}. \tag{7.71}$$

Of course, because the flow is incompressible, the identity

$$s_{33} = -(s_{11} + s_{22})$$

allows us to express general criterion (7.71) solely in terms of the strain-matrix components orthogonal to the columnar filament. Obviously, a positive axial-strain component is a necessary condition for the columnar filament to remain stable. We remark that the inequality in criterion (7.71) does not involve the transverse rotation components for the background flow field ω from Eq. (7.60).

We interpret criterion (7.71) in the special case in which the strain matrix is diagonal so that $s_{12} = 0$, $s_{11} = \gamma_1$, $s_{22} = \gamma_2$, and $s_{33} = \gamma_{33}$, with $\gamma_1 + \gamma_2 + \gamma_3 = 0$. Without loss of generality, we assume that $\gamma_1 \ge \gamma_2$; then stability criterion (7.71) becomes

$$\gamma_2 < 0, \qquad \gamma_1 < \frac{3}{5}|\gamma_2|. \tag{7.72}$$

The regime of criteria (7.72), in which γ_1 and γ_2 satisfy

$$\gamma_2 < 0, \qquad \frac{1}{2}|\gamma_2| < \gamma_1 < \frac{3}{5}|\gamma_2|, \tag{7.73}$$

is very interesting for qualitative comparison with the computational results mentioned earlier. With conditions (7.73) we also have the following conditions:

(1) The axial component of the strain matrix is positive but γ_3 is the intermediate eigenvalue of the strain matrix.

(2) The strongest strain effect is the flattening of the vortex core by the compressive strain γ_2 orthogonal to the filament.
(3) Despite the structure in (1) and (2) and also rapid planar rotation from ω, the columnar vortex is stable in time on the τ time scale according to criteria (7.72).

Conditions (1) and (2) are the ones typically observed for the nearly columnar vortices that emerge in the numerical experiments cited earlier that are conditioned on the large-vorticity sets. Here we have provided supporting analytic evidence for the stability of such columnar vortices to suitable short-wavelength perturbations on appropriate time scales. We have utilized the special case in criteria (7.72) and (7.73) to elucidate this structure in a transparent fashion. Clearly, from the more general criterion (7.71), we can even allow suitable temporal rotation of the planar-strain axis in time and fulfill all three of conditions (1)–(3).

Linearized Theory for Planar-Strain Flows

Here we develop the linearized theory for Eq. (7.60) in the important special case in which the background flow field is a simple planar strain in directions orthogonal to the unperturbed filament. This case has some physical importance as it occurs naturally in the development of secondary 3D instabilities in mixing layers and other 2D basic flows. Also, the stability criteria just discussed are not satisfied; the unperturbed filament is linearly unstable in the presence of these background flows.

With constant planar-strain flows as the background field, filament equation (7.60) is given by

$$\left(\frac{1}{i}\right)\psi_\tau = \psi_{\sigma\sigma} + \epsilon^2\left(\frac{1}{2}|\psi|^2\psi - I[\psi] - i\gamma\overline{\psi}\right), \tag{7.74}$$

where γ is the strain rate. To see the effect of the strain alone, we ignore all of the other terms and arrive at the ODE

$$\psi_\tau = \epsilon^2\gamma\overline{\psi}. \tag{7.75}$$

The simple ODE (7.75) has both damping and growing solutions; thus the background planar-strain flow can be a source of instability for filament equation (7.74) at suitable wavelengths.

We linearize Eq. (7.74) at the unperturbed filament $\psi = 0$; thus the linearized problem is given by

$$\left(\frac{1}{i}\right)\psi_\tau = \psi_{\sigma\sigma} + \epsilon^2(-I[\psi] - i\gamma\overline{\psi}). \tag{7.76}$$

We write the exact solution of problem (7.76) by using the Fourier transform

$$\widehat{\psi}(k,\tau) = (2\pi)^{-1/2}\int_{-\infty}^{\infty} e^{ik\sigma}\psi(\sigma,\tau)d\sigma. \tag{7.77}$$

The effect of the straining term involving $\overline{\psi}$ in problem (7.76) is that the two modes k and $-k$ are coupled together so that the Fourier-transformed equation is no longer local. In fact, we calculate that the pair $\widehat{\psi}(k,\tau)$, $\overline{\widehat{\psi}}(-k,\tau)$ satisfies the coupled equation

$$\left\{\begin{bmatrix} \widehat{\psi}(k,\tau) \\ \overline{\widehat{\psi}}(-k,\tau) \end{bmatrix}\right\}_\tau = i\left\{\widehat{\mathcal{L}}(k)\begin{bmatrix} 1 & 0 \\ 0 & -1 \end{bmatrix} - i\epsilon^2\gamma\begin{bmatrix} 0 & 1 \\ 1 & 0 \end{bmatrix}\right\} \times \left\{\begin{bmatrix} \widehat{\psi}(k,\tau) \\ \overline{\widehat{\psi}}(-k,\tau) \end{bmatrix}\right\}. \tag{7.78}$$

Here, $\widehat{\mathcal{L}}(k)$ is given by

$$\widehat{\mathcal{L}}(k) = -k^2 - \epsilon^2 \hat{I}(k), \tag{7.79}$$

with the Fourier symbol for the nonlocal operator I from Eqs. (6.45) given by

$$\hat{I}(k) = -k^2 \ln|k| + C_0 k^2,$$

where $C_0 = \frac{1}{2} - \gamma = -0.0772$ and γ is Euler's constant.

Given coupled equation (7.78) we explicitly calculate the solution:

$$\left\{\begin{bmatrix} \widehat{\psi}(k,\tau) \\ \overline{\widehat{\psi}}(-k,\tau) \end{bmatrix}\right\} = \{\cos[\tau p(k)]I + i\sin[\tau p(k)]P\} \times \begin{bmatrix} \widehat{\psi}(k,0) \\ \overline{\widehat{\psi}}(-k,0) \end{bmatrix}, \tag{7.80}$$

where

$$p(k) = [\widehat{\mathcal{L}}^2(k) - \epsilon^4\gamma^2]^{1/2}, \tag{7.81}$$

I is the 2×2 identity matrix, and

$$P = [p(k)]^{-1}\left\{i\epsilon^2\gamma\begin{bmatrix} 0 & 1 \\ 1 & 0 \end{bmatrix} + \widehat{\mathcal{L}}(k)\begin{bmatrix} 1 & 0 \\ 0 & -1 \end{bmatrix}\right\}. \tag{7.82}$$

The functions $\cos(z)$ and $\sin(z)$ are evaluated for complex variables z in solution (7.80). In particular, we have oscillatory behavior for the larger wave numbers,

$$k \quad \text{with} \quad \widehat{\mathcal{L}}^2(k) > \epsilon^4\gamma^2,$$

and unstable growth for the sufficiently low wave numbers,

$$\widehat{\mathcal{L}}^2(k) < \epsilon^4\gamma^2,$$

in the behavior of the solutions of linearized equation (7.76). To summarize, we see that there are two effects of the planar-strain flow as regards linearized theory: first, to introduce instability at low wave numbers; second, to couple the modes k and $-k$ in a nonlocal fashion in Fourier space.

7.2.5. The Creation of Kinks, Folds, and Hairpins in Vortex Tubes by Means of the Filament Equations

Vortex tubes that develop kinks, folds, and hairpins appear prominently in both numerical solutions (Chorin, 1982) and experiments in flows such as shear layers and trailing wakes (Corcos and Lin, 1984; Van Dyke, 1982). The simplified asymptotic equations developed in Eqs. (7.39) and (7.60) arise from initial short-wavelength perturbations of the form of Eq. (7.33). In Subsections 7.2.2 and 7.2.4 we studied properties of solutions of the linear operators in the equations. We might conjecture that the nonlinearity in these equations can interact with these linear operators to produce the incipient formations of kinks and hairpins in the vortex filaments seen in experiments. Here we present numerical solutions of asymptotic equations (7.39) and (7.60), confirming this behavior.

First we consider spatially periodic solutions of the filament equations with self-stretch in Eq. (7.39), without the effect of a background flow field. Equations of the form of Eq. (7.39) have exact oscillatory solutions of the form

$$\psi_H(\sigma, \tau) = A \exp(i[\xi\sigma + \Omega\tau]), \tag{7.83}$$

provided that $\Omega(\xi, A, \varepsilon)$ is given by

$$\Omega(\xi, A, \varepsilon) = -\xi^2 - \varepsilon^2 \hat{I}(\xi) + \frac{1}{2}\varepsilon^2 A^2, \tag{7.84}$$

where, $\hat{I}(\xi)$ is given explicitly in Eq. (7.51). We claim that the exact solutions in Eq. (7.83) of the asymptotic equations correspond to helical vortex filaments. To see this, we recall Hasimoto transform (7.38); we have the general formulas

$$\begin{aligned} |\psi|(\sigma, \tau) &= \text{curvature}, \\ \arg \psi &= \int_0^\sigma T(\sigma, \tau) d\sigma, \end{aligned} \tag{7.85}$$

where $T(\sigma, \tau)$ the torsion. From Eqs. (7.85), the exact solutions in Eq. (7.83) have constant curvature and constant torsion and thus are necessarily helical vortex filaments.

Because the exact solutions in Eq. (7.83) are linearly unstable to suitable long-wavelength perturbations (Klein and Majda, 1991b), it is very interesting to see the nonlinear consequences of perturbations of such solutions. Consider spatially periodic initial data describing a suitable small perturbation of the helical filament with the value of $\varepsilon^2 = 0.25$. Figure 7.1 shows the evolving curvature of the filament in plots of $|\psi|$. These plots indicate that a narrow, highly localized spike in the curvature is developing in time. Figure 7.2 shows the corresponding spatial curves that produce a periodic array of strong kinks in the vortex filament. Thus solutions of the asymptotic equations for vortex filaments with self-stretch produce strong curvature spikes that correspond to the spontaneous generation of kinks. Such behavior is completely absent in corresponding solutions of nonlinear Schrödinger equation (7.23) that arises from the self-induction approximation without the self-stretch terms (Klein and Majda, 1991b).

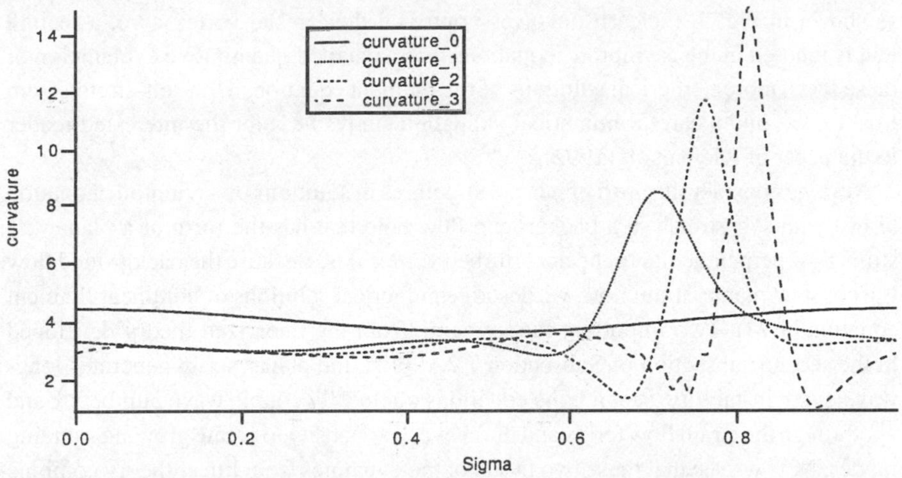

Figure 7.1. Time series of curvature distributions, $|\psi|$ vs. σ, for perturbation of the helix solution. Background helix: $|\psi H| = A = 4.0$, $\xi H = 2.0$; perturbation mode: $|\psi H|b = 0.4$, $\xi H + \beta = 3.0$; $\epsilon^2 = 0.25$. solid curve, $t = 0$; dotted curve, $t = 1.6$; short-dashed curve, $t = 1.75$; long-dashed curve, $t = 1.9$.

Although the solution indicated in Figs. 7.1 and 7.2 continues to exist for all times, it loses asymptotic validity as a physical solution of the vortex-filament equations precisely near the final time shown in Fig. 7.1. As the kink in the vortex filament forms, the solution simultaneously generates large amplitudes at wavelengths of the scale of the vortex-core thickness. Thus the solution violates implicit assumptions (7.35) that perturbations have wavelengths larger than the core thickness. This makes physical sense because the subsequent evolution of the kink in the vortex filament,

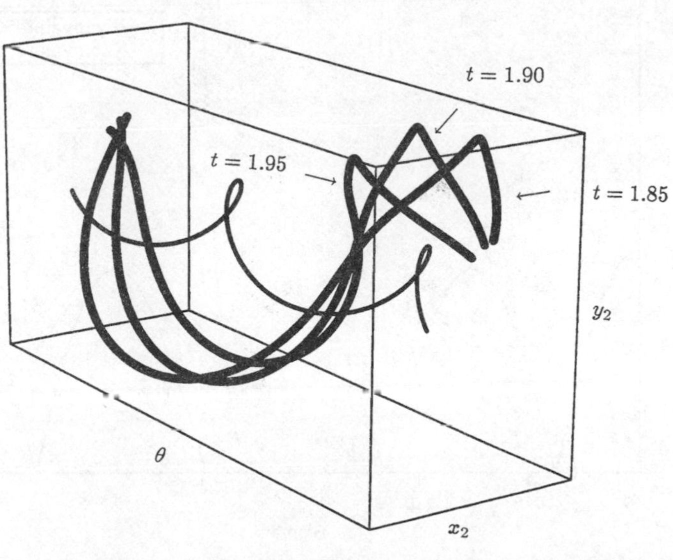

Figure 7.2. Time sequence of 3D plot of the filament curve showing the initial helix, followed by times just before, just at, and just after the peak in maximum stretch rate. Viewed from $\theta = 3$, $x_2 = 2$, $y_2 = 1$. Note the formation of the kink structure in the vortex filament.

as shown in Fig. 7.2, clearly involves nontrivial flow in the vortex core, a feature that is ignored in the asymptotic equations. For a detailed quantitative explanation of these facts, more numerical solutions of the filament equations with self-stretch from Eqs. (7.39), and a careful numerical validation study, we refer the interested reader to the paper of Klein et al. (1992).

Next we consider the birth of hairpin structures in solutions of asymptotic equation (7.60) with self-stretch in a background flow field that has the form of a planar 2D strain flow perpendicular to the unperturbed vortex axis. Because the background flow is a constant planar-strain flow, we describe numerical solutions of nonlinear filament equation (7.74). Two important facts emerge from the linearized theory developed in the second subsection of Subsection 7.2.4: First, the planar strain generates long-wavelength instability; second, the equations nonlocally couple wave numbers k and $-k$ through the strain flow terms and thus have a propensity to create growing standing modes. Next we see that these two facets of the dynamics from linear theory combine with nonlinearity to produce the birth of hairpin vortices.

Figure 7.3 shows the initial curvature magnitude and the curvature magnitude at a later time for a numerical solution of Eq. (7.74) with strain rate $\gamma = 2$ and $\varepsilon^2 = 0.25$ for suitable long-wavelength standing-mode initial data. The plot at the later time shows the emergence of two narrow localized curvature spikes. Because each narrow curvature spike generates a kink in the vorticity and two kinks are required for creating a hairpin vortex filament, we might anticipate that this solution of Eq. (7.74) displays the birth of a hairpin. The graph in Fig. 7.4 of the actual filament curve at a time later than that of Fig. 7.3 confirms the formation of a spatially periodic array of localized hairpins. The later time solution shown in Figs. 7.3 and 7.4 is near the time at which

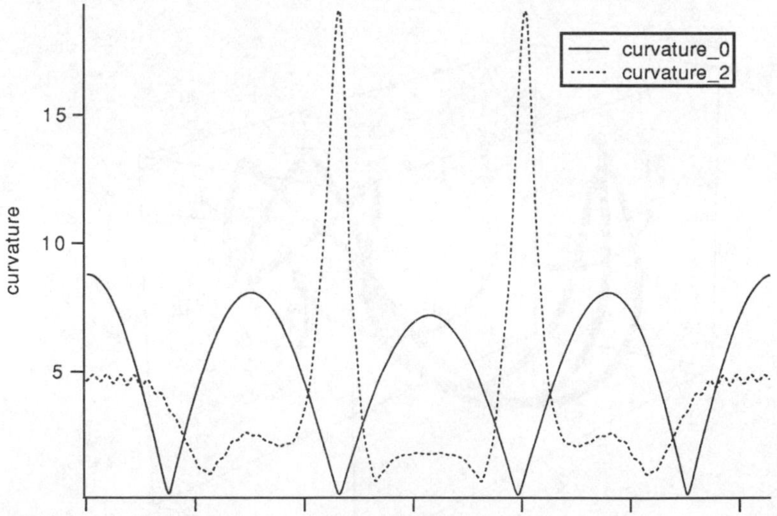

Figure 7.3. Plots of the spatial curvature distribution at times $t = 0.0$ and $t = 1.66$ for perturbed standing-mode initial data $\psi(\sigma, 0) = 4.0[\exp(4\pi i\sigma) + \exp(-4\pi i\sigma)] + 0.4[\exp(6\pi i\sigma) + \exp(-6\pi i\sigma)]$ and weak strain $\gamma = -2.0 + i0.0$.

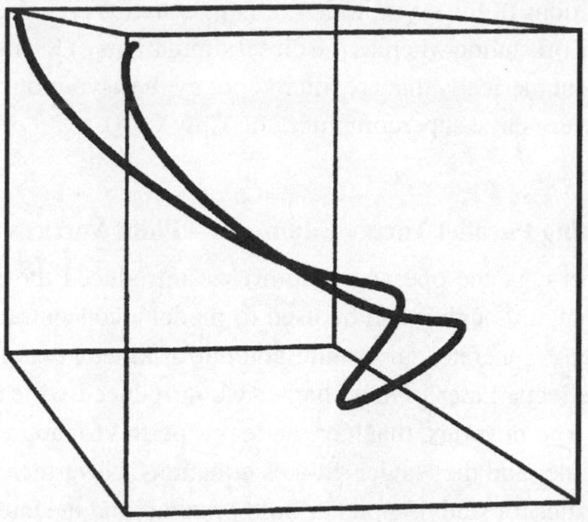

Figure 7.4. 3D representation of the filament geometry for the solution of Fig. 7.3 with viewpoint (3.0, −0.4, 0.2).

the solution of asymptotic equation (7.74) loses its validity for the same reasons mentioned earlier – the creation of the hairpin generates wavelengths of the order of the core thickness. The interested reader can consult the paper by Klein et al. (1992) for more details and further numerical simulations.

The Finite Effects of ε in the Asymptotic Theories

Despite the intuitive appeal and elegant mathematical underpinning related to physical effects in solutions of the asymptotic filament equations discussed throughout Section 7.2, one can criticize this theory at the outset as being unrealistic because the requirements in assumptions (7.35) simultaneously require that

$$\varepsilon \cong (|\ln(R_e)|)^{-\frac{1}{2}}, \qquad \varepsilon \ll 1. \tag{7.86}$$

Thus the appearance of ε^2 in simplified equations (7.39) and (7.60) might imply that these equations study transcendentally small effects at ultrahigh Reynolds numbers. Furthermore, the formation of kinks and hairpins in solutions of these asymptotic equations involved the moderate values of $\varepsilon^2 = 0.25$ that seem unrealistically large.

Are such criticisms of the theory real or illusory? One way to address this is to develop detailed numerical simulations of the complete incompressible fluid equations for initial data like those from Eq. (7.33), which are compatible with the asymptotic theory. Recently, in an important paper, Klein and Knio developed direct simulations of the incompressible fluid equations, utilizing computational vortex methods in three dimensions (see Chaps. 2 and 6) to address these issues. They emphatically confirm the quantitative predictions of the asymptotic filament equations described in this section by their direct numerical simulations, even for the moderate values of $\varepsilon^2 = 0.25$! This provides dramatic evidence that the appealing physical asymptotic theory for vortex filaments developed in this section has quantitative and qualitative validity beyond the strict regime of asymptotic validity listed in requirements (7.35) and (7.86).

Numerical solutions of the asymptotic filament equations require only a few minutes on a desktop workstation whereas the direct simulations of Klein and Knio need to resolve stiff terms numerically that are filtered out by the asymptotics, which requires many hours on a very large supercomputer, the Cray C-90.

7.3. Interacting Parallel Vortex Filaments – Point Vortices in the Plane

In Chapter 6, (see, e.g., the opening section) we introduced the notion of a point vortex and saw that, although it can be used to model a concentrated region of vorticity, the fact that it generates an infinite amount of kinetic energy can sometimes lead to spurious effects. Later in that chapter we introduced vortex blobs, a smooth approximation to a point vortex, that have better properties for approximating smooth solutions of the Euler and the Navier–Stokes equations. Nevertheless, point vortices are useful diagnostics for studying planar fluid systems, and the interested reader can consult Chorin and Marsden (1990) for more information on the motion of collections of point vortices in the plane. Here, for the purposes of motivating an important generalization presented in Section 7.4, we regard a collection of point vortices in the plane as equivalently a collection of exactly parallel vortex filaments with no structural variation along the x_3 axis. For point vortices, the vorticity $\omega(x, t)$ has the postulated form as a superposition of δ functions for all times:

$$\begin{aligned} \omega(x, t) &= \sum_{j=1}^{N} \Gamma_j \delta[x - X_j(t)], \\ X_j(t)|_{t=0} &= X_j^0, \end{aligned} \tag{7.87}$$

where Γ_j is the circulation of the jth vortex. Recall from Eq. (7.1) that the formal induced velocity associated with vorticity (7.87) is

$$v(x, t) = \sum_{j=1}^{N} \Gamma_j \frac{(x - X_j)^{\perp}}{|x - X_j|^2}, \tag{7.88}$$

where

$$x^{\perp} = \begin{pmatrix} -x_2 \\ x_1 \end{pmatrix}.$$

Ignoring the fact that the velocity of a point vortex is infinite at its center, intuitively as in the exact radial eddies described in Example 2.1, Chap. 2, we find that a point vortex induces no motion at its center. Thus, by using the particle-trajectory equations

$$\frac{dX_j}{dt} = v(X_j, t), \qquad X_j|_{t=0} = X_j^0, \tag{7.89}$$

and the above formal fact, we arrive at the *equations for N Interacting Exactly Parallel Vortex Filaments:*

$$\begin{aligned} \frac{dX_j}{dt} &= \sum_{k \neq j} \Gamma_k \frac{(X_j - X_k)^{\perp}}{|X_j - X_k|^2}, \\ X_j|_{t=0} &= X_j^0. \end{aligned} \tag{7.90}$$

It is worth noting here that such dynamic equations as those of Eqs. (7.90) have a self-consistent derivation through formal asymptotic expansions as the high Reynolds number limit of suitable solutions of the 2D Navier–Stokes equations. The interested reader can consult the book by Ting and Klein (1991) for a detailed discussion.

It is well known that the equations for point vortices are a Hamiltonian system, i.e., dynamical equations (7.90) can be rewritten in the generalized Hamiltonian form

$$\Gamma_j \frac{d}{dt} X_j = J \nabla_{X_j} H_p, \tag{7.91}$$

where J is the usual skew-symmetric matrix

$$J = \begin{bmatrix} 0 & -1 \\ 1 & 0 \end{bmatrix}. \tag{7.92}$$

For the motion of point vortices in dynamical equations (7.90), the reader can readily verify that the Hamiltonian is

$$H_p = 2 \sum_{j<k} \Gamma_j \Gamma_j \ln |X_j - X_k|. \tag{7.93}$$

Conserved Quantities

In general, according to Noether's Theorem (Arnold, 1989), symmetries in a Hamiltonian lead to conserved quantities. First, the translational symmetry $H(X_1 + \mathbf{Y}, X_2 + \mathbf{Y}, \ldots, X_N + \mathbf{Y}) = H(X_1, \ldots, X_N)$ for all $\mathbf{Y}$ leads to the conserved quantity for the dynamics:

$$M = \sum_{j=1}^{N} \Gamma_j X_j(t). \tag{7.94}$$

Second, the rotational symmetry of the Hamiltonian,

$$H[\mathcal{Q}(\theta)X_1, \mathcal{Q}(\theta)X_2, \ldots, \mathcal{Q}(\theta)X_N] = H(X_1, \ldots, X_N)$$

for any rotation matrix $\mathcal{Q}(\theta)$ in the plane, yields the conserved quantity

$$A = \sum_{j=1}^{N} \Gamma_j |X_j(t)|^2. \tag{7.95}$$

We leave the verification that Eqs. (7.94) and (7.95) are conserved quantities for the dynamics of point vortices as an exercise for the interested reader. However, we remark here that M and A are discrete versions of continuous fluid impulse (1.74), $\int_{\mathbb{R}^2} (x_2, -x_1)^t \omega \, dx$, and moment of fluid impulse (1.75), $\int_{\mathbb{R}^2} |x|^2 \omega \, dx$, both conserved quantities for the 2D Euler equations (see Section 1.7 in Chap. 1).

Elementary Exact Solutions

It is well known (Lamb, 1932) that the motion of two exactly parallel vortex filaments is exactly solvable; here we simply record these exact solutions in coordinates convenient for the discussion in Section 7.4. For a pair of parallel vortex filaments, we can always rescale time so that one of the filaments has circulation 1 and the other filament has circulation Γ, where Γ satisfies $-1 \leq \Gamma \leq 1$ with $\Gamma \neq 0$; thus Γ represents the circulation ratio. We set $X_j(t) = (x_j, y_j)$ and introduce the complex coordinate for each filament,

$$\psi_j = x_j + iy_j \qquad \text{for} \quad j = 1, 2, \tag{7.96}$$

with $i = (-1)^{1/2}$.

With these simplifications, the equations for a pair of point vortices have the form

$$\begin{aligned} \frac{1}{i}\frac{\partial \psi_1}{\partial t} &= 2\Gamma \frac{\psi_1 - \psi_2}{|\psi_1 - \psi_2|^2}, \\ \frac{1}{i}\frac{\partial \psi_2}{\partial t} &= -2\frac{\psi_1 - \psi_2}{|\psi_1 - \psi_2|^2}. \end{aligned} \tag{7.97}$$

Because the nonlinear term in Eqs. (7.97) is a function of only one variable, we introduce the coordinates $\psi = \psi_1 - \psi_2$ and $\varphi = \psi_1 + \psi_2$ so that

$$\psi_1 = \frac{1}{2}(\varphi + \psi), \qquad \psi_2 = \frac{1}{2}(\varphi - \psi). \tag{7.98}$$

Then Eqs. (7.97) become

$$\begin{aligned} \frac{1}{i}\varphi_t &= 2(\Gamma - 1)\frac{\psi}{|\psi|^2}, \\ \frac{1}{i}\psi_t &= 2(\Gamma + 1)\frac{\psi}{|\psi|^2}. \end{aligned} \tag{7.99}$$

With the form of Eqs. (7.99) it is easy to write down the exact solutions describing the motion of two point vortices. We choose the origin of coordinates so that at time $t = 0$, $\psi_1 = \left(\frac{1}{2}d, 0\right)$ and $\psi_2 = \left(-\frac{1}{2}d, 0\right)$, where d is the separation distance. Then Eqs. (7.99) have the explicit exact solutions for $\Gamma \neq -1$,

$$\begin{aligned} \varphi^0(t) &= d\left\{1 - \exp\left[\frac{2i(1+\Gamma)t}{d^2}\right]\right\}\frac{1-\Gamma}{1+\Gamma}, \\ \psi^0(t) &= d\exp\left[\frac{2i(1+\Gamma)t}{d^2}\right] \end{aligned} \tag{7.100}$$

and for $\Gamma = -1$,

$$\begin{aligned} \varphi^0(t) &= -\frac{4i}{d^2}t, \\ \psi^0(t) &= (d, 0). \end{aligned} \tag{7.101}$$

The case of $\Gamma = -1$ is called the *antiparallel vortex pair* because the two vortices have equal and opposite circulations. For the antiparallel pair the motion of the two vortices is uniform translation at a velocity related to the initial separation distance.

In the general case, with $\Gamma \neq -1$, the two vortices simply rotate about their common center of mass at a fixed angular velocity. These *motions of exactly parallel vortex filaments are clearly stable within other exactly parallel vortex-filament perturbations with the same fixed circulation ratio,* although the antiparallel pair with $\Gamma = -1$ exhibits weak instability involving mild $O(t)$ growth in time, as is clear from Eqs. (7.101).

7.4. Asymptotic Equations for the Interaction of Nearly Parallel Vortex Filaments

Nearly parallel interacting vortex filaments with large strength and narrow cross section are called fluid-mechanical structures in mixing layers (Corcos and Lin, 1984) and trailing wakes (Van Dyke, 1982). A prominent and important example is the antiparallel pair shed by the wingtips of an aircraft: The hazards of flying smaller aircraft into the turbulent wake generated by larger aircraft through the interaction of an antiparallel vortex pair is a notable practical example of such an interaction.

How can we develop a simplified theory for the interaction of nearly parallel collections of vortex filaments? Here we take the opposite approach from that of the discussion in Sections 7.1 and 7.2 on asymptotic models for individual vortex filaments. We make an educated guess for an attractive plausible qualitative model for the interaction of nearly parallel vortex filaments and then indicate the fashion in which such simplified equations arise as a self-consistent asymptotic limit.

Consider N vortex filaments that are all nearly parallel to the x_3 axis with perturbed centerlines described by pairs of coordinates

$$X_j(\sigma, t) = [x_j(\sigma, t), y_j(\sigma, t)], \qquad 1 \leq j \leq N, \tag{7.102}$$

where σ parameterizes the x_3 axis. Equations (7.29) from Section 7.1 present the simplest theory of self-interaction for a nearly parallel vortex filament, involving the linearized self-induction equation for each parallel vortex perturbation:

$$\frac{\partial X_j}{\partial t}(\sigma, t) = J\Gamma_j \frac{\partial^2 X_j}{\partial \sigma^2}(\sigma, t), \qquad 1 \leq j \leq N. \tag{7.103}$$

On the other hand, the simplest theory for mutual interaction of exactly parallel vortex filaments is given by the equations for N interacting point vortices described in Section 7.3. Combining these two effects gives *simplified equations for the interaction of nearly parallel vortex filaments:*

$$\frac{\partial X_j}{\partial t} = J\left[\Gamma_j \frac{\partial^2 X_j}{\partial \sigma^2}\right] + J\left[\sum_{k \neq j}^{N} 2\Gamma_k \frac{(X_j - X_k)}{|X_j - X_k|^2}\right], \qquad 1 \leq j \leq N, \tag{7.104}$$

where J is the skew-symmetric matrix in Eq. (7.92). In Eqs. (7.104), the perturbed vortex filament curves have mutual interaction through the point-vortex interaction in a layered fashion for each σ and simultaneously have self-interaction in the

σ variable through the linearized self-induction of each filament as described in Eq. (7.103). Solutions of these equations have remarkable mathematical and physical properties, which we describe briefly below.

On the other hand, intuition suggests that these two physical effects in simplified equations (7.104) might dominate for nearby interacting almost parallel vortex filaments in which

(1) the wavelength of perturbations is much longer than the separation distance,
(2) the separation distance is much larger than the core thickness. (7.105)

Recently Klein et al. (1995) developed a systematic asymptotic expansion of the Navier–Stokes equations in a suitable distinguished limit in which conditions (7.105) are satisfied and the simplified asymptotic equations for the interaction of nearly parallel vortex filaments (7.104) emerge as the leading-order limit equations. Next, we briefly describe the precise asymptotic conditions needed in the derivation.

First, the centerline of each nearly parallel vortex filament has an assumed asymptotic form,

$$X_j^\varepsilon = s\mathbf{e}_3 + \varepsilon^2 X_j\left(\frac{s}{\varepsilon}, \frac{t}{\varepsilon^4}\right) + \mathcal{O}(\varepsilon^2), \qquad 1 \le j \le N, \tag{7.106}$$

and each vortex filament has a cross-sectional core radius satisfying the relations

$$\varepsilon^2 = \ln\left(\frac{1}{\delta}\right), \qquad \delta = R_e{}^{-1/2}, \quad \delta \ll 1, \tag{7.107}$$

which are already familiar to the reader from assumptions (7.2) and conditions (7.35). These conditions guarantee that (2) from conditions (7.105) is automatically satisfied. From Eq. (7.106) we see that

the separation distance between vortices is $\mathcal{O}(\varepsilon^2)$ whereas
the wavelength of perturbations is $\mathcal{O}(\varepsilon)$. (7.108)

Thus in the very precise sense from conditions (7.108) for $\varepsilon \ll 1$, the wavelengths of perturbations are much longer than the separation distance and (1) from conditions (7.105) is automatically satisfied.

We also note that, as in the discussion above conditions (7.35) in Section 7.2, the unit length scale for Eq. (7.106) is defined by the typical radius of curvature of each individual filament, that is, $\mathcal{O}(1)$. This means that the vortices must be close within $\mathcal{O}(\varepsilon^2)$ on this length scale; in practical terms the interacting filaments can be fairly distant provided that the vortex perturbations have large radii of curvature. We also remark that the time scale in Eq. (7.106) is a more rapid time scale than the one utilized in Eq. (7.33) of Section 7.2 for the self-stretch of an individual filament. In fact, the self-stretching mechanisms of an individual vortex filament studied extensively in Section 7.2 are higher-order asymptotic corrections with the ansatz in Eq. (7.106); the interested reader can find a detailed comparison and discussion of regimes for asymptotic vortex theories in Section 6 of the paper of Klein et al. (1995). With the ansatz in Eq. (7.106) under conditions (7.107), the asymptotic equations in (7.104) emerge from a detailed asymptotic expansion of Biot-Savart law (7.1) under these assumptions in a similar fashion as described earlier in Sections 7.1 and 7.2. The details are developed in Section 2 of the paper by Klein et al. (1995).

7.4.1. Hamiltonian Structure and Conserved Quantities

We write the simplified equations for interacting nearly parallel vortex filaments from simplified equations (7.104) in the form

$$\Gamma_j \frac{\partial X_j}{\partial t} = J\left[\Gamma_j^2 \frac{\partial^2}{\partial \sigma^2} X_j\right] + J\left[\sum_{k \neq j}^{N} 2\Gamma_j \Gamma_k \frac{(X_j - X_k)}{|X_j - X_k|^2}\right]. \tag{7.109}$$

To find the Hamiltonian for this system of equations, we need to find a functional $\mathcal{H}$ of the N vortex filaments such that

$$\Gamma_j \frac{\partial X_j}{\partial t} = J \frac{\delta \mathcal{H}}{\delta X_j}, \qquad 1 \leq j \leq N. \tag{7.110}$$

In Eq. (7.110), $\delta\mathcal{H}/\delta X_j$ denotes the functional or variational derivative with respect to the curve $X_j(\sigma)$ computed through the L^2 inner product for curves:

$$(\Phi, \Psi)_0 = \int \Phi(\sigma) \cdot \psi(\sigma) d\sigma. \tag{7.111}$$

In Eq. (7.111) we assume that the filament curves are either periodic in σ so that the integration range is over a period interval or that the filament curve perturbations vanish sufficiently rapidly together with their derivatives in the situation in which the range of integration is the entire line.

To find the Hamiltonian satisfying Eq. (7.110) we write $\mathcal{H}$ in the form

$$\mathcal{H} = \mathcal{H}_s + \mathcal{H}_p, \tag{7.112}$$

where

$$\frac{\delta \mathcal{H}_s}{\delta X_j} = \Gamma_j^2 \frac{\partial^2}{\partial \sigma^2} X_j, \tag{7.113}$$

$$\frac{\delta \mathcal{H}_p}{\delta X_j} - \sum_{k \neq j}^{N} 2\Gamma_j \Gamma_k \frac{(X_j - X_k)}{|X_j - X_k|^2} \tag{7.114}$$

for $1 \leq j \leq N$. The functional $\mathcal{H}_s$ satisfying Eq. (7.113) is

$$\mathcal{H}_s = -\sum_{j=1}^{N} \frac{\Gamma_j^2}{2} \int \left|\frac{\partial X_j}{\partial \sigma}\right|^2 d\sigma \tag{7.115}$$

and the functional $\mathcal{H}_p$ satisfying Eq. (7.114) is merely the integral over σ of the familiar N point-vortex Hamiltonian discussed earlier in Eq. (7.93), i.e.,

$$\mathcal{H}_p = 2\sum_{j<k}^{N} \int \Gamma_j \Gamma_k \ln|X_j(\sigma) - X_k(\sigma)| d\sigma. \tag{7.116}$$

With Eqs. (7.112)–(7.116), we conclude that the Hamiltonian satisfying Eq. (7.110) is given by

$$\mathcal{H} = -\sum_{j=1}^{N} \frac{\alpha_j \Gamma_j^2}{2} \int \left|\frac{\partial X_j}{\partial \sigma}\right|^2 d\sigma + 2\sum_{j<k}^{N} \int \Gamma_j \Gamma_k \ln|X_j(\sigma) - X_k(\sigma)| d\sigma. \tag{7.117}$$

Of course, it follows immediately from Eq. (7.110) that *the Hamiltonian* $\mathcal{H}$ *is conserved in time for solutions of the asymptotic filament equations in the form of Eq. (7.109).*

Other Conserved Quantities

Equations (7.109) have other symmetries that lead to integrated analogs of the conservation of the center of vorticity and angular momentum for 2D point vortices from conserved quantities (7.94) and (7.95).

The Hamiltonian remains unchanged under the transformations $X_j \mapsto X_j + \mathbf{Z}_0$, where $\mathbf{Z}_0$ is an arbitrary two-vector. Thus we define the *mean center of vorticity* by

$$\mathbf{M} = \int \sum_{j=1}^{N} \Gamma_j X_j(\sigma) d\sigma. \tag{7.118}$$

An elementary calculation establishes that $(d\mathbf{M}/dt) = 0$ for solutions of Eqs. (7.109) so that *the mean center of vorticity* $\mathbf{M}$ *is conserved in time for solutions.* Similarly, the Hamiltonian remains invariant under the transformations $X_j \mapsto \mathcal{O}(\theta) X_j$, where $\mathcal{O}(\theta)$ is an arbitrary rotation matrix. We define the *mean angular momentum* by

$$A = \int \sum_{j=1}^{N} \Gamma_j |X_j(\sigma)|^2 d\sigma, \tag{7.119}$$

and elementary calculations establish that *the mean angular momentum A defined in Eq. (7.119) is conserved in time.*

There is another conserved quantity that arises for solutions of the interacting filament equations and that is not derived as a direct analog in integrated form from the equations for 2D point vortices. The Hamiltonian in Eq. (7.117) remains invariant under the translation $X_j(\sigma) \mapsto X_j(\sigma + h)$, where h is arbitrary, so that Noether's theorem guarantees that there is another conserved quantity. By following the procedure for Noether's theorem in the form stated in Arnold (1989), we claim that the quantity

$$W = \int \sum_{j=1}^{N} \Gamma_j [J X_j(\sigma)] \cdot \frac{\partial X_j(\sigma)}{\partial \sigma} d\sigma \tag{7.120}$$

is conserved in time for solutions of filament equations (7.109). To verify this, we calculate that

$$\begin{aligned}\frac{dW}{dt} &= 2\int\sum_{j=1}^{N}\Gamma_j J\frac{dX_j}{dt}\cdot\frac{\partial X_j}{d\sigma}d\sigma\\ &= \left[-2\int\left(\sum_{j=1}^{N}\Gamma_j^2\frac{\partial^2 X_j}{\partial\sigma^2}\cdot\frac{\partial X_j}{\partial\sigma}\right)d\sigma\right]\\ &\quad+\left\{-4\int\left[\sum_{j<k}\Gamma_j\Gamma_k\frac{X_j-X_k}{|X_j-X_k|^2}\cdot\frac{\partial}{\partial\sigma}(X_j-X_k)\right]d\sigma\right\}\\ &= \{1\}+\{2\}. \end{aligned}\tag{7.121}$$

We claim that all of the integrands in both terms {1} and {2} are perfect derivatives so that all contributions on the right-hand side of Eq. (7.121) vanish. Because

$$2\frac{\partial^2 X_j}{\partial\sigma^2}\cdot\frac{\partial X_j}{\partial\sigma}=\frac{\partial}{\partial\sigma}\left(\left|\frac{\partial X_j}{\partial\sigma}\right|^2\right)$$

the integrand in {1} is a perfect derivative, and

$$\frac{X_j-X_k}{|X_j-X_k|^2}\cdot\frac{\partial}{\partial\sigma}(X_j-X_k)=\frac{\partial}{\partial\sigma}\ln|X_j-X_k|,$$

so that the integrand in {2} is also a perfect derivative. Thus the quantity W is conserved by solutions of the filament equations.

The Average Distance Functional

We briefly consider a generalized distance functional I defined by

$$I=\frac{1}{2}\int\sum_{j,k}\Gamma_j\Gamma_k|X_j-X_k|^2\,d\sigma. \tag{7.122}$$

We show below that I is conserved for special configurations consisting of identical vortex filaments. Through integration by parts, we compute in general that

$$\begin{aligned}\frac{dI}{dt} &= -\int\sum_{j,k}^{N}\left(\Gamma_j X_j\cdot\Gamma_k\frac{dX_k}{dt}+\Gamma_k X_k\cdot\Gamma_j\frac{dX_j}{dt}\right)d\sigma\\ &= -\sum_{j,k}^{N}\Gamma_j\Gamma_k(\Gamma_k-\Gamma_j)\int\left(\frac{\partial x_j}{\partial\sigma}\frac{\partial y_k}{\partial\sigma}-\frac{\partial x_k}{\partial\sigma}\frac{\partial y_j}{\partial\sigma}\right)d\sigma,\end{aligned}$$

where $X_j(\sigma,t)=[x_j(\sigma,t),y_j(\sigma,t)]$. From the above identity, we observe that *the average distance functional I is conserved in time for solutions provided that, for all the filaments,* $\Gamma_j=\Gamma_k$. In particular, for corotating filament pairs with the same structure in the vortex core, the distance functional I is conserved.

7.4.2. The Equations for Pairs of Interacting Filaments

Here we study solutions of the simplified equations for interacting filaments (7.104) in the important special case involving two interacting nearly parallel vortex filaments. As in Section 7.3, without loss of generality we assume that one of the filaments has circulation $\Gamma_1 = 1$ and that the other vortex filament has circulation $\Gamma_2 = \Gamma$, where Γ satisfies $-1 \leq \Gamma \leq 1$, with $\Gamma \neq 0$; thus Γ represents the circulation ratio of the two interacting filaments. For the pair of interacting filaments with $X_j(\sigma, t) = (x_j, y_j)$, it is convenient to introduce the complex coordinates for each filament,

$$\psi_j = x_j(\sigma, t) + i y_j(\sigma, t), \tag{7.123}$$

for $j = 1, 2$, with $i = \sqrt{-1}$.

Equations (7.104), when specialized to the case of two interacting filaments, have the form

$$\begin{aligned} \frac{1}{i}\frac{\partial \psi_1}{\partial t} &= \frac{\partial^2 \psi_1}{\partial \sigma^2} + 2\Gamma \frac{\psi_1 - \psi_2}{|\psi_1 - \psi_2|^2}, \\ \frac{1}{i}\frac{\partial \psi_2}{\partial t} &= \Gamma \frac{\partial^2 \psi_2}{\partial \sigma^2} - 2\frac{\psi_1 - \psi_2}{|\psi_1 - \psi_2|^2}. \end{aligned} \tag{7.124}$$

From Subsection 7.4.1 these equations have the following conserved quantities:

$$\begin{aligned} H &= -\frac{1}{2}\int \left|\frac{\partial \psi_1}{\partial \sigma}\right|^2 d\sigma - \frac{\Gamma^2}{2}\int \left|\frac{\partial \psi_2}{\partial \sigma}\right|^2 d\sigma \\ &\quad + 2\left(\int \Gamma \ln|\psi_1 - \psi_2| d\sigma\right), \\ M &= \int \psi_1\, d\sigma + \int \Gamma \psi_2\, d\sigma, \\ A &= \int |\psi_1|^2 d\sigma + \int \Gamma |\psi_2|^2 d\sigma, \\ W &= i\int \psi_1 \frac{\partial \overline{\psi_1}}{\partial \sigma} + \Gamma \psi_2 \frac{\partial \overline{\psi_2}}{\partial \sigma} d\sigma. \end{aligned} \tag{7.125}$$

Furthermore, for corotating pairs so that $\Gamma = 1$, we have the additional conserved quantity

$$I = \int |\psi_1 - \psi_2|^2 d\sigma \tag{7.126}$$

To make the nonlinear term in Eqs. (7.124) a function of only one variable, as in Section 7.3, we introduce the coordinates $\psi = \psi_1 - \psi_2$ and $\varphi = \psi_1 + \psi_2$ such that

$$\begin{aligned} \psi_1 &= \frac{(\varphi + \psi)}{2}, \\ \psi_2 &= \frac{(\varphi - \psi)}{2}. \end{aligned} \tag{7.127}$$

Equations (7.124) have the following form in the new variables, *the equivalent equation for a nearly parallel filament pair:*

$$\frac{1}{i}\varphi_t = \frac{(1+\Gamma)}{2}\varphi_{\sigma\sigma} + \frac{(1-\Gamma)}{2}\left[\psi_{\sigma\sigma} - 4\frac{\psi}{|\psi|^2}\right],$$
$$\frac{1}{i}\psi_t = \frac{(1-\Gamma)}{2}\varphi_{\sigma\sigma} + \frac{(1+\Gamma)}{2}\left[\psi_{\sigma\sigma} + 4\frac{\psi}{|\psi|^2}\right]. \tag{7.128}$$

7.4.3. Corotating Filament Pairs

For the special case of corotating filament pairs, we set $\Gamma = 1$ and Eqs. (7.128) completely decouple into two separate scalar equations given by

$$\frac{1}{i}\varphi_t = \varphi_{\sigma\sigma},$$
$$\frac{1}{i}\psi_t = \psi_{\sigma\sigma} + \frac{4\psi}{|\psi|^2}. \tag{7.129}$$

The first equation in Eqs. (7.129) is the linear Schrödinger equation, and the second equation is a nonlinear Schrödinger equation with an unusual nonlinearity.

These equations have a dispersive wavelike behavior. We demonstrate this by writing some elementary exact solutions. The complex function ψ is a nonlinear plane-wave solution of the form

$$\psi = Be^{i(k\sigma+\omega t)},$$

provided that

$$\omega = \frac{4}{B^2} - k^2.$$

With the exact solutions, we obtain general wavelike solutions in the original complex filament coordinates given by

$$\psi_1 = \varphi(\sigma, t) + \frac{B}{2}e^{i(k\sigma+\omega t)},$$
$$\psi_2 = \varphi(\sigma, t) - \frac{B}{2}e^{i(k\sigma+\omega t)}, \tag{7.130}$$

where φ is an arbitrary solution of the linear Schrödinger equation.

7.4.4. Linearized Stability for the Filament Pair

It should be evident to the reader that any collection of exactly parallel vortex filaments that satisfies N point-vortex equations (7.90) is automatically an exact solution of Eqs. (7.104) for N interacting nearly parallel filaments. Thus one can begin to understand the behavior of solutions of Eqs. (7.104) by studying the linearized stability of the point-vortex solutions within the class of general 3D filament curve perturbations.

In Eqs. (7.100) and (7.101) from Section 7.3, we wrote the well-known exact solutions for a pair of exactly parallel vortex filaments. As we remarked earlier in Section 7.3, such solutions are essentially always stable to exactly parallel filament perturbations. Here, as the most important special case of the strategy outlined in the preceding paragraph, we study the linearized stability of these exact solutions within general filament pair perturbations satisfying Eqs. (7.128). We find the remarkable fact that *these exactly parallel configurations are always unstable at suitably long wavelengths for any negative-circulation ratio,* $-1 \leq \Gamma \leq 0$, *and are always* (neutrally) *stable for any positive-circulation ratio,* $0 < \Gamma \leq 1$.

We linearize Eqs. (7.128) about the exact solutions φ^0, ψ^0 from Eqs. (7.100) and (7.101) and obtain the linearized equations for perturbations, still denoted by φ and ψ, given by

$$\begin{aligned}\frac{\partial \varphi}{\partial t} &= i\frac{(1+\Gamma)}{2}\varphi_{\sigma\sigma} + i\frac{(1-\Gamma)}{2}\left[\psi_{\sigma\sigma} + 4\frac{\overline{\psi}}{\overline{\psi}_0^2}\right],\\ \frac{\partial \psi}{\partial t} &= i\frac{(1-\Gamma)}{2}\varphi_{\sigma\sigma} + i\frac{(1+\Gamma)}{2}\left[\psi_{\sigma\sigma} - 4\frac{\overline{\psi}}{\overline{\psi}_0^2}\right],\end{aligned} \tag{7.131}$$

where

$$\psi_0 = d e^{\frac{2i(1+\Gamma)t}{d^2}}$$

and an overbar indicates complex conjugation. To remove the time dependence that enters through $\overline{\psi}_0^2$, we go to a coordinate frame rotating with the vortex filaments, through the variables χ and Θ defined by

$$\begin{aligned}\chi(\sigma, t) &= \psi(\sigma, t)e^{\frac{-2i(1+\Gamma)t}{d^2}},\\ \Theta(\sigma, t) &= \varphi(\sigma, t)e^{\frac{-2i(1+\Gamma)t}{d^2}}.\end{aligned} \tag{7.132}$$

With this transformation, the linearized equations of motion and the equations for the corresponding complex conjugates are

$$\begin{aligned}\frac{1}{i}\chi_t &= \frac{(1+\Gamma)}{2}\left[\chi_{\sigma\sigma} - \frac{4}{d^2}\chi\right] + \frac{(1-\Gamma)}{2}\Theta_{\sigma\sigma} - \frac{2(1+\Gamma)}{d^2}\overline{\chi},\\ \frac{1}{i}\Theta_t &= \frac{(1-\Gamma)}{2}\chi_{\sigma\sigma} + \frac{(1+\Gamma)}{2}\left[\Theta_{\sigma\sigma} - \frac{4}{d^2}\Theta\right] + \frac{2(1-\Gamma)}{d^2}\overline{\chi},\\ \frac{1}{i}\overline{\chi}_t &= -\frac{(1+\Gamma)}{2}\left[\overline{\chi}_{\sigma\sigma} - \frac{4}{d^2}\overline{\chi}\right] - \frac{(1-\Gamma)}{2}\overline{\Theta}_{\sigma\sigma} + \frac{2(1+\Gamma)}{d^2}\chi,\\ \frac{1}{i}\overline{\Theta}_t &= -\frac{(1-\Gamma)}{2}\overline{\chi}_{\sigma\sigma} - \frac{(1+\Gamma)}{2}\left[\overline{\Theta}_{\sigma\sigma} - \frac{4}{d^2}\overline{\Theta}\right] - \frac{2(1-\Gamma)}{d^2}\chi.\end{aligned} \tag{7.133}$$

We solve this 4×4 system with the Fourier transform. The coupling of χ and Θ to their complex conjugates $\overline{\chi}$ and $\overline{\Theta}$ in real space means that $\widehat{\chi}(\xi, t)$ is coupled to $\widehat{\overline{\chi}}(-\xi, t)$ in wave-number space, where ξ is the wave number, and the same is true for Θ. Similar behavior also occurred earlier in Subsection 7.2.3. In wave-number space

Eqs. (7.133) reduce to the 4 × 4 matrix ODE

$$\frac{d}{dt}\begin{bmatrix}\widehat{\chi}(\xi,t)\\ \widehat{\Theta}(\xi,t)\\ \overline{\widehat{\chi}}(-\xi,t)\\ \overline{\widehat{\Theta}}(-\xi,t)\end{bmatrix} = iA(\xi)\begin{bmatrix}\widehat{\chi}(\xi,t)\\ \widehat{\Theta}(\xi,t)\\ \overline{\widehat{\chi}}(-\xi,t)\\ \overline{\widehat{\Theta}}(-\xi,t)\end{bmatrix}, \tag{7.134}$$

where

$$A(\xi) = \begin{bmatrix} -\frac{(1+\Gamma)}{2}\left(\xi^2+\frac{4}{d^2}\right) & -\frac{(1-\Gamma)}{2}\xi^2 & -\frac{2(1+\Gamma)}{d^2} & 0 \\ -\frac{(1-\Gamma)}{2}\xi^2 & -\frac{(1+\Gamma)}{2}\left(\xi^2+\frac{4}{d^2}\right) & \frac{2(1-\Gamma)}{d^2} & 0 \\ \frac{2(1+\Gamma)}{d^2} & 0 & \frac{(1+\Gamma)}{2}\left(\xi^2+\frac{4}{d^2}\right) & \frac{(1-\Gamma)}{2}\xi^2 \\ -\frac{2(1-\Gamma)}{d^2} & 0 & \frac{(1-\Gamma)}{2}\xi^2 & \frac{(1+\Gamma)}{2}\left(\xi^2+\frac{4}{d^2}\right) \end{bmatrix}. \tag{7.135}$$

The matrix $A(\xi)$ has the eigenvalues $\lambda^1_\pm = \pm\frac{1}{2}\sqrt{\mathcal{R}+2\sqrt{P}}$ and $\lambda^2_\pm = \pm\frac{1}{2}\sqrt{\mathcal{R}-2\sqrt{P}}$, where

$$\mathcal{R} = 8\frac{a^2}{d^4} + 8\frac{a^2\xi^4}{d^2} + a^2\xi^4 + b^2\xi^4,$$

$$\mathcal{P} = 16\frac{a^4}{d^8} + 32\frac{a^2b^2\xi^2}{d^6} + 20\frac{a^2b^2\xi^4}{d^4} + 4\frac{b^4\xi^4}{d^4} + 8\frac{a^2b^2\xi^6}{d^2} + a^2b^2\xi^8, \tag{7.136}$$

$a = 1+\Gamma$, and $b = 1-\Gamma$. We have growing modes and linearized instability whenever the $\lambda_\pm$ are imaginary. From the form of the $\lambda_\pm$ it is easy to see that because both $\mathcal{R}$ and $\mathcal{P}$ are positive real numbers from Eqs. (7.136), $\lambda^1_\pm$ is always a real number. Thus only $\lambda^2_\pm$ that has a minus sign under the radicand can assume imaginary values and yield instability. It is useful to define the growth factor $\mathcal{G}$, which depends on the signs of the radicands in $\lambda^2_\pm$:

$$\mathcal{G} = \text{sgn}[-(\mathcal{R}-2\sqrt{P})]\left|\sqrt{\mathcal{R}-2\sqrt{\mathcal{P}}}\right|. \tag{7.137}$$

Whenever the $\lambda^2_\pm$ are real, $\mathcal{G}$ is negative and the perturbations are *neutrally stable* with the oscillation frequency given by the absolute value of $\mathcal{G}$. If, however, any one of the $\lambda^2_\pm$ is imaginary, $\mathcal{G}$ is positive, and we have either exponentially growing and damped solutions, with the growth/damping rate given by the absolute value of $\mathcal{G}$. In general, the growing solution dominates and we have an instability. From the definition of $\mathcal{G}$, we have $\mathcal{G} > 0$ and instability if and only if $\mathcal{R}^2 < 4\mathcal{P}$. Therefore we consider the quantity $\mathcal{R}^2 - 4\mathcal{P}$ given explicitly from Eqs. (7.136) by

$$\mathcal{U} = \mathcal{R}^2 - 4\mathcal{P} = [a^2-b^2]\left[128\frac{a^2\xi^2}{d^6} + 80\frac{a^2\xi^4}{d^4} + 16\frac{b^2\xi^4}{d^2} + 16\frac{a^2\xi^6}{d^2} + (a^2-b^2)\xi^8\right], \tag{7.138}$$

where, as before, $a = 1+\Gamma$ and $b = 1-\Gamma$.

First we consider the situation with a positive-circulation ratio so that $1 \geq \Gamma > 0$; in this case we have the inequality $a > b$, guaranteeing by Eq. (7.138) that $\mathcal{R}^2 - 4\mathcal{P}$ is positive for all wave numbers ξ and that there are no growing modes. Thus we have linearized (neutral) stability for the vortex pair in this situation.

Next we consider the situation with negative-circulation ratios Γ with $-1 \leq \Gamma < 0$, implying that $a < b$. From Eq. (7.138), in this case the only positive term in $\mathcal{U} = \mathcal{R}^2 - 4\mathcal{P}$ is $(b^2 - a^2)b^2\xi^8$. At long wavelengths, $\xi^2 \ll 1$, this positive term is dominated in magnitude by the negative contributions to $\mathcal{U}$ of the order of ξ^2, ξ^4, and ξ^6; thus there is always long-wavelength instability for the straight-line vortex pair for any negative-circulation ratio. On the other hand, at short wavelengths, $\xi^2 \gg 1$, $\mathcal{R}^2 - 4\mathcal{P}$ is dominated by the positive factor $[a^2 - b^2]^2\xi^8$ and there is always short-wavelength stability.

We summarize our analysis presented in the previous two paragraphs in the following fashion:

For simplified vortex-filament equations (7.124), straight-line point-vortex pairs have linearized long-wavelength instability for arbitrary negative-circulation ratios and linearized (neutral) stability for arbitrary positive-circulation ratios. (7.139)

For any fixed negative-circulation ratio, the graph of the stability function $\mathcal{G}$ from Eq. (7.137) is positive over a finite interval of wave numbers extending from zero where $\mathcal{G}$ vanishes. For each negative-circulation ratio there is a unique wave number with the largest growth rate.

7.4.5. Finite-Time Collapse and Wavelike Behavior for Nearly Parallel Pairs of Vortex Filaments

The linear stability results of summary (7.139) suggest very different behavior of interacting pairs of filaments for negative-circulation ratio with strong instability at long wavelengths compared with the case for a positive-circulation ratio in which wave like neutral stability without growth occurs. The special structure of the filament pair equations for the corotating case with $\Gamma = 1$ and presented in Subsection 7.4.3 also suggests that wavelike behavior dominates for positive-circulation ratios.

Next we present numerical solutions of filament pair equations (7.124). For negative-circulation ratios, we see that the linearized instability and nonlinearity conspire to produce a finite-time local collapse of the two vortex filaments in a distinctive fashion for each different circulation ratio Γ, with $-1 \leq \Gamma < 0$. On the other hand, the numerical solutions always indicate wavelike behavior without any finite-time collapse for any positive-circulation ratio.

7.4.6. Finite-Time Collapse for Negative-Circulation Ratios

First we consider the antiparallel pair with the circulation ratio, $\Gamma = -1$. This configuration corresponds to two trailing-wake vortices shed by the wingtips of an aircraft.

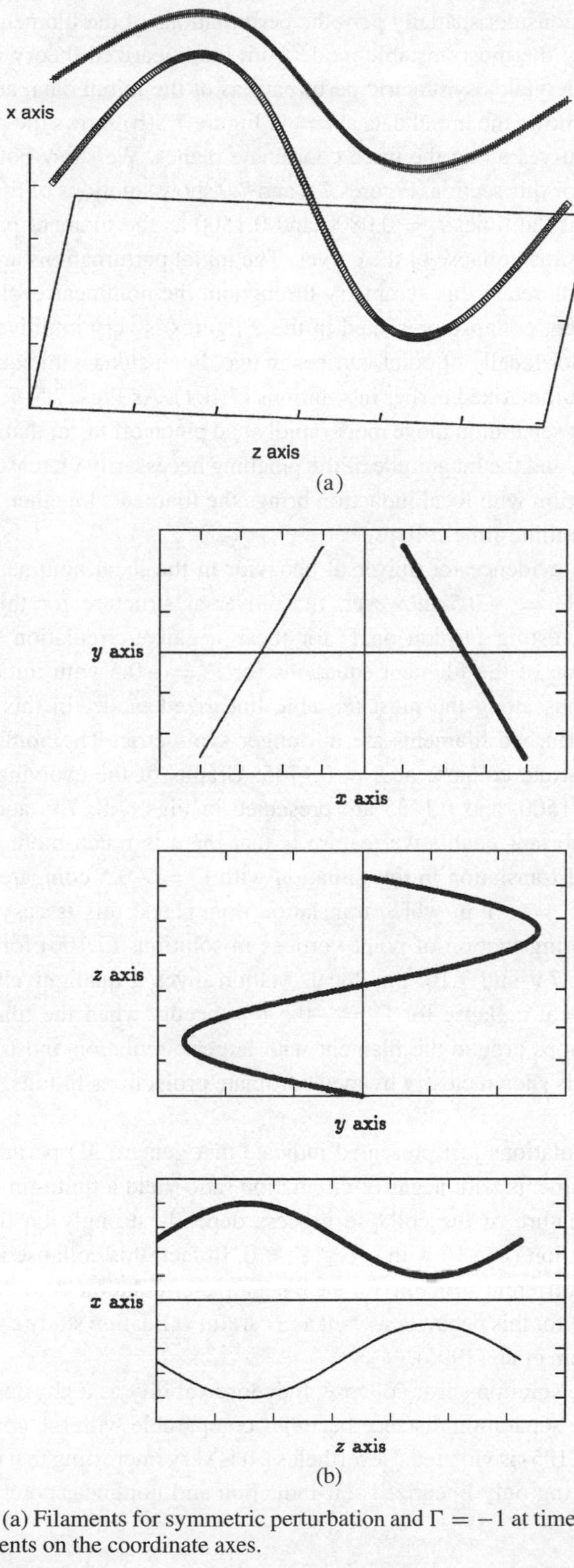

Figure 7.5. (a) Filaments for symmetric perturbation and $\Gamma = -1$ at time $t = 0$, (b) projection of the filaments on the coordinate axes.

For initial data we consider spatially periodic perturbations of the filament pair with a structure given by the most unstable mode from the linearized theory in Subsection 7.4.4; this mode yields symmetric perturbations of the initial data, as shown in Fig. 7.5(a), which shows the initial data at $t = 0$. Figure 7.5(b) shows the projections of these filament curves along the three coordinate planes. We show both types of data in the figures for this section. Figures 7.6 and 7.9 show solutions of filament pair equations (7.124) at the times $t = 0.0800$ and 0.1500 as the filament pair evolves toward local finite-time collapse of the curves. The initial perturbations are symmetric, and the solutions retain this symmetry throughout the nonlinear evolution. The nature of the filament collapse presented in these figures is very intuitive when we consider the behavior locally of point vortices in two dimensions with equal and opposite circulation summarized earlier in solutions (7.101). As Figs. 7.5–7.7 indicate, the points of closest separation move more rapidly and pinch off faster than the rest of the vortex filament, and the magnitude of the pinching necessarily increases substantially as the interaction with local induction brings the filaments together. The result of the interaction is finite-time collapse.

Next we present evidence for universal behavior in the local nonlinear collapse of solutions with $\Gamma = -0.5$; however, the universal structure for this collapse depends in an interesting fashion on Γ for these negative-circulation ratios. We consider the solution of the filament equations for $\Gamma = -0.5$ with initial data involving perturbations along the most unstable linearized mode. In this case with $\Gamma = -0.5$, the perturbed filaments are no longer symmetric. The nonlinear solution exhibits finite-time collapse at $t = 0.1735$. Graphs of the evolving solutions at times $t = 0$, 0.1500, and 0.1735 are presented in Figs. 7.8, 7.9, and 7.10, respectively. An important qualitative feature is that there is much more local rotation without spatial translation in this situation with $\Gamma = -0.5$ compared with the earlier case with $\Gamma = -1$ in which translation dominates; this is easy to understand with the rotating motion of point vortices in solutions (7.100) for $\Gamma \neq -1$. As shown in Figs. 7.9 and 7.10, this local rotation gives a qualitatively different structure for the local collapse for $\Gamma = -0.5$ that occurs when the filament with smaller vorticity loops around the filament with larger circulation and is sucked in by it. This process is shown clearly by the coordinate projections in Figs. 7.9(b) and 7.10(b).

Both of the calculations just presented indicate that general 3D perturbations of a pair of vortex filaments with negative-circulation ratio yield a finite-time collapse. Furthermore, the nature of the collapse process depends strongly on the specific value of the circulation ratio Γ, with $-1 \leq \Gamma < 0$. In fact, this collapse is probably self-similar with a different structure for each negative-circulation ratio. Much more numerical evidence for this behavior as well as a careful validation study can be found in the paper by Klein et al. (1995).

Of course, as the solutions near collapse, they lose validity as a physical approximation because the separation distance becomes comparable with the core size and (2) in conditions (7.105) is violated. Nevertheless, it is very interesting that the asymptotic solutions, having only linearized self-induction and nonlinear potential vortex

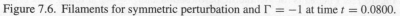

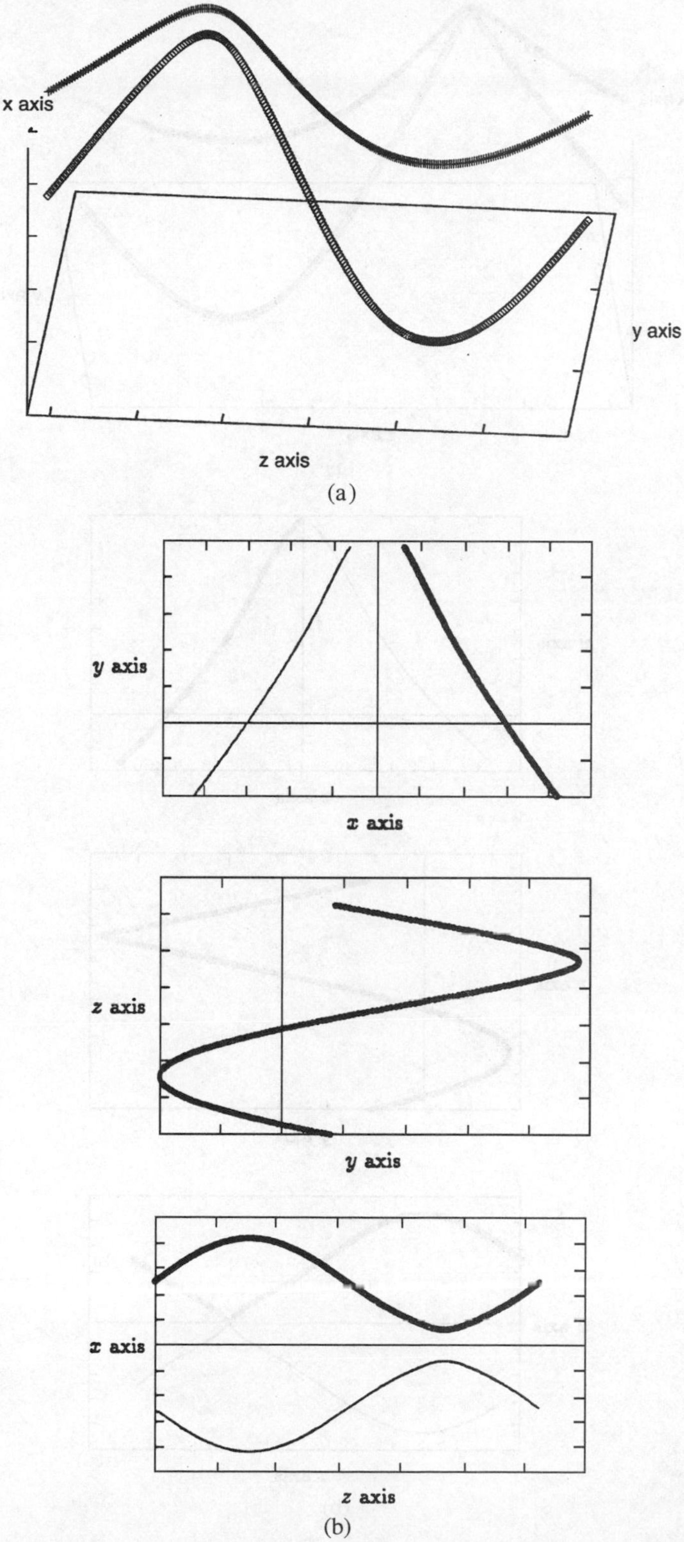

Figure 7.6. Filaments for symmetric perturbation and $\Gamma = -1$ at time $t = 0.0800$.

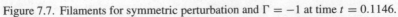

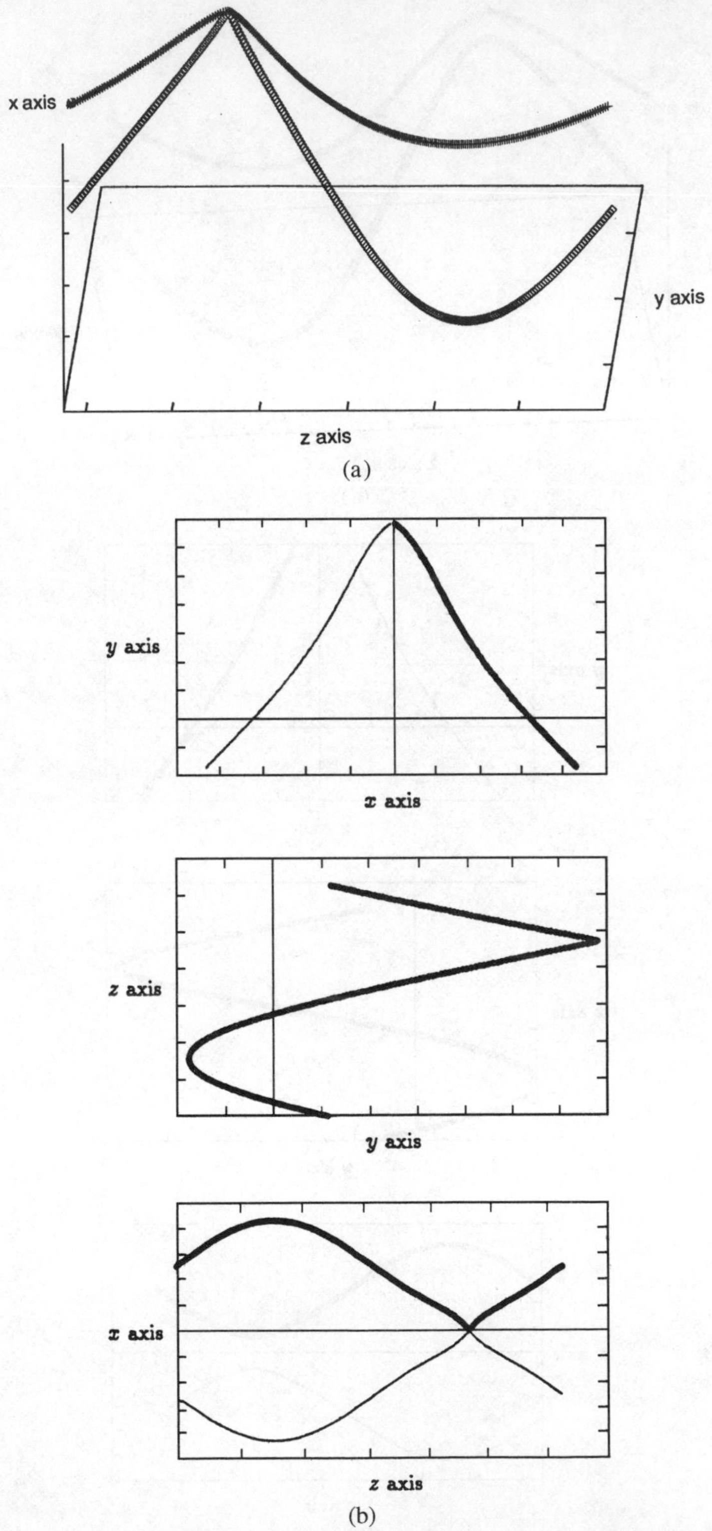

Figure 7.7. Filaments for symmetric perturbation and $\Gamma = -1$ at time $t = 0.1146$.

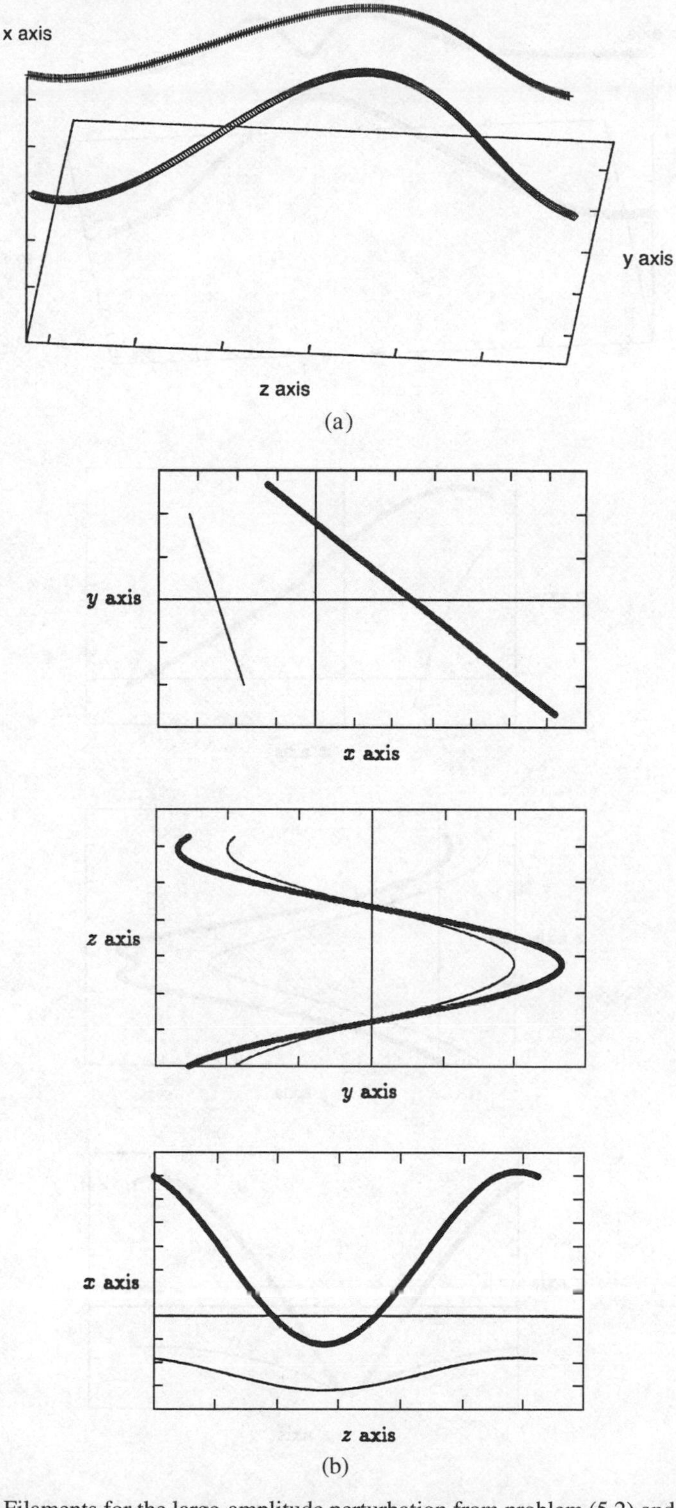

Figure 7.8. Filaments for the large-amplitude perturbation from problem (5.2) and $\Gamma = -0.5$ at time $t = 0$.

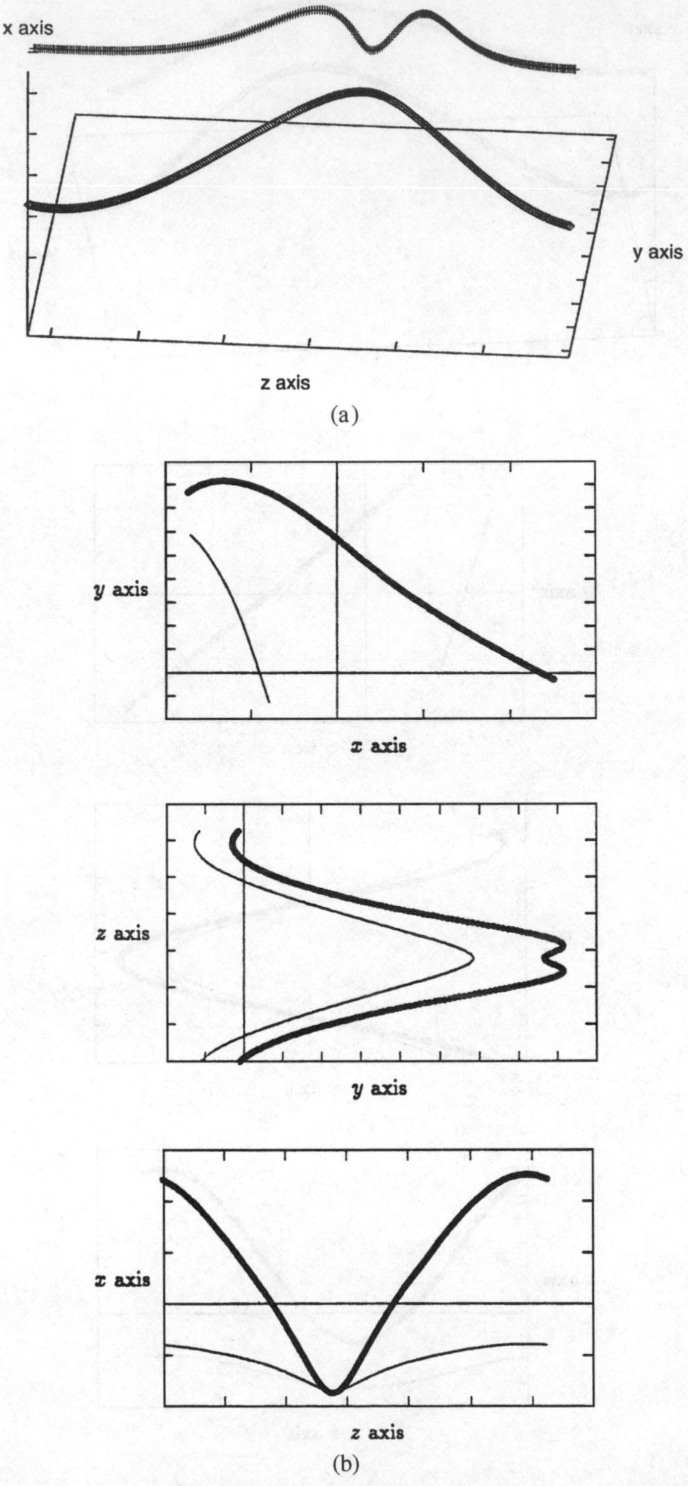

Figure 7.9. Filaments for the large-amplitude perturbation from problem (5.2) and $\Gamma = -0.5$ at time $t = 0.1500$.

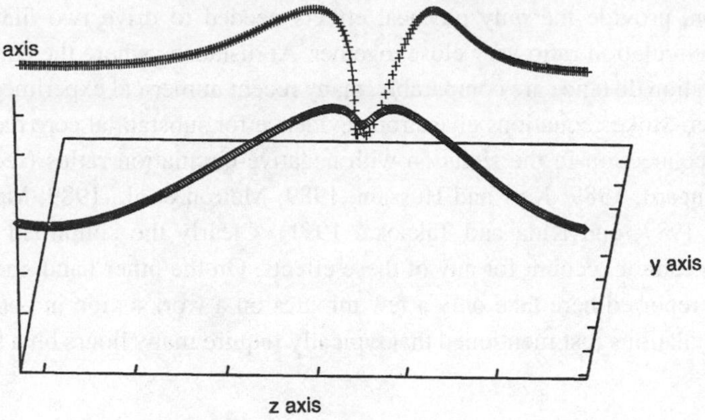

(a)

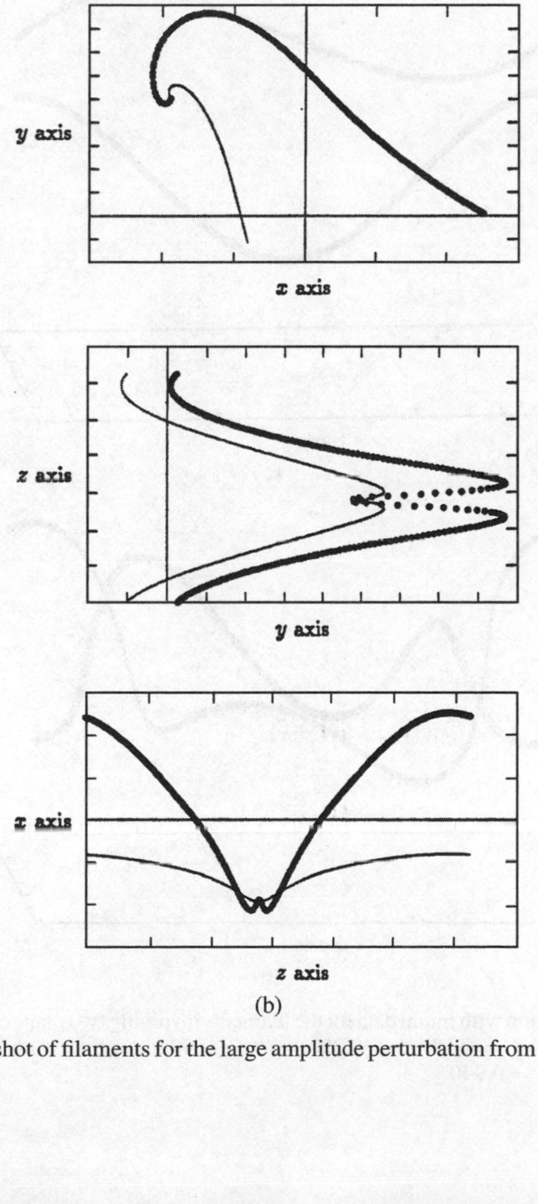

(b)

Figure 7.10. Snapshot of filaments for the large amplitude perturbation from (5.2) and $\Gamma = -.5$ at time $t = .1735$.

interaction, provide the only physical effects needed to drive two filaments with negative-circulation ratio very close together. At distances where the core thickness and separation distance are comparable, many recent numerical experiments with the full Navier–Stokes equations give strong evidence for substantial core flattening and vortex reconnection in the situation with negative-circulation ratios (see Anderson and Greengard, 1989; Kerr and Hussain, 1989; Meiron et al., 1989; Melander and Zabusky, 1987; and Kida and Takaoka, 1991). Clearly the simplified asymptotic equations cannot account for any of these effects. On the other hand, the numerical solutions reported here take only a few minutes on a workstation in contrast to the direct simulations just mentioned that typically require many hours on a large supercomputer.

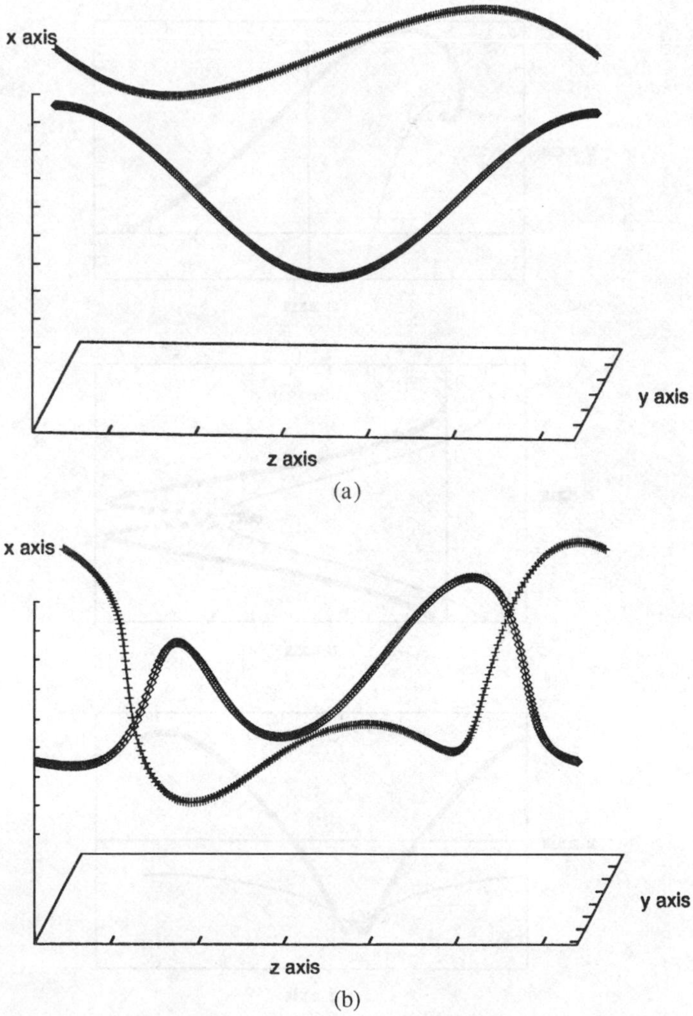

Figure 7.11. Solution with initial data for the filaments involving two plane curves in orthogonal planes at times (a) $t = 0$; (b) $t = 0.300$; (c) $t = 0.908$, the time with smallest separation distance; and (d) $t = 0.940$.

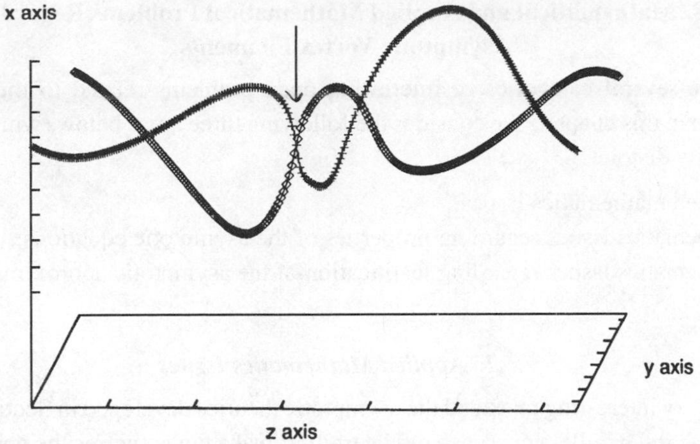

(c)

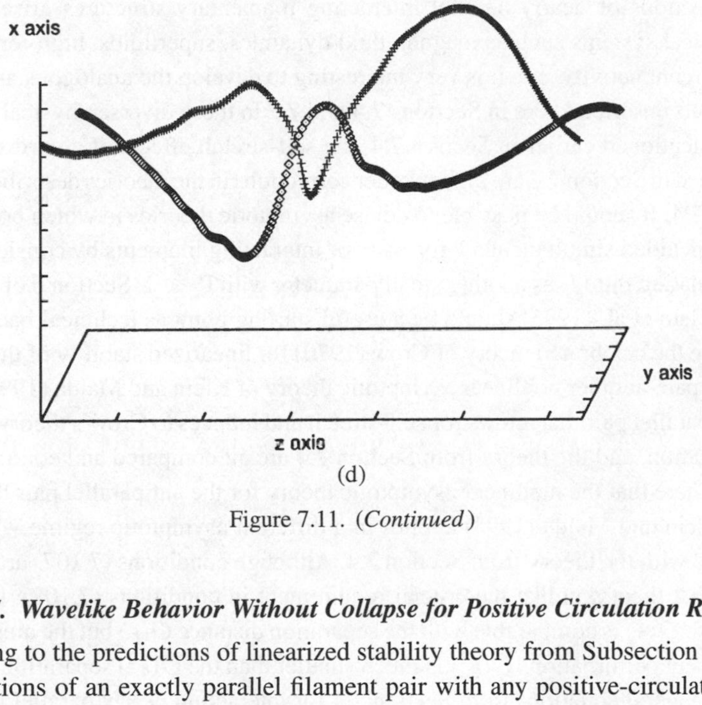

(d)

Figure 7.11. (*Continued*)

7.4.7. Wavelike Behavior Without Collapse for Positive Circulation Ratios

According to the predictions of linearized stability theory from Subsection 7.4.4, all perturbations of an exactly parallel filament pair with any positive-circulation ratio are neutrally stable and wavelike at all wave numbers. Numerical experiments for a fixed positive-circulation ratio Γ, with $0 < \Gamma \leq 1$, show that the fully nonlinear solutions of filament pair equations (7.124) continue to remain wavelike throughout their evolution without any finite-time collapse.

We illustrate this typical wavelike behavior through the numerical solutions shown in Figs. 7.11(a), 7.11(b), and 7.11(c) at t times $t = 0$, 0.300, and 0.908, respectively. The wavelike evolution of the solution is evident from these snapshots; we have singled out the time $t = 0.908$ because this is the time of closest approach of the filament pair throughout the entire time history. Clearly there is no evidence of collapse as occurred in the situation with negative-circulation ratios described earlier.

7.5. Mathematical and Applied Mathematical Problems Regarding Asymptotic Vortex Filaments

There are several categories of interesting open problems related to the material described in this chapter. We consider the following three areas below (which are not necessarily disjoint!).

(1) applied mathematics issues,
(2) mathematics issues regarding properties of the asymptotic equations,
(3) mathematics issues regarding justification of the asymptotic approximation.

7.5.1. Applied Mathematics Issues

(1) It is very interesting to extend the asymptotic theories developed in Section 7.4 for interacting parallel vortices to model trailing-wake flows such as the von Kármán vortex street.
(2) Collections of nearly parallel interacting filamentary structures arise in many physical systems such as magnetofluid dynamics, superfluids, high-temperature superconductivity, etc. It is very interesting to develop the analogous asymptotic models described here in Sections 7.4 and 7.2 in these diverse physical contexts.
(3) As mentioned earlier in Section 7.4, the self-stretch effects of a vortex filament studied in Section 7.2 are higher-order corrections in the theories described in Section 7.4. It should be possible to devise asymptotic theories in which both effects are included simultaneously for pairs of interacting filaments by considering the circulation ratio Γ as another small parameter with $\Gamma \ll 1$. Section 6 of the paper by Klein et al. (1995) should be a useful starting point as technical background. There the celebrated theory of Crow (1970) for linearized stability of the antiparallel pair, another nonlinear asymptotic theory of Klein and Majda (1993) for the antiparallel pair that allows for self-stretch and reduces to Crow's theory after linearization, and the theory from Section 7.4 are all compared and contrasted. We note here that the nonlinear asymptotic theory for the antiparallel pair that is due to Klein and Majda (1993) applies in a different asymptotic regime when compared with the theory from Section 7.4. Although conditions (7.107) are satisfied for that theory, unlike the crucial requirement in conditions (7.108), the wavelength $O(\varepsilon)$ is comparable with the separation distance $\mathcal{O}(\varepsilon)$ but the amplitude of filament perturbations $\mathcal{O}(\varepsilon^2)$ is much smaller than the $\mathcal{O}(\varepsilon^2)$ separation distance.
(4) The general equations from Section 7.4 for interacting nearly parallel vortex filaments, with their concise Hamiltonian structure and conserved quantities, are appealing candidates for simplified statistical theories of nearly parallel vortex filaments, an interpolant between 2D and 3D statistical theories for vortices (see Chorin, 1988, 1994). Lions and Majda (2000) have recently made some progress on this topic but much more work remains to be done.

7.5.2. Mathematical Issues Regarding Properties of the Asymptotic Equations

(1) As regards the vortex-filament pair discussed in Subsection 7.4.2, it would be very interesting to develop mathematical theorems proving finite-time collapse

for negative-circulation ratios with appropriate initial data and disproving collapse for positive-circulation ratios with arbitrary initial data. The conserved quantities from Subsection 7.4.1 should be useful in this regard. The novel nonlinear Schrödinger equation described in Subsection 7.4.2 for $\Gamma = 1$ is an obvious starting point for showing that collapse does not occur.

(2) Are filament pair equations (7.124) a completely integrable Hamiltonian system for at least some special values of the circulation ratio such as $\Gamma = 1$ or $\Gamma = -1$?

(3) It is well known (see Aref, 1983) that the motion of three point-vortices in the plane is integrable but can exhibit finite-time collapse whereas the motion of four-point vortices can be chaotic. How do genuine 3D filament perturbations of these exactly parallel solutions modify this behavior?

7.5.3. Mathematical Issues Regarding Justification of the Asymptotic Approximation

The simplest situations to begin justification of the asymptotic approximations occur in Sections 7.3 and 7.1.

(1) In Chap. 2 of their book, Ting and Klein (1991) present a detailed formal asymptotic derivation of the point-vortex equations in Section 7.3 from solutions of the Navier–Stokes equations. Can this formal work be combined with estimates for the 2D Navier–Stokes equations to rigorously justify this approximation?

(2) The same issues and references for the self-induction equations in Section 7.1.

Notes for Chapter 7

Crow (1970) pioneered the cutoff approximation to compute the finite part of a Biot–Savart integral, and this device has been developed further by Moore and Saffman, Widnall, and others. Callegari and Ting (1978) present a succinct critique of this cutoff approximation. One can apply our asymptotic techniques in Section 7.2 involving the Hasimoto transform as well as our techniques in Section 7.4 in conjunction with the cutoff approximation to obtain the same asymptotic equations developed here. We have chosen to emphasize the methods of Callegari and Ting (1978) here because they are less ad hoc and apply to Navier–Stokes solutions.

On heuristic ground, Zakharov (1988) wrote an equation describing the special case of symmetric perturbations of the antiparallel pair. Under these very special circumstances, the theory from Section 7.4 recovers a slightly simplified version of Zakharov's equation in a quantitative fashion (see Klein et al., 1992).

References for Chapter 7

Ablowitz, M. and Segur, H., "Solitons and the inverse scattering transform," *SIAM Stud. Appl. Math.* **4**, 1981.

Anderson, H. and Greengard, C., "The vortex ring merger problem at infinite Reynolds numbers," *Commun. Pure Appl. Math.* **42**, 1123–1139, 1989.

Aref, H., "Integrable, chaotic and turbulent vortex motion in two-dimensional flows," *Ann. Rev. Fluid Mech.* **15**, 345–389, 1983.

Arms, R. J. and Hama, F. R., "Localized-induction concept on a curved vortex and motion of an elliptic vortex ring," *Phys. Fluids* **8**, 553–559, 1965.

Arnold, V. I., *Mathematical Methods of Classical Mechanics*, 2nd ed., Springer, New York, 1989.

Ashurst, W., Kerstein, A., Kerr, R., and Gibson, C., "Alignment of vorticity and scalar gradient with strain rate in simulated Navier–Stokes turbulence," *Phys. Fluids* **30**, 2343–2353, 1987.

Callegari, A. J. and Ting, L., "Motion of a curved vortex filament with decaying vortical core and axial velocity," *SIAM J. Appl. Math.* **35**, 148–175, 1978.

Chorin, A. J., "Evolution of a turbulent vortex," *Commun. Math. Phys.* **83**, 517–535, 1982.

Chorin, A. J., "Spectrum, dimension and polymer analogies in fluid turbulence," *Phys. Rev. Lett.* **60**, 1947–1949, 1988.

Chorin, A. J. *Vorticity and Turbulence*, Springer, New York, 1994.

Chorin, A. J. and Marsden, J. E., *A Mathematical Introduction to Fluid Mechanics*, Springer, New York, 1993.

Corcos, G. and Lin, S., "The mixing layer: deterministics models of a turbulent flow. Part 2: the origin of three-dimensional motion," *J. Fluid Mech.* **139**, 67–95, 1984.

Crow, S., "Stability theory for a pair of trailing vortices," *AIAA J.* **8**, 2172–2179., 1970.

do Carmo, M. P., *Differential Geometry of Curves and Surfaces*, Prentice-Hall, Englewood Cliffs, NJ, 1976.

She, Z. S., Jackson, E., and Orszag, S., "Intermittent vortex structures in homogeneous isotropic turbulence," *Nature (London)* **344**, 226–228, 1990.

Kerr, R. M. and Hussain, A. K. M. F., "Simulation of vortex reconnection," *Physica D* **37**, 474, 1989.

Kida, S. and Takaoka, M., "Breakdown of frozen motion fields and vorticity reconnection," *J. Phys. Soc. Jpn.* **60**, 2184–2196, 1991.

Klein, R. and Knio, O. M., "Asymptotic vorticity structure and numerical simulation of slender vortex filaments," *J. Fluid Mech.* **284**, 275–321, 1995.

Klein, R. and Majda, A., "Self-stretching of a perturbed vortex filament I. The asymptotic equations for deviations from a straight line," *Physica D* **49**, 323–352, 1991a.

Klein, R. and Majda, A., "Self-stretching of perturbed vortex filaments II. Structure of solutions," *Physica D* **53**, 267–294, 1991b.

Klein, R. and Majda, A., "An asymptotic theory for the nonlinear instability of anti-parallel pairs of vortex filaments," *Phys. Fluids A* **5**, 369–387, 1993.

Klein, R., Majda, A., and Damodaran, K., "Simplified equations for the interaction of nearly parallel vortex filaments," *J. Fluid Mech.*, **288**, 201–248, 1995.

Klein, R., Majda, A., and McLaughlin, R. M., "Asymptotic equations for the stretching of vortex filaments in a background flow field," *Phys. Fluids A* **4**, 2271–2281, 1992.

Knio, O. M. and Klein, R., "Improved thin tube methods for slender vortex simulations," *J. Comp. Phys.* **162**, 1–15, 2000.

Lamb, G. L., *Elements of Soliton Theory*, Wiley-Interscience, New York, 1980.

Lamb, H., *Hydrodynamics*, 6th ed., Cambridge Univ. Press, Cambridge, U.K., 1932.

Lions, P. L. and Majda, A., "Equilibrium Statistical Theory for Nearly Parallel Vortex Filaments" *Comm. Pure Appl. Math.* **53**(1), 76–142, 2000.

Majda, A., "Vorticity, turbulence and acoustics in fluid flow," *SIAM Rev.* **33**, 349–388, 1991.

Meiron, D., Shelley, M., Ashurst, W., and Orszag, S., "Numerical studies of vortex reconnection," in R. Caflisch, ed., *Mathematical Aspects of Vortex Dynamics*, Society for Industrial and Applied Mathematics, Philadelphia. PA, 1989, pp. 183–194.

Melander, M. V. and Zabusky, N., "Interaction and "apparent" reconnection of vortex tubes via direct numerical simulations," *Fluid Dynamics Research 3*: 247–250, 1988.

Ting, L. and Klein, R., *Viscous Vortical Flows*, Springer-Verlag, Berlin, 1991.

Van Dyke, M., *An Album of Fluid Motion*, Parabolic, Stanford, CA, 1982.

Zakharov, V. E., "Wave collapse," *Usp. Fiz. Nauk* **155**, 529–533, 1988.

8
Weak Solutions to the 2D Euler Equations with Initial Vorticity in L^∞

So far we have discussed classical smooth solutions to the Euler and the Navier–Stokes equations. In the first two chapters we discussed elementary properties of the equations and exact solutions, including some intuition for the difference between 2D and 3D and the role of vorticity. In Chaps. 3 and 4 we established the global existence of smooth solutions from smooth initial data in two dimensions (e.g., Corollary 3.3) and global existence in three dimensions, provided that the maximum of the vorticity is controlled (see, e.g., Theorem 3.6 for details). However, many physical problems possess localized, highly unstable structures whose complete dynamics cannot be described by a simple smooth model.

The remaining chapters of this book deal with mathematical issues related to nonsmooth solutions of the Euler equations. This chapter addresses a type of weak solution appropriate for modeling an isolated region of intense vorticity, such as what one might use to model the evolution of a hurricane. In particular, we consider problems that have vorticity that is effectively discontinuous, exhibiting a strong eddylike motion in one region while being essentially irrotational in an adjacent region. To treat this problem mathematically, we must derive a formulation of the Euler equation that makes sense when the vorticity is discontinuous but bounded. We also assume that vorticity can be decomposed by means of a radial-energy decomposition (Definition 3.1) and in particular that it has a globally finite integral. The prototypical example is the vortex patch that has vorticity localized to a bounded region in the plane. Recall from Chap. 1, Corollary 1.2, that, in two dimensions, vorticity is conserved along particle trajectories. Hence, if the vorticity is the characteristic function of a bounded domain Ω,

$$\omega(x) = \begin{cases} \omega_0, & x \in \Omega \\ 0, & x \notin \Omega \end{cases},$$

then it remains like this with only the region Ω evolving in time. Patches of constant vorticity are a special case of weak solutions with vorticity in $L^p(\mathbb{R}^2) \cap L^1(\mathbb{R}^2)$. The L^p spaces on the unbounded domain $\mathbb{R}^2$ are not ordered; hence there is no relation between these weak solutions and those in L^∞. In Chap. 10 we show that for any $1 < p < \infty$, the class $L^1 \cap L^p$ is strong enough to guarantee existence but not necessarily uniqueness of solutions.

Mathematically, we cannot directly use PDEs to describe solutions with discontinuities. Instead we use analogous integral identities that come from integration by parts on a test function. We begin by considering weak versions of the vorticity-stream form of the *Euler* equation.

Later chapters of this book address even weaker solutions than that of the vortex patch. In particular, a *vortex sheet* is a structure whose vorticity is concentrated as a delta function on a 1D curve in the plane. They are a special class of measure valued weak solutions $\omega(\cdot, t) \in \mathcal{M}(\mathbb{R}^2) \cap H^{-1}_{\mathrm{loc}}(\mathbb{R}^2)$. Chapter 9 provides an introduction to basic properties of vortex sheets and some of the mathematical issues connected with their theoretical understanding. Chapter 11 provides detailed mathematical results concerning solutions with vortex-sheet initial data, including the recent result that such solutions exist for all time when the initial data have vorticity of distinguished sign.

This chapter presents the mathematical theory of weak solutions with vorticity in $L^\infty(\mathbb{R}^2) \cap L^1(\mathbb{R}^2)$. In Section 8.1 we discuss in detail a class of exact solutions in which the vorticity is constant inside an elliptcal region that rotates in time. In Section 8.2 we prove existence and uniqueness of general weak solutions of this type, a result first derived by Yudovich (1963). There we prove an estimate (Lemma 8.2) that shows that the particle paths are at least Hölder continuous with a decay in the time exponent. Last, in Section 8.3 we discuss the special case of vortex patches, in which the vorticity is a constant multiple of a characteristic function of a time-evolving domain. We derive contour dynamics equation (CDE) (8.57) for the evolution of the boundary of the patch. However, the estimates from Lemma 8.2 suffice only to guarantee Hölder continuity of the boundary of the patch as it evolves in time, whereas the derivation of CDE (8.57) assumes at least Lipschitz continuity. In Section 8.3 we use CDE (8.57) to prove that the boundary, if initially smooth, stays smooth for all time.

8.1. Elliptical Vorticies

We begin by introducing a family of exact solutions to the Euler equation for which the vorticity ω is discontinuous and has compact support. These examples are significant as simple models of actual physical phenomena and for their use in testing numerical methods.

We consider an elliptical columnar vortex in two situations. In the first case, without the presence of an external flow, the ellipse is a uniformly rotating solution to the Euler equation; it simply rotates with a constant angular velocity without changing shape. Imposing a strain flow causes the ellipse to retain elliptical geometry while changing its shape, as time evolves, through a change in its aspect ratio. We consider, in general, 3D strains; however, this example remains inherently 2D as in the case of the strained shear layers (Example 1.7 in Chap. 1).

We assume that the elliptical columnar vortex is infinite in the z direction and has a cross section described by an ellipse $E(t)$ with the semiaxes $a(t)$ and $b(t)$ and a rotation angle $\theta(t)$ (Fig. 8.1). Moreover, we assume that the vorticity has support inside $E(t)$ and aligns with the z axis with uniform strength $\omega(t)$. Now we show

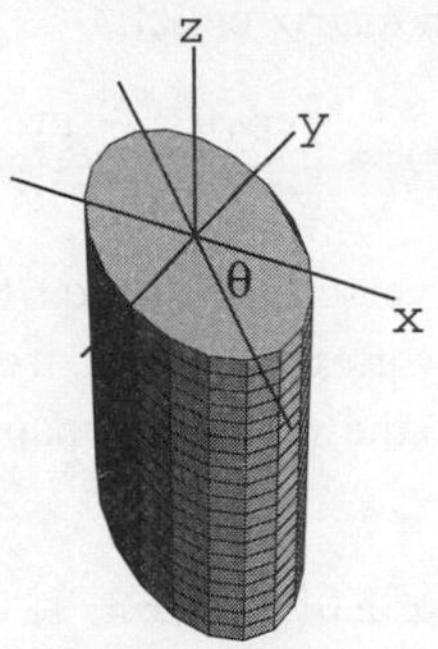

Figure 8.1. An elliptical columnar vortex, infinite in the z direction with a cross section described by an ellipse $E(t)$ with semiaxes $a(t)$ and $b(t)$ and a rotation angle $\theta(t)$.

that such an elliptical columnar vortex evolves as a solution to the Euler equation by simply rotating without a change of shape, so that its evolution is fully specified by the parameter $\theta(t)$.

Consider the stream function $\psi(x', t), x' = (x_1, x_2)^t$ and corresponding velocity $v^c = (\psi_{x_2}, -\psi_{x_1})^t$ self-induced by the elliptical columnar vortex. Recall from Chap. 2, Eq. (2.8), that the stream function ψ satisfies

$$-\Delta_{x'}\psi(x', t) = \begin{cases} \omega & \text{inside } E(t) \\ 0 & \text{otherwise} \end{cases}, \tag{8.1}$$

with the continuity of v^c across $E(t)$ as the boundary condition. If the rotation angle $\theta(t)$ is zero, the calculation of the velocity field from potential theory (see Lamb, 1945) gives

$${}^t\left(v_1^c, v_2^c\right) = U(a, b)x',$$

where

$$U(a, b) = \begin{cases} \dfrac{\omega}{a+b}\begin{bmatrix} 0 & -a \\ b & 0 \end{bmatrix} & \text{inside } E(t) \\ \dfrac{\omega}{\sqrt{a^2+\lambda^2}+\sqrt{b^2+\lambda^2}}\begin{bmatrix} 0 & -\sqrt{a^2+\lambda} \\ \sqrt{b^2+\lambda} & 0 \end{bmatrix} & \text{otherwise} \end{cases}, \tag{8.2}$$

and the parameter $\lambda > 0$ satisfies

$$\frac{x_1^2}{a^2+\lambda} + \frac{x_2^2}{b^2+\lambda} = 1. \tag{8.3}$$

Because solutions to the Euler equation have rotation symmetry (recall Proposition 1.1), for any angle $\theta(t)$ the self-induced velocity v^c is

$$\left(v_1^c, v_2^c\right)^t = U(a, b, \theta)x', \qquad x' = (x_1, x_2)^t, \tag{8.4}$$

where

$$U(a, b, \theta) = R(\theta)U(a, b)R(-\theta), \tag{8.5}$$

and $R(\theta)$ is the standard rotation matrix on $\mathbb{R}^2$:

$$R(\theta) = \begin{bmatrix} \cos\theta & -\sin\theta \\ \sin\theta & \cos\theta \end{bmatrix}. \tag{8.6}$$

Note that inside the ellipse the velocity v^c in Eq. (8.4) depends linearly on x' and hence preserves the shape of the vortex. Moreover, it does not depend on z, so material lines inside E that are parallel to the x_3 axis are mapped onto lines parallel to x_3.

The equation of the ellipse is

$$\boldsymbol{x}^t E(a, b, \theta)\boldsymbol{x} = 1, \tag{8.7}$$

where

$$E(a, b, \theta) = R(\theta) \begin{bmatrix} 1/a^2 & 0 \\ 0 & 1/b^2 \end{bmatrix} R(-\theta). \tag{8.8}$$

Let $\mathbf{x}(t) = \mathbf{x}(t)\mathbf{x} + y(t)\mathbf{y}$ denote a particle path on the boundary of the ellipse. At all times $\boldsymbol{x}$ satisfies Eq. (8.7); hence we can differentiate it to obtain

$$\dot{\boldsymbol{x}}^t E(a, b, \theta)\boldsymbol{x} + \boldsymbol{x}^t \dot{E}(a, b, \theta)\boldsymbol{x} + \boldsymbol{x}^t E(a, b, \theta)\dot{\boldsymbol{x}} = 0. \tag{8.9}$$

Because $\dot{\boldsymbol{x}} = U(a, b, \theta)\boldsymbol{x}$ we have

$$\boldsymbol{x}^T(U^T E + \dot{E} + EU)\boldsymbol{x} = 0, \tag{8.10}$$

and hence

$$U^T E + \dot{E} + EU = 0. \tag{8.11}$$

Plugging in the exact forms of U and E gives the following example.

Example 8.1. Kirchhoff's Steady Elliptical Columnar Vortex. The velocity v^c from Eq. (8.4) gives evolution equations for $a(t)$, $b(t)$, and $\theta(t)$:

$$\begin{aligned} \frac{da}{dt}(t) &= 0, \\ \frac{db}{dt}(t) &= 0, \\ \frac{d\theta}{dt}(t) &= \omega\frac{ab}{(a+b)^2} \equiv \frac{\Gamma}{\pi(a+b)^2}, \end{aligned} \tag{8.12}$$

where $\Gamma = \omega\pi ab$ is the total circulation of the vortex (Γ is conserved in time – see Proposition 1.11 on Kelvin's conservation of circulation). Hence a and b remain fixed while the elliptical columnar vortex rotates (without a change of shape) with angular velocity $\{\Gamma/[\pi(a+b)^2]\}$.

We now generalize Example 8.1 by assuming that the elliptical columnar vortex is subjected to an external strain velocity v^s:

$$v^s = (\gamma_1 x_1, -\gamma_2 x_2, \gamma_3 x_3)^t, \tag{8.13}$$

where $\gamma_1 - \gamma_2 + \gamma_3 = 0$ by the incompressibility condition. The resulting total velocity

v is a superposition of the strain velocity v^s and the self-induced velocity v^c:

$$(v_1, v_2)^t = U(a, b, \theta)x' + \begin{bmatrix} \gamma_1 & 0 \\ 0 & -\gamma_2 \end{bmatrix} x', \; x' = (x_1, x_2)^t. \tag{8.14}$$

Again, the velocity v in Eq. (8.14) depends linearly on x'. We leave it as an elementary exercise for the reader to show that elliptical columnar vortex preserves its shape, and its evolution is specified by the parameters $a(t)$, $b(t)$, and $\theta(t)$. Using these observations, Neu (1984) derived the following example.

Example 8.2. Neu's Unsteady Elliptical Columnar Vortex in an Imposed Strain flow. The external velocity v in Eq. (8.14) implies evolution equations for $a(t), b(t)$, and $\theta(t)$:

$$\begin{aligned} &\frac{da}{dt}(t) + (\gamma_2 \sin^2\theta - \gamma_1 \cos^2\theta)a = 0, \\ &\frac{db}{dt}(t) + (\gamma_2 \cos^2\theta - \gamma_1 \sin^2\theta)b = 0, \\ &\frac{d\theta}{dt}(t) - \omega\frac{ab}{(a+b)^2} + \frac{1}{2}(\gamma_1 + \gamma_2)\frac{a^2+b^2}{a^2-b^2}\sin 2\theta = 0. \end{aligned} \tag{8.15}$$

Defining the aspect ratio $\eta = (a/b)$ and the dimensionless time τ related to t by

$$d\tau = \omega\frac{\eta^2}{\eta^2 - 1}dt,$$

Neu rewrote the above system in the Hamiltonian form:

$$\begin{aligned} \frac{d\eta}{d\tau} &= -\frac{dH}{d\theta}, \\ \frac{d\theta}{d\tau} &= \frac{dH}{d\eta}, \end{aligned} \tag{8.16}$$

where the Hamiltonian H is

$$H = \ln\frac{(1+\eta)^2}{\eta} - \frac{1}{2}\frac{\gamma_1+\gamma_2}{\omega}\left(\eta - \frac{1}{2}\right)\sin 2\theta. \tag{8.17}$$

In general, H may also depend on τ by the vorticity ω, so that system (8.16) may not be autonomous. Because Eqs. (8.15) give

$$\frac{d}{dt}(ab) = (\gamma_1 - \gamma_2)ab,$$

the magnitude of the vorticity satisfies

$$\omega = \frac{\Gamma}{\pi ab} = \frac{\Gamma}{\pi a_0 b_0}e^{(\gamma_2-\gamma_1)t}. \tag{8.18}$$

System (8.16) is autonomous provided that $\gamma_1 = \gamma_2 \equiv \gamma$. This implies that $\gamma_3 \equiv 0$, so there is no stretching along the x_3 axis. Moreover, level curves of the Hamiltonian H define possible trajectories on a phase plane whose polar coordinates are (η, θ). In lieu of using the η-θ coordinates, we follow Bertozzi (1988) and use the $\theta \to \theta + \pi$ symmetry as well as the $\eta \to 1/\eta$, $\theta \to \theta + \pi/2$ symmetry of the ellipse. Hence we capture the entire dynamics with the polar coordinate variables $r = \log \eta$ and $\varphi = 2\theta$. In real time, the evolution equations are

$$\dot{r} = (\gamma + \gamma') \cos \varphi,$$
$$\dot{\varphi} = \frac{2\omega e^r}{(e^r + 1)^2} - (\gamma + \gamma') \frac{e^{2r} + 1}{e^{2r} - 1} \sin \varphi.$$

The reader can verify (see also Bertozzi, 1988) that if r and φ are viewed polar coordinates the above set of equations indeed defines a 2D dynamical system with a vector field continuous at the origin. Figure 8.2 shows the phase portraits for different values of γ/ω [regimes (1) (2) and (3)]. When $(\gamma/\omega) = 0$ (no strain), $\dot{r} = 0$ and the ellipse rotates at a constant speed (this is Example 8.1). When $(\gamma/\omega) = 0 < 0.1227$ [regime (1)] there are three possible behaviors for the ellipse: oscillation, rotation, or elongation. Figure 8.2, regime (1) shows $(\gamma/\omega) = 0.1$. There is a homoclinic orbit containing periodic orbits and fixed points. The periodic orbit intersecting the origin separates the rotation modes from the oscillation modes. The orbits pictured divide the phase portrait up into regions corresponding to the three modes of motion: a rotation, an oscillation, and an elongation. The trajectories inside the inner periodic orbit all correspond to oscillating ellipses. The trajectories outside the periodic orbit but inside the homoclinic orbit correspond to rotating ellipses. All other trajectories describe elongating ellipses.

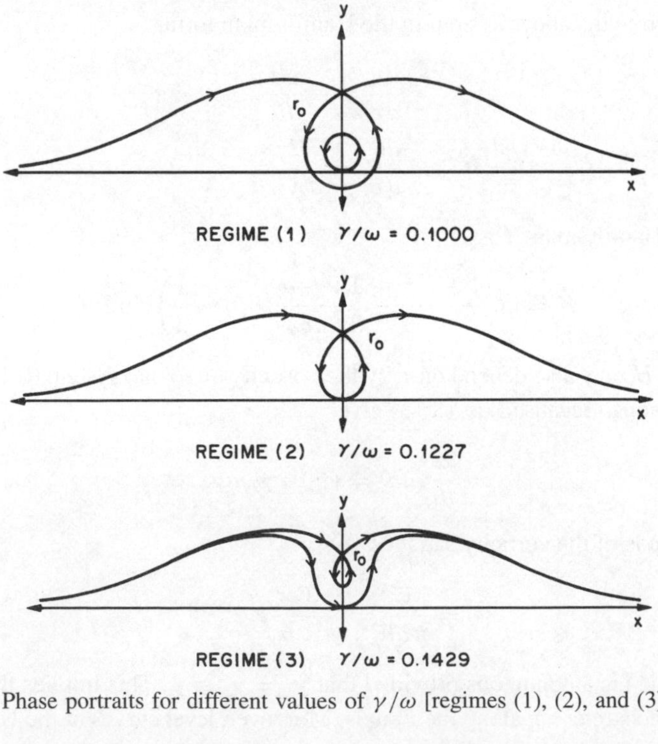

Figure 8.2. Phase portraits for different values of γ/ω [regimes (1), (2), and (3)] (Bertozzi, 1988).

At $(\gamma/\omega) = 0.1227$ the homoclinic orbit crosses the origin, eliminating the rotation mode [regime (2)]. The phase diagram undergoes a bifurcation. The homoclinic orbit crosses the x axis, eliminating the possibility for a rotation mode.

When $0.1227 < (\gamma/\omega) < 0.15$ there is still a homoclinic orbit with interior trajectories of the dynamical system that correspond to oscillation modes of the ellipse and trajectories outside that correspond to elongation modes. Figure 8.2, regime (3), shows the case $(\gamma/\omega) = 0.1429$. The trajectories inside the homoclinic orbit correspond to oscillating ellipses. Outside the loop, the motion described is elongation of the ellipse. For $(\gamma/\omega) \geq 0.15$, the oscillation mode vanishes because of the lack of a homoclinic orbit in the dynamical system. Hence there is only an elongation mode of motion.

If the vortex undergoes a stretching with $\gamma_3 > 0$, then $\gamma_1 \neq \gamma_2$ so that by Eq. (8.18) the vorticity inside the ellipse $E(t)$ grows exponentially with time and Hamiltonian system (8.16) is not autonomous. For $0 < |\frac{\gamma_1 - \gamma_2}{\omega}| \ll 1$, Neu (1984) analyzed the evolution of the elliptical columnar vortex by means of adiabatic invariants.

Another possibility is for γ_i to be time dependent. The case of γ_3 of the form $\epsilon\gamma''(t)$ with $\gamma''(t)$ time periodic was first analyzed by Bertozzi (1988), who used the Melnikov theory for perturbed homoclinic orbits. This work shows that simple, small oscillations in the γ_i typically introduce chaotic dynamics into the phase portrait of the evolution equation for the ellipse. Thus, in general, the motion of columnar vortex is extremely sensitive to small changes in an external strain field.

8.2. Weak L^∞ Solutions to the Vorticity Equation

The elliptical vortices of Section 8.1 are just one example of a class of weak solutions to the vorticity-stream form of the 2D Euler equation in which the vorticity is discontinuous but bounded. Such solutions are not included in the existence and uniqueness theory from Chaps. 3 and 4, in which we required the initial data ω_0 to be Hölder continuous. In this section we present the mathematical theory of general 2D $L^\infty \cap L^1$ weak solutions to the vorticity-stream formulation [Eq. (2.5) with $\nu = 0$] of the Euler equation. In Section 8.3 we discuss the special case of vortex patches. This section is organized as follows: In Subsection 8.2.2 we discuss the existence of weak solutions, in Subsection 8.2.4 their uniqueness, and in Subsection 8.2.5 the propagation of regularity for such weak solutions.

8.2.1. A Weak Vorticity-Stream Formulation

In Chap. 2 we analyzed the vorticity-stream formulation for smooth 2D flows [Eq. (2.5)]:

$$\frac{D\omega}{Dt} = 0,$$
$$\omega|_{t=0} = \omega_0, \tag{8.19}$$

where $(D/Dt) = (\partial/\partial t) + v \cdot \nabla$ and the velocity v is determined from the vorticity

ω by

$$v(x,t) = \int_{\mathbb{R}^2} K(x-y)\omega(y,t)dy, \tag{8.20}$$

with the kernel

$$K(x) = \frac{1}{2\pi}\left(\frac{-x_2}{|x|^2}, \frac{x_1}{|x|^2}\right)^t. \tag{8.21}$$

Recall from Chap. 2 that, in two dimensions, the vorticity stays constant on particle trajectories. We use this idea to formulate a concept of weak solutions to the 2D Euler equation with initial vorticity $\omega_0 \in L^1(\mathbb{R}^2) \cap L^\infty(\mathbb{R}^2)$. To do this, we look for an equivalent expression for vorticity-stream formulation (8.19) that makes sense for vorticities $\omega(\cdot,t)$ that are not smooth but merely in $L^1(\mathbb{R}^2) \cap L^\infty(\mathbb{R}^2)$. A correct form comes from transport formula (1.15):

$$\frac{d}{dt}\int_{\mathbb{R}^2} \varphi\omega\, dx = \int_{\mathbb{R}^2} \frac{D\varphi}{Dt}\omega\, dx + \int_{\mathbb{R}^2} \varphi \frac{D\omega}{Dt}\, dx,$$

where ω and φ are smooth functions vanishing as $|x| \nearrow \infty$. Integrating this equation in time, we have

$$\int_{\mathbb{R}^2} \varphi(x,T)\omega(x,T)dx - \int_{\mathbb{R}^2} \varphi(x,0)\omega(x,0)dx$$
$$= \int_0^T \int_{\mathbb{R}^2} \frac{D\varphi}{Dt}\omega\, dxdt + \int_0^T \int_{\mathbb{R}^2} \varphi \frac{D\omega}{Dt}\, dxdt. \tag{8.22}$$

Suppose that the vorticity ω is a smooth solution (so that the velocity $v = K * \omega$ is divergence free and $\omega = \text{curl } v$). Then by formulation (8.19), $(D\omega/Dt) = 0$, and we get the identity

$$\int_{\mathbb{R}^2} \varphi(x,T)\omega(x,t)dx - \int_{\mathbb{R}^2} \varphi(x,0)\omega_0(x)dx = \int_0^T \int_{\mathbb{R}^2} \frac{D\varphi}{Dt}\omega\, dxdt,$$

which is valid for all test functions $\varphi \in C^1([0,T] \times \mathbb{R}^2)$ with compact support $\text{supp}\, \varphi(\cdot,t) \subset \{x : |x| \le R\}$. Note that this identity is also well defined for a larger class of solutions ω such that ω and the product $v\omega[\equiv (K*\omega)\omega]$ are locally integrable, $\omega, v\omega \in L^1_{\text{loc}}([0,T] \times \mathbb{R}^2)$, where v is defined by Eq. (8.20). We make the following definition.

Definition 8.1. *Given $\omega_0 \in L^1(\mathbb{R}^2) \cap L^\infty(\mathbb{R}^2)$, (v,ω) is a weak solution to the vorticity-stream formulation of the 2D Euler equation with initial data $\omega_0(x)$, provided that*

(i) $\omega \in L^\infty\{[0,T]; L^1(\mathbb{R}^2) \cap L^\infty(\mathbb{R}^2)\}$,
(ii) $v = K * \omega$ *and* $\omega = \text{curl } v$,
(iii) for all $\varphi \in C^1\{[0,T]; C_0^1(\mathbb{R}^2)\}$

$$\int_{\mathbb{R}^2} \varphi(x,T)\omega(x,T)dx - \int_{\mathbb{R}^2} \varphi(x,0)\omega_0(x)dx = \int_0^T \int_{\mathbb{R}^2} \frac{D\varphi}{Dt}\omega\, dxdt. \tag{8.23}$$

In the following subsections we prove the existence and the uniqueness of these weak solutions. First, in order for this definition to be a good one, we require that a weak solution that is C^1 smooth be a classical solution. Indeed, we have the following proposition.

Proposition 8.1. *(i) Every smooth solution to the 2D Euler equation is a weak solution in the sense of Definition 8.1. (ii) Conversely, if the above weak solution is additionally C^1 smooth, then it is a classical solution.*

Proof of Proposition 8.1. We have already proved part (i) by construction. To prove (ii), suppose that (v, ω) is a C^1-smooth weak solution. Combining (8.22) and (8.23) gives

$$\int_0^T \int_{\mathbb{R}^2} \varphi \frac{D\omega}{Dt} \, dx dt = 0, \qquad \forall \varphi \in C^1\{[0, T]; C_0^1(\mathbb{R}^2)\},$$

which, by the continuity of $(D\omega/Dt)$, implies that $(D\omega/Dt) = 0$ pointwise. □

8.2.2. The Existence of Weak Solutions

In the previous subsection we formulated Definition 8.1 of a weak solution to the 2D Euler equation with initial data in $L^1 \cap L^\infty$. We showed that smooth solutions satisfy this definition, so that we know the class is nonempty. In fact, in this section we prove the following theorem.

Theorem 8.1. *The Existence of Weak Solutions. Let the initial vorticity $\omega_0 \in L^1(\mathbb{R}^2) \cap L^\infty(\mathbb{R}^2)$. Then for all time there exists a weak solution (v, ω) to the vorticity-stream formulation of the 2D Euler equation in the sense of Definition 8.1.*

These weak solutions indeed form a much broader class than the class of smooth solutions. In Section 8.1 we introduced the Kirchoff elliptical vortices, the simplest example of nonsmooth weak solutions in the sense of Definition 8.1. The Kirchoff ellipse has a jump discontinuity in the vorticity across the boundary of the ellipse and is an example of a "patch of constant vorticity." Despite the fact that the evolution described by the elliptical vorticies in Section 8.1 is so simple, general vortex-patch solutions can develop more complex structures. However, as we show in Subsection 8.3.3, the boundary of a vortex patch, if initially smooth, stays smooth for all time.

The existence (and the uniqueness) of weak solutions with $\omega_0 \in L^\infty$ in bounded domains was first proved by Yudovich (1963). Below we give a simpler proof by using a strategy similar to the one used in Chap. 3 to construct solutions to the Euler and the Navier–Stokes equations with smooth initial data. The strategy has two steps. First we smooth the initial data so that Corollary 3.3 guarantees the existence of a smooth solution for all time. Then we pass to the limit in the regularization parameter.

Our first step is to mollify $\omega_0 \in L^1 \cap L^\infty$ to obtain a family of smooth initial data (ω_0^ϵ), $\epsilon > 0$. We recall mollification operator J_ϵ (3.34) from Chap. 3. First we choose a function $\rho \in C_0^\infty(\mathbb{R}^2),\ \rho \geq 0,\ \int_{\mathbb{R}^2} \rho\, dx = 1$. We then define the mollification of ω by

$$\omega_0^\epsilon(x) = J_\epsilon \omega_0 = \epsilon^{-2} \int_{\mathbb{R}^2} \rho\left(\frac{x-y}{\epsilon}\right) \omega_0(y) dy, \qquad \epsilon > 0.$$

Recall from Lemma 3.5 that mollifiers satisfy the following inequalities:

$$\begin{aligned} \left|\left|\omega_0^\epsilon\right|\right|_{L^\infty} &\leq ||\omega_0||_{L^\infty}, \\ \left|\left|\omega_0^\epsilon\right|\right|_{L^1} &\leq ||\omega_0||_{L^1}, \\ \lim_{\epsilon \to 0} \left|\left|\omega_0^\epsilon - \omega_0\right|\right|_{L^1} &= 0. \end{aligned} \tag{8.24}$$

To complete the construction of the approximate solutions we use Corollary 3.3 from Chap. 3 to obtain a unique global smooth solution $\omega^\epsilon(x, t)$, $\omega^\epsilon(x, 0) = \omega_0^\epsilon(x)$, $v^\epsilon = K * \omega^\epsilon$ to the 2D Euler equation. Being smooth solutions, these solutions also solve identity (8.23) that defines weak solutions:

$$\int_{\mathbb{R}^2} \varphi(x, T)\omega^\epsilon(x, T) dx - \int_{\mathbb{R}^2} \varphi(x, 0)\omega_0^\epsilon(x) dx = \int_0^T \int_{\mathbb{R}^2} \frac{D^\epsilon \varphi}{Dt} \omega^\epsilon\, dx dt \tag{8.25}$$

for all $\varphi \in C^1\{[0, T]; C_0^1(\mathbb{R}^2)\}$, where $(D^\epsilon/Dt) = (\partial/\partial t) + v^\epsilon \cdot \nabla$.

The next step of our strategy is to extract subsequences $(\omega^{\epsilon'})$ and $(v^{\epsilon'})$, then to show their convergence to limits ω and v, and finally to show that these limits are a weak solution in the sense of Definition 8.1. To do this we need bounds for ω^ϵ and v^ϵ independent of ϵ. These bounds involve the following norm,

$$|||\omega_0||| = \|\omega_0\|_{L^1} + \|\omega_0\|_{L^\infty}, \tag{8.26}$$

and are derived by use of potential theory.

Proposition 8.2. *Let the initial vorticity $\omega_0 \in L^1(\mathbb{R}^2) \cap L^\infty(\mathbb{R}^2)$, and let ω^ϵ, v^ϵ be a smooth solution to the regularized initial data ω_0^ϵ on a time interval $[0, T]$. Then, for all $t \in [0, T]$,*

(i) ω^ϵ and v^ϵ are uniformly bounded, and

$$||v^\epsilon(\cdot, t)||_{L^\infty} \leq c_1 |||\omega^\epsilon(\cdot, t)||| \leq c_2 |||\omega_0|||,$$

(ii) there exist functions $\omega(\cdot, t) \in L^1(\mathbb{R}^2) \cap L^\infty(\mathbb{R}^2)$ and $v = K * \omega$ such that $\forall t \in [0, T]$

$$\begin{aligned} \omega^\epsilon(\cdot, t) &\to \omega(\cdot, t) \text{ in } L^1(\mathbb{R}^2), \\ v^\epsilon(\cdot, t) &\to_x v(\cdot, t) \text{ locally.} \end{aligned}$$

Here $\to_x$ denotes uniform convergence in the space variables. We prove this proposition in the next section.

Proof of Theorem 8.1. Assuming Proposition 8.2, it remains to show that the limit functions ω and v are weak solutions in the sense of Definition 8.1. We must show that identity (8.25),

$$\int_{\mathbb{R}^2} \varphi(x,T)\omega^\epsilon(x,T)dx - \int_{\mathbb{R}^2} \varphi(x,0)\omega_0^\epsilon(x)dx = \int_0^T \int_{\mathbb{R}^2} \frac{D^\epsilon \varphi}{Dt}\omega^\epsilon \, dxdt,$$

converges to

$$\int_{\mathbb{R}^2} \varphi(x,T)\omega(x,T)dx - \int_{\mathbb{R}^2} \varphi(x,0)\omega_0(x)dx = \int_0^T \int_{\mathbb{R}^2} \frac{D\varphi}{Dt}\omega \, dxdt$$

as the mollification parameter $\epsilon \searrow 0$.

Because φ has compact support, $v^\epsilon \rightarrow_x v$ locally, $\omega^\epsilon \rightarrow \omega$ in L^1, and ω^ϵ is uniformly bounded, we obtain the convergence of the nonlinear term by the dominated convergence theorem:

$$\int_0^T \int_{\mathbb{R}^2} v^\epsilon \cdot \nabla\varphi\omega^\epsilon \, dxdt \rightarrow \int_0^T \int_{\mathbb{R}^2} v \cdot \nabla\varphi\omega \, dxdt.$$

All the linear terms converge by Proposition 8.2. In the following subsection we provide some key potential theory estimates, thus proving Proposition 8.2 and concluding the proof of Theorem 8.1. □

8.2.3. Potential Theory Estimates

Our goal in this section is to prove Proposition 8.2, the necessary ingredient for completing the proof of THEOREM 8.1. We recall Proposition 8.2.

Proposition 8.2. *Let the initial vorticity $\omega_0 \in L^1(\mathbb{R}^2) \cap L^\infty(\mathbb{R}^2)$, and let ω^ϵ, v^ϵ be a smooth solution, on a time interval $[0, T]$, with the regularized initial data ω_0^ϵ. Then, for all $t \in [0, T]$,*

(i) ω^ϵ and v^ϵ are uniformly bounded, and

$$||v^\epsilon(\cdot,t)||_{L^\infty} \leq c_1|||\omega^\epsilon(\cdot,t)||| \leq c_2|||\omega_0|||, \tag{8.27}$$

(ii) there exist functions $\omega(\cdot,t) \in L^1(\mathbb{R}^2) \cap L^\infty(\mathbb{R}^2)$ and $v = K * \omega$ such that

$$\omega^\epsilon(\cdot,t) \rightarrow \omega(\cdot,t) \text{ in } L^1, \tag{8.28}$$
$$v^\epsilon(\cdot,t) \rightarrow_x v(\cdot,t) \text{ locally}. \tag{8.29}$$

Proof of Proposition 8.2, part (i). The proof uses potential theory. First note that the vorticity ω^ϵ satisfies $|||\omega^\epsilon(\cdot,t)||| \leq c|||\omega_0|||$ because of the properties of mollifiers (8.24) and the fact that vorticity is conserved along particle trajectories for 2D smooth solutions.

Now we show that the velocity v^ϵ is uniformly bounded. Using the Biot–Savart law, we split v^ϵ into two parts:

$$\begin{aligned} v^\epsilon(x,t) &\\ = \int_{\mathbb{R}^2} \rho(x-x')K_2(x-x')\omega^\epsilon(x',t)dx' &+ \int_{\mathbb{R}^2} [1-\rho(x-x')]K_2(x-x')\omega^\epsilon(x',t)dx' \\ = v_1^\epsilon(x,t) + v_2^\epsilon(x,t), & \end{aligned}$$

where $\rho \in C_0^\infty(\mathbb{R}^2)$ is a cutoff function such that $\rho(x) \equiv 1$ for $|x| \le 1$ and $\rho(x) \equiv 0$ for $|x| \ge 2$. Recall from Proposition 2.1 that the kernel K is homogeneous of degree -1, $|K_2(x)| \le c|x|^{-1}$. Hence Young's inequality gives

$$\begin{aligned} ||v^\epsilon(\cdot,t)||_{L^\infty} &\le ||\rho K||_{L^1}||\omega^\epsilon||_{L^\infty} + ||(1-\rho)K||_{L^\infty}||\omega^\epsilon||_{L^1} \\ &\le c|||\omega^\epsilon(\cdot,t)|||. \end{aligned}$$

This concludes the proof of part (i) of Proposition 8.2. □

Before we prove the second part of Proposition 8.2, first we note that, although the family ω^ϵ is composed of smooth solutions, we cannot define a function $\omega(\cdot,t) \in L^1(\mathbb{R}^2) \cap L^\infty(\mathbb{R}^2)$ by a direct passing to the limit $\omega^\epsilon \to \omega$ in $L^1 \cap L^\infty$ (as the space L^∞ is not separable). Instead we construct the limiting vorticity $\omega \in L^1 \cap L^\infty$ by means of particle trajectories. In Chap. 1, Eq. (1.53), we saw that smooth solutions to the 2D Euler equation satisfy

$$\omega(x,t) = \omega_0(X^{-t}(x)),$$

where X^{-t} is the inverse (at time t) of the particle-trajectory map X, satisfying

$$\frac{dX}{dt} = v(X(\alpha,t),t), \qquad X(\alpha,t)|_{t=0} = \alpha.$$

It is not obvious that these formulas are also valid for the initial vorticity $\omega_0 \in L^1 \cap L^\infty$. They suggest, however, the following strategy of constructing the weak solution $\omega(\cdot,t) \in L^1 \cap L^\infty$. Mollifying the initial data ω_0 for all time $t > 0$, we have smooth solutions

$$\omega^\epsilon(x,t) = \omega_0^\epsilon\left(X_\epsilon^{-t}(x)\right),$$

where the (smooth) particle trajectories X_ϵ satisfy

$$\frac{dX_\epsilon}{dt} = v^\epsilon\left(X_\epsilon^t(\alpha,t),t\right), \qquad X_\epsilon(\alpha,t)|_{t=0} = \alpha,$$

and the velocity $v^\epsilon = K * \omega^\epsilon$. We expect that the method of characteristics should work for weak solutions. By passing to a subsequence, we can take a uniform limit $X_{\epsilon'} \to_x X$ and then define the weak solutions by

$$\omega(x,t) = \omega_0(X^{-t}(x)), \tag{8.30}$$

$$v(x,t) = \int_{\mathbb{R}^2} K(x-x')\omega(x',t)dx'. \tag{8.31}$$

To prove the second part of Proposition 8.2, we use the following potential theory estimates for the velocity v^ϵ and the particle trajectories X_ϵ^t independent of ϵ.

Lemma 8.1. *Potential Theory Estimates for a Velocity. Let the initial vorticity $\omega_0 \in L^1(\mathbb{R}^2) \cap L^\infty(\mathbb{R}^2)$ and let ω^ϵ, v^ϵ be a smooth solution on a time interval $[0, T]$ with the regularized initial data ω_0^ϵ. Then $v^\epsilon(\cdot, t)$ is quasi-Lipschitz continuous:*

$$\sup_{0\le t\le T} |v^\epsilon(x^1, t) - v^\epsilon(x^2, t)| \le c|||\omega_0||| \, |x^1 - x^2|(1 - \ln^- |x^1 - x^2|), \tag{8.32}$$

where $\ln^- a = \ln a$ for $0 < a < 1$ and $\ln^- a = 0$ for $a \ge 1$.

Lemma 8.2. *Potential Theory Estimates for Particle Trajectories. Let the assumptions of Lemma 8.1 be satisfied. Then for any $T > 0$ there exist $c > 0$ and the exponent $\beta(t) = \exp(-c|||\omega_0|||t)$ such that for all $\epsilon > 0$, $0 \le t \le T$,*

$$\left|X_\epsilon^{-t}(x^1) - X_\epsilon^{-t}(x^2)\right| \le c|x^1 - x^2|^{\beta(t)}, \tag{8.33}$$

$$|X_\epsilon(\alpha^1, t) - X_\epsilon(\alpha^2, t)| \le c|\alpha^1 - \alpha^2|^{\beta(t)}, \tag{8.34}$$

and for all $0 \le t_1, t_2 \le T$,

$$\left|X_\epsilon^{-t_1}(x) - X_\epsilon^{-t_2}(x)\right| \le c|t_1 - t_2|^{\beta(t)}, \tag{8.35}$$

$$|X_\epsilon(\alpha, t_1) - X_\epsilon(\alpha, t_2)| \le c|t_1 - t_2|^{\beta(t)}. \tag{8.36}$$

Assuming these potential theory estimates, we give the remaining proof of Proposition 8.2.

Proof of Proposition 8.2, part (ii). First we prove the existence of the particle trajectories $X^{-t}(x)$ such that $X_\epsilon^{-t}(x) \to_x X^{-t}(x)$ locally. Estimate (8.27) provides a uniform bound for $\{v^\epsilon\}$ and hence the uniform bound

$$\left|X_\epsilon^{-t}(x) - x\right| = |X_\epsilon(\alpha, t) - \alpha| = \left|\int_0^t v^\epsilon(X_\epsilon(\alpha, \tau), \tau) d\tau\right| \le cT,$$

so the family (X_ϵ^{-t}) is uniformly bounded for all $0 \le t \le T$. Also, by potential theory estimates (8.33) and (8.35), for all $\eta > 0$ there exists $\delta > 0$ such that if

$$c|x^1 - x^2|^\beta + c|t_1 - t_2|^\beta < \delta \qquad \text{for} \quad (x^j, t_j) \in \{x : |x| \le R\} \times [0, T]$$

then

$$\left|X_\epsilon^{-t_1}(x^1) - X_\epsilon^{-t_2}(x^2)\right| < \eta,$$

so that the family (X_ϵ^{-t}) is equicontinuous on $\{x : |x| \le R\} \times [0, T]$. Likewise, the family (X_ϵ) is equicontinuous on $\{\alpha : |\alpha| \le R\} \times [0, T]$. Thus by the Arzela–Ascoli theorem there exists a subsequence $(X_{\epsilon'}^{-t})$ so that $X_{\epsilon'}^{-t}(x) \to_x X^{-t}(x)$ uniformly on $\{x : |x| \le R\} \times [0, T]$. We need the uniform limit to exist in order to guarantee that X^{-t} is a measure-preserving transformation:

Claim: *For all $t \in [0, T]$, X^{-t} is a measure-preserving map from $\mathbb{R}^2$ to $\mathbb{R}^2$ and for any $f \in L^1(\mathbb{R}^2)$, $\int f(X^{-t}(x))dx = \int f(x)dx$.*

Proof of Claim: We first note that the claim holds for all continuous, compactly supported f. This is because the X_ϵ^{-t} are measure preserving and converge uniformly to X^{-t}. Thus $f(X_\epsilon^{-t})$ converges pointwise to $f(X^{-t})$ and by Lebesgue dominated convergence, we have $\int f(X^{-t}(x))dx = \lim \int f(X_\epsilon^{-t}(x))dx = \int f(x)dx$. By the fact that C_0 is dense in L^1 we obtain the result. By the Riesz representation theorem for C_0 functions, we see that in fact the measures m and $\mu = m \circ (X^{-t})^{-1}$ are equivalent on the Borel sets, where m denotes Lebesgue measure. Thus X^{-t} is in fact a measure-preserving transformation. □

We need this claim in the following arguments.

Given the above limiting particle trajectories $X^{-t}(x)$, we define the vorticity ω by solution (8.30), $\omega(x, t) = \omega_0(X^{-t}(x))$, and the velocity v by solution (8.31), $v = K * \omega$.

To finish the proof of part (ii) of Proposition 8.2 we show that $\omega^\epsilon(\cdot, t) \to \omega(\cdot, t)$ in L^1 and that $v^\epsilon(\cdot, t) \to_x v(\cdot, t)$ locally. We have

$$||\omega^\epsilon(\cdot, t) - \omega(\cdot, t)||_{L^1} \le \left|\left|\omega_0^\epsilon(X_\epsilon^{-t}) - \omega_0(X_\epsilon^{-t})\right|\right|_{L^1} + \left|\left|\omega_0(X_\epsilon^{-t}) - \omega_0(X^{-t})\right|\right|_{L^1}$$

Because the particle trajectories X_ϵ^{-t} are volume preserving, by the properties of mollifiers (8.24)

$$\left|\left|\omega_0^\epsilon(X_\epsilon^{-t}) - \omega_0(X_\epsilon^{-t})\right|\right|_{L^1} = \left|\left|\omega_0^\epsilon - \omega_0\right|\right|_{L^1} \to 0 \qquad \text{as} \quad \epsilon \searrow 0.$$

We need to show that $||\omega_0(X_\epsilon^{-t}) - \omega_0(X^{-t})||_{L^1} \to 0$ as $\epsilon \to 0$.

Because ω_0 is in $L^\infty \cap L^1$, standard real analysis (Royden, 1968) arguments give the existence of $\omega_0^n \in C_0(\mathbb{R}^2)$ such that

$$\left|\left|\omega_0 - \omega_0^n\right|\right|_{L^1} \le 1/n.$$

Consider the bound

$$\begin{aligned} \left|\left|\omega_0(X_\epsilon^{-t}) - \omega_0(X^{-t})\right|\right|_{L^1} &\le \left|\left|\omega_0(X_\epsilon^{-t}) - \omega_0^n(X_\epsilon^{-t})\right|\right|_{L^1} \\ &\quad + \left|\left|\omega_0^n(X_\epsilon^{-t}) - \omega_0^n(X^{-t})\right|\right|_{L^1} \\ &\quad + \left|\left|\omega_0^n(X^{-t}) - \omega_0(X^{-t})\right|\right|_{L^1} \\ &= (1) + (2) + (3). \end{aligned}$$

Given $\delta > 0$, we want to show that there exists $\epsilon_0 > 0$ such that for all $\epsilon < \epsilon_0$, $(1) + (2) + (3) \le \delta$. We choose $n > 3/\delta$. Using the fact that X_ϵ^{-t} and X^{-t} are measure-preserving transformations, we have that $(1) + (3) \le 2\delta/3$. For this fixed n, we have that ω_0^n is continuous, and thus $\omega_0^n(X_\epsilon^{-t}) \to \omega_0^n(X^{-t})$ pointwise. Therefore, by the Lebesgue dominated convergence theorem, $\omega_0^n(X_\epsilon^{-t}) \to \omega_0^n(X^{-t})$ in L^1 as $\epsilon \to 0$. Thus there exists $\epsilon_0(n)$ such that, if $\epsilon \le \epsilon_0(n)$,

$$\left|\left|\omega_0^n(X_\epsilon^{-t}) - \omega_0^n(X^{-t})\right|\right|_{L^1} \le \delta/3.$$

Putting this all together, we have shown that $||\omega_0(X_\epsilon^{-t}) - \omega_0(X^{-t})||_{L^1} \to 0$ as $\epsilon \to 0$.

Finally we show that $v^\epsilon(\cdot,t) \to_x v(\cdot,t)$ locally. We take $\rho \in C_0^\infty$, $\rho(x) = 1$ for $|x| \le 1$, $\rho(x) = 0$ for $|x| \ge 2$, and we define the cutoff function $\rho_\delta(x) = \rho(x/\delta)$. By the definition of v^ϵ and v we have

$$
\begin{aligned}
|v^\epsilon(x,t) - v(x,t)| \le& \left|(\rho_\delta K) * \left\{\omega_0^\epsilon\left(X_\epsilon^{-t}(x)\right) - \omega_0(X^{-t}(x))\right\}\right| \\
&+ \left|[(1-\rho_\delta)K] * \omega_0^\epsilon\left(X_\epsilon^{-t}(x)\right) - \omega_0(X^{-t}(x))\right| = I_1 + I_2.
\end{aligned}
$$

The first term, I_1, we estimate as

$$
\begin{aligned}
I_1 &\le ||\rho_\delta K||_{L^1}\left[\left|\left|\omega_0^\epsilon\left(X_\epsilon^{-t}\right)\right|\right|_{L^\infty} + ||\omega_0(X^{-t})||_{L^\infty}\right] \\
&\le 2||\omega_0||_{L^\infty}\int_{|x|\le 2\delta} |\rho_\delta(x)K(x)|dx \le c\int_{|x|\le 2\delta}|x|^{-1}dx \\
&\le c\delta.
\end{aligned}
$$

The second term, I_2, contains the smooth kernel $(1-\rho_\delta)K_2$, so we estimate

$$
\begin{aligned}
I_2 &\le ||(1-\rho_\delta)K_2||_{L^\infty}\left|\left|\omega_0^\epsilon\left(X_\epsilon^{-t}\right) - \omega_0(X^{-t})\right|\right|_{L^1} \\
&\le \frac{c}{\delta}\left|\left|\omega_0^\epsilon\left(X_\epsilon^{-t}\right) - \omega_0(X^{-t})\right|\right|_{L^1},
\end{aligned}
$$

so that finally we have

$$
|v^\epsilon(x,t) - v(x,t)| \le c\delta + \frac{c}{\delta}||\omega^\epsilon(\cdot,t) - \omega(\cdot,t)||_{L^1}.
$$

Now for any $\eta > 0$ we take $\delta = (\eta/2c)$. Because $||\omega^\epsilon(\cdot,t) - \omega(\cdot,t)||_{L^1} \to 0$ as $\epsilon \searrow 0$, we can pick $\epsilon_0 > 0$ such that for all $0 < \epsilon \le \epsilon_0$, $2c^2||\omega^\epsilon(\cdot,t) - \omega(\cdot,t)||_{L^1} \le \eta^2$. Thus for any $\eta > 0$ there exists $\epsilon_0 > 0$ such that for all $0 < \epsilon \le \epsilon_0$, $|v^\epsilon(\cdot,t) - v(\cdot,t)| \le \eta$, so that $v^\epsilon \to_x v$. □

Remark: In Proposition 8.2 we do not have convergence of $\omega^\epsilon(\cdot,t) \to \omega(\cdot,t)$ in $L^1 \cap L^\infty$. This is because the operation of translating L^∞ functions is not continuous in the L^∞ norm.

Now we give proofs of the potential theory estimates in Lemmas 8.1 and 8.2.

Proof of Lemma 8.1. Potential Theory Estimates for the Velocity. Let $d \equiv |x^1 - x^2| < 1$ and let $B(x,r)$ denote a ball centered at x and with a radius r. Split the function $v^\epsilon(x^1,t) - v^\epsilon(x^2,t)$ as

$$
\begin{aligned}
|v^\epsilon(x^1,t) - v^\epsilon(x^2,t)| \le& \left[\int_{\mathbb{R}^2 - B(x^1,2)} + \int_{B(x^1,2)-B(x^1,2d)} + \int_{B(x^1,2d)}\right] \\
&\times |K_2(x^1 - \tilde{x}) - K_2(x^2 - \tilde{x})|\,|\omega^\epsilon(\tilde{x},t)|d\tilde{x} \\
\equiv& I_1 + I_2 + I_3.
\end{aligned}
$$

Now estimate the successive terms I_j. For all $x, y \in \mathbb{R}^2$,

$$
\left|\frac{x}{|x|^2} - \frac{y}{|y|^2}\right|^2 = \frac{|x|^2 - 2(x,y) + |y|^2}{|x|^2|y|^2} = \frac{|x-y|^2}{|x|^2|y|^2};
$$

thus

$$|K_2(x) - K_2(y)| = \frac{1}{2\pi}\left(\left|\frac{y_2}{|x|^2} - \frac{x_2}{|y|^2}\right|^2 + \left|\frac{x_1}{|x|^2} - \frac{y_1}{|y|^2}\right|^2\right)^{1/2} \le \frac{1}{\pi}\frac{|x-y|}{|x||y|}.$$

Using this estimate, we find that the first term I_1 is bounded by

$$I_1 \le \frac{1}{\pi}|x^1 - x^2| \int_{\mathbb{R}^2 - B(x^1,2)} |\omega^\epsilon(\tilde{x},t)| \frac{d\tilde{x}}{|x^1 - \tilde{x}||x^2 - \tilde{x}|} \le c||\omega_0||_{L^1}|x^1 - x^2|.$$

Now for $\tilde{x} \in B(x^1,2) - B(x^1,2d)$ the mean-value theorem implies that

$$|K_2(x^1 - \tilde{x}) - K_2(x^2 - \tilde{x})| \le \sup_{0\le\theta\le 1} |\nabla K_2[x^1 - \tilde{x} + \theta(x^2 - x^1)]|\, |x^1 - x^2| \le c\frac{|x^1 - x^2|}{|x^1 - \tilde{x}|^2},$$

so the second term, I_2, is bounded by

$$I_2 \le c|x^1 - x^2|\, ||\omega^\epsilon(\cdot,t)||_{L^\infty} \int_{B(x^1,2)-B(x^1,2d)} \frac{d\tilde{x}}{|x^1 - \tilde{x}|^2} \le c||\omega_0||_{L^\infty}|x^1 - x^2| \int_{2d}^{2} \frac{dr}{r}$$
$$\le c||\omega_0||_{L^\infty}|x^1 - x^2|(1 - \ln^-|x^1 - x^2|).$$

Finally, we estimate the third term, I_3, as

$$I_3 \le c||\omega^\epsilon(\cdot,t)||_{L^\infty}\left[\int_{B(x_1,2d)} \frac{d\tilde{x}}{|x^1 - \tilde{x}|} + \int_{B(x_1,2d)} \frac{d\tilde{x}}{|x^2 - \tilde{x}|}\right]$$
$$\le c||\omega||_{L^\infty}\left(\int_0^{2d} dr + \int_0^{3d} dr\right) \le c||\omega_0||_{L^\infty}|x^1 - x^2|.$$

Combining these estimates gives estimate (8.32),

$$|v^\epsilon(x^1,t) - v^\epsilon(x^2,t)| \le c|||\omega_0|||\, |x^1 - x^2|(1 - \ln^-|x^1 - x^2|),$$

so that $v^\epsilon(\cdot,t)$ is quasi-Lipschitz continuous uniformly in ϵ. □

Finally we give the proof of Lemma 8.2.

Proof of Lemma 8.2. Potential Theory Estimates for Particle Trajectories. First we prove estimate (8.33) for the backward particle trajectories. Let $Y_\epsilon(x,t;\tau)$ denote the backward particle trajectories such that $Y_\epsilon(x,t;\tau)|_{\tau=t} = X_\epsilon^{-t}(x)$. We denote $\rho(\tau) = |Y_\epsilon(x^1,t;\tau) - Y_\epsilon(x^2,t;\tau)|$. Because Y_ϵ satisfies the ODE

$$\frac{dY_\epsilon}{d\tau}(x^j,t;\tau) = -v^\epsilon[Y_\epsilon(x^j,t;\tau), t-\tau], \qquad Y_\epsilon(x^j,t;\tau)|_{\tau=0} = x^j,$$

the quasi-Lipschitz continuity of v^ϵ estimate (8.32) implies that

$$\frac{d}{d\tau}\rho(\tau) \le c|||\omega_0|||\rho(\tau)[1 - \ln^-\rho(\tau)].$$

By the substitution $z(\tau) = \ln \rho(\tau)$, this inequality reduces to the linear inequality

$$\frac{dz}{d\tau}(\tau) \leq c|||\omega_0|||[1 - z(\tau)], \qquad z(\tau)|_{\tau=0} = \ln^-|x^1 - x^2|.$$

Solving for $z(\tau)$ bounds $\rho(\tau)$ by

$$\rho(\tau) \leq e\rho(0)^{\exp(-c|||\omega_0|||\tau)}, \qquad \forall\, 0 \leq \tau \leq T^* \leq T,$$

provided that $\rho(\tau) \leq 1$. Imposing the restriction $\rho_0 \leq \exp(-\exp c|||\omega_0|||T)$, we find that the above estimate for $\rho(\tau)$ is valid for all $0 \leq \tau \leq T$. Taking, in particular, $\tau = t$, we get

$$\left|X_\epsilon^{-t}(x^1) - X_\epsilon^{-t}(x^2)\right| \leq e|x^1 - x^2|^{\exp(-c|||\omega_0|||T)}, \qquad \forall\, 0 \leq t \leq T,$$

which concludes the proof of Hölder estimate (8.33) independent of ϵ. To obtain estimate (8.34) for the forward particle trajectories, we apply the same technique as above to the forward-trajectory equation.

Finally we prove Hölder estimates (8.35) and (8.36). Let $0 \leq t_1 \leq t_2 \leq T$, so that, denoting $\alpha^* = Y_\epsilon(x, t_2; t_2 - t_1)$, we have $|X_\epsilon^{-t_1}(x) - X_\epsilon^{-t_2}(x)| = |X_\epsilon^{-t_1}(x) - X_\epsilon^{-t_1}(\alpha^*)|$. Thus

$$\begin{aligned}\left|X_\epsilon^{-t_1}(x) - X_\epsilon^{-t_2}(x)\right| &\leq e|x - \alpha^*|^{\exp(-c|||\omega_0|||T)} \\ &= e\left|\int_{t_1}^{t_2} v^\epsilon\left[X_\epsilon^\tau(\alpha^*), \tau\right] d\tau\right|^{\exp(-c|||\omega_0|||T)} \\ &\leq c|||\omega_0|||\; |t_1 - t_2|^{\exp(-c|||\omega_0 T)}. \end{aligned} \tag{8.37}$$

Estimate (8.36) follows similarly. □

8.2.4. The Uniqueness of Weak Solutions

In Subsections 8.2.2 and 8.2.3 we proved the existence of weak solutions with vorticity

$$\omega \in L^\infty[0, \infty; L^1(\mathbb{R}^2) \cap L^\infty(\mathbb{R}^2)].$$

The uniqueness of these solutions does not follow from the construction discussed in Subsections 8.2.2 and 8.2.3 – although the limit particle trajectory $X^t(\cdot)$ is unique, they may be other weak solutions ω obtained for example by different regularizations. Now we prove uniqueness of weak solutions $\omega \in L^\infty[0, \infty; L^1(\mathbb{R}^2) \cap L^\infty(\mathbb{R}^2)]$ for an initial vorticity ω_0 with a compact support. Recall from the previous subsection that the particle trajectories have bounded speed and are measure preserving. Hence any such weak solution $\omega(x, t)$ retains compact support with the measure of the support fixed. Because the $L^p(\Omega)$ spaces are ordered on bounded domains Ω, $L^{p_2}(\Omega) \hookrightarrow L^{p_1}(\Omega)$ if $p_1 \leq p_2$, we actually have $\omega(\cdot, t) \in L_0^p(\mathbb{R}^2)$ for all $1 \leq p \leq \infty$.

Our goal in this section is to prove Theorem 8.2.

Theorem 8.2. *The Uniqueness of Weak Solutions. Let the initial vorticity* $\omega_0 \in L_0^\infty(\mathbb{R}^2)$ *have compact support* $\text{supp}\omega_0 \subset \{x : |x| \le R_0\}$*. Then the weak solution* $\omega \in L^\infty[0, \infty; L_0^\infty(\mathbb{R}^2)]$ *is unique.*

Uniqueness also holds for weak solutions $\omega \in L^\infty[0, \infty; L^1(\mathbb{R}^2) \cap L^\infty(\mathbb{R}^2)]$ without compact support. The assumption supp $\omega_0 \subset \{x : |x| \le R_0\}$ simplifies the proof. As in the strategy used by Yudovich (1963),we use ordinary differential inequalities with nonunique solutions.

Lemma 8.3. *Let a weak solution (in the sense of Definition 8.1) to the 2D Euler equation* $\omega(\cdot, t) \in L_0^\infty(\mathbb{R}^2)$ *have compact support* supp $\omega(\cdot, t) \subset \{x : |x| \le R(t)\}$*. Then, for all* $1 < p < \infty$,

$$||\nabla v(\cdot, t)||_{L^p} \le c(||\omega_0||_{L^\infty})p. \tag{8.38}$$

We prove this lemma at the end of this section.

Proof of Theorem 8.2. First we show that any weak solution $\omega(x, t)$ in the sense of Definition 8.1 with the initial data $\omega_0(x)$, supp $\omega_0 \subset \{x : |x| \le R_0\}$, obeys

$$\int_{\mathbb{R}^2} \omega(x, t)dx = \int_{\mathbb{R}^2} \omega_0(x)dx. \tag{8.39}$$

Any solution $\omega(\cdot, t) \in L^1 \cap L^\infty$ has uniformly bounded velocity v, $||v(\cdot, t)||_{L^\infty} \le c|||\omega_0|||$. Because the vorticity is convected by the particle trajectories, the support has finite propagation speed so that there exists a uniformly increasing bounded function $R(t)$ such that supp $\omega(\cdot, t) \subset \{x : |x| \le R(t)\}$. Consider identity (8.23), which defines weak solutions $\omega(\cdot, t) \in L^1 \cap L^\infty$:

$$\int_{\mathbb{R}^2} \omega(x, T)\varphi(x, T)dx - \int_{\mathbb{R}^2} \omega_0(x)\varphi(x, 0)dx = \int_0^T \int_{\mathbb{R}^2} \frac{D\varphi}{Dt}\omega\, dxdt, \tag{8.40}$$

where $\varphi \in C^1\{[0, T]; C_0^1(\mathbb{R}^2)\}$. Because supp $\omega(\cdot, t) \subset \{x : |x| \le R(T)\}$ for all $0 \le t \le T$, any test function φ satisfying $\varphi(x, t) = 1$ for all $0 \le t \le T$ and $|x| \le R(T)$ gives zero for the right-hand side of identity (8.40), implying Eq. (8.39).

Now we suppose that there are two weak solutions $\omega_j, v_j = K_2 * \omega_j$, with the same initial vorticity $\omega_0 \in L_0^\infty(\mathbb{R}^2)$, supp $\omega_0 \subset \{x : |x| \le R_0\}$. The velocities v_j solve the 2D Euler equations in a distribution sense:

$$\frac{\partial}{\partial t}v_j + v_j \cdot \nabla v_j = -\nabla p_j,$$
$$\nabla \cdot v_j = 0.$$

We note that although each velocity v_j does not have a finite energy, their difference $w = v_1 - v_2$ does:

$$E(t) \equiv \int_{\mathbb{R}^2} (w(x, t))^2 dx < \infty. \tag{8.41}$$

We used this same idea in Chap. 3 to construct a radial-energy decomposition (Definition 3.1) for smooth solutions to the Euler and the Navier–Stokes equation.

Because supp $\omega_j(\cdot, t) \subset \{x : |x| \leq R_j(T)\}$, by using an asymptotic expansion for the kernel K_2 we have

$$v_j(x,t) = \frac{c}{|x|}\int_{\mathbb{R}^2} \omega_j(y,t)dy + \mathcal{O}\big(|x|^{-2}\big) \qquad \text{for} \quad |x| \geq 2R_j(T),$$

so that by Eq. (8.39) the integral terms in $v_1 - v_2$ cancel and $w(x,t) = \mathcal{O}(|x|^{-2})$ for $|x| \geq 2\max_j R_j(T)$. This implies Eq. (8.41).

Because $w = v_1 - v_2$ has finite energy E and satisfies

$$w_t + v_1 \cdot \nabla w + w \cdot \nabla v_2 = -\nabla(p_1 - p_2)$$

in the sense of distributions, taking the L^2 inner product of this equation with w and integrating by parts gives

$$\frac{1}{2}\frac{d}{dt}E(t) - \int_{\mathbb{R}^2} w^2\, \nabla \cdot v_1\, dx + \int_{\mathbb{R}^2} (w \cdot \nabla v_2) w\, dx = \int_{\mathbb{R}^2} (p_1 - p_2)\, \nabla \cdot w\, dx.$$

The Hölder inequality implies that

$$\begin{aligned}\frac{d}{dt}E(t) &\leq 2\int_{\mathbb{R}^2} w^2 |\nabla v_2| dx\\ &\leq 2||\nabla v_2||_{L^p}\left(\int_{\mathbb{R}^2} |w|^{2p/(p-1)} dx\right)^{(p-1)/p}\\ &\leq 2||\nabla v_2||_{L^p}\left(||w(\cdot,t)||_{L^\infty}^{2/(p-1)}\int_{\mathbb{R}^2} |w|^2 dx\right)^{(p-1)/p}.\end{aligned}$$

Estimate (8.38) applied to $||\nabla v_2||_{L^p}$ and the fact that the velocities v_j are uniformly bounded yields

$$\frac{d}{dt}E(t) \leq pM\, E(t)^{1-1/p}, \tag{8.42}$$

where $M = c(||\omega_0||_{L^\infty})||\omega_0||_{L^\infty}$.

We want to conclude that $E(t) \equiv 0$ for all $t > 0$. Because $E(0) = 0$, $E(T) = 0$ is a trivial solution to inequality (8.42). However, this inequality does not have unique solutions (see, e.g., Hartman 1982, p. 33, exercise 6.4). However, the maximal solution $\bar{E}(t)$ to (2.24) is $\bar{E}(t) = (Mt)^p$ and any solution $E(t)$ satisfies $E(t) \leq \bar{E}(t)$.

Now we take an interval $[0, T^*]$ such that $MT^* \leq \frac{1}{2}$. Passing to the limit as $p \nearrow \infty$ we have

$$E(t) \leq \left(\frac{1}{2}\right)^p \searrow 0 \qquad \text{as} \quad p \nearrow \infty, \tag{8.43}$$

so $E(t), \equiv 0$, $\forall\, 0 \leq t \leq T^*$. Repeating these arguments, we conclude that $E(t) = 0$, $\forall\, 0 \leq t \leq T$, so that $v_1 = v_2$ almost everywhere. □

Finally, it remains to prove the crucial potential theory estimate in Lemma 8.3. From Proposition 2.17 on differentiating kernels of degree $1 - N$, we see that

$$\nabla v(x) = \text{PV}\int_{\mathbb{R}^2} \nabla K_2(x-y)\omega(y)dy + c\omega(x) \equiv P\omega(x) + c\omega(x), \tag{8.44}$$

where P is a (matrix) SIO, $P \in$ SIO and PV$\int$ denotes the principal-value integral. In our case the vorticity $\omega \in L_0^\infty(\mathbb{R}^2)$ has a compact support supp $\omega \subset \{x : |x| \leq R\}$. We can immediately prove a rough version of Lemma 8.3 by using the Calderon–Zygmund inequality

$$||P\omega||_{L^p} \leq c_p ||\omega||_{L^p}, \qquad 1 < p < \infty, \tag{8.45}$$

with the rough estimate of the constant c_p given by (see Stein, 1970)

$$c_p \leq \begin{cases} c/(p-1), & 1 < p < 2 \\ cp, & 2 \leq p < \infty \end{cases}.$$

We see that from relation (8.43) that we need Lemma 8.3 for only sufficiently high p to prove the uniqueness of weak solutions with compact support. Thus the Calderon–Zygmund estimate is sufficient for our purposes here. For completeness and also to gain some more insight into the behavior of weak solutions we now prove a sharper estimate by using specific properties of SIOs.

Below we present arguments that explicitly use the properties of SIOs on the function space L^∞. This case is not included in Calderon–Zygmund inequality (8.45).

First we introduce the space of functions of a bounded-mean oscillation (BMO) as defined by John and Nirenberg (1961). Suppose that f is defined on $\mathbb{R}^N$. We say that $f \in$ BMO if there exists a constant M_f such that

$$\frac{1}{\text{meas } Q} \int_Q |f(x) - f_Q| dx \leq M_f \tag{8.46}$$

for every cube $Q \subset \mathbb{R}^N$, where f_Q is the mean value of f over Q and $f_Q = \int_Q f(x)dx$.

The boundedness of f is not necessary for the boundedness of its mean oscillation, for example, $\ln|x| \in$ BMO. Let f be any (locally) integrable function with the property that we can associate with every subcube Q a value f_Q such that

$$m_f(Q, \lambda) \equiv \text{meas}\{x \in Q : |f(x) - f_Q| \geq \lambda\} \leq c_1 e^{-c_2 \lambda} \text{ meas } Q,$$

where $\lambda > 0$. Then

$$\int_Q |f(x) - f_Q| dx = \int_0^\infty m_f(Q, \lambda) d\lambda \leq \frac{c_1}{c_2} \text{ meas } Q,$$

so that the mean oscillation of f does not exceed c_1/c_2.

The above result is also true in the opposite direction, and we have the following lemma.

Lemma 8.4. *Given $f \in$ BMO, let $m_f(Q, \lambda)$ be the measure of the set of points $x \in Q$, where $|f(x) - f_Q| \geq \lambda$:*

$$m_f(Q, \lambda) = \text{meas}\{x \in Q : |f(x) - f_Q| \geq \lambda\}. \tag{8.47}$$

There exist constants $c_1, c_2 > 0$ independent of f such that

$$m_f(Q,\lambda) \le c_1 \text{ meas } Q \exp\left(-c_2 \frac{\lambda}{M_f}\right) \tag{8.48}$$

for every cube $Q \subset \mathbb{R}^N$.

The proof of this lemma is based on a decomposition of integrable functions (see John and Nirenberg, 1961).

The space BMO is very important in the theory of SIOs because it often arises as a substitute for the space L^∞. For preciseness, we define the SIO to be the set of operators of the form

$$P(f) = \text{PV}\int_{\mathbb{R}^n} K_2(x-y)f(y)dy,$$

where PV$\int$ denotes a principal-value integral centered at the origin. Here we require the kernel $K_2(y)$ to be of the form $\Omega(y)/|y|^n$, where Ω is homogeneous of degree zero and smooth. We then have (see Stein, 1967) Lemma 8.5.

Lemma 8.5. *Let P be a* SIO, *$P \in$* SIO. *Then*

$$P : L_0^\infty \to \text{BMO}. \tag{8.49}$$

Proof of Lemma 8.5. Given $f \in L_0^\infty$, let R be large enough so that supp $f \in B_R(0)$. We fix a cube Q that we may assume is centered at the origin. We split $f = f_1 + f_2$, where

$$f_1(x) = \begin{cases} f(x), & |x| \le 2 \text{ diam } Q \\ 0, & \text{otherwise} \end{cases}.$$

The Hölder and the Calderon–Zygmund inequalities imply that

$$\int_Q |Pf_1(x)|dx \le (\text{meas } Q)^{1/2}||Pf_1||_{L^2(Q)} \le c(\text{meas } Q)^{1/2}||f_1||_{L^2(Q)}.$$

However, $||f_1||_{L^2(Q)} \le ||f_1||_{L^\infty}^{1/2}(\text{meas } Q)^{1/2}$, so that

$$\frac{1}{\text{meas } Q}\int_Q |Pf_1(x)|dx \le c||f_1||_{L^\infty}^{1/2}.$$

Next we set

$$f_Q = -\int_{\mathbb{R}^2} P(y)f_2(y)dy.$$

This integral exists in the view of our assumptions: P is locally integrable away from the origin, and f_2 is supported away from the origin. Therefore

$$|Pf_2(x) - f_Q| \le \int_{\mathbb{R}>|y|\ge 2 \text{ diam } Q} |P(x-y) + P(y)|\,|f_2(y)|dy.$$

Claim.

$$\int_{\mathbb{R}>|y|\geq 2|x|} |P(x-y)+P(y)|dy \leq c, \tag{8.50}$$

so that

$$|Pf_2(x) - f_Q| \leq c||f_2||_{L^\infty} \qquad \text{if} \quad x \in Q.$$

Combining the above estimates gives

$$\frac{1}{\text{meas } Q}\int_Q |Pf(x) - f_Q|dx \leq c||f||_{L^\infty} + c||f||_{L^\infty}^{1/2},$$

which concludes the proof, provided that we can prove relation (8.50). P is a smooth function, homogeneous of degree $-n$, so that we can write $P = p(x)/|x|^n$, where p is smooth and homogeneous of degree zero. We have

$$\begin{aligned} P(x-y)+P(y) &= \frac{p(x-y)+p(y)}{|x-y|^n} - p(y)\left[\frac{1}{|x-y|^n} - \frac{1}{|y|^n}\right] \\ &= r_1(x,y) + r_2(x,y). \end{aligned}$$

Note that if $|y| \geq |2x|$,

$$\begin{aligned} |p(x-y)+p(y)| &= \left|p\left(\frac{x-y}{|x-y|}\right) + p\left(\frac{y}{|y|}\right)\right| \\ &\leq c\left|\frac{x-y}{|x-y|} - \frac{y}{|y|}\right| \\ &\leq c\left|\frac{x}{y}\right|. \end{aligned} \tag{8.51}$$

Thus $r_1(x, y)$ is bounded by a constant for $|y| \geq |2x|$. Note that $\int_{|y|\geq|2x|} |r_2(x,y)|dy$ is bounded in magnitude by

$$\int_{\mathbb{R}>|y|\geq 2|x|} \left|\frac{1}{|x-y|^n} - \frac{1}{|y|^n}\right|dy,$$

which is also bounded by a constant. □

Finally, by combining Lemmas 8.4 and 8.5, we are ready to give the proof of Lemma 8.3.

Proof of Lemma 8.3. Recalling Eq. (8.44), we have

$$\nabla v(x) = \text{PV}\int_{\mathbb{R}^2} \nabla K_2(x-y)\omega(y)dy + c\omega(x) \equiv \nabla v^{(1)}(x) + \nabla v^2(x), \tag{8.52}$$

where $\text{supp } \omega \subset \{x : |x| \leq R\}$. The second term $\nabla v^2(x)$ satisfies the crude estimate

$$||\nabla v^2||_{L^p} \leq c||\omega||_{L^p} \leq cp||\omega||_{L^\infty}, \qquad 1 < p < \infty.$$

For the first term, $\nabla v^{(1)}(x)$, we split the L^p norm into

$$||\nabla v^{(1)}||^p_{L^p} = ||\nabla v^{(1)}||^p_{L^p(|x|\le 2R)} + ||\nabla v^{(1)}||^p_{L^p(|x|\ge 2R)}.$$

First observe that if $f \in \text{BMO}$; then, by the definition of $m_f(Q, f)$ in Eq. (8.47),

$$m_f(Q, f) = \text{meas}\{x \in Q : |f(x) - f_Q| \ge \lambda\},$$

and by integration by parts,

$$||f||^p_{L^p(Q)} \equiv \int_Q |f(x)|^p\, dx = -\int_0^\infty \lambda^p dm_f(Q, \lambda) = p\int_0^\infty \lambda^{p-1} m_f(Q, \lambda) d\lambda.$$

Lemma 8.5 implies that $\nabla K_2 : L_0^\infty \to \text{BMO}$, so that the characterization of BMO in Lemma 8.4 gives

$$m_{\nabla K_2 \omega}(Q, \lambda) \le c_1 \text{ meas } Q \exp\left(-c_2 \frac{\lambda}{M_{\nabla K_2 \omega}}\right).$$

Taking a fixed cube $Q \supset \{x : |x| \le 2R\}$, we estimate

$$||\nabla v^{(1)}||^p_{L^p(|x|\le 2R)} \le ||\nabla v^{(1)}||^p_{L^p(Q)} \le c_1 p \text{ meas } Q \int_0^\infty \lambda^{p-1} \exp\left(-c_2 \frac{\lambda}{M_{\nabla K_2 \omega}}\right) d\lambda.$$

The last integral looks like the $\Gamma(p)$ function, so by using the fact that $\Gamma(1) = 1$, $\Gamma(p+1) = p!$, we get the crude estimate

$$p\int_0^\infty \lambda^{p-1} \exp\left(-\frac{c_2 \lambda}{M_{\nabla K_2 \omega}}\right) d\lambda \le cp!,$$

and

$$||\nabla v^{(1)}||^p_{L^p(|x|\le 2R)} \le c(R)p!. \tag{8.53}$$

Because $|\nabla K_2(x)| \le c|x|^{-N}$,

$$|\nabla v(x)| \le \frac{c}{|x|^{-2}} \int_{\mathbb{R}^2} |\omega(y)| dy + 0(|x|^{-3}), \qquad |x| \ge 2R,$$

$$||\nabla v^{(1)}||^p_{L^p(|x|\ge 2R)} \le c||\omega||^p_{L^1} \int_{|x|\ge 2R} |x|^{-2p}\, dx \le \frac{c(R)}{2(p-1)} ||\omega||^p_{L^1}.$$

Thus

$$||\nabla v^{(1)}||^p_{L^p(|x|\ge 2R)} \le cp! ||\omega||^p_{L^\infty}. \tag{8.54}$$

Combining relations (8.53) and (8.54) yields

$$||\nabla v^{(1)}||_{L^p} \le c(||\omega_0||_{L^\infty})(p!)^{1/p} \le c(||\omega_0||_{L^\infty})p. \qquad \square$$

8.2.5. Propagation of Regularity

So far we have proved the existence and the uniqueness of weak solutions $\omega \in L^\infty[0, \infty; L^1(\mathbb{R}^2) \cap L^\infty(\mathbb{R}^2)]$. Now we want to consider the following problem: Suppose that the initial vorticity ω_0 is smooth on some open set $\Omega(0) \subset \operatorname{supp} \omega_0$. Does the weak solution $\omega(\cdot, t)$ remain smooth on the set $\Omega(t) = \{X^t(\alpha) : \alpha \in \Omega(0)\}$? This is in fact the case.

Proposition 8.3. *Let the initial vorticity $\omega_0 \in L^1(\mathbb{R}^2) \cap L^\infty(\mathbb{R}^2)$, and $\omega_0|_{\Omega(0)} \in C^{m,\gamma}[\bar{\Omega}(0)]$, $m \geq 0$, $0 < \gamma < 1$, where $\Omega(0)$ is a bounded open subset of $\mathbb{R}^2$. Denote $\Omega(t) = \{X^t(\alpha) : \alpha \in \Omega(0)\}$. Then for all $t \geq 0$*

$$\omega(\cdot, t)|_{\Omega(t)} \in C^{m,\gamma}_{\text{loc}}[\Omega(t)],$$
$$v(\cdot, t)|_{\Omega(t)} \in C^{m+1,\gamma}_{\text{loc}}[\Omega(t)].$$

Throughout this subsection we use $C^{m,\gamma}(\Omega)$ to indicate the space of functions $f \in C^m(\Omega)$ satisfying

$$\sup_{x \neq y,\ x, y \in \Omega} \left| \frac{f^m(x) - f^m(y)}{x - y} \right| \leq \infty,$$

and $C^{m,\gamma}_{\text{loc}}(\Omega)$ to indicate the space of functions $f \in C^{m,\gamma}(K_2)$ for all compact $K_2 \subset \Omega$. These Hölder spaces were introduced in Chap. 4 for the study of the existence and the uniqueness of smooth Euler solutions by means of particle-trajectory methods.

Proof of Proposition 8.3. We proceed inductively in m, starting with $m = 0$.

Step 1: $\omega \in C^{0,\gamma\beta}[\bar{\Omega}(t)]$.

Suppose that $\omega_0|_{\Omega(0)} \in C^{0,\gamma}[\bar{\Omega}(0)]$, $0 < \gamma < 1$. Because the potential theory estimates in Lemma 8.2 are uniform in ϵ, the limit particle trajectory $X^{-t}(x)$ also satisfies estimates (8.33) for all $t \in [0, T]$:

$$|X^{-t}(x^1) - X^{-t}(x^2)| \leq c|x^1 - x^2|^\beta,$$

so that for $t \in [0, T]$, $X^{-t}(\cdot) \in C^{0,\beta(T)}[\bar{\Omega}(t)]$. Here the exponent $0 < \beta < 1$ is rapidly deteriorating with time, $\beta(T) = \exp(-c|||\omega_0|||T)$. Now, by using the Hölder continuity of $\omega_0|_{\Omega(0)}$, we estimate for $x^1, x^2 \in \bar{\Omega}(t)$, $t \in [0, T]$,

$$\begin{aligned} |\omega(x^1, t) - \omega(x^2, t)| &= |\omega_0[X^{-t}(x^1)] - \omega_0[X^{-t}(x^2)]| \leq c|X^{-t}(x^1) - X^{-t}(x^2)|^\gamma \\ &\leq c|x^1 - x^2|^{\gamma\beta}. \end{aligned}$$

Thus for all $t \in [0, T]$, $\omega \in C^{0,\gamma\beta}[\bar{\Omega}(t)]$.

Step 2: $v \in C^{1,\gamma\beta}_{\text{loc}}[\Omega(t)]$.

We now show that $v(\cdot, t)|_{\Omega(t)} \in C^{1,\gamma\beta}[\Omega(t)]$ for all $t \in [0, T]$. We note that there exists a stream function $\Psi(t)$ such that in $\Omega(t)$, $v = \nabla^\perp \Psi$ and $\omega = \Delta\Psi(t)$ (Ψ is just the Newtonian potential of ω in $\mathbb{R}^2$). We apply the following proposition on the elliptic regularity of Hölder spaces.

Proposition 8.4. *Let* $\Psi \in C^0(\mathbb{R}^2)$ *satisfy*

$$\Delta\Psi = \omega$$

where $\omega \in C^{m,\alpha}_{\text{loc}}(\Omega)$ *for some* $\alpha > 0$; Ω *is an open subset of* $\mathbb{R}^n$. *Then* $\Psi \in C^{m+2,\alpha}_{\text{loc}}(\Omega)$. *Furthermore, if* $v = \nabla^\perp\Psi$, *then*

$$|v|_{C^{m+1,\gamma}(\Omega_{2\epsilon})} \leq C(\epsilon, m, \gamma)\left[|v|_{C^{m-1}(\Omega_{2\epsilon})} + |\omega|_{C^{m,\gamma}(\Omega_\epsilon)}\right],$$

where

$$\Omega_\epsilon = \{x \in \Omega | B_\epsilon(x) \in \Omega\}.$$

This is a direct result of Theorems 4.6 and 6.13 in Gilbarg and Trudinger (1998). Proposition 8.4 tells us that, in general, $\omega \in C^{n,\gamma\beta}[\bar{\Omega}(t)]$ implies that $v \in C^{n+1,\gamma\beta}_{\text{loc}}[\Omega(t)]$. In particular, for this case, for all $t \in [0, T]$, $v(\cdot, t) \in C^{1,\beta\gamma}_{\text{loc}}[\Omega(t)]$. We note, however, that this proposition is not valid when $\alpha = 0$.

Step 3: $X(\alpha, t) \in C^1[\Omega(0)]$.

We now show that $v \in C^{1,\gamma\beta}[\Omega(t)]$ for all $t \in [0, T]$ implies that $X(\alpha, t) \in C^1[\Omega(0)]$ for all $t \in [0, T]$. We use the following proposition that is a special case of more general standard ODE theorems (see Hartman, 1982).

Proposition 8.5. *Let* $f(\cdot, t) \in C\{[0, T); C^n[\Omega(t)]\}$, *where* $\Omega(t)$ *is an open subset of* R^n. *We also assume that* $\{(x, t) \in \mathbb{R}^n \times [0, T)|x \in \Omega(t)\}$ *has a continuous boundary in* $R^n \times [0, \infty)$. *Let* $X(\cdot, t)$ *be the unique solution of*

$$\frac{dX(\alpha, t)}{dt} = f(X(\alpha, t), t)$$

on the time interval $[0, T)$, *with initial data* $X(\alpha, 0) = \alpha$. *Furthermore, assume that* $X(\alpha, t) \in \Omega(t)$ *for all* $\alpha \in \Omega(0)$. *Then* $X(\cdot, t) \in C\{[0, \infty); C^n[\Omega(t)]\}$.

We apply this proposition to the case $n = 1$ with $f(x, t) = v(x, t)$. We know from relation (8.37) that $X(\alpha, t)$ is continuous in t; hence $\omega(x, t) = \omega_0(X^{-t}(x))$ is continuous in t and thus $v(x, t)$ is continuous in t. Also, we have that $\{(x, t) \in \mathbb{R}^n \times [0, \infty)|x \in \Omega(t)\}$ has a continuous boundary. Thus Proposition 8.5 gives us $X(\alpha, t) \in C^1[\Omega(0)]$ for all $t \in [0, T]$.

Step 4: $\omega \in C^\gamma[\Omega(t)]$, $v \in C^{1,\gamma}_{\text{loc}}[\Omega(t)]$.

Using the backward particle-trajectory equations, we obtain the above result for $X^{-t}(x)$, that is, $X^{-t}(x) \in C^1[\Omega(t)]$. Hence, for all $t \in [0, T]$, $\omega(x, t) \in C^\gamma_{\text{loc}}[\Omega(t)]$, and thus, by Proposition 8.4, $v(x,t) \in C^{1,\gamma}_{\mathrm{loc}}[\Omega'(t)]$ for all $t \in [0,T]$ and for any $\Omega'(t)$ with closure in $\Omega(t)$; hence also $v(x,t) \in C^{1,\gamma}_{\mathrm{loc}}[\Omega(t)]$.

The above arguments work for any $T > 0$, with of course a different β in the proof; hence $\omega \in C^{\gamma}_{\mathrm{loc}}[\Omega(t)]$ for all $t > 0$, $v \in C^{1,\gamma}_{\mathrm{loc}}[\Omega(t)]$ for all $t > 0$.

Remark: We note that, in the Proof of Proposition 8.5, the particle trajectories are in $C^1[\bar{\Omega}'(0)]$ for $\bar{\Omega}' \subset \Omega$. However, in general, we do not have $X(\alpha,t) \in C^1[\bar{\Omega}(0)]$. That is, we do not gain any information about regularity of the boundary other than what we already know from the Hölder continuity of the particle trajectories. In particular, the above regularity results do not yield any new information about the behavior of the boundary of a vortex patch. In Subsection 8.3.3 we derive estimates for the tangent vector to the patch boundary to prove global regularity of the boundary given an initially smooth boundary.

We now consider the case $m = 1$.

We first follow the arguments in step 4 to see that $\omega(x) = \omega_0(X^{-t}(x))$ combined with the fact that $\omega_0 \in C^{1,\gamma}[\Omega(0)]$ implies that $\omega \in C^1[\Omega(t)]$ for all time. Unfortunately, Proposition 8.4 does not hold for $\alpha = 0$ and we cannot immediately show that $v \in C^2[\Omega(t)]$. We must first show that $\omega \in C^{1,\gamma\beta}_{\mathrm{loc}}[\Omega(t)]$.

To obtain this result, we proceed by showing that the backward particle trajectories X^{-t} are in $C^{1,\gamma\beta}_{\mathrm{loc}}[\Omega(t)]$. To do this, we recall the mollified initial data ω_0^ϵ, the resulting smooth solutions $\omega^\epsilon(x,t)$, $v^\epsilon(x,t)$, and the corresponding particle trajectories X^t_ϵ.

Claim 1. *$|\omega_0^\epsilon|_{C^{1,\gamma}(\Omega_0^\epsilon)} \le |\omega_0|_{C^{1,\gamma}(\Omega_0)}$, where $\Omega_0^\epsilon = \{x \in \Omega(0) | B_\epsilon(x) \in \Omega(0)\}$ and we have taken one to be the radius of the support of the function ρ defining the mollification of the initial data:*

$$\omega_0^\epsilon = \frac{1}{\epsilon^2}\rho\left(\frac{x}{\epsilon}\right) * \omega_0.$$

This is an elementary calculus exercise, and the proof is left to the reader.

Claim 2. *For all ϵ_0 there exists a uniform bound in $\epsilon < \epsilon_0$ and $t \in [0,T]$ on the following quantities:*

$$|v^\epsilon|_{C^{1,\gamma\beta}[\Omega^{2\epsilon_0}(t)]},$$
$$|X^\epsilon(\alpha,t)|_{C^1[\Omega^{2\epsilon_0}(t)]},$$

where $\Omega^{\epsilon_0}(t) = X^t_\epsilon(\Omega_0^{\epsilon_0})$.

Proof of Claim 2. We know from the above that for all $\epsilon < \epsilon_0$, $|\omega_0^\epsilon|_{C^{1,\gamma}(\Omega_0^{\epsilon_0})} \le |\omega_0|_{C^{1,\gamma}(\Omega_0)}$. Furthermore, from Lemma 8.2 we have a uniform bound on $|X^t_\epsilon|_{C^{0,\beta(T)}(\mathbb{R}^2)}$ on the time interval $[0,T]$. Combining this with Claim 1, we obtain a uniform bound on $|\omega^\epsilon|_{C^{0,\gamma\beta}[\Omega^{\epsilon_0}(t)]}$ and from Proposition 8.4 a uniform bound on $|v^\epsilon|_{C^{1,\gamma\beta}[\Omega^{2\epsilon_0}(t)]}$. To obtain the uniform bound on X_ϵ we use the fact that $(\partial X_\epsilon/dt) = v^\epsilon(X_\epsilon)$ combined with the above estimates. □

Claim 3. *$|X_\epsilon(\alpha, t)|_{C^{1,\beta\gamma}[\Omega^{2\epsilon_0}(0)]}$ is uniformly bounded for all $\epsilon < \epsilon_0$ and $t \in [0, T]$.*

Proof of Claim 3. We have

$$\begin{aligned}\frac{d}{dt}|X_\epsilon(\alpha, t)|_{C^{1,\gamma\beta}[\Omega^\epsilon(0)]} &\le |v^\epsilon(X_\epsilon(\alpha, t), t)|_{C^{1,\gamma\beta}[\Omega^\epsilon(0)]}\\ &\le |v^\epsilon(x, t)|_{C^1[\Omega^\epsilon(t)]}|X_\epsilon(\alpha, t)|_{C^1[\Omega(t)]}\\ &\quad + |v^\epsilon(x, t)|_{C^{1,\beta\gamma}[\Omega^\epsilon(t)]}|X_\epsilon(\alpha, t)|^2_{C^1[\Omega(0)]}\\ &\quad + |v^\epsilon(x, t)|_{C^1[\Omega^\epsilon(t)]}|\nabla_\alpha X_\epsilon(\alpha, t)|_{C^{1,\beta\gamma}[\Omega(0)]}\\ &\le C_1|x_\epsilon(\alpha, t)|_{C^{1,\gamma\beta}[\Omega^\epsilon(0)]} + C_2.\end{aligned}$$

Thus $|X_\epsilon(\alpha, t)|_{C^{1,\beta\gamma}[\Omega^\epsilon(0)]}$ grows at most exponentially with a uniform growth rate. □

We now define $\Omega_\epsilon(t) = \{x \in \Omega(t)|B_\epsilon(x) \in \Omega(t)\}$. We recall that we have a subsequence $X_\epsilon^{-t}(x)$ that converges uniformly on $\{(x, t)|x \in \Omega(t)\}$ to $X^{-t}(x)$. We can redefine the parameter ϵ so that $|X_\epsilon^{-t}(x) - X^{-t}(x)| \le \epsilon$. By combining this estimate with Lemma 8.2, we see that, for all ϵ,

$$\Omega^\epsilon(t) \subset \Omega_{\varphi(\epsilon)}(t) \subset \Omega^{\varphi^2(\epsilon)}(t),$$

where $\varphi(\epsilon) = 2(\epsilon/C)^{[1/\beta(T)]}$. Given this fact, a similar argument to that described in Claim 3 reveals that we also have, for fixed ϵ_0, a uniform bound {in $\epsilon < \epsilon_0$ and $t \in [0, T]$} for $|X_\epsilon^{-t}(\cdot)|_{C^{1,\beta\gamma}[\Omega_\epsilon(t)]}$ and a uniform bound [in $\epsilon < \epsilon_0$ and $x \in \Omega_{\epsilon_0}(t)$] on $|\nabla_x X_\epsilon^{-t}(x)|_{C^{\gamma\beta}([0,T])}$. The upshot is that the family $\{\nabla X_\epsilon^{-t}\}_{\epsilon \le \epsilon_0}$ forms an equicontinuous family on $\{(x, t)|x \in \Omega^{\epsilon_0}(t)\}$. Thus by the Arzela–Ascoli theorem, there exists a subsequence that converges uniformly to some limit. It is a simple analysis exercise to show that that limit must indeed be $\nabla_x X^{-t}$. As in the Hölder estimates for the particle trajectories (Lemma 8.2), the above uniform bounds also carry through to the limit and we have for all ϵ, $X^t \in C^{1,\gamma\beta(T)}(\Omega_0^\epsilon)$, $X^{-t} \in C^{1,\gamma\beta(T)}[\Omega_\epsilon(t)]$. Because this is true for all ϵ, we have for all $t \in [0, T]$, $\omega(x) \in C^{1,\gamma\beta(T)}[\Omega(t)]$. We now follow parallel arguments to steps 2–4 of the case $m = 0$ to obtain the final result for the case $m = 1$.

A bootstrapping argument proves Proposition 8.3 for any m. □

8.3. Vortex Patches

In Section 8.2 we introduced the notion of a 2D L^∞ weak solution. As in the case of smooth solutions, the weak solutions have well-defined particle trajectories along which the vorticity is preserved. If we start at some time t_0 with a vorticity field that is constant inside some region of the fluid, the region evolves in time, keeping the same constant vorticity inside. This fact leads to a natural class of weak solutions known as *patches of constant vorticity*. The Kirchoff ellipse (Example 8.1) introduced in Section 8.1 is an example of a vortex patch. In general, a vortex patch has an L^∞

vorticity distribution of the form

$$\begin{aligned} \omega(x,t) &= \omega_0, & x &\in \Omega(t), \\ &= 0, & x &\notin \Omega(t), \end{aligned} \tag{8.55}$$

where $\Omega(t) = \{X^t(\alpha) : \alpha \in \Omega_0\}$ is the "patch" that moves with the flow and X^t are the particle trajectories.

The extensive study of vortex patches over the past two decades has led to a number of interesting observations regarding the structure of the boundary of a vortex patch. The uniformly rotating Kirchoff ellipses of Section 8.1 are very special cases. In general, vortex-patch boundaries can deform quite dramatically despite the conservation of the area of the patch. For example, the length or curvature of the boundary may grow rapidly (Alinhac, 1991; Constantin and Titi, 1988), Majda (1986) proposed the vortex-patch problem, in contour dynamics form (below), as a model for the inviscid, incompressible creation of small scales. Motivated by analogy with the stretching of vorticity in three dimensions and by a simple model (Constantin et al., 1985), he suggested the possibility of finite-time singularities. In other words, some smooth initial contours might, in finite time, lead to loss of regularity (infinite length, corners, or cusps, for instance). At the end of this chapter we prove that such a possibility cannot occur.

If the boundary of the patch is sufficiently smooth, then the velocity field induced by the patch can be completely determined by the position of the boundary. This follows from the Green's theorem, and we shall prove it in detail in Subsection 8.3.1. In the event that the boundary is at least piecewise C^1, the motion of the boundary of the patch satisfies the CDE (first derived in Zabusky et al., 1979):

$$\frac{dz}{dt}(\alpha,t) = -\frac{\omega_0}{2\pi}\int_0^{2\pi} \text{In}|z(\alpha,t) - z(\alpha',t)|z_\alpha(\alpha',t)d\alpha', \tag{8.56}$$

$$z(\alpha,t)|_{t=0} = z_0(\alpha). \tag{8.57}$$

Here $z(\alpha,t)$ is a Lagrangian parameterization of the curve at time t, that is, $z(\alpha,t) = X^t[z(\alpha,0)]$, where X^t is that particle-trajectory map for the flow. In addition to deriving the CDE, Subsection 8.3.1 contains a proof of the fact that any solution to the CDE defines a unique weak L^∞ solution to the 2D Euler equation. We note that Eq. (8.56) is valid only for patches that are simply connected. The equation for a nonsimply connected patch has the same form as that of Eq. (8.56) with a sum of integrals on the right-hand side. The results of this section can be directly extended to the case of a patch that is not simply connected and even to the case of multiple patches. They are also true for patches in domains other than $\mathbb{R}^2$, such as vortex patches on the surface of a sphere and vortex patches with periodic boundary conditions. For simplicity, we consider a single simply connected domain in our analysis.

In Subsection 8.3.1 we use the material from Section 8.2 on general weak L^∞ solutions to formulate some basic facts about vortex patches. Also, we show there that the CDE, as an integrodifferential equation, parallels the Euler equation of 3D incompressible flow. The results from Section 8.2 merely ensure that the boundary of a patch of constant vorticity stays Hölder continuous with a decaying Hölder exponent. A natural question to ask is whether the boundary of such a patch stays smooth for all time or for even a finite time if initially smooth. In Subsection 8.3.2, by using the contour dynamics formulation, we show that an initially C^∞ boundary stays C^∞ for a finite time. The strategy is similar to that of Chap. 4, in which we proved local existence of smooth solutions to the 3D Euler equation (see Theorem 4.3). We use the CDE here in the same way that we used the integrodifferential equations for the particle trajectories of Eq. (4.6) in Chap. 4. We then show in Subsection 8.3.3 that an argument similar to that used to prove Theorem 4.3 proves global-in-time smoothness of the boundary of the vortex patch.

The CDE is just one example of a self-deforming curve equation. There are numerous other processes with representations as self-deforming curve equations in which the physics present is drastically different from that of the vortex patch. For example, plasma clouds in the ionosphere can be modeled by constant-density regions whose evolution is governed by the motion of the boundary through a self-deforming curve equation (Berk and Roberts, 1970). A simple model is that of the "electron patches" for the 1D Vlasov–Poisson equation. These are layers of constant electron density in a noncollisional plasma of charged particles. For smooth-enough boundaries, we can derive a contour dynamics formulation for the evolution of these constant-density regions. We discuss briefly the "electron-patch" solutions (Subsection 13.1.2) and in detail analogous "electron sheets" in Chap. 13. Other examples of self-deforming curve problems not discussed in this text include Hele–Shaw flows (Meiburg and Homsy, 1988) that describe the displacement of a less viscous fluid by a more viscous fluid in which surface tension plays a critical role. For such problems the dynamics of topological transitions in the fluid is an ongoing subject of research (Almgren, 1996; Almgren et al., 1996; Goldstein et al., 1993, 1995; Shelley, 1995; Constantin et al., 1993).

8.3.1. Derivation of the Contour Dynamics Equation and Elementary Properties of Vortex Patches

Derivation of the CDE

We begin with a derivation of CDE (8.57) for the evolution of the boundary of a vortex patch. We follow arguments similar to those used in Section 2.5 of Chap. 2 to derive integrodifferential equations (2.118) and (2.119) for the particle trajectories. The main difference is that the simple form of the vorticity allows us to use the Green's theorem to rewrite the integral over all of $\mathbb{R}^2$ as an integral over the boundary of the patch. Let $\Omega_0 \in \mathbb{R}^2$ be a bounded, simply connected region with smooth boundary $\partial\Omega_0$. We consider the weak solution $\omega(\cdot, t) \in L^1(\mathbb{R}^2) \cap L^\infty(\mathbb{R}^2)$ with initial vorticity

$$\begin{aligned} \omega_0(x) &= \omega_0, && x \in \Omega_0, \\ &= 0, && x \notin \Omega_0. \end{aligned} \tag{8.58}$$

The results of Section 8.2 guarantee a unique weak solution in the sense of Definition 8.1 with initial data (8.58). For all time $t > 0$, the vorticity ω satisfies

$$\begin{aligned} \omega(x, t) &= \omega_0, && x \in \Omega(t), \\ &= 0, && x \notin \Omega(t), \end{aligned} \tag{8.59}$$

where the patch $\Omega(t)$ moves with the flow,

$$\Omega(t) = \{X(\alpha, t) : \alpha \in \Omega_0\},$$

and X are the particle trajectories corresponding to v.

Note that because the vorticity has such a simple form, the solution $\omega(\cdot, t)$ is completely determined by the evolution of the boundary $\partial\Omega(t)$. At a given time $t > 0$, $\partial\Omega(t) = \{X(\alpha, t) : \alpha \in \partial\Omega_0\}$. Hence a natural choice of parameterization for the boundary of the patch is the Lagrangian one, $z(\alpha, t) = X[z_0(\alpha), t]$, $0 \le \alpha < 2\pi$ on $\partial\Omega(t)$. We assume for now that $z(\cdot, t)$ is C^1 smooth. We make further restrictions on $z_0(\alpha)$ in later subsections as they become necessary. As in the case of smooth 2D Euler flows, the stream function $\psi(x, t)$ satisfies $v = (-\psi_{x_2}, \psi_{x_1})^t$ and

$$\begin{aligned} \Delta\psi &= \omega_0, \qquad x \in \Omega(t), \\ &= 0, \qquad x \notin \Omega(t). \end{aligned}$$

The Biot–Savart law gives the explicit formula

$$v(x, t) = \frac{\omega_0}{2\pi} \int_{\Omega(t)} \left(-\frac{\partial}{\partial x_2}, \frac{\partial}{\partial x_1} \right)^t \ln|x - x'| dx'. \tag{8.60}$$

Using the Green's formula

$$\int_\Omega u v_{x_j}\, dx = -\int_\Omega u_{x_j} v\, dx + \int_{\partial\Omega} u v n_j\, ds,$$

we find that Eq. (8.60) becomes

$$\begin{aligned} v(x, t) &= \frac{\omega_0}{2\pi} \int_{\partial\Omega(t)} \ln|x - x'| [-n_2(x'), n_1(x')]^t ds(x') \\ &= -\frac{\omega_0}{2\pi} \int_0^{2\pi} \ln|x - z(\alpha', t)| z_\alpha(\alpha', t) d\alpha'. \end{aligned} \tag{8.61}$$

Here $z_\alpha = (dz/d\alpha)$ and the curve $\partial\Omega(t)$ is parameterized with a clockwise orientation. Expression (8.61) is a single-layer potential, that is, continuous on $\mathbb{R}^2$ (Folland, 1995, p. 72). Thus we have a valid expression for the velocity on the boundary $\partial\Omega(t)$. Replacing $v(x, t)$ with $(dz/dt)(\alpha, t)$ in Eq. (8.61), we obtain the CDE

$$\begin{aligned} \frac{dz}{dt}(\alpha, t) &= -\frac{\omega_0}{2\pi} \int_0^{2\pi} \ln|z(\alpha, t) - z(\alpha', t)| z_\alpha(\alpha', t) d\alpha', \\ z(\alpha, t)|_{t=0} &= z_0(\alpha). \end{aligned} \tag{8.62}$$

Indeed, this equation contains all the information we need to have a weak solution to the 2D Euler equation. We have the following proposition.

Proposition 8.6. *Let $z_0(\alpha)$ be a smooth map from the circle S^1 to $\mathbb{R}^2$. Let $z(\alpha, t)$ be a $C^1([-T, T]; C[0, 2\pi]) \cap C([-T, T]; C^1[0, 2\pi])$ solution to the CDE:*

$$\begin{aligned} \frac{dz}{dt}(\alpha, t) &= -\frac{\omega_0}{2\pi} \int \ln|z(\alpha, t) - z(\alpha', t)| z_\alpha(\alpha', t) d\alpha', \\ z(\alpha, t)|_{t=0} &= z_0(\alpha). \end{aligned}$$

Define the velocity field v by

$$v(x,t) = -\frac{\omega_0}{2\pi}\int \ln|x - z(\alpha',t)| z_\alpha(\alpha',t)d\alpha',$$

and the vorticity (in the sense of distributions) by $\omega(x,t) = \text{curl } v$. *Then* v, ω *form a 2D Euler weak solution of the vortex-patch type [as defined in initial data (8.58) and (8.59)].*

Proposition 8.6 is analogous to Proposition 2.23 for the integrodifferential equations for particle trajectories.

We note that by the derivation of the CDE, if v is a weak solution to the 2D Euler equation corresponding to a vortex patch with a sufficiently smooth boundary, then $z(\alpha,t) = X[z_0(\alpha),t]$ solves the CDE. Here we prove that the converse is true.

Proof. We want to show that $z(\alpha,t)$, a solution to the CDE, defines a 2D weak solution in the sense of Definition 8.1. We begin by showing the existence of $\omega(x,t)$ and $v(x,t)$, satisfying

$$\begin{aligned} \text{curl } v &= \omega \\ \omega(x,t) &= \omega_0, \qquad x \in \Omega(t) \\ &= 0, \qquad x \notin \Omega(t), \end{aligned}$$

where the derivatives are taken in a distributional sense. Given a solution $z(\alpha,t)$ to the CDE, we define

$$v(x,t) = -\frac{\omega_0}{2\pi}\int \ln|x - z(\alpha,t)| z_\alpha(\alpha',t)d\alpha'.$$

Then, for $\mathbf{x} \notin \{z(\alpha,t)|\alpha \in S^1\}$, we can differentiate v under the integral sign to obtain

$$\frac{\partial v}{\partial x_1} = -\frac{\omega_0}{2\pi}\int \frac{x_1 - z_1(\alpha',t)}{|x - z(\alpha',t)|^2} z_\alpha(\alpha',t)d\alpha', \tag{8.63}$$

$$\frac{\partial v}{\partial x_2} = -\frac{\omega_0}{2\pi}\int \frac{x_2 - z_2(\alpha',t)}{|x - z(\alpha',t)|^2} z_\alpha(\alpha',t)d\alpha'. \tag{8.64}$$

If we view $\mathbb{R}^2$ as the complex plane, then we can combine Eqs. (8.63) and (8.64) to obtain, by means of standard one-variable residue theory (Ahlfors, 1979),

$$\begin{aligned} \text{curl } v = \frac{\partial v_2}{\partial x_1} - \frac{\partial v_1}{\partial x_2} &= \text{Re}\frac{\omega_0}{2\pi i}\int_{\partial\Omega(t)} \frac{1}{z - x}dz \\ &= \omega_0, \qquad x \in \Omega(t) \\ &= 0, \qquad x \notin \bar{\Omega}(t). \end{aligned}$$

Here $\Omega(t)$ is simply the interior of the curve defined parametrically by $z(\alpha,t)$. Because z satisfies the CDE, $\partial z/\partial t = v$ and Ω must be convected by the velocity v. This uniquely defines ω everywhere in $\mathbb{R}^2$ but on the boundary $\partial\Omega$. Because we are interested in obtaining only $\omega \in L^\infty$, we can define it arbitrarily on a set of measures zero in $\mathbb{R}^2$ without changing its value in L^∞. We show later in this chapter that a natural choice of ω on the boundary is $\omega(\mathbf{x}) = \omega_0/2, \mathbf{x} \in \partial\Omega$.

Now we show that ω and v satisfy the condition for a weak solution as defined in Definition 8.1. We must show that, for all $\varphi \in C^1\{\mathbb{R}^2 \times [0, T)\}$,

$$\int_{\mathbb{R}^2} \omega(x, T)\varphi(x, T)dx - \int_{\mathbb{R}^2} \varphi(x, 0)\omega_0(x)dx = \int_0^T \int_{\mathbb{R}^2} \frac{D\varphi}{Dt}\omega\, dxdt. \quad (8.65)$$

Because ω is zero outside $\Omega(t)$, the right-hand side of condition (8.65) is

$$\begin{aligned}\int_0^T \int_{\Omega(t)} \frac{D\varphi}{Dt}\omega_0\, dxdt &= \omega_0 \int_0^T \int_{\Omega(t)} \frac{D\varphi}{Dt} dxdt \\ &= \omega_0 \int_0^T \frac{d}{dt} \int_{\Omega(t)} \varphi\, dxdt \\ &= \omega_0 \left[\int_{\Omega(T)} (x, T) - \int_{\Omega(0)} \varphi(x, 0) \right] \\ &= \int_{\mathbb{R}^2} \omega(x, T)\varphi(x, t) - \int_{\mathbb{R}^2} \omega(x, 0)\varphi(x, 0).\end{aligned}$$

In the preceding equation, we use transport formula (1.15) to pull the time derivative out of the inner integral. □

Elementary Properties of Patches of Constant Vorticity

The existence and uniqueness theory from Section 8.2 has direct application to the vortex-patch problem. Recall that the particle trajectories $X^t : \mathbb{R}^2 \to \mathbb{R}^2$ for the weak solutions $\omega(\cdot, t) \in L^1(\mathbb{R}^2) \cap L^\infty(\mathbb{R}^2)$ are one to one and onto. This immediately implies that patches of constant vorticity do not merge or reconnect for all time $t > 0$. The equivalence of contour dynamics to the weak solution theory implies that a smooth solution to the CDE describes a curve evolving in time that never crosses or touches itself.

From estimate (8.33) we also know that the particle trajectories $X^t(\cdot)$ are Hölder continuous for all time intervals [0,T]:

$$|X(\alpha_1, t) - X(\alpha_2, t)| \le c|||\omega_0|||\,|\alpha_1 - \alpha_2|^{\beta(t)}, \qquad \forall 0 \le t \le T, \quad (8.66)$$

with the Hölder exponent $\beta(t) = \exp(-c|||\omega_0|||t)$. This bound guarantees only Hölder continuity (with decaying exponent) of the boundary. However, for the CDE to make sense, we require that the boundary of the vortex patch be at least piecewise C^1 smooth, so that z_α is well defined almost everywhere. Hence a natural question to ask is whether such a curve, if initially Lipschitz, remains so for some positive time. The answer to this question is still not known. In this chapter we prove global regularity for the stronger condition of a $C^{1,\gamma}$-smooth boundary; that is, solutions with an initially $C^{1,\gamma}$-smooth boundary remain $C^{1,\gamma}$ smooth for all time. A standard induction argument combined with higher-derivative estimates, analogous to

Chap. 4, Section 4.4, on higher-derivative control of smooth solutions to the Euler equation, gives global smoothness for a patch boundary that is initially smooth.

We now give some insight into the importance of the role of the tangent vector in the CDE. In studying the evolution equation of the tangent vector, we find that there is a direct link between the CDE and the stretching of vorticity in 3D Euler flows (Majda, 1986). Differentiating CDE (8.62) with respect to α yields Lemma 8.6.

Lemma 8.6. *Let $z(\alpha, t) \in C^1\{[0, T^*); C^{1,\gamma}(S^1)\}$ be a solution to the CDE:*

$$\frac{dz(\alpha, t)}{dt} = -\frac{\omega_0}{2\pi} \int \ln|z(\alpha, t) - z(\alpha', t)| z_\alpha(\alpha', t) d\alpha'.$$

Then the derivative of z with respect to α (in the sense of distributions) satisfies

$$\frac{dz_\alpha}{dt}(\alpha, t) = -\frac{\omega_0}{2\pi} \mathrm{PV} \int_{\alpha-\pi}^{\alpha+\pi} \frac{z_\alpha(\alpha, t) \cdot [z(\alpha, t) - z(\alpha', t)]}{|z(\alpha, t) - z(\alpha', t)|^2} z_\alpha(\alpha', t) d\alpha'. \tag{8.67}$$

Here PV$\int$ denotes the principal value centered at α. The proof of Eq. (8.67) is standard; the details are presented in the appendix.

If we define the following matrix of SIOs,

$$\tilde{\mathcal{V}}(\omega) = -\frac{\omega_0}{2\pi} \mathrm{PV} \int_{\alpha-\pi}^{\alpha+\pi} \frac{[z(\alpha, t) - z(\alpha', t)]_j}{|z(\alpha, t) - z(\alpha', t)|^2} \omega_i(\alpha', t) d\alpha', \tag{8.68}$$

then Eq. (8.67) becomes

$$\frac{dz_\alpha}{dt}(\alpha, t) = \tilde{\mathcal{V}}(z_\alpha) z_\alpha(\alpha, t). \tag{8.69}$$

$\tilde{\mathcal{V}}(\omega)$, is analogous to $\mathcal{D}(\omega)$, the deformation matrix for 3D flows (see Corollary 2.1 from Chap. 2), that is, $\tilde{\mathcal{V}}(\omega)$ is a SIO with tr $\tilde{\mathcal{V}}(\omega) = 0$. We see that Eq. (8.69) is analogous to vorticity equation (2.109) for 3D flows:

$$\frac{D\omega}{dt} = \mathcal{D}(\omega)\omega.$$

The simplest equation of this form is the scalar equation

$$\omega_t = (H\omega)\omega; \quad \omega(x, t) : \mathbb{R} \times [0, \infty) \to \mathbb{R}, \tag{8.70}$$

where H is the Hilbert transform:

$$Hf(x) = \int \frac{f(y)}{x - y} dy.$$

This equation, viewed as a model problem for the preceding, more complicated equations, was analyzed by Constantin, et al. (1985). They showed that, for certain initial data, the solution ω to Eq. (8.70) blows up in finite time. In light of this fact, it is remarkable that solutions to the CDE with smooth initial data do not develop singularities in finite time. In Subsection 8.3.3 we present a proof of this result that shows in particular that we cannot rule out the possibility of superexponential (infinite-time) growth of quantities associated with the tangent vector.

To prove global regularity, we first need to prove local existence and to link the smoothness of the boundary with quantities associated with the tangent vector. A natural framework for the local existence theory is that of contour dynamics.

8.3.2. Local Existence and Uniqueness of Smooth Solutions to the CDE

We know from relation (8.66) that the boundary of a patch of vorticity stays continuous for all time. In this section, we use the equivalence of solutions to the CDE with the weak solutions of Section 8.2 to prove local regularity of the boundary of a vortex patch. Following the method of Chap. 4, we apply the Picard theorem on a Banach space to the CDE to prove the local (-in-time) existence and uniqueness of solutions to the CDE. We begin by showing that if the boundary starts out $C^{1,\gamma}$ smooth it remains so for a finite time. Using the fact that the CDE is an autonomous ODE on a Banach space, we can determine sufficient conditions for continuing the $C^{1,\gamma}$ solution for all time. In the next subsection we derive an a priori bound for the controlling quantity, thus proving global regularity.

We start by showing the local existence of $C^{1,\gamma}$ solutions of the CDE

$$\frac{dz(\alpha,t)}{dt} = -\frac{\omega_0}{2\pi}\int \ln|z(\alpha,t)-z(\alpha',t)|z_\alpha(\alpha',t)d\alpha', \tag{8.71}$$
$$z(\alpha,t)|_{t=0} = z_0(\alpha).$$

We define $C^{1,\gamma}(S^1;\mathbb{R}^2)$ to be the space of maps $z : S^1 \to \mathbb{R}^2$ with Hölder continuous first derivatives. We wish to know for a given γ whether a solution $z(\alpha,t)$ remains in $C^{1,\gamma}$ for some fixed-time interval $(-T,T)$. To answer this question, we recall Picard Theorem 4.1 from Chap. 4.

Theorem 8.3. *Picard Theorem on a Banach Space. Let $O \subseteq \mathbf{B}$ be an open subset of a Banach space $\mathbf{B}$ and let $F(X)$ be a nonlinear operator satisfying the following criteria:*

(i) $F(X)$ maps O to $\mathbf{B}$,
(ii) $F(X)$ is locally Lipschitz continuous, i.e., for any $X \in O$ there exists $L > 0$ and an open neighborhood $U_X \subset O$ of X such that

$$\|F(\tilde{X}) - F(\hat{X})\|_{\mathbf{B}} \le L\|\tilde{X} - \hat{X}\|_{\mathbf{B}} \qquad \text{for all} \quad \tilde{X}, \hat{X} \in U_X.$$

Then for any $X_0 \in O$, there exists a time T such that the ODE

$$\frac{dX}{dt} = F(X), \qquad X|_{t=0} = X_0 \in O, \tag{8.72}$$

has a unique (local) solution $X \in C^1[(-T,T);O]$.

We now apply Theorem 8.3 to the study of solutions to CDE (8.71). Consider CDE (8.71) as an ODE on a Banach space **B**:

$$dz/dt = F(z),$$
$$F(z) = -\frac{\omega_0}{2\pi} \int \ln|z(\alpha, t) - z(\alpha', t)| z_\alpha(\alpha', t) d\alpha',$$
$$z(\alpha, t)|_{t=0} = z_0(\alpha) \in \mathbf{B}.$$

Consider $z(\alpha, t)$, for fixed t, as belonging to the space $\mathbf{B}^{n,\gamma}(S^1)$ equipped with the norm

$$|||z|||_{n,\gamma} = \sum_{i=0}^{n} \left[\left| \frac{\partial^i z(\alpha, t)}{\partial \alpha^i} \right|_0 + \left| \frac{\partial^i z(\alpha, t)}{\partial \alpha^i} \right|_\gamma \right], \qquad 0 < \gamma \le 1,$$
$$= \sum_{i=0}^{n} \left\| \frac{\partial^i z(\alpha, t)}{\partial \alpha^i} \right\|_\gamma . \tag{8.73}$$

Here $|\ |_0$ denotes the supremum norm and $|\ |_\gamma$ denotes the Hölder γ seminorm:

$$|z|_\gamma = \max_{\theta_1 \neq \theta_2} \frac{|z(\theta_1) - z(\theta_2)|}{|\theta_1 - \theta_2|^\gamma}. \tag{8.74}$$

$\mathbf{B}^{n,\gamma}$ is equivalent to $C^{n,\gamma}$, the space of C^n functions with Hölder continuous nth derivatives. We use the notation $\mathbf{B}^{n,\gamma}$ to emphasize the fact that it is a Banach space. We discuss here the case $n = 1$. We then state a result for higher derivatives that shows that low-order quantities associated with the tangent vector control all higher derivatives. The proof of this is left for the reader and can be found in Bertozzi (1991). The ideas are quite similar to those used in Chap. 4 for the 3D Euler equation.

To apply Theorem 8.3, we need an open set

$$O^M = \{z \in B^{1,\gamma} \,|\, |z|_* > 1/M, |z_\alpha|_0 < M\}, \tag{8.75}$$

where

$$|z|_* = \inf_{\theta_1 \neq \theta_2} \frac{|z(\theta_1) - z(\theta_2)|}{|\theta_1 - \theta_2|}. \tag{8.76}$$

We neglect the (n, γ) dependence of O^M as it is made clear by the context.

Proposition 8.7. *For all $M > 0$, $0 < \gamma \le 1$, $n \in \mathbb{N}$, the set*

$$O^M = \{z \in B^{n,\gamma} \,|\, |z|_* > 1/M, |z_\alpha|_0 < M\}$$

is nonempty, open, and consists only of one-to-one mappings of S^1 to $\mathbb{R}^2$.

Proof of Proposition 8.7. O^M is nonempty because it contains the identity map of S^1 to the unit circle in $\mathbb{R}^2$. Also, the maps $|z|_* : B^{n,\gamma} \to [0, \infty)$ and $|z_\alpha|_0 : B^{n,\gamma} \to [0, \infty)$ are continuous. To see that $|\cdot|_*$ defines continuous maps from $B^{n,\gamma}$ to $\mathbb{R}$, we note that, for $z_1, z_2 \in B^{n,\gamma}$,

$$\begin{aligned}||z_1|_* - |z_2|_*| &= \left| \inf_{\theta_1 \neq \theta_2} \frac{|z_1(\theta_1) - z_1(\theta_2)|}{|\theta_1 - \theta_2|} - \inf_{\theta_1 \neq \theta_2} \frac{|z_2(\theta_1) - z_2(\theta_2)|}{|\theta_1 - \theta_2|} \right| \\ &\leq \sup_{\theta_1 \neq \theta_2} \left| \frac{|z_1(\theta_1) - z_1(\theta_2)|}{|\theta_1 - \theta_2|} - \frac{|z_2(\theta_1) - z_2(\theta_2)|}{|\theta_1 - \theta_2|} \right| \\ &\leq \sup_{\theta_1 \neq \theta_2} \frac{|z_1(\theta_1) - z_2(\theta_1) - [z_1(\theta_2) - z_2(\theta_2)]|}{|\theta_1 - \theta_2|} \\ &\leq |z_1 - z_2|_{n,\gamma}.\end{aligned}$$

Because O^M is the intersection of two preimages of open subsets of $\mathbb{R}$, it is open. Last, the condition $|z|_* > 1/M$ requires that O^M contain only one-to-one maps. This is because if z is not one to one, then $z(\theta_1^*) = z(\theta_2^*)$ for some $\theta_1^* \neq \theta_2^*$. Then,

$$|z|_* = \inf_{\theta_1 \neq \theta_2} \frac{|z(\theta_1) - z(\theta_2)|}{|\theta_1 - \theta_2|} = \frac{|z(\theta_1^*) - z(\theta_2^*)|}{|\theta_1^* - \theta_2^*|} = 0. \tag{8.77}$$

We note that a lower bound on the $|\cdot|_*$ norm tells us two things: There is a lower bound on $|z_\alpha(\alpha)|$ and the curve does not cross itself. □

Theorem 8.4. *Local Existence of $C^{1,\gamma}(S^1)$ Solutions. Let $z(\alpha, t)$ be a solution to the CDE*

$$\begin{aligned}\frac{dz(\alpha, t)}{dt} &= -\frac{\omega_0}{2\pi} \int \ln|z(\alpha, t) - z(\alpha', t)| z_\alpha(\alpha', t) d\alpha', \\ z(\alpha, t)|_{t=0} &= z_0(\alpha),\end{aligned}$$

with initial data $z_0(\alpha) \in B^{n,\gamma}(S^1)$. Then there exists $T(M)$ such that $z(\alpha, t) \in O^M$ $\forall t \in (-T(M), T(M))$.

If the boundary is initially $C^{n,\gamma}$ smooth, it stays $C^{n,\gamma}$ smooth for some finite-time interval. This theorem follows directly from Picard Theorem 8.3, provided that we can prove the following proposition.

Proposition 8.8. *Let O^M and $B^{n,\gamma}$ be defined as in open set (8.75) and norm (8.73), respectively. Let $F : O^M \to B^{n,\gamma}$ be defined by*

$$F[z(\alpha)] = -\omega_0/2\pi \int \ln|z(\alpha) - z(\alpha')| z_\alpha(\alpha') d\alpha'.$$

Then F satisfies the assumptions of Picard Theorem 8.3, that is, F is a locally Lipschitz continuous operator from O^M to $B^{n,\gamma}$.

Proposition 8.8 is analogous to Proposition 4.2 from Chap. 4. To prove it, we need to show the following two things, that F maps O^M to $B^{n,\gamma}$ and that F is locally Lipschitz continuous. We prove here the case $n = 1$. Higher derivatives can be proved analogously, following the same ideas used in Chap. 4, Section 4.4, to prove higher regularity of Euler solutions. The interested reader can read Bertozzi (1991) for details.

The case $n = 1$ requires differentiation of $F(z)$ once with respect to the variable α. We note that although $\ln|z(\alpha) - z(\alpha')|z_\alpha(\alpha)$ is an L^1 function of α', its derivative with respect to α,

$$\frac{z_\alpha(\alpha) \cdot [z(\alpha) - z(\alpha')]}{|z(\alpha) - z(\alpha')|^2} z_\alpha(\alpha'), \tag{8.78}$$

is not in L^1 so we cannot a priori differentiate F under the integral sign. What is true, however, is that $dF/d\alpha$ can be expressed as the principal-value integral of expression (8.78). We have the following lemma.

Lemma 8.7. *Let $z \in O^M$, where*

$$O^M = \{z \in B^{1,\gamma} ||z|_* > 1/M, \ |z_\alpha|_0 < M\}.$$

Let

$$F(z) = -\frac{\omega_0}{2\pi} \int_0^\pi \ln|z(\alpha) - z(\alpha')|z_\alpha(\alpha')d\alpha'.$$

Then,

$$\frac{dF}{d\alpha}[z(\alpha)] = -\frac{\omega_0}{2\pi} \mathrm{PV} \int \frac{z_\alpha(\alpha) \cdot [z(\alpha) - z(\alpha')]}{|z(\alpha) - z(\alpha')|^2} z_\alpha(\alpha')d\alpha', \tag{8.79}$$

where PV $\int$ *represents the principal-value integral.*

Lemma 8.7 follows directly from Lemma 8.6.

We note that Eq. (8.79) represents $dF/d\alpha$ as a singular integral, where the integrand blows up like $z(\alpha)/(\alpha - \alpha')$ near $\alpha = \alpha'$. To simplify the analysis, we desingularize $dF/d\alpha$ by putting it in an alternative form.

Lemma 8.8. *Let $z \in O^M$, where*

$$O^M = \{z \in B^{1,\gamma} ||z|_* > 1/M, |z_\alpha|_0 < M\}. \tag{8.80}$$

Then,

$$\frac{dF}{d\alpha}[z(\alpha)] = -\frac{\omega_0}{2\pi} \mathrm{PV} \int \frac{z_\alpha(\alpha) \cdot [z(\alpha) - z(\alpha')]}{|z(\alpha) - z(\alpha')|^2} z(\alpha')d\alpha'$$

can be written as the sum of the following four integrals:

$$I_1 = \frac{\omega_0}{2\pi} \int_{-\pi}^{\pi} \frac{z_\alpha(\alpha) \cdot [z(\alpha - \theta) - z(\alpha) - \theta z_\alpha(\alpha)]}{|z(\alpha) - z(\alpha - \theta)|^2} z_\alpha(\alpha - \theta)d\theta,$$

$$I_2 = \frac{\omega_0}{2\pi} \int_{-\pi}^{\pi} \frac{|z_\alpha(\alpha)|^2 \theta [z_\alpha(\alpha - \theta) - z_\alpha(\alpha)]}{|z(\alpha) - z(\alpha - \theta)|^2} d\theta$$

$$I_3 = \frac{\omega_0}{2\pi} \int_{-\pi}^{\pi} \frac{z_\alpha(\alpha) \cdot [z(\alpha - \theta) - z(\alpha) - \theta z_\alpha(\alpha)]}{|z(\alpha) - z(\alpha - \theta)|^2} z_\alpha(\alpha)d\theta$$

$$I_4 = \frac{\omega_0}{2\pi} \int_{-\pi}^{\pi} \frac{[z(\alpha) - z(\alpha - \theta)] \cdot [z(\alpha - \theta) - z(\alpha) - \theta z_\alpha(\alpha)]}{\theta |z(\alpha) - z(\alpha - \theta)|^2} z_\alpha(\alpha)d\theta,$$

where it is not necessary to use the principal-value form of the integral.

The proof is left for the reader.

The Hölder spaces satisfy various calculus inequalities. We introduced these properties in Chap. 4 and recall several that we use throughout the remainder of this section.

Lemma 8.9. *Calculus Inequalities of the Hölder Spaces. Let f and g be Hölder continuous functions belonging to the space C^γ. In addition, let f be differentiable. Then the product fg satisfies*

$$|fg|_\gamma \leq (|f|_0|g|_\gamma + |f|_\gamma|g|_0), \tag{8.81}$$

the composition $g \circ f$ satisfies

$$|g \circ f|_\gamma \leq |f'|_0^\gamma |g|_\gamma, \tag{8.82}$$

and, if $g \in C^{1,\gamma}$, then

$$|g(x) - g(x-\theta) - g'(x)\theta| \leq |\theta|^{1+\gamma}|g'|_\gamma. \tag{8.83}$$

For proofs, we refer the reader to the appendix of Chap. 4. The last inequality, (8.83), is not explicitly stated there, but follows from identical arguments.

To prove Proposition 8.8 we first show that F is a bounded operator from O_M to $B^{1,\gamma}$, that is, we prove Lemma 8.10.

Lemma 8.10. *Let F and $z \in O^M$ be as defined in Eqs. (8.79) and (8.80). Then there exists $C(M,\gamma)$ such that, for all $0 \leq \delta \leq 1$,*

$$||F||_\gamma \leq C(|z|_0 + 1), \tag{8.84}$$

$$\left|\left|\frac{dF}{d\alpha}\right|\right|_\gamma \leq C(||z_\alpha||_\gamma + 1). \tag{8.85}$$

Proof of Lemma 8.10. We have

$$F[z(\alpha)] = -\frac{\omega_0}{\pi}\int \ln|z(\alpha) - z(\alpha')|z_\alpha(\alpha')d\alpha'.$$

To show estimate (8.84) we must first estimate $|F|_0$. We note that, for $|z| < 1$,

$$\ln|z(\alpha) - z(\alpha')| \leq G\left|\ln\left|\frac{\alpha - \alpha'}{M}\right|\right|.$$

If $|z| > 1$, then

$$\begin{aligned}\ln|z(\alpha) - z(\alpha')| &\leq C\left[\left|\ln\frac{|\alpha - \alpha'|}{M}\right| + |\ln(2|z|_0)|\right]\\ &\leq C\left(\left|\ln\frac{|\alpha - \alpha'|}{M}\right| + |z|_0| + 1\right).\end{aligned}$$

Thus

$$|F|_0 \leq |z_\alpha|_0 C(M)(1+|z|_0). \tag{8.86}$$

To estimate $|F|_\gamma$, we fix $|\alpha - \bar{\alpha}| = \epsilon$ and write

$$\begin{aligned}
&\frac{2\pi}{\omega_0}\{F[z(\alpha)] - F[z(\bar{\alpha})]\} \\
&\quad = \int \ln|z(\alpha) - z(\alpha')| z_\alpha(\alpha') d\alpha' - \int \ln|z(\bar{\alpha}) - z(\alpha')| z_\alpha(\alpha') d\alpha' \\
&\quad = \int_{|\alpha-\alpha'|\leq 2\epsilon} [\ln|z(\alpha) - z(\alpha')| z_\alpha(\alpha') d\alpha' - \ln|z(\bar{\alpha}) - z(\alpha')| z_\alpha(\alpha')] d\alpha' \\
&\quad\quad + \int_{|\alpha-\alpha'|>2\epsilon} [\ln|z(\alpha) - z(\alpha')| z_\alpha(\alpha') d\alpha' - \ln|z(\bar{\alpha}) - z(\alpha')| z_\alpha(\alpha')] d\alpha' \\
&\quad = I_1 + I_2.
\end{aligned}$$

To estimate I_1 we observe that

$$\frac{|z(\alpha) - z(\alpha')|}{|\alpha - \alpha'|} \leq CM,$$

$$1/M \leq \frac{|z(\alpha) - z(\alpha')|}{|\alpha - \alpha'|}.$$

Therefore

$$\frac{C}{M^2}\frac{|\alpha - \alpha'|}{|\bar{\alpha} - \alpha'|} \leq \frac{|z(\alpha) - z(\alpha')|}{|z(\bar{\alpha}) - z(\alpha')|} \leq M^2 \frac{|\alpha - \alpha'|}{|\bar{\alpha} - \alpha'|}.$$

Thus

$$I_1 = \int_{|\alpha-\alpha'|\leq 2\epsilon} \left[\ln \frac{|z(\alpha) - z(\alpha')|}{|z(\bar{\alpha}) - z(\alpha')|} z_\alpha(\alpha')\right] d\alpha',$$

$$|I_1| \leq C(M) \int_{|\theta|\leq 2\epsilon} |\ln|\theta|| d\theta \leq C(M)\epsilon \log \epsilon. \tag{8.87}$$

To estimate I_2, we note that for $|\alpha - \alpha'| > 2\epsilon$, the mean-value theorem gives us

$$\ln|z(\alpha) - z(\alpha')| - \ln|z(\bar{\alpha}) - z(\alpha')| = \epsilon \frac{z_\alpha(\alpha_0) \cdot [z(\alpha_0) - z(\alpha')]}{|z(\alpha_0) - z(\alpha')|^2}$$

where $\alpha_0 \in [\alpha, \alpha]$. Thus

$$\begin{aligned}
|I_2| &\leq C\epsilon M^4 \int_{|\theta|>2\epsilon} \frac{1}{|\theta|} d\theta \\
&\leq \epsilon M^4(|\ln \epsilon|).
\end{aligned} \tag{8.88}$$

If we combine relations (8.87) and (8.88) we have

$$|F[z(\alpha)] - F[z(\bar{\alpha})]| \leq C(M)|\alpha - \alpha'| \log|\alpha - \alpha'|.$$

Hence $F(\alpha)$ is quasi Lipschitz continuous and, in particular, is Hölder–γ continuous for all $\gamma < 1$. This, combined with the estimate for $|F|_0$, gives us the desired result.

To prove estimate (8.85), we use the form of F_α from Lemma 8.8. We eliminate many details because the arguments are straightforward and similar to those used in Chap. 4. For example, consider the first integral I_1. The bound on $|I_1|_0$ follows from a direct application of the calculus inequalities and the definition of O_M to the integrand in I_1; that is,

$$I_1 = \frac{\omega_0}{2\pi}\int_{-\pi}^{\pi} \frac{z_\alpha(\alpha)\cdot[z(\alpha-\theta)-z(\alpha)-\theta z_\alpha(\alpha)]}{|z(\alpha)-z(\alpha-\theta)|^2} z_\alpha(\alpha-\theta)d\theta, \tag{8.89}$$

$$|I_1| \le \frac{\omega_0}{2\pi}\int_{-\pi}^{\pi} \frac{|z_\alpha|_0^2|z_\alpha|_\gamma\theta^{1+\gamma}}{|z|_*^2\theta^2}d\theta \tag{8.90}$$

$$\le C(\omega_0,\gamma)\frac{|z_\alpha|_0^2|z_\alpha|_\gamma}{|z|_*^2}. \tag{8.91}$$

A bound for $|I_1|_\gamma$ follows from a combination of arguments used in the derivation of the above bound for $|I_2|_0$ and the arguments used to compute analogous bounds for $|\nabla F|_\gamma$ in the proof of local existence of solutions to the Euler equation from Chap. 3. The other integrals are computed similarly. We leave the details to the reader. A complete discussion can be found in Bertozzi (1991). □

To finish the proof of Proposition 8.8 we must show that F is locally Lipschitz continuous on O_M. We prove this fact by proving a sufficient condition, that the derivative $F'(z)$ is bounded as a linear operator from O^M to $B^{1,\gamma}$, $||F'(z)|| < \infty$, $\forall z \in O^M$. Then, by the mean-value theorem, we have

$$|F(z_1)-F(z_2)|_{1,\gamma} = \left|\int_0^1 \frac{d}{d\epsilon}F[z_1+\epsilon(z_2-z_1)]d\epsilon\right|_{1,\gamma}$$
$$\le \int_0^1 ||F'[z_1+\epsilon(z_2-z_1)]||d\epsilon|z_1-z_2|_{1,\gamma}. \tag{8.92}$$

So F is locally Lipschitz continuous on O^M. We compute $F'(z)$ as

$$\begin{aligned} F'(z)y &= -\frac{d}{d\epsilon}F(z-\epsilon y)|_{\epsilon=0} \\ &= \frac{d}{d\epsilon}\left\{\frac{\omega_0}{2\pi}\int_0^{2\pi} \ln|z(\alpha)+\epsilon y(\alpha)-z(\alpha')-\epsilon y(\alpha')| \right. \\ &\quad \left.\left. \times [z_\alpha(\alpha')+\epsilon y_\alpha(\alpha')]d\alpha'\right\}\right|_{\epsilon=0} \\ &= \frac{-\omega_0}{2\pi}\int_0^{2\pi} \ln|z(\alpha)-z(\alpha')|y_\alpha(\alpha')d\alpha' \\ &\quad -\frac{\omega_0}{2\pi}\int_0^{2\pi} \frac{[y(\alpha)-y(\alpha')]\cdot[z(\alpha)-z(\alpha')]}{|z(\alpha)-z(\alpha')|^2}z_\alpha(\alpha')d\alpha' \\ &= G_1(z)y+G_2(z)y. \end{aligned} \tag{8.93}$$

We need to obtain estimates for the $|\cdot|_{1,\gamma}$ norms of $G_1(z)y$ and $G_2(z)y$. As in Lemma 8.7 we can differentiate under the integral sign as long as we consider the resulting expressions as principal-value integrals.

Lemma 8.11. *Let $G_1(z)y$ and $G_2(z)y$ be as defined in Eq. (8.93). Then,*

$$\frac{d}{d\alpha}[G_1(z)y(\alpha)] = -\frac{\omega_0}{2\pi}\mathrm{PV}\int \frac{z_\alpha(\alpha)\cdot[z(\alpha)-z(\alpha')]}{|z(\alpha)-z(\alpha')|^2}y_\alpha(\alpha')d\alpha',$$

$$\begin{aligned}\frac{d}{d\alpha}[G_2(z)y(\alpha)] = {}& -\frac{\omega_0}{2\pi}\mathrm{PV}\int \frac{y_\alpha(\alpha)\cdot[z(\alpha)-z(\alpha')]}{|z(\alpha)-z(\alpha')|^2}z_\alpha(\alpha')d\alpha' \\ & -\frac{\omega_0}{2\pi}\mathrm{PV}\int \frac{z_\alpha(\alpha)\cdot[y(\alpha)-y(\alpha')]}{|z(\alpha)-z(\alpha')|^2}z_\alpha(\alpha')d\alpha' \\ & +\frac{\omega_0}{\pi}\mathrm{PV}\int \frac{z_\alpha(\alpha)\cdot[z(\alpha)-z(\alpha')][y(\alpha)-y(\alpha')]\cdot[z(\alpha)-z(\alpha')]}{|z(\alpha)-z(\alpha')|^4} \\ & \times z_\alpha(\alpha')d\alpha'.\end{aligned}$$

The proof of Lemma 8.11 is essentially identical to that of Lemma 8.7, proved in the appendix.

Furthermore, we can write $(d/d\alpha)[G_1(z)y]$ and $(d/d\alpha)[G_2(z)y]$ in the form of Lemma 8.8. For example,

$$\begin{aligned}&\frac{d}{d\alpha}[G_1(z)y] \\ &\quad = -\frac{\omega_0}{2\pi}\int_{-\pi}^{\pi} \frac{z_\alpha(\alpha)\cdot[z(\alpha-\theta)-z(\alpha)-\theta z_\alpha(\alpha)]}{|z(\alpha)-z(\alpha-\theta)|^2}y_\alpha(\alpha-\theta)d\theta \\ &\quad + \frac{\omega_0}{2\pi}\int_{-\pi}^{\pi} \frac{|z_\alpha(\alpha)|^2\theta[y_\alpha(\alpha-\theta)-y_\alpha(\alpha)]}{|z(\alpha)-z(\alpha-\theta)|^2}d\theta \\ &\quad - \frac{\omega_0}{2\pi}\int_{-\pi}^{\pi} \frac{z_\alpha(\alpha)\cdot[z(\alpha-\theta)-z(\alpha)-\theta z_\alpha(\alpha)]}{|z(\alpha)-z(\alpha-\theta)|^2}y_\alpha(\alpha)d\theta \\ &\quad - \frac{\omega_0}{2\pi}\int_{-\pi}^{\pi} \frac{[z(\alpha)-z(\alpha-\theta)]\cdot[z(\alpha-\theta)-z(\alpha)-\theta z_\alpha(\alpha)]}{\theta|z(\alpha)-z(\alpha-\theta)|^2}y_\alpha(\alpha)d\theta.\end{aligned} \tag{8.94}$$

The analogous expression for $[G_2(z)y]$ has more terms than G_1 but is similar. Standard singular integral estimates analogous to the ones used to prove Lemma 4.10 of Chap. 4 give Proposition 8.9.

Proposition 8.9. *Let $G_1(z)y$ and $G_2(z)y$ be as defined in Eq. (8.93). Then there exists $C(M, |z|_{1,\gamma})$ such that*

$$|G_i(z)y|_{1,\gamma} \le C(M, |z|_{1,\gamma})|y|_{1,\gamma} \qquad i = 1, 2. \tag{8.95}$$

Thus F is a locally Lipschitz operator on $O^M \subset B^{1,\gamma}$.

This concludes the proof of Proposition 8.8 and therefore the proof of the local existence of solutions $z(\alpha, t) \in B^{1,\gamma}$.

Now that we have proved local existence and uniqueness, we can use the continuation part of the Picard Theorem (Theorem 4.4 on page 181) to see that as long as we have control over M we have control over $|||z|||_{1,\gamma}$. We recall this result from Chap. 4.

Theorem 8.5. *Continuation of an Autonomous ODE on a Banach Space. Let $O \subset \mathbf{B}$ be an open subset of a Banach space $\mathbf{B}$ and let $F : O \to \mathbf{B}$ be a locally Lipschitz continuous operator. Then the unique solution $X \in C^1\{[0, T); O\}$ to the autonomous ODE,*

$$\frac{dX}{dt} = F(X), \qquad X|_{t=0} = X_0 \in O,$$

either exists globally in time or $T < \infty$ and $X(t)$ leaves the open set O as $t \nearrow T$.

In particular, this shows that solutions can be continued in time provided we have good control of $|z|_{1,\gamma}$ and $1/|z|_*$. We now show that this is sufficient to control higher derivatives. Repeated differentiation of $F[z(\alpha)]$ with respect to α combined with arguments similar to those discussed above gives the following proposition.

Proposition 8.10. *Let F and $z \in O^M \cap B^{n,\gamma}$ be as defined in Eqs. (8.73) and (8.75). Then there exists $C(M, |z|_{n-1,\gamma})$ such that*

$$|||F(z)|||_{n,\gamma} \leq C(M, |z|_{n-1,\gamma})|||z|||_{n,\gamma}.$$

We recall Grönwall's Lemma (Lemma 3.1 in Chap. 3).

Lemma 8.12. *Grönwall's Lemma. If u, q, and $c \geq 0$ are continuous on $[0, t]$, c is differentiable, and*

$$q(t) \leq c(t) + \int_0^t u(s)q(s)ds,$$

then

$$q(t) \leq c(0) \exp \int_0^t u(s)ds + \int_0^t c'(s) \left[\exp \int_s^t u(\tau)d\tau \right] ds.$$

This shows that Proposition 8.10 provides an a priori bound for higher derivatives in terms of lower ones; that is, the continuation property of ODEs on a Banach space gives the following proposition.

Proposition 8.11. *Let $z(\alpha, t)$ be an exact solution to*

$$\frac{dz}{dt} = \frac{\omega_0}{2\pi} \int \ln|z(\alpha, t) - z(\alpha', t)| z_\alpha(\alpha', t)d\alpha'$$

with initial data $z(\alpha, 0) \in B^{n,\gamma}$ for $n \geq 2$. To guarantee that $z(\alpha, t)$ remains bounded in $B^{n,\gamma}$ on any time interval $[0, T]$, it suffices to to control the norm in $B^{n-1,\gamma}$ and $1/|z|_*$ on that time interval.

Using an induction argument on n, we finally obtain the following theorem for C^∞ curves.

Theorem 8.6. *Let $z(\alpha, t)$ be an exact solution to*

$$\frac{dz}{dt} = \frac{\omega_0}{2\pi} \int \ln|z(\alpha, t) - z(\alpha', t)| z_\alpha(\alpha', t) d\alpha'$$

with initial data $z(\alpha, 0) \in C^\infty(S^1)$. Then in order to guarantee that $z(\alpha, t) \in C^\infty(S^1)$ for all time it is sufficient to control $|z_\alpha(t)|_{1,\gamma}$ and $|z(t)|_*$.

In the next subsection we prove a priori bounds for these quantities. Note that $dz/dt = v(z)$ implies that

$$\frac{d}{dt}[z(\alpha, t) - z(\alpha', t)] = v[z(\alpha) - v(z(\alpha')]$$

and hence

$$\frac{d}{dt} \frac{|\alpha - \alpha'|}{|z(\alpha) - z(\alpha')|} \leq \frac{|v(z\alpha) - v[z(\alpha')]||\alpha - \alpha'|}{|z(\alpha) - z(\alpha')|^2} \leq |\nabla v|_{L^\infty} \frac{|\alpha - \alpha'|}{|z(\alpha) - z(\alpha')|}$$

or, taking a supremum,

$$\frac{d}{dt} \frac{1}{|z|_*} \leq |\nabla v|_{L^\infty} \frac{1}{|z|_*}.$$

Grönwall's lemma then shows that an a priori bound on $\int_0^T |\nabla v|_{L^\infty} dt$ provides an a priori bound for $1/|z|_*$ on the time interval $[0, T]$. A similar argument shows that $|z_\alpha|_{L^\infty}$ is also controlled a priori by $\int_0^T |\nabla v|_{L^\infty} dt$. Furthermore, if we define an Eulerian vector field $\boldsymbol{W}(x, t)$ to satisfy $\boldsymbol{W}[z(\alpha, t), t] = z_\alpha(\alpha, t)$ on the patch boundary, then a bound for $|\boldsymbol{W}(\cdot, t)|_{C^\gamma}$ and $|z_\alpha|_{L^\infty}$ produces a bound on $|z_\alpha(\cdot, t)|_\gamma$ by calculus inequality (8.82). We introduce such an Eulerian vector field in the next subsection by considering the perpendicular gradient $\nabla^\perp$ of a defining function φ for the patch boundary $\Omega = \{x | \varphi(x) > 0\}$. We derive a bound for $\int_0^T |\nabla v|_{L^\infty} dt$ and $|\boldsymbol{W}(x, t)|_\gamma = |\nabla^\perp \varphi|_\gamma$ on an arbitrary time interval in terms of the initial data alone. Consequently a $C^{1,\gamma}$-smooth patch boundary remains $C^{1,\gamma}$ smooth for all time.

8.3.3. Global Regularity for Vortex Patches

This question of global regularity of the boundary of the patch has been the subject of some debate in the computational literature (Buttke, 1989; Dritschel and McIntyre, 1990). Recently Chemin (1993), using machinery from paradifferential calculus, proved that smooth contours stay smooth for all time. The simpler proof presented here is due to Bertozzi and Constantin (1993) and uses standard techniques from harmonic analysis, related to the arguments from Chap. 4 and from the previous local-existence theory for vortex-patch boundaries.

We switch now from a Lagrangian framework as used in the previous subsection to the Eulerian framework of Section 8.2. The important quantity is ∇v. Using

Proposition 2.17 we explicitly compute

$$\nabla v(x) = \frac{\omega_0}{2\pi}\mathrm{PV}\int_\Omega \frac{\sigma(x-y)}{|x-y|^2}dy + \frac{\omega_0}{2}\begin{bmatrix} 0 & -1 \\ 1 & 0 \end{bmatrix}\chi_\Omega(x), \tag{8.96}$$

where PV$\int$ stands for Cauchy's principal-value integral. The characteristic function χ_Ω we take to be 1 in Ω, $\frac{1}{2}$ on $\partial\Omega$, and 0 outside $\bar{\Omega}$. The 2×2 symmetric matrix $\sigma(z)$ has the explicit form

$$\sigma(z) = \frac{1}{|z|^2}\begin{bmatrix} 2z_1 z_2 & z_2^2 - z_1^2 \\ z_2^2 - z_1^2 & -2z_1 z_2 \end{bmatrix}. \tag{8.97}$$

As in previous discussions (see, for example, Chap. 2, Subsection 2.4.2)involving singular integrals, we do not use the explicit form of σ but in fact only the following three properties:

(1) it is a smooth function, homogeneous of degree 0,
(2) it has zero mean on the unit circle, and
(3) it is symmetric with respect to reflections,

$$\sigma(-z) = \sigma(z). \tag{8.98}$$

Formula (8.96) indicates that ∇v is discontinuous across the boundary of Ω, no matter how smooth this boundary might be. However, we show that $|\nabla v(x)|$ is bounded across smooth ($C^{1,\gamma}$) boundaries. We show this to be true by using the reflection symmetry of σ, which in particular gives that the mean of σ on half-circles is zero. Near the boundary of Ω the intersection of Ω with a small circle looks very much like a half-circle. We explicitly exploit this fact in the Geometric Lemma below. Following the arguments used in the proof of Lemma 8.8 we could prove the boundedness of $|\nabla v|$ in Lagrangian coordinates. However, in Eulerian coordinates we obtain a simple logarithmic bound analogous to bound (4.50) 15 in Chap. 4:

$$|\nabla v|_{L^\infty} \le C_\gamma|\omega_0|\{1 + (\log[\Delta_\gamma])\}, \tag{8.99}$$

where Δ_γ is related to the γ–Hölder modulus of continuity of the tangent to the boundary (Proposition 8.12).

A second key point is that, despite its discontinuity, ∇v is nevertheless continuous in the tangential direction. More precisely, we show that the formula

$$\nabla v(x)\boldsymbol{W} = \frac{\omega_0}{2\pi}\mathrm{PV}\int_\Omega \frac{\sigma(x-y)}{|x-y|^2}[\boldsymbol{W}(x) - \boldsymbol{W}(y)]dy. \tag{8.100}$$

holds for divergence-free vector fields $\boldsymbol{W}$ that are tangent to $\partial\Omega$. We apply this idea to the vector field

$$\boldsymbol{W} = \nabla^\perp\varphi, \tag{8.101}$$

where φ is a $C^{1,\gamma}$ function in $\mathbb{R}^2$ that defines the patch by

$$\Omega = \{x \in \mathbb{R}^2 | \varphi(x) > 0\}. \tag{8.102}$$

We combine these two ideas to obtain an a priori bound for $\nabla\varphi$.

In the previous section on contour dynamics, we showed that, for sufficiently smooth boundaries, the vortex-patch evolution was equivalent to a solution to the CDE. In this section we reformulate the vortex-patch problem in terms of a scalar function $\varphi(x,t)$ that defines the patch boundary by $\{\varphi(x,t) > 0\} = \Omega(t)$ and is convected with the flow by

$$\frac{\partial \varphi}{\partial t} + v \cdot \nabla \varphi = 0, \tag{8.103}$$

$$\varphi(x,0) = \varphi_0(x). \tag{8.104}$$

Equation (8.104) guarantees that $\varphi(x,t)$ defines the patch boundary at later times if it initially satisfies $\{\varphi(x,0) > 0\} = \Omega_0$. We can express the entire evolution in terms of φ by means of the Biot–Savart law

$$\begin{aligned} v(x,t) &= \frac{\omega_0}{2\pi} \int_\Omega \nabla_x^\perp \ln|x-y| dy, \\ \Omega &= \{x \in \mathbb{R}^2 | \varphi(x) > 0\}. \end{aligned} \tag{8.105}$$

We assume that

$$\Omega_0 = \{x \in \mathbb{R}^2 | \varphi_0(x) > 0\} \tag{8.106}$$

is bounded and has a smooth boundary. We also assume that

$$\inf_{x \in \partial\Omega_0} |\nabla \varphi_0(x)| \geq m > 0 \tag{8.107}$$

and that $\varphi_0 \in C^{1,\gamma}(\mathbb{R}^2)$. In view of law (8.105), the vorticity $\omega(x) = \partial_2 v_1(x) - \partial_1 v_2(x)$ is

$$\omega(x,t) = \omega_0 \chi_{\Omega(t)}(x) \tag{8.108}$$

with characteristic function $\chi_\Omega(x) = 1$ if $x \in \Omega$, $\frac{1}{2}$ if $x \in \partial\Omega$, and 0 if $x \notin \bar{\Omega}$.

The existence of a solution to Eqs. (8.103) and (8.104) is guaranteed by the weak solution theory of Section 8.2; that is, it guarantees a unique solution in the sense of Definition 8.1 for an initial vorticity of the form $\omega_0(x) = \omega_0 \chi_{\Omega_0}$. If $X(x,t)$ are the particle trajectories associated with this solution, then a solution to Eqs. (8.103) and (8.104) is simply

$$\varphi(x,t) = \varphi_0(X^{-t}(x,t)). \tag{8.109}$$

Note that taking $\nabla^\perp$ of Eq. (8.103) combined with the divergence-free condition on v gives an equation for $\boldsymbol{W}(x,t) = \nabla^\perp \varphi(x,t)$:

$$\frac{\partial \boldsymbol{W}}{\partial t} + v \cdot \nabla \boldsymbol{W} = \nabla v \boldsymbol{W}. \tag{8.110}$$

In particular we see that if $z(\alpha,t)$ is a Lagrangian parameterization of the patch boundary with $z_\alpha(\alpha,0) = \boldsymbol{W}(z(\alpha,0),0)$ at $t=0$, then $z_\alpha(\alpha,t) = \boldsymbol{W}[z(\alpha,t),t]$ at a later time t. Hence it is sufficient to prove appropriate bounds on the Eulerian field $\boldsymbol{W}(x,t)$ in order to control the Lagrangian tangent vector $z_\alpha(\alpha,t)$.

We use the following notation during the rest of this chapter: The area of $\Omega(t)$ is constant in time and defines a unit of length,

$$L^2 = \text{area}(\Omega_0) = \text{area}[\Omega(t)]. \tag{8.111}$$

We also use the notation

$$|\nabla\varphi(\cdot)|_{\inf} = \inf_{x\in\partial\Omega} |\nabla\varphi(x)| = \inf_{x,\varphi(x)=0} |\nabla\varphi(x)|,$$

$$|\nabla\varphi(\cdot)|_\gamma = \sup_{x\neq x'} \frac{|\nabla\varphi(x) - \nabla\varphi(x')|}{|x-x'|^\gamma}, \qquad 0<\gamma<1.$$

Our goal in this section is to prove global regularity for the patch boundary by the equivalence of the evolution of φ with vortex-patch evolution. Hence the main result is one of regularity of φ.

Theorem 8.7. *Given $\omega_0 \neq 0$, Ω_0 bounded, and $\varphi_0 \in C^{1,\gamma}(\mathbb{R}^2)$ satisfying assumptions (8.106) and (8.107), there exists a constant C that depends on only $|\omega_0|$, L, $|\nabla\varphi_0|_\gamma$, $|\nabla\varphi_0|_{L^\infty}$, and $|\nabla\varphi_0|_{\inf}$ so that problems (8.103)–(8.106) have a unique solution $\varphi(x,t)$ defined for all $x\in\mathbb{R}^2$ and $t\in\mathbb{R}$ and satisfy*

$$|\nabla v(\cdot,t)|_{L^\infty} \leq |\nabla v(\cdot,0)|_{L^\infty} e^{C|t|}, \tag{8.112}$$

$$|\nabla\varphi(\cdot,t)|_\gamma \leq |\nabla\varphi(\cdot,0)|_\gamma \exp[(C_0+\gamma)e^{C|t|}]. \tag{8.113}$$

$$|\nabla\varphi(\cdot,t)|_{L^\infty} \leq |\nabla\varphi(\cdot,0)|_{L^\infty} \exp(e^{C|t|}). \tag{8.114}$$

$$|\nabla\varphi(\cdot,t)|_{\inf} \geq |\nabla\varphi(\cdot,0)|_{\inf} \exp(-e^{C|t|}). \tag{8.115}$$

C_0 is a fixed constant. From the previous section we have that relations (8.112)–(8.115) are sufficient for higher regularity. That is, we have Theorem 8.8.

Theorem 8.8. *Consider $\omega_0 \neq 0$, Ω_0 bounded, and $\varphi_0 \in C^{k,\gamma}(\mathbb{R}^2)$ satisfying assumptions (8.106) and (8.107). Let $\varphi(x,t)$ be as defined in solution (8.109). A sufficient condition for $\varphi(x,t)\in C^{k,\gamma}(\mathbb{R}^2)$ for all $t\in[0,T]$ is that $|\nabla\varphi(\cdot,t)|_\gamma$, $|\nabla\varphi(\cdot,t)|_{L^\infty}$, and $1/(|\nabla\varphi(\cdot,t)|_{\inf})$ remain bounded for all $t\in[0,T]$.*

In particular the above theorem is true for $C^\infty = \cap_{k=1}^\infty C^{k,\gamma}$.

Proposition 8.12. *Assume that v is given by Biot–Savart formula (8.105) and φ is related to Ω by assumption (8.106). Then*

$$|\nabla v|_{L^\infty} \leq \left(\frac{9}{2}+\frac{2}{\gamma}\right)|\omega_0|\{1+(\log[\Delta_\gamma])\}, \tag{8.116}$$

where

$$\Delta_\gamma = \frac{|\nabla\varphi(\cdot)|_\gamma L^\gamma}{|\nabla\varphi(\cdot)|_{\inf}}.$$

Recall from Eq. (8.96) that the velocity gradient can be decomposed into symmetric and antisymmetric parts,

$$\nabla v(x) = \frac{\omega_0}{2\pi} \mathrm{PV} \int_\Omega \frac{\sigma(x-y)}{|x-y|^2} dy + \frac{\omega_0}{2} \begin{bmatrix} 0 & -1 \\ 1 & 0 \end{bmatrix} \chi_\Omega(x), \tag{8.117}$$

and we need only to estimate the symmetric part of $(\nabla v)(x_0)$. The only nontrivial case is when x_0 is close to the boundary of Ω. Let us denote by

$$d(x_0) = \inf_{x \in \partial\Omega} \{|x - x_0|\} \tag{8.118}$$

the distance from x_0 to $\partial\Omega$. We take a cutoff distance $0 < \delta \leq \infty$ defined by

$$\delta^\gamma = \frac{|\nabla\varphi|_{\inf}}{|\nabla\varphi|_\gamma} \tag{8.119}$$

and consider the set of points x_0 so that

$$d(x_0) < \delta. \tag{8.120}$$

For every ρ, $\rho \geq d(x_0)$, we consider the directions z so that $x_0 + \rho z \in \Omega$:

$$S_\rho(x_0) = \{z | |z| = 1, x_0 + \rho z \in \Omega\}. \tag{8.121}$$

Also, we choose a point $\tilde{x} \in \partial\Omega$ so that $|x_0 - \tilde{x}| = d(x_0)$ and consider the semicircle

$$\Sigma(x_0) \equiv \{z | |z| = 1, \qquad \nabla_x \varphi(\tilde{x}) \cdot z \geq 0\}. \tag{8.122}$$

As $d(x_0)$ approaches zero, the symmetric difference,

$$R_\rho(x_0) = [S_\rho(x_0) \setminus \Sigma(x_0)] \cup [\Sigma(x_0) \setminus S_\rho(x_0)], \tag{8.123}$$

becomes negligible. More precisely we have the following lemma.

Geometric Lemma. *If $R_\rho(x_0)$ is the symmetric difference in Eq. (8.123) and H^1 denotes the Lebesgue measure on the unit circle then*

$$H^1[R_\rho(x_0)] \leq 2\pi \left[(1 + 2^\gamma) \frac{d(x_0)}{\rho} + 2^\gamma \left(\frac{\rho}{\delta} \right)^\gamma \right] \tag{8.124}$$

holds for all $\rho \geq d(x_0)$, $1 > \gamma > 0$ and x_0 so that $d(x_0) < \delta = (|\nabla\varphi|_{\inf} / |\nabla\varphi|_\gamma)^{1/\gamma}$.

Proof of Geometric Lemma. Recall that

$$S_\rho(x_0) = \{z | |z| = 1, x = x_0 + \rho z \in \Omega\}, \tag{8.125}$$

$$\Sigma(x_0) = \{z | |z| = 1 \quad [\nabla_x \varphi(\tilde{x}) \cdot (z)] \geq 0\}, \tag{8.126}$$

$$R_\rho(x_0) = [S_\rho(x_0) \setminus \Sigma(x_0)] \cup [\Sigma(x_0) \setminus S_\rho(x_0)]. \tag{8.127}$$

Let $\theta(z)$ be the angle in $R_\rho(x_0)$ that corresponds to the point z in $R_\rho(x_0)$. Parameterize R_ρ by $\theta(z)$, given by

$$\sin\theta(z) = \frac{\nabla\varphi(\tilde{x}) \cdot (z)}{|\nabla\varphi(\tilde{x})||z|} = \frac{\nabla\varphi(\tilde{x}) \cdot (\tilde{x} - x_0)}{|\nabla\varphi(\tilde{x})|\rho} + \frac{\nabla\varphi(\tilde{x}) \cdot (x_0 + \rho z - \tilde{x})}{|\nabla\varphi(\tilde{x})|\rho}.$$

If $z \in R_\rho(x_0)$ then either $\{\sin\theta(z) > 0$ and $\varphi(x_0+\rho z) < 0\}$ or $\{\sin\theta(z) < 0$ and $\varphi(x_0+\rho z) > 0\}$. In either case, because $\varphi(\tilde{x}) = 0$ and $\nabla\varphi(\tilde{x})$ is parallel to $x_0 - \tilde{x}$,

$$\begin{aligned}
|\sin\theta(z)| &\le \frac{d(x_0)}{\rho} + \left| \frac{\nabla\varphi(\tilde{x})\cdot(x_0+\rho z - \tilde{x})}{|\nabla\varphi(\tilde{x})|\rho} - \frac{\varphi(x_0+\rho z) - \varphi(\tilde{x})}{|\nabla\varphi(\tilde{x})|\rho} \right| \\
&\le \frac{d(x_0)}{\rho} + \frac{|\nabla\varphi|_\gamma |x_0 + \rho z - \tilde{x}|^{1+\gamma}}{\rho|\nabla\varphi|_{\inf}} \\
&\le \frac{d(x_0)}{\rho} + \frac{|\nabla\varphi|_\gamma}{\rho|\nabla\varphi|_{\inf}}[d(x_0)+\rho]^{1+\gamma} \\
&\le \frac{d(x_0)}{\rho} + 2^\gamma \frac{|\nabla\varphi|_\gamma}{\rho|\nabla\varphi|_{\inf}}[d(x_0)^{1+\gamma} + \rho^{1+\gamma}].
\end{aligned}$$

Hence

$$\begin{aligned}
H^1[R_\rho(x_0)] &\le 2\pi \left\{ \frac{d(x_0)}{\rho} + 2^\gamma \frac{|\nabla\varphi|_\gamma}{\rho|\nabla\varphi|_{\inf}}[d(x_0)^{1+\gamma} + \rho^{1+\gamma}] \right\} \\
&\le 2\pi \left\{ (1+2^\gamma)\left[\frac{d(x_0)}{\rho}\right] + 2^\gamma \left(\frac{\rho}{\delta}\right)^\gamma \right\}. \qquad \square
\end{aligned}$$

Proof of Proposition 8.12. Recall that we need only an estimate for the symmetric part of ∇v, which we write as the sum of two integrals, I_1 and I_2. I_2 is the integral

$$I_2(x_0) = \frac{\omega_0}{2\pi}\int_{\Omega\cap\{|x_0-y|\ge\delta\}} \frac{\sigma(x_0-y)}{|x_0-y|^2}dy,$$

where δ is defined in Eq. (8.119). It follows that

$$|I_2| \le |\omega_0|\left[1 + \log\left(\frac{\delta}{L}\right)\right].$$

On the other hand,

$$I_1(x_0) = \frac{\omega_0}{2\pi}\int_{\Omega\cap\{|x_0-y|<\delta\}} \frac{\sigma(x_0-y)}{|x_0-y|^2}dy$$

vanishes if $d(x_0) > \delta$, so we may assume that $d(x_0) \le \delta$. Passing to polar coordinates centered at x_0 and using $\int_{\Sigma(x_0)} \sigma\, dH^1(z) = 0$, we obtain

$$|I_1(x_0)| \le \frac{|\omega_0|}{2\pi}\int_{d(x_0)}^{\delta} \frac{d\rho}{\rho} H^1[R_\rho(x_0)].$$

Applying the Geometric Lemma and integrating yields

$$|I_1| \le |\omega_0|\left[(1+2^\gamma) + \frac{2^\gamma}{\gamma}\right]. \qquad \square$$

Proposition 8.13. *If v is given by Biot–Savart formula (8.105) and W is a divergence-free vector field tangent to $\partial\Omega$, then*

$$\nabla v(x) W = \frac{\omega_0}{2\pi}\mathrm{PV}\int_\Omega \frac{\sigma(x-y)}{|x-y|^2}[W(x) - W(y)]dy. \qquad (8.128)$$

Proof of Proposition 8.13. We use the fact that

$$\nabla_y[\nabla_y^\perp \log|x-y|] = \frac{\sigma(x-y)}{|x-y|^2}$$

to rewrite Eq. (8.96). Integration by parts and the fact that $\boldsymbol{W}$ is divergence free and tangent to $\partial\Omega$ imply that

$$\begin{aligned}
&\frac{1}{2\pi}\mathrm{PV}\int_\Omega \nabla[\nabla_y^\perp \log|x-y|]\boldsymbol{W}(y)dy \\
&= -\lim_{\delta\to 0}\frac{\omega_0}{2\pi}\int_{|x-y|=\delta, y\in\Omega}\left[\boldsymbol{W}(y)\cdot\left(\frac{x-y}{\delta}\right)\right]\nabla_y^\perp \log|x-y|dy \\
&= -\frac{1}{2}\chi_\Omega(x)\begin{bmatrix} 0 & -1 \\ 1 & 0 \end{bmatrix}\boldsymbol{W}(x).
\end{aligned}$$

Substitution of the left-hand side of the above equation for the second term in Eq. (8.96) gives the desired result. □

A corollary to Proposition 8.13 is Corollary 8.1

Corollary 8.1. *There exists a constant C_0 such that if v is given by Biot–Savart formula (8.105) and if $\boldsymbol{W}$ is a divergence-free vector field $\boldsymbol{W} \in C^\gamma(\mathbb{R}^2, \mathbb{R}^2)$ tangent to $\partial\Omega$, then*

$$|\nabla v \boldsymbol{W}|_\gamma \le C_0 |\nabla v|_{L^\infty} |\boldsymbol{W}|_\gamma. \tag{8.129}$$

The proof is straightforward and is given in the appendix.

We use this bound to prove the first estimate in Proposition 8.14.

Proposition 8.14. *We assume that φ is a solution of problems (8.103)–(8.106) on some interval of time $|t| < T$. We also assume that $\varphi(\cdot, t) \in C^{1,\gamma}(\mathbb{R}^2)$ and that $|\nabla\varphi(\cdot, t)|_{\inf} > 0$ for $|t| \le T$. Then*

$$|\nabla\varphi(\cdot,t)|_\gamma \le |\nabla\varphi(\cdot,0)|_\gamma \exp\left[(C_0+\gamma)\int_0^t |\nabla v(\cdot,s)|_{L^\infty} ds\right], \tag{8.130}$$

$$|\nabla\varphi(\cdot,t)|_{L^\infty} \le |\nabla\varphi(\cdot,0)|_{L^\infty} \exp\left[\int_0^t |\nabla v(\cdot,s)|_{L^\infty} ds\right], \tag{8.131}$$

$$|\nabla\varphi(\cdot,t)|_{\inf} \ge |\nabla\varphi(\cdot,0)|_{\inf} \exp\left[-\int_0^t |\nabla v(\cdot,s)|_{L^\infty} ds\right]. \tag{8.132}$$

We conclude that, by using Proposition 8.12 in conjunction with Proposition 8.14, the estimates in the Theorem 8.7 are true as long as we have a $C^{1,\gamma}$ solution; that is, plugging bounds (8.130)–(8.132) into bound (8.116) from Proposition 8.12 produces the following estimate:

$$|\nabla v(\cdot,t)|_{L^\infty} \le C(\omega_0)\left[1+\int_0^t |\nabla v(\cdot,s)|_{L^\infty} ds\right],$$

where $C(\omega_0)$ is a constant that depends on only the initial data. Grönwall's Lemma 8.12 implies an a priori bound for $|\nabla v(\cdot,t)|_{L^\infty}$ on any time interval. Recall from the remark at the end of the preceding section that this is sufficient to control $1/|z|_*$ in the Lagrangian variables. Furthermore, by Proposition 8.14 we have an a priori bound for $|\nabla\varphi(\cdot,t)|_\gamma$, $|\nabla\varphi(\cdot,t)|_{L^\infty}$, and $|\nabla\varphi(\cdot,t)|_{\text{inf}}$. The equivalence of the φ evolution problem to the CDE (in particular we can estimate $|z|_{1,\gamma}$ in terms of $|\varphi_{1,\gamma}$ and vice versa) combined with Theorem 8.6 of the previous section proves global regularity of the boundary of the vortex patch.

Using the same approach, we can prove that several disjoint patches retain the smoothness of their boundaries. Also, under appropriate assumptions, the result holds for a patch of nonconstant vorticity.

To prove Proposition 8.14, we recall that $\mathbf{W}=\nabla^\perp\varphi$ satisfies

$$\frac{\partial \mathbf{W}}{\partial t}+v\cdot\nabla\mathbf{W}=\nabla v\mathbf{W}. \tag{8.133}$$

Let $X(\alpha,t)$ be the particle-trajectory map associated with v. We can obtain bounds (8.131) and (8.132) directly by writing Eq. (8.133) in Lagrangian coordinates and computing pointwise bounds for $|\mathbf{W}|$; that is,

$$\mathbf{W}(X(\alpha,t),t)=\mathbf{Z}(\alpha,t)$$

so that $\mathbf{Z}$ satisfies

$$\begin{aligned}
\frac{d\mathbf{Z}(\alpha,t)}{dt} &= \nabla v(X(\alpha,t),t)\mathbf{Z}(\alpha,t)\\
\frac{d\ln|\mathbf{Z}(\alpha,t)|}{dt} &\le |\nabla v(X(\alpha,t),t)|\\
e^{-\int_0^t |\nabla v(\cdot,s)|_{L^\infty} ds} \le \frac{|\mathbf{Z}(\alpha,t)|}{|\mathbf{Z}(\alpha,0)|} &\le e^{-\int_0^t |\nabla v(\cdot,s)|_{L^\infty} ds},
\end{aligned}$$

which certainly implies bounds (8.131) and (8.132).

We present the details of bound (8.130). The proof of this bound is almost identical to the proof of bound (4.48) from Lemma 4.8 in Chap. 4. The particle trajectories and the definition of material derivative imply that Eq. (8.133) is equivalent to the integral equation

$$\mathbf{W}(x,t)=\mathbf{W}_0(X^{-t}(x))+\int_0^t \nabla v\mathbf{W}(X^{s-t}(x),s)ds.$$

Hence the difference

$$\begin{aligned}
|\mathbf{W}(x,t)-\mathbf{W}(x',t)| &\text{ satisfies } |\mathbf{W}(x,t)-\mathbf{W}(x',t)|\\
&\le |\mathbf{W}_0(X^{-t}(x))-\mathbf{W}_0(X^{-t}(x'))|\\
&\quad +\left|\int_0^t \nabla v\mathbf{W}(X^{s-t}(x),s)-\nabla v\mathbf{W}(X^{s-t}(x'),s)ds\right|
\end{aligned}$$

$$\leq |W_0|_\gamma |\nabla X^{-t}|^\gamma_{L^\infty}|x-x'|^\gamma + \int_0^t |\nabla v W(\cdot,s)|_\gamma |\nabla X^{s-t}|^\gamma_{L^\infty}|x-x'|^\gamma ds$$

$$\leq |W_0|_\gamma \exp\left[\gamma \int_0^t |\nabla v(\cdot,s)|_{L^\infty} ds\right]|x-x'|^\gamma$$

$$+ \int_0^t |\nabla v W(\cdot,s)|_\gamma \exp\left[\gamma \int_s^t |\nabla v(s',\cdot)|_{L^\infty} ds'\right]|x-x'|^\gamma ds,$$

where we use calculus inequality (8.82) and the fact that $\nabla_x X^{-t}$ satisfies

$$|\nabla X^{s-t}|_{L^\infty} \leq \exp\left[\int_s^t |\nabla v(s',\cdot)|_{L^\infty} ds'\right].$$

Thus

$$|W(\cdot,t)|_\gamma \leq |W_0|_\gamma \exp\left[\gamma \int_0^t |\nabla v(\cdot,s)|_{L^\infty} ds\right]$$

$$+ \int_0^t |\nabla v W(\cdot,s)|_\gamma \exp\left[\gamma \int_s^t |\nabla v(\cdot,s')|_{L^\infty} ds'\right] ds.$$

Writing $Q(s) = |\nabla v(\cdot,s)|_{L^\infty}$ and using estimate (8.129) from Corollary 8.1 we have

$$|W(\cdot,t)|_\gamma \leq |W_0|_\gamma \exp\left[\gamma \int_0^t Q(s) ds\right]$$

$$+ C \int_0^t Q(s)|W(\cdot,s)|_\gamma \exp\left[\gamma \int_s^t Q(s')ds'\right] ds.$$

Multiplying both sides by $\exp[-\gamma \int_0^t Q(s')ds']$, we obtain

$$|W(\cdot,t)|_\gamma \exp\left[-\gamma \int_0^t Q(s')ds'\right] \leq |W_0|_\gamma$$

$$+ C_0 \int_0^t Q(s)|W(\cdot,s)|_\gamma \exp\left[-\gamma \int_0^s Q(s')ds'\right] ds,$$

so that $|W(\cdot,t)|_\gamma \exp[-\gamma \int_0^t Q(s')ds'] = G(t)$ satisfies

$$G(t) \leq |W_0|_\gamma + C_0 \int_0^t Q(s)G(s)ds$$

and thus by Grönwall's Lemma satisfies

$$G(t) \leq |W_0|_\gamma \exp\left[C_0 \int_0^t Q(s)ds\right],$$

which gives

$$|W(\cdot,t)|_\gamma \leq |W_0|_\gamma \exp\left[(C_0+\gamma) \int_0^t |\nabla v(\cdot,s)|_{L^\infty} ds\right].$$

8.4. Appendix for Chapter 8

8.4.1. Differentiation of the CDE

Below we present the proof of Lemma 8.6, on the differentiation of the CDE.

Lemma 8.6. *Let $z(\alpha, t) \in C^1\{[0, T^*); C^{1,\gamma}(S^1)\}$ be a solution to the CDE*

$$\frac{dz(\alpha,t)}{dt} = -\frac{\omega_0}{2\pi}\int \ln|z(\alpha,t) - z(\alpha',t)| z_\alpha(\alpha',t)d\alpha';$$

then the derivative of z with respect to α (in the sense of distributions) satisfies

$$\frac{dz_\alpha}{dt}(\alpha,t) = -\frac{\omega_0}{2\pi}\mathrm{PV}\int_{\alpha-\pi}^{\alpha+\pi}\frac{z_\alpha(\alpha,t)\cdot[z(\alpha,t)-z(\alpha',t)]}{|z(\alpha,t)-z(\alpha',t)|^2}z_\alpha(\alpha',t)d\alpha'. \tag{8.134}$$

Proof of Lemma 8.6. Without loss of generality, we omit the $\omega_0/2\pi$. By the definition of the distribution derivative, for all $\Phi \in C^\infty(S^1)$,

$$\begin{aligned}
\left(\frac{d}{d\alpha}\frac{dz}{dt}, \Phi\right) &= -\left(\frac{dz}{dt}, \Phi_\alpha\right)\\
&= \int_{S^1}\int_{S^1}\Phi_\alpha(\alpha)\ln|z(\alpha,t)-z(\alpha',t)|z_\alpha(\alpha',t)d\alpha' d\alpha\\
&= \lim_{\epsilon\to 0}\int\int_{S^1\times S^1/\{|\alpha-\alpha'|>\epsilon\}}\Phi_\alpha(\alpha)\ln|z(\alpha,t)-z(\alpha',t)|z_\alpha(\alpha',t)d\alpha' d\alpha\\
&= \lim_{\epsilon\to 0}\int_{\alpha'\in S^1}\int_{|\alpha-\alpha'|>\epsilon}\Phi_\alpha(\alpha)\ln|z(\alpha,t)-z(\alpha',t)|z_\alpha(\alpha',t)d\alpha d\alpha'\\
&= \lim_{\epsilon\to 0}\int_{\alpha'\in S^1}[\Phi(\alpha'-\epsilon)\ln|z(\alpha'-\epsilon,t)-z(\alpha',t)|\\
&\quad -\Phi(\alpha'+\epsilon)\ln|z(\alpha'+\epsilon,t)-z(\alpha',t)|]z_\alpha(\alpha',t)d\alpha'\\
&\quad +\lim_{\epsilon\to 0}\int_{\alpha'\in S^1}\int_{|\alpha-\alpha'|>\epsilon}\Phi(\alpha)\frac{z_\alpha(\alpha,t)\cdot[z(\alpha,t)-z(\alpha',t)]}{|z(\alpha,t)-z(\alpha',t)|^2}d\alpha d\alpha'\\
&= (1)+(2).
\end{aligned}$$

We now show that $(1) = 0$. We show that the integrand is bounded by a constant times ϵ^γ as $\epsilon \to 0$. Indeed, $|\Phi(\alpha'-\epsilon)\ln|z(\alpha'-\epsilon,t)-z(\alpha',t)| - \Phi(\alpha'+\epsilon)\ln|z(\alpha'+\epsilon,t)-z(\alpha',t)||$

$$\begin{aligned}
&< |[\Phi(\alpha'-\epsilon)-\Phi(\alpha'+\epsilon)]|\ln|z(\alpha'-\epsilon,t)-z(\alpha',t)||\\
&\quad +|\Phi(\alpha'+\epsilon)[\ln|z(\alpha'-\epsilon,t)-z(\alpha',t)| - \ln|z(\alpha'+\epsilon,t)-z(\alpha',t)|]|\\
&= (1a)+(1b)
\end{aligned}$$

$$(1a) \le 2\epsilon|\Phi_\alpha(\alpha)|_0 \ln\epsilon \ln|z|_*.$$

By the mean-value theorem,

$$(1b) \le |\Phi|_0\left|\ln\left|\frac{z(\alpha'-\epsilon,t)-z(\alpha',t)}{\epsilon}\right| - \ln\left|\frac{z(\alpha'+\epsilon,t)-z(\alpha',t)}{\epsilon}\right|\right|$$

$$\leq |\Phi|_0 \ln \frac{|[z_{1\alpha}(\alpha_1,t), z_{2\alpha}(\alpha_2,t)]|}{|[z_{1\alpha}(\alpha_3,t), z_{2\alpha}(\alpha_4,t)]|}$$
$$\leq |\Phi|_0 \left| \ln \left\{ 1 + \frac{|[z_{1\alpha}(\alpha_1,t), z_{2\alpha}(\alpha_2,t)]| - |[z_{1\alpha}(\alpha_3,t), z_{2\alpha}(\alpha_4,t)]|}{|[z_{1\alpha}(\alpha_3,t), z_{2\alpha}(\alpha_4,t)]|} \right\} \right|$$
$$\leq C|\Phi_0||z_\alpha|_\gamma \epsilon^\gamma / |z|_*.$$

Thus, using the dominated convergence theorem,

$$\left(\frac{d}{d\alpha}\frac{dz}{dt}, \Phi\right) = \lim_{\epsilon \to 0} \int_{\alpha \in S^1} \int_{|\alpha - \alpha'| > \epsilon} \Phi(\alpha) \frac{z_\alpha(\alpha,t) \cdot [z(\alpha,t) - z(\alpha',t)]}{|z(\alpha,t) - z(\alpha',t)|^2} d\alpha' d\alpha$$
$$= \int_{\alpha \in S^1} \lim_{\epsilon \to 0} \int_{|\alpha - \alpha'| > \epsilon} \Phi(\alpha) \frac{z_\alpha(\alpha,t) \cdot [z(\alpha,t) - z(\alpha',t)]}{|z(\alpha,t) - z(\alpha',t)|^2} d\alpha' d\alpha$$

we obtain the desired result. □

8.4.2. Singular Integral Estimate

We prove Corollary 8.1 in a more general form.

Lemma. *Let K be a Calderon–Zygmund kernel, homogeneous of degree $-n$, with mean zero on spheres, satisfying $|\nabla K(x)| \leq C|x|^{-n-1}$. There exists a constant C_0 so that all $f \in C^\epsilon(\mathbb{R}^n)$ and $\omega \in L^\infty(\mathbb{R}^n)$ satisfy*

$$|G|_\epsilon \leq C_0(\epsilon, n)|f|_\epsilon(|K * \omega|_{L^\infty} + |\omega|_{L^\infty})$$

where

$$G(x) = \mathrm{PV}\int_{\mathbb{R}^n} K(x-y)[f(x) - f(y)]\omega(y)dy.$$

Proof. We write

$$\begin{aligned}
&G(x) - G(x+h) \\
&= \mathrm{PV}\int K(x-y)[f(x) - f(y)]\omega(y)dy \\
&\quad - \mathrm{PV}\int K(x+h-y)[f(x+h) - f(y)]\omega(y)dy \\
&= \mathrm{PV}\int_{|x-y|<2h} K(x-y)[f(x) - f(y)]\omega(y)dy \\
&\quad - \mathrm{PV}\int_{|x-y|<2h} K(x+h-y)[f(x+h) - f(y)]\omega(y)dy \\
&\quad + \mathrm{PV}\int_{|x-y|\geq 2h} K(x-y)[f(x) - f(x+h)]\omega(y)dy \\
&\quad + \mathrm{PV}\int_{|x-y|\geq 2h} [K(x-y) - K(x+h-y)][f(x+h) - f(y)]\omega(y)dy \\
&= (1) + (2) + (3) + (4).
\end{aligned}$$

Clearly $|(1)|, |(2)| \leq C_\epsilon |f|_\epsilon h^\epsilon |\omega|_{L^\infty}$. Also, we have

$$|(4)| \leq \int_{|x-y|\geq 2h} h \frac{C}{|x-y|^{n+1-\epsilon}} |f|_\epsilon |\omega|_{L^\infty} dy \leq C_\epsilon h^\epsilon |f|_\epsilon |\omega|_{L^\infty}.$$

We obtain a bound for (3) by using a lemma that is due to Cotlar (see Torchinsky, 1986, p. 291):

$$|(3)| \leq |f|_\epsilon h^\epsilon \frac{1}{2\pi} \left| \mathrm{PV} \int_{|x-y|\geq 2h} K(x-y)\omega(y)dy \right| \leq C_0 |f|_\epsilon h^\epsilon (|K * \omega|_{L^\infty} + |\omega|_{L^\infty}).$$

Combining these four estimates gives the desired result. □

Notes for Chapter 8

The weak L^∞ solution theory dates back to the 1963 work of Yudovich.

In 1979, Zabusky et al. (1979) showed that the vortex-patch solutions are useful from a computational point of view. By restricting the problem to a patch of constant vorticity, they introduced the CDE as a means of reducing the dimensions of the equation from two to one. Since that time, there has been quite a vast literature illustrating the rapid development of structure on the boundary on a vortex patch (Zabusky and Overman, 1983; Zou et al., 1988; Dritschel, 1989; Dritschel and McIntyre, 1990; Buttke, 1989). Furthermore, there are several works addressing an exact, steady solution like the Kirchoff ellipse discussed in Section 8.1. Examples include uniformly rotating vortex solutions with corners (Overman, 1986; Wu et al., 1984) and uniformly translating pairs of vorticies (Pierrehumbert, 1980; Saffman and Tanveer, 1982).

The work of Chemin (1983) and Bertozzi and Constantin (1993) on the regularity of the boundary of a vortex patch has been recently extended to the case of a vortex patch in a bounded domain by Morgulis (1993) and to the case of a patch with corners by Danchin (1997). Constantin and Wu (1996) address the behavior at the vortex-patch boundary for the vanishing viscosity limit of the Navier–Stokes equations.

References for Chapter 8

Ahlfors, L. V., *Complex Analysis*, McGraw-Hill, New York, 1979.

Alinhac, S., "Remarques sur l'instabilité du problème des poches de tourbillon," *J. Funct. Anal.* **98**(2), 361–379, 1991.

Almgren, R. "Singularity formation in Hele–Shaw bubbles," *Phys. Fluids* **8**, 344–352, 1996.

Almgren, R., Bertozzi, A. L., and Brenner, M. P., "Stable and unstable singularities in the unforced Hele–Shaw cell," *Phys. Fluid* **8**, 1354–1370, 1996.

Berk, H. and Roberts, K., "The water-bag model," *Methods Comput. Phys.* **9**, 87–134, 1970.

Bertozzi, A. L., "Heteroclinic orbits and chaotic dynamics in planar fluid flows," *SIAM J. Math. Anal.* **19**, 1271–1294, 1988.

Bertozzi, A., "Existence, uniqueness, and a characterization of solutions to the contour dynamics equation," Ph.D. dissertation, Princeton University, Princeton , NJ, 1991.

Bertozzi, A. L. and Constantin, P. "Global regularity for vortex patches," *Commun. Math. Phys.* **152**, 19–28, 1993.

Buttke, T. F., "The observation of singularities in the boundary of patches of constant vorticity," *Phys. Fluids A* **1**, 1283–1285, 1989.

Chemin, J.-Y., "Persistance de structures geometriques dans les fluides incompressibles bidimensionnels," *Ann. Ec. Norm. Supér.* **26**(4), 1–16, 1993.

Constantin, P. Dupont, T. F., Goldstein, R. E., Kadanoff, L. P., Shelley, M. J., and Zhou, S.-M., "Droplet breakup in a model of the Hele-Shaw cell," *Phys. Rev. E* **47**, 4169–4181, 1993.

Constantin, P. Lax, P. D., and Majda, A., "A simple one dimensional model for the three dimensional vorticity equation," *Commun. Pure Appl. Math.* **38**, 715–724, 1985.

Constantin, P. and Titi, E. S. "On the evolution of nearly circular vortex patches," *Commun. Math. Phys.* **119**, 177–198, 1988.

Constantin, P. and Wu, J., "The inviscid limit for non-smooth vorticity," *Indiana Univ. Math J.* **1**, 67–81, 1996.

Danchin, R., "Evolution temporelle d'une poche de tourbillon singulière," *Commun. Partial. Diff. Eqns.* **22**, 685–721, 1997.

Dritschel, D. G., "Contour dynamics and contour surgery: numerical algorithms for extended, high-resolution modeling of vortex dynamics in two-dimensional, inviscid, incompressible flows," *Comput. Phys. Rep.* **10**, 77–146, 1989.

Dritschel, D. G. and McIntyre, M. E., "Does contour dynamics go singular?" *Phys. Fluids A* **2**, 748–753, 1990.

Folland, G. B., *Introduction to Partial Differential Equations*, Princeton Univ. Press, Princeton, NJ, 1995.

Gilbarg, D. and Trudinger, N. S., "*Elliptic Partial Differential Equations of Second Order*, Springer-Verlag, Berlin, revised third printing.

Goldstein, R. E., Pesci, A. I., and Shelley, M. J., "Topology transitions and singularities in viscous flows," *Phys. Rev. Lett.* **70**, 3043–3046, 1993.

Goldstein, R. E., Pesci, A. I., and Shelley, M. J., "An attracting manifold for a viscous topology transition," *Phys. Rev. Lett.* **75**, 3665–3668, 1995.

Hartman, P., *Ordinary Differential Equations*, Birkhäuser, Boston, 1982.

John, F. and Nirenberg, L., "On functions of bounded mean oscillation," *Commun. Pure Appl. Math.* **14**, 415–426, 1961.

Lamb, H., *Hydrodynamics*, Dover, New York, 1945.

Majda, A., "Vorticity and the mathematical theory of incompressible fluid flow," *Commun. Pure Appl. Math.* **39**, 5187–5220, 1986.

Meiburg, E. and Homsy, G. M., "Nonlinear unstable viscous fingerings in Hele–Shaw flows. ii. Numerical simulation.," *Phys. Fluids* **31**, 429–439, 1988.

Morgulis, A. B., "Global regularity of perfect incompressible weakly discontinuous flows in a bounded 2D domain," Preprint 47, Rostov State University, Rostov, Russia, 1993.

Neu, J. C., "The dynamics of a columnar vortex in an imposed strain," *Phys. Fluids* **27**, 2397–2402, 1984.

Overman II, E. A., "Steady-state solutions of the Euler-equations in two dimensions. II. Local analysis of limiting V-states," *SIAM J. Appl. Math.* **46**, 765–800, 1986.

Pierrehumbert, R. T., "A family of steady translating vortex pairs with distributed vorticity," *J. Fluid Mech.* **99**, 129–144, 1980.

Royden, H. L., *Real Analysis*, Macmillan, New York, 1968.

Saffman, P. G. and Tanveer, S., "The touching pair of equal and opposite uniform vorticities," *Phys. Fluids* **25**, 1929–1930, 1982.

Shelley, M. J., "Dynamical interfaces in fluid dynamics," Course Notes, Courant Institute of Mathematics and Science, New York, NY, 1995.

Stein, E. M., "Singular integrals, harmonic functions, and differentiability properties of functions of several variables," in *Proceedings of the Symposium on Pure Mathematics*, American mathematical society, Providence, RI, 1967, Vol. 10, pp. 316–335.

Stein, E. M., *Singular Integrals and Differentiability Properties or Functions*, Princeton Univ. Press., Princeton, NJ, 1970.

Torchinsky, A., *Real Variable Methods in Harmonic Analysis*, Academic, New York, 1986.

Wu, H. M., Overman II, E. A., and Zabusky, N. J., "Steady-state solutions of the Euler equations in two dimensions: rotating and translation v-states with limiting cases: I. Numerical algorithms and results," *J. Comput. Phys.* **53**, 42–71, 1984.

Yudovich, V. I., "Non-stationary flow of an incompressible liquid," *Zh. Vychisl. Mat. Mat. Fiz.* **3**, 1032–1066, 1963.

Zabusky, N., Hughes, M. H., and Roberts, K. V., "Contour dynamics for the Euler equations in two dimensions," *J. Comput. Phys.* **48**, 96–106, 1979.

Zabusky, N. J. and Overman II, E. A., "Regularization of contour dynamical algorithms. I. Tangential Regularization," *J. Comput. Phys.* **52**, 351–373, 1983.

Zou, Q., Overman II, E. A., Wu, H.-M., and Zabusky N. J., "Contour dynamics for the Euler equations: curvature controlled initial node placement and accuracy," *J. Comput. Phys.* **78**, 350–368, 1988.

9

Introduction to Vortex Sheets, Weak Solutions, and Approximate-Solution Sequences for the Euler Equation

Many physical phenomena possess strong, irregular, and typically unstable fluctuations with intense activity on small scales. The small-scale dynamics can sometimes induce large-scale coherent structures and affect the overall behavior of a system. Small space or time scales or a small parameter in a nonlinear model provide the mathematical infrastructure for understanding such problems. A current challenge of modern applied mathematics is to develop new tools to understand such singular limits in nonlinear equations and their physical significance. Fully developed turbulence (Avellaneda and Majda, 1994) is one such exciting research area and provides one motivation for the material in this and subsequent chapters. Some other examples in which modern analysis is playing such a role include the focusing of laser beams (Merle, 1992), contact line motion in coating flows (Bertozzi and Pugh, 1996), phase transition and homogenization in materials (James and Kinderlehrer, 1989), and statistical physics (Lions, 1991). Our main focus in the next four chapters (Chaps. 9–12) is the development of analytical tools to address these questions for the 2D and the 3D Euler equations. In Chap. 13 we summarize a recent body of research (Zheng and Majda, 1994; Majda et al., 1994a, 1994b) on an analogous problem, that of the 1D Vlasov–Poisson equations.

The Navier–Stokes equations (1.1)–(1.3) of Chap. 1, can be rewritten as

$$\frac{\partial v^\epsilon}{\partial t} + \operatorname{div}(v^\epsilon \otimes v^\epsilon) = -\nabla p^\epsilon + \epsilon \Delta v^\epsilon, \qquad \operatorname{div} v^\epsilon = 0, \tag{9.1}$$

where ϵ denotes the kinematic viscosity and the notation $u \otimes v$, denoting the matrix product $u_i v_j$, provides a convenient way of introducing a weak solution theory. We are interested in properties of this equation in the limit as $\epsilon \to 0$ or, equivalently, as the Reynolds number goes to infinity. Fully developed turbulence, present in many high Reynolds number flows, is characterized by the fact that the parameter ϵ in Eq. (9.1) is extremely small. Mathematicians can contribute to the scientific understanding of turbulence by addressing natural questions associated with the Navier–Stokes equations. For example, given a sequence of solutions $\{v^\epsilon\}$ to Eq. (9.1), what kinds of structures persist in the limit as $\epsilon \to 0$? Moreover, do we recover a solution to the Euler equation in this limit? What physically motivated constraints must a family of solutions $\{v^\epsilon\}$ (approximate or exact) to the Euler equations satisfy in order for a weak limit to satisfy the Euler equation? As we saw from Chap. 3, given fixed smooth initial data for which the Euler problem possesses a smooth solution, the zero-viscosity limit of the Navier–Stokes solutions satisfies the Euler equation. However, for less smooth initial data the answer becomes less clear. The study of such small-scale dynamics of this kind was recently addressed by a comprehensive body of research by (DiPerna and Majda, 1987a, 1987b, 1988). Chapters 10–12 address this and related work.

In the previous chapter on vortex patches, we introduced a weak formulation of the vorticity-stream form of the 2D Euler equation. Using the fact that vorticity is conserved along particle paths in two dimensions, we showed that initial data with vorticity bounded in $L^\infty \cap L^1(\mathbb{R}^2)$ is sufficient to guarantee the existence of such a weak solution to the Euler equation. Furthermore, arguments from harmonic analysis show that such a limit is unique. An integral part of this analysis involved constructing approximating solutions and showing that a limit satisfies the weak form of the equation.

In the remaining chapters of this book we address the same questions for more singular solutions to the Euler equation. In physical examples of fluids with small viscosity, flows tend to separate from rigid walls and sharp corners.† Mathematically, the most natural way to model this is by a solution to the *inviscid* Euler equation in which the velocity changes sign discontinuously across a streamline. Such structures are known to be extremely unstable, leading to coherent vortex structures often found in mixing layers, jets, and wakes (see Van Dyke, 1982). A recent numerical study by Nitsche and Krasny (1994) shows that the solution to the inviscid mathematical model agrees quite well with the high Reynolds number experiments of Didden (1979) for a fluid forced by a moving piston. A velocity discontinuity in an inviscid flow is called a *vortex sheet*. Unlike the vortex patch, in which ω is pointwise bounded, a vortex sheet has vorticity concentrated as a measure (delta function) along a surface of codimension one. In two-space dimensions, this is a curve in the plane. In three-space dimensions, the vorticity lives on a 2D sheet or surface in $\mathbb{R}^3$. Although some of the results in this section can be extended to the 3D case, for simplicity we consider the 2D case in this chapter.

Provided that the vorticity remains smoothly distributed along the sheet, the solution retains the property of *locally finite kinetic energy*, that is, for every bounded set $\Omega \in \mathbb{R}^2$, there exists a constant $C(\Omega)$ such that

$$\int_\Omega |v|^2 dx \leq C(\Omega).$$

However, the sheet becomes rapidly unstable, generating small-scale structures instantaneously.

Our purpose in this chapter is to introduce the reader to the kinds of complex behavior associated with very weak solutions of the Euler equation. We first review the classical theory of vortex sheets and then describe some of the phenomena observed in computing vortex sheets by means of more regular approximations. The chapter is organized as follows. Based on the primitive-variable form of the equation, in Section 9.1 a weaker form of the Euler equation than the formulation from Chap. 8,

† See, e.g., Birkhoff (1960, Chap. 3).

based on the vorticity-stream form, is derived. In Section 9.2 we derive the classical theory of vortex sheets and the Birkhoff–Rott (B-R) equation. In Section 9.3 we discuss the linear instability associated with the B-R equation, typically referred to as the Kelvin–Helmholtz instability. In Section 9.4 we discuss computational methods for vortex sheets and the various structures observed in these simulations. Section 9.5 serves as an introduction to the general theory of approximate-solution sequences, concentrations, and oscillations, which are the main topics of Chaps. 10–12.

9.1. Weak Formulation of the Euler Equation in Primitive-Variable Form

The discontinuity of the velocity in the vortex sheet necessitates a weaker reformulation of the Euler equation. The vorticity, being the curl of the discontinuous velocity, forms a delta function on a curve in $\mathbb{R}^2$. This means, for example, that the weak formulation of the Euler equation, used in Chap. 8 for vortex-patch initial data in two dimensions, is still too strong for such a solution to make sense. It is even too strong to consider the case of vorticity $\omega \in L^p \cap L^1(\mathbb{R}^2)$ with $1 < p < \infty$. We reformulate the Euler equation in such a way that it makes sense for both this class of initial data and the weaker case $\omega \in H^{-1}(\mathbb{R}^2) \cap \mathcal{M}(\mathbb{R}^2)$, where $\mathcal{M}(\mathbb{R}^2)$ denotes a measure in $\mathbb{R}^2$. The latter case includes the case of classical vortex-sheet initial data in which the vorticity is concentrated on a smooth curve in the plane.

To make sense of vorticity concentrated as a measure, we consider a distribution form of the primitive-variable form of the equation

$$\frac{Dv}{Dt} = -\nabla p, \qquad \frac{D}{Dt} = \frac{\partial}{\partial t} + v \cdot \nabla, \qquad\qquad (9.2)$$
$$\operatorname{div} v = 0, \qquad v(x,0) = v_0.$$

Let $\Phi = (\Phi_1, \Phi_2) \in C_0^\infty([0,T] \times \mathbb{R}^2)$; then transport formula (1.15) of Chap. 1 gives

$$\begin{aligned}\frac{d}{dt}\int_{\mathbb{R}^2} \Phi \cdot v\, dx &= \int_{\mathbb{R}^2} \frac{D}{Dt}(\Phi \cdot v) dx,\\ &= \int_{\mathbb{R}^2} \left(\frac{D\Phi}{Dt} \cdot v + \Phi \cdot \frac{Dv}{Dt}\right) dx, \qquad\qquad (9.3)\\ &= \int_{\mathbb{R}^2} \left(\frac{D\Phi}{Dt} \cdot v - \Phi \cdot \nabla p\right) dx.\end{aligned}$$

Assuming $\operatorname{div} \Phi = 0$ implies that $\int_{\mathbb{R}^2} \Phi \cdot \nabla g\, dx = 0$ for all scalar $g \in C^1(\mathbb{R}^2)$. Thus for all $\Phi \in C_0^\infty$ with $\nabla \cdot \Phi = 0$

$$\frac{d}{dt}\int_{\mathbb{R}^2} \Phi \cdot v\, dx = \int_{\mathbb{R}^2} \frac{D\Phi}{Dt} \cdot v\, dx. \qquad\qquad (9.4)$$

Integrating Eq. (9.4) with respect to t on the time interval $[0,T]$ yields the following definition.

Definition 9.1. *Weak Solution of the Euler Equation in Primitive-Variable Form. A velocity field $v(x,t)$ with initial data v_0 is a weak solution of Euler equation in primitive-variable form provided that*

(i) $v \in L^1([0,T] \times B_R)$ *for any* $R > 0$, $B_R = \{x \in \mathbb{R}^2, |x| \le R\}$,
(ii) $v \otimes v = (v_i v_j) \in L^2([0,T] \times B_R)$,
(iii) $\operatorname{div} v = 0$ *in the sense of distributions,*

$$\int \nabla\varphi \cdot v = 0, \qquad \forall \varphi \in C\{[0,T], C_0^1(\mathbb{R}^2)\}, \tag{9.5}$$

(iv) for any $\Phi = (\Phi_1, \Phi_2) \in C^1\{[0,T], C_0^1(\mathbb{R}^2)\}$ *with* $\operatorname{div}\Phi = 0$,

$$\int \Phi(x,T) \cdot v(x,T)dx - \int \Phi(x,0) \cdot v_0(x)dx = \int_0^T \int (\Phi_t \cdot v + \nabla\Phi : v \otimes v)dxdt, \tag{9.6}$$

where $\nabla\Phi = (\frac{\partial}{\partial x_i}\Phi_j)$, $A : B = \sum_{i,j=1}^2 A_{ij}B_{ij}$.

Requirements (i) and (ii) are such that the time term and the nonlinear term of the right-hand side of Eq. (9.6) are meaningful, (iii) is the "weak" incompressibility condition, and (iv) is the "weak" Euler equation with the pressure eliminated by a special restriction ($\operatorname{div}\Phi = 0$) on the test functions. Note that whereas the vorticity-stream weak formulation is valid for only the 2D fluid equations, the above formulation is valid for both the 2D and the 3D cases.

The reader can verify by going back through steps (9.4)–(9.6) that for smooth solutions to the Euler equations, the above definition is equivalent to the Euler equation in primitive-variable form (9.2).

We showed in Chaps. 3 and 4 that for initial data that are sufficiently smooth, there is a unique solution to the classical form of the Euler equation. We proved global existence of such a solution in two dimensions and local existence in three dimensions. Furthermore, in Chap. 8 we rederived the well-known Yudovich theory for weak solutions of the 2D Euler equation with vorticity in $L^1 \cap L^\infty(\mathbb{R}^2)$ and proved a uniqueness result for this weaker class of solutions to the Euler equation. In particular, these results show that there is a unique solution for all time to the Euler equation with *vortex-patch* initial data.

A natural question to ask is if, for the weak form in Definition 9.1, we can prove an existence result for the Euler equation with weaker initial data such as *vortex-sheet* initial data and, if so, if such a solution is unique. In Chap. 10 we show that for the case of the 2D Euler equation with initial vorticity in $L^p(\mathbb{R}^2) \cap L^1(\mathbb{R}^2)$ such a solution exists for all time. In Chap. 11 we show that under the assumption that the vorticity is initially of fixed sign we can also prove a global-existence result for the vortex-sheet problem. In either case, the question of uniqueness is still unanswered. In Chap. 12 we introduce the notion of a Young measure. Here we show that even when a sequence of solutions does not have a limit that satisfies the equations in a classical weak sense, the limit is nevertheless a measure-valued solution of the Euler equation. All of these chapters make heavy use of the concepts of concentrations and oscillations in solution sequences. We describe these ideas and some elementary examples later in this chapter (Section 9.5). In Chap. 13 we provide an analogy with the 1D Vlasov–Poisson equation and make some conjectures about the uniqueness of solutions to the Euler equation with vortex-sheet initial data.

In Sections 9.2–9.4 we present elementary properties of vortex sheets and also present some results of numerical simulations. In particular, in Section 9.3 we show that the behavior of the vortex sheet is much more unstable than that of its cousin, the vortex patch. We derive a self-deforming curve equation, called the *Birkhoff-Rott* equation, for the evolution of the sheet. This equation is analogous to CDE (8.56) derived in Chap. 8 for the evolution of the boundary of the vortex patch. Recall that the CDE has globally smooth solutions for smooth initial data. We show that, unlike the CDE, the B-R equation is linearly ill posed and can be solved in general only on short time intervals for the case of analytic initial data.

9.2. Classical Vortex Sheets and the Birkhoff–Rott Equation

The classical vortex sheet is a special case of a solution to the 2D Euler equation with vorticity in $\mathcal{M}(\mathbb{R}^2) \cap H^{-1}(\mathbb{R}^2)$. In Chap. 11 we prove a general existence theorem for such initial data for the case in which the vorticity has a distinguished sign.

Consider a solution to the incompressible 2D Euler equation in which the vorticity is concentrated on a smooth hypersurface S and the velocity field is discontinuous across the surface (see Fig. 9.1). The surface S divides the plane into two regions, Ω_+ and Ω_-, with normal vector $\mathbf{n}$ pointing into Ω_+. Let $z(\alpha, t)$ denote a parameterization of the sheet at time t. In addition, we specify the vortex-sheet strength $\gamma(\alpha, t)$ along the curve. At a fixed time t, if the vorticity is in $\mathcal{M}(\mathbb{R}^2)$ then the sheet strength satisfies for all spatial test functions $\varphi(x) \in C_0^\infty(\mathbb{R}^2)$

$$\int_{\mathbb{R}^2} \varphi(x) d\omega(x, t) = \int_{-\infty}^{\infty} \varphi(z(\alpha, t))\gamma(\alpha, t)|z_\alpha| d\alpha.$$

On either side of the sheet, $\Omega_\pm$, the velocity $v_\pm$ is smooth and hence satisfies the

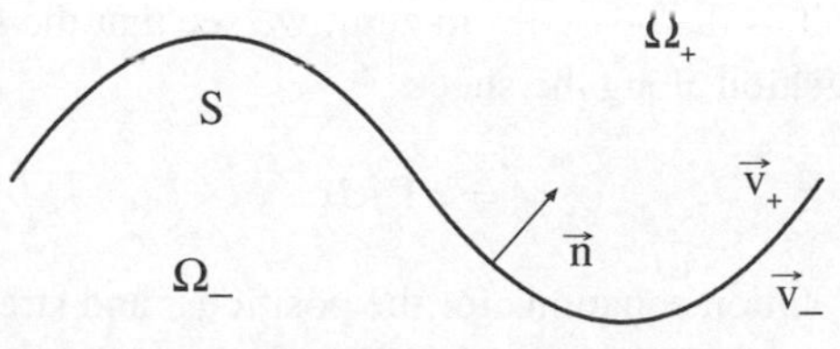

Figure 9.1. A 2D vortex sheet. The vorticity is concentrated along the curve S with strength γ. Off the sheet, the flow is irrotational and smooth.

strong form of the Euler equations:

$$\partial_t v_\pm + v_\pm \cdot \nabla v_\pm + \nabla p_\pm = 0,$$
$$\nabla \cdot v_\pm = 0.$$

On the sheet the velocity field $\mathbf{v}$ satisfies the weak form of incompressibility condition (9.5),

$$\int_{\mathbb{R}^2} \mathbf{v} \cdot \nabla\varphi \, dx = 0$$

for all smooth test functions φ. We denote by $[\mathbf{v}] = \mathbf{v}_+ - \mathbf{v}_-$ the jump in $\mathbf{v}$ across the sheet.

Now we consider test functions φ whose support crosses S. Integrating the weak incompressibility condition by parts, we see that

$$0 = \int \mathbf{v} \cdot \nabla\varphi = -\int_{\Omega_-} \operatorname{div} \mathbf{v}\varphi - \int_{\Omega_+} \operatorname{div} \mathbf{v}\varphi + \int_S \varphi[\mathbf{v}] \cdot \mathbf{n} \, dS.$$

On the right-hand side, the first two terms are zero by strong incompressibility; hence the last term must vanish. Because φ is arbitrary this implies that

$$[\mathbf{v}] \cdot \mathbf{n} = 0,$$

meaning that the jump in $\mathbf{v}$ must occur along a path that is tangent to the field $\mathbf{v}$.

Furthermore we can compute the vortex-sheet strength γ in terms of $[\mathbf{v}]$ by noting that

$$\begin{aligned} \int_{\mathbb{R}^2} \varphi \, d\omega &= \int_S \varphi(z(\alpha, t))\gamma(\alpha, t)|z_\alpha| d\alpha \\ &= \int_{\mathbb{R}^2} \mathbf{v} \times \nabla\varphi \, dx = \int_{\Omega^+ \cup \Omega^-} (\nabla \times v)\varphi \, dx + \int [\mathbf{v}] \times \mathbf{n}\varphi \, dS. \end{aligned} \tag{9.7}$$

The vorticity is zero off the sheet; hence $\int_{\Omega^+\cup\Omega^-} (\nabla \times v)\varphi \, dx = 0$ and Eq. (9.7) is satisfied for all test functions φ provided that

$$[\mathbf{v}] \times \mathbf{n} = [\mathbf{v}] \cdot \mathbf{s} = \gamma. \tag{9.8}$$

Thus the jump in $\mathbf{v}$ across the vortex-sheet determines the vortex-sheet strength γ. Furthermore, if we take a small box around the sheet and compute the line integral and then take the limit as the box goes to zero, we see that the sheet strength is just the incremental circulation along the sheet:

$$\gamma = d\Gamma/ds.$$

Now we derive evolution equations for the position z and strength γ of the vortex sheet. Recall that for smooth solutions of the Euler equation, the vorticity is conserved along particle paths in two dimensions. We saw that this was also true for weak solutions with vorticity in $L^1(\mathbb{R}^2) \cap L^\infty(\mathbb{R}^2)$. However, the particle paths may not be well defined for the case of a vortex sheet because the velocity field is at best in L^∞, and this is not sufficient to guarantee uniqueness of solutions to the particle-trajectory equation $dX/dt = v(X,t)$ (Coddington and Levinson, 1955). Nonetheless, if we can find consistent equations of motion for the sheet S itself and the sheet strength γ, because the vorticity is zero off the sheet, we have determined the evolution for the whole flow. Recall that the weak form of the incompressibility condition implies that the jump in $\mathbf{v}$ is in only the tangent component to the sheet. It is convenient to choose the average velocity $V = \frac{1}{2}(\mathbf{v}_+ + \mathbf{v}_-)$ on the sheet. Following the derivation of the particle-trajectory equations (see Section 2.5 of Chap. 2) for smooth solutions of the Euler equations, we see that the Biot–Savart law gives

$$\mathbf{v}(x) = \int_{\mathbb{R}^2} K_2(x - x')\omega(x')dx' = \int_S \gamma(\alpha')K_2[x - z(\alpha')]|z_\alpha|d\alpha',$$

where, as we recall from Chap. 2, Eq. (2.11), the kernel $K_2(x) = (1/2\pi)\nabla^\perp \log|x|$. The reader can verify by means of standard complex variables (or see Caflisch and Li, 1992, Appendix A) that the limit as x approaches a point on S from either side gives (ignoring the time dependence)

$$\mathbf{v}_\pm[z(\alpha)] = \mathrm{PV}\int_S K_2[z(\alpha) - z(\alpha')]|z_\alpha(\alpha')|\gamma[z(\alpha')]d\alpha' \pm \frac{1}{2}\gamma[z(\alpha)]\mathbf{s}[z(\alpha)],$$

where $\mathbf{s}[z(\alpha)]$ is the unit tangent to the sheet. It follows that the average velocity can be expressed as a principal-value integral along the sheet

$$V[z(\alpha)] = \mathrm{PV}\int_S \gamma[z(\alpha'),t]K_2[z(\alpha) - z(\alpha')]|z_\alpha(\alpha')|d\alpha'.$$

Advecting the sheet by this averaged velocity field gives the evolution equation for the position of the sheet:

$$\partial_t z(\alpha,t) = \mathrm{PV}\int_S \gamma[z(\alpha'),t]K_2[z(\alpha) - z(\alpha')]|z_\alpha(\alpha')|d\alpha'.$$

To complete the derivation, we need to determine γ. To do this we use some ideas from potential theory. Note that, off the sheet, the velocity field $v_\pm$ is irrotational and incompressible so that div $v_\pm$ and curl $v_\pm$ both vanish. This implies that there exist harmonic potentials $\Psi_\pm$ so that $\nabla\Psi_\pm = \mathbf{v}_\pm$. Formula (9.8) implies that the vortex sheet strength can be expressed in terms of $[\Psi] = \Psi_+ - \Psi_-$:

$$\gamma = \mathbf{s}\cdot\nabla[\Psi]. \tag{9.9}$$

Because curl $v_\pm = 0$ off the sheet, $v_\pm \cdot \nabla v_\pm = \frac{1}{2}\nabla|v_\pm|^2$. This leads to the well-known potential equation called *Bernoulli's law*:

$$\partial_t\Psi_\pm + \frac{1}{2}|\nabla\Psi_\pm|^2 + p_\pm = 0. \tag{9.10}$$

Note that

$$|\nabla\Psi_+|^2 - |\nabla\Psi_-|^2 = (\nabla\Psi_+ + \nabla\Psi_-)\cdot(\nabla\Psi_+ - \nabla\Psi_-)$$
$$= 2V\cdot\nabla[\Psi].$$

By subtracting the two equations in law (9.10) and using $p_+ = p_-$ across the sheet, we obtain

$$\partial_t[\Psi] + V\cdot\nabla[\Psi] = 0.$$

Thus $[\Psi]$ is conserved along particle paths that move with the flow V. Note that the vortex-sheet strength $\gamma = [\mathbf{v}]\cdot\mathbf{s} = [\nabla\Psi]\cdot\mathbf{s} = (d[\Psi]/ds) = d\Gamma/ds$. Thus, in particular, $d\Gamma = \gamma ds$ is conserved along particle paths of the flow

$$\gamma(z(\alpha,t),t)|z_\alpha(\alpha,t)|d\alpha = \gamma(z(\alpha,0),0)|z_\alpha(\alpha,0)|d\alpha$$

for a Lagrangian parameter α.

Recall that

$$K_2(x) = \frac{1}{2\pi}\nabla^\perp \log|x| = \frac{1}{2\pi}\mathbf{x}^\perp/|x|^2,$$

and that the sheet strength $\gamma = d\Gamma/ds$, where Γ is the total circulation. Using the circulation as a parameter along the sheet and combining the above expressions gives the B–R equation for the vortex sheet:

$$\frac{\partial}{\partial t}z^*(\Gamma,t) = \frac{1}{2\pi i}\mathrm{PV}\int_S \frac{d\Gamma'}{z(\Gamma,t) - z(\Gamma',t)}, \qquad (9.11)$$

where the circulation Γ is itself a local Langrangian parameterization of the sheet and z^* denotes the complex conjugate.

The B-R equation is analogous to the CDE

$$\partial_t z(\alpha,t) = \frac{1}{2\pi}\int \log|z(\alpha,t) - z(\alpha',t)|z_\alpha(\alpha',t)d\alpha'$$

from Chap. 8 for the boundary of a vortex patch. Both are self-deforming curve equations for the position z of the curve by use of a Lagrangian parameterization. There is, however, one very important difference. Note that the kernel in the B-R equation is of degree -1. This is in contrast to the kernel in the CDE that is of degree zero. This more singular kernel implies an *ill posedness* for the B-R equation that was not present in the case of the CDE. In fact, we showed in Chap. 8 that $C^{k,\alpha}$ initial data yield a unique solution to the CDE for all time. In contrast, the B-R equation is in general solvable for only short times with analytic initial data. The ill posedness means that the solution does not depend continuously on the initial data in topologies based on usual norms. We discuss in more detail the ill posedness of the B-R equation in the next section on *linear stability*.

9.3. The Kelvin–Helmholtz Instability

The instability of a vortex sheet has been known for over a century. In the 1860's, Helmholtz (1868) made the qualitative observation that boundaries of jets from organ pipes "rolled up" into periodically spaced spirals. Kelvin (1894; see also Rayleigh, 1896, Chaps. XX–XXI) provided the mathematical analysis of this instability. He considered a sinusoidal interface separating two fluids moving with different velocities u and u'.

We derive this linear stability theory by considering the perturbation of a flat, constant solution of the B-R equation:

$$z(\Gamma, t) = \Gamma. \tag{9.12}$$

We denote the perturbed sheet by

$$z(\Gamma, t) = \Gamma + \zeta(\Gamma, t). \tag{9.13}$$

In the derivation of the linearization below, we omit the time dependence from the notation but assume it implicitly. We have

$$\begin{aligned}\frac{\partial \zeta^*}{\partial t} &= \frac{1}{2\pi i}\text{PV}\int_{-\infty}^{\infty} \frac{d\Gamma'}{\Gamma - \Gamma' + \zeta(\Gamma) - \zeta(\Gamma')} \\ &= \frac{1}{2\pi i}\text{PV}\int_{-\infty}^{\infty} \left\{\frac{1}{1 + [\zeta(\Gamma) - \zeta(\Gamma')]/(\Gamma - \Gamma')}\right\}\frac{d\Gamma'}{\Gamma - \Gamma'}.\end{aligned}$$

Because ζ is a perturbation, ζ' is uniformly small and hence so is $[\zeta(\Gamma) - \zeta(\Gamma')]/(\Gamma - \Gamma')$. To leading order, the above integral is

$$\frac{1}{2\pi i}\text{PV}\int_{-\infty}^{\infty}\left[\frac{d\Gamma'}{\Gamma - \Gamma'} - \frac{\zeta(\Gamma) - \zeta(\Gamma')}{(\Gamma - \Gamma')^2}d\Gamma'\right]$$

The first integral vanishes by the cancellation property of the kernel. The second term simplifies by means of integration by parts to be $-(2\pi i)^{-1}$ times:

$$\lim_{\epsilon \to 0}\Big|_{\Gamma' = \Gamma - \epsilon}^{\Gamma' = \Gamma + \epsilon} \frac{\zeta(\Gamma) - \zeta(\Gamma')}{\Gamma - \Gamma'} + \text{PV}\int_{-\infty}^{\infty} \frac{\zeta'(\Gamma')}{\Gamma' - \Gamma}d\Gamma'.$$

The first term vanishes because both sides approach $\zeta'(\Gamma)$, and we are left with

$$\frac{\partial \zeta^*}{\partial t} = \frac{1}{2}\mathcal{H}\zeta', \tag{9.14}$$

where $\mathcal{H}$ is the Hilbert transform:

$$\mathcal{H}f(x) = \frac{1}{\pi i}\text{PV}\int_{-\infty}^{\infty} \frac{f(y)dy}{y - x}. \tag{9.15}$$

The Hilbert Transform

Suppose that $f(z)$ is a complex function that is analytic in the upper half-plane and on the real axis and satisfies $f(z) \to 0$ as $\operatorname{Im} z \to +\infty$. Then, for real x, $f(z)/(z-x)$ is analytic in the upper half-plane and on the real axis except for x. Suppose also that f decays fast enough at infinity on the real line so that the Hilbert transform makes sense. We then have

$$\mathcal{H}f(x) = \frac{1}{\pi i}\mathrm{PV}\int_{-\infty}^{\infty}\frac{f(y)dy}{y-x} = f(x)$$

(this is the Cauchy integral theorem).

Similarly, if f is analytic in the lower half-plane, including the real axis, and if f decays sufficiently rapidly as $\operatorname{Im} z \to -\infty$, then

$$\mathcal{H}f(x) = -f(x).$$

We summarize these two facts by writing

$$\mathcal{H}f = f_+ - f_-,$$

where f_+ is the upper analytic part of f and f_- is the lower analytic part. The constant term is omitted because the Hilbert transform of a constant is zero. Any function that is analytic in a strip containing the real line can be decomposed in such a fashion. In particular, $f(x)$ can be rewritten in terms of its Fourier transform: Consider k real; then

$$f(x) = \int e^{ikx}\hat{f}(k)dk,$$

in which case

$$\mathcal{H}f = \int_{k>0} e^{ikx}\hat{f}(k)dk - \int_{k<0} e^{ikx}\hat{f}(k)dk.$$

We now look at the stability of a solution to linearized equation (9.14) by looking for solutions of the form

$$\zeta(\Gamma, t) = A_k(t)e^{ik\Gamma} + B_k(t)e^{-ik\Gamma}, \qquad k > 0.$$

Then

$$\mathcal{H}\zeta' = ikA_k(t)e^{ik\Gamma} + ikB_k(t)e^{-ik\Gamma}.$$

Plugging this into Eq. (9.14) gives equations for A and B:

$$\dot{B}_k = -\frac{1}{2}ikA_k^*,$$

$$\dot{A}_k = -\frac{1}{2}ikB_k^*,$$

with the solutions

$$A_k(t) = A_k^+ e^{kt/2} + A_k^- e^{-kt/2}, \qquad B_k(t) = B_k^+ e^{kt/2} + B_k^- e^{-kt/2}, \tag{9.16}$$

where $A_k^+ = -iB_k^{+*}$, $A_k^- = iB_k^{-*}$.

The upshot is that the kth Fourier mode has a component that grows like $e^{|k|t/2}$. This instability is called the *Kelvin–Helmholtz* instability (Birkhoff, 1962; Saffman and Baker, 1979). This growth rate implies that the linear evolution problem is *ill posed.* This means that, given an initial condition, the solution at later times does not depend continuously on the initial data in any Sobolev norm. The only way to be guaranteed of a solution even on a short time interval is to take analytic initial data. To see this, consider the following lemma.

Lemma. *Suppose that the initial data $\zeta_0(\Gamma)$ are analytic in a strip of half-width Γ_0 around the real axis. Then the Fourier transform*

$$\hat{\zeta}_0(k) = \int_{-\infty}^{\infty} e^{-ik\Gamma} \varsigma_0(\Gamma) d\Gamma$$

decays at least as fast as

$$|\hat{\zeta}_0(k)| < ce^{-|k|\Gamma_0}.$$

The proof follows by use of the formula for a Fourier transform, deformation of the path of integration within the strip of analyticity, and then application of a straightforward estimate.

Note that if the initial data satisfy the above estimate, then the solution to linearized equation (9.14) at later times (which we construct by computing the solution for each mode as described above) satisfies

$$|\hat{\zeta}(k,t)| < Ce^{\frac{1}{2}|k|(t-2\Gamma_0)}.$$

In this case, $s(\Gamma, t)$ is analytic in a strip of half-width $\Gamma(t) = \Gamma_0 - \frac{1}{2}t$ so that singularities cannot form before time $t_* = 2\Gamma_0$.

We now show that there exists a family of exact solutions to *linearized* equation (9.14) that are analytic initially and have a singularity at a finite time.

They take the form

$$\zeta(\Gamma, t) = (1-i)\left[\left(1 - e^{-\frac{1}{2}t - i\Gamma}\right)^{1+\nu} - \left(1 - e^{-\frac{1}{2}t + i\Gamma}\right)^{1+\nu}\right], \qquad \nu > 0,$$

which is finite with finite derivatives at all times and has a singularity in the curvature $\zeta_{\Gamma\Gamma}$ at $t = 0$. The solution is smooth and decays to zero as $t \to \infty$. Finally this solution is analytic in the strip $|\text{Im}\,\Gamma| < \frac{1}{2}t$. By use of the time-reversal symmetry $\zeta(\Gamma, t) \to \zeta^*(\Gamma, -t)$, this gives a solution that forms a singularity in finite time.

Birkhoff (1962) conjectured that the full nonlinear problem (9.11) with analytic initial data is well posed, at least for a finite time. This is proved in several forms (Caflisch and Orellana, 1986; Sulem et al., 1981). Caflisch and Orellana (1989) prove nonlinear ill posedness of the B-R equation in all Sobolev spaces H^n with $n > 3/2$. Duchon and Robert (1988) show that for a very special class of initial conditions there exists a global analytic solution for all time.

9.4. Computing Vortex Sheets

There has been a tremendous amount of work done on the numerical computation of vortex sheets. In this section we present some of the useful techniques used and describe some of the recent results in the context of the issues of interest in the following chapters. Difficulties arise in computing vortex-sheet motion that is due to the Kelvin–Helmholtz instability described in Section 9.3.

The B-R equation

$$\partial_t z^*(\Gamma, t) = \int_S K[z(\Gamma, t) - z(\Gamma', t)] d\Gamma', \qquad K = \frac{1}{2\pi i z}, \tag{9.17}$$

can be solved only for analytic initial data, and careful numerics shows that typically finite-time singularities in the sheet occur in which a spiral roll-up forms. Birkhoff conjectured that the nonlinearity would cause a singularity to form in finite time (Birkhoff, 1962; Birkhoff and Fisher, 1959). An asymptotic analysis that is due to Moore (1979, 1984) suggests that an $\alpha^{3/2}$ branch point forms in the vortex sheet at a finite critical time. Whether or not this is the correct form of the singularity has been the subject of debate in the literature. Meiron et al. (1982) examine a Taylor series approximation of z and obtain results consistent with Moore's. Shelley (1992) observes that Moore's analysis is valid at early times but that, close to singularity formation, although the real part of the solution remains consistent with Moore's analysis, the imaginary part shows a weaker behavior.

Regardless of what the exact form of the scaling in the initial singularity is, it is clear that the onset of a singularity is characterized by the appearance of a very weak nonanalyticity in the sheet, that of a blowup in the curvature while the tangent to the sheet remains continuous. However, immediately following this initial weak singularity, computations of slightly smoothed solutions of the equations indicate that the vortex sheet undergoes a roll-up with a tight spiral occurring at the point where the initial weak singularity occurred.

In Chap. 11 we prove that when the initial vorticity has a fixed sign, there is a solution to the Euler equation describing the evolution of the vortex sheet for all time, including after the initial singularity. It is still an open question as to whether or not such solutions are unique. The issue of nonuniqueness may be approached by examination of different well-posed approximations to the vortex sheet. Three such methods have been addressed in the computational literature.

9.4.1. The Point-Vortex Approximation

In Chap. 6 we introduced several *vortex methods* for computing smooth solutions to the Euler equation. Vortex methods use a discretization of the vorticity-stream form of the Euler equation combined with a smoothing of the Biot–Savart kernel. As we saw in

Chap. 6, this smoothing was necessary to remove the infinite energy associated with a *point-vortex* discretization of the equations, the natural discretization that arises from using directly the unsmoothed Biot–Savart kernel. Using point vortices for numerical simulations is a delicate issue and requires some understanding of the underlying mechanics to avoid spurious behavior sometimes seen in simulations (Rosenhead, 1932).

Point vortices have been used successfully to compute the short time evolution of vortex sheets with analytic initial data. The point-vortex approximation describes the vorticity by a discrete set of point vortices. In the case of a vortex sheet, the point-vortex approximation can give a simple discretization of the B-R equation. Discretize the sheet by point vortices $z_j(t)$, $j = 1, \ldots, N$. The next step involves the best way to approximate the B-R integral. The most basic point-vortex method, used originally by Rosenhead in 1931 (Rosenhead, 1932) to study periodic vortex sheets, is to solve the ODEs:

$$\frac{dz_j^*}{dt} = \sum_{k \neq j} K(z_j - z_k) N^{-1}, \qquad K(z) = \frac{1}{2i} \cot \pi z. \tag{9.18}$$

Both these and higher-order quadrature rules for the B-R integral (Higdon and Pozrikidis, 1985; van de Vooren, 1980; Shelley, 1992) all suffer from the same problem: No matter how good the order of accuracy in approximating the B-R integral, some level of noise [usually from round-off error (Krasny, 1986b)] is always introduced *at the frequency of the grid*. Figure 9.2 shows the evolution of a periodic sheet (as in Krasny, 1986b) taken initially to have a uniform strength of one with shape $z(x) = \Gamma + 0.01(1 - i)\sin[2\pi(\Gamma)]$. Here the time is $t = 0.37$, right before the singularity occurs at $t \sim 0.375$. Shown are two periods of the solution, one calculated with 100 points in single precision and the other with 100 points in double precision. Note the huge reduction of the noise gained by the higher precision calculation. From solutions (9.16) we see that for near-planar sheets this noise tends to grow as $e^{kt/2}$, where k is the frequency associated with the grid spacing. For an evenly spaced grid with N points this means that the noise generated from round-off error grows at a rate of $e^{Nt/2}$. The result is that the finer the resolution desired to resolve a structure, the faster the noise is amplified by the Kelvin–Helmholtz instability. It is important to note here that this is a feature of the underlying equation itself as opposed to an instability of the numerical method.

A clever idea that is due to Krasny (1986b) is the implementation of a Fourier filter to halt the amplification of grid-scale noise. Note that the function $\xi(\Gamma, t) = z(\Gamma, t) - \Gamma$

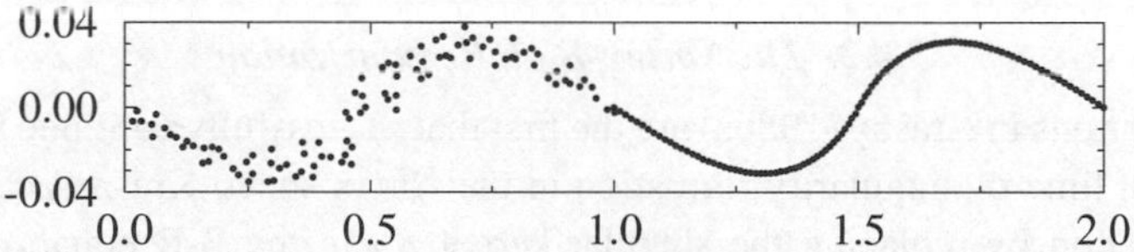

Figure 9.2. The analytic vortex sheet calculated with the point-vortex method. The time shown is $t = 0.37$. The period $[0, 1]$ shows data from a calculation done in single precision and the period $[1, 2]$ shows data from a calculation done in double precision. Both computations use 100 points on one interval.

is periodic. Thus we can compute the discrete Fourier transform

$$\hat{\xi}(t) = N^{-1/2} \sum_{j=1}^{N} \xi_j(t) e^{-2\pi i k \Gamma_j}.$$

Krasny showed that the source of the fluctuations in Fig. 9.2 is from the round-off error's swamping the initially exponentially small high-frequency modes and being amplified by the Kelvin–Helmholtz instability. He proposed a filtering technique whereby a threshold value δ is chosen (usually based on machine precision) and on each time step any value of $\hat{\xi}(t)$ that is less than the threshold value is set to zero. Krasny showed by numerical computations that this method can produce results similar to those produced by higher-order precision in the numerical arithmetic alone without the added expense that typically accompanies higher-order precision computation. Although point vortices are useful for computing very smooth vortex-sheet structures, in which small-scale noise can be controlled on relatively short time scales, other methods must be used to analyze longer time and more singular behavior in the sheet. Convergence of the filtering technique was recently proved in Caflisch et al. (1999).

Perhaps the most exciting work in the field of vortex-sheet computation is the results for the behavior of the sheet after the initial weak singularity. There are two main issues of interest. First, for the case of fixed-sign vorticity initial data, is there a unique way to continue the evolution? Second, for the case of mixed-sign vorticity initial data, does the weak Euler evolution exist for all time or do coherent structures in the form of *concentrations* develop from the interaction of vorticity of opposite sign? Neither of these questions has been answered analytically, and both pose interesting problems for detailed future numerical and analytical research.

After the singularity time the B-R equation breaks down, and the only recourse is to look for a solution to the Euler equation in weak form (9.6) by means of the limit of a smooth or less singular approximation.

There are a number of different regularizations for vortex-sheet initial data. They include approximating the vortex sheet by smoothed-out vortex blobs (Krasny, 1986a), approximating the sheet by a thin vortex patch (Baker and Shelley, 1990), approximating the Euler solution by a high Reynolds number Navier–Stokes solution combined with a smoothing of the initial data (Tryggvason et al., 1991) and stabilizing the instability on the sheet by use of surface tension (Hou et al., 1994). We describe briefly these techniques below.

9.4.2. *The Vortex-Blob Regularization*

The calculations in Krasny (1986a) are the first that successfully continue beyond the first critical time of singularity formation in the vortex sheet. Krasny accomplishes this calculation by replacing the singular kernel K_2 in the B-R equation with the smoothed kernel

$$K_\delta(x) = K(x)\frac{|x|^2}{|x|^2 + \delta^2} = \frac{x^\perp}{|x|^2 + \delta^2}. \tag{9.19}$$

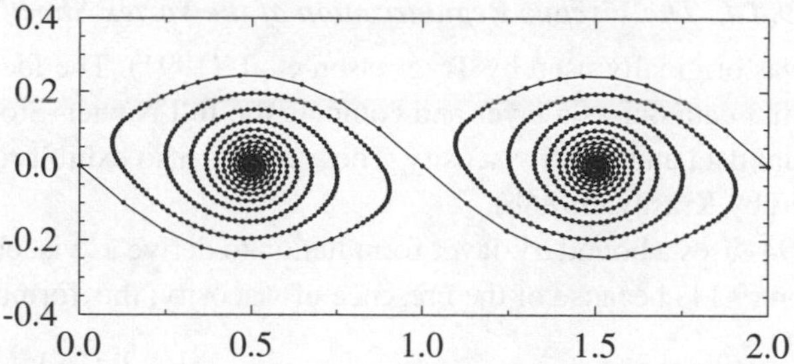

Figure 9.3. Computation of the periodic vortex sheet after singularity by means of a desingularization of the B-R kernel. Here the smoothing parameter $\delta = 0.25$, and 512 points are used for the calculation. The computation is shown at $t = 3.9$, well after the initial singularity. Note the approximation to a double-branched spiral.

Recall from Chap. 6, Eqs. (6.45), that this is one of the possible kernels for the *vortex-blob* method. This choice of desingularization changes the Kelvin–Helmholtz instability so that the dispersion relation for the growth rate $e^{\beta(k)t}$ satisfies

$$\beta^2 = \frac{k\left[1 - e^{-k\cosh^{-1}(1+\delta^2)}\right] e^{-k\cosh^{-1}(1+\delta^2)}}{4\delta(2+\delta^2)^{1/2}}.$$

In particular, for fixed δ, $\beta(k) \to 0$ as $k \to \infty$, so the evolution is linearly well posed. Also, for fixed k as $\delta \to 0$, we recover $\omega^2 = k^2/4$.

The important key step in computing the vortex sheet lies in the order of the limits $\delta \to 0$ and $h \to 0$, where h is the discretization or mesh size. If N points are used to describe the initial data, then h is L/N, where L is the size of the domain. Fixing δ and letting $h \to 0$ produces a smooth solution to the Euler equation with the initial data of a "regularized" sheet. Taking the limit $\delta \to 0$ after $h \to 0$ yields the dynamics of the vortex sheet. Figure 9.3 shows a computation of the periodic vortex sheet that uses the same regularization as that in Krasny (1986a) with $\delta = 0.25$ and 512 points initially evenly distributed along the periodic interval. The time shown in the figure is $t = 3.9$, well after the initial singularity. Note the spiral roll-up pattern shown here. As the regularization parameter δ decreases, the spiral at any fixed time t shows more turns. By extrapolation, the limit as $\delta \to 0$ appears to produce a spiral with an infinite number of turns. In this calculation we show a roll-up of a vortex sheet with a fixed sign. The computations are consistent with Pullin's conjecture (Pullin, 1989) that the solution beyond the critical time of singularity formation is locally a double-branched spiral with an infinite number of turns.

9.4.3. The Vortex-Patch Regularization

Another regularization of the vortex sheet, used by Baker and Shelley (1990), is an approximation by a layer with mean finite thickness h and uniform vorticity inside. If the uniform vorticity scales as $1/h$, then the $h \to 0$ limit yields the measure-valued initial condition of a vortex sheet. An interesting facet of their calculations is the appearance of an approximate Kirchoff ellipse during the roll-up process. In all runs, the ellipse always has the same aspect ratio, with an area scaling like $O(h^{1.6})$ as $h \to 0$.

9.4.4. The Viscous Regularization of the Vortex Sheet

This method was originally used by Trygvasson et al. (1991). The idea is simply to smooth the initial data out as a layer and compute the full Navier–Stokes equations with those initial data and small viscosity. These results also exhibit roll-up, similar to that observed by Krasny (1986b).

Dhanak (1994) uses a boundary-layer formulation to derive a "viscous correction" to B-R equation (9.11) because of the presence of viscosity; this formulation is

$$\frac{\partial}{\partial t} z^*(\Gamma, t) = \frac{1}{2\pi i} \mathrm{PV} \int_S \frac{d\Gamma'}{z(\Gamma, t) - z(\Gamma', t)} + \nu \frac{1}{|z_\Gamma|^3} \frac{\partial |z_\Gamma|}{\partial \Gamma} \frac{\partial z^*}{\partial \Gamma}. \tag{9.20}$$

For this modified B-R equation, the dispersion relation is

$$\beta(k) = \frac{1}{2}\left[(k^4\nu^2 + k^2)^{1/2} - k^2\nu\right]. \tag{9.21}$$

Note that as $k \to \infty$, $\beta \to (4\nu)^{-1}$. As $\nu \to 0$ we recover the dispersion relation for the B-R equation.

In Chap. 11 we prove a result concerning the viscous regularization of the vortex-sheet: For vortex-sheet initial data with distinguished sign, the Navier–Stokes regularization of the problem produces a weak solution to the inviscid Euler equation in the sense of Eq. (9.6) in the limit as the viscosity goes to zero. In fact, for the distinguished-sign case, the method of proof in Chap. 11 shows that all of the above regularizations produce a solution to the Euler equation in the limit as the regularization parameter goes to zero.

9.4.5. Regularizing with Surface Tension

A recent paper by Hou et al. (1997) examines a regularization of the vortex sheet by means of surface tension. This would be the correct model to study two different immiscible fluids separated by an interface. However, the method does not directly apply to a single fluid with a sharp transition in velocity, as we consider in this book. For the sake of completeness, we mention this method. Surface tension modifies the dynamics in a simple way. Recall from the derivation of B-R equation (9.11) that we used the fact that the pressure $p^\pm$ is continuous across the sheet. In the presence of surface tension, the pressure has a jump discontinuity across the sheet that is equal to the surface tension σ times the local curvature κ of the sheet. The result is a modified B-R equation:

$$\begin{aligned} \frac{\partial}{\partial t} z^*(\Gamma, t) &= \frac{1}{2\pi i} \mathrm{PV} \int_S \frac{\tilde{\gamma}(\alpha) d\alpha'}{z(\alpha, t) - z(\alpha', t)}, \\ \frac{\partial \tilde{\gamma}}{\partial t} &= \sigma \kappa_\alpha, \\ \tilde{\gamma} &= |z_\alpha| \gamma = |z_\alpha| [v] \cdot \mathbf{s}. \end{aligned} \tag{9.22}$$

As in the case of vortex-blob regularization, surface tension serves to damp the exponential growth in the high modes and turns the problem into one that is linearly well posed. This can be seen immediately in the fact that dispersion relation for (9.22) is

$$\beta(k)^2 = \frac{k^2}{4} - \frac{\sigma}{2}|k|^3.$$

Hou et al. (1997) consider a slightly perturbed single-sign vortex sheet and shows that the roll-up occurs, not as a tight spiral, but as a winding of two "bubbles" of the different fluids with an initial singularity that corresponds to a pinch-off or breaking of the sheet. This kind of singularity has not been observed in other simulations without surface tension. However, it is typical of interface motion in which surface tension dominates (Bertozzi et al., 1994).

A particularly interesting case study is one in which the sign of the initial vorticity is not fixed.

9.4.6. The Case of Mixed-Sign Initial Data

An illuminating calculation performed by Krasny (1987) involving mixed-sign vorticity is that of the elliptically loaded wing. He again uses the *vortex-blob* approach with desingularized kernel (9.19). He demonstrates the convergence of the algorithm in the limit as $\delta \to 0$, especially in the vicinity of the tips where the roll-up occurs. As further evidence for the validity of his approach, Krasny compares the numerical solution with Kaden's self-similar spiral (Kaden, 1931). A second calculation reported by Krasny considers initial data with a vortex-sheet strength that changes sign three times. Although the initial condition is a flat finite sheet, the moderate time behavior shown in Fig. 9.4 illustrates the roll-up at six different locations along the sheet. Figure 9.5 shows a close-up of one of the sides at a later-time. Note the incredibly small-scale complexity generated by large-scale coherent structures on the sheet that drive their development. A mathematical framework designed to address such complexity is briefly outlined in the next section and discussed in detail in the subsequent Chapters.

9.5. The Development of Oscillations and Concentrations

Vortex sheets suggest a natural framework within which to build a mathematical theory. Despite the singularity in vorticity, they retain the physically significant feature of *finite kinetic energy*. As recent numerical experiments show (Nitsche and Krasny, 1994), vortex sheets are good models for real physical objects in which the vorticity is actually finite but nonetheless sharply peaked along a surface. As parameters such as the Reynolds number in certain experiments are increased, the data more closely align with the idealized model of an inviscid-vortex sheet.

This connection suggests a mathematical family of problems: Consider a sequence of exact or approximate solutions v^ϵ to the Euler equation that satisfies a uniform bound on local kinetic energy:

$$\int_\Omega |v^\epsilon|^2 dx \le C(\Omega) \tag{9.23}$$

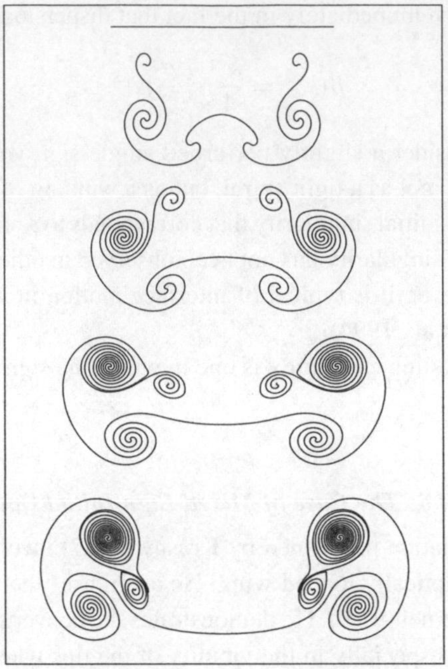

Figure 9.4. Intermediate time behavior of the sheet. Six vortices emerge during the time $1 \leq t \leq 4$.

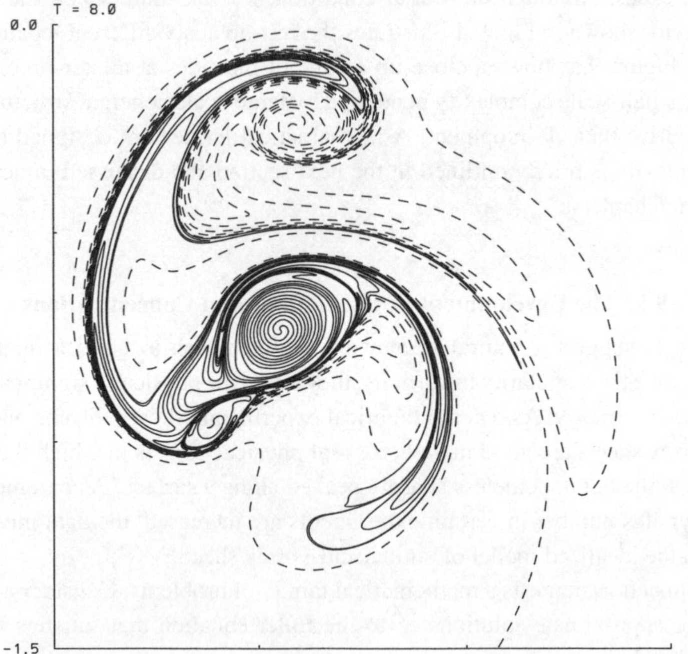

Figure 9.5. Late time evolution of the roll-up ($t = 8$). Pictured is a close-up of the right-hand side. The solid (dashed) curves correspond to negative (positive) vorticity. Note the complex intertwining of positive and negative vorticity.

(here C is independent of ϵ). Uniform bound (9.23) implies that there is a function v and a subsequence that converges weakly in L^2_{loc}:

$$\forall \text{ bounded } \Omega, \qquad v^\epsilon \rightharpoonup v \in L^2(\Omega).$$

Does the limit v satisfy the Euler equation? What kinds of defects are formed in the limiting process? We consider the heuristic (to be made rigorous later) decomposition

$$v^\epsilon = \bar{v}^\epsilon + v^\epsilon_{\text{osc}} + v^\epsilon_{\text{conc}}.$$

Here $\bar{v}^\epsilon$ denotes the "part" of v^ϵ that converges strongly in L^2, so that there is a $\bar{v}$ satisfying

$$\int_\Omega |\bar{v}^\epsilon - \bar{v}|^2 dx \to 0 \qquad \text{as} \quad \epsilon \to 0.$$

The function v^ϵ_{osc} measures the small-scale oscillations in the limit in the sense that v^ϵ_{osc} does not converge strongly in L^2; however,

$$\left|v^\epsilon_{\text{osc}}(x)\right|_{L^\infty} \leq C, \tag{9.24}$$

and there exists $v_{\text{osc}} \in L^\infty$ such that

$$\int_\Omega \varphi v^\epsilon_{\text{osc}}\, dx \to \int_\Omega \varphi v_{\text{osc}}\, dx \qquad \text{for all functions} \quad \varphi \in C_0(\Omega).$$

A prototypical example of a sequence with small-scale oscillations in $\mathbb{R}$ is $v^\epsilon_{\text{osc}} = v(x/\epsilon)$, where v is a fixed smooth periodic function.

The function v^ϵ_{conc}, which measures concentrations in the limit, also satisfies the uniform L^2 bound that implies that a subsequence converges weakly to some limit function $v_{\text{conc}} \in L^2$. Unlike v^ϵ_{osc}, v^ϵ_{conc} does not satisfy a uniform L^∞ bound but does converges pointwise almost everywhere to a function v_{conc}. The convergence is not strong in L^2. A simple example of concentration in $\mathbb{R}^N$ is the sequence $v^\epsilon_{\text{conc}} = \epsilon^{-N/2}\rho(x/\epsilon)$, where ρ is a smooth positive function $\int \rho^2 = 1$. As $\epsilon \to 0$, ρ^2 converges to a delta function, but v^ϵ_{conc} converges weakly in L^2 and pointwise almost everywhere to 0. Until approximately 15 years ago, mathematicians studying nonlinear PDEs mainly considered problems in which a priori estimates guarantee strong convergence (so that quantities like v_{osc} and v_{conc} are zero). Many exciting current research efforts consider situations with both fine-scale oscillations and concentrations. It is our hope that a greater understanding of the underlying mathematics will shed more light on the physics at hand.

The Young measure, introduced by Tartar in the context of the nonlinear PDE (Tartar, 1979, 1983, 1986), is an effective tool for studying a sequence $\{v^\epsilon_{\text{osc}}\}$ satisfying an L^∞ bound. If the vector field v maps $\Omega \subset \mathbb{R}^N$ to $\mathbb{R}^M$, the Young measure theorem asserts that there is a family of probability measures on $\mathbb{R}^M$ $\{\nu_y\}$ indexed by $y \in \Omega$ so that, after passing to a subsequence, weak limits of composite nonlinear maps $g(v^\epsilon)$

satisfy

$$\lim_{\epsilon \to 0} \int_\Omega \varphi(y) g\left(v_{\text{osc}}^\epsilon\right) dy = \int_\Omega \varphi(y) \int_{\mathbb{R}^M} g(\lambda) d\nu_y(\lambda) dy. \qquad (9.25)$$

The sequence v_{osc}^ϵ converges strongly to v_{osc} on Ω if and only if the probability measure ν_y is given by the Dirac mass, $\delta_{v_{\text{osc}}(y)}$ for each $y \in \Omega$. To attack problems from mathematical physics in which both fine-scale oscillations and concentrations develop in the limit, DiPerna and Majda (1987b) developed a very general version of the Young measure theorem for sequences with only a uniform L^2 or L^p bound. Problems involving the zero-viscosity limit of solutions of Navier–Stokes equations with finite kinetic energy (9.1) provide motivation for this work. In general, the limiting solution is a measure-valued solution of the Euler equation with both oscillations and concentrations. We discuss this work in detail in Chap. 12. Unlike Eq. (9.25), the generalized Young measures in this context are often continuous with respect to Lebesgue measure, reflecting the development of concentrations.

The simplest way to generate examples of approximate-solution sequences and measure-valued solutions for the 3D Euler equation is to take a sequence of smooth exact solutions and examine the generalized Young measure in the limit. The reader can verify that any sequence of smooth velocity fields

$$v^\epsilon = \left[0, 0, v_3^\epsilon(x_1, x_2)\right], \qquad |v^\epsilon|_{L^2}(\mathbb{R}^3) \leq C, \qquad (9.26)$$

defines a sequence of exact smooth solutions of the 3D Euler equation converging to a measure-valued solution of the 3D Euler equation. By following the constructions above, one can generate solution sequences with both oscillations and concentrations. If $v_3^\epsilon \rightharpoonup v_3(x_1, x_2)$ in L^2, then $v = [0, 0, v_3(x_1, x_2)]$ defines a weak solution of the 3D Euler equation, despite the fact that typically $(v_3^\epsilon)^2$ does not converge weakly to v_ϵ^2.

Recall from Chap. 2, Proposition 2.7, that every 2D flow always generates a large family of 3D flows with vorticity. That is, let $\tilde{v}(\tilde{x}, t), \tilde{p}(\tilde{x}, t), \tilde{x} \in \mathbb{R}^2$, be a solution to the 2D Euler equation and let $v^3(\tilde{x}, t)$ be a solution to the linear scalar equation $(\widetilde{D}/Dt)v^3 = 0$, $v^3|_{t=0} = v_0^3$, where $(\widetilde{D}/Dt) = (\partial/\partial t) + \tilde{v} \cdot \nabla_{\tilde{x}}$. Then $v = (\tilde{v}, v^3)$ solves the 3D Euler equation provided that the velocity $v^3(\tilde{x}, t)$ does not depend on the x_3 variable. A special class of such "$2\frac{1}{2}$D" flows is generated by taking as the base 2D flow a shear flow (e.g., Example 1.5 from Chap. 1) that depends on two scales x_2/ϵ and x_2:

$$v^\epsilon = \{v(x_2/\epsilon, x_2), 0, u[x_1, -v(x_2/\epsilon, x_2)t, x_2, x_2/\epsilon]\}, \qquad (9.27)$$

where v and u are smooth bounded functions with $v(y, x_2)$ periodic in y. Although this sequence satisfies the uniform L^∞–bound $|v^\epsilon|_{L^\infty} \leq C$, the weak limit of Eq. (9.27) is an explicit smooth function that *does not* satisfy the 3D Euler equations. This is an explicit example of a non-trivial measure-valued solution showing explicitly that new phenomena can occur in three dimensions by means of persistence of oscillations in the limiting process. We discuss this (see Example 12.3) and other examples in greater detail in Chap. 12.

When concentration occurs without the development of oscillations, it is natural to measure the size of the concentration set. Two types of defect measures exist for this purpose, the weak* defect measure (Lions, 1984) and the reduced defect measure (DiPerna and Majda, 1988). Given uniform L^2 bound (9.23) and a weak limit $v^\epsilon \rightharpoonup v$, there is a nonnegative Radon measure σ, called the *weak* defect measure*, such that

$$\int \varphi d\sigma = \lim_{\epsilon\to 0} \int_\Omega \varphi |v^\epsilon - v|^2 dy \tag{9.28}$$

for all $\varphi \in C^0(\Omega)$. In particular, if σ is zero on an open set, then v^ϵ converges strongly on that set. Motivated by problems in the calculus of variations with critical Sobolev exponents, Lions (1984) proved a beautifully simple theorem guaranteeing that if, in addition to uniform bound (9.23), one also has a uniform bound on the L^1 norm of the gradient

$$\int_\Omega |v^\epsilon|^2 dx + \int_{\mathbb{R}^2} |\nabla v^\epsilon| \le C(\Omega) \tag{9.29}$$

then the weak* defect measure lives on, at most, a countable number of points $\{y_j\}_{j=1}^\infty$ with associated weights $\alpha_j \ge 0$. Furthermore, $\sum \alpha_j^{1/2} < \infty$ such that

$$\sigma = \sum_j \alpha_j \delta_{y_j},$$

where δ_z denotes Dirac mass at z. The reduced defect measure is a finitely subadditive outer measure θ defined by

$$\theta(E) = \limsup \int_E |v^\epsilon - v|^2 dx.$$

The reduced defect measure has the property that $\theta(E)$ vanishes for any set E if and only if there is strong convergence on E. Furthermore, for any closed set F,

$$\theta(F) \le \sigma(F),$$

so that θ can be very small even though $\sigma(F)$ is large. We make heavy use of these measures in Chaps. 11 and 12.

The radial eddies (Example 2.1)

$$v = (-x_2/r, x_1/r) \int_0^r s\omega(s) ds \tag{9.30}$$

provide the basic building blocks of several elementary examples of exact solution sequences for the 2D Euler equation that exhibits concentration in the limit.

Example 9.1. Given a *positive* function $\omega(r) \ge 0$ with bounded support, define

$$v^\epsilon = (\log 1/\epsilon)^{-1/2} \epsilon^{-1} v(x/\epsilon)$$

with v determined from ω by means of radial eddies (9.30) above.

Example 9.2. Consider a function $\omega(r)$ with compact support and vorticity of zero mean, $\int_0^\infty s\omega(s)ds = 0$, implying that the circulation is zero outside the support of ω. Define v^ϵ by $\epsilon^{-1}v(x/\epsilon)$, with v defined by radial eddies (9.30).

The scaling in both examples guarantees uniform local kinetic-energy bound (9.23). In both examples, $v^\epsilon \rightharpoonup 0$, but the nonlinear terms in the Euler equation $v \otimes v$ satisfy

$$v^\epsilon \otimes v^\epsilon \rightharpoonup C \begin{bmatrix} 1 & 0 \\ 0 & 1 \end{bmatrix} \delta(x)$$

with weak convergence in the sense of measures. We discuss these and other interesting examples that demonstrate concentration in Chap. 11.

Notes for Chapter 9

The B-R equation has many derivations. The form presented here is similar to that of Caflisch and Li (1992).

We acknowledge useful discussions with Mike Shelley and Robert Krasny on the computational side of vortex-sheet motion and also the lecture notes of Almgren (1995) for the exposition of the Kelvin–Helmholtz instability.

The calculations of Krasny (1987) are not the first to address these issues. Indeed, the earlier work of Chorin and Bernard (1973) on the same problem confirms the trends observed by Krasny. For a nice review article on these and other computations see Krasny (1991).

For a more analytical treatment of the diffusing vortex sheet, the reader is referred to Dhanak (1994). For a more extensive description of interfaces with surface tension and numerical methods for such problems see Hou et al., (1994).

For a proof that the vortex-patch regularization of the vortex sheet converges to a solution of the B-R equation for analytic initial data and short times, see Benedetto and Pulvirenti (1992).

References for Chapter 9

Almgren, R., "Interface motion," Lecture notes from graduate course at University of Chicago, 1995.

Avellaneda, M., and Majda, A. J., "Simple examples with features of renormalization for turbulent transport," *Philos. Trans. R. Soc. London A* **346**, 205–233, 1994.

Baker, G. R. and Shelley, M. J., "On the connection between thin vortex layers and vortex sheets," *J. Fluid Mech.* **215**, 161–194, 1990.

Benedetto, D. and Pulvirenti, M., "From vortex layers to vortex sheets," *SIAM J. Appl. Math.* **52**, 1041–1056, 1992.

Bertozzi, A. L., Brenner, M. P., Dupont, T. F., and Kadanoff, L. P., "Singularities and similarities in interface flow," in L. Sirovich, ed., *Trends and Perspectives in Applied Mathematics*, Vol. 100 of Applied Mathematical Series, Springer-Verlag, New York, 1994, pp. 155–208.

Bertozzi, A. L. and Pugh, M., "The lubrication approximation for thin viscous films: regularity and long time behavior of weak solutions," *Commun. Pure Appl. Math.* **49**, 85–123, 1996.

Birkhoff, G., *Hydrodynamics: "A Study in Logic, Fact and Similitude,"* Princeton Univ. Press, Princeton, NJ, 1960.

Birkhoff, G., "Helmholtz and Taylor instability," in *Hydrodynamics Instability*, Proceedings of the Symposium on Applied Mathematics, American Mathematical Society, Providence, RI, 1962, Vol. 13, pp. 55–76.

Birkhoff, G. and Fisher, J., "Do vortex sheets roll up?" *Rend. Circ. Math. Palermo* **8**, 77–90, 1959.

Caflisch, R. E., Hou, T. Y., and Lowengrub, J., "Almost optimal convergence of the point vortex method for vortex sheets using numerical filtering," *Math. Comput.* **68**, 1465–1496, 1999.

Caflisch, R. E. and Li, X.-F., "Lagrangian theory for 3D vortex sheets with axial or helical symmetry," *Transp. Theory Stat. Phys.* **21**, 559–578, 1992.

Caflisch, R. E. and Orellana, O. F., "Long time existence for a slightly perturbed vortex sheet," *Commun. Pure Appl. Math.* **39**, 807–838, 1986.

Caflisch, R. E. and Orellana, O. F., "Singular solutions and ill-posedness for the evolution of vortex sheets," *SIAM J. Math. Anal.* **20**, 293–307, 1989.

Chorin, A. J. and Bernard, P. J., "Discretization of vortex sheet with and example of roll-up," *J. Comput. Phys.* **13**, 423–428, 1973.

Coddington, E. A. and Levinson, N., *Theory of Ordinary Differential Equations*, McGraw-Hill, New York, 1955.

Dhanak, M. R., "Equations of motion of a diffusing vortex sheet," *J. Fluid Mech.* **269**, 265–281, 1994.

Didden, N., "On the formation of vortex rings: rolling-up and production of circulation," *Z. Angew. Math. Phys.* **30**, 101–116, 1979.

DiPerna, R. J. and Majda, A. J., "Concentrations in regularizations for 2-D incompressible flow," *Commun. Pure Appl. Math.* **40**, 301–345, 1987a.

DiPerna, R. J. and Majda, A. J., "Oscillations and concentrations in weak solutions of the incompressible fluid equations," *Commun. Math. Phys.* **108**, 667–689, 1987b.

Duchon, J. and Robert, R., "Global vortex sheet solutions of Euler equations in the plane," *J. Diff. Eqns.* **73**, 215–224, 1988.

Higdon, J. J. L. and Pozrikidis, C., "The self-induced motion of vortex sheets," *J. Fluid Mech.* **150**, 203–231, 1985.

Hou, T. Y. and Lowengrub, J. S., and Shelley, M. J., "Removing the stiffness from interfacial flows with surface tension," *J. Comput. Phys.* **114**, 312–338, 1994.

Hou, T. Y., Lowengrub, J. S., and Shelley, M. J., "The long-time motion of vortex sheets with surface tension," *Phys. Fluids*. **9**, 1933–1954, 1997.

James, R. D. and Kinderlehrer, "Theory of diffusionless phase transitions," in: PDEs and continuum models of phase transitions (Nice 1988) 51–84. Lecture Notes in Physics 344, Springer Berlin 1989.

Kaden, H., "Aufwicklung einer unstabilen Unstetigkeitsfläche," *Ing. Arch. Berlin* **2**, 140–168, 1931.

Lord Kelvin, *Nature (London)* **50**, 524, 549, 573, 1894.

Krasny, R., "Desingularization of periodic vortex sheet foll-up," *J. Comput. Phys.* **65**, 292–313, 1986a.

Krasny, R., "A study of singularity formation in a vortex sheet by the point vortex approximation," *J. Fluid Mech.* **167**, 292–313, 1986b.

Krasny, R., "Computation of vortex sheet foll-up in the Treffitz plane," *J. Fluid Mech.* **184**, 123–155, 1987.

Krasny, R., "Computing vortex sheet motion," in *Proceedings of International Congress of Mathematicians, Kyoto 1990*, Math. Soc. Japan, Tokyo, 1991, Vol. 2, pp. 1573–1583.

Lions, P.-L., "The concentration–compactness principle in the calculus of variations. The locally compact case. II, *Ann. Inst. H. Poincaré Anal. Non Linéaire*, Annales de 1'Institut Henri Poincaré. Analyse Non Linéaire, **1**(4), 223–283, 1984.

Lions, P.-L., "The concentration–compactness principle in the calculus of variations. The locally compact case. I, *Ann. Inst. H. Poincaré Anal. Non Linéaire*, Annales de 1'Institut Henri Poincaré. Analyse Non Linéaire, **1**(2), 109–145, 1984.

Lions. P. L. "On kinetic equations." in *Proceedings of International Congress of Mathematicians, Kyoto 1990*, Math. Soc. Japan, Tokyo, New York, pp. 1173–1185, 1991.

Majda, A., Majda, G., and Zheng, Y., "Concentrations in the one-dimensional Vlasov–Poisson equations, i: temporal development and nonunique weak solutions in the single component case," *Physica D*, **74**, 268–300, 1994a.

Majda, A., Majda, G., and Zheng, Y., "Concentrations in the one-dimensional Vlasov–Poisson equations, ii: screening and the necessity for measure-valued solutions in the two component case," *Physica D*, **79**, 41–76, 1994b.

Meiron, D. I., Baker, G. R., and Orszag, S. A., "Analytic structure of vortex sheet dynamics 1. Kelvin–Helmholtz instability," *J. Fluid Mech.* **114**, 283–298, 1982.

Merle, F., "On uniqueness and continuation properties after blow-up time of self-similar solutions of nonlinear Schrödinger equation with critical exponent and critical mass," *Commun. Pure Appl. Math.* **45**, 203–254, 1992.

Moore, D. W., "The spontaneous appearance of a singularity in the shape of an evolving vortex sheet," *Proc. R. Soc. London A* **365**, 105–119, 1979.

Moore, D. W., "Numerical and analytical aspects of Helmholtz instability, in F. I. Niordson and Olhoff, eds.," *Theoretical and Applied Mechanics*, (Lyngby, 1984), North-Holland, Amsterdam, 1985, pp. 263–274.

Nitsche, M. and Krasny, R., "A numerical study of vortex ring formation at the edge of a circular tube," *J. Fluid Mech.* **276**, 139–161, 1994.

Pullin, D. I., "On similarity flows containing two-branched vortex sheets," in R. Caflisch, ed., *Mathematical Aspects of Vortex Dynamics*, Society for Industrial and Applied Mathematics, Philadelphia, 1989.

Lord Rayleigh, *Theory of Sound*, Printed by Dover in 1945. Original produced in 2 volumes in 1877–8.

Rosenhead, L. "The point vortex approximation of a vortex sheet or the formation of vortices from a surface of discontinuity-check," *Proc. R. Soc. London A* **134**, 170–192, 1932.

Saffman, P. and Baker, G., "Vortex interactions," *Ann. Rev. Fluid Mech.* **11**, 95–122, 1979.

Shelley, M. J., "A study of singularity formation in vortex sheet motion by a spectrally accurate vortex method," *J. Fluid Mech.* **244**, 493–526, 1992.

Sulem, C., Sulem, P. L., Bardos, C., and Frisch, U., "Finite time analyticity for the two and three dimensional Klevin–Helmholtz instability," *Commun. Math. Phys.* **80**, 485–516, 1981.

Tartar, L., "Compensated compactness and applications to partial differential equations," in *Nonlinear Analysis and Mechanics: Heriot–Watt Symposium, Vol. IV*, Vol. 39 of Research Notes in Mathematics Series, Pitman, Boston, 1979, pp. 136–212.

Tartar, L., "The compensated compactness method applied to systems of conservation," in J. Ball, ed., *Systems of Nonlinear Partial Differential Equations*, Vol. 111 of NATO Advanced Study Institute Series C, Reidel, Dordrecht, The Netherlands, 1983, pp. 263–285.

Tartar, L., "Oscillations in nonlinear partial differential equations: compensated compactness and homogenization," in *Nonlinear Systems of Partial Differential Equations in Applied Mathematics*, Vol. 23 of Lectures in Applied Mathematics Series, American Mathematical Society, Providence, RI, 1986, pp. 243–266.

Tryggvason, G., Dahm, W. J. A., and Sbeih, K., "Fine structure of vortex sheet rollup by viscous and inviscid simulation," *J. Fluids Eng.* **113**, 31–36, 1991.

van de Vooren, A. I. "A numerical investigation of the rolling up of vortex sheets," *Proc. R. Soc. London A* **373**, 67–91, 1980.

Van Dyke, M., *An Album of Fluid Motion*, Parabolic, Stanford, CA, 1982.

von Helmholtz, H., "Uber discontinuierliche Flussigkeitsbewgungen," *Monatsber. Dtsch. Akad. Wiss. Berlin* **23**, 215–228, 1868.

Zheng, Y. and Majda, A., "Existence of global weak solutions to one component Vlasov–Poisson and Fokker–Planck–Poisson systems in one space dimension with initial data of measures," *Commun. Pure Appl. Math.* **47**, 1365–1401, 1994.

10

Weak Solutions and Solution Sequences in Two Dimensions

The last five chapters of this book address the mathematical theory connected with small-scale structures and dynamics in high Reynolds number and inviscid fluid flow. In Chap. 8 we introduced the notion of a vortex patch, a solution to the 2D Euler equation in which the vorticity is constant in a region, with a discontinuiuty across a boundary. To study these objects, we had to reformulate the Euler equations in a weaker form, essentially the weak form of the vorticity-stream form of the Euler equation (see Definition 8.1). Despite this weaker form for the equations, we showed that the mathematical model is well posed, that is, for initial vorticity $\omega_0 \in L^\infty \cap L^1$, there exists for all time a unique solution to the equation (Theorems 8.1 and 8.2).

In Chap. 9 we introduced the notion of a vortex sheet, used to describe coherent structures seen in mixing layers, jets, and wakes. The mathematical theory for these highly unstable objects requires the notion of a very weak solution to the Euler equation in primitive-variable form (see Definition 9.1 in Chap. 9). This very weak definition is necessary in order to make sense of solutions for which the vorticity is unbounded, for example,

$$\omega_0 \in L^p(\mathbb{R}^2) \cap L^1(\mathbb{R}^2), \qquad p > 1, \tag{10.1}$$

and especially for vortex-sheet initial data in which

$$\omega_0 \in \mathcal{M}(\mathbb{R}^2) \cap H^{-1}_{\text{loc}}(\mathbb{R}^2). \tag{10.2}$$

There is a wide applied mathematics literature based largely on numerical calculations and formal asymptotic methods that analyze solutions to the Euler equations with such singular initial data. As we saw in Section 9.2 in the analysis of the Birkhoff–Rott equation, vortex sheets define a classical ill-posed problem. Although very special initial data (Duchon and Robert, 1988) gives a global analytic solution to the vortex-sheet problem, numerical simulations, including the ones discussed in Section 9.4, show that generically even analytic initial data lead to finite-time nonanalyticities corresponding to double-branched spirals and more complex coherent structures in the sheet.

The ill posedness of vortex sheets suggests an interesting plan of attack for the mathematical theory describing such phenomena. In real-life problems, the actual fluid possesses some residual viscosity and complex structures are usually smoothed on a very fine scale. We are really interested in capturing the essence of the dynamics in a limiting process. This chapter begins the development of a rigorous framework within which to study *approximate-solution sequences* of the 2D Euler equation. To construct a solution with such weak initial data, typically we must approximate the weak solution by a family of smooth solutions and pass to the limit in the approximating parameter. In Section 9.5 we briefly discussed the motivation for these ideas and showed examples of some approximate-solution sequences that had the interesting features of *concentrations* and *oscillations* in the limit. The general study of approximate-solution sequences is extremely useful in understanding the complex limiting process associated with constructing very weak solutions to the Euler and the Navier–Stokes equations. In this chapter and in Chaps. 11 and 12 we address this general theory of approximate-solution sequences for the Euler equation. In Chap. 11 we use some of the techniques developed here to prove that vortex-sheet initial data in the class of expression (10.2) with a distinguished sign produce a weak solution to the Euler equations for all time.

Our goal of this chapter is to understand convergence results for approximate solution sequences to the 2D Euler equation,

$$\frac{\partial v}{\partial t} + v \cdot \nabla v = -\nabla p, \qquad \operatorname{div} v = 0, \quad x \in \mathbb{R}^2,$$

in which the vorticity satisfies certain L^p constraints. In particular, the case of L^1 vorticity control is satisfied by many approximating sequences for vortex sheets, including the numerical *vortex-blob regularization* described in Subsection 9.4.2.

A useful feature of 2D incompressible inviscid flow is the conservation of vorticity along particle trajectories. Let $X(\alpha, t)$ denote the particle-trajectory map satisfying

$$\frac{dX(\alpha, t)}{dt} = v(X(\alpha, t), t), \qquad \operatorname{div} v = 0$$
$$X(\alpha, 0) = \alpha, \qquad \det|\nabla_\alpha X| = 1.$$

As we saw in Chap. 2, for the 2D Euler equation, vorticity is conserved along particle paths:

$$\omega(x, t) = \omega_0(X^{-t}(x)), \qquad X^{-t}(x) = (X(\alpha, t))^{-1}. \tag{10.3}$$

We explicitly use this fact here to show that L^p norms ($p \geq 1$) are conserved and hence arise as a natural constraint in the study of approximate-solution sequences to the 2D Euler equation.

In Section 10.1 we define the rigorous notion of an approximate-solution sequence for the Euler equation and derive some elementary results concerning these sequences. In Section 10.2 we prove some convergence results for 2D approximate-solution sequences with L^1 and L^p vorticity control. Two important results appear here in this section. The first is a general convergence theorem for approximate-solution sequences with L^1 vorticity control. In this case we show that there is a subsequence for which the velocity converges *strongly* in L^1. This is a direct result of the conservation of vorticity in two dimensions, a feature not found in three dimensions. As a corollary to this theorem we obtain the fact that oscillations are not possible for approximate-solution sequences to the 2D Euler equation with L^1 vorticity control. Thus, in two dimensions, the only possible singular behavior in the limit is *concentration*. The second main result of this chapter is the fact that when the approximate-solution sequence has additional L^p vorticity control, there is strong convergence of the velocity in L^2_{loc} to a limit that satisfies the Euler equations. As a consequence we obtain the existence of weak solutions to the 2D Euler equation with initial vorticity in $L^1 \cap L^p(\mathbb{R}^2)$. This is the first weak existence result for the equations in primitive-variable form. Uniquess, however, is still an open problem. The fact that we obtain existence for vorticities in L^p for all $p > 1$ implies that the existence theorem also holds for vorticity in L^∞. By Uniqueness Theorem 8.2 in Subsection 8.2.4, however, it can be shown that in this special case of L^∞ vorticity control, the solution is unique.

In Chap. 11 we introduce the notion of *concentration–cancellation*, in which the velocities v^ϵ do not converge strongly in L^2 yet their L^2 weak limit does satisfy the Euler equation. We present some elementary examples of this phenomenon and show this behavior to occur in general for approximate-solution sequences derived from either smoothing of vortex-sheet initial data with a distinguished sign (as in Example 10.1 below), or smoothing of the equation (as in Example 10.2 below). This leads to the *global existence* of solutions to the Euler equation with such initial data. In Chap. 12 we introduce the notion of a *measure-valued solution to the Euler equation* that can include such phenomena as oscillations and concentrations in which cancellation does not occur.

10.1. Approximate-Solution Sequences for the Euler and the Navier–Stokes Equations

To begin, we give two natural examples of approximate-solution sequences to the Euler equation. Both of these examples were discussed in Chap. 9 in the context of computing vortex sheets. An interesting question concerns the possible structures, of the form of *oscillations* and *concentrations* that can develop in the limit.

The following example is motivated by the vortex-blob method (Subsection 9.4.2) for computing vortex sheets.

Example 10.1. Smoothing (Mollification) of Initial Data. Given $v_0 \in L^p(\mathbb{R}^N)$ and $\omega_0 = \text{curl } v_0$, not necessarily smooth, define for $\epsilon > 0$

$$
\begin{aligned}
v_0^\epsilon &= J_\epsilon v_0 = \epsilon^{-N} \int_{\mathbb{R}^N} \rho\left(\frac{x-y}{\epsilon}\right) v_0(y)dy, \\
\omega_0^\epsilon &= \text{curl } v_0^\epsilon,
\end{aligned}
\tag{10.4}
$$

where ρ is any function in $C_0^\infty(\mathbb{R}^N)$, $\rho \geq 0$, $\int \rho(x)dx = 1$. Recall from Chap. 3 that such smooth initial data v_0^ϵ, ω_0^ϵ give smooth solutions $v^\epsilon(x,t)$, $\omega^\epsilon(x,t)$ for all $t > 0$ in the case of two dimensions and for at least a finite time $t > 0$ in the case of three dimensions.

Another alternative is to regularize the equation itself by adding diffusion (see, e.g., Subsection 9.4.4).

Example 10.2. Smoothing Initial Data and Equation. As in Example 10.1, smooth (mollify) the initial data v_0 to obtain v^ϵ and solve the Navier–Stokes equation with viscosity ϵ:

$$\frac{\partial v^\epsilon}{\partial t} + v^\epsilon \cdot \nabla v^\epsilon = -\nabla p^\epsilon + \epsilon \Delta v^\epsilon, \qquad \operatorname{div} v^\epsilon = 0. \tag{10.5}$$

This gives us another solution sequence v^ϵ.

Both kinds of approximate-solution sequences are natural regularizations of vortex sheets. The numercal results of Krasny (1986) (vortex-blob regularization) and Trygvasson et al. (1991) (viscous regularization) suggest that for the roll-up of a single-sign sheet, the approximations in Examples 10.1 and 10.2 have the same limit. However, it is unclear if the other approximations described in Section 9.4 also have the same limit. Moreover, the case of mixed-sign initial data is even more complex as demonstrated by the example presented in Subsection 9.4.6. After developing some elementary theories of approximate-solution sequences, we show in Chap. 11 that, for the case of 2D vortex-sheet initial data, both regularizations in Examples 10.1 and 10.2 lead to a weak solution to the Euler equation in primitive-variable form.

10.1.1. 2D Euler and Navier–Stokes Equations and A Priori Vorticity Bounds

To make either strategy useful and to obtain useful information in the limit as $\epsilon \to 0$, we require estimates for the solutions that are independent of the smoothing parameter ϵ. The results of this chapter address solution sequences for 2D Euler and Navier–Stokes equations. These results exploit the conservation of vorticity along particle paths for the Euler equation in two dimensions to produce a priori bounds associated with various norms of the vorticity. An important result that distinguishes the case of 2D sequences from 3D sequences is the strong L^1 convergence of a subsequence of velocity profiles for an approximate-solution sequence with the L^1 vorticity bound. This result explicitly uses the conservation of vorticity along particle paths in two dimensions and has as a direct corollary the fact that the *oscillations* can not occur for such 2D approximate-solution sequences. In Chap. 12 we show some explicit examples of oscillations occurring in 3D approximate-solution sequences to the Euler equation.

First, in the case of the 2D Euler equation, recall from Chap. 3 that although solutions in general do not have globally finite energy, any initial data v_0 with velocity in $L^2_{\text{loc}}(\mathbb{R}^2)$ and vorticity in L^1 have a *radial-energy decomposition* (see Definition 3.1) of the form $v_0 = \bar{v}(r) + \tilde{v}_0$, where $\bar{v}$ is a smooth radial function. This decomposition yields a solution at later times $v = \bar{v}(r) + \tilde{v}$ with the same radial part $\bar{v}$ and an a priori L^2 energy bound for the time-dependent nonradial part [from relation (3.25)]:

$$\sup_{0 \le t \le T} \|\tilde{v}\|_0 \le \|\tilde{v}_0(\cdot, 0)\|_{L^2} + C(|\nabla \bar{v}|_{L^\infty}, T). \tag{10.6}$$

To produce more refined estimates on the vorticity we introduce some useful background materials from real analysis.

Distribution Function: Let $m(E)$ denote the Lebesgue measure of a measurable set $E \subset \mathbb{R}^M$ and let f be a measurable function. The *distribution function* of f, $m_f(\cdot)$ is

$$m_f(\lambda) = m\{x \in \mathbb{R}^M : |f(x)| \geq \lambda\}. \tag{10.7}$$

For $f \in L^p(\mathbb{R}^M)$, $0 < p < \infty$, we can write the L^p norm in terms of the distribution function:

$$\|f\|_{L^p}^p = -\int_0^\infty \lambda^p dm_f(\lambda). \tag{10.8}$$

The justification of Eq. (10.8) follows by the approximation of f by simple functions and application of the monotone convergence theorem.

For smooth solutions of the 2D Euler equation, conservation of vorticity along particle trajectories yields the following very strong conservation property.

Proposition 10.1. *Let $\omega(\cdot, t)$ be the vorticity associated with a smooth solution of the 2D Euler equation with initial data ω_0; then*

$$m_{\omega_0}(\lambda) = m_{\omega(\cdot,t)}(\lambda) \qquad \text{for any} \quad 0 \leq t < \infty. \tag{10.9}$$

Conservation of the distribution function of vorticity implies that any function of vorticity determined by the distribution function is also conserved. For example, we have the following corollary.

Corollary 10.1. *Let $\omega(\cdot, t)$ be the vorticity corresponding to a smooth solution of the 2D Euler equation with initial vorticity ω_0. Then we have*

$$\|\omega(\cdot, t)\|_{L^p} = \|\omega_0\|_{L^p}, \qquad 1 \leq p \leq \infty, \quad 0 \leq t < \infty. \tag{10.10}$$

For $g(s)$, $g \geq 0$, let $G(f) = \int_0^\infty g(s) dm_f(\lambda)$. Then

$$C_T[\omega(\cdot, t)] = G[\omega_0(\cdot)], \qquad 0 \leq t < \infty. \tag{10.11}$$

Note that Eq. (10.10) for $p = \infty$ does not follow directly from Proposition 10.1 but instead from the fact that, for smooth solutions, vorticity is (pointwise) conserved along the particle trajectories.

Proof of Proposition 10.1. Let X denote the particle-trajectory map satisfying

$$\frac{dX(\alpha, t)}{dt} = v(X(\alpha, t), t), \qquad \text{div}\, v = 0$$
$$X(\alpha, 0) = \alpha, \qquad \det|\nabla_\alpha X| = 1.$$

Recall that for the 2D Euler equation, vorticity is conserved along particle paths:

$$\omega(x, t) = \omega_0(X^{-t}(x)), \qquad X^{-t}(x) = (X(\alpha, t))^{-1}. \tag{10.12}$$

Given an arbitrary Borel measurable set $B \subset \mathbb{R}^M$, denote $X(B,t) = \{x = X(\alpha,t) \mid \alpha \in B\}$. The *generalized version* of the charge of variables formula (see Rudin, 1987, Chap. 9) and the incompressibility of the velocity fields implies that

$$\int_{X(B,t)} dx = \int_B d\alpha, \qquad \text{i.e.,} \quad m(X(B,t)) = m(B). \tag{10.13}$$

Conservation of vorticity in two dimensions along particle trajectories then gives

$$\{x \mid |\omega(x,t)| > \lambda\} = \{X(\alpha,t) \mid |\omega_0(\alpha)| > \lambda\} = X(B,t),$$

where $B = \{\alpha \mid |\omega_0(\alpha)| > \lambda\}$.

Hence,

$$\begin{aligned} m_{\omega(\cdot,t)}(\lambda) &= m\{x \mid |\omega(x,t)| > \lambda\} \\ &= m(X(B,t)) = m(B) \\ &= m\{\alpha \mid |\omega_0(\alpha)| > \lambda\} = m_{\omega_0}(\lambda). \end{aligned}$$

□

The 2D Navier–Stokes equation does not have conservation of vorticity along particle paths because of the diffusion resulting from nonzero viscosity. Instead, the vorticity for smooth solutions to the Navier–Stokes equation satisfies the following inequalities. These estimates do not depend on high derivatives or viscosity and are valid for smooth solutions to the Euler equations ($\nu = 0$) also.

Proposition 10.2. *Let ω^ν be a smooth solution to the vorticity-stream formulation of the 2D Navier–Stokes equation,*

$$\frac{\partial \omega}{\partial t} + v \cdot \nabla \omega = 0 \quad |x| \in \mathbb{R}^2. \tag{10.14}$$

Then for any smooth convex function $\Phi(s)$, $s \in \mathbb{R}$, with $\Phi''(s) \geq 0$ and $\Phi(0) = 0$,

$$\int \Phi(\omega^\nu(x,t))dx \leq \int \Phi(\omega_0(x))dx. \tag{10.15}$$

Furthermore, and as a consequence, for all $1 \leq p \leq \infty$ we have decay *of the L^p norm of vorticity:*

$$\left[\int |\omega^\nu(x,t)|^p \, dx\right]^{1/p} \leq \left(\int |\omega_0(x)|^p \, dx\right)^{1/p}. \tag{10.16}$$

Proof of Proposition 10.2. For $2 \leq p < \infty$, relation (10.16) is a direct consequence of relation (10.15) when $\Phi(s) = |s|^p$ is chosen. Although $\Phi(s) = |s|^p$ is not smooth for $1 \leq p < 2$, approximating $\Phi(s)$ by a smooth function by means of mollification and standard density arguments yields relation (10.16) from relation (10.15) for this case. The case $p = \infty$ in relation (10.16) follows from the parabolicity of the vorticity-stream form of the Navier-Stokes equation, which implies the existence of a maximum principle.† Alternatively, because relation (10.16) is true for all $p \in [1, \infty)$

† See, e.g., Gilbarg and Trudinger, (1998).

we obtain the case $p = \infty$ by taking the limit $p \to \infty$ and using the fact that $\|\cdot\|_p \to \|\cdot\|_\infty$ as $p \to \infty$. Now, it remains to give the proof of relation (10.15).

Let $\Phi(s)$ be a function satisfying the hypothesis of Proposition 10.2. Then,

$$\begin{aligned}\frac{d}{dt}\int \Phi(\omega^\nu(x,t))dx &= \int \Phi'(\omega^\nu(x,t))\frac{\partial \omega^\nu}{\partial t}\,dx\\ &= -\int \Phi'(\omega^\nu(x,t))v^\nu \cdot \nabla\omega^\nu\,dx\\ &\quad + \nu\int \Phi'(\omega^\nu(x,t))\Delta\omega^\nu dx = \{1\} + \{2\},\end{aligned}$$

$$\{1\} = -\int \operatorname{div}\,(\Phi(\omega^\nu)v^\nu)dx = -\lim_{R\to\infty}\int_{|x|=R} \Phi(\omega^\nu)v^\nu \cdot \hat{n}\,d\sigma = 0.$$

The limit in the above equation goes to zero because v^ν vanishes in the far field like $1/R$ and and $\Phi(\omega^\nu) \to 0$ as $R \to \infty$:

$$\begin{aligned}\{2\} &= \nu\sum_{i=1}^{2}\int \Phi'(\omega^\nu)\frac{\partial^2\omega^\nu}{\partial x_i^2}\,dx\\ &= -\nu\int \Phi''(\omega^\nu)|\nabla\omega^\nu|^2 dx < 0.\end{aligned}$$

Hence, by integrating with respect to t, we obtain

$$\begin{aligned}\int \Phi(\omega^\nu(x,t))dx &= \int \Phi(\omega_0(x)) - \nu\int_0^T\int \nu\Phi''(\omega^\nu)|\nabla\omega^\nu|^2 dx\\ &\le \int \Phi(\omega_0(x)).\end{aligned}$$
□

Note that relation (10.15) holds regardless of the sign of Φ itself. However, in order for it to be useful, as in the case of relation (10.16), it should be a nonnegative function of its argument.

Recall from Chap. 8 that we constructed weak solutions to the 2D Euler equation with vorticity in $L^\infty(\mathbb{R}^2) \cap L^1(\mathbb{R}^2)$. We showed that such solutions were unique and studied a special class of these weak solutions known as vortex patches. In this chapter and the next two chapters we examine two weaker classes of solutions to the 2D Euler equations: ones with initial vorticity $\omega_0 \in L^p(\mathbb{R}^2) \cap L^1(\mathbb{R}^2)$ $(1 < p < \infty)$ and ones with $\omega_0 \in \mathcal{M}(\mathbb{R}^2)$, the space of bounded Borel measures on $\mathbb{R}^2$. In the latter case we require a condition that the sign of the vorticity remain fixed in order to prove existence.

10.1.2. Definition of Weak Solutions and Approximate-Solution Sequences

In Chap. 8 we considered weak $L^1 \cap L^\infty$ solutions of 2D Euler equations by means of vorticity-stream formulation (10.14). For weaker initial data, we need a definition that makes sense, $v \in L^2$ and ω only L^p, $1 < p < \infty$ or worse, so that, in particular, the nonlinearity in the vorticity-stream form $v\omega$ may not make sense. In the previous chapter, we derived the meaning of a weak solution based on the primitive-variable form for the Euler equation (Definition 9.1):

$$\begin{cases} \dfrac{Dv}{Dt} = -\nabla p, & \dfrac{D}{Dt} = \dfrac{\partial}{\partial t} + v \cdot \nabla, \\ \operatorname{div} v = 0, & v(x,0) = v_0. \end{cases} \tag{10.17}$$

We recap that construction here.

Let $\Phi = (\Phi_1, \Phi_2) \in C_0^\infty([0,T] \times \mathbb{R}^2)$; then by transport formula (1.15) from Chap. 1 we have

$$\begin{aligned} \frac{d}{dt}\int_{\mathbb{R}^2} \Phi \cdot v\, dx &= \int_{\mathbb{R}^2} \frac{D}{Dt}(\Phi \cdot v) dx, \\ &= \int_{\mathbb{R}^2} \left(\frac{D\Phi}{Dt} \cdot v + \Phi \cdot \frac{Dv}{Dt} \right) dx, \\ &= \int_{\mathbb{R}^2} \left(\frac{D\Phi}{Dt} \cdot v - \Phi \cdot \nabla p \right) dx. \end{aligned}$$

Assume that div $\Phi = 0$ so that $\int_{\mathbb{R}^2} \Phi \cdot \nabla g\, dx = 0$ for all scalar C^1 functions g. This implies that, for all $\Phi \in C_0^\infty([0,T] \times \mathbb{R}^2)$ with div $\Phi = 0$,

$$\frac{d}{dt}\int_{\mathbb{R}^2)} \Phi \cdot v\, dx = \int_{\mathbb{R}^2} \frac{D\Phi}{Dt}. \tag{10.18}$$

Integrating Eq. (10.18) with respect to t on the time interval $[0, T]$ and using a test function $\Phi \in C^1\{[0,T], C_0^1(\mathbb{R}^2)\}$ to include the initial condition in the formulation gives the following definition.

Definition 10.1. *Weak Solution in Primitive-Variable Form. A velocity field $v(x,t)$, with initial data v_0, is a weak solution of the Euler equation in primitive-variable form provided that*

(i) $v \in L^1([0,T] \times B_R)$ *for any* $R > 0$, $B_R = \{x \in \mathbb{R}^2, |x| \le R\}$,
(ii) $v \otimes v = (v_i v_j) \in L^1([0,T] \times B_R)$,
(iii) div $v = 0$ in the sense of distributions, i.e.,

$$\int \nabla\varphi \cdot v = 0 \qquad \forall \varphi \in C\{[0,T], C_0^1(\mathbb{R}^2)\},$$

(iv) for any $\Phi = (\Phi_1, \Phi_2) \in C^1\{[0,T], C_0^1(\mathbb{R}^2)\}$ *with* div $\Phi = 0$,

$$\begin{aligned} &\int \Phi(x,T) \cdot v(x,T) dx - \int \Phi(x,0) \cdot v_0(x) dx \\ &\quad = \int_0^T \int (\Phi_t \cdot v + \nabla\Phi : v \otimes v) dx dt, \end{aligned} \tag{10.19}$$

where

$$\nabla\Phi = \left(\frac{\partial}{\partial x_i}\Phi_j \right), \qquad A : B = \sum_{i,j=1}^{2} A_{ij} B_{ij}.$$

Note that the requirements (i) and (ii) are for the time term and the nonlinear term of the right-hand side of Eq. (10.19) to be meaningful, (iii) is the "weak" incompressibility condition, and (iv) is the "weak" Euler equation with the pressure eliminated by a special restriction (div $\Phi = 0$) on the test functions. Note also that although the vorticity-stream weak formulation is valid for only 2D fluid equations, the above formulation is valid for both 2D and 3D cases. To consider the existence problem for the primitive-variable formulation by means of sequences of approximate solutions we make the following precise definition.

Definition 10.2. *Approximate-Solution Sequence for the Euler Equation. A sequence of functions* $v^\epsilon \in C\{[0,T], L^2_{\text{loc}}(\mathbb{R}^2)\}$ *is an* approximate solution sequence *for the 2D Euler equation if*

(i) for all $R, T > 0$, $\max_{0\le t\le T}\int_{|x|\le R}|v^\epsilon(x,t)|^2 dx \le C(R)$, *independent of* ϵ,
(ii) $\operatorname{div} v^\epsilon = 0$ *in the sense of distributions,*
(iii) (weak consistency with the Euler equation),

$$\lim_{\epsilon\to 0}\int_0^T\int_{\mathbb{R}^2}(v^\epsilon\cdot\Phi_t + \nabla\Phi : v^\epsilon\otimes v^\epsilon)dxdt = 0$$

for all test function $\Phi \in C_0^\infty([0,T]\times\mathbb{R}^2)$ *with* $\operatorname{div}\Phi = 0$.

Note that the conditions (i)–(iii) are extremely flexible (*minimal* in some sense). The natural and very interesting questions to ask are (1) does every approximate-solution sequence converge in some sense to a weak solution of Euler equation? and (2) does any new phenomena happen during the limiting process? The purpose of our study in this and the following chapters is to answer the above questions. Definition 10.2 applies to both the 2D and the 3D case. For the 2D case, we just derived some a priori bounds on the vorticity that depend on only the initial vorticity. This naturally leads to the following additional constraint for a 2D approximate-solution sequence.

Definition 10.3. *A 2D approximate-solution sequence [satisfying (i), (ii), and (iii) of Definition 10.2] has* L^1 **vorticity control** *if (iv)* $\max_{0\le t\le T}\int|\omega^\epsilon(x,t)|dx \le C(T)$, $\omega^\epsilon = \operatorname{curl} v^\epsilon$.

Definition 10.4. *An approximate-solution sequence with* L^1 *vorticity control [satisfying (i)–(iv) above] has* L^p **vorticity control** $p > 1$ *if*

$$\text{(v)}\qquad \max_{0\le t\le T}\int|\omega^\epsilon(x,t)|^p\,dx \le C(T), \qquad \omega^\epsilon = \operatorname{curl} v^\epsilon.$$

We now verify that Examples 10.1 and 10.2, with appropriate constraints on the initial vorticity, form an approximate-solution sequence of the 2D Euler equation as defined above.

Example 10.1. Smoothing Data. Consider initial data $v_0 \in L^2_{\text{loc}}(\mathbb{R}^2)$ and the mollification $v_0^\epsilon = J_\epsilon * v_0$, where J_ϵ is the standard mollifier defined in Eqs. (10.4). Let $v^\epsilon(x,t)$

be the smooth global-in-time solution given by the energy method (see Chap. 3). Then estimate (10.6) obtained by the radial-energy decomposition implies that

$$\int_{|x|\le R} |v^\epsilon(x,t)|^2 dx \le \int_{|x|\le R} \left|\bar{v}_0^\epsilon(x)\right|^2 dx + \|\tilde{v}_0(\cdot,0)\|_{L^2}$$
$$\le C \int_{|x|\le R} |v_0(x)|^2 \, dx < \infty, \tag{10.20}$$

and, moreover, the vorticity $\omega^\epsilon = \text{curl } v^\epsilon$ satisfies

$$\int |\omega^\epsilon(x,t)|^p dx \le \int \left|\omega_0^\epsilon(x)\right|^p dx \tag{10.21}$$
$$\le \int |\omega_0(x)|^p dx, \qquad 1 \le p < \infty. \tag{10.22}$$

The other conditions [(ii) and (iii)] are trivially satisfied by the v^ϵ's. These observations lead to the following result.

If $\omega_0 \in L^1 \cap L^p(\mathbb{R}^2)$, then $v^\epsilon(x,t)$, the solutions associated with the smooth initial data $v_0^\epsilon = J_\epsilon * v_0$, curl $v_0 = \omega_0$, generate an approximate-solution sequence with $L^p(1 \le p < \infty)$ vorticity control.

Example 10.4. Zero Diffusion Limit of the Navier-Stokes Equation. This example takes smooth solutions v^ϵ of the Navier–Stokes equation with viscosity ϵ and mollified initial data:

$$\begin{cases} \dfrac{Dv^\epsilon}{Dt} = -\nabla p^\epsilon + \epsilon \Delta v^\epsilon, & \text{div } v^\epsilon = 0 \\ v^\epsilon|_{t=0} = J^\epsilon v_0, & v_0 \in L^2(\mathbb{R}^2). \end{cases} \tag{10.23}$$

Then we know from the energy estimate and the viscosity-independent estimate (10.16) that relations (10.20) and (10.22) are true. Hence, to show that the sequence of solutions of Navier-Stokes equations (10.23) generates an approximate-solution sequence with or without vorticity control, it remains to verify the weak consistency with the Euler equation, condition (iii). It is easy to see that the weak consistency from the estimates,

$$\epsilon \left| \int_0^T \int_{\mathbb{R}^2} \Phi \cdot \Delta v^\epsilon \, dx dt \right| \le \epsilon \int_0^t \int_{\mathbb{R}^2} |\Delta \Phi \cdot v^\epsilon| dx dt$$
$$\le \epsilon C(T) \left[\int_{|x|\le R_0} |v_0^\epsilon(x)| dx \right]^{1/2}$$
$$\le \epsilon C(T) \left[\int_{|x|\le R_0} |v_0(x)|^2 dx \right]^{1/2},$$

where sup $\Phi \subset \{|x| \le R_0\}$.

In addition to conditions (i) – (v) in the definition of an approximate-solution sequence, we need one more technical condition for the approximate-solution sequences to give an interpretation of the weak sense in which the initial data are assumed at time $t = 0$ in the limiting process. Before stating this condition, we review some background information and notation. We denote by Ω a locally compact Hausdorff space and by $\mathcal{M}(\Omega)$ the space of the Radon measures on Ω with $|\mu|(\Omega) < \infty$. $C_0(\Omega)$ is the space of continuous functions on Ω that vanish at infinity. For $\varphi \in C_0(\Omega)$ the L^∞ norm $|\varphi|_{L^\infty}$ is equal to $\max_{x\in\Omega} |\varphi(x)|$; hence

$$\left| \int_\Omega \varphi \, d\mu \right| \leq |\varphi|_{L^\infty} |\mu| (\Omega) \qquad \forall \varphi \in C_0(\Omega), \quad \mu \in \mathcal{M}(\Omega).$$

Thus Radon measures $\mathcal{M}(\Omega)$ are embedded in $C_0^*(\Omega)$, the dual of $C_0(\Omega)$. Actually the *Riesz representation theorem* (Rudin, 1987) says that $\mathcal{M}(\Omega)$ is isometrically isomorphic to $C_0^*(\Omega)$.

By the notion $\mu_j \rightharpoonup \mu$ in $\mathcal{M}(\Omega)$ we mean that

$$\int_\Omega \varphi \, d\mu_j \to \int_\Omega \varphi \, d\mu \qquad \text{for any} \quad \varphi \in C_0(\Omega).$$

From the fact that $[C_0(\Omega), |\cdot|]$ is a Banach space and by use of the Banach–Alaoglu theorem we have the following lemma,

Lemma 10.1. *Compactness Lemma for Measures. The family $\{\mu_j\} \subset \mathcal{M}(\Omega)$ with $|\mu_j|(\Omega) \leq C$ is weak-* sequentially compact, i.e., there is a subsequence $\{\mu_{ji}\}$ and $\mu \in \mathcal{M}(\Omega)$ such that*

$$\mu_{ji} \rightharpoonup \mu \qquad \text{in} \quad \mathcal{M}(\Omega).$$

We now recall some basic properties of Sobolev spaces. These properties were explained in detail in Chap. 3.

Let $S(\mathbb{R}^N)$ be the Schwartz space of rapidly decreasing smooth functions on $\mathbb{R}^N$; for $f \in S^*(\mathbb{R}^N)$ let $\hat{f}(\xi) = \int_{\mathbb{R}^N} e^{-ix\cdot\xi} f(x)dx$ be its Fourier transform. Recall from Subsection 3.2.1 for $s \in \mathbb{R}$ the Sobolev space

$$H^s(\mathbb{R}^N) = \left\{ f \in S^*(\mathbb{R}^N) | \, \|f\|_s^2 = \int_{\mathbb{R}^N} |\hat{f}(\xi)|^2 \left(1 + |\xi|^2\right)^2 d\xi < \infty \right\}.$$

The notation $\|f\|_s$ denotes the Sobolev s norm.

Proposition 10.3. *The function spaces $H^{-s}(\mathbb{R}^N)$ and $H^s(\mathbb{R}^N)$ are the duals to each other through the $L^2(\mathbb{R}^N)$ inner product. In particular,*

$$| < f, g > | \leq \|f\|_{-s} \|g\|_s \qquad \forall f \in H^{-s}(\mathbb{R}^N) \quad \textit{and} \quad g \in H^s(\mathbb{R}^N).$$

Let $\alpha = (\alpha_1, \ldots, \alpha_N)$ be a multi-index with $|\alpha| = \sum \alpha_i = L$. Then, from the fact that $|D^{\hat{\alpha}} f|(\xi) \leq |\xi|^L |\hat{f}(\xi)|$, we easily obtain

$$D^\alpha : H^s(\mathbb{R}^N) \to H^{s-L}(\mathbb{R}^N).$$

Another important tool from Chap. 3 is the fact that the spaces $H^{s+k}(\mathbb{R}^N)$ are continuously embedded in $C^k(\mathbb{R}^N)$ for $s > N/2$ (Lemma 3.3). We use the C_0 version of this result here in this chapter.

Lemma 10.2. *Sobolev's Lemma. For $s > (N/2)$ the inclusion mapping $i : H^s(\mathbb{R}^N) \to C_0(\mathbb{R}^N)$ is continuous, i.e., there exists a constant C_s such that*

$$|f|_{L^\infty} \leq C_s \|f\|_s.$$

Using the fact that, from elementary functional analysis, if some mapping between the two Banach spaces is continuous then its transpose mapping is also continuous, we obtain, as a corollary of Sobolev's lemma, the fact that all Radon measures are in a negative-order Sobolev space.

Proposition 10.4. *Let $s > (N/2)$; then for any $\rho \in C_0^\infty(\mathbb{R}^N)$ there exists a constant C_s such that*

$$\|\rho\mu\|_{-s} \leq C_s |\rho\mu|(\mathbb{R}^N) \qquad \forall \mu \in \mathcal{M}(\mathbb{R}^N).$$

We have a useful characterization of $H^s_{\text{loc}}(\mathbb{R}^N)$:

Proposition 10.5. *For any $s \in \mathbb{R}$, $f \in H^s_{\text{loc}}(\mathbb{R}^N)$ if and only if $\rho f \in H^s(\mathbb{R}^N)$ for any $\rho \in C_0^\infty(\mathbb{R}^N)$.*

We also need the following compactness lemma.

Lemma 10.3. *Rellich Compactness Lemma. (Folland, 1955, p. 255) Let $\{f_j\}$ be a sequence in $H^s(\mathbb{R}^N)$ with $\|f_j\|_s \leq C$; then there exists a subsequence f_{ji} and $f \in H^s(\mathbb{R}^N)$ such that for any $\rho \in C_0^\infty(\mathbb{R}^N)$, $\rho f_{ji} \to \rho f$ in $H^r(\mathbb{R}^N)$ for any $r < s$.*

We also need the following version of the Lions–Aubin lemma.

Lemma 10.4. *Lions–Aubin Lemma. Let $\{f^\epsilon(t)\}$ be a sequence in $C\{[0, T], H^s(\mathbb{R}^N)\}$ such that*

(i) $\max_{0\leq t\leq T} \|f^\epsilon(t)\|_s \leq C$,
(ii) for any $\rho \in C_0^\infty(\mathbb{R}^N)$, $\{\rho f^\epsilon\}$ is uniformly in $\text{Lip}\{[0, T], H^M(\mathbb{R}^N)\}$, *i.e.,*

$$\|\rho f^\epsilon(t_1) - \rho f^\epsilon(t_2)\|_M \leq C_M |t_1 - t_2|, \qquad 0 \leq t_1, \quad t_2 \leq T$$

for some constant C_M; then there exists a subsequence $\{f^{\epsilon_j}\}$ and $f \in C\{[0, T], H^s(\mathbb{R}^N)\}$ such that for all $R \in (M, S)$ and $\rho \in C_0^\infty(\mathbb{R}^N)$

$$\max_{0\leq t\leq T} \|\rho f^{\epsilon_j}(t) - \rho f(t)\|_R \to 0 \qquad \text{as} \quad j \to \infty.$$

Now we are prepared to state the final technical condition for the approximate-solution sequences.

Condition (vi): An approximate-solution sequence $\{v^\epsilon\}$ of the 2D Euler equation [hence satisfying (i)–(iii)] should satisfy, for some constant $C > 0$,

$$||\rho v^\epsilon(t_1) - \rho v^\epsilon(t_2)||-L \leq C|t_1 - t_2|,$$

$$0 \leq t_1, t_2 \leq T \qquad \forall L > 0, \quad \forall \rho \in C_0^\infty(\mathbb{R}^N), \tag{10.24}$$

i.e., $\{v^\epsilon\}$ is uniformly bounded in Lip$\{[0, T], H_{\text{loc}}^{-L}(\mathbb{R}^N)\}$.

For illustration we check condition (vi) for a model equation:

$$\frac{\partial v}{\partial t} + v \cdot \nabla v = \nu \Delta v, \qquad 0 \leq \nu \leq 1. \tag{10.25}$$

Observe that, in contrast with the Navier–Stokes equations, in Eq. (10.25) we do not have the pressure term that introduces nonlocality to the equations. We assume that the exact-solution sequence $\{v^\epsilon(x, t)\}$ satisfies the condition of locally finite energy [condition (i)]. Then for any cutoff function $\rho \in C_0^\infty(\mathbb{R}^N)$ with $\sup \rho \subset \{|x| \leq R\}$, we have

$$\|\rho(v^\epsilon \otimes v^\epsilon)(t)\|_{L^1} \leq C(R, T). \tag{10.26}$$

For any $s > (N/2)$ by the dual version of Sobolev's lemma,

$$\begin{aligned}
&\|\rho \operatorname{div}(v^\epsilon \otimes v^\epsilon)(t)\|_{H^{-s-1}(\mathbb{R}^N)} \\
&\quad \leq \|\operatorname{div}[\rho(v^\epsilon \otimes v^\epsilon)(t)]\|_{H^{-s-1}(\mathbb{R}^N)} + \|(v^\epsilon \otimes v^\epsilon)(t) : \nabla\rho\|_{H^{-s-1}(\mathbb{R}^N)} \\
&\quad \leq C_1\|\rho(v^\epsilon \otimes v^\epsilon)(t)\|_{H^{-s}(\mathbb{R}^N)} \leq C_2\|\rho(v^\epsilon \otimes v^\epsilon)(t)\|_{L^1}
\end{aligned}$$

Hence, from Eq. (10.25),

$$\|\rho \operatorname{div}(v^\epsilon \otimes v^\epsilon)(t)\|_{H^{-s-1}(\mathbb{R}^N)} \leq C(R, T). \tag{10.27}$$

Clearly, from the local finiteness of the kinetic energy,

$$\|\rho\nu\Delta v^\epsilon(t)\|_{H^{-2}(\mathbb{R}^N)} \leq C(R, T) \tag{10.28}$$

for a suitable constant $C(R, T)$. Now we set $-L = \min_{s>(N/2)}\{-2, -s - 1\}$; then from Eq. (10.25) and estimates (10.27) and (10.28) we have

$$\begin{aligned}
\|\rho v^\epsilon(t_1) - \rho v^\epsilon(t_2)\|_{-L} &= \left\|\rho \int_{t_1}^{t_2} \frac{\partial v^\epsilon}{\partial t} dt\right\|_{-L} \\
&\leq |t_2 - t_1| \max_{0 \leq t \leq T} \left\|\rho \frac{\partial v^\epsilon}{\partial t}(t)\right\|_{-L} \\
&\leq |t_2 - t_1|[\|\rho(v^\epsilon \cdot \nabla v^\epsilon)(t)\|_{-L} + \|\rho\nu\Delta v^\epsilon(t)\|_{-L}] \\
&\leq |t_2 - t_1|\left[\|\rho(v^\epsilon \otimes v^\epsilon)(t)\|_{H^{-s-1}(\mathbb{R}^N)} + \|\rho\nu\Delta v^2(t)\|_{H^{-2}(\mathbb{R}^N)}\right] \\
&\leq C(R, T)|t_2 - t_1|
\end{aligned}$$

for all $0 \leq t_1, t_2 \leq T$. Hence condition (vi) is satisfied by the exact-solution sequences $\{v^\epsilon(x,t)\}$ satisfying the condition of uniform finiteness of local energy, (i). The proof of the fact that an approximate-solution sequence of 2D Euler equation satisfies condition (vi) is similar to the above argument; for more details see Appendix A of DiPerna and Majda, 1987a.

10.2. Convergence Results for 2D Sequences with L^1 and L^p Vorticity Control

Our goal in this section is to prove two important convergence results for approximate-solution sequences in two dimensions. First we prove a general theorem about the convergence of approximate-solution sequences with L^1 vorticity control. We show that the L^1 bound in two dimensions implies that in addition to weak convergence in L^2 for a subsequence of velocities (this results directly from the energy bound) these fields converge strongly in $L^1_{\text{loc}}(\mathbb{R}^2)$. This result has two important consequences. The first obvious one is that it forbids oscillations to occur in approximate-solution sequences in two dimensions. The second important consequence is that it allows us to prove that an approximate-solution sequence with additional L^p vorticity control has a subsequence of velocities converging *strongly* in L^2_{loc}. This fact directly implies the existence of a weak solution to the primitive-variable form of the 2D Euler equation with initial vorticity in $L^1 \cap L^p$, $p > 1$.

In Chaps. 11 and 12 we address what happens to the solution sequences when the L^p control no longer holds. In particular, in Chap. 11, we address the case of *vortex-sheet* initial data in which the vorticity is a measure approximated by smooth vorticities with only a uniform L^1 bound. In this very weak case we can prove a global-existence theorem when the vorticity has a distinguished sign. This result directly uses the phenomenon *concentration-cancellation*, which we introduce in the next chapter.

10.2.1. Convergence of Approximate-Solution Sequences with L^1 Vorticity Control

We begin by proving the following theorem for approximate-solution sequences with L^1 vorticity control. It has the important consequence that oscillations cannot occur in the limit.

Theorem 10.1. *Convergence of Approximate-Solution Sequences with L^1 Vorticity Control. Let $\Omega_T = [0,T] \times B_R(\cdot)$, $B_R(\cdot) = \{x \in \mathbb{R}^2 | \ |x| \leq R\}$. Given an approximate-solution sequence v^ϵ with L^1 vorticity control [hence satisfying conditions (i)–(iv) and (vi)], there exists a function v with*

$$\max_{0 \leq t \leq T} \int_{|x| \leq R} |v|^2 dx \leq C(R,T), \qquad \operatorname{div} v = 0$$

and a subsequence v^ϵ such that

$$\int\int_{\Omega} |v^\epsilon - v| dx dt \to 0 \quad \text{as} \quad \epsilon \to 0 \tag{10.29}$$

and

$$\operatorname{curl} v^\epsilon = \omega^\epsilon \rightharpoonup \omega = \operatorname{curl} v \textit{ weak* in } \mathcal{M}(\Omega_T). \tag{10.30}$$

The natural question to ask is, does the limit v above satisfy the weak form of the Euler equation? As we show in Theorem 10.2, this is true for solution sequences with the additional L^p vorticity control.

Proof of Theorem 10.1. From condition (iv) on uniform L^1 vorticity control and the compactness lemma 10.1 we know there exists a subsequence $\{\omega^\epsilon\}$ and $\omega \in \mathcal{M}(\Omega)$ such that Eq. (10.30) holds. From technical condition (vi) we have, for all $L > 0$, $\rho \in C_0^\infty(\mathbb{R}^2)$, and $0 \le t_1, t_2 \le T$

$$\|\rho\omega^\epsilon(t_1) - \rho\omega^\epsilon(t_2)\|_{H^{-L-1}(\mathbb{R}^2)} \le C|t_1 - t_2|. \tag{10.31}$$

The dual of Sobolev's lemma and the condition of uniform L^1 vorticity control imply that

$$\max_{0\le t\le T} \|\rho\omega^\epsilon(t)\|_{H^{-s}(\mathbb{R}^2)} \le C_1 \max_{0\le t\le T} \|\rho\omega^\epsilon(t)\|_{L^1(\mathbb{R}^2)} \le C_2, \qquad s > 1. \tag{10.32}$$

From relations (10.31) and (10.32) and Lions-Aubin Lemma 10.4 there exists $\omega \in C\{[0,T], H_{\text{loc}}^{-s}(\mathbb{R}^2)\}$ such that

$$\max_{0\le t\le T} \|\rho\omega^\epsilon(t) - \rho\omega(t)\|_{H^{-s}(\mathbb{R})} \to 0 \qquad \forall s > 1. \tag{10.33}$$

To show convergence (10.29) it is sufficient to show that $\{v^\epsilon\}$ is a Cauchy sequence in $L^1(\Omega)$. The Biot–Savart law implies that

$$v^\epsilon(x,t) = \int_{\mathbb{R}^2} K(x-y)\omega^\epsilon(y,t)dy, \; |K(x)| \le \frac{C}{|x|}. \tag{10.34}$$

Let

$$\rho(|x|) \in C_0^\infty(\mathbb{R}) \qquad \text{with} \quad \rho(x) = \begin{cases} 1 & |x| \le 1 \\ 0 & |x| > 2 \end{cases} \tag{10.35}$$

and $\rho_\delta(|x|) = \rho(|x|/\delta)$. Then

$$\begin{aligned}(v^{\epsilon_1} - v^{\epsilon_2})(x,t) &= \left\{\int_{\mathbb{R}^2} \rho_\delta(|x-y|)K(x-y)(\omega^{\epsilon_1} - \omega^{\epsilon_2})(y,t)dy\right\} \\ &\quad + \left\{\int_{\mathbb{R}^2} (\rho_{\bar{R}} - \rho_\delta)(|x-y|)K(x-y)(\omega^{\epsilon_1} - \omega^{\epsilon_2})(y,t)dy\right\} \\ &\quad + \left\{\int_{\mathbb{R}^2} (1 - \rho_{\bar{R}})(|x-y|)K(x-y)(\omega^{\epsilon_1} - \omega^{\epsilon_2})(y,t)dy\right\} \\ &= \{1\} + \{2\} + \{3\}.\end{aligned}$$

Hence

$$\int_0^T \int_{\mathbb{R}^2} |(v^{\epsilon_1} - v^{\epsilon_2})(x,t)| dx dt$$
$$\leq \|\{1\}\|_{L^1([0,T]\times\mathbb{R}^2)} + \|\{2\}\|_{L^1([0,T]\times\mathbb{R}^2)} + \|\{3\}\|_{L^1([0,T]\times\mathbb{R}^2)}.$$

The terms {1} and {3} are easy to treat as follows: For term {1} we use the fact that a small neighborhood of the origin contributes an amount to the integral that is proportional to its radius:

$$\|\{1\}\|_{L^1([0,T]\times\mathbb{R}^2)} = \int_0^T \int_{\mathbb{R}^2} \left| \int_{\mathbb{R}^2} \rho_\delta(|x-y|) K(x-y) dx (\omega^{\epsilon_1} - \omega^{\epsilon_2})(y,t) dx \right| dy dt$$
$$\leq \|\rho_\delta K\|_{L^1(\mathbb{R}^2)} \|\omega^{\epsilon_1} - \omega^{\epsilon_2}\|_{L^1([0,T]\times\mathbb{R}^2)}$$
$$\leq C\delta.$$

For term {3} we use the fact that the kernel decays like $1/R$ in the far field, that is

$$(1-\rho_{\tilde{R}})K \in L^\infty(\mathbb{R}^2),$$
$$\omega^{\epsilon_1} - \omega^{\epsilon_2} \in L^\infty\{[0,T], L^1(\mathbb{R}^2)\},$$
$$\{3\} = [(1-\rho_{\tilde{R}})K] * (\omega^{\epsilon_1} - \omega^{\epsilon_2}) \in L^\infty([0,T]\times\mathbb{R}^2).$$

Hence

$$\|\{3\}\|_{L^1([0,T]\times\mathbb{R}^2)} \leq \|\{3\}\|_{L^\infty([0,T]\times\mathbb{R}^2)} \leq C\|(1-\rho_{\tilde{R}})K\|_{L^\infty(\mathbb{R}^2)} \leq C\tilde{R}^{-1}.$$

Because δ and $\tilde{R}^{-1}$ can be made arbitrarily small, it is sufficient to show that $\|\{2\}\|_{L^1(\Omega)} \to 0$ as $\epsilon_1, \epsilon_2 \to 0$. Here we exploit the fact that we have convergence of ω^ϵ in H^{-s} for $s < 0$. For this purpose we use the dominated convergence theorem. Let

$$\sigma^{\epsilon_1,\epsilon_2}(x,t) = \int_{\mathbb{R}^2} [(\rho_{\tilde{R}} - \rho_\delta)K](x-y)(\omega^{\epsilon_1}(y,t) - \omega^{\epsilon_2}(y,t)dy (= \{2\}).$$

Then, for fixed $x, |x| < R$, $[(\rho_{\tilde{R}} - \rho_\delta)K](x-y) \in H_0^s(\mathbb{R})$ for any $s \in R$. [Actually, it is in $C_0^\infty(\mathbb{R}^2)$.] Now, for $\tilde{\rho} = \rho_{2(\tilde{R}+R)}$, we have from convergence (10.33)

$$|\sigma^{\epsilon_1,\epsilon_2}(x,t)| \leq \|(\rho_{\tilde{R}} - \rho_\delta)K\|_{H^s(\mathbb{R}^2)} \|\tilde{\rho}\omega^{\epsilon_1}(t) - \tilde{\rho}\omega^{\epsilon_2}(t)\|_{H^{-s}(\mathbb{R}^2)}$$
$$\leq \|(\rho_{\tilde{R}} - \rho_\delta)K\|_{H^s(\mathbb{R}^2)} \max_{0\leq t\leq T} \|\tilde{\rho}\omega^{\epsilon_1}(t) - \tilde{\rho}\omega^{\epsilon_2}(t)\|_{H^{-s}(\mathbb{R}^2)}$$
$$\to 0 \text{ as } \epsilon_1, \epsilon_2 \to 0 \text{ for all } s > 1,$$

which implies both the uniform boundedness and the pointwise convergence of $\sigma^{\epsilon_1,\epsilon_2}(\cdot)$ in $\Omega_T (= [0,T] \times B_R)$. Hence, by the dominated convergence theorem, we conclude the proof. □

Note that, in proving convergence (10.29), the condition of uniform L^1 vorticity control was crucial. Without this condition the result would be false: For example, recall the steady inviscid periodic solution from Chap. 2 (Example 2.3). We can create a solution sequence by using decreasingly smaller lattices for the periodic pattern:

$$v^\epsilon(x) = \begin{bmatrix} \sin\left(\frac{2\pi x_1}{\epsilon}\right) & \sin\left(\frac{2\pi x_2}{\epsilon}\right) \\ \cos\left(\frac{2\pi x_1}{\epsilon}\right) & \cos\left(\frac{2\pi x_2}{\epsilon}\right) \end{bmatrix}. \tag{10.36}$$

This sequence satisfies all the conditions (i)–(iii) of the approximate-solution sequences except the condition of uniform vorticity control. For this sequence there is *no* $v \in L^1(B_R)$ such that v^ϵ converges *strongly* to v in $L^1(B_R)$, i.e., convergence (10.29) does not hold. Case (10.36) is one of the simplest examples of a sequence with *oscillations*, that is, there is no subsequence that converges pointwise almost everywhere to a limit, but nonetheless the sequence converges weakly to zero. For the case of a 2D approximate-solution sequence with L^1 vorticity control, strong convergence (10.29) implies the following corollary.

Corollary 10.2. *Let v^ϵ be an approximate-solution sequence with L^1 vorticity control. Then* oscillations *cannot occur in the limiting process and the only possible singular behavior is* concentration.

Proof of Corollary 10.2. The strong convergence in L^1, convergence (10.29), implies that the convergence also must be satisfied pointwise almost everywhere, and hence no oscillation can occur. □

10.2.2. Existence of a Weak Solution to the 2D Euler Equation with Vorticity in $L^p \cap L^1$

Now we show that the additional constraint of L^p vorticity control implies that a subsequence of v^ϵ converges *strongly* in L^2. Applying this result to Example 10.1 on smoothing of the initial data implies, by passing to the limit, the existence of a weak solution when the vorticity $\omega_0 \in L^p(\mathbb{R}^{2)}$.

First we recall some useful facts.

Lemma 10.5. *Interpolation Lemma for L^p Spaces. If $0 < p < q < r \leq \infty$, then $L^p \cap L^r \subset L^q$ and*

$$\|f\|_{L^q} \leq \|f\|_{L^p}^{\lambda} \|f\|_{L^r}^{1-\lambda}, \tag{10.37}$$

where $(1/q) = \{(\lambda/p) + [(1-\lambda)/r]\}$.

Proof of Lemma 10.5. For $r = \infty$, $\int |f|^q dx \leq \|f\|_{L^\infty}^{q-p} \int |f|^p dx$, so

$$\|f\|_{L^q} \leq \|f\|_{L^p}^{p/q} \|f\|_{L^\infty}^{1-p/q} = \|f\|_{L^p}^{\lambda} \|f\|_{L^\infty}^{1-\lambda}.$$

For $r < \infty$, by Hölder's inequality,

$$\int |f|^q\, dx = \int |f|^{\lambda q} |f|^{(1-\lambda)q} dx \le \left\| |f|^{\lambda q} \right\|_{L^{p/\lambda q}} \left\| |f|^{(1-\lambda)q} \right\|_{L^{r/(1-\lambda)q}}$$

$$= \left(\int |f|^p dx \right)^{\lambda q/p} \left(\int |f|^r dx \right)^{(1-\lambda)q/r}$$

$$= \|f\|_{L^p}^{\lambda q} \|f\|_{L^r}^{(1-\lambda)q}.$$

Taking the qth root gives relation (10.37) □

Recall from Chap. 8, Subsection 8.2.4, Calderon–Zygmund inequality (8.45).

Proposition 10.6. *Calderon-Zygmund Inequality. Let $1 < p < \infty$ and P be a SIO on $L^p(\mathbb{R}^N)$, i.e.,*

$$(Pf)(x) = \mathrm{PV} \int p(x-y) f(y) dy \qquad \text{for} \quad f \in L^p(\mathbb{R}^N),$$

with

$$p(\lambda x) = \lambda^{-N} p(x) \quad \text{and} \quad \int_{S^{N-1}} p(\omega) d\omega = 0.$$

Then there exists a constant C_p such that

$$\|Pf\|_{L^p(\mathbb{R}^N)} \le C_p \|f\|_{L^p(\mathbb{R}^N)}. \tag{10.38}$$

As $p \searrow 1$, $C_p \nearrow \infty$, and result (10.38) is false for $p = 1$. Also recall the Sobolev inequality.

Proposition 10.7. *Sobolev Inequality. Let $p < N$; then for $v \in W_0^{1,p}(\mathbb{R}^N)$, there exists a constant C such that*

$$\|v\|_{L^{p'}(\mathbb{R}^N)} \le C \|\nabla v\|_{L^p(\mathbb{R}^N)} \tag{10.39}$$

where $p' = [Np/(N-p)]$.

We now state and prove our main theorem of this section.

Theorem 10.2. *Existence of Weak Solutions to the 2D Euler Equation with Vorticity in $L^1 \cap L^p$. Consider an approximate-solution sequence $\{v^\epsilon\}$ with uniform L^p vorticity control $(p > 1)$ [hence satisfying conditions (i)–(vi)]; then there exists $v \in L^2(\Omega_T)$, $\Omega_T = [0, T] \times B_R$, such that*

$$\|v^\epsilon - v\|_{L^2(\Omega_T)} \to 0 \qquad \text{as} \quad \epsilon \to 0, \tag{10.40}$$

and v is a weak solution for the 2D Euler equation.

Because the smooth solutions of the 2D Navier-Stokes equation with initial vorticity $\omega_0^\epsilon = J_\epsilon * \omega_0,\ \omega_0 \in L^p(\mathbb{R}^2)$ and the smooth solutions of 2D Euler equation with the same initial data ω_0^ϵ form approximate-solution sequences with uniform L^p vorticity control, we have the following immediate corollary of the above theorem.

Corollary 10.3. *Let $\{v^\epsilon\}$ be a sequence of smooth solutions of the 2D Navier–Stokes equations with viscosity ϵ and initial data $\omega_0^\epsilon = J_\omega * \omega_0,\ \omega_0 \in L^p(\mathbb{R}^2)$. Alternatively, let $\{v^\epsilon\}$ be a sequence of smooth solutions of the 2D Euler equations with initial data ω_0^ϵ; then there exists $v \in L^2(\Omega)$ such that $v^\epsilon \to v$ in $L^2(\mathbb{R}^2)$ and v is a weak solution of 2D Euler equation.*

Proof of Theorem 10.2. First recall the identity introduced in Chaps. 2–4:

$$\nabla v^\epsilon(x,t) = C_0\omega^\epsilon(x,t) + (P\omega^\epsilon)(x,t), \tag{10.41}$$

where C_0 is some constant and

$$(P\omega^\epsilon)(x,t) = \mathrm{PV}\int P_2(x-y)\omega^\epsilon(y,t)dy,$$

where

$$P(\lambda x) = \lambda^{-2}P_2(x), \quad \int_0^{2\pi} P_2(\sigma)d\sigma = 0$$

is a SIO. The Calderon-Zygmund inequality (Proposition 10.6) and the condition of uniform L^p vorticity control combined with the decomposition of ∇v in identity (10.41) imply that

$$\|\nabla v^\epsilon(t)\|_{L^p(\mathbb{R}^2)} \le C_p(T) \tag{10.42}$$

for some constant $C_p(T)$.

Let $\rho \in C_0^\infty(\mathbb{R}^2)$ be any cutoff function. We write

$$\begin{aligned} v^\epsilon(t) &= (K * \omega^\epsilon)(t) = (\rho K) * \omega^\epsilon(t) + [(1-\rho)K] * \omega^\epsilon(t) \\ &= v_1^\epsilon(t) + v_2^\epsilon(t). \end{aligned}$$

Because $\rho K \in L^1(\mathbb{R}^2),\ \|\omega^\epsilon(t)\|_{L^p(\mathbb{R}^2)} \le C_1(T)$, Young's inequality gives

$$\left\|v_1^\epsilon(t)\right\|_{L^p(\mathbb{R}^2)} \le C_1(T). \tag{10.43}$$

By a similar argument, $(1-\rho)K \in L^\infty(\mathbb{R}^2)$ and $\omega^\epsilon(t) \in L^1(\mathbb{R}^2)$; hence

$$\left\|v_2^\epsilon(t)\right\|_{L^\infty(\mathbb{R}^2)} \le C_2(T). \tag{10.44}$$

Combining relations (10.43) and (10.44) we have

$$\|v^\epsilon(t)\|_{L^p(|x|\le R)} \le C \tag{10.45}$$

for some constant $C = C(T,R)$.

By Sobolev's inequality (Proposition 10.7), for some constant C we have for $p' = 2p/(2-p) > 2$

$$\begin{aligned}\|\rho v^\epsilon(t)\|_{L^{p'}(\mathbb{R}^2)} &\le C_3\|\nabla[\rho v^\epsilon(t)]\|_{L^p(\mathbb{R}^2)}\\ &\le C_3\|(\nabla\rho)\cdot v^\epsilon(t)\|_{L^p(\mathbb{R}^2)} + C_3\|\rho\nabla v^\epsilon(t)\|_{L^p(\mathbb{R}^2)}.\end{aligned}$$

Inequalities (10.45) and (10.42) yield

$$\|\rho v^\epsilon(t)\|_{L^{p'}(\mathbb{R}^2)} \le C(T). \tag{10.46}$$

We now apply Lemma 10.5, the interpolation lemma for L^p spaces, choosing $q = 2$, $r = 1$, $p = p'$, and $\Omega = \Omega_T = [0, T] \times B_R$. This gives

$$\|\rho[v^\epsilon(t) - v(t)]\|_{L^2(\Omega_T)} \le \|\rho[v^\epsilon(t) - v(t)]\|^{1-\lambda}_{L^1(\Omega_T)}\|\rho[v^\epsilon(t) - v(t)]\|^{\lambda}_{L^{p'}(\Omega_T)}.$$

From Theorem 10.1, $\|\rho[v^\epsilon(t) - v(t)]\|_{L^1(\Omega-T)} \to 0$ as $\epsilon \to 0$, and, by inequality (10.46), $\|\rho[v^\epsilon(t) - v(t)]\|_{L^{p'}(\Omega_T)} \le C(T)$; hence convergence (10.40) is true:

$$\|\rho[v^\epsilon(t) - v(t)]\|_{L^2(\Omega_T)} \to 0 \quad \text{as} \quad \epsilon \to 0. \tag{10.47}$$

Furthermore, this strong convergence implies that, for $\Phi \in [C_0^\infty([0, T] \times \mathbb{R}^2)]^2$,

$$\text{(A)} \qquad \int\int_\Omega \Phi_t \cdot v^\epsilon\, dxdt \to \int\int_\Omega \Phi_t \cdot v\, dxdt \qquad \text{as} \quad \epsilon \to 0,$$

$$\text{(B)} \qquad \int\int_\Omega \nabla\Phi : v^\epsilon \otimes v\, dxdt \to \int\int_\Omega \nabla\Phi : v \otimes v\, dxdt \qquad \text{as} \quad \epsilon \to 0.$$

The decomposition

$$\begin{aligned}&\int\int_\Omega (\nabla\Phi : v^\epsilon \otimes v^\epsilon - \nabla\Phi : v \otimes v) dxdt\\ &\quad = \int\int_\Omega \nabla\Phi : (v^\epsilon - v) \otimes v^\epsilon\, dxdt + \int\int_\Omega \nabla\Phi : v \otimes (v^\epsilon - v) dxdt\end{aligned}$$

with result (10.47) shows that, from (B), we have

$$\text{(C)} \qquad \int\int_\Omega \nabla\Phi : v^\epsilon \otimes v^\epsilon\, dxdt \to \int\int_\Omega \nabla\Phi : v \otimes v\, dxdt \;\; \epsilon \to 0.$$

(A) combined with (C) implies that v is a weak solution of the Euler equation. □

From the constructions in this proof, we see that whenever we have strong convergence in L^2 as in result (10.47) the limit v of the approximate-solution sequence will satisfy the Euler equation in the weak sense of Definition 10.1. If for all subsequences strong convergence in L^2 does not occur, we say that *concentration* happens. In the next chapter we show several examples for which concentration occurs and yet the limit still satisfies the Euler equation. We call this phenomenon *concentration–cancellation*. It is still an open problem whether or not concentration–cancellation always occurs for general 2D approximate-solution sequences with only L^1 vorticity control. In

Chap. 11 we can prove concentration–cancellation for the special case in which the vorticity has a distinguished sign.

Another interesting open problem is the question of uniqueness of the solution from Corollary 10.3. In the special case in which the initial data have vorticity in L^∞, the solution inherits this bound, and by the Uniqueness Theorem 8.2 in Subsection 8.2.4, is in fact unique.

Notes for Chapter 10

As we saw in Chap. 9, vortex-sheet evolution is a classically ill-posed problem. One research program has attacked this problem by means of classical existence theorems for analytic vortex sheets for finite-time intervals through nonlinear Cauchy–Kovaleski theorems (see, e.g., Caflisch and Orellana, 1986; Caflisch, 1989; Sulem et al., 1981). The material from this chapter comes from the first paper of DiPerna and Majda (1987a) in a series of works including those of Diperna and Majda (1987b, 1988) that address the behavior of such complex inviscid dynamics by means of modern analysis. The material in these two additional papers forms the framework for Chaps. 11, 12, and 13.

We remark here that the question of existence of a solution with compactly supported initial data $\omega_0 \in L^1(\mathbb{R}^2)$ without a sign constraint was solved recently by Vecchi and Wu (1993). They follow the arguments of Delort (1991) for the case of vortex-sheet initial data with a distinguished sign and use the fact that the stronger assumption of L^1 initial data allows one to apply the Dunford–Pettis theorem to prove the required convergence in the proof.

This chapter assumes knowledge of basic analysis including measure theory, distributions, and some potential theory. The books of Rudin (1987), Gilbarg and Trudinger (1998), and Folland (1995) contain most of the material needed to understand this chapter. In addition, we recommend *Real Analysis* by Folland (1984) for details on measure theory; e.g., for details on the distribution function $m_f(\lambda)$ see Chap. 6 of this text.

For the proof of the Sobolev inequality (Proposition 10.7) and related results see Gilbarg and Trudinger (1998).

References for Chapter 10

Caflisch, R., *Mathematical Aspects of Vortex Dynamics*. Society for Industrial and Applied Mathematics, Philadelphia, 1989.

Caflisch, R. E. and Orellana O. F., "Long time existence for a slightly perturbed vortex sheet," *Commun. Pure Appl. Math.* **39**, 807–838, 1986.

Delort, J. M., "Existence de nappes de tourbillon en dimension deux," *J. Am. Math. Soc.* **4**, 553–586, 1991.

DiPerna, R. J. and Majda, A. J., "Concentrations in regularizations for 2-D incompressible flow," *Commun. Pure Appl. Math.* **40**, 301–345, 1987a.

DiPerna, R. J. and Majda, A. J., "Oscillations and concentrations in weak solutions of the incompressible fluid equations," *Commun. Math. Phys.* **108**, 667–689, 1987b.

DiPerna, R. J. and Majda, A. J., "Reduced Hausdorff dimension and concentration–cancellation for two-dimensional incompressible flow," *J. Am. Math. Soc.* **1**, 59–95, 1988.

Duchon, J. and Robert, R., "Global vortex sheet solutions of Euler equations in the plane," *J. Diff. Eqns.* **73**, 215–224, 1988.

Folland, G. B., *Real Analysis: Modern Techniques and Their Applications*, Wiley, New York, 1984.

Folland, G. B., *Introduction to Partial Differential Equations*, Princeton Univ. Press, Princeton, NJ, 1995.

Gilbarg, D. and Trudinger, N. S., *Elliptic Partial Differential Equations of Second Order*, Springer-Verlag, Berlin, 1998, revised 3rd printing.

Krasny, R., "A study of singularity formation in a vortex sheet by the point vortex approximation," *J. Fluid Mech.* **167**, 292–313, 1986.

Rudin, W., *Real and Complex Analysis*, 3rd ed., McGraw-Hill, New York, 1987.

Sulem, C., Sulem, P. L., Bardos, C., and Frisch, U., "Finite time analyticity for the two and three dimensional Kelvin–Helmholtz instability," *Commun. Math. Phys.* **80**, 485–516, 1981.

Tryggvason, G., Dahm, W. J. A., and Sbeih, K., "Fine structure of vortex sheet rollup by viscous and inviscid simulation," *J. Fluids Eng.* **113**, 31–36, 1991.

Vecchi, I. and Wu, S., "On L^1 vorticity for incompressible flow," *Manuscr. Math.* **78**, 403–412, 1993.

11

The 2D Euler Equation: Concentrations and Weak Solutions with Vortex-Sheet Initial Data

In the first half of this book we studied smooth flows in which the velocity field is a pointwise solution to the Euler or the Navier–Stokes equations. As we saw in the introductions to Chaps. 8 and 9, some of the most interesting questions in modern hydrodynamics concern phenomena that can be characterized only by nonsmooth flows that are inherently only weak solutions to the Euler equation. In Chap. 8 we introducted the vortex patch, a 2D solution of a weak form of the Euler equation, in which the vorticity has a jump discontinuity across a boundary. Despite this apparent singularity, we showed that the problem of vortex-patch evolution is *well posed* and, moreover, that such a patch will retain a smooth boundary if it is initially smooth.

In Chap. 9 we introduced an even weaker class of solutions to the Euler equation, that of a *vortex sheet*. Vortex sheets occur when the *velocity* field forms a jump discontinuity across a smooth boundary. Unlike its cousin, the vortex patch, the vortex sheet is known to be so unstable that it is in fact an ill-posed problem. We saw this expicitly in the derivation of the Kelvin–Helmholtz instability for a flat sheet in Section 9.3. This instability is responsible for the complex structure observed in mixing layers, jets, and wakes. We showed that an analytic sheet solves self-deforming curve equation (9.11), called the Birkhoff–Rott equation. However, because even analytic sheets quickly develop singularities (as shown in, e.g., Fig. 9.3), analytic initial data are much too restrictive for practical application.

Later in Chap. 9 we showed that a more natural class of initial data, which we refer to here as **vortex-sheet initial data**, is that of a 2D incompressible velocity field $v_0 = (v_1^0, v_2^0)$ with vorticity $\omega_0 = \operatorname{curl} v_0$ satisfying the condition that ω_0 is a nonnegative Radon measure

$$\omega_0 \in \mathcal{M}(\mathbb{R}^2), \qquad \omega_0 \geq 0, \tag{11.1}$$

and v_0 has locally finite kinetic energy:

$$\int_{|x|<R} |v_0|^2 dx \leq C(R) < \infty \qquad \text{for any} \quad R > 0. \tag{11.2}$$

For simplicity we assume here that ω_0 has compact suport. Note that condition (11.2) excludes the case of point vorticies but includes vorticities concentrated as a smooth distribution on a curve in the plane.

At first glance, because of the ill posedness of the simple flat sheet, it seems quite difficult to formulate a rigorous theory for vortex-sheet evolution. To understand the dynamics of a high Reynolds number fluid, we must find a way to describe the evolution of the sheet with very general initial data and on long time intervals. In particular, we would like to be able to show that a solution exists for all time, starting from very general constraints on the initial data,

$$v_0 \in L^2_{\mathrm{loc}}(\mathbb{R}^2), \qquad \omega \in \mathcal{M}(\mathbb{R}^2), \tag{11.3}$$

that is, the data have finite kinetic energy and the vorticity is initially a Radon measure. From now on we refer to *vortex-sheet initial data* as those that satisfy conditions (11.3). One important goal of this chapter is to show that, for the special case of vortex-sheet initial data with *vorticity of distinguished sign*, such an existence theorem holds. The general question of uniquness and that of existence for mixed-sign vortex sheets is still an open question.

A natural framework, motivated both by the physics and numerical methods, is to consider a sequence of smooth solutions and pass to a limit in a parameter. The hope is that the resulting limit will be a very weak solution to the Euler equation that will possess a certain degree of structure. Some examples of approximate-solution sequences were introduced in Chaps. 9 and 10. They include smoothing initial data (10.4), smoothing the equation by means of viscosity (10.5), and computational vortex-blob regularization (9.19). In Chap. 10 we defined precisely what we mean by an approximate-solution sequence to the 2D Euler equation. We used it to proved some more elementary properties including the existence of weak solutions with initial vorticity in $L^p(\mathbb{R}^2)$ $(p > 1)$, in which the vorticity could have mixed sign. Our goal in this chapter is to understand properties of approximate-solution sequences without the L^p vorticity constraint. Ultimately we use these properties to prove the existence of solutions with vortex-sheet initial data of distinguished sign.

This chapter is organized as follows. In Section 11.1 we introduce the *weak-** and *reduced defect* measures associated with the limiting process of an approximate-solution sequence. In Section 11.2 we present some examples in which concentration–cancellation occurs. In particular, in Subsection 11.2.2 we give an example that is due to Greengard and Thomann (1988) with concentration–cancellation in which the weak-* defect measure is the Lebesgue measure on the unit square. This example emphasizes the fact that the weak-* defect measure is not a particularly good choice for studying the concentration set of an approximate-solution sequence. In Subsection 11.2.3 we show, by example, that the reduced defect measure is not countably subadditive. In Chap. 12 we show by means of the Greengard–Thomann example that, despite this fact, the reduced defect measure can be much better behaved than the weak* defect measure. In Section 11.3 we introduce the *vorticity maximal function*, which plays an important role in the analysis. Finally, in Section 11.4 we prove an important result of this chapter. When the initial vorticity has a *distinguished sign*, there exists a solution to the 2D Euler equation with vortex-sheet initial data $\omega_0 \in \mathcal{M}(\mathbb{R}^2) \cap H^{-1}_{\mathrm{loc}}(\mathbb{R}^2)$. The proof uses a result from the previous section on the decay rate of the vorticity maximal function when vorticity has a distinguished sign, combined with some ideas from the theory of approximate-solution sequences. The original proof of this result is due to Delort (1991) and our proof (Majda, 1993) makes use of an insightful lemma from that paper.

We begin now by recalling the definition of an approximate-solution sequence, and reviewing some facts from Chap. 10.

Definition 11.1. An approximate-solution sequence *for the Euler equation is a sequence of functions* $v^\epsilon \in C\{[0, T], L^2_{\text{loc}}(\mathbb{R}^2)\}$ *satisfying*

(i) $\max_{0\le t\le T} \int_{|x|\le R} |v^\epsilon(\cdot, t)|^2 dx \le C(R)$,
(ii) $\operatorname{div} v^\epsilon = 0$ *in the sense of distributions,*
(iii) (weak consistency with the Euler equation):

$$\lim_{\epsilon\to 0} \int_0^T \int_{\mathbb{R}^2} (v^\epsilon \cdot \Phi_t + \nabla\Phi : v^\epsilon \otimes v^\epsilon) dx dt = 0$$

for all test functions $\Phi \in C_0^\infty([0, T] \times \mathbb{R}^2)$ *with* $\operatorname{div} \Phi = 0$ *with an additional property of* L^1 *vorticity control,*
(iv) $\max_{0\le t\le T} \int |\omega^\epsilon(\cdot, t)| dx \le C(T)$, $\omega^\epsilon = \operatorname{curl} v^\epsilon$, *and the technical condition that for some constant* $C > 0$ *and sufficiently large* $L > 0$,

$$\|\rho v^\epsilon(t_1) - \rho v^\epsilon(t_2)\|_{-L} \le C|t_1 - t_2|, \quad 0 \le t_1, t_2 \le T, \quad \forall \rho \in C_0^\infty(\mathbb{R}^N), \tag{11.4}$$

i.e., $\{v^\epsilon\}$ *is uniformly bounded in* $\text{Lip}\{[0, T], H^{-L}_{\text{loc}}(\mathbb{R}^N)\}$.

We use the notation $\Omega_T = [0, T] \times B_R(0)$ for the time-dependent case and $\Omega = B_R(0)$ for the time-independent, case where $B_R(0) = \{x \in \mathbb{R}^2 |\ |x| \le R\}$. For any such approximate-solution sequence the Banach–Alaoglu theorem implies that there exists $v \in L^2(\Omega_T)$ with $\operatorname{div} v = 0$ such that

$$v^\epsilon \rightharpoonup v \text{ in } L^2(\Omega_T) \qquad \text{as} \quad \epsilon \to 0. \tag{11.5}$$

In Theorem 10.1 we proved that

$$v^\epsilon \to v \text{ in } L^1(\Omega_T) \qquad \text{as} \quad \epsilon \to 0. \tag{11.6}$$

Here we use "$\rightharpoonup$" to denote weak L^2 convergence and "$\to$" to denote strong L^2 convergence. In particular, strong convergence (11.6) implies that $v^\epsilon \to v$ pointwise almost everywhere. By definition, this means that *oscillations cannot occur in approximate-solution sequences of the 2D Euler equation.* In Chap. 12 we show that such behavior *is* possible for the case of approximate-solution sequences in three dimensions. The proof of convergence (11.6) used the a priori L^1 control of vorticity, which is a property of the equations in two but not in three dimensions. Because of weak convergence (11.5) it is clear that the limit v is a weak solution of the 2D Euler equation provided that, for all $\Phi \in C_0^\infty(\Omega_T)$,

$$\iint_{\Omega_T} \nabla\Phi : v^\epsilon \otimes v^\epsilon \, dxdt \rightarrow \iint_{\Omega_T} \nabla\Phi : v \otimes v \, dxdt \quad \text{as} \quad \epsilon \to 0. \tag{11.7}$$

This condition is guaranteed whenever

$$v^\epsilon \to v \quad \text{in} \quad L^2(\Omega_T). \tag{11.8}$$

For the case of 2D approximate-solution sequences, there are still interesting phenomena associated with weak convergence (11.5). When the convergence is not strong in $L^2(\Omega_T)$ we say that *concentration* occurs. In this chapter we present some examples of concentration in the limiting process and introduce two measures designed for examining this phenomenon. The examples we study all have the property of *concentration–cancellation*, which means that although convergence (11.5) is not strong, the limit v still satisfies the Euler equation.

In Theorem 10.2 we proved that, with the additional constraint of L^p vorticity control,

$$\text{(v)} \qquad \max_{0\le t\le T} \int |\omega^\epsilon(\cdot, t)|^p dx \le C(T), \qquad \omega^\epsilon = \text{curl } v^\epsilon,$$

we have strong convergence $v^\epsilon \to v$ in $L^2(\Omega_T)$ and hence condition (11.7) holds. The proof used the following two steps:

(i) *The Calderon–Zygmund inequality* implies, that, for $1 < p < \infty$,

$$\|\nabla v^\epsilon\|_{L^p} \le C_p \|\omega^\epsilon\|_{L^p} \quad \text{with} \quad C_p \nearrow \infty \quad \text{as} \quad p \searrow 1. \tag{11.9}$$

(ii) *Sobolev's inequality:* For any $p \in C_0^\infty(\mathbb{R})$

$$||\rho v^\epsilon||_{L^{p'}} \le C||\nabla v^\epsilon||_{L^p} \qquad \text{with} \quad p > 1, \quad p' = \frac{2p}{2-p} > 2.$$

To summarize the argument, uniform L^p control of vorticity for $p > 1$ implies a uniform full gradient control of velocity in $L^p(p > 1)$. This, in turn, implies the uniform control of the $L^{p'}$ norm of the velocity for $p' > 2$. We then use an interpolation inequality to establish strong convergence (11.8). The requirement that $p > 1$ for the uniform L^p vorticity control in Theorem 10.2 is sharp. Indeed, even if we have uniform L^1 control of the full velocity gradient, the best we can hope for is uniform L^2 control of the velocity (uniform kinetic energy), which is not good enough to use the interpolation inequality to prove strong L^2 convergence. In this chapter we show that the condition of uniform L^1 vorticity control is actually worse than that of uniform L^1 velocity gradient control; in the former case the limiting process can create a much more complex structure.

11.1. Weak-* and Reduced Defect Measures

As we showed in Chap. 10, the a priori bound on the kinetic energy of the approximate-solution sequence guarantees that for any Ω_T there exists a subsequence† $\{v^\epsilon\}$ and a limit v such that $v^\epsilon \rightharpoonup v$ weakly in $L^2(\Omega_T)$. We now introduce two tools to measure the degree to which this convergence is strong. These tools will be further examined in Chap. 12 in which we study the Hausdorff dimension of the set on which concentration occurs.

First we view weak convergence (11.5) by using the concept of a Radon measure. We define

$$d\mu^\epsilon = |v^\epsilon|^2 dxdt, \tag{11.10}$$

that is, for all Borel sets $E \subset \Omega_T$,

$$\mu^\epsilon(E) = \int_E |v^\epsilon|^2 dxdt.$$

The condition of uniform control of kinetic energy,

$$\max_{0\le t\le T} \int_{|x|\le R} |v^\epsilon|^2 dxdt \le C(R, T), \tag{11.11}$$

implies that $\mu^\epsilon(\Omega_T) \le C(R, T)$; hence by Compactness Lemma 10.1 there exists a subsequence $\{\mu^\epsilon\}$ and $\mu \in \mathcal{M}(\Omega_T)$ such that

$$\mu^\epsilon \rightharpoonup \mu \quad \text{in} \quad \mathcal{M}(\Omega_T) \qquad \text{as} \quad \epsilon \to 0, \tag{11.12}$$

i.e., for all $\varphi \in C_0(\Omega_T)$,

$$\iint_{\Omega_T} \varphi |v^\epsilon|^2 dxdt \to \iint_{\Omega_T} \varphi\, d\mu \qquad \text{as} \quad \epsilon \to 0. \tag{11.13}$$

This viewpoint is more flexible as the following proposition shows.

Proposition 11.1. *Let $d\mu$ be as defined as in Eq. (11.12) and let $v \in L^2(\Omega_T)$ satisfy*

$$v^\epsilon \rightharpoonup v \quad \text{in} \quad L^2(\Omega_T). \tag{11.14}$$

If $d\mu = |v|^2 dxdt$, then

$$v^\epsilon \to v \quad \text{in} \quad L^2(\Omega_T). \tag{11.15}$$

Proof of Proposition 11.1. From $\iint_{\Omega_T} |v^\epsilon|^2 dxdt \to \iint_{\Omega_T} \varphi |v|^2 dxdt$,

$$(v^\epsilon, v^\epsilon) \to (v, v), \tag{11.16}$$

† We use the same notation as the original sequence both here and throughout this chapter.

and

$$(v^\epsilon - v, v^\epsilon - v) = (v^\epsilon, v^\epsilon) - 2(v^\epsilon, v) + (v, v).$$

Weak convergence (11.14) implies that $(v^\epsilon, v) \to (v, v)$. Combining this with strong convergence (11.16) yields the desire result. □

In general, $v^\epsilon \rightharpoonup v$ in $L^2(\Omega_T)$ implies that

$$\begin{aligned} 0 \le \lim(v^\epsilon - v, v^\epsilon - v) &= \lim[(v^\epsilon, v^\epsilon) - 2(v^\epsilon, v) + (v, v)] \\ &= \lim[(v^\epsilon, v^\epsilon) - (v, v)]. \end{aligned}$$

Hence we always have the inequality relation

$$\iint_{\Omega_T} |v|^2 dx dt \le \lim \iint_{\Omega_T} |v^\epsilon|^2 dx dt. \tag{11.17}$$

Proposition 11.1 says that if equality holds in relation (11.17), then strong convergence holds, $v^\epsilon \to v$ in $L^2(\Omega_T)$.

Consider a Radon measure $\sigma \in \mathcal{M}(\Omega_T)$ defined on all Borel sets $E \subset \Omega_T$:

$$\sigma(E) = \mu(E) - \iint_E |v|^2 dx dt. \tag{11.18}$$

We now show that for all $\varphi \in C_0(\Omega)$, $\varphi \ge 0$,

$$\int_{\Omega_T} \varphi \, d\mu - \iint_{\Omega_T} \varphi |v|^2 dx dt \ge 0, \tag{11.19}$$

which implies that for all Borel sets $E \in \Omega_T$

$$\sigma(E) \ge 0. \tag{11.20}$$

For $\varphi \in C_0(\Omega_T)$, $\varphi \ge 0$ we have the general inequality

$$\begin{aligned} 0 &\le \lim[\varphi^{1/2}(v^\epsilon - v),\ \varphi^{1/2}(v^\epsilon - v)] \\ &= \lim[(\varphi v^\epsilon, v^\epsilon) - 2(\varphi v, v^\epsilon) + (\varphi v, v)] \\ &= \lim(\varphi v^\epsilon, v^\epsilon) - (\varphi v, v) \\ &= \iint_{\Omega_T} \varphi \, d\mu - \iint_{\Omega_T} |v|^2 dx dt, \end{aligned}$$

which shows relation (11.19). We call σ, defined above in Eq. (11.18), the **weak-* defect measure.**

We now introduce another related (outer) measure θ defined on Borel sets $E \subset \Omega_T$:

$$\theta(E) = \limsup_{\epsilon \to 0} \int_E |v^\epsilon - v|^2 dx dt. \tag{11.21}$$

One important fact from this definition is

$$\theta(E) = 0 \Leftrightarrow v^\epsilon \to v \quad \text{strongly in} \quad L^2(E). \tag{11.22}$$

In general, however, the vanishing of σ on E is not equivalent to strong convergence on E. The partial relation between them is stated in the following proposition.

Proposition 11.2. *If F is a closed set, then*

$$\sigma(F) = 0 \Rightarrow \lim_{\epsilon \to 0} \iint_F |v^\epsilon - v|^2 dx dt = 0. \tag{11.23}$$

$$\sigma(F) \geq \theta(F). \tag{11.24}$$

Proof of Proposition 11.2. Equation (11.23) follows immediately from relations (11.24) and (11.22). We prove relation (11.24) here. Because σ is a Radon measure, σ is outer regular. Thus for any $\delta > 0$, there exists an open set G with $F \subset G$ and

$$\sigma(F) + \delta \geq \sigma(G).$$

By Urysohn's lemma, there exists $f \in C_0(\mathbb{R}^N)$, $0 \leq f \leq 1$, and

$$f(x) = \begin{cases} 1, & x \in F \\ f(x) = 0, & x \in G^c \end{cases}.$$

Therefore

$$\begin{aligned} \sigma(F) + \delta \geq \sigma(G) \geq \int_G f\, d\sigma &= \lim_{\epsilon \to 0} \iint_G f|v^\epsilon|^2 dx dt - \iint_G f|v|^2 dx dt \\ &= \lim_{\epsilon \to 0} \iint_G f\left[(v^\epsilon)^2 - 2v^\epsilon \cdot v + v^2\right] dx dt \\ = \lim_{\epsilon \to 0} \iint_G f|v^\epsilon - v|^2 dx dt &\geq \limsup_{\epsilon \to 0} \sup_{\epsilon > 0} \int\int_F |v^\epsilon - v|^2 dx dt = \theta(F). \end{aligned}$$

□

The outer measure θ above is called the **reduced defect measure**.

The inequality $\sigma(F) \geq \theta(F)$ on closed sets F implies that θ might be small on a closed set on which σ is large. In fact, we show Section 11.2.2 by a concrete example that $\theta(F)$ could be extremely small on a closed set F on which σ becomes a full Lebesgue measure. Before presenting this example, we illustrate the phenomenon of concentration in some elementary approximate-solution sequences of the 2D Euler equations.

11.2. Examples with Concentration

In this section we describe several different examples of approximate-solution sequences with concentrations in the limit as $\epsilon \to 0$. First we present some elementary examples in which concentration occurs at a point. Then we present an example that is due to Greengard and Thomann (1988) in which the weak-* defect measure is the Lebesgue measure on the unit square. In all of these examples, concentration–cancellation occurs and the weak limit satisfies the Euler equation.

All of the examples below are motivated by a combination of elementary exact solutions to the Euler equation and natural desingularization methods as discussed in Chap. 6 on vortex methods and in Section 9.4 of Chap. 9. Although these examples are very simple, they serve to illustrate some of the subtle complex behavior that can occur in limiting process.

11.2.1. Elementary Examples

The basic building block for all of the examples in this section is the radial eddy (Example 2.1) from Chap. 2. Let ω_0 be a radially symmetric function. Recall then that

$$v_0(x) = {}^t\left(-\frac{x_2}{r^2}, \frac{x_1}{r^2}\right)\int_0^r s\omega_0(s)ds \tag{11.25}$$

is an exact steady solution to the 2D Euler equation. In this example, all streamlines of the flow are circles, and the fluid rotates depending on the sign of ω_0.

Example 11.1. Concentration of Phantom Vortices. This first example of an approximate-solution sequence takes a family of eddies in which the total vorticity $\int w\,dx = 0$. Such objects have zero circulation outside the support of the vorticity (hence the name "phantom vortices"). Furthermore, in addition to having locally finite kinetic energy, as we saw in Definition 3.1 (Radial-Energy Decomposition) of Chap. 3, such eddies have globally finite kinetic energy. Choose a velocity field

$$v(x) = r^{-2}\begin{pmatrix} -x_2 \\ x_1 \end{pmatrix}\int_0^r s\omega(s)ds, \tag{11.26}$$

satisfying $\operatorname{supp}\omega \subset \{|x| \leq 1\}$ and $\int_0^1 s\omega(s)ds = 0$ [hence $v(x) = 0$ for $|x| > 1$]. Set $v^\epsilon(x) = \epsilon^{-1}v(x/r^\epsilon)$; then by the scale symmetry of the Euler equation, v^ϵ is an exact solution for each $\epsilon > 0$. The sequence $\{v^\epsilon\}$ is an approximate-solution sequence. In particular, we have

$$\int |v^\epsilon(x)|^2 dx = \int \epsilon^{-2}\left|v\left(\frac{x}{\epsilon}\right)\right|^2 dx \left(\frac{x}{\epsilon} = x'\right)$$

$$= \int |v(x')|^2 dx' = C_1,$$

$$\int |\nabla v^\epsilon(x)| dx = \int \epsilon^{-2}\left|\nabla v\left(\frac{x}{\epsilon}\right)\right| dx \left(\frac{x}{r\epsilon} = x'\right)$$

$$= \int |\nabla v(x')| dx' = C_2,$$

and the L^1 vorticity control

$$\int |\omega^\epsilon(x)|dx \le \int |\nabla v^\epsilon(x)|dx = C_3.$$

Because $\int |v^\epsilon(x)|dx = \int \epsilon^{-1} v(x/\epsilon)dx = \epsilon \int v(x')dx' \to 0$, as $\epsilon \to 0$, $v^\epsilon \to 0$ in $L^1(\Omega)$, $\Omega = \{x \in \mathbb{R}^2 |x| \le 1\}$. Thus

$$v^\epsilon \rightharpoonup 0 \quad \text{in} \quad L^2(\Omega) \quad \text{as} \quad \epsilon \to 0.$$

Below, we calculate the weak limit of

$$v^\epsilon \otimes v^\epsilon \qquad \text{in} \quad \mathcal{M}(\Omega).$$

For any $\varphi \in C_0(\Omega)$ we have

$$\begin{aligned}
\lim_{\epsilon\to 0} \int \varphi(x) v_i^\epsilon(x) v_j^\epsilon(x) dx &= \lim_{\epsilon\to 0} \int \varphi(x)\epsilon^{-2} v_i\left(\frac{x}{\epsilon}\right) v_j\left(\frac{x}{\epsilon}\right) dx \\
&= \lim_{\epsilon\to 0} \int \varphi(\epsilon x') v_i(x') v_j(x') dx' \\
&= \varphi(0) W_{ij}, \qquad W_{ij} = \int v_i v_j \, dx.
\end{aligned}$$

In fact,

$$\begin{aligned}
W_{11} &= W_{22} + \frac{1}{2} \int_0^\infty \frac{1}{r^2} \left[\int_0^r s\omega(s) ds \right]^2 dr \equiv C_0, \\
W_{12} &= W_{21} = 0.
\end{aligned}$$

Thus

$$\lim_{\epsilon\to 0} \int \varphi |v^\epsilon(x)|^2 dx = \varphi(0)(W_{11} + W_{22}) = 2C_0\varphi(0).$$

To summarize,

$$v^\epsilon \otimes v^\epsilon \rightharpoonup C_0 \begin{pmatrix} \delta_0 & 0 \\ 0 & \delta_0 \end{pmatrix} \quad \text{in} \quad \mathcal{M}(\Omega) \quad \text{as} \quad \epsilon \to 0. \tag{11.27}$$

Because

$$\begin{aligned}
\int \varphi \, d\mu &= \lim_{\epsilon\to 0} \int \varphi |v^\epsilon|^2 dx = 2C_0\varphi(0), \\
\sigma &= \mu - \int |v|^2 dx = \mu = 2C_0\delta_0.
\end{aligned}$$

In this example, a finite amount of kinetic energy $(2C_0)$ concentrates at the origin during the weak limit process.

Example 11.2. Concentration in Positive Vorticity. In this example we consider a family of eddies in which the vorticity has a fixed sign. Unlike the phantom vortices, these eddies have locally but not globally finite kinetic energy. Thus, in order to satisfy property

(i) of an approximate solution sequence (kinetic-energy bound) we must choose a different rescaling than in the previous example. They are

$$v^\epsilon(x) = \left[\log\left(\frac{1}{\epsilon}\right)\right]^{-\frac{1}{2}} \epsilon^{-1} v\left(\frac{x}{\epsilon}\right), \qquad \epsilon > 0. \tag{11.28}$$

In this case

$$\int |v^\epsilon(x)|^2 dx \leq C \tag{11.29}$$

$$\int |\omega^\epsilon(x)| dx \to 0 \quad \text{as} \quad \epsilon \to 0. \tag{11.30}$$

In particular, we have uniform vorticity control. On the other hand,

$$\int |\nabla v^\epsilon(x)| dx = 0 \left[\log\left(\frac{1}{\epsilon}\right)\right]^{1/2} \to \infty \qquad \text{as} \quad \epsilon \to 0, \tag{11.31}$$

$$v^\epsilon \otimes v^\epsilon \rightharpoonup C_1 \begin{bmatrix} \delta_0 & 0 \\ 0 & \delta_0 \end{bmatrix} \quad \text{in} \quad \mathcal{M}(\Omega) \tag{11.32}$$

with $\Omega = \{x \in \mathbb{R}^2 |\, |x| \leq R\}$, $R > 0$, and $\sigma = 2C_1\delta_0$.

The form of concentration in this example is the same as the previous one except the constant C_1 is different.

Example 11.3. Time-Dependent Concentration. Recall the uniformly rotating Kirchoff elliptical vortex from Chap. 8, Section 8.1. Given v, the velocity field corresponding to the Kirchoff rotating elliptical vortex, define the sequence $\{v^\epsilon\}$ to be

$$v^\epsilon(x,t) = \left[\log\left(\frac{1}{\epsilon}\right)\right]^{-1/2} \epsilon^{-1} v\left\{\frac{x}{r\epsilon}, \left[\log\left(\frac{1}{\epsilon}\right)\right]^{-1/2} \frac{t}{\epsilon^2}\right\}. \tag{11.33}$$

The example also satisfies

$$v^\epsilon \otimes v^\epsilon \rightharpoonup C_2 \begin{bmatrix} \delta_0 & 0 \\ 0 & \delta_0 \end{bmatrix} \qquad \text{in} \quad \mathcal{M}(\Omega_T), \tag{11.34}$$

$\Omega_T = [0, T] \times B_R(0)$.

Note that in all of the above examples

$$v^\epsilon \rightharpoonup 0 \quad \text{in} \quad L^2(\Omega_T), \tag{11.35}$$

but

$$v^\epsilon \otimes v^\epsilon \rightharpoonup C\begin{bmatrix}\delta_0 & 0\\ 0 & \delta_0\end{bmatrix} \quad \text{in} \quad \mathcal{M}(\Omega_T), \tag{11.36}$$

i.e., for all $\varphi \in C_0^\infty(\Omega_T)$

$$\iint_{\Omega_T} \varphi v^\epsilon \otimes v^\epsilon dxdt \to \iint_{\Omega_T} \varphi v \otimes v\, dxdt.$$

In spite of this fact, for $\Phi \in C_0^\infty(\Omega_T)$ with $\operatorname{div}\Phi = 0$,

$$\begin{aligned}\iint_{\Omega_T} \nabla\Phi : v^\epsilon \otimes v^\epsilon dxdt \to c\iint_{\Omega_T} \nabla\Phi : \begin{bmatrix}\delta_0 & 0\\ 0 & \delta_0\end{bmatrix} \\ = C\left(\left\langle \frac{\partial\varphi_1}{\partial x_2}, \delta_0\right\rangle + \left\langle \frac{\partial\Phi_2}{\partial x_2}, \delta_0\right\rangle\right) \\ = C\left(\frac{\partial\Phi_1}{\partial x_1} + \frac{\partial\Phi_2}{\partial x_2}\right)(0) = 0.\end{aligned}$$

Thus, because of the special property of test functions ($\operatorname{div}\Phi = 0$) imposed to remove the pressure term of the Euler equation, the nonlinear term is insensitive to the phenomena of concentration and the weak limit of the approximate solution becomes a weak solution of the Euler equation. We address this very subtle phenomena, called concentration–cancellation in detail in Chap. 12.

The form of concentration at the level of the weak-* defect measure depends greatly on the behavior of the full gradient in the limiting process. The following theorem by Lions (1984a, 1984b) says that the full gradient control strongly restricts the concentration mode.

Theorem 11.1. *Let $\Omega = \{x \in \mathbb{R}^2\}|\ |x| \leq R\}$ and $\{v^\epsilon\}$ be a sequence in $L^2_{\text{loc}}(\mathbb{R}^2)$ satisfying*

$$\int_\Omega |v^\epsilon(x)|^2 dx \leq C_1, \tag{11.37}$$

$$\int_\Omega |\nabla v^\epsilon(x)| dx \leq C_2; \tag{11.38}$$

then

$$v^\epsilon \rightharpoonup v \quad \text{in} \quad L^2(\Omega), \qquad v^\epsilon \to v \quad \text{in} \quad L^2(\Omega),$$

and there exist at most a countable number of points $\{Z_j\}_{j=1}^\infty \in \Omega$ and a sequence $\{\alpha_j\}_{j=1}^\infty$ of positive numbers with $\sum_{j=m}^\infty \alpha_j^{1/2} < \infty$ such that

$$\sigma = \sum_{j=1}^\infty \alpha_j \delta(x - Z_j). \tag{11.39}$$

The theorem says that the condition of full gradient control restricts the concentration set to be at most countable number of points. Because the sequence of scaled phantom vortices (Example 11.1) satisfies the full gradient control, we can apply the above theorem directly in the following exercise.

Exercise 11.1. *Take an exact solution sequence of the type*

$$v^\epsilon(x) = \sum_{j=1}^{\infty} \Gamma_{j,\epsilon} v_{ph}^{\epsilon(j)}(x - z_j),$$

where v_{ph}^ϵ is the velocity field of phantom vortices given in Example 11.1.

(i) Using these vortices, construct an approximate-solution sequence that satisfies conditions (11.37) and (11.38) such that

$$\sigma(x) = \sum_{j=1}^{\infty} \alpha_j \delta(x - z_j).$$

(ii) What is the weak limit of $v^\epsilon \otimes v^\epsilon$ in $\mathcal{M}(\Omega)$ as $\epsilon \to 0$?

11.2.2. Arbitrary Behavior for the Weak-* Defect Measure: An Example Constructed by Greengard and Thomann

Although the above theorem tells us that full velocity gradient control severely restricts the form of the weak-* defect measure, a slightly weaker condition allows essentially arbitrary behavior for the weak-* defect measure. We present an example with uniform L^1 vorticity control in which the reduced defect measure is the Lebesgue measure on the unit square (essentially a worse case). This example implies that the weak-* defect measure is a bad tool for measuring the "bad set" on which strong L^2 convergence fails. We show that the reduced defect measure is indeed a finer tool than the weak-* defect measure.

Theorem 11.2. *(Greengard and Thomann, 1988.) There exists a sequence $\{v^m\}$ of exact steady solutions of the 2D Euler equation satisfying all the requirements of a approximate-solution sequence such that*

$$\int |v^m|^2 dx \le C, \qquad \int |\omega^\epsilon| dx \le C, \tag{11.40}$$

$$\int |\nabla v^\epsilon| dx \to \infty \qquad as \quad \epsilon \to 0, \tag{11.41}$$

$$\int |v^m| dx \to 0 \qquad as \quad m \to \infty \text{ (hence, } v^\epsilon \rightharpoonup 0). \tag{11.42}$$

The associated reduced defect measure σ satisfies

$$\sigma(E) = m(E \cap [0,1] \times [0,1]) \; \forall \text{ Borel sets } E \subset \mathbb{R}^2, \tag{11.43}$$

i.e., σ is the Lebesgue measure on unit square and

$$v^m \otimes v^m \rightharpoonup 2Cm(E \cap [0,1] \times [0,1]). \tag{11.44}$$

This theorem has the following generalization.

Corollary 11.1. *Given an arbitrary positive Radon measure σ, there exists an exact solution sequence $\{v^\epsilon\}$ for the 2D Euler equations satisfying all the conditions of an approximate-solution sequence so that σ is the weak-* defect measure associated with $\{v^\epsilon\}$.*

Exercise 11.2. *Prove Corollary 11.1.*

Because the weak limit $v = 0$ [see convergence (11.42)] is trivially a weak solution of the Euler equation, we observe that even in this very bad case concentration–cancellation happens.

Proof of Theorem 11.2. We explicitly construct the exact-solution sequence in the theorem. We exploit the special property of phantom vorticies that they have zero velocity and circulation outside the support of the vorticity. This means that, unlike general solutions to the Euler equations, two or more phantom vorticies with nonintersecting vorticities sum together to form another solution to the Euler equation.

Consider a dyadic lattice:

$$\Lambda^m = \left\{(x,y) | x = \frac{1}{2^m}\left(k_1 - \frac{1}{2}\right), y = \frac{1}{2^m}\left(k_2 - \frac{1}{2}\right), \; 1 \le k_i \le 2^m, i = 1,2\right\}.$$

For each point in Λ^m consider a phantom vortex with two scales (Fig. 11.1) given by

$$\omega_m(x) = \chi_{A_m}(x)\Omega_m^+ + \chi_{B_m}(x)\Omega_m^-,$$

where A_m is a circle of radius δ_m and B_m is an annulus, $\delta_m^{\frac{1}{2}} \le r \le R_m$.

The values of the constants Ω_m^+, Ω_m^-, δ_m and R_m are specified below. We impose the phantom-vortex condition

$$\int_{B_{R_m}} \omega_m(x) dx = 0. \tag{11.45}$$

The corresponding velocity field V_m is given by

$$V_m(x) = \begin{pmatrix} -x_2 \\ x_1 \end{pmatrix} \frac{1}{r^2} \int_0^r s\omega_m(s) ds, \tag{11.46}$$

where condition (11.45) implies that $V_m \equiv 0$ outside B_{R_m}. Consider the superposition of phantom vortices ω_m given by

$$\omega^m(x) = \sum_{z_j \in \Lambda^m} \omega_m(x - z_j). \tag{11.47}$$

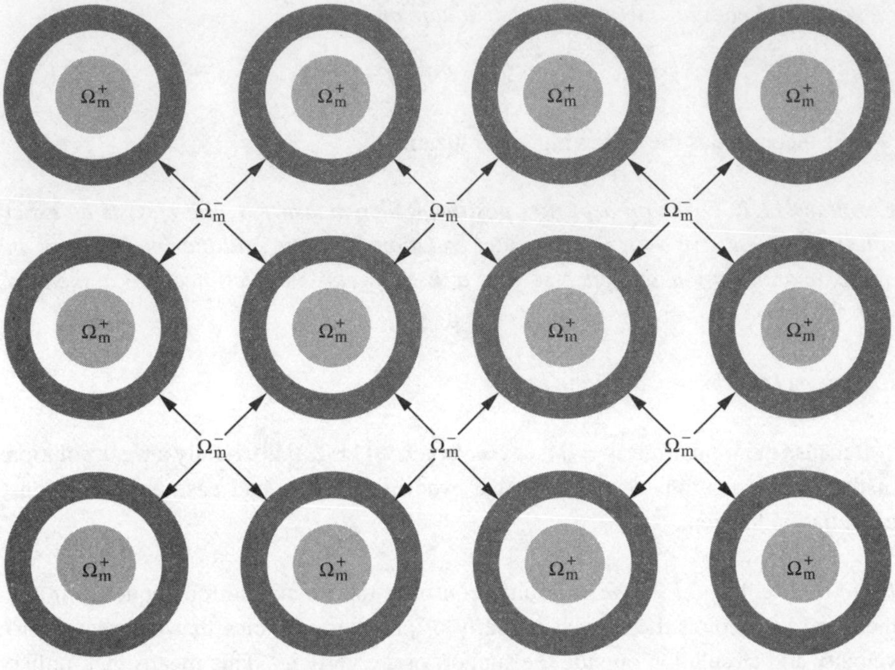

Figure 11.1. Phantom vorticies on a dyadic lattice make up the solutions in the approximate solution sequence of Greengard and Thomann. This sequence has a weak* defect measure equal to the Lebesgue measure on the unit square.

The vortex ω_m generates an exact-steady-solution sequence $\{v^m\}$

$$v^m(x) = \sum_{z_j \in \Lambda^m} V_m(x - z_j), \tag{11.48}$$

provided that each phantom vortex on the lattice points does not overlap. This condition is satisfied when

$$R_m \le \left(\frac{1}{2}\right)^{m+1}. \tag{11.49}$$

Constraint (11.45) that ω_m has mean zero implies that

$$\Omega_m^+ \delta_m^2 + \Omega_m^-(R_m - \delta_m) = 0. \tag{11.50}$$

Moreover, the L^1 norm of the vorticity satisfies

$$\begin{aligned} \int_0^1 \int_0^1 |\omega^m(x)| dx_1 dx_2 &\le 4^m \int |\omega_m(x)| ds \\ &\le \pi 4^m \left[\Omega_m^+ \delta_m^2 + |\Omega_m^-|(R_m - \delta_m)\right] = 2\pi \Omega_m^+ \delta_m^2 4^m, \end{aligned}$$

where we eliminate Ω_m^- by using Eq. (11.50). The uniform L^1 control of vorticity results by setting

$$\Omega_m^+ \delta_m^2 = 4^{-m}. \tag{11.51}$$

To obtain the uniform kinetic energy control we see directly from formula (11.43) that

$$|V_m(r)| = \begin{cases} \frac{1}{2}\Omega_m^+ r, & 0 \le r \le \delta_m \\ \frac{1}{2}\Omega_m^+ \delta_m^2 \frac{1}{r}, & \delta_m \le r \le \delta_m^{\frac{1}{2}}. \\ \frac{1}{2}|\Omega_m^-| \frac{R_m - r^2}{r}, & \delta_m^{\frac{1}{2}} \le r \le R_m \end{cases} \tag{11.52}$$

The kinetic energy is computed easily:

$$\begin{aligned} \int_0^1 \int_0^1 |v^m|^2 &\le 4^m 2\pi \int_0^{R_m} |V_m(r)|^2 r\, dr \\ &= 2\pi 4^m \left(\int_0^{\delta_m} |V_m|^2 r\, dr + \int_{\delta_m}^{\delta_m^{\frac{1}{2}}} |V_m|^2 r\, dr + \int_{\delta_m^{\frac{1}{2}}}^{R_m} |V_m|^2 r\, dr \right) \\ &= \{1\} + \{2\} + \{3\}. \end{aligned}$$

The term {2} contributes most of the kinetic energy. From Eqs. (11.52) and (11.51),

$$\{2\} = \frac{\pi}{2} 4^m \left(\Omega_m^+ \delta_m^2\right)^2 [-\ln(\delta_m)] = \frac{\pi}{2} \cdot 4^{-m} [-\ln(\delta_m)].$$

Thus, by choosing δ_m to be

$$\delta_m = \exp(-4^{m+1}/2\pi), \tag{11.53}$$

we have $\{2\} = 1$. Using Eqs. (11.52) and (11.51) we have

$$\{1\} = \frac{1}{4} \cdot \Omega_m^+ \delta_m^2 = 4^{-m-1} \le 4^{-m},$$

$$\{3\} \le \frac{\pi}{4} \cdot 4^{-m} \frac{1}{\delta_m} (R_m - \delta_m).$$

So, by choosing R_m to be

$$R_m = \frac{5}{4} \delta_m, \tag{11.54}$$

we have $\{3\} \le 4^{-m}$. Combining these gives

$$1 \le \int_0^1 \int_0^1 |v^m(x)|^2 dx \le 1 + C4^{-m}. \tag{11.55}$$

We now show convergence (11.42). From Eqs. (11.52), (11.51), and (11.53) we have

$$\begin{aligned} \int |v^m| dx &= 4^m \pi \left(\Omega_m^+ \delta_m^2\right) \left(\delta_m^{\frac{1}{2}} - \delta_m\right) \\ &\le C \left(\delta_m^{\frac{1}{2}} - \delta_m\right) \le C e^{-4m} \to 0 \quad \text{as} \quad m \to \infty. \end{aligned}$$

Next we show convergence (11.41). From Eq. (11.52) we have

$$|\nabla V_m| = \left|\frac{\partial V_m}{\partial r}\right| = \frac{1}{2}\Omega_m^+\delta_m^2\frac{1}{r^2}, \qquad \delta_m \le r \le \delta_m^{\frac{1}{2}}.$$

$$\begin{aligned}\int_{[0,1]\times[0,1]} |\nabla v^m|dx &\ge C4^m\Omega_m^+\delta_m^2\int_{\delta_m}^{\delta_m^{\frac{1}{2}}}\frac{1}{r}dr\\ &\ge C4^m\end{aligned}$$

[by Eqs. (11.51) and (11.53)]. Finally, we calculate the weak-* defect measure associated with the sequence $\{v^m\}$. The same calculation as in relations (11.52)–(11.55) for the rectangle

$$R = \left\{(x,y)|\frac{k_{11}}{2^{\ell_1}} \le x \le \frac{k_{21}}{2^{\ell_1}},\ \frac{k_{21}}{2^{\ell_2}} \le y \le \frac{k_{22}}{2^{\ell_2}}\right\}$$

for $m > \ell_1, \ell_2$ leads to

$$m(R) \le \int\int_R |v^m|^2 dx \le m(R)(1 + C4^{-m}). \tag{11.56}$$

Thus,

$$\int\int_R |v^m|^2 dx \to m(R) \text{ as } m \to \infty.$$

Because any Borel set E can be approximated by a union of $R's$, for all continuous test functions φ,

$$\lim_{m\to\infty}\int\int \varphi(x)|v^m(x)|^2 dx = \int\int \varphi\, dx$$

and σ, the weak* limit of $|v^m|^2$, is the Lebesgue measure on the unit square. □

If the reduced defect measure θ were countably subadditive, then in this example, by the above argument, θ would be equal to σ. However, in general, and in particular for the Greengard–Thomann example, the reduced defect measure is not countably subadditive and in fact it can concentrate on a set of much smaller dimension than that of the weak-* defect measure.

The reduced defect measure θ is finitely subadditive:

$$\theta\left(U_{r=1}^n E_r\right) \le \sum_{r=1}^n \theta(E_r) \text{ for measurable sets } \{E_r\}_{r=1}^n. \tag{11.57}$$

We show below with a simple concrete counterexample that θ is not countable subadditive, i.e., relation (11.57) does not hold for $n = \infty$ in general.

11.2.3. Counterexample to the Countable Subadditivity of θ

Let $\{v^\epsilon\}$ be the sequence of scaled phantom vortices of Example 11.1. We know from previous calculations that

$$\int |v^\epsilon|^2 dx = C_1 > 0, \tag{11.58}$$

$$v^\epsilon \rightharpoonup 0 \quad \text{in} \quad L^2(\Omega), \qquad |v^\epsilon|^2 \rightharpoonup 2C_1\delta_0 \quad \text{in} \quad \mathcal{M}(\Omega) \tag{11.59}$$

with $\Omega = \{|x| \leq 1\}$, or $\Omega = \mathbb{R}^2$. Hence,

$$\theta(\Omega) = \lim_{\epsilon\to 0} \int |v^\epsilon|^2 dx = C_1 > 0. \tag{11.60}$$

Let $F_N = \{x \in \mathbb{R}|\, |x| > \frac{1}{N}\}$; then, because supp $|v^\epsilon| \subseteq \{x \in \mathbb{R}^2|\, |x| \leq \epsilon\}$, for each $N \leq 1$ we have

$$\theta(F_N) = \lim_{\epsilon\to 0} \int_{F_N} |v^\epsilon| dx = \lim_{\epsilon\to 0} \int_{|x|>\frac{1}{N}} |v^\epsilon|^2 dx = 0.$$

Consider $\mathbb{R}^2/\{0\} = \cup_{N=1}^{\infty} F_N$. If θ is countable subadditive then

$$\theta(\mathbb{R}^2/\{0\}) \leq \sum_{N=1}^{\infty} \theta(F_N) = 0. \tag{11.61}$$

However,

$$\theta(\mathbb{R}/\{0\}) = \lim_{\epsilon\to 0} \int_{\mathbb{R}-\{0\}} |v^\epsilon - v|^2 dx = \lim_{\epsilon\to 0} \int_{\mathbb{R}} |v^\epsilon - v|^2 dx = \theta(\mathbb{R}) = C_1 > 0$$

from Eq. (11.60).

As we remarked before, $\theta(F)$ can be very small even though $\sigma(F)$ is large in general for a closed set F. We show this explicitly in Chap. 12, in which we use the example that is due to Greengard and Thomann (1988) in Theorem 11.2. In the next section we show that the *vorticity maximal function* plays a special role in determining the nature of the convergence of a general approximate-solution sequence.

11.3. The Vorticity Maximal Function: Decay Rates and Strong Convergence

Recall that the total circulation in a region Ω is

$$\omega(\Omega) = \int_\Omega \omega(x) dx.$$

In this section we show that the circulation associated with the absolute value of the vorticity plays an important role in the study of concentration–cancellation. We call this quantity

$$|\omega|(B_R(x)) \equiv \int_{B_R(x)} |\omega(y)| dy \tag{11.62}$$

the *vorticity maximal function*.

In this section, we address some relationships among uniform *decay rates* $|\omega|[B_R(x)] \leq f(R),\ f > 0$, of the vorticity maximal functions, strong convergence, and the sign of the vorticity. In particular, we present two important results for the vorticity maximal functions. Recall from Subsection 9.4.3 that Baker and Shelley showed numerically that the vortex-patch regularization of a fixed-sign vortex sheet exhibits a Kirchoff ellipse-type roll-up with uniform aspect ratio and an area scaling like $\Omega(h^{1.6})$ as $h \to 0$. In particular, the magnitude of the total circulation of this ellipse scales like $h^{0.6} \to 0$ as $h \to 0$. One might hope that if the circulation inside a small region is uniformly controlled for all elements of the approximate-solution sequence, and moreover, that this value uniformly decays to zero as the size of the ball decreases, then no concentration can occur in the limit. We prove rigorously here that a simple sufficient condition on $f(R)$ is sufficient to guarantee that no concentration occurs in the limit.

Theorem 11.3. *Assume that $\{v^\epsilon\}$ is an approximate-solution sequence to the 2D Euler equations satisfying*

$$\max_{x_0\in\mathbb{R}^2, 0\leq t\leq T} |\omega^\epsilon|(B_R(x_0), t) \leq C \log\left(\frac{1}{R}\right)^{-\beta} \tag{11.63}$$

for all $R \leq R_0,\ 0 < \epsilon \leq \epsilon_0$, where $\beta > 1$, and

$$\max_{0\leq t\leq T} \int_{\mathbb{R}^2} |\log^+(|x|)|\, |\omega^\epsilon(x,t)| dx \leq C. \tag{11.64}$$

Then, given any Ω_T as defined before, there exists a subsequence $\{v^\epsilon\}$ and a limit $v \in L^2(\Omega_T)$ such that $v^\epsilon \to v$ strongly in $L^2(\Omega_T)$ and v is a classical weak solution for the 2D Euler equations (i.e., no concentration occurs during the limit process.)

It is interesting to note that the scaling law observed by Baker and Shelley satisfies the condition of this theorem. Therefore their calculations suggest that concentration does not occur in the limit as the approximating parameter goes to zero.

In fact, the discrepancy between weak and strong convergence is very small. We show, by means of an example from Section 11.2, that a slightly weaker decay rate than that in condition (11.63) allows concentration to occur. However, in the case of positive-sign vorticity, we can show that this weaker decay rate is the worst that can happen. We then use this result in Section 11.4 to prove the existence of solutions to the Euler equation with vortex-sheet initial data of distinguished sign.

We use the following notation in condition (11.63):

$$\log^+(s) = \begin{cases} \log s, & s \geq 1 \\ 0, & s < 1 \end{cases},$$

$$\log^-(s) = \log s - \log^+(s).$$

If $\omega \geq 0$, then the Stokes theorem implies that

$$|\omega^\epsilon|[B_R(x)] = \int_{C_R(x)} v^\epsilon \cdot \mathbf{t}\, ds,$$

where $C_R(x) = \{y \in \mathbb{R}^2 |\ |y - x| = R\}$ and $\mathbf{t}$ is a limit tangent vector $C_R(x)$. In this case, condition (11.63) represents the bound for uniform decay rate for the net circulations. This theorem is in some sense sharper than the analogous result proved in Chap. 10 for approximate-solution sequences with L^p vorticity control. Theorem 11.3 applies to a family of solutions resulting from the smoothing of vortex-sheet initial data (which have only the L^1 control of vorticity).

In the case of vorticity with positive sign the following example shows that condition (11.63) is nearly optimal.

Example 11.4. Take $\omega \in C_0^\infty(\mathbb{R}^2)$, $\omega \geq 0$, supp $\omega \subseteq \{x \in \mathbb{R}^2 |\ |x| \leq 1\}$. This vorticity field has corresponding velocity

$$v(x) = \begin{pmatrix} -x_2 \\ x_1 \end{pmatrix} \frac{1}{r^2} \int_0^r s\omega(s) ds. \tag{11.65}$$

Consider the scaled sequence of exact solutions $\{v^\epsilon\}$ defined by

$$v^\epsilon(x) = \left[\log\left(\frac{1}{\epsilon}\right)\right]^{-\frac{1}{2}} \epsilon^{-1} v\left(\frac{x}{\epsilon}\right).$$

The corresponding sequence of vorticities is

$$\begin{aligned} \omega(x) = \frac{\partial v_2^\epsilon}{\partial x_1} - \frac{\partial v_1^\epsilon}{\partial x_2} &= \left[\log\left(\frac{1}{\epsilon}\right)\right]^{-\frac{1}{2}} \frac{\partial}{\partial x_1}\left[\frac{x_1}{r^2}\int_0^{\frac{r}{\epsilon}} s\omega(s) ds\right] \\ &+ \left[\log\left(\frac{1}{\epsilon}\right)\right]^{-\frac{1}{2}} \frac{\partial}{\partial x_2}\left[\frac{x_2}{r^2}\int_0^{\frac{r}{\epsilon}} s\omega(s) ds\right] \\ &= \left[\log\left(\frac{1}{\epsilon}\right)\right]^{-\frac{1}{2}} \epsilon^{-2}\omega\left(\frac{|x|}{\epsilon}\right). \end{aligned}$$

The vorticity maximal function is

$$\begin{aligned} |\omega|[B_R(0)] &= \left[\log\left(\frac{1}{\epsilon}\right)\right]^{-\frac{1}{2}} \epsilon^{-2} \int_0^R 2\pi\omega\left(\frac{r}{\epsilon}\right) r\, dr \\ &= 2\pi \left[\log\left(\frac{1}{\epsilon}\right)\right]^{-\frac{1}{2}} \int_0^{\frac{R}{\epsilon}} \omega(r) r\, dr. \end{aligned}$$

Define

$$f(\epsilon, R) \equiv \frac{|\omega|[B_R(0)]}{\left[\log\left(\frac{1}{R}\right)\right]^{-\frac{1}{2}}} = 2\pi \left[\log\left(\frac{1}{R}\right) \log\left(\frac{1}{\epsilon}\right)\right]^{\frac{1}{2}} \int_0^{\frac{R}{\epsilon}} \omega(r) r\, dr.$$

Then $f(\epsilon, R)$ is continuous and positive in

$$(0, \epsilon_0] \times (0, R_0], \qquad 0 < \epsilon_0 < 1, \quad 0 < R_0 < 1.$$

Because

$$\lim_{\epsilon\to 0} f(\epsilon, R) = \lim_{R\to 0} f(\epsilon, R) = 0,$$

the lower bound cannot be improved beyond 0. For the upper bound of f, we denote $(R/\epsilon) = t$; then

$$\begin{aligned}
|f|^2 &= 4\pi^2 \left[\frac{\log\left(\frac{1}{\epsilon t}\right)}{\log\left(\frac{1}{\epsilon}\right)}\right] \left[\int_0^t \omega(r) r\, dr\right]^2 \\
&= 4\pi^2 \left[1 + \frac{\log\left(\frac{1}{t}\right)}{\log\left(\frac{1}{\epsilon}\right)}\right] \left[\int_0^t \omega(r) r\, dr\right]^2 \\
&\le C\left\{ |\omega|_{L^1}^2 + \left[\frac{1}{\log\left(\frac{1}{\epsilon_0}\right)}\right] \log\left(\frac{1}{t}\right) \left[\int_0^t \omega(r) r\, dr\right]^2 \right\}
\end{aligned}$$

[if $\log(1/t) \ge 0$]. Now the function $g(t) = \log(\frac{1}{t})[\int_0^t \omega(r) r\, dr]^2$ is nonnegative only in $(0, 1]$, and has a continuous extension

$$\tilde{g}(t) = \begin{cases} g(t) & t \in (0, 1] \\ \lim_{t\to 0} g(t) = 0, & t = 0 \end{cases}.$$

Because $\tilde{g} \in C([0, 1])$ we have

$$\sup_{t\in(0,1]} |g(t)| \le \max_{t\in[0,1]} |\tilde{g}(t)| < \infty.$$

Thus there exists $C < \infty$ such that $0 \le f \le C$, i.e.,

$$|\omega|[B_R(0)] \le C \left[\log\left(\frac{1}{R}\right)\right]^{-\frac{1}{2}}. \tag{11.66}$$

Even with this slightly slower decay than that in condition, (11.63) we know from the previous chapter that the approximate-solution sequence $\{v^\epsilon\}$ exhibits concentration during the limiting process. The decay rate in condition (11.66) is the worst possible for the positive-vorticity case. This case includes that of Kirchoff elliptical vortices with the same scalings as those in Example 11.4.

In contrast to the above remark, in the case of vorticities that change sign, the vorticity maximal functions do not decay at all in general.

Example 11.5. Consider the phantom vorticies $\{v_\epsilon^{\text{ph}}\}$, $v_\epsilon^{\text{ph}} = \epsilon^{-1} v^{\text{ph}}(x/\epsilon)$:

$$v^{\text{ph}}(x) = \begin{pmatrix} -x_2 \\ x_1 \end{pmatrix} \frac{1}{r^2} \int_0^r s\omega^{\text{ph}}(s) ds, \quad \int_0^1 s\omega^{\text{ph}}(s) ds = 0,$$
$$\operatorname{supp} \omega^{\text{ph}} \subset \{x \in \mathbb{R}^2 |\, |x| \le 1\}$$

from Example 11.2 in Section 11.2.

For $\epsilon < R$

$$|\omega_\epsilon^{\mathrm{ph}}|[B_R(0)] = \int_{|x|\leq R} \epsilon^{-2}\left|\omega^{\mathrm{ph}}\left(\frac{x}{\epsilon}\right)\right| dx = \int_{|x|\leq \frac{R}{\epsilon}} |\omega^{\mathrm{ph}}(x)|dx$$
$$= \|\omega^{\mathrm{ph}}\|_{L^1(|x|\leq 1)},$$

which is a constant independent of R and ϵ.

As promised, we now show that for positive vorticity the decay rate in condition (11.66) is general.

Lemma 11.1. *Pick a family of smooth velocity fiel ds $\{v_0^\epsilon\}$ such that the vorticities $\omega_0^\epsilon = \mathrm{curl}\, v_0^\epsilon$ satisfy*

$$\omega_0^\epsilon \geq 0,$$
$$\int |x|^2 \omega_0^\epsilon(x) \equiv M_0^\epsilon \leq M_0 < \infty,$$
$$\int \omega_0^\epsilon(x)dx \equiv C_0^\epsilon \leq C_0 < \infty,$$
$$H_0^\epsilon \leq H_0 < \infty, \tag{11.67}$$

where $H_0^\epsilon \equiv H(\omega_0)$ and $H(t) \equiv H[\omega(t)]$ is the pseudoenergy defined below in definition (11.71). Then there exists a constant $C_1 = C_1(H_0, M_0, C_0)$ so that for any $R \leq R_0$, $T > 0$ we have

$$\max_{0\leq t\leq \frac{T}{x_0}\in\mathbb{R}^2} |\omega^\epsilon|(B_R(x_0), t) \leq C_1 \left[\log\left(\frac{1}{R}\right)\right]^{-\frac{1}{2}}, \tag{11.68}$$

where C_1 is a constant independent of ϵ.

Recall from previous sections that, given smooth initial data, the 2D Euler equation has a global-in-time unique smooth solution. Moreover, the pointwise conservation of vorticity along particle trajectories ensures the preservation of the sign of the vorticity:

$$\omega(\cdot, t) \geq 0 \qquad \forall t > 0 \qquad \text{if} \quad \omega_0 \geq 0.$$

Recall from Chap. 1 that smooth solutions of the 2D Euler equation preserve the following quantities:

(i) $\int \omega(x,t)dx = \int \omega_0(x)dx,$

(ii) $\int \mathbf{x}\omega(x,t)dx = \int \mathbf{x}\omega_0(x)dx,$

(iii) $\int |\mathbf{x}|^2\omega(x,t)dx = \int |\mathbf{x}|^2\omega_0(x)dx$,

(iv) $\int |v(x,t)|^2 dx = \int |v_0(x)|^2 dx$.

However, the energy of quantity (iv) is conserved only if the initial energy satisfies

$$\int |v_0(x)|^2 dx < \infty. \tag{11.69}$$

As we saw in Chap. 3, when the vorticity has a fixed sign, the global energy is unbounded for 2D flows. Fortunately, even in the case in which the initial energy is not finite, there is a related conserved bounded quantity called the **pseudoenergy**.

For a smooth 2D velocity field, recall that the stream satisfies

$$v = \nabla^\perp\psi, \qquad \Delta\psi = \omega, \quad \omega = \text{curl } v. \tag{11.70}$$

For fields with globally finite kinetic energy, integration by parts and the Biot–Savart law gives

$$\begin{aligned} E(t) = \int |v(x,t)|^2 dx &= \int \nabla^\perp\psi\cdot\nabla^\perp\psi \, dx \\ &= -\int \psi\Delta\psi \, dx \\ &= -\int \omega\psi \, dx \\ &= -\frac{1}{2\pi}\iint \log(|x-y|)\omega(x,t)\omega(y,t)dxdy. \end{aligned}$$

This last quantity is finite for a general 2D smooth velocity field with vorticity of compact support. Hence the pseudoenergy,

$$H(t) = \frac{1}{2\pi}\iint \log(|x-y|)\omega(x,t)\omega(y,t)dxdy, \tag{11.71}$$

is a useful quantity, especially for dealing with cases in which the energy is unbounded. Solutions to the Euler equations have

$$H(t) = H(0) \equiv H_0, \tag{11.72}$$

where $H(t) = -E(t)$ in the case $E(t) < \infty$. We can easily see conservation of pseudoenergy by differentiating H with respect to time and by using the vorticity-stream form of the Euler equation.

We now prove Lemma 11.1 by using conservations (i) and (iii) and Eq. (11.72).

Proof of Lemma 11.1. First we claim that there exists a constant $C = C(H_0, M_0, C_0)$ such that

$$\iint \log^-(|x-y|)\omega(x,t)\omega(y,t)dxdy \le C. \tag{11.73}$$

Proof of the Claim: Because $\log^-(s) = \log s - \log^+(s)$ and $\log^+(|x-y|) \le |x-y|^2 \le (|x|^2 + |y|^2)$, we obtain

$$\begin{aligned}
&\iint |\log^-(|x-y|)|\omega(x,t)\omega(y,t)dxdy \\
&\le |H^\epsilon(t)| + \int\int \log^+(|x-y|)\omega(x,t)\omega(y,t)dxdy \\
&\le H_0^\epsilon + \int\int (|x|^2 + |y|^2)\omega(x,t)\omega(y,t)dxdy \\
&\le H_0^\epsilon + 2M^\epsilon(t)C^\epsilon(t) \\
&\le H_0^\epsilon + 2M_0^\epsilon C_0^\epsilon \le H_0 + 2M_0C_0 \equiv C(H_0, M_0, C_0),
\end{aligned}$$

thus proving condition (11.73).

From conditions (11.67) and the preservation of the sign of vorticity we have $\omega(0,t) \ge 0$ for all $t \in [0,T]$. Hence, for any $R \le R_0 < 1$, $x_0 \in \mathbb{R}^2$

$$\begin{aligned}
&\iint |\log^-(|x-y|)|\omega(x,t)\omega(y,t)dxdy \\
&\ge \int_{B_R(x_0)}\int_{B_R(x_0)} |\log^-(|x-y|)|\omega(x,t)\omega(y,t)dxdy \\
&\text{Note:} \quad |\log^-(|x-y|)| \ge -\log(2R) \text{ for } x, y \in B_R(x_0) \\
&\ge \log(2R)^{-1}\int_{B_R(x_0)}\int_{B_R(x_0)} \omega(x,t)\omega(y,t)dxdy \\
&= \log(2R)^{-1}\{|\omega|[B_R(x_0), t]\}^2 .
\end{aligned}$$

Combining this with condition (11.73) we obtain relation (11.68). □

We now prove Theorem 11.3. During the course of this proof, we keep in mind the identity $\int |v|^2 dx = -\int \psi\omega\, dx$ ($\Delta\psi = \omega$, $\nabla^\perp\psi = v$). Thus it is sufficient to show that

$$\int \psi^\epsilon \omega^\epsilon dx \to \int \psi\omega\, dx. \tag{11.74}$$

We can do this by showing the strong convergence

$$\psi^\epsilon \to \psi \text{ in } C_0(\mathbb{R}^2)$$

by using the decay estimates of the vorticity maximal function and the weak convergence of measures,

$$\omega^\epsilon \rightharpoonup \omega \text{ weakly in } \mathcal{M}(\Omega_T), \qquad \Omega_T = B_R(0) \times [0,T],$$

by the a proiri L^1 bound for the vorticity. This idea for the proof is carried out concisely below.

Proof of Theorem 11.3. To set up the analysis, we make the following claim:

Claim. *To prove $v^\epsilon \to v$ strongly in $L^2(\Omega_T)$ it is sufficient to show that for all $\rho \in C_0^\infty(\mathbb{R}^2)$*

$$\lim_{\epsilon\to 0}\int_0^T\int_{\mathbb{R}^2}\rho|v^\epsilon|^2 dxdt = \int_0^T\int \rho|v|^2 dxdt. \tag{11.75}$$

We postpone the proof of this auxilliary result and continue with the proof of the theorem. Integration by parts gives

$$\int_0^T\int \rho|v^\epsilon|^2 dxdt = \int_0^T\int \rho|\nabla^\perp\psi^\epsilon|^2 dxdt$$

$$= -\int_0^T\int \rho\psi^\epsilon\omega^\epsilon\, dxdt + \int \psi^\epsilon\nabla\rho\cdot\nabla\psi^\epsilon\, dxdt.$$

Thus it is sufficient to show that

$$\lim_{\epsilon\to 0}\int_0^T\int \rho\psi^\epsilon\omega^\epsilon\, dxdt = \int_0^T\int \rho\psi\omega\, dxdt, \tag{11.76}$$

$$\lim_{\epsilon\to 0}\int_0^T\int \psi^\epsilon\nabla\rho\cdot\nabla\psi^\epsilon\, dxdt = \int_0^T\int \psi\nabla\rho\cdot\nabla\psi\, dxdt. \tag{11.77}$$

Limit (11.76) is the crux of matter, and limit (11.77) is much easier, as we will see below. In fact limit (11.76) is the rigorous version of the convergence (11.74) in the formal argument outlined above.

Step 1: We begin with the easy part of the proof, limit (11.77).

We know from the kinetic-energy bound for an approximate-solution sequence that

$$\nabla^\perp\psi^\epsilon \rightharpoonup v = \nabla^\perp\psi \tag{11.78}$$

in $L^2(\Omega_T)$. We now show that $\psi^\epsilon \to \psi$ *strongly* in $L^2(\Omega_T)$. By the duality of L^2, this proves the result. To show the necessary strong convergence, we use a combination of potential theory and a compactness lemma. Weak convergence (11.78) implies that the vorticities ω^ϵ are uniformly bounded in and converge weakly to ω in $L^2\{[0,T]; H^{-1}(B_R)\}$. Recall that an approximate-solution sequence also satisfies a uniform bound $\omega^\epsilon \in \mathrm{Lip}\{[0,T]; H^{-L-1}(B_R)\}$. We now recall the Lions–Aubin Lemma (Lemma 10.4) from Chap. 10.

The Lions–Aubin Compactness Lemma. *Let $\{f^\epsilon(t)\}$ be a sequence in $C\{[0,T], H^s(\mathbb{R}^N)\}$ such that*

(i) $\max_{0\le t\le T}\|f^\epsilon(t)\|_s \le C,$
(ii) for any $\rho \in C_0^\infty(\mathbb{R}^N)$, $\{\rho f^\epsilon\}$ is uniformly in $\mathrm{Lip}\{[0,T], H^M(\mathbb{R}^N)\}$, *i.e.,*

$$\|\rho f^\epsilon(t_1) - \rho f^\epsilon(t_2)\|_M \le C_M|t_1 - t_2|,\ 0 \le t_1, t_2 \le T$$

for some constant C_M, then there exists a subsequence $\{f^{\epsilon_j}\}$ and $f \in C\{[0,T], H^s(\mathbb{R}^N)\}$ such that for all $R \in (M, S)$ and $\rho \in C_0^\infty(\mathbb{R}^N)$

$$\max_{0 \le t \le T} \|\rho f^{\epsilon_j}(t) - \rho f(t)\|_R \to 0 \text{ as } j \to \infty.$$

This lemma implies that $\omega^\epsilon \to \omega$ *strongly* in $L^2\{[0,T]; H^{-1-s}(B_R)\}$ for all $s > 0$. Because $\psi^\epsilon = (1/2\pi)\log * \omega^\epsilon$, by standard elliptic compactness results for the Laplacian (Folland, 1995), $\psi^\epsilon \to \psi$ strongly in $L^2\{[0,T]; H^{1-s}(B_R)\}$ for all $s > 0$. Taking $s = 1$ gives the desired result.

Step 2: This step is the key part of the proof. We prove the first convergence (11.76), assuming that the following lemma is true.

Lemma 11.2. *We assume that $\omega(x)$ is smooth and satisfies for some $\beta > 1$ and all $R \le R_0$*

$$\max_{x \in \mathbb{R}^2} |\omega|[B_R(x)] \le C\left[\log\left(\frac{1}{R}\right)\right]^{-\beta},$$

$$\int |\log^+(r)|\,|\omega| dx \le C. \tag{11.79}$$

If $\psi = (1/2\pi)\log(r) * \omega$ is the corresponding logarithmic potential, then for any $R > 0$

$$|\psi(x) - \psi(y)| \le C_R \left|\log\left(\frac{1}{|x-y|}\right)\right|^{1-\beta} \tag{11.80}$$

for

$$|x|,\ |y| \le R,\ |x-y| \le \frac{1}{8}.$$

Proof of the first convergence (11.76): Lemma 11.2 implies that

$$\rho\psi^\epsilon \text{ is uniformly bounded in } L^\infty\left[0, T; C_0^\theta(\mathbb{R}^2)\right] \tag{11.81}$$

where $C^\theta(\mathbb{R}^2)$ is the Hölder space of functions with modulus of continuity defined by

$$\theta(|x-y|) = \left|\log\left(\frac{1}{|x-y|}\right)\right|^{1-\beta}. \tag{11.82}$$

Because the injection $C_0^\theta(\mathbb{R}^2) \to C_0(\mathbb{R}^2)$ is compact by the Arzela–Ascoli theorem and $\rho\psi^\epsilon$ is uniformly Lipschitz in $H^{-L+1}(\mathbb{R}^2)$ [This comes from the previous fact that the velocity $v^\epsilon = \nabla^\perp\psi^\epsilon$ is uniformly Lipschitz in $H^{-L}(\mathbb{R}^2)$], by an application of the Lions–Aubin lemma above,

$$\rho\psi^\epsilon \to \rho\psi \qquad \text{uniformly in} \quad L^\infty[0, T : C_0(\mathbb{R}^2)]$$

for some subsequence $\{\rho\psi^{\epsilon}\}$. On the other hand, because $\omega^{\epsilon} \rightharpoonup \omega$ in $\mathcal{M}(\Omega_T)$‡ we conclude that convergence (11.76) is true because it contains the dual pairing $\rho\psi^{\epsilon}, \omega$.

It remains to prove Lemma 11.2, thus concluding the whole proof of the theorem.

Step 5: Proof of Lemma 11.2. Because condition (11.79) is translational invariant it is sufficient to estimate

$$|\psi(0) - \psi(y)| = \frac{1}{2\pi}\left|\int [\log(|z|) - \log(|z-y|)]\omega(z)dz\right| \tag{11.83}$$

for $|y| < (1/8)$. Define $m(R) = \sup_{y\in\mathbb{R}^2} \int_{|z|\le R} |\omega(y+z)|dz$.

Decompose the integral in Eq. (11.83) into the near-field ($|z| \le 2|y|$) integral and the far-field ($|z| > 2|y|$) integral. For the near field integral,

$$\begin{aligned}
&\frac{1}{2\pi}\int_{|z|\le 2|y|} [\log|z| - \log(|z-y)]|\omega(z)|dz \\
&\le -\frac{1}{2\pi}\int_{|z|\le 2|y|} \log|z|\omega(z)|dz - \frac{1}{2\pi}\int_{|z|\le 1|y|} \log|z-y|\,|\omega(z)|dz \\
&\le -\frac{1}{2\pi}\int_0^{2|y|} \log(r)dm(r) - \frac{1}{2\pi}\int_{|z+y|\le 2|y|} \log|z|\,|\omega(z+y)|dz \\
&\le -\frac{1}{2\pi}\int_0^{2|y|} \log(r)dm(r) - \frac{1}{2\pi}\int_{|z|\le 4|y|} \log|z|\,|\omega(z+y)|dz \\
&\le -\frac{1}{2\pi}\int_0^{2|y|} \log(r)dm(r) - \frac{1}{2\pi}\int_0^{4|y|} \log(r)dm(r) \\
&\le -\frac{1}{\pi}\int_0^{4|y|} \log(r)dm(r) \\
&\le \frac{1}{\pi}\left|\int_0^{4|y|} \frac{m(r)}{r}dr\right| + \frac{1}{\pi}m(4|y|)\log(|y|)\Bigg|.
\end{aligned}$$

Because $m(r) \le C\lambda[\log(\frac{1}{r})]^{-\beta}$, from condition (11.79),

$$\begin{aligned}
\left|\int_0^{4|y|} \frac{m(r)}{r}dr\right| &\le C\left|\int_0^{4|y|} \frac{1}{r}\left[\log\left(\frac{1}{r}\right)\right]^{-\beta} dr\right| \\
&= C\int_{\log\left(\frac{1}{4|y|}\right)}^{\infty} u^{-\beta}du \qquad \left[\log\left(\frac{1}{r}\right) \equiv u\right] \\
&\le C\left|\log\left(\frac{1}{|y|}\right)\right|^{1-\beta}.
\end{aligned}$$

‡ This comes from $\max_{0\le t\le T} \int |\omega(x,t)|dx \le C$.

Thus the near-field integral satisfies

$$\frac{1}{2\pi}\int_{|z|\le 2|y|} \big|\log|z| - \log|z-y|\,\big|\,|\omega(z)|dz \\ \le C\left|\log\left(\frac{1}{|y|}\right)\right|^{1-\beta}. \qquad (11.84)$$

For the far-field integral, when $|\log|z| - \log|z-y|\,| \le C(|y|/|z|)$ is used for $|z| > 2|y|$,

$$\begin{aligned}
\frac{1}{2\pi}\int_{|z|>2|y|} &\big|\log|z| - \log|z-y|\,\big|\,|\omega(z)|dz \\
&\le C|y|\int_{|z|>2|y|}\frac{1}{|z|}|\omega(z)|dz \\
&= C|y|\int_{2|y|}^{R}\frac{1}{r}dm(r) \\
&\le C|y|\left|\int_{2|y|}^{R}\frac{m(r)}{r^2}dr\right| + \left|C|y|\left[\frac{m(r)}{r}\right]_{r=2|y|}^{r=R}\right| \\
&= \{1\} + \{2\}, \\
\{1\} &\le C|y|\left|\int_{2|y|}^{R}\left|\frac{1}{r^2}\right|\log\left(\frac{1}{r}\right)\right|^{-\beta}dr\right|.
\end{aligned}$$

Denote

$$u = \log\left(\frac{1}{r}\right), \quad \lambda = \log\left(\frac{1}{|y|}\right).$$

Then

$$\begin{aligned}
\{1\} &\le C\int_{\log(\frac{1}{R})}^{\lambda} e^{u-\lambda}u^{-\beta}\,du \le C\int_{\frac{\lambda}{2}}^{\lambda} u^{-\beta}\,du + Ce^{-\frac{\lambda}{2}}\int_{\log(\frac{1}{R})}^{\frac{\lambda}{2}} u^{-\beta}\,du \\
&\le C_1(R)(1-\beta)^{-1}\left\{\left[\log\left(\frac{1}{|y|}\right)\right]^{1-\beta} + |y|\right\}, \\
\{2\} &\le C\frac{m(R)}{R}|y| + m(2|y|) \\
&\le C_2(R)\left[|y| + \left|\log\left(\frac{1}{|y|}\right)\right|^{-\beta}\right].
\end{aligned}$$

Thus, for the far-field integral, because $|y| < (1/8)$,

$$\frac{1}{2\pi}\int_{|z|>2|y|}(|\log|z| - \log|z-y||)|\omega(z)|dz \le C(R)\left|\log\left(\frac{1}{|y|}\right)\right|^{1-\beta} \quad (11.85)$$

Combining conditions (11.84) and (11.85) concludes the proof.

Finally, for completeness, we prove the elementary claim stated in the beginning of this proof. To prove this claim consider $\rho_\delta \in C_0^\infty(\mathbb{R}^2)$ defined by

$$\rho_\delta(x) = \rho_\delta(|x|) = \begin{cases} 1, & |x| \le R \\ \text{smoothly decrease}, & R < |x| < R+\delta. \\ 0, & |x| \ge R+\delta \end{cases}$$

Then

$$\limsup_{\epsilon\to 0}\int_{\Omega_T}|v^\epsilon|^2 dxdt \le \limsup_{\epsilon\to 0}\int_0^T\int_{\mathbb{R}}\rho_\delta|v^\epsilon|^2 dxdt = \int_0^T\int_{\mathbb{R}}\rho_\delta|v|^2 dxdt.$$

Let $\delta \to 0$; then, by the dominated convergence theorem, we have $\limsup_{\epsilon\to 0}\int_{\Omega_T}|v^\epsilon|^2 dxdt \le \int_0^T\int_{B_R(0)}|v|^2 dxdt$, i.e., $\limsup_{\epsilon\to 0}\|v^\epsilon\|^2_{L^2(\Omega_T)} \le \|v\|^2_{L^2(\Omega_T)}$. Combining this with the weakly lower semicontinuity of the L^2 norm,

$$\|v\|^2_{L^2(\Omega_T)} \le \lim_{\epsilon\to 0}\|v^\epsilon\|^2_{L^2(\Omega_T)} \qquad \text{for} \quad v^\epsilon \rightharpoonup v \quad \text{weakly in} \quad L^2(\Omega_T)$$

we obtain

$$\|v\|^2_{L^2(\Omega_T)} = \lim_{\epsilon\to 0}\|v^\epsilon\|^2_{L^2(\Omega_T)}.$$

This completes the proof of the claim. □

11.4. Existence of Weak Solutions With Vortex-Sheet Initial Data of Distinguished Sign

An extremely important problem, unsolved until recently, is the question of existence of a solution for *vortex-sheet* initial data. As motivated in Chap. 9, such questions are relevant to understanding the instabilities created by jets, wakes, and mixing layers in high Reynolds number fluids. An exciting recent result, which is due to Delort (1991), is the existence of a solution to the Euler equation with vortex-sheet initial data when the vorticity has a *distinguished sign*. The proof we present (Majda, 1993) is motivated by our framework on approximate-solution sequences for the 2D Euler equation. As in the case of data in L^p, $1 \le p < \infty$, uniqueness of the solution is not known, nor are results known for the case of vorticity with mixed sign.

The key simplification over Delort's original proof is the use of the sharp a priori estimate for the decay of the vorticity maximal function, expressed in Lemma 11.1, valid only when the vorticity has a distinguished sign. This existence proof has enough flexibility to establish, as a corollary, that the high Reynolds number limit of Navier–Stokes solutions with vortex-sheet initial data with a distinguished sign converges to a weak solution to the 2D Euler equations.

For simplicity in exposition, we assume throughout this section that the initial vorticity ω_0 has compact support,

$$\operatorname{supp} \omega_0 \subseteq \{x \mid |x| < R\}, \tag{11.86}$$

for some fixed $R > 0$.

In the case of distinguished-sign ω_0, all fluid elements spin in the same direction so that subtle cancellation and screening effects at small scales (Majda, 1988) cannot occur. Our goal in this section is to translate this physical intuition into quantitative a priori estimates that confine the complexity that can occur in the limiting process of a family of approximate solutions.

First we review the notion of a weak solution in primitive-variable form (Definitions 9.1 and 10.1 in Chaps. 9 and 10). The velocity field $v(x,t) \in L^\infty\{[0,T], L^2_{\text{loc}}(\mathbb{R}^2)\}$ is a *weak solution* to the 2D Euler equations with the initial data $v_0(x)$ provided that

(1) for all test functions $\Phi \in C_0^\infty(\mathbb{R}^2 \times \mathbb{R}^+)$ with $\operatorname{div} \Phi = 0$,

$$\iint (\Phi_t \cdot v + \nabla\Phi : v \otimes v) dx dt = 0, \tag{11.87}$$

(2) the velocity $v(x,t)$belongs to Lip $\{[0,T], H^{-L}_{\text{loc}}(\mathbb{R}^2)\}$ for some $L > 0$ and $v(0) = v_0$ in $H^{-L}_{\text{loc}}(\mathbb{R}^2)$.

In the weak form (11.87) of evolution equation, $v \otimes v = (v_i v_j)$, $\nabla\Phi = (\partial\Phi_i/\partial x_j)$, *1:2* denotes the matrix product $\sum_{i,j} a_{ij} b_{ij}$, and $H^s_{\text{loc}}(\mathbb{R}^2)$ denotes the local L^2 Sobolev space with s derivatives. Recall that technical condition (2) gives a precise meaning to the sense in which the initial data are achieved. We also recall the Biot–Savart law (2.94) and (2.95) that relates the velocity and vorticity by means of a stream function ψ. The equations

$$\begin{aligned} \Delta\psi &= \omega \\ v = \nabla^\perp\psi &= \begin{pmatrix} -\psi_{x_2} \\ \psi_{x_1} \end{pmatrix} \end{aligned} \tag{11.88}$$

imply that

$$v(x) = \int_{\mathbb{R}^2} K(x-y)\omega(y) dy, \tag{11.89}$$

with

$$K(y) = \left[2\pi\left(y_1^2 + y_2^2\right)\right]^{-1} \begin{pmatrix} -y_2 \\ y_1 \end{pmatrix} = \nabla^\perp \left(\frac{1}{2\pi} \log(|y|) \right).$$

Recall that vortex-sheet initial data mean a 2D incompressible velocity field $v_0 = (v_1^0, v_2^0)$ with vorticity $\omega_0 = \text{curl } v_0$ satisfying the condition that ω_0 is a nonnegative Radon measure,

$$\omega_0 \in \mathcal{M}(\mathbb{R}^2), \qquad \omega_0 \geq 0 \tag{11.90}$$

and v_0 has locally finite kinetic energy

$$\int_{|x|<R} |v_0|^2 dx \leq C(R) < \infty \qquad \text{for any} \quad R > 0. \tag{11.91}$$

For simplicity we assume that ω_0 has compact suport.

Given vortex-sheet initial data v_0 satisfying condition (11.90) and (11.91) with $\omega_0 \geq 0$, we now construct an *approximate-solution sequence* that has the solution v as its limit. First, smooth the initial data by mollification, $v_0^\epsilon = \mathcal{J}_\epsilon \omega = \rho_\epsilon * v_0$ $\int \rho = 1, \rho \geq 1, \rho \in C_0^\infty(\mathbb{R}^2), \rho_\epsilon = \frac{1}{\epsilon^2}\rho(x/\epsilon)$.

As we showed in Chap. 3, this standard mollification, described above, satisfies the following conditions:

(1) The smoothed vorticity ω^ϵ satisfies

$$\omega_0^\epsilon = \text{curl } v_0^\epsilon,$$
$$\omega_0^\epsilon \geq 0,$$
$$\text{supp } \omega_0^\epsilon \subseteq \{x| \, |x| < \rho\}.$$

(2) For any $R > 0$,

$$\int_{|x|\leq R} |v_0^\epsilon|^2 dx + \int \omega_0^\epsilon \, dx \leq C(R^2). \tag{11.92}$$

(3) The smoothed vorticities ω_0^ϵ converge strongly to ω_0 in $H^{-1}_{\text{comp}}(\mathbb{R}^2)$ and v_0^ϵ converges strongly to v_0 in $L^2_{\text{loc}}(\mathbb{R}^2)$.

We now construct an approximate-solution sequence that has the solution v as its limit. Using the smoothed initial data v_0^ϵ, we let v^ϵ be the unique smooth solution to the (Navier–Stokes or Euler) equations:

$$\begin{aligned} \frac{\partial v^\epsilon}{\partial t} + v^\epsilon \cdot \nabla v^\epsilon &= -\nabla p^\epsilon + \nu_\epsilon \Delta v^\epsilon, \\ \text{div } v^\epsilon &= 0, \\ v^\epsilon|_{t=0} &= v_0^\epsilon. \end{aligned} \tag{11.93}$$

Recall from Chap. 3 that the solution v^ϵ exists and remains smooth for all positive times for any value of the viscosity ν_ϵ satisfying $\nu_\epsilon \geq 0$. Taking the limit of these approximations gives the main result of this section.

Theorem 11.4. *Existence of Vortex Sheets with Distinguished Sign. Consider v_0, vortex-sheet initial data, such that the vorticity ω_0 has a distinguished sign. Assume* *that the viscosity ν_ϵ is either identically zero (Euler limit) or that ν_ϵ converges to zero as $\epsilon \to 0$ in some arbitrary fashion (Navier–Stokes limit). Then, after passing to a subsequence, the smooth solutions $v^\epsilon(x,t)$ of the Navier–Stokes (Euler) equations (11.93) converge weakly in $L^2_{\text{loc}}([0,T] \times \mathbb{R}^2)$ to an incompressible velocity field $v(x,t)$, which is a weak solution of the 2D Euler equations satisfying evolution equation (11.87) with the initial data $v_0(x)$.*

This theorem generalizes Delort's result (Delort, 1991) to the case of approximations by means of solutions to the Navier–Stokes equation. There is strong evidence (Zheng and Majda, 1994; Majda et al., 1994a) through explicit examples for an analogous but simpler model problem involving the 1D Vlasov–Poisson equations that the weak solutions in the theorem are probably not unique and depend on the regularization even though the vorticity has a distinguished sign. We discuss this analogy in Chap. 13. Next, we present a brief summary of the structure of the proof. The proof of the theorem has three key elements:

(1) Through the *special nonlinear structure of the 2D Euler equations* it is sufficient to guarantee weak convergence for the special nonlinear *functions* $v_1 v_2$, $v_1^2 - v_2^2$ in order to satisfy requirement (11.87) for a weak solution. We discuss this special nonlinear structure in more detail in Chap. 12 (DiPerna and Majda, 1987b).

(2) An a priori decay estimate for the vorticity maximal functions when *the vorticity has a distinguished sign.* We show that the approximate-solution sequence defined in Eqs. (11.93) satisfies the conditions of Lemma 11.1 and hence satisfies the a priori bound on the vorticity maximal functions derived there.

(3) *Delort's key observation regarding the special nonlinearities* $v_1 v_2$ and $v_1^2 - v_2^2$ and *iterated mildly singular integrals.* In the following subsections, we present these three ingredients in detail and combine them to yield Theorem 11.4.

11.4.1. Step 1: Special Nonlinear Structure of the 2D Euler Equation

The first step exploits the fact that the vector test functions Φ satisfy div $\Phi = 0$. An equivalent condition is to substitute for Φ

$$\Phi = \nabla^\perp \eta = \begin{pmatrix} -\eta_{x_2} \\ \eta_{x_1} \end{pmatrix},$$

$$\eta \in C_0^\infty(\mathbb{R}^2 \times [0,T]) \tag{11.94}$$

into Eq. (11.87).

The equation then reduces to

$$\iint (\eta_{x_2 t} v_1 - \eta_{x_1 t} v_2) dx dt$$

$$= \iint \left[\eta_{x_1 x_2}\left(v_2^2 - v_1^2\right) + (\eta_{x_2 x_2} - \eta_{x_1 x_1})(v_1 v_2)\right] dx dt. \tag{11.95}$$

Thus an incompressible velocity field $v(x,t)$ defines a weak solution provided that identity (11.95) holds.

To establish the theorem, we need to show that the regularized solutions $v^\epsilon(x,t)$ defined in Eqs. (11.93) converge to a weak solution as $\epsilon \to 0$. The family $v^\epsilon(x,t)$ forms an approximate-solution sequence for the 2D Euler equation with vortex-sheet initial data in the precise sense of Definition 11.1. Therefore the functions v^ϵ and $\omega^\epsilon = \text{curl } v^\epsilon$ satisfy the uniform bounds

$$\max_{0\le t\le T}\left(\int_{\mathbb{R}^2}|\omega^\epsilon|dx + \int_{|x|\le R}|v^\epsilon|^2 dx\right) \le C_{R,T}, \tag{11.96}$$

with constants $C_{R,T}$ uniform in ϵ for every $R, T > 0$. Also, recall from Theorem 10.1 of Chap. 10 that, by passing to a subsequence, there exists $v \in L^\infty\{[0,T], L^2_{\text{loc}}(\mathbb{R}^2)\}$ with curl $v = \omega \in L^\infty\{[0,T], \mathcal{M}(\mathbb{R}^2)\}$ satisfying

$$\begin{aligned} v^\epsilon &\rightharpoonup v \quad \text{weakly in} \quad L^2(\Omega_T),\\ \omega^\epsilon &\rightharpoonup \omega \quad \text{weakly in} \quad L^\infty\{[0,T], \mathcal{M}(\mathbb{R}^2)\}, \end{aligned} \tag{11.97}$$

and

$$v^\epsilon \to v \quad \text{strongly in} \quad L^1(\Omega_T), \tag{11.98}$$

for any $T > 0$ and any space–time open set $\Omega_T = \{(x,t)|\,|x| < R,\ 0 < t < T\}$. Furthermore, v satisfies the weak regularity condition in time required in the definition of an approximate-solution sequence with $v(0) = v_0$ so that v assumes the required vortex-sheet initial data. If we had the additional constraint of L^p, $p > 1$, vorticity control, we could show, as in Chap. 10, that no concentration occurs in the limit and v satisfies the weak form of Euler equation (11.87). However, as Example 11.4 in Section 11.3 shows, uniform bound (11.96) is insufficient to guarantee that no concentration occurs. Instead, we exploit the fact that all members of the approximate-solution sequence have vorticity of fixed sign to prove that concentration–cancellation occurs in the limit.

It follows from Eqs. (11.93) that ω^ϵ satisfies the *vorticity-stream form* of the Euler/Navier–Stokes equation:

$$\begin{aligned} \frac{\partial \omega^\epsilon}{\partial t} + v^\epsilon \cdot \nabla\omega^\epsilon &= \nu_\epsilon \Delta\omega^\epsilon,\\ \omega^\epsilon|_{t=0} &= \omega_0^\epsilon. \end{aligned} \tag{11.99}$$

Thus, if the initial vorticity has a distinguished sign, $\omega_0^\epsilon \ge 0$, it follows from the maximum principle for $t \ge 0$ that

$$\omega^\epsilon(x,t) \ge 0 \qquad \omega(x,t) \ge 0, \tag{11.100}$$

so both $\omega^\epsilon(x,t)$ and the limiting vorticity measure ω have a distinguished sign for any later time.

Smooth solutions satisfying Eqs. (11.93) are trivially weak solutions of either the Euler or the Navier–Stokes equations. This fact combined with weak convergences (11.97) and the special structure for the weak form in identity (11.95), shows that the limiting velocity, v, is a weak solution for 2D Euler provided that the special nonlinear functions $v_2^2 - v_1^2$ and $v_1 v_2$ satisfy

$$\iint \eta_{x_1 x_2}(v_2^2 - v_1^2)dxdt = \lim_{\epsilon \to 0} \iint \eta_{x_1 x_2}\left[(v_2^\epsilon)^2 - (v_1^\epsilon)^2\right]dxdt, \quad (11.101)$$

$$\iint (\eta_{x_2 x_2} - \eta_{x_1 x_1})(v_1 v_2)dxdt = \lim_{\epsilon \to 0} \iint (\eta_{x_2 x_2} - \eta_{x_1 x_1})v_1^\epsilon v_2^\epsilon dxdt. \quad (11.102)$$

The Navier–Stokes (Euler) equations are rotationally invariant in that if $v^\epsilon(x,t)$ satisfies these equations and estimates, $O(\theta)v^\epsilon[{}^t O(\theta)x, t]$ satisfies these same equations and estimates for any rotation matrix $O(\theta)$; through rotation by 45°, the term $(v_2^\epsilon)^2 - (v_1^\epsilon)^2$ becomes $v_1^\epsilon v_2^\epsilon$ in the rotated reference frame (see DiPerna and Majda, 1988). Thus, in order to prove the theorem that v is a weak solution for the 2D Euler equation, it is sufficient to establish the weak convergence of $v_1^\epsilon v_2^\epsilon$ to $v_1 v_2$, i.e.,

$$\iint \varphi v_1 v_2 \, dxdt = \lim_{\epsilon \to 0} \iint \varphi v_1^\epsilon v_2^\epsilon \, dxdt \quad (11.103)$$

for all test functions $\varphi \in C_0^\infty(\mathbb{R}^2 \times [0, T])$. This reduction completes the first key element of the proof of Theorem 11.4.

11.4.2. Step 2: Uniform Decay of the Vorticity Maximal Function

The circulation around any point x_0 enclosed inside a circle of radius R is given by

$$\int_{|x - x_0| \le R} \omega^\epsilon(x,t)dx. \quad (11.104)$$

In Section 11.3 we showed that if the *vorticity maximal functions*

$$\int_{|x - x_0| \le R} |\omega^\epsilon(x,t)|dx, \quad (11.105)$$

which are identical to the circulation for positive vorticity, decay at a sufficiently rapid uniform rate as $R \to 0$ independent of ϵ then the limiting behavior of v^ϵ in L^2 can be controlled. Theorem 11.3 showed that an explicit bound on the decay rate is sufficient to guarantee strong convergence and hence no concentration in the limit. In Example 11.4, we showed that, in general, approximate-solution sequences for vortex-sheet initial data do not satisfy the more stringent bound of Theorem 11.3. Nevertheless, as we say in Lemma 11.1, all solution sequences possessing a uniform L^1 bound, energy bound, and bounded second moment satisfy a slightly weaker decay estimate for the vorticity maximal function. This decay estimate is what ultimately allows us to prove that concentration–cancellation occurs in the limit for the distinguished-sign vortex-sheet problem. The first step in this process is to show that, for our chosen approximate-solution sequence, the assumptions of Lemma 11.1 are satisfied. The result is then the following proposition.

Proposition 11.3. *Uniform Decay of the Vorticity Maximal Function. Assume that* ω_0^ϵ *and* v_0^ϵ *satisfy relations (11.86) and (11.92) and that* ω_0^ϵ *is nonnegative,* $\omega_0^\epsilon \geq 0$*. Then the vorticity maximal functions*

$$|\omega^\epsilon|(B_R(x)) \equiv \int_{B_R(x)} |\omega(y)| dy \tag{11.106}$$

have an a priori uniform decay rate and satisfy for $R < (1/2)$

$$\max_{0 \leq t \leq T, x_0 \in \mathbb{R}^2, 0 < \epsilon \leq \epsilon_0} \int_{|x - x_0| \leq R} \omega^\epsilon(x, t) dx \leq C_{T, R_0} |\log(2R)|^{-\frac{1}{2}}, \tag{11.107}$$

where C_{T,R_0} *is a fixed constant depending on* T, R_0 *from support (11.86) and uniform bounds (11.92) and (11.96).*

The same estimate in condition (11.107) also applies to the limiting vorticity measure, $\omega = \text{curl } v$ from weak convergence (11.97); in fact, for $R < (1/4)$,

$$\max_{0 \leq t \leq T, x_0 \in \mathbb{R}^2} \int_{|x - x_0| \leq R} d\omega(x, t) \leq C_{T, R_0} |\log(4R)|^{-\frac{1}{2}}, \tag{11.108}$$

As we showed in Section 11.3 the uniform a priori estimate in estimate (11.107) is sharp for initial measures with a distinguished sign, $\omega_0^\epsilon \geq 0$, and false for general initial vorticity distributions without a distinguished sign.

To prove this proposition, we must show that the approximate-solution sequence satisfies the condition of Lemma 11.1 from Section 11.3.

Lemma 11.1. *Pick a family of smooth velocity fields* $\{v_0^\epsilon\}$ *such that the vorticities* $\omega_0^\epsilon = \text{curl } v_0^\epsilon$ *satisfy*

$$\begin{gathered} \omega_0^\epsilon \geq 0, \\ \int |x|^2 \omega_0^\epsilon(x) \equiv M_0[\omega^\epsilon(x)] \leq M_0 < \infty, \\ \int \omega_0^\epsilon(x) dx \equiv C_0 \leq C_0 < \infty, \\ H_0^\epsilon \leq H_0 < \infty, \end{gathered} \tag{11.109}$$

where $H_0^\epsilon \equiv H(\omega_0^\epsilon)$ *and* $H(t) \equiv H[\omega(t)]$ *is the pseudoenergy*

$$H(\omega) = \frac{1}{2\pi} \iint \log|x - y| \omega(x) \omega(y) dx dy.$$

Then there exists a constant $C = C(H_0, M_0, C_0)$ *so that for any* $R \leq R_0$, $T > 0$, *we have*

$$\max_{0 \leq t \leq T, x_0 \in \mathbb{R}^2} |\omega^\epsilon|[B_R(x_0), t] \leq C_1 \left[\log\left(\frac{1}{R}\right)\right]^{-\frac{1}{2}}, \tag{11.110}$$

where C_1 is a constant independent of ϵ and

$$|\omega^\epsilon|(B_R(x_0), t) = \int_{B_R(x_0)} |\omega^\epsilon(x, t)| dx \tag{11.111}$$

is the vorticity maximal function at time t.

Next, we show how the above proposition directly yields the proof of Proposition 11.3 for the special case in which the approximating solutions v^ϵ from Eqs. (11.93) are smooth solutions to the 2D Euler equations so that $\nu_\epsilon \equiv 0$ in Eq. (11.93).[§] Recall from Proposition 1.12 and from the conservation of the pseudoenergy H, defined in Definition (11.71), that for the 2D Euler equations, the quantities $\int_{\mathbb{R}^2} \omega(t) dx$, $H[\omega(t)]$, and $M_0[\omega(t)]$ are all exactly conserved for all times $t \geq 0$ and have the same value as defined at time $t = 0$. Thus, to deduce estimate (11.107) in the lemma, it is sufficient to obtain uniform boun ds independent of ϵ in the initial data for the functionals $\int_{\mathbb{R}^2} \omega_0^\epsilon(x) dx$, $M_0(\omega_0^\epsilon)$, and $|H(\omega_0^\epsilon)|$. Given the regularization procedure, the ϵ-independent bounds for $\int_{\mathbb{R}^2} \omega_0^\epsilon(x) dx$ and $M_0(\omega_0^\epsilon)$ are obvious. In particular, the approximations ω_0^ϵ satisfy

$$\int_{\mathbb{R}^2} \omega_0^\epsilon(x) dx = \int_{\mathbb{R}^2} d\omega_0 \equiv \Gamma_0. \tag{11.112}$$

The uniform bound for $|H(\omega_0^\epsilon)|$ is slightly more tedious and involves similar considerations as were utilized in Chap. 3 to handle the fact that $v^\epsilon \notin L^2(\mathbb{R}^2)$ when ω_0^ϵ is nonnegative. We sketch the argument for completeness. Recall from Chap. 3, Definition 3.1, that any solution to the 2D Euler equation has a radial-energy decomposition. Choose a fixed radial function $\overline{\omega}_0(r) \in C_0^\infty(\mathbb{R}^2)$ with supp $\overline{\omega}_0(r) \subseteq \{|x| < R_0\}$ and

$$\Gamma_0 = 2\pi \int_0^\infty r\overline{\omega}_0(r) dr. \tag{11.113}$$

The induced velocity

$$\overline{v}_0(x) = |x|^{-2} \begin{pmatrix} -x_2 \\ x_1 \end{pmatrix} \int_0^{|x|} s\overline{\omega}(s) ds \tag{11.114}$$

is the corresponding radial eddy defining an exact steady solution of the 2D Euler equations with $\overline{\psi}(x)$ the corresponding stream function. It is an elementary exercise to show that $\overline{\psi}(x)$ satisfies the bound

$$|\overline{\psi}(x)| \leq C[1 + \log^+(|x|)]. \tag{11.115}$$

Consider the velocity field $\tilde{v}_0^\epsilon = v_0^\epsilon - \overline{v}$; with the condition (11.112) and (11.113), a standard far-field calculation refining (11.112) yields the uniform estimate,

$$\int_{|x|>2R_0} |\tilde{v}_0^\epsilon|^2 \leq C\left(\int_{|x|\leq R_0} |\tilde{\omega}_0^\epsilon| dx\right)^2 \leq 4C\Gamma_0^2, \tag{11.116}$$

§ This is the only case treated by Delort in Delort (1991).

so that relations (11.92), (11.114), and (11.116) guarantee the uniform L^2 velocity bound

$$\max_{\epsilon\le 1}\int_{\mathbb{R}^2}\left|\tilde{v}_0^\epsilon\right|^2 dx \le C. \tag{11.117}$$

To establish the required uniform bound on $H(\omega_0^\epsilon)$, we decompose $H(\omega_0^\epsilon)$ as follows:

$$\begin{aligned} H(\omega_0^\epsilon) &= H(\tilde{\omega}_0^\epsilon) + H(\overline{\omega}_0) - 2\int_{\mathbb{R}^2}(\tilde{\omega}_0^\epsilon\overline{\psi})dx \\ &= -\int_{\mathbb{R}^2}\left|\tilde{v}_0^\epsilon\right|^2 dx + H(\overline{\omega}_0) - 2\int_{\mathbb{R}^2}(\tilde{\omega}_0^\epsilon\overline{\psi})dx. \end{aligned} \tag{11.118}$$

The second identity in Eq. (11.118) uses the fact that for $\tilde{v} \in L^2$ with vorticity $\tilde{\omega}$ rapidly decreasing,

$$\int_{\mathbb{R}^2}|\tilde{v}|^2 = -\int_{\mathbb{R}^2}\tilde{\psi}\tilde{\omega}. \tag{11.119}$$

For bound (11.117) the first two terms on the right-hand side of Eq. (11.118) are uniformly bounded, whereas bound (11.115) implies that the remaining term satisfies

$$\left|\int_{\mathbb{R}^2}\tilde{\omega}_0^\epsilon\overline{\psi}\,dx\right| \le C\Gamma_0. \tag{11.120}$$

Thus, $|H(\omega_0^\epsilon)| \le \overline{H}$ as required in the lemma.

Exercise 11.3. *Verify that because $\omega^\epsilon \rightharpoonup \omega$ in $\mathcal{M}(\mathbb{R}^2)$, as shown in relation (11.97), the measure ω automatically inherits the slightly weaker bound (11.108) from the uniform bound on the vorticity maximal functions for ω^ϵ in estimate (11.107).*

This completes the proof of the proposition in the case in which $v^\epsilon(x,t)$ is a sequence of smooth solutions satisfying the 2D Euler equations.

To apply the lemma to an approximating sequence v^ϵ involving smooth solutions of the Navier–Stokes equations, it is necessary to establish ϵ-independent bounds on the three functionals $\int_{\mathbb{R}^2}\omega^\epsilon(x,t)dx$, $M_0[\omega^\epsilon(t)]$, and $H[\omega^\epsilon(t)]$ on the fixed-time interval, $0 \le t \le T$ for an arbitrary $T > 0$. However, as we say in the earlier chapters of this book, both $M_0(\omega)$ and $\int\omega$ are controlled on any time interval by the initial data for solutions of the 2D Navier–Stokes equation. Because $M_0(\omega_0^\epsilon)$ and $\int\omega_0^\epsilon$ are uniformly bounded, this guarantees the required ϵ-independent bounds for $\int_{\mathbb{R}^2}\omega^\epsilon(x,t)dx$ and $M_0(\omega^\epsilon(t))$ on any finite-time interval. To establish the ϵ-independent bound for $H(\omega^\epsilon(t))$, we use the radial-energy decomposition from Chap. 3 for the Navier–Stokes equation. Let $\overline{\omega}^\epsilon(|x|,t)$ be the radial solution of the diffusion equation

$$\begin{aligned} \frac{\partial\overline{\omega}^\epsilon}{\partial t} &= \nu_\epsilon\Delta\overline{\omega}^\epsilon, \\ \overline{\omega}^\epsilon|_{t=0} &= \overline{\omega}_0(|x|), \end{aligned} \tag{11.121}$$

with the initial data $\overline{\omega}_0(|x|)$ defined above in condition (11.113). It follows (see Chap. 3) that

$$\overline{v}^\epsilon(x,t) = |x|^{-2}\begin{pmatrix}-x_2\\ x_1\end{pmatrix}\int_0^{|x|} s\overline{\omega}^\epsilon(s,t)ds \qquad (11.122)$$

defines an exact solution of the Navier–Stokes equations with

$$\frac{\partial \overline{v}^\epsilon}{\partial t} = \nu_\epsilon \Delta \overline{v}^\epsilon, \qquad \overline{v}^\epsilon \cdot \nabla \overline{v}^\epsilon = -\nabla \overline{p}^\epsilon = 0 \qquad (11.123)$$

Furthermore, the stream functions $\overline{\psi}^\epsilon$ and velocity fields $\overline{v}^\epsilon$ satisfy the uniform bounds,

$$\max_{0\le t\le T, 0<\epsilon\le 1} \left|\overline{\psi}^\epsilon(|x|,t)\right| \le C(1+\log^+|x|) \qquad (11.124a)$$

$$\max_{0\le t\le T, 0<\epsilon\le 1} |\nabla \overline{v}^\epsilon(t)|_{L^\infty} \le C. \qquad (11.124b)$$

Consider $\tilde{v}^\epsilon = v^\epsilon(x,t) - \overline{v}^\epsilon(x,t)$; because both $\omega^\epsilon(x,t)$ and $\overline{\omega}^\epsilon(x,t)$ are rapidly decreasing and the total vorticity is conserved in Navier–Stokes solutions, it follows by similar far-field arguments as presented earlier that $\tilde{v}^\epsilon(x,t) \in L^2(\mathbb{R}^2)$ for any fixed ϵ. With Eqs. (11.123) and (11.93) the function $\tilde{v}^\epsilon$ satisfies the equations

$$\frac{D\tilde{v}^\epsilon}{Dt} + (\tilde{v}^\epsilon \cdot \nabla)\overline{v}^\epsilon(x,t) = -\nabla \tilde{p}^\epsilon + \nu_\epsilon \Delta \tilde{v}^\epsilon,$$

$$\operatorname{div} \tilde{v}^\epsilon = 0. \qquad (11.125)$$

Recall from Chap. 3 on energy methods that Eqs. (11.125) combined with the bound (11.124b) establish that

$$\max_{0\le t\le T}\int_{\mathbb{R}^2} |\tilde{v}^\epsilon|^2(x,t)dx \le C(T)\int_{\mathbb{R}^2} |\tilde{v}_0^\epsilon|^2 dx \le \tilde{C}(T). \qquad (11.126)$$

Thus, inequalities (11.126) together with bound (11.117) guarantee an ϵ-independent bound for the energy of the velocity fields, $\tilde{v}^\epsilon(x,t)$. As in Eq. (11.118), we have

$$H(\omega^\epsilon(t)) = \int_{\mathbb{R}^2} |\tilde{v}^\epsilon|^2(t)dx + H(\overline{\omega}^\epsilon(t)) - 2\int_{\mathbb{R}^2} \omega^\epsilon(t)\overline{\psi}^\epsilon(t)dx. \qquad (11.127)$$

With estimates (11.124a) and (11.124b), it follows that

$$\left|\int_{\mathbb{R}^2} \omega^\epsilon(x,t)\overline{\psi}^\epsilon(x,t)dx\right|$$

$$\le C\int_{\mathbb{R}^2}(1+|x|^2)\omega^\epsilon(x,t)dx \le C[\Gamma_0 + L_0 + \nu_\epsilon t N(\omega_0)]. \qquad (11.128)$$

The end result is the ϵ-independent bound,

$$\max_{0\le t\le T, 0<\epsilon\le 1} |H[\omega^\epsilon(t)]| \le C(T)$$

as required for applying the lemma. This fact completes the proof of Proposition 11.3.

11.4.3. Step 3: Delort's Key Observation and Completion of the Proof

To pass to the limit as $\epsilon \to 0$ and obtain a weak solution of the 2D Euler equation, it follows from Eq. (11.103) that we need to understand the expression

$$\int_{\mathbb{R}^2} \varphi v_1 v_2 \, dx \qquad \text{for any} \quad \varphi \in C_0^\infty(\mathbb{R}^2).$$

The Biot–Savart law implies that we can write

$$\int_{\mathbb{R}^2} \varphi v_1 v_2 \, dx = \int\int_{\mathbb{R}^2 \times \mathbb{R}^2} H_\varphi \omega(x)\omega(y) dx dy, \tag{11.129}$$

where $H_\varphi(x, y)$ is the function (distribution) on $\mathbb{R}^2 \times \mathbb{R}^2$ given by the formula

$$H_\varphi(x, y) = -\frac{1}{4\pi^2} \frac{\partial^2}{\partial x_1 \partial x_2} \int_{\mathbb{R}^2} \log|x - z| \log|y - z| \varphi(z) dz. \tag{11.130}$$

In general, when kernels that are homogeneous of degree -1 in $\mathbb{R}^2$ are iterated to build a function such as $H_\varphi(x, y)$, logarithmic singularities occur along the diagonal, $x = y$. Delort's (1991) *key new observation* (see Proposition 1.2.3) is the important fact that with the *special nonlinearities* $v_1 v_2$, such iterated singular integrals result in a *function*, H_φ, that remains *bounded along the diagonal.* The detailed statement of his result is the following lemma.

Delort's Lemma: *The function*

$$H_\varphi(x, y) = -\frac{1}{4\pi^2} \frac{\partial^2}{\partial x_1 \partial x_2} \int_{\mathbb{R}^2} \log|x - z| \log|y - z| \varphi(z) dz$$

is a bounded function on $\mathbb{R}^2 \times \mathbb{R}^2$*, continuous on the complement of the diagonal, and tends to zero at infinity. In fact,*

$$H_\varphi(x, y) = \frac{1}{2}[\varphi(x) + \varphi(y)] h(x - y) + r(x, y) \tag{11.131}$$

where $h(\omega) = [(\omega_1 \cdot \omega_2)/(4\pi|\omega|^2)]$ *and* $r(x, y)$ *is a bounded continuous function.*

We save the proof of this lemma for the end of this section. Next, with the three main facts in relations (11.103), (11.107), (11.108), and (11.129)–(11.131), we pass to the limit in Eq. (11.103) and verify that v is a weak solution of the 2D Euler equation. Through a standard density argument, it is no loss of generality to verify Eq. (11.103) for only the special product test functions $\varphi(x, t) = \psi(t)\varphi(x)$, with $\psi(t) \in C_0^\infty(\mathbb{R}^+)$, $\varphi(x) \in C_0^\infty(\mathbb{R}^2)$. Let $\rho(|x|) \in C_0^\infty(\mathbb{R}^2)$ be a fixed positive cutoff function with $\rho \equiv 1$ for $|x| < 1$ and $\rho \equiv 0$ for $|x| > 2$. Expression (11.129) implies that

$$\begin{aligned}\int\int \varphi(x,t)v_1^\epsilon v_2^\epsilon\,dxdt &= \int_0^\infty \int_{\mathbb{R}^2\times\mathbb{R}^2}\int \psi(t)\left[1-\rho\left(\frac{|x-y|}{\delta}\right)\right] H_\varphi(x,y)\,\omega^\epsilon(x,t)\omega^\epsilon(y,t)dxdydt\\ &+ \int_0^\infty \int_{\mathbb{R}^2\times\mathbb{R}^2}\int \psi(t)\rho\left(\frac{|x-y|}{\delta}\right) H_\varphi(x,y)\,\omega^\epsilon(x,t)\,\omega^\epsilon(y,t)dxdydt.\end{aligned} \tag{11.132}$$

For any fixed δ with $1 > \delta > 0$, Delort's lemma implies that

$$\psi(t)\left[1-\rho\left(\frac{|x-y|}{\delta}\right)\right] H_\varphi(x,y) \in C_0(\mathbb{R}^2). \tag{11.133}$$

From weak convergences (11.97) it follows directly that

$$\omega^\epsilon(x,t)\otimes\omega^\epsilon(y,t) \rightharpoonup d\omega(x,t)\otimes d\omega(y,t) \tag{11.134}$$

weakly in measures, $L^\infty\{[0,T], M(\mathbb{R}^2\times\mathbb{R}^2)\}$, for any $T > 0$. The continuity and convergence in relations (11.133) and (11.134) combine to yield

$$\begin{aligned}\lim_{\epsilon\to 0}\int_0^\infty \int\int_{\mathbb{R}^2\times\mathbb{R}^2} g(|x-y|,t)H_\varphi(x,y)\omega^\epsilon(x,t)\omega^\epsilon(y,t)dxdydt\\ = \int_0^\infty dt \int\int_{\mathbb{R}^2\times\mathbb{R}^2} g(|x-y|,t)H_\varphi(x,y)d\omega(x,t)\otimes d\omega(y,t).\end{aligned}$$

$$g(|x-y|,t) = \psi(t)\left[1-\rho\left(\frac{|x-y|}{\delta}\right)\right]. \tag{11.135}$$

Next, we use the uniform decay rate for the vorticity maximal function in estimate (11.107) together with Delort's lemma to get uniform control on the second term on the right-hand side of Eq. (11.132), resulting in the estimates

$$\begin{aligned}&\left|\int_0^\infty \int_{\mathbb{R}^2\times\mathbb{R}^2}\int \psi(t)\rho\left(\frac{|x-y|}{\delta}\right) H_\varphi(x,y)\omega^\epsilon(x,t)\omega^\epsilon(y,t)dxdydt\right|\\ &\le C_{\varphi,\psi}\int_0^T \int_{\mathbb{R}^2\times\mathbb{R}^2} \rho\left(\frac{|x-y|}{\delta}\right)\omega^\epsilon(x,t)\omega^\epsilon(y,t)dxdydt\\ &\le C_{\varphi,\psi}C_{R_0,T}\left[\log\left(\frac{1}{2\delta}\right)\right]^{-\frac{1}{2}}\int_0^T\int_{\mathbb{R}^2}\omega^\epsilon(y,t)dydt\\ &\le C\left[\log\left(\frac{1}{2\delta}\right)\right]^{-\frac{1}{2}}.\end{aligned} \tag{11.136}$$

The first inequality in estimates (11.136) uses Delort's lemma and the nonnegativity of ω^ϵ whereas the second inequality applies the uniform decay in estimate (11.107) for the vorticity maximal function. The final inequality in Eq. (11.135) uses the uniform control on the mass of the measures ω^ϵ guaranteed by uniform bounds (11.96). The decay rates of the vorticity maximal functions, a priori bounds (11.96), and weak convergence of measures (11.97) imply that the same uniform estimates (11.136) hold with ω^ϵ replaced with the limiting vorticity measure ω. Thus Eq. (11.135) and estimates (11.136) guarantee that

$$\left| \iint \varphi(x,t) v_1 v_2 \, dx dt - \lim_{\epsilon \to 0} \iint \varphi v_1^\epsilon v_2^\epsilon \, dx dt \right| \leq C_\varphi \left[\log \left(\frac{1}{4\delta} \right) \right]^{-\frac{1}{2}}. \tag{11.137}$$

Because δ in relation (11.137) can be made arbitrarily small, requirement (11.103) is satisfied and v is a weak solution of the 2D Euler equations with the required vortex-sheet initial data, $v_0(x)$.

For completeness, we conclude with the proof of Delort's lemma (Delort, 1991).

Proof of Delort's Lemma. The proof we present here is different from the one in Delort (1991). It uses directly the ideas from singular integral theory from Chap. 4.

First note that, off the diagonal,

$$H_\varphi(x,y) = -\frac{1}{4\pi^2} \int \frac{x_1 - z_1}{|x-z|^2} \frac{y_2 - z_2}{|y-z|^2} \varphi(z) dz.$$

It is easy to see that H_φ decays to zero at infinity, because φ has compact support. We leave the details of this as an exercise for the reader. Rewrite this expression:

$$\begin{aligned} H_\varphi(x,y) &= -\frac{1}{4\pi^2} \int \frac{x_1 - z_1}{|x-z|^2} \frac{y_2 - z_2}{|y-z|^2} \left\{ \varphi(z) - \frac{1}{2}[\varphi(x) + \varphi(y)] \right\} dz \\ &\quad - \frac{1}{4\pi^2} \int \frac{x_1 - z_1}{|x-z|^2} \frac{y_2 - z_2}{|y-z|^2} [\varphi(x) + \varphi(y)] dz. \\ &= H_1(x,y) + H_2(x,y) \end{aligned}$$

The integrals are well defined as $|z| \to \infty$ by assuming a principal-value integral in the far field, that is, $\lim_{R\to\infty} \int_{|z|=R} \cdots dz$. This is due to the cancellation property of the leading-order behavior of integrands for large $|z|$. Our proof of Delort's lemma has two parts:

(1) show that H_1 is a continuous, bounded function on all of $\mathbb{R}^2 \times \mathbb{R}^2$,
(2) show that H_2 has the explicit form

$$H_2 = \frac{w_1 w_2}{4\pi |w|^2} [\varphi(x) + \varphi(y)], \qquad w = x - y.$$

Step 1 is the harder of the two and requires some ideas from the theory of singular integrals. Step 2 is a simple exercise involving the Fourier transform.

Step 1: Note that on the diagonal, $H_1(x,x)$ has the simple form

$$H_1 = -\frac{1}{4\pi^2}\int [\varphi(x) - \varphi(z)]\frac{(x_1 - z_1)(x_2 - z_2)}{|x - z|^4}dz, \tag{11.138}$$

where a principal-value integral near $z = x$ is not necessary because of the differentiability of φ at x.

To prove continuiuty of H_1 in both x and y, for simplicity we assume that $x \neq y$; however, the reader can verify that all steps of the proof hold in the event that $x = y$. The ideas presented below are similar to those used to prove potential theory Lemma 4.6 of Chap. 4. First we rewrite

$$\begin{aligned}-4\pi^2 H_1 = &\frac{1}{2}\int [\varphi(x) - \varphi(z)]\frac{x_1 - z_1}{|x - z|^2}\frac{y_2 - z_2}{|y - z|^2}dz\\ &+\frac{1}{2}\int [\varphi(y) - \varphi(z)]\frac{x_1 - z_1}{|x - z|^2}\frac{y_2 - z_2}{|y - z|^2}dz.\end{aligned}$$

Without loss of generality, we prove boundedness and continuity in x and y of the first term,

$$\frac{1}{2}\int [\varphi(x) - \varphi(z)]\frac{x_1 - z_1}{|x - z|^2}\frac{y_2 - z_2}{|y - z|^2}dz, \tag{11.139}$$

in the above decomposition. The second part of H_1 is proved continuous in an identical manner.

Boundedness is a simple exercise that the reader can check by using estimate (11.141) below. To show continuity in x, we write

$$\begin{aligned}&\int [\varphi(x) - \varphi(z)]\frac{x_1 - z_1}{|x - z|^2}\frac{y_2 - z_2}{|y - z|^2}dz\\
&\quad - \int [\varphi(x+h) - \varphi(z)]\frac{x_1 + h_1 - z_1}{|x + h - z|^2}\frac{y_2 - z_2}{|y - z|^2}dz\\
&= \int_{|x-z|<2h} [\varphi(x) - \varphi(z)]\frac{x_1 - z_1}{|x - z|^2}\frac{y_2 - z_2}{|y - z|^2}dz\\
&\quad - \int_{|x-z|<2h} [\varphi(x+h) - \varphi(z)]\frac{x_1 + h_1 - z_1}{|x + h - z|^2}\frac{y_2 - z_2}{|y - z|^2}dz\\
&\quad + \int_{|x-z|>2h, |y-z|<2h}\left\{\frac{[\varphi(x) - \varphi(z)](x_1 - z_1)}{|x - z|^2}\right.\\
&\quad \left. - \frac{[\varphi(x+h) - \varphi(z)](x_1 + h_1 - z_1)}{|x + h - z|^2}\right\}\frac{y_2 - z_2}{|y - z|^2}dz\\
&\quad + \int_{|x-z|>2h, |y-z|<2h}\left\{\frac{[\varphi(x) - \varphi(z)](x_1 - z_1)}{|x - z|^2}\right.\\
&\quad \left. - \frac{[\varphi(x+h) - \varphi(z)](x_1 + h_1 - z_1)}{|x + h - z|^2}\right\}\frac{y_2 - z_2}{|y - z|^2}dz\\
&= T1 + T2 + T3 + T4.\end{aligned}$$

Because $[\varphi(x)-\varphi(z)][(x_1 - z_1)/(|x - z|^2)] < M$ for all $x, z \in \mathbb{R}^2$, $|T1|$ and $|T2|$ are both bounded by $M \int_{|x-z|<2h} dz/|y - z| \leq M \int_{|y-z|<4h} dz/|y - z| = 8\pi Mh$. On $|x - z| \geq 2h$,

$$\left\{ \frac{[\varphi(x) - \varphi(z)](x_1 - z_1)}{|x - z|^2} - \frac{[\varphi(x + h) - \varphi(z)](x_1 + h_1 - z_1)}{|x + h - z|^2} \right\} \leq \frac{hk_1}{|x - z|},$$

which means that

$$|T3| \leq k1 \int_{|y-z|<2h} dz/|y - z| \leq k_1 4h\pi.$$

Finally,

$$\begin{aligned} |T4| &\leq k_1 h \int_{|x-z|,|y-z|\geq 2h} \frac{dz}{|z - x||z - y|} \\ &\leq k_1 h \int_{|x-z|,|y-z|\geq 2h} \left(\frac{dz}{2|z - x|^2} + \frac{dz}{2|z - y|^2} \right) \qquad (11.140) \\ &\leq k_1 2\pi h \log h. \end{aligned}$$

We now must show continuity in y. To do this, we use the fact that

$$[\varphi(x) - \varphi(z)]\frac{x_1 - z_1}{|x - z|^2} \equiv M_x(z) < \frac{M}{1 + |x - z|}. \qquad (11.141)$$

We now compare the function evaluated at (x, y) and $(x, y+h)$. Write this difference as

$$\begin{aligned} &\int_{|y-z|<2h} M_x(z)\frac{y_2 - z_2}{|y - z|^2}dz \\ &\quad - \int_{|y-z|<2h} M_x(z)\frac{y_2 + h_2 - z_2}{|y + h - z|^2}dz \\ &\quad - \int_{|y-z|>2h} M_x(z)\left(\frac{y_2 - z_2}{|y - z|^2} - \frac{y_2 + h_2 - z_2}{|y + h - z|^2}\right)dz = T1 + T2 + T3. \end{aligned}$$

$|T1|$ and $|T2|$ are both bounded by $M2\pi h$. $|T3|$ is likewise bounded by $M2\pi h \log h$. We can see this by following the same ideas used to obtain bound (11.140) above. We have shown that H_1 is continuous in all of $\mathbb{R}^2 \times \mathbb{R}^2$ with modulus of continuiuty $\rho(h) = C(k_1, k_2, M)h \log h$. This completes step 1 of the proof.

We can easily show *Step 2* by noting that, for $w = x - y$,

$$\int \frac{x_1 - z_1}{|x - z|^2}\frac{y_2 - z_2}{|y - z|^2}dz = \int \frac{w_1 - z_1}{|w - z|^2}\frac{z_2}{|z|^2}dz = (\partial_{w_1} \log|w|) * (\partial_{w_2} \log|w|).$$

The Fourier transform of this expression is

$$\frac{2\pi i\xi_1 \cdot 2\pi i\xi_2}{(2\pi)^2|\xi|^2} = -\frac{\xi_1}{|\xi|^2}\frac{\xi_2}{|\xi|^2} = \frac{1}{2}\partial_{\xi_1}\partial_{\xi_2}(\log|\xi|)$$

which is in turn the Fourier transform of $-\pi[(w_1 w_2)/|w|^2]$. The reader unfamiliar with the Fourier transform of these expressions can easily check this by noting (see, e.g., Folland, 1995, Chap. 2, Section B) that the transform of $\log|w|$ in $\mathbb{R}^2$ is $-|\xi|^{-2}/2\pi$. □

We mention here that the distinguished sign for the vorticity is used in only two places in the proof of Theorem 11.4: to derive the proposition on uniform decay of the vorticity maximal function and in the second step in inequality (11.136).

One important ingredient in this version the proof of Theorem 11.4 is the uniform decay rate for the vorticity maximal functions in the approximating problem. The proof of this result makes use of the pseudoenergy H for 2D fluid flow. Computational vortex-blob methods, which we discussed in detail in Chap. 6, exactly conserve a discrete version of pseudoenenergy $H(\omega)$ (11.71) [see DiPerna and Majda, 1987a, p. 332, Eq. (2.31)]; and this discrete function serves as the Hamiltonian for smoothed-core discrete vortex dynamics. Because vortex-blob algorithms converge for smooth solutions of the Euler equations (see Majda, 1988), we can pass to the continuous limit and verify the conserved quantity in pseudoenergy (11.71). This conservation suggests the following problem: Let ω_h^δ be a vortex-blob regularization of a vortex sheet. (Here δ denotes the smoothing parameter of the vortex-blob regularization and h denotes the size of the discretization.) For an appropriately chosen limit as $\delta, h \to 0$, show that there is a weak limit that is a solution to the Euler equation in the sense of Definition 11.1. This problem was solved recently for the case of initial data with vorticity of fixed sign (Liu and Xin, 1995).

Another regularization of vortex-sheet initial data discussed in Chap. 9 was the vortex-patch regularization with a smooth boundary. Recall from Chap. 8 that such a regularization has a global-in-time solution that possesses the same conserved quantities as the smooth solution of the Euler equations.

Exercise 11.4. *Consider the case of initial vorticity of distinguished sign* $\omega_0 = \gamma(x)\delta_S$*, where* S *is a smooth curve in the plane and* γ *is a* C^∞ *function. Let* ω_0^ϵ *be a vortex-patch regularization of the sheet so that* $\omega_0^\epsilon \rightharpoonup \omega_0$ *in the sense of measures. Show that the unique solutions of the vortex-patch approximating problems* $\omega^\epsilon(x,t)$ *coverge to a weak solution of the Euler equation in the sense of Definition 11.1 with vortex-sheet initial data* ω_0.

As mentioned in Chap. 9, both these and numerical calculations indicate much more complex behavior in the 2D vortex-sheet evolution without a distinguished sign. Such behavior suggests the possibility of a nontrivial measure-valued solution to the Euler equations in the case of mixed-sign vortex-sheet initial data. In fact, the Majda conjectured early on in Majda (1988) that perhaps such measure-valued solutions are necessary in some circumstances to describe vortex-sheet evolution when the vorticity changes sign. The 1D Vlasov–Poisson equations for a two-component plasma have an analogous structure to the behavior of the 2D Euler equations when the vorticity changes sign (see Majda et al., 1984a, 1984b). In Majda et al. (1994b), simple explicit examples with dynamically evolving concentrations are constructed in which *perturbed weak solutions dynamically converge to a measure-valued solution for the 1D Vlasov–Poisson equations, which is not a weak solution*. These examples provide further supporting evidence regarding the possibility of a significant role for measure-valued solutions in limiting procedures for 2D vortex sheets with vorticity that changes sign. Obviously it is a major open problem to understand the behavior of weak solutions and measure-valued solutions for the 2D Euler equation and general 2D vortex sheets without a distinguished sign. We discuss this analogy problem in detail in Chap. 13

Notes for Chapter 11

The simple yet illustrative theoretical examples presented in this chapter come from Krasny (1991), Majda (1988), DiPerna and Majda (1987a, 1987b, 1988), and Greengard and Thomann (1988).

Evans and Müller (1994) developed another "harmonic analyst's" proof of Delort's theorem by using the detailed properties of special quadratic functions operating on Hardy spaces. The proof in Evans and Müller (1994), although more complex than Delort's original proof, serves to elucidate some of the general features of the relevant nonlinear structure interacting with the requirement of nonnegative vorticity.

The first key element in the proof of Theorem 11.4 was the special nonlinear structure of the 2D Euler equations, which made it sufficient to guarantee weak convergence for the special nonlinear functions $v_1 v_2$ and $v_1^2 - v_2^2$ in order to satisfy the requirement for a weak solution. This fact was first developed and used by Diperna and Majda (1988, p. 71).

References for Chapter 11

Delort, J. M., "Existence de nappes de tourbillon en dimension deux," *J. Am. Math. Soc.* **4**, 553–586, 1991.

DiPerna, R. J. And Majda, A. J., "Concentrations in regularizations for 2-D incompressible flow," *Commun. Pure Appl. Math.* **40**, 301–345, 1987a.

DiPerna, R. J. and Majda, A. J., "Oscillations and concentrations in weak solutions of the incompressible fluid equations," *Commun. Math. Phys.* **108**, 667–689, 1987b.

DiPerna, R. J. and Majda, A. J., "Reduced Hausdorff dimension and concentration–cancellation for two-dimensional incompressible flow," *J. Am. Math. Soc.* **1**, 59–95, 1988.

Evans, L. C. and Müller, S., "Hardy spaces and the two-dimensional Euler equations with nonnegative vorticity," *J. Am. Math. Soc.* **1**, 199–219, 1994.

Folland, G. B., *Introduction to Partial Differential Equations*, Princeton Univ. Press, Princeton, NJ, 1995.

Greengard, C. and Thomann, E., "On DiPerna–Majda concentration sets for two-dimensional incompressible flow," *Commun. Pure Appl. Math.* **41**, 295–303, 1988.

Krasny, R., "Computing vortex sheet motion," in *Proceedings of International Congress of Mathematicians, Kyoto 1990*, Springer-Verlag, New York, 1991, Vol. 2, pp. 1573–1583.

Lions, P. L., "The concentration–compactness principle in the calculus of variations, the limit case, part i," *Riv. Mat. Iberoam.* **1**, 145–201, 1985.

Lions, P. L., "The concentration–compactness principle in the calculus of variations. *Ann. Inst. Henri Poincaré* Vol. 1, Part I, pp. 109–145, Part II, pp. 223–283, 1984b.

Lions, P. L., "The concentration–compactness principle in the calculus of variations, the limit case, part ii," *Riv. Mat. Iberoam.* **1**, 45–121, 1985.

Liu, J.-G. and Xin, Z., "Convergence of vortex methods for weak solutions to the 2-D Euler equations with vortex sheet data," *Commun. Pure Appl. Math.* **48**, 611–628, 1995.

Majda, A., "Vortex dynamics: numerical analysis, scientific computing, and mathematical theory," in *ICIAM '87: Proceedings of the First International Conference on Industrial and Applied Mathematics*, Society for Industrial and Applied Mathematics, Philadelphia, 1988, pp. 153–182.

Majda, A. J., "Remarks on weak solutions for vortex sheets with a distinguished sign," *Indiana Univ. Math. J.* **42**, 921–939, 1993.

Majda, A. J., Majda, G., and Zheng, Y., "Concentrations in the one-dimensional Vlasov–Poisson equations, i: temporal development and nonunique weak solutions in the single component case," *Physica D* **74**, 268–300, 1994a.

Majda, A., Majda, G., and Zheng, Y., "Concentrations in the one-dimensional Vlasov–Poisson equations, ii: screening and the necessity for measure-valued solutions in the two component case," *Physica D*, **79**, 41–76, 1994b.

Zheng, Y. and Majda, A., "Existence of global weak solutions to one component Vlasov–Poisson and Fokker–Flanck–Poisson systems in one space dimension with initial data of measure," *Commun. Pure Appl. Math.* **47**, 1365–1401, 1994.

12

Reduced Hausdorff Dimension, Oscillations, and Measure-Valued Solutions of the Euler Equations in Two and Three Dimensions

In Chaps. 10 and 11 we introduced the notion of an approximate-solution sequence to the 2D Euler equations. The theory of such sequences is important in understanding the kinds of small-scale structures that form in the zero-viscosity limiting process and also for modeling the complex phenomena associated with jets and wakes. One important result of Chap. 11 was the use of the techniques developed in this book to prove the existence of solutions to the 2D Euler equation with vortex-sheet initial data when the vorticity has a fixed sign.

To understand the kinds of phenomena that can occur when vorticity has mixed sign and is in three dimensions, we address three important topics in this chapter. First, we analyze more closely the case of *concentration* by devising an effective way to measure the set on which concentration takes place. In Chap. 11 we showed that a kind of "concentration–cancellation" occurs for solution sequences that approximate a vortex sheet when the vorticity has distinguished sign. This cancellation property yielded the now-famous existence result for vortex sheets of distinguished sign (see Section 11.4). In this chapter we show that for *steady* approximate-solution sequences with L^1 vorticity control, concentration–cancellation occurs even in the case of *mixed-sign* vorticity (DiPerna and Majda, 1988).

We go on to discuss what kinds of phenomena can occur when L^1 vorticity control is not known. This topic is especially relevant to the case of 3D Euler solutions in which no a priori estimate for L^1 vorticity control is known. We show that without this control *oscillations* can occur. We present some concrete examples of 3D Euler approximate-solution sequences whose limit does not satisfy the Euler equation in a classical weak sense.

Finally we show that even when concentrations and oscillations do take place, there still exists a very weak notion of the equations in the sense of a *Young measure*. We define the notion of a measure-valued solution the Euler equation and discuss the previous examples of concentrations and oscillations within this context.

For review, we recall that an approximate-solution sequence for the 2D Euler equation is a sequence of functions $v^\epsilon \in C\{[0, T], L^2_{\mathrm{loc}}(\mathbb{R}^2)\}$ satisfying

(1) a local kinetic-energy bound, $\max_{0\le t\le T} \int_{|x|\le R} |v^\epsilon(\cdot, t)|^2 dx \le C(R)$,
(2) incompressibility $\operatorname{div} v^\epsilon = 0$ in the sense of distributions,

(3) (weak consistency with the Euler equation):

$$\lim_{\epsilon \to 0} \int_0^T \int_{\mathbb{R}^2} (v^\epsilon \cdot \Phi_t + \nabla \Phi : v^\epsilon \otimes v^\epsilon) dx dt = 0$$

for all test function $\Phi \in C_0^\infty([0, T] \times \mathbb{R}^2)$ with div $\Phi = 0$, with an additional property of L^1 vorticity control

(4) $\max_{0 \le t \le T} \int |\omega^\epsilon(\cdot, t)| dx \le C(T)$, $\omega^\epsilon = \text{curl } v^\epsilon$, the *optional* property of L^p vorticity control

(5) $\max_{0 \le t \le T} \int |\omega^\epsilon(\cdot, t)|^p dx \le C(T)$ and the technical condition that for some constant $C > 0$,

$$\|\rho v^\epsilon(t_1) - \rho v^\epsilon(t_2)\|_{-L} \le C|t_1 - t_2|,$$
$$0 \le t_1, \quad t_2 \le T \quad \forall L > 0, \quad \forall \rho \in C_0^\infty(\mathbb{R}^N)$$

i.e., $\{v^\epsilon\}$ is uniformly bounded in Lip$\{[0, T], H_{\text{loc}}^{-L}(\mathbb{R}^N)\}$. This last condition dictates how the solution attains its initial data in the $t \to 0$ limit.

In Chap. 11 we introduced two important measures associated with the limiting process associated with this sequence. The a priori L^2 bound in (1) implies the existence of a $v \in L^2_{\text{loc}}(\mathbb{R}^2)$ so that $v^\epsilon \rightharpoonup v$ in $L^2_{\text{loc}}(\mathbb{R}^2)$. The *weak* defect measure* associated with this sequence is then a measure σ such that $(v^\epsilon)^2 dx dt - v^2 dx dt \rightharpoonup \sigma$ converges in the weak* topology.

On the other hand, the *reduced defect measure* θ, a finitely subadditive measure, is defined for each Borel set E as $\theta(E) = \limsup_{\epsilon \to 0} \int \int_E [(v^\epsilon)^2 - v^2] dx dt$.

The reduced defect measure is not in general countably subadditive as the example in Subsection 11.2.3 showed. Nonetheless it provides a more useful tool than the weak* defect measure for understanding the nature of the limit v. To see this in more detail, in Section 12.1 we introduce the reduced Hausdorff dimension for measuring the dimension of the set on which θ concentrates. One important result of this analysis is that if the concentration set has a Hausdorff dimension strictly less than one, then no concentration occurs in the limit. Moreover, any approximate-solution sequence of *steady* solutions to the Euler equation, regardless of the sign of the vorticity, has concentration–concentration in the limit. In Section 12.2 we discuss the case of approximate-solution sequences to the 3D Euler equation. In addition to the phenomena observed in the case of 2D approximate-solution sequences, we show that in three dimensions oscillations can occur in which no subsequence converges pointwise almost everywhere and, furthermore, the weak limit is not a solution to the Euler equation. In Section 12.3 we introduce the notion of a *measure-valued solution* of the Euler equation and show that even when oscillations or concentrations occur, the resulting limit still satisfies this very weak notion of a solution to the equations. Finally in Section 12.4 we discuss the possibility of a single approximate-solution sequence's yielding both oscillation and concentration in the limit.

12.1. The Reduced Hausdorff Dimension

First, we recall the notion of Hausdorff measure. For $r > 0$, E a Borel set, let $\{B_{r_i}(x_i)\}$ be a countable cover of E by open balls, $E \subseteq \cup_{i=1}^{\infty} B_{r_i}(x_i)$. The *Hausdorff premeasure of the order of* γ is

$$H_r^\gamma(E) = \inf_{r_i \le r} \sum_{i=1}^{\infty} r_i^\gamma, \qquad E \subseteq \bigcup_{i=1}^{\infty} B_{r_i}(x_i). \tag{12.1}$$

The *Hausdorff measure of the order of* γ is defined as

$$H^\gamma(E) = \lim_{r \to 0} H_r^\gamma(E). \tag{12.2}$$

Instead of $h(r) = r^\gamma$ we can use a different strictly increasing function $h(r)$ satisfying $h(0) = 0$ to define the Hausdorff measure. In this case the Hausdorff premeasure and Hausdorff measure become, respectively,

$$H_r^h(E) = \inf_{r_i \le r} \sum_{i=1}^{\infty} h(r_i), \qquad E \subseteq \bigcup_{i=1}^{\infty} B_{r_i}(x_i), \tag{12.3}$$

$$H^h(E) = \lim_{r \to 0} H_r^h(E). \tag{12.4}$$

For example, we could use

$$h(r) = \left(\log \frac{1}{r}\right)^{-\beta}, \qquad 0 < r \le 1. \tag{12.5}$$

We remark that for integer values of γ, the Hausdorff measure coincides with the Lebesgue measure up to multiplication by a constant. As promised we present an example in which the weak-* defect measure is huge but the reduced defect measure is arbitrarily small on some closed sets.

Claim. *There exists a family of closed sets* $\{F_m\}$ *with* $F_m \subset [0, 1] \times [0, 1]$ *for all* $m \ge 1$ *such that*

(i) $\theta(F_m) = 0$ *for all* $m \ge 1$,
(ii) $H_{r_m}^\gamma(F_m^c) \le C$ *for all* $\gamma > 0$, *where* C *is independent of* γ, m,
(iii) $\sigma(F_m) = m(F_m)$ *(= Lebesgue measure of* F_m*) with*
(iv) $m(F_m) \ge 1 - \epsilon_m$, $\epsilon_m \to 0$ *as* $m \to \infty$.

Proof of Claim. Our argument resides in the example of Greengard and Thomann from the proof of Theorem 11.2. Recall in this example that we consider $\{\Lambda^m\}$ a $2^m \times 2^m$ dyadic lattice with velocity $v^m = \sum_{z_j \in \Lambda^m} V^m(x - z_j)$ where V^m denotes a special phantom vortex with outer radius R_m and inner radii δ_m and $\delta_m^{1/2}$, respectively. (See Subsection 11.2.2 for details and Fig. 10.1 therein). Recall that R_m and δ_m satisfy the equation

$$R_m = \left(\frac{5}{4}\delta_m\right)^{1/2} = \frac{\sqrt{5}}{2} e^{-\frac{4m+1}{2\pi}}. \tag{12.6}$$

Define F_m such that

$$F_m^c = \bigcup_{j\geq m} \bigcup_{z_k \in \Lambda^j} B_{R_j}(Z_k). \tag{12.7}$$

Because $v^j \equiv 0$ on F_m for any $j \geq m$, trivially

$$v^j \rightharpoonup 0 \text{ in } L^2(F_m) \quad \text{as} \quad j \to \infty. \tag{12.8}$$

Hence $\theta(F_m) = \lim_{j\to\infty} \int_{F_m} |v^j - v|^2 dx = \lim_{j\to\infty} \int_{F_m} |v^j|^2 dx = 0$ for all $m \geq 1$, thus proving (i).

To prove (ii) we take $r_m = e^{-4^m/2\pi}$; then by comparing with Eq. (12.6) we have $R_j \leq R_m < r_m \; \forall j \geq m$. Thus, from covering (12.3), we have

$$H_{r_m}^{\gamma}\left(F_m^C\right) \leq \sum_{j=m}^{\infty} 4^j R_j^{\gamma} \leq \sum_{j=m}^{\infty} 4^j e^{-\frac{4^j}{2\pi}\gamma} < \sum_{j=1}^{\infty} 4^j e^{-\frac{4^j}{2\pi}\gamma} \equiv S(\gamma).$$

Because $S(\gamma)$ is a convergent series for any $\gamma > 0$ and $S(\gamma)$ is monotonically decreasing as $\gamma \nearrow \infty$, we have (ii). Part (iii) is shown in Subsection 11.2.2. (Note: $F_m \subset [0,1] \times [0,1] \forall m \geq 1$). We now verify (iv):

$$\begin{aligned} m(F_m) &= m([0,1]\times[0,1]) - m\left(F_m^c\right) \\ &\geq 1 - \sum_{j=m}^{\infty} 4^j R_j^2 \geq 1 - \sum_{j=m}^{\infty} 4^j e^{-\frac{4^j}{\pi}} \equiv 1 - \epsilon_m. \end{aligned}$$

Because $\epsilon_m = \sum_{j=m}^{\infty} 4^j e^{-\frac{4^j}{\pi}}$ is the tail of the convergent series $S(2)$ above, $\epsilon_m \to 0$ as $m \to \infty$. □

Note that even if we use the Hausdorff premeasure defined with another monotone function of type (12.5), for example, $h(r) = (\log \frac{1}{4})^{-(1+\delta)}$, $\delta > 0$, we arrive at the same conclusion as (ii) of the above claim, as can be seen in the following estimates:

$$\begin{aligned} H_{r_m}^h\left(F_m^c\right) &\leq \sum_{j=m}^{\infty} 4^j h(R_j) \leq \sum_{j=m}^{\infty} 4^j (r^{-j})^{1+\delta} = \sum_{j=m}^{\infty} (4^{-\delta})^j \\ &\leq C(\delta) \text{ with } C(\delta) = \frac{4^{-\delta}}{1-4^{-\delta}} < \infty, \quad \forall \delta < 0. \end{aligned}$$

We are interested in determining the set on which θ, a finitely subadditive set function, concentrates.

Definition 12.1. *We say θ that* concentrates inside a set *with Hausdorff dimension p if, given $\delta > 0$, $r > 0$, there exists a family of closed sets $\{F_r\}$ such that*

(i) $\theta(F_r) = 0$

(ii) $H_r^{p+\delta}(F_r^c) \leq C$ with C independent of r and δ.

In the Greengard–Thomann example, the above claim shows that θ concentrates inside a set with Hausdorff dimension 0 (strikingly in contrast to the fact that σ concentrates on a set with Hausdorff dimension 2). Later we prove the following theorem, illustrating that this is typical of the steady case.

Theorem 12.1. *(DiPerna and Majda, 1988.) Let $\{v^\epsilon\}$ be a steady (elliptic) sequence in $L^2_{\text{loc}}(\mathbb{R}^2)$ with*

(i) $\operatorname{div} v^\epsilon = 0$,
(ii) $\operatorname{curl} v^\epsilon = \omega^\epsilon$,
(iii) $\int_{|x|\le R} |v^\epsilon|^2 dx + \int_{|x|\le R} |\omega^\epsilon| dx \le C(R)$.

Then θ concentrates inside a set with Hausdorff dimension 0.

We recall that concentration–cancellation for the sequence $\{v^\epsilon\}$ in $L^2_{\text{loc}}(\mathbb{R}^2)$ with $\int_{|x|\le R} |v^\epsilon|^2 \le C(R)$ means that the weak $L^2_{\text{loc}}(\mathbb{R}^2)$ limit v of v^ϵ is a weak solution of the 2D Euler equations even though we may have $v^\epsilon \otimes v^\epsilon \not\rightharpoonup v \otimes v$ (i.e., even if concentration occurs). For the case of of *steady* sequences, we use Theorem 12.1 to prove that concentration–cancellation always occurs.

Theorem 12.2. *(DiPerna and Majda, 1988.) Consider an approximate-solution sequence $\{v^\epsilon\}$ for the 2D Euler equations defined by a sequence of steady exact solutions with (optional) external force f^ϵ, $f^\epsilon \to f$ in $L^1_{\text{loc}}(\mathbb{R}^2)$; then concentration–cancellation occurs.*

For the time-dependent case, we have the following result, proving strong convergence when the concentration set is small enough.

Theorem 12.3. *(DiPerna and Majda, 1988.) Consider an approximate-solution sequence $\{v^\epsilon\}$ for the 2D Euler equation and assume θ concentrates on a space-time set with a Hausdorff dimension strictly less than 1. Then concentration does not occurs, and in fact we have strong convergence in $L^2_{\text{loc}}(\mathbb{R}^2)$ for the sequence $\{v^\epsilon\}$.*

DiPerna and Majda (1988) show that, for the time-dependent case, θ always concentrates inside a space–time set of a *cylindrical* Hausdorff dimension less than or equal to 1.† Motivated by this and related results, Zheng (1991) showed that, for a stronger notion of a concentration set, concentration–callcellation always occurs when the "strong-concentration set" has a Hausdorff dimension of less than or equal to 1. These results suggest that the case of a space–time concentration set of Hausdorff dimension 1 may be critical, in which it is not known whether concentration–cancellation will occur. Nussenzveig Lopes (1994, 1997) has achieved some refined estimates for the true Hausdorff dimension of the concentration set. In Chap. 11, we showed that concentration–cancellation occurs for sequences with positive vorticity. However, it

† This is not the same sense as in measures (12.1) and (12.2) above.

may be true that concentration–cancellation does not occur for some time-dependent Euler sequences with both positive and negative vorticity.

In the next chapter we introduce an analogy problem, that of the 1D Vlasov–Poisson system, which has a mathematical structure much like that of the 2D Euler equations. We show that the same kinds of questions can be formulated for this system and, because of its simpler nature, can actually be answered much more precisely.

The proof of Theorem 12.1 is somewhat technical, so we save it for a separate subsection. We first prove Theorem 12.3 following Greengard and Thomann (1988).

Proof of Theorem 12.3. For $F \subseteq \mathbb{R}^2 \times [0, \infty)$, denote by π the projection onto the time axis

$$\pi F = \{t \in \mathbb{R} |\ \exists (x, t) \in F\}.$$

By hypothesis there exists a number γ with $0 < \gamma < 1$ and a family of closed sets $\{F_r\}$ such that

$$\theta(F_r) = 0 \ \forall r > 0, \tag{12.9}$$

$$H_r^\gamma \left(F_r^c\right) \leq C, \qquad C \text{ independent of } \gamma \text{ and } r. \tag{12.10}$$

Decompose

$$\begin{aligned}\lim_{\epsilon \to 0} \int_0^T \int_{|x| \leq R} |v^\epsilon - v|^2 \, dx dt &= \left\{ \lim_{\epsilon \to 0} \int \int_{F_r} |v^\epsilon - v|^2 dx dt \right\} \\ &+ \left\{ \lim_{\epsilon \to 0} \int \int_{F_r^c} |v^\epsilon - v|^2 dx dt \right\} = \{1\} + \{2\}.\end{aligned} \tag{12.11}$$

Then

$$\{1\} \leq \lim_{\epsilon \to 0} \int_{F_r} \int |v^\epsilon - v|^2 dx dt = \theta(F_r) = 0.$$

We now control $\{2\}$. From the definition of an approximate-solution sequence,

$$\max_{0 \leq t \leq T} \int_{|x| \leq R} \int |v^\epsilon|^2 dx \leq C(T, R),$$

which implies that

$$\max_{0 \leq t \leq T} \int_{|x| \leq R} |v|^2 dx \leq C(T, R).$$

Thus

$$\max_{0 \leq t \leq T} \int_{|x| \leq R} |v^\epsilon - v|^2 dx \leq C(T, R). \tag{12.12}$$

The bound on Hausdorff premeasure (12.10) implies that there exists a countable covering $\{B_{r_i}[\cdot(x_i, t_i)]\}$ of F_r^c with $r_i \leq r$ that gets within a factor of 2 of the correct

Hausdorff premeasure. That is,

$$F_r^c \subset \bigcup_{i=1}^{\infty} B_{r_i}(x_i, t_i), \qquad r_i \le r, \qquad \sum_{i=1}^{\infty} r_i^\gamma \le 2H_r^\gamma.$$

This implies that the Lebesgue measure in time of $\pi F_r^c \subset \cup_{i=1}^{\infty} \pi B_{r_i}(x_i, t_i)$ satisfies

$$\begin{aligned} m\left(\pi F_r^c\right) &\le m\left[\bigcup_{i=1}^{\infty} \pi B_{r_i}(x_i, t_i)\right] \le \sum_{i=1}^{\infty} m[\pi B_{r_i}(x_i, t_i)] \\ &\le \sum_{i=1}^{\infty} r_i \le r^{1-\gamma} \sum_{i=1}^{\infty} r_i^\gamma \le r^{1-\gamma} 2H_r^\gamma\left(F_r^c\right). \end{aligned}$$

Hence from premeasure (12.10) we have

$$m\left(\pi F_r^c\right) \le C r^{1-\gamma}, \qquad 0 < \gamma < 1. \tag{12.13}$$

From this and relation (12.12) we obtain

$$\begin{aligned} \int\int_{F_r^c} |v^\epsilon - v|^2 dx dt &\le \int_{\pi(F_r^c)} \left(\int_{|x| \le R} |v^\epsilon - v|^2 dx \right) dt \\ &\le C(T, R) m\left(\pi F_r^c\right) \le C'(T, R) r^{1-\gamma}. \end{aligned}$$

This combined with limit (12.11) gives

$$\lim_{\epsilon \to 0} \int_0^T \int_{|x| \le R} |v^\epsilon - v|^2 dx dt \le C'(T, R) r^{1-\gamma}, \qquad 0 < \gamma < 1. \tag{12.14}$$

Because r is arbitrary the left-hand side is zero. Hence we have strong convergence:

$$v^\epsilon \to v \quad \text{in} \quad L^2\{[0, T] \times B_R(0)\}.$$ □

The proof actually shows that the locally finite kinetic-energy bound implies that if there exists a set E that can be covered by a union U of balls with $\theta(U^c) = 0$ and whose total projected time coordinates are of arbitrarily small Lebesgue measure in time, then the convergence of the sequence v^ϵ is strong. Thus concentration sets must have time coordinates of positive Lebesgue measure. Furthermore, in this proof, the Euler equation itself is not used directly. The time axis is singled out because of the nature of the a priori kinetic-energy bound (sup in time, L^2_{loc} in space). We express this remark as a corollary.

Corollary 12.1. *If there exists a family of closed sets $\{F_r\}$ such that*

(i) $\theta(F_r) = 0\ \forall r > 0$,

(ii) $m(\pi F_r^c) \to 0$ as $r \to 0$, where m is the 1D Lebesque measure and $\{v^\epsilon\}$ is an approximate-solution sequence with $v^\epsilon \rightharpoonup v$ in $L^2\{[0, T] \times B_R(0)\}$, then

$$v^\epsilon \to v \text{ in } L^2\{[0, T] \times B_R(0)\}.$$

Next, we prove Theorem 12.2, assuming Theorem 12.1.‡ Before proving the theorem we prepare some facts and lemmas. First, when we defined an approximate-solution sequence of the 2D Euler equations, we did not include the requirement that the v^ϵ vanish uniformly as $|x| \to \infty$; these were not needed in our discussion until now. However, imposing these requirements on an approximate-solution sequence $\{v^\epsilon\}$ implies that we can legitimately truncate the v^ϵ's, i.e., for some fixed R_0 we can use only the v^ϵ's with supp $v^\epsilon \subset \{x \in \mathbb{R}^2 | \, |x| \leq R_0\}$. This point is discussed in detail in the appendix (DiPerna and Majda, 1988). We use this fact for the proof of Theorem 12.2. Also, we need the following lemmas.

Lemma 12.1. *For any $\eta \in C_0^\infty(\mathbb{R}^N)$ with supp $\eta \subset \{x \in \mathbb{R}^N | \, |x| \leq R_0\}$ we can approximate η uniformly by a finite sum of plane waves. Specifically, given L and $\epsilon > 0$, there exists $\{a_m\}_{m=1}^M$, $\varphi_m \in S^{N-1}(|\varphi_m| = 1)$ such that*

$$\sup_{|x| \leq R_0} \sum_{|\alpha| \leq L} \left| D^\alpha \left[\eta(x) - \sum_{m=1}^{M} a_m \bar{\eta}(\varphi_m, x \cdot \varphi_m) \right] \right| < \epsilon \tag{12.15}$$

where $\bar{\eta}(\varphi, y) \in C^\infty(S^{N-1} \times \mathbb{R})$.

Below we sketch the proof of this lemma.

Proof of Lemma 12.1. Sketch. We simply rewrite the Fourier transform formula

$$\begin{aligned} \eta(x) &= \int_{\mathbb{R}^N} e^{ix\cdot\xi} \hat{\eta}(\xi) d\xi \\ &= \int_{S^{N-1}} \left[\int_0^\infty e^{i\lambda x \cdot \varphi} \hat{\eta}(\lambda, \varphi) \lambda^{N-1} \, d\lambda \right] d\varphi \\ &= \int_{S^{N-1}} \bar{\eta}(\varphi, x \cdot \varphi) d\varphi, \bar{\eta} = \int_0^\infty e^{i\lambda x \cdot \varphi} \hat{\eta}(\lambda, \varphi) \lambda^{N-1} \, d\lambda. \end{aligned}$$

Now, approximating the integral over S^{N-1} by the Riemann sum, we obtain result (12.15) for $|\alpha| = 0$.

We can apply the same idea for the case for any α, $|\alpha| \leq 1$. □

Recall from Chap. 11 that the Euler equations have a covariant structure. This means that the equation is preserved under rotations by an orthogonal transformation.

Lemma 12.2. *Covariant Structure of Euler Equation. Let $\mathcal{O}$ be any rotation matrix in $\mathbb{R}^N (s\mathcal{O}, \mathcal{O}^t = \mathcal{O}^{-1})$; then for any scalar test function $\eta \in C_0^\infty(\mathbb{R}^N)$ we have the following identities for the nonlinear term:*

$$\begin{aligned} &\int_{\mathbb{R}^N} (\nabla_y \nabla_y^\perp)\eta : (\mathcal{O}^t v \otimes \mathcal{O}^t v)(\mathcal{O}y, t) dy \\ &\quad = \int_{\mathbb{R}^N} (\nabla_x \nabla_x^\perp)\eta : v \otimes v(x, t) dx. \end{aligned} \tag{12.16}$$

‡ Actually, a weaker version of Theorem 12.1 is all that is needed; we require concentration on a set with a Hausdorff dimension of less than 1.

Proof of Lemma 12.2. We put $\mathcal{O}y = x$; then, by using $\nabla_y = \mathcal{O}^t \nabla_x$, $\nabla_y^\perp = \mathcal{O}^t \nabla_x^\perp$, $\det(\mathcal{O}^t) = 1$, we have

$$\begin{aligned}
&\int_{\mathbb{R}^N} (\nabla_y \nabla y^\perp)\eta : (\mathcal{O}^t v \otimes \mathcal{O}^t v)(\mathcal{O}y, t)dy \\
&\quad = \int_{\mathbb{R}^N} (\mathcal{O}^t \nabla_x \mathcal{O}^t \nabla_x^\perp)\eta(x) : (\mathcal{O}^t v \otimes \mathcal{O}^t v)(x, t)\det(\mathcal{O}^t)dx \\
&\quad = \int_{\mathbb{R}^N} \sum \mathcal{O}^t_{i\ell}\mathcal{O}^t_{jm}\mathcal{O}^t_{ip}\mathcal{O}^t_{jq}(\nabla_x)_\ell(\nabla_x^\perp)_m\eta(x)(v_p v_q)(x, t)dx \\
&\quad = \int_{\mathbb{R}^N} \sum \delta_{\ell p}\delta_{mq}(\nabla_x)_\ell(\nabla_x^\perp)_m\eta(x)(v_p v_q)(x, t)dx \\
&\quad = \int_{\mathbb{R}^N} \sum (\nabla_x)_\ell(\nabla_x^\perp)_m\eta(x)(v_\ell v_m)(x, t)dx \\
&\quad = \int_{\mathbb{R}^N} (\nabla_x \nabla_x^\perp \eta)(x) : (v \otimes v)(x, t)dx.
\end{aligned}$$

□

In addition, we have the following straightforward result concerning the rotational invariance of the dimension of concentrations.

Lemma 12.3. *Rotational Invariance of the Dimension of Concentrations. If the reduced defect measure θ associated with the sequence $\{v^\epsilon\}$, $v^\epsilon \rightharpoonup v$ in $L^2(\mathbb{R}^2)$ concentrates inside a set of Hausdorff dimension p, then so does the defect measure $\tilde{\theta}$ associated with $\{\tilde{v}^\epsilon\}$, where $\tilde{v}^\epsilon(x) = \mathcal{O}^t v^\epsilon(\mathcal{O}x)$, $\mathcal{O}^t = \mathcal{O}^{-1}$.*

Proof of Lemma 12.3. By hypothesis there exists a family $\{F_r\}$ of closed sets such that

$$\theta(F_r) = 0 \; \forall r > 0, \tag{12.17}$$

$$H_r^{p+\delta}(F_r^c) \le C \; \forall \delta > 0, \quad r > 0. \tag{12.18}$$

Let $\tilde{F}_r = \{x \in \mathbb{R}^2 | \, x = \mathcal{O}y, \, y \in F_r\}$.

Claim. *The family $\{\tilde{F}_r\}$ has the same properties (12.17) and (12.18) as $\{F_r\}$ for the same number p and C.*

Obviously, from $v^\epsilon \rightharpoonup v$ we have $\tilde{v}^\epsilon \rightharpoonup \tilde{v}$, where $\tilde{v}(x) = \mathcal{O}^t v(\mathcal{O}x)$. Hence,

$$\theta(\tilde{F}_r) = \lim_{\epsilon \to 0} \int_{\tilde{F}_r} |\tilde{v}^\epsilon - \tilde{v}|^2 dx = \lim_{\epsilon \to 0} \int_{F_r} |v^\epsilon - v|^2 dy = 0.$$

Inequality (12.18) for $\{\tilde{F}_r\}$ follows from the definition of the Hausdorff premeasure by use of the rotational invariance of the covering property and the radius of each ball in the covering family. □

Now we are ready to prove one of our main theorems in this section.

Proof of Theorem 12.2. Let $\{v^\epsilon\}$ be the sequence of exact solutions and $v^\epsilon \rightharpoonup v$ in $L^2_{\text{loc}}(\mathbb{R}^2)$. We use the test function of the form

$$\Phi = \nabla^\perp \eta = \begin{pmatrix} -\eta_{x_2} \\ \eta_{x_1}, \end{pmatrix} \eta \in C_0^\infty(\mathbb{R}^2). \tag{12.19}$$

We must show that

$$\int\!\!\int_{\mathbb{R}^2} \nabla\nabla^\perp \eta : v \otimes v \, dx = \int\!\!\int_{\mathbb{R}^2} f \cdot \nabla^\perp \eta \, dx, \tag{12.20}$$

which means that v is the steady weak solution of the 2D Euler equation with external force f. Now,

$$\begin{aligned} \nabla\nabla^\perp \eta : v \otimes v &= \partial_1(\nabla^\perp \eta)_1 v_1 v_1 + \partial_1(\nabla^\perp \eta)_2 v_1 v_2 \\ &\quad + \partial_2(\nabla^\perp \eta)_1 v_2 v_1 + \partial_2(\nabla^\perp \eta)_2 v_2 v_2 \\ &= \frac{\partial^2 \eta}{\partial x_1 \partial x_2}(v_2^2 - v_1^2) + \left(\frac{\partial^2 \eta}{\partial x_2^2} - \frac{\partial^2 \eta}{\partial x_2^2}\right) v_1 v_2. \end{aligned}$$

Hence, we can rewrite Eq. (12.20) as

$$\begin{aligned} &\int\!\!\int_{\mathbb{R}^2} \left[\frac{\partial^2 \eta}{\partial x_1 \partial x_2}(v_2^2 - v_1^2) + \left(\frac{\partial^2 \eta}{\partial x_2^2} - \frac{\partial^2 \eta}{\partial x_1^2}\right) v_1 v_2 \right] dx \\ &\quad = \int\!\!\int_{\mathbb{R}^2} \left(f_2 \frac{\partial \eta}{\partial x_1} - f_1 \frac{\partial \eta}{\partial x_2} \right) dx. \end{aligned} \tag{12.21}$$

From the remarks following Corollary 12.1 we assume that there exists an R_0 so that, for all $\epsilon > 0$, supp $v^\epsilon \subset B_{R_0}(0)$ without loss of generality. This implies that we may assume supp $\eta \subset \{x \in \mathbb{R}^2 | \, |x| \le R_0\}$. Lemma 12.1 implies that because Eq. (12.20) is linear in η, it is sufficient to show Eq. (12.20) for a single plane wave $\bar{\eta}(\varphi, x \cdot \varphi)$, $\varphi \in S^1$. The covariant structure of the Euler equation (Lemma 12.2) implies that we can make a change of variables $x = \mathcal{O}y$, $\mathcal{O}^t = \mathcal{O}^{-1}$, $\tilde{v}(y) = \mathcal{O}^t v(\mathcal{O}y)$ and rewrite the equation as

$$\begin{aligned} &\int_{\mathbb{R}^2}\!\!\int \left[\frac{\rho^2 \tilde{\eta}}{\rho y, \rho y_2}(\tilde{v}_2^2 - \tilde{v}_1^2) + \left(\frac{\rho^2 \tilde{\eta}}{\rho y_2^2} - \frac{\rho^2 \tilde{\eta}}{\rho y_1^2}\right) \tilde{v}_2 \tilde{v}_2 \right] dy \\ &\quad = \int_{\mathbb{R}^2}\!\!\int \left(\tilde{f}_2 \frac{\rho \tilde{\eta}}{\rho y_1} - \tilde{f}_1 \frac{\rho \tilde{\eta}}{\rho y_2} \right) dy. \end{aligned} \tag{12.22}$$

where $\tilde{f}$ and $\tilde{\eta}$ denote

$$\tilde{f}(y) = (\mathcal{O}^t f)(\mathcal{O}y), \qquad \tilde{\eta}(y) = \bar{\eta}(\varphi, y \cdot \varphi)$$

We need to show only that the rotated velocity $\tilde{v}$ and force $\tilde{f}$ satisfy Eq. (12.22) for a rotated scaler test function $\tilde{\eta}$. Clearly $\tilde{v}^\epsilon \rightharpoonup \tilde{v}$ in $L^2(|x| \le R)$ and $\tilde{f}^\epsilon \to \tilde{f}$ in $L^1_{\text{loc}}(\mathbb{R}^2)$ for "rotated" $\tilde{v}^\epsilon$ and $\tilde{f}^\epsilon$. The reduced defect measure associated with $\{\tilde{v}^\epsilon\}$, $\tilde{\theta}(F) = \limsup_{\epsilon \to 0} \int_F |\tilde{v}^\epsilon - \tilde{v}|^2 dy$, has the same dimension of concentration set as θ by Lemma 12.3. From now on, even if we work in the new rotated coordinate system, for simplicity we use the same notation for variables, velocities, forces, etc., as in the old coordinate system. In particular, consider Eq. (12.20). If we choose the coordinate system so that $\varphi = (1,0)$ and denote $\eta[(1,0), x_1] = h(x_1)$, Eq. (12.20) transforms to

$$-\int\int_{\mathbb{R}^2} h''(x_1)v_1v_2\,dx = \int\int_{\mathbb{R}^2} f_2h'(x_1)dx. \tag{12.23}$$

An approximate- (exact-) solution sequence satisfies

$$-\int\int_{\mathbb{R}^2} h''(x_1)v_1^\epsilon v_2^\epsilon\,dx = \int\int_{\mathbb{R}^2} f_2^\epsilon h'(x_1)dx. \tag{12.24}$$

For a steady approximate-solution sequence, Theorem 12.1 implies that there exists a family $\{F_r\}$ of closed sets such that for all $r > 0$

$$\theta(F_r) = \overline{\lim}_{\epsilon\to 0}\int_F |v^\epsilon - v|^2 dx = 0 \tag{12.25}$$

and a constant C so that for all $r > 0, 0 < p$,

$$H_r^p\left(F_r^c\right) \le C. \tag{12.26}$$

Our strategy below is to first pass to limit (12.24) by using truncated test functions that vanish on the set where the convergence is not strong and then to remove the truncation.

Let P_1F_r denote the projection of F_r onto the x_1 axis. Then, given a covering $\{B_{r_i}(x_i)\}$, $r_i \le r$ of F_r^c, that approximates its H_r^p premeasure to within a factor of 2 we have

$$m\left(P_1F_r^c\right) \le \sum_{i=1}^{\infty} m[P_1B_{r_i}(x_i)] \le \sum_{i=1}^{\infty} r_i \le r^{1-p}\sum_{i=1}^{\infty} r_i^p \le 2Cr^{1-p}. \tag{12.27}$$

Hence,

$$m\left(P_1F_r^c\right) \to 0 \quad \text{as} \quad r \to 0. \tag{12.28}$$

Consider a rough test function $h_r \in W^{2,N}(R)$ defined by§

$$h_r'' = \chi_{(P_1F_r^c)^c}h''. \tag{12.29}$$

Then convergence (12.28) implies that $h_r'' \to h''$ in measure. By choosing a subsequence we have

$$h_r'' \to h'' \text{ boundedly almost everywhere in } |x_1| \le R. \tag{12.30}$$

This, in turn, implies that (when the dominated convergence theorem is applied)

$$h_r' \to h' \quad \text{in} \quad L^\infty(|x_1| \le R_0). \tag{12.31}$$

§ Here χ denotes the characteristic function.

From Eq. (12.25) for each r

$$\lim_{\epsilon\to 0}\int_{F_r} |v^\epsilon - v|^2 dx \to 0 \qquad \text{as} \quad \epsilon \to 0. \tag{12.32}$$

Because $\pi^{-1}\{[\pi(F_r^c)]^c\} \subset F_r$,

$$v^\epsilon \to v \qquad \text{(strongly) in} \quad L^2(\pi^{-1}\{[\pi(F_r^c)]^c\}).$$

Hence

$$\int\int_{\mathbb{R}^2} h''{}_r(x_1)v_1^\epsilon v_2^\epsilon\, dx = \int\int_{\mathbb{R}} \chi_{(P_1F_r^c)^c} h''(x_1)v_1^\epsilon v_2^\epsilon\, dx$$

$$= \int_{P_1^{-1}\{[P_1(F_r^c)]^c\}}\int h''(x_1)v_1^\epsilon v_2^\epsilon\, dx \to \int_{P_1^{-1}[(F_r^c)^c]} h''(x_1)v_1v_2\, dx$$

$$= \int_{\mathbb{R}}\int h''{}_r(x_1)v_1v_2\, dx, \text{ i.e.,}$$

$$\int\int_{\mathbb{R}^2} h''{}_r(x_1)v_1^\epsilon v_2^\epsilon\, dx \to \int\int_{\mathbb{R}} h_r''(x_1)v_1v_2\, dx. \tag{12.33}$$

Also, $f^\epsilon \to f$ in $L^1_{\text{loc}}(\mathbb{R}^2)$ (hypothesis) and the assumption on h_r implies that

$$-\int\int_{\mathbb{R}^2} h_r'(x_1)f_2^\epsilon\, dx \to \int\int_{\mathbb{R}^2} h_r'(x_1)f_2\, dx. \tag{12.34}$$

Combining convergences (12.33) and (12.34) with limit (12.24), we have for each $r > 0$

$$-\int\int_{\mathbb{R}^2} h''{}_r(x_1)v_1v_2\, dx = \int\int_{\mathbb{R}^2} h_r'(x_1)f_2\, dx. \tag{12.35}$$

Now, convergences (12.30) and (12.31) allow us to use the dominated convergence theorem to pass to a limit $r \to 0$ on both sides of Eq. (12.35), from which we obtain convergence (12.23). □

Note that in the proof we require relation (12.27) to hold only for some $0 \le p < 1$, which is true provided that concentration occurs on a set of Hausdorff measure of strictly less than 1.

12.1.1. Proof of Theorem 12.1 on 2D Elliptic Sequences

A first proof of this result is contained in DiPerna and Majda (1988). The proof we present here uses capacity theory (Evans, 1990). Given $v^\epsilon \in L^2_{\text{loc}}(\mathbb{R}^2)$, $\operatorname{div} v^\epsilon = 0$, $\operatorname{curl} v^\epsilon = \omega^\epsilon$, $\int_{|x|<R} |\omega^\epsilon| dx < C$, our goal is to show that θ concentrates on a set of Hausdorff dimension 0.

Let $v^\epsilon \rightharpoonup v$ weakly in $L^2(\mathbb{R}^2)$. For this sequence $\{v^\epsilon\}$, the *reduced defect measure* is defined as

$$\theta(E) = \limsup_{\epsilon\to 0}\int_E |v^\epsilon - v|^2 dx.$$

Recall also that θ *concentrates on a set of* H^s(s–dim Hausdorff) *measure* 0 if there exist a family of closed sets $\{E_\delta\}_{\delta>0}$ with

(1) $\theta(E_\delta) = 0 \ \forall \delta > 0$,
(2) $H^s_{r(\delta)}(\mathbb{R}^2 \sim E_\delta) \to 0$ as $\delta \to 0$, where $r(\delta) \searrow 0$ as $\delta \searrow 0$.

Theorem 12.1 can be restated with stream functions as Theorem 12.1A.

Theorem 12.1A. *Let* $\{\psi^\epsilon\}$ *be a sequence of smooth functions in* $\mathbb{R}^2$ *such that*

$$\int |\nabla\psi^\epsilon|^2 dx \le C, \qquad \Delta\psi^\epsilon = \omega^\epsilon, \quad v^\epsilon = \nabla^\perp\psi^\epsilon, \tag{12.36}$$

$$\int |\omega^\epsilon| dx \le C. \tag{12.37}$$

Then there exists $\psi \in H^1_{\text{loc}}(\mathbb{R}^2)$ *and a subsequence (denoted also by* ψ^ϵ*) such that* $\nabla\psi^\epsilon \rightharpoonup \nabla\psi$ *weakly in* $L^2(\mathbb{R}^2)$ *and the corresponding reduced defect measure* θ *concentrates on a set of* H^γ *measure 0 for any* $\gamma > 0$.

Recall that a priori bounds (12.36) and (12.37) imply that

$$\omega^\epsilon \rightharpoonup \omega \qquad \text{weakly in} \quad \mathcal{M}(\mathbb{R}^2), \tag{12.38}$$

$$\nabla\psi^\epsilon \to \nabla\psi \in L^p(\mathbb{R}^2), \qquad 1 \le p < 2. \tag{12.39}$$

The strategy of the proof of Theorem 12.1 is in the following steps,

Step 1: Refinement of Egoroff's theorem by use of the capacity for functions in W^{1-p}, $p \le 2$.

Step 2: Estimates (12.36) and (12.37) combined with step 1 yield a family of closed sets $\{E_\delta\}$ such that

$$\int_{E_\delta} |\nabla\psi^\epsilon - \nabla\psi|^2 dx \to 0 \qquad \text{as} \quad \epsilon \to 0$$

which implies that $\theta(E_\delta) = 0$ for all $\delta > 0$.

Let $\mathcal{F}_A = \{f \in C_0^\infty(\mathbb{R}^N) |\ f \ge 1 \text{ on } A \subset \mathbb{R}^N\}$; then for $p \le N$, the capacity $\text{Cap}_p(A)$ of A is defined as

$$\text{Cap}_p(A) = \inf_{f\in\mathcal{F}_A} \int_{\mathbb{R}^N} |\nabla f|^p dx. \tag{12.40}$$

Step 3: We use step 1 to show $\text{Cap}_p(\mathbb{R}^2 \sim E_\delta) \to 0$ as $\delta \to 0$. Estimating the Hausdorff premeasure by Cap_p, we finish the proof. [$\text{Cap}_p(\cdot)$ is defined below.]

Step 1: Egoroff's theorem says that if $m(\Omega) < \infty$ for $\Omega \subset \mathbb{R}^N$ and $\psi^\epsilon \to \psi$ almost everywhere in Ω then for each $\delta > 0$ there exists a set E_δ such that (1) $\psi^\epsilon \to \psi$ uniformly on E_δ and (2) $m(\Omega \sim E_\delta) \le \delta$. We require the following refinement of Egoroff's theorem.

Theorem 12.4. *Let $\{\rho\}$ be a uniformly bounded family in $W^{1,q}_{\text{loc}}(\mathbb{R}^N)$, $q \le N$; then there exists a subsequence $\{\rho\}$ and a function ψ in $W^{1,q}_{\text{loc}}(\mathbb{R}^N)$ with the following properties:*

For each p with $1 \le p < q$ and $\delta > 0$ there exist closed sets $E_\delta \subset \mathbb{R}^N$ such that

$$\psi^\epsilon \to \psi \qquad \textit{uniformly on} \quad E_\delta, \tag{12.41}$$

$$\text{Cap}_p(\mathbb{R}^N \sim E_\delta) \le \delta. \tag{12.42}$$

Theorem 12.4 is true. The reader is referred to Chap. 1 of Evans (1990) for a proof of Theorem 12.4.

Energy bounds (12.36) allow us to apply Theorem 12.4 with $q = 2$ so that, for any $\delta > 0$, $1 \le p < 2$, there exist closed sets E_δ such that

$$\psi^\epsilon \to \psi \qquad \text{uniformly on} \quad E_\delta, \tag{12.43}$$

$$\text{Cap}_p(\mathbb{R}^2 \sim E_\delta) \le \delta. \tag{12.44}$$

Step 2: We require the following proposition.

Proposition 12.1. *Let ψ^ϵ and ω^ϵ satisfy estimates (12.36), (12.37), and (12.43). Then for any $\sigma > 0$ there exists $\epsilon_0 = \epsilon_0(\sigma) > 0$ such that $\forall \epsilon_1, \epsilon_2 < \epsilon_0$,*

$$\int_{E_\delta} |\nabla\psi^{\epsilon_1} - \nabla\psi^{\epsilon_2}|^2 dx \le C\sigma, \tag{12.45}$$

where

$$C = \tilde{C} \sum_{j=1}^{2} \left[\|\omega^{\epsilon_j}\|_{L^1(\mathbb{R}^2)} + r\|\nabla\psi^{\epsilon_j}\|_{L^2(\mathbb{R}^2)}\right] \qquad \textit{for some constant } \tilde{C}.$$

Estimates (12.36), (12.37), and (12.43) combined with Proposition 12.1 yield

$$\int_{E_\delta} |\nabla\psi^\epsilon - \nabla\psi|^2 dx \to 0 \qquad \text{as} \quad \epsilon \to 0, \tag{12.46}$$

from which we get

$$\begin{aligned}\theta(E_\delta) &= \overline{\lim}_{\epsilon\to 0} \int_{E_\delta} |\nabla^\perp\psi^\epsilon - \nabla^\perp\psi|^2 dx \\ &= \overline{\lim}_{\epsilon\to 0} \int_{E_\delta} |\nabla\psi^\epsilon - \nabla\psi|^2 dx = 0\ \forall\delta > 0.\end{aligned}$$

The step is complete if we prove Proposition 12.1.

Proof of Proposition 12.1. Consider a cutoff function $\beta_\sigma(s)$ with $|\beta_\sigma| \le \sigma$ defined as

$$\beta_\sigma(s) = \begin{cases} \sigma, & s > \sigma \\ s, & -\sigma \le s \le \sigma. \\ -\sigma, & s < -\sigma \end{cases}$$

From estimate (12.43) there exists $\epsilon_0 = \epsilon_0(\sigma)$ such that $\epsilon_1, \epsilon_2 < \epsilon_0$ implies that

$$|\psi^{\epsilon_1} - \psi^{\epsilon_2}| \leq \frac{\sigma}{2} \quad \text{on} \quad E_\delta. \tag{12.47}$$

Thus by the definition of β_σ

$$\beta_\sigma\,(\psi^{\epsilon_1} - \psi^{\epsilon_2}) = \psi^{\epsilon_1} - \psi^{\epsilon_2} \quad \text{on} \quad E_\delta. \tag{12.48}$$

The chain rule for the weak derivative (see Gilbarg and Trudinger, 1988, p. 146) of the piecewise smooth function β_σ gives

$$\nabla\beta_\sigma(\psi^{\epsilon_1} - \psi^{\epsilon_2}) = \begin{cases} 0, & |\psi^{\epsilon_1} - \psi^{\epsilon_2}| \geq \sigma \\ \nabla\psi^{\epsilon_1} - \nabla\psi^{\epsilon_2}, & |\psi^{\epsilon_1} - \psi^{\epsilon_2}| < \sigma \end{cases}.$$

Consider $\rho \in C_0^\infty(\mathbb{R}^2)$ with $\rho \equiv 1$ on E_δ. (Existence of such a function ρ is guaranteed by Urysohn's lemma because E_δ is *closed.*) Then we have the estimates

$$\begin{aligned}
\int_{E_\delta} |\nabla\psi^{\epsilon_1} - \nabla\psi^{\epsilon_2}|^2 dx &= \int_{E_\delta} \nabla(\psi^{\epsilon_1} - \psi^{\epsilon_2}) \cdot \nabla(\psi^{\epsilon_1} - \nabla\psi^{\epsilon_2}) dx \\
&\leq \int_{\mathbb{R}^2} \rho\nabla(\psi^{\epsilon_1} - \psi^{\epsilon_2}) \cdot \nabla(\beta_\sigma(\psi^{\epsilon_1} - \psi^{\epsilon_2})) dx \\
&\quad \text{(integration by parts)} \\
&\leq \left| \int_{\mathbb{R}^2} \rho\Delta(\psi^{\epsilon_1} - \psi^{\epsilon_2})\beta_\sigma(\psi^{\epsilon_1} - \psi^{\epsilon_2}) dx \right| \\
&\quad + \left| \int_{\mathbb{R}^2} \nabla\rho \cdot \nabla(\psi^{\epsilon_1} - \psi^{\epsilon_2})\beta_\sigma(\psi^{\epsilon_1} - \psi^{\epsilon_2}) dx \right| \\
&\equiv \{1\} + \{2\}, \\
\{1\} &= \left| \int_{\mathbb{R}^2} \rho(\omega^{\epsilon_1} - \omega^{\epsilon_2})\beta_\sigma(\psi^{\epsilon_1} - \psi^{\epsilon_2}) dx \right| \\
&\leq \sigma \int_{\mathbb{R}^2} |\omega^{\epsilon_1} - \omega^{\epsilon_2}| dx = \sigma \sum_{j=1}^{2} \|\omega^{\epsilon_1}\|_{L^1(\mathbb{R}^2)}
\end{aligned}$$

where we used $|\beta_\sigma| \leq \sigma$. Similarly

$$\begin{aligned}
\{2\} &\leq \sigma \int_{\mathbb{R}^2} |\nabla\rho \cdot \nabla(\psi^{\epsilon_1} - \psi^{\epsilon_2})| dx \\
&\leq \sigma \left(\int |\nabla\rho \cdot \nabla\psi^{\epsilon_1}| dx + \int |\nabla\rho \cdot \nabla\psi^{\epsilon_2}| dx \right) \\
&\leq \sigma \|\nabla\rho\|_{L^2(\mathbb{R}^2)} \sum_{j=1}^{2} \|\nabla\psi^{\epsilon_j}\|_{L^2(\mathbb{R}^2)} \leq \sigma\tilde{C} \sum_{j=1}^{2} \|\nabla\psi^{\epsilon_j}\|_{L^2(\mathbb{R})}
\end{aligned}$$

with $\tilde{C} = \max\big[1, \|\nabla\rho\|_{L^2(\mathbb{R}^2)}\big]$.

Thus we obtain relation (12.45). □

This finishes Step 2.

Step 3: In this step we bound the H^γ premeasure by capacity. Accomplishing this will finish the whole proof [except the proof of Theorem 12.4 by condition (12.44)].

Proposition 12.2. *Given V_δ, an open set with $\mathrm{Cap}_p(V_\delta) = \delta$, and $\gamma = N - p + \epsilon$, $\epsilon > 0$, there exists a constant $C = C(N, p, \epsilon)$ such that*

$$H^\gamma_{r(\delta)}(V_\delta) \leq C\ \mathrm{Cap}_p(V_\delta)$$

where $r(\delta) = \tilde{C}\delta^{[p/(N-p)]}$, $\tilde{C}$ = a constant.

We prove this proposition as a corollary of the following Chebyshev inequality (Theorem 12.5 below). Recall that the classical Chebyshev inequality states that the Lebesgue distribution function σ of an element f of $L^p(\mathbb{R}^2)$ is dominated by an algebraic function of the order of p:

$$\sigma(\lambda) = H^n\{x : |f| \geq \lambda\} \leq \lambda^{-p} \int |f|^p dx. \tag{12.49}$$

For functions with compact support, the essential behavior is associated with large values of λ.

If $f \in W^{1,p}$ then the n–dimensional Lebesgue measure can be replaced with a finer set function, namely a γ– order Hausdorff premeasure where $\gamma = n - p + \epsilon$ and ϵ is arbitrarily small.

Theorem 12.5. *(DiPerna and Majda, 1988.) If $\epsilon > 0$ and $p < N$, then there exists a constant $C = C(N, p, \epsilon)$ such that*

$$H^\gamma_{r(\theta)}(\{x|\,|f| \geq \lambda\}) \leq C\theta^p \tag{12.50}$$

for all f in $W^{1,p(\mathbb{R}^N)}$ where

$$\theta = \alpha/\lambda, \qquad \alpha = \left\{\int_{\mathbb{R}^N} |\nabla f|^p dx\right\}^{1/p}, \qquad r(\theta) = C\theta^{q/N},$$

$$q = \frac{Np}{N-p}.$$

Proof of Theorem 12.5. Claim. The interval set

$$A = \{x : |f| \geq \lambda\}$$

is contained in a set of the form

$$D = \left\{x : r^{-\gamma} \int_{B(r,x)} |\nabla f|^p dy \geq M \text{ for some } r \text{ with } 0 \leq r \leq r_0\right\}$$

if the parameters r_0 and M are chosen appropriately. In short, A is contained in a set in which the gradient fails to exhibit uniform local decay of a specific order: The normalized local maximal function of the gradient remains above M for some ball of radius $r \leq r_0$. The choice of r_0 and M are given below.

We use the following lemma.

Lemma 12.4. *Covering Lemma (Stein, 1970). If J is an arbitrary family of balls B_α contained in a bounded set of $\mathbb{R}^n$, then there exists a countable disjoint collection B_{r_j} whose fivefold expansion covers J,*

$$\cup_J B_\alpha \subset \cup B_{5r_j}.$$

We see that there exists B_{r_j}, disjoint balls, with $r_j < r_0$ such that $D \subset \cup B_{5r_j}$ and

$$r_j^{-\gamma} \int_{B_{r_j}} |\nabla f|^p dy \geq M.$$

Because the balls B_{r_j} are disjoint,

$$H_{5r_0}(D) \leq \sum_j (5r_j)^\gamma \leq 5^\gamma \sum_j \frac{1}{M} \int_{B_{r_j}} |\nabla f|^p dy \leq \frac{5^\gamma}{M} \int |\nabla f|^p dy,$$

that is,

$$H^\gamma_{5r_0}(D) \leq CM^{-1} \int |\nabla f|^p dy$$

with a universal constant C. Because A is contained in D a similar inequality holds for A and yields the desired result.

The proof of the claim is based on the fact that the pointwise values of a function are dominated by the normalized maximal function of the derivative. At a Lebesgue point f may be expressed as a telescoping sum of the dyadic local averages:

$$f(x) = f(x_0) + \sum_{j\geq 1} f_{j+1}(x) - f_j(x) \tag{12.51}$$

where

$$f_j = {}_{\text{avg}}\int_{B_j} f(y)dy$$

and $B_j = B_j(r_j, x)$, $r_j = 2^{-j} r_0$. The notation

$${}_{\text{avg}}\int_E f\,dy = \frac{1}{m(E)} \int_E f\,dy.$$

denotes integration with respect to normalized Lebesgue measure. The Poincaré inequality reveals the dependence on the local behavior of the derivative

$$|f| \leq |f_0| + 2^n \sum r_j \left\{ {}_{\text{avg}}\int_{B_j} |\nabla f|^p dy \right\}^{1/p},$$

because the difference of the two consecutive averages is bounded by the average oscillation,

$$|f_{j+1} - f_j| \leq 2^n {}_{\text{avg}}\int_{B_j} |f - f_j| dy.$$

Suppose that A is not contained in D; choose a Lebesgue point z in A such that

$$\operatorname{avg}\int_{B(r,z)} |\nabla f|^p dy \leq M r^{-p+\epsilon} \qquad \text{if} \quad 0 < r \leq r_0.$$

If follows that

$$\lambda \leq |f_0(z)| + 2^n M^{1/p} \sum r_j^{\epsilon/p}, \tag{12.52}$$

because z lies in A and $r_j \leq r_0$. The strategy is to select M and r_0 so that condition (12.52) is violated. For this purpose the first term on the right of sum (12.51) can be estimated with the Sobolev inequality

$$\begin{aligned} |f_0(z)| &= \left|\operatorname{avg}\int_{B_0} f\, dy\right| \leq \left\{\operatorname{avg}\int_{B_0} |f|^q dy\right\}^{1/q} \\ &\leq r_0^{-n/q}\left\{\int_{R^n} |f|^q dy\right\}^{1/q} \leq c_1 r_0^{-n/q}\alpha, \end{aligned}$$

where c_1 is a universal constant and $B_0 = B(r_0, z)$. Letting

$$c_2 = 2^n \sum_j r_j^{\epsilon/p}$$

yields the inequality

$$\lambda \leq c_1 r_0^{-n/q}\alpha + c_2 M^{1/p} \tag{12.53}$$

as a consequence of condition (12.52). We conclude that A is contained in D if r_0 and M are chosen so that each term of the right-hand side of inequality (12.53) equals $\lambda/3$:

$$r_0 = (3c_1\theta)^{q/n}, \qquad M = (3c_2)^{-p}\lambda^p.$$

The value of c in the conclusion of the theorem is easily derived from c_1 and c_2. □

Proof of Proposition 12.2. By definition of capacity (12.40) there exists a function $f \in C_0^\infty(\mathbb{R}^2)$, $f \geq 1$ on V_δ such that

$$\int_{\mathbb{R}^N} |\nabla f|^p dx \leq 2\,\mathrm{Cap}_p(V_\delta). \tag{12.54}$$

From the fact that $V_\delta \subset \{x \in \mathbb{R}^N \mid |f| \geq \frac{1}{2}\}$ we have, by application of Theorem 12.3 with $\lambda \equiv \frac{1}{2}$, $\theta \equiv \delta = \mathrm{Cap}_p(V_\delta)$,

$$\begin{aligned} H^\gamma_{r(\delta)}(V_\delta) &\leq H^\gamma_{r(\delta)}\left(\left\{x \mid |f| \geq \frac{1}{2}\right\}\right) \leq 2^p \tilde{C}\int_{\mathbb{R}^N} |\nabla f|^p dx \\ &\leq 2^{p+1}\tilde{C}\,\mathrm{Cap}_p(V_\delta) \equiv C\,\mathrm{Cap}_p(V_\delta) \\ &\quad \text{with} \quad r(\delta) = \tilde{C}\delta^{\frac{p}{N-p}},\ \tilde{C} = \tilde{C}(N, p, \epsilon). \end{aligned}$$

□

The completion of Step 3 follows by setting $N = 2$, $V_\delta = \mathbb{R}^2 \sim E_\delta$ in proposition (12.37), thus obtaining

$$H^{\gamma}_{r(\delta)}(\mathbb{R}^2 \sim E_\delta) \le C\ \mathrm{Cap}_p(V_\delta) \le C\delta \to 0$$

as

$$\delta \to 0 \qquad \text{for all} \quad \gamma = 2 - p + \epsilon > 0.$$

This concludes the proof of Theorem 12.1 assuming that Theorem 12.4 is true.

We remark that the above proof uses condition (12.36) crucially in condition (12.45) of Step 2 whereas the original proof (DiPerna and Majda, 1988) relies on the weaker condition

$$\nabla\psi^\epsilon \to \nabla\psi \text{ in } L^p(\mathbb{R}^2), \qquad 1 \le p < 2.$$

12.1.2. Improved Velocity Estimates

Recall from Section 11.3 that the vorticity maximal function $|\omega|(\cdot,\cdot)$ of a finite Borel measure ω is

$$|\omega(s, x)| \equiv \int_{|x-y|\le s} d|\omega(y)| = \int_{B_s(x)} d|\omega| = |\omega|[B_s(x)]. \tag{12.55}$$

In this section we prove some results on the decay of the vorticitly maximal functions associated with an approximate-solution sequence. This result yields some improved velocity estimates for v^ϵ.

Let $\beta(\cdot)$ be any monotone increasing function satisfying for some constant K

$$\beta(5s) \le K[\beta(s)]. \tag{12.56}$$

For example $\beta(s) = [\log(\frac{1}{s})]^{-1+\gamma}$ and s^γ with $0 < \gamma < 1$ are in this class.

Proposition 12.3. *Suppose that ω is a nonnegative measure with finite total mass and* $\mathrm{supp}|\omega| \subset \{|x| \le R\}$*; then the closed set*

$$E_r = \left\{ x \in \mathbb{R}^N \,\middle|\, |\omega|(s, x) \le \beta(s),\ 0 \le s \le \frac{r}{5} \right\}$$

satisfies

$$H^{\beta}_r\left(E^c_r\right) \le K|\omega|(R, 0). \tag{12.57}$$

Note that $\{E_r\}$ $r > 0$ is the family of "good" sets in terms of the decay of the radial distribution $\omega(\cdot,\cdot)$; $\{E^c_r\}$ is the family of "bad" sets.

Proof of Proposition 12.3. By hypothesis we have

$$E^c_r = \left\{ x \in \mathbb{R}^N \,\middle|\, |\omega|(s, x) > \beta(s) \text{ for some } s,\ 0 \le s \le \frac{r}{5} \right\}.$$

We observe that $E_r^c \subset \cup_{x \in E_r^c} B_{s(x)}(x)$. Let $\{B_j\},\ B_j \equiv B_{s_j(x_j)}(x_j)$ be the subcovering described in the covering lemma (Lemma 12.4); then

$$E^c \subset U_{j=1}^{\infty} B_j^* \qquad B_j^* = B_{5s_j}(x_j).$$

Hence

$$H_r^\beta\left(E_r^c\right) \le \sum_{j=1}^{\infty} \beta(5s_j) \le K \sum_{j=1}^{\infty} \beta(s_j) \le K \sum_{j=1}^{\infty} |\omega|(s_j, x_j)$$
$$= K\omega\left(\bigcup_{j=1}^{\infty} B_j\right) \le K|\omega|(R, 0). \qquad \square$$

The full uniformization theorem concerns the *uniform* radial decay of mass distributions for a *family* of measures.

Theorem 12.6. *Uniformization Theorem. Consider a family $\{\omega^\epsilon\}$ of finite, nonnegative measures on $\mathbb{R}^N$. Fix parameters δ, γ such that $\delta > 0$ and $\gamma > n\delta$, $n =$ dimension of space. Then there exists a subsequence $\{\omega^k\}$ with the following property. For every $r > 0$ there exists a closed set F_r such that*

$$F_r \subset \left\{x \in \mathbb{R}^N | \omega^k(s, x) \le K s^\delta,\ 0 \le s \le 1,\ k \ge \frac{1}{r}\right\}, \tag{12.58}$$

$$H_r^\gamma\left(F_r^c\right) \le C(\delta, \gamma), \tag{12.59}$$

where

$$K = C + Cr^{-\delta}, \tag{12.60}$$

(C is a universal constant).

The proof of the full uniformization theorem can be found in DiPerna and Majda (1988). We use the uniformization theorem to prove the following improved velocity estimates.

Theorem 12.7. *Improved Velocity Estimates. Assume the same conditions as those in Theorem 12.6; then the family of sets $\{F_r\}$ satisfies*

$$\int_{F_r} |v^k|^{p'} dx \le C(\delta, r) \qquad \textit{for some} \quad p' > 2 \tag{12.61}$$

for all $k \ge (1/r)$, where curl $v^k = \omega^k$.

We now prove Theorem 12.7 on improved velocity estimates.

We introduce the velocity distribution function $\sigma_F(\lambda)$ associated with the Marcinkiewicz space M^p defined by

$$\sigma_F(\lambda) = m\{x \in F|\ |v(x)| > \lambda\}. \tag{12.62}$$

We say that $v \in M^p$ if $\sigma_F(\lambda) \le c\lambda^{-p}$. Note that this definition generalizes Eq. (12.49) of the previous section.

We now show that the improved velocity estimates follow from the fact that if a radial distribution function $|w|$ decays on a set F with order δ then the associated velocity field $v = \int K_2(x-y)dw(y)$ lies in the Marcinkiewicz space M^p where $p = 2 + \delta/(\delta - 1)$. We state this fact as a theorem.

Theorem 12.8. *Suppose that ω is a Borel measure on $\mathbb{R}^2$ with finite total mass. Assume for all $x \in F$ that*

$$|\omega|(s, x) \le K s^\delta \qquad for \quad 0 \le s \le 1. \tag{12.63}$$

Then there exists a constant λ_0 such that for all $\lambda \ge \lambda_0$ the velocity $v = \int K_2(x-y)\, dw(y)$ has a distribution function satisfying

$$\sigma_F(\lambda) \le C(K, \delta)\lambda^{-p}, \qquad p = 2 + \frac{\delta}{1-\delta}. \tag{12.64}$$

This theorem says that the L^∞ norm of $|\omega|$ controls the M^p norm of v.

The following corollary follows from the fact that the M^p norm is controlled by the L^p norm when $p > p'$.

Corollary 12.2. *With the same assumptions as those of Theorem 12.8, we have for any p' with $2 < p' < p = 2 + \frac{\delta}{1-\delta}$,*

$$\int_F |v|^{p'} dx \le C(K, \delta). \tag{12.65}$$

Proof of Corollary 12.2.

$$\begin{aligned}
\|v\|^{p'}_{L^{p'}(F)} &= \int_0^\infty \lambda^{p'} d\sigma_F(\lambda) \\
&= \left\{ \int^{\lambda_0} \lambda^{p'} d\sigma_F(\lambda) \right\} + \left\{ \int_{\lambda_0}^\infty \lambda^{p'} d\sigma_F(\lambda) \right\} = \{1\} + \{2\}, \\
|\{1\}| &= \int_{|v| \le \lambda_0} |v|^{p'} dx \le \lambda_0^{p'} \int_{|x| \le 1} dx \le C, \\
|\{2\}| &= \lambda_0^{p'} \sigma_F \left(\lambda_0 - p' \int_{\lambda_0}^\infty \lambda^{p'-1} \sigma_F(\lambda) d\lambda \right) \\
&\le C(K, \delta) \left(\lambda_0^{p'-p} - p' \int_{\lambda_0}^\infty \lambda^{p'-p-1} d\lambda \right) \le C(K, \delta) p' \lambda_0^{p'-p} \le C'(K, \delta),
\end{aligned}$$

where we integrated by parts and used the fact that $p' - p - 1 < -1$. □

This corollary implies directly the improved velocity estimates.

It remains to prove Theorem 12.8. Recall that for a general vorticity measure $\omega \in \mathcal{M}(\mathbb{R}^2)$ the velocity field satifies

$$v(x) = \int_{\mathbb{R}^2} K(x-y)d\omega(y),$$

$$K(z) = \frac{1}{2\pi|z|^2}\begin{pmatrix} -z_2 \\ z_1 \end{pmatrix} \tag{12.66}$$

where $|K(z)| \le C|z|^{-1}$ for some constant C. Without loss of generality we assume that

$$\operatorname{supp} w \subset B_1(0), \tag{12.67}$$

$$|\omega|(1,0) = 1. \tag{12.68}$$

We define $A_k = \{x \in \mathbb{R}^2 |\, \omega(s,x) \le ks^{1+\delta} \text{ if } 0 \le s \le \frac{1}{k}\}$. Our strategy is

(1) to show that $x \in A_k \cap F$ implies $|v(x)| \le f(k)$ for some function $f(k)$,
(2) then to choose k so that $f(k) = \lambda$ if $\lambda > \lambda_0$ for some constant λ_0,
(3) From (2) $\{x \in F |\, |v(x)| > \lambda\} \subset A_k^c$; hence $\sigma_F(\lambda) \le m(A_k^c)$.

Because A_k^c is the "badly" decaying set, we can control its Hausdorff premeasure by using Proposition 12.3, thus establishing the theorem.

Step 1: If $x \in A_k \cap F$, then

$$\begin{aligned}
|v(x)| &\le C\int_{|x-y|} |x-y|^{-1} d\omega(y) = C\int_0^1 s^{-1} d\omega(s,x) \\
&\le C\omega(1,x) + C\int_0^1 s^{-2}\omega(s,x)ds \\
&= \{1\} + \{2\}, \\
\{1\} &= C|\omega|(1,0) = C, \\
\{2\} &= C\int_0^{1/k} s^{-2}\omega(s,x)ds + C\int_{1/k}^1 s^{-2}\omega(s,x)ds \\
&\le Ck\int_0^{1/k} s^{-1+\delta} ds + C\int_{1/k}^1 ks^{-2+\delta}\, ds \\
&= \left(\frac{C}{\delta} + CK\right)k^{1-\delta}.
\end{aligned}$$

Thus if $x \in A_k \cap F$

$$|v(x)| \le \left(\frac{C}{\delta} + CK\right)k^{1-\delta} + C. \tag{12.69}$$

Step 2: We choose k so that

$$\left(\frac{C}{\delta} + CK\right) k^{1-\delta} + C = \lambda; \tag{12.70}$$

then relation (12.69) implies that $\{x \in F \mid |v(x)| > \lambda\} \subset A_k^c$.

Step 3: From step 2 we have

$$\sigma_F(\lambda) \le m\left(A_k^c\right). \tag{12.71}$$

From Proposition 12.3 [choosing $\beta(s) = ks^{1+\delta}$]

$$H_{5/k}^{1+\delta}\left(A_k^c\right) \le C_1 |\omega|(1,0) = C_1. \tag{12.72}$$

Now relation (12.72) implies that there exists a family of countable balls $s\{B_j\}$ with radii $r_j \le (5/k)$ so that $A_k^c \subset \cup_{j=1}^{\infty} B_j$ and $\sum r_j^{1+\delta} \le \frac{C}{k}$ for some constant C. Hence,

$$\begin{aligned} m\left(A_k^c\right) &\le \sum_{j=1}^{\infty} m(B_j) = \pi \sum_{j=1}^{\infty} r_i^2 = \pi \sum_{j=1}^{\infty} r_i^{1-\delta} r_i^{1+\delta} \\ &\le \pi \left(\frac{5}{k}\right)^{1-\delta} \sum_{j=1}^{\infty} r_i^{1+\delta} \le \pi \left(\frac{5}{k}\right)^{1-\delta} \frac{C}{k} \le C_2 k^{\delta-2}. \end{aligned}$$

From Eq. (12.70)

$$k = \left(\frac{\lambda - C}{\frac{C}{\delta} - CK}\right)^{\frac{1}{1-\delta}} \le C_3 \lambda^{\frac{1}{1-\delta}} \qquad \text{for} \quad \lambda \ge \lambda_0$$

with sufficiently large λ_0, i.e.,

$$m\left(A_k^C\right) \le C_4 \lambda^{\frac{\delta-2}{1-\delta}} = C_4 \lambda^{-p}, \qquad p = 2 + \frac{\delta}{1-\delta}.$$

Combining this with relation (12.71) concludes the proof.

12.2. Oscillations for Approximate-Solution Sequences without L^1 Vorticity Control

Recall that one feature of approximate-solution sequences $\{v^\epsilon\}$ for vortex-sheet initial data in two dimensions is that they always possess uniform L^1 vorticity control that, as we showed in Chap. 10, implies the existence of a subsequence that converges pointwise almost everywhere to the weak limit v. Hence oscillations are not possible for 2D approximate-solution sequences with uniform L^1 vorticity control. In three dimensions, however, the L^1 norm of the vorticity is not a conserved quantity; therefore we lose this a priori bound for an approximate-solution sequence. Our purpose in this section is to study the limiting phenomena of approximate-solution sequences *without* L^1 vorticity control. It is interesting to think about this problem for both 2D and 3D approximate-solution sequences.

As we show in this section, this can lead to other phenomena's occurring in the limit. For a sequence $\{v^\epsilon\}$ in $L^\infty(\mathbb{R}^N)$ with $|v^\epsilon|_{L^\infty} \le C$ there is a subsequence, which we denote also by $\{v^\epsilon\}$, such that

$$v^\epsilon \rightharpoonup v \text{ weak-* in } L^\infty(\mathbb{R}^N) \qquad \text{as} \quad \epsilon \to 0 \tag{12.73}$$

{i.e. for all $\rho \in L^1(\mathbb{R}^N)$, $\int_{\mathbb{R}^N} \rho v^\epsilon dx \to \int_{\mathbb{R}^N} \rho v\, dx$}. A sequence $\{v^\epsilon\}$ is said to exhibit oscillation in the limit when there is no subsequence $\{v^{\epsilon'}\}$ of $\{v^\epsilon\}$ such that

$$v^{\epsilon'} \to v \qquad \text{almost everywhere in} \quad \mathbb{R}^N, \tag{12.74}$$

even if we have weak convergence (12.73). As a preliminary to more complex examples involving solutions of the Euler equations, we present the following proposition on the weak L^∞ limit of a sequence of functions of two scales.

Proposition 12.4. *Let $u(x, y)$ be a smooth, bounded function on $\mathbb{R}^N \times \mathbb{R}^N$ periodic with period p in the y vector,*

$$u(x, y + pe_j) = u(x, y), \qquad j = 1, \ldots, n.$$

Then the sequence $\{u^\epsilon\}$, $u^\epsilon(x) = u(x, \frac{x}{\epsilon})$, satisfies

$$u^\epsilon \rightharpoonup u \qquad \textit{weak-*in} \quad L^\infty(\mathbb{R}^N), \tag{12.75}$$

where

$$\bar{u}(x) = \frac{1}{p^n} \int_0^p \cdots \int_0^p u(x, y)\, dy^1 \cdots dy^n. \tag{12.76}$$

Proof of Proposition 12.4. Step 1: We first prove the proposition for $u(x, y) = \varphi(x)T(y)$, where φ is smooth, bounded in $\mathbb{R}^N$, and T is a trigonometric function given by $T(x) = \sum_{|k|\le L} \hat{T}_k e[(2\pi i)/p]\langle x, k\rangle$. For any $\rho \in L^1(\mathbb{R}^N)$ we have

$$\begin{aligned}
\int \rho(x) u^\epsilon(x) dx &= \int \rho(x)\varphi(x) T\left(\frac{x}{\epsilon}\right) dx \\
&= \left\{ \int \rho(x)\varphi(x)\bar{T} \right\} + \left\{ \sum_{1\le|l|\le L} \hat{T}_k \int \rho(x)\varphi(x) e^{\frac{2\pi i}{p}\langle \frac{x}{\epsilon}, k\rangle} dx \right\} \\
&= \{1\} + \{2\},
\end{aligned}$$

where

$$\bar{T} = \frac{1}{p^n} \int_0^p \cdots \int_0^p T(y) dy \cdots dy^n. \tag{12.77}$$

Now,

$$\{2\} = \sum_{1\le|k|\le L} \hat{T}_k \widehat{\rho\varphi}\left(-\frac{k}{\epsilon}\right).$$

Because $\psi\varphi \in L^1$, by the Riemann–Lebesgue lemma (Rudin, 1987)

$$\widehat{\rho\varphi}\left(-\frac{k}{\epsilon}\right) \to 0 \qquad \text{as} \quad \epsilon \to 0.$$

Thus $\{2\} \to 0$ as $\epsilon \to 0$.

Hence we have shown that, given $\delta > 0$, $\rho \in L^1(\mathbb{R}^N)$, there exists $\epsilon = \epsilon(\delta)$ such that

$$\left| \int \left[\rho(x)\varphi(x)T\left(\frac{x}{\epsilon}\right) - \rho(x)\overline{\varphi T}(x) \right] dx \right| < \delta. \tag{12.78}$$

Step 2: We extend the result of Step 1 to the case $u(x, y) = \varphi(x)\psi(y)$, where ψ is any smooth, p-periodic function on $\mathbb{R}^N$. We know that, given $\delta > 0$, there exists a trigonometric polynomial T_δ such that

$$|\psi(y) - T_\delta(y)|_{L^\infty(\mathbb{R}^N)} < \delta, \tag{12.79}$$

where

$$T_\delta(y) = \sum_{|\kappa| \le L(\delta)} \hat{T}_k(\delta) e^{\frac{2\pi i}{p}\langle y,k\rangle}.$$

We use the standard $(\epsilon/3)$ argument as follows:
Given $\rho \in L^1(\mathbb{R}^N)$, then

$$\begin{aligned}
&\left| \int \left[\rho(x)\varphi(x)\psi\left(\frac{x}{\epsilon}\right) - \rho(x)\overline{\varphi\psi}(x) \right] dx \right| \\
&\quad \le \left| \int \left[\rho(x)\varphi(x)\psi\left(\frac{x}{\epsilon}\right) - \int \rho(x)\varphi(x)T_\delta\left(\frac{x}{\epsilon}\right) \right] dx \right| \\
&\quad\quad + \left| \int \left[\rho(x)\varphi(x)T_\delta\left(\frac{x}{\epsilon}\right) - \int \rho(x)\overline{\varphi T_\delta}(x) \right] dx \right| \\
&\quad\quad + \left| \int \left[\rho(x)\overline{\varphi T}_\delta(x) - \int \rho(x)\overline{\varphi\psi}(x) \right] dx \right| \\
&\quad = \{1\} + \{2\} + \{3\}.
\end{aligned}$$

Estimates (12.78) and (12.79) imply that

$$\begin{aligned}
&\{1\} \le |\rho\varphi|_{L^1(\mathbb{R}^N)} |\psi - T_\delta|_{L^\infty(\mathbb{R}^N)} \le |\rho\varphi|_{L^1(\mathbb{R}^N)} \le |\rho\varphi|_{L^1(\mathbb{R}^N)}\delta, \\
&\{2\} < \delta, \\
&\{3\} \le |\rho\varphi|_{L^1(\mathbb{R}^N)} \frac{1}{p^n} \int_0^p \cdots \int_0^p |T_\delta(y) - \psi(y)| dy^1 \cdots dy^n, \\
&\quad \le |\rho\varphi|_{L^1(\mathbb{R}^N)}\delta.
\end{aligned}$$

Hence we have shown that, given $\delta > 0$, $\rho \in L^1(\mathbb{R}^N)$, there exists $\epsilon = \epsilon(\delta)$ such that

$$\left| \int \rho(x)\varphi(x)\psi\left(\frac{x}{\epsilon}\right)dx - \int \rho(x)\overline{\varphi\psi}(x)dx \right| < \delta \tag{12.80}$$

for any smooth, periodic function ψ on $\mathbb{R}^N$.

Step 3: We extend the result of step 2 to the function $u(\cdot, \cdot)$ given in the proposition. We know that, given $\delta > 0$, there exists a finite family of functions $\{\varphi_i(x)\psi_i(y)\}_{i=1}^{L(\delta)}$ with φ_i smooth, ψ_i smooth such that

$$\left| u(x, y) - \sum_{i=1}^{L(\delta)} \varphi_i(x)\psi_i(y) \right|_{L^\infty(\mathbb{R}^N \times \mathbb{R}^N)} < \delta \tag{12.81}$$

for any smooth function $u(\cdot, \cdot)$ on $\mathbb{R}^N \times \mathbb{R}^N$.

Let $\rho \in L^1(\mathbb{R}^N)$ be given; then

$$\begin{aligned}
&\left| \int \left[\rho(x)u\left(x, \frac{x}{\epsilon}\right) - \rho\bar{u}(x) \right] dx \right| \\
&\quad \le \left| \int \left[\rho(x)u\left(x, \frac{x}{\epsilon}\right) - \rho \sum_{i=1}^{L(\delta)} \varphi_i(x)\psi_i\left(\frac{x}{\epsilon}\right) \right] dx \right| \\
&\qquad + \left| \int \left[\rho \sum_{i=1}^{L(\delta)} \varphi_i(x)\psi_i\left(\frac{x}{\epsilon}\right) - \rho \sum_{i=1}^{L(\delta)} \overline{\varphi\psi}(x) \right] dx \right| \\
&\qquad + \left| \int \rho \left[\sum_{i=1}^{L(\delta)} \bar{\varphi\psi} - \bar{u} \right] (x)dx \right| = \{1\} + \{2\} + \{3\},
\end{aligned}$$

$$\{1\} \le |\rho|_{L^1(\mathbb{R}^N)} \left| u - \sum_{i=1}^{L} \varphi_i\psi_i \right|_{L^\infty(\mathbb{R}^N \times \mathbb{R}^N)} \le |\rho|_{L^1(\mathbb{R}^N)}\delta,$$

$$\{2\} \le \delta,$$

$$\begin{aligned}
\{3\} &= \left| \int \rho(x) \left\{ \frac{1}{p^n} \int_0^p \cdots \int_0^p \left[\sum_{i=1}^{L} \varphi_i(x)\psi_i(y) - u(x, y) \right] dy^1 \cdots dy^n \right\} dx \right| \\
&\le \int |\rho(x)| \left\{ \frac{1}{p^n} \int_0^p \cdots \int_0^p dy^1 \cdots dy^n \right\} dx \left| \sum_{i=1}^{L} \varphi_i\psi_i - \omega \right|_{L^\infty(\mathbb{R}^N \times \mathbb{R}^N)} \\
&\le |\rho|_{L^1(\mathbb{R}^N)}\delta.
\end{aligned}$$

□

The following proposition provides the other aspect of strong convergence in $L^2_{\text{loc}}(\mathbb{R}^N)$, which is actually the content of the claim in Step 1 of the Proof of Theorem 11.3.

Proposition 12.5. *Consider a sequence $\{v^\epsilon\}$ in $L^2_{\text{loc}}(\mathbb{R}^N)$ that is uniformly bounded, i.e., for any $R > 0$*

$$\int_{B_R(0)} |v^\epsilon|^2 dx \leq C(R).$$

Furthermore, let us assume that

$$v^\epsilon \rightharpoonup v \quad \textit{weakly in} \quad L^2_{\text{loc}}(\mathbb{R}^N).$$

Then

$$v^\epsilon \to v \quad \textit{strongly in} \quad L^2_{\text{loc}}(\mathbb{R}^N)$$

if and only if

$$|v^\epsilon|^2 \rightharpoonup |v|^2 \quad \textit{weakly in} \quad \mathcal{M}_{\text{loc}}(\mathbb{R}^N).$$

Now consider a function $u(\cdot, \cdot)$ satisfying the conditions of Proposition 12.4. Then we know that for $u^\epsilon(x) = u(x, \frac{x}{\epsilon})$,

$$u^\epsilon(x) \rightharpoonup \bar{u}(x) = \frac{1}{p^m} \int_0^p \cdots \int_0^p u(x, y) dy^1 \cdots dy^n \tag{12.82}$$

weak-* in $L^\infty(\mathbb{R}^N)$ from the previous proposition. Also, the same proposition tells us that, for $(u^\epsilon)^2(x)$,

$$[u^\epsilon(x)]^2 \rightharpoonup \overline{u^2}(x) = \frac{1}{p^n} \int_0^p \cdots \int_0^p u^2(x, y) dy^1 \cdots dy^n \tag{12.83}$$

weak-* in $L^\infty(\mathbb{R}^N)$,

However, Proposition 12.5 combined with convergence (12.82) says that $u^\epsilon \to u$ strongly if and only if

$$(u^\epsilon)^2(x) \rightharpoonup \overline{u^2}(x) \qquad \text{weakly in} \quad \mathcal{M}_{\text{loc}}(\mathbb{R}^N). \tag{12.84}$$

Thus, from convergences (12.83) and (12.84) we have $u^\epsilon \to u$ strongly if and only if $\overline{u}^2(x) = \overline{u^2}(x)\ \forall x \in \mathbb{R}^N$. However, the Cauchy–Schwarz inequality implies in general that $\overline{u}^2(x) \leq \overline{u^2}(x)$ with equality if and only if $u(x, y)$ is constant in y. Thus we have proved Proposition 12.6.

Proposition 12.6. *Let u, u^ϵ, and $\bar{u}$ be as given in Proposition 12.4; then $u^\epsilon \to u$ in $L^2_{\text{loc}}(\mathbb{R}^2)$ if and only if*

$$u(x, y) \equiv u(x) \ (\textit{constant in } y).$$

Below we present some examples of sequences of exact solutions of the Euler equation that exhibit oscillation in the limiting process.

Example 12.1. Recall from Chap. 2 that, for any smooth solution ψ of the nonlinear elliptic equation, $\Delta\psi = F(\psi)$ defines an exact steady solution of the 2D Euler equation by $v = {}^t(-\psi_{x_2}, \psi_{x_1})$. Here we take ψ as the solution of the eigenvalue equation of the Laplacian

$$\Delta\psi = -4\pi^2 K^2 \psi$$

given by

$$\psi(x) = \sum_{|\mathbf{k}|^2=K^2} a_{\mathbf{k}} e^{2\pi i <\mathbf{k}\cdot x>}, \qquad a_{\mathbf{k}} = a_{-\mathbf{k}}. \tag{12.85}$$

This stream function defines a steady 2π-periodic solution of the 2D Euler equation with the explicit form given by

$$v(x) = \sum_{|\mathbf{k}|^2=K^2} a_{\mathbf{k}} \begin{pmatrix} -k_2 \\ k_1 \end{pmatrix} e^{2\pi i <\mathbf{k}\cdot x>}, \qquad a_{\mathbf{k}} = a_{-\mathbf{k}}. \tag{12.86}$$

We define $v^\epsilon(x) \equiv v(x/\epsilon)$; then we have

$$|v^\epsilon|_{L^\infty(\mathbb{R}^2)} \le C, \tag{12.87}$$

$$|\mathrm{curl}\, v^\epsilon|_{L^1(\mathbb{R}^2)} \ge \mathcal{O}(\epsilon^{-1}). \tag{12.88}$$

Thus we do not have uniform L^1 control of vorticity.

By applying Proposition 12.4 we obtain

$$v^\epsilon \rightharpoonup \bar{v} = a_0 = \mathbf{0} \qquad \textit{weak-* in} \quad L^\infty(\mathbb{R}^2). \tag{12.89}$$

By the averaging theorem,

$$v^\epsilon \otimes v^\epsilon \rightharpoonup \mathrm{avg}(v \otimes v). \tag{12.90}$$

However,

$$v \otimes v = \sum_{|\mathbf{k}^i|^2=K^2 p, |\mathbf{k}^j|^2=K^2} a_{\mathbf{k}^i} a_{\mathbf{k}^j} \begin{pmatrix} -k_2^i \\ k_1^i \end{pmatrix} \otimes \begin{pmatrix} -k_2^j \\ k_1^j \end{pmatrix} e^{2\pi i <\mathbf{k}^i + \mathbf{k}^j, x>}$$

$$= -\sum_{|\mathbf{k}|^2=K^2} |a_{\mathbf{k}}|^2 \mathbf{k}^\perp \otimes \mathbf{k}^\perp, \ \mathbf{k}^\perp \equiv \begin{pmatrix} -k_2 \\ k_1 \end{pmatrix}.$$

Thus

$$v^\epsilon \otimes v^\epsilon \rightharpoonup -\sum_{|\mathbf{k}|^2=K^2} |a_{\mathbf{k}}|^2 \mathbf{k}^\perp \otimes \mathbf{k}^\perp \neq 0.$$

We note that even in this case the weak limit $\rightharpoonup 0$ is trivially a weak solution of Euler equation.

Example 12.2. Shear Layer. Let $v(x, y)$ be a smooth, bounded function in $\mathbb{R}^2$ which is 1−periodic in y and define $v^\epsilon(x_1, x_2) = [v(x_2, x_2/\epsilon), 0]$. Then Proposition 12.4 gives

$$v^\epsilon \rightharpoonup \bar{v}(x) = \left[\int_0^1 v(x_2, y)dy, 0\right]$$

*weak-** in $L^\infty(\mathbb{R}^2)$.

We now ask

Question 1: If $\{v^\epsilon\}$ is a sequence of exact solutions to the 3D Euler equation with $|v^\epsilon|_{L^\infty(\mathbb{R}^3)} \leq C$ and $v^\epsilon \rightharpoonup v$ weak-* in $L^\infty(\mathbb{R}^3)$, is v a solution of the 3D Euler equation?

The following example shows that the answer to the question is, in general, no.

Example 12.3. Recall from Chap. 2, Proposition 2.7, that if $\tilde{v} = [v_1(\tilde{x}, t), v_2(\tilde{x}, t)]$, $\tilde{x} = (x_1, x_2)$ solves the 2D Euler equation; then the function $v = [\tilde{v}, u(x_1, x_2, t)]$ is an exact solution of the 3D Euler equation provided that u satisfies

$$\frac{\tilde{D}}{\tilde{D}t}u = 0, \qquad \frac{\tilde{D}}{\tilde{D}t} \equiv \frac{\partial}{\partial t} + \tilde{v}_1\frac{\partial}{\partial x_1} + \tilde{v}_2\frac{\partial}{\partial x_2}.$$

Let $v \in C^\infty(\mathbb{R}^2) \cap L^\infty(\mathbb{R}^2)$, $u \in C^\infty(\mathbb{R}^3) \cap L^\infty(\mathbb{R}^3)$, and $v(x, y)$ be 1-periodic in y and let $u(x, y, z)$ be 1-periodic in z. Furthermore we assume that the average $\bar{v} = 0$.

Consider $\tilde{v} = [v_1(x_2, x_2/\epsilon), 0]$, which is an exact (steady shear layer) solution of the 2D Euler equation. Then because $u[x_1 - v_1(x_2, x_2/\epsilon)t, x_2, x_2/\epsilon]$ is a solution of the equation

$$\frac{\partial}{\partial t}u + v_1\left(x_1, \frac{x_2}{\epsilon}\right)\frac{\partial u}{\partial x_1} = 0$$

with initial data $u|_{t=0} = u(x_1, x_2, x_2/\epsilon)$, by application of the above fact to $(\tilde{v}, u)$, the sequence of velocities $\{v^\epsilon\}$ defined by

$$v^\epsilon(x) = \left\{\begin{matrix} v_1\left(x_2, \frac{x_2}{\epsilon}\right) \\ 0 \\ u\left[x_1 - v_1\left(x_2, \frac{x_2}{\epsilon}\right)t, x_2, \frac{x_2}{\epsilon}\right] \end{matrix}\right\}$$

is a sequence of the exact solutions of the 3D Euler equation. By the assumptions on v and u we have $|v^\epsilon|_{L^\infty(\mathbb{R}^3)} \leq C$; thus by application of Proposition 12.4 we have

$$v^\epsilon \rightharpoonup \begin{bmatrix} 0 \\ 0 \\ \int_0^1 u(x_1 - v_1(x_2, y)t, x_2, y)dy \end{bmatrix} \equiv \begin{pmatrix} 0 \\ 0 \\ \bar{u} \end{pmatrix}$$

weak-* in $L^\infty(\mathbb{R}^3)$.

To answer Question 1, i.e., to check if v is a solution of the 3D Euler equation or not, we need to look at only the third component of v. The velocity field v is a solution of the 3D Euler equation if and only if there exists a pressure p such that

$$\frac{\partial \bar{u}}{\partial t}(x_1, x_2) = \frac{\partial p}{\partial x_3}(x_1, x_2) \equiv 0. \tag{12.91}$$

However,

$$\frac{\partial \bar{u}}{\partial t} = \int_0^1 \left\{ v_1(x_2, y) \frac{\partial u}{\partial x_1}[x_1 - v_1(x_2, y)t, y] \right\} dy,$$

$$\frac{\partial^2 \bar{u}}{\partial t^2} = \frac{\partial^2 u}{\partial x_1^2} \int_0^1 |v_1(x_2, y)|^2 dy \neq 0.$$

In particular if we take initial data $\partial_{x_1}^2 u|_{t=0} \neq 0$ and $v_1(x_2, \cdot)$ not identically zero then $\frac{\partial^2 \bar{u}}{\partial t^2}|_{t=0}$ will be nonzero, necessarily violating condition (12.91).

In general the weak limit v of $\{v^\epsilon\}$ is not a solution of the 3D Euler equation even if v^ϵ is so for any $\epsilon > 0$.

Because we lose vorticity control in three dimensions the following question is natural to ask.

Question 2: Suppose $\{v^\epsilon\}$ is a sequence of exact solutions of the 2D Euler equation with uniform $L^p(\mathbb{R}^2)$ bound for any given $p \in [1, \infty]$ such that $v^\epsilon \rightharpoonup v$ weakly in $L^p(\mathbb{R}^2)$. Is it possible (by an example) that v is not a solution to the 2D Euler equation?

We do not know any example to answer to the above question yet. The concentration–cancellation phenomena we discussed in Chap. 11 and Example 12.3 above suggest that the answer may be no.

12.3. Young Measures and Measure-Valued Solutions of the Euler Equations

In this section we introduce the notion of a measure-valued solution of the Euler equation. This very weak definition allows us to consider a broader class of phenomena than allowed by the distribution Definition 10.1. In particular we can analyze measure-valued solutions that are the weak limits of approximate-solution sequences with oscillations and concentrations. The classical Young measure is expressed by the following theorem.

Theorem 12.9. *Young Measure Theorem. Let $\{u^\epsilon\}$ be a sequence of mappings from $O \subset \mathbb{R}^p$ to $\mathbb{R}^M$ such that*

$$|\omega^\epsilon(x)| \leq C \qquad \textit{uniformly in} \quad x \in \Omega \quad \textit{and} \quad \epsilon > 0. \tag{12.92}$$

Then there exists a subsequence $\{u^\epsilon\}$ *and a weakly measurable mapping, called the Young measure,* $x \mapsto \nu_x$ *from* Ω *to* prob $[\mathcal{M}(\mathbb{R}^M)]$ *such that*

$$u^\epsilon \rightharpoonup u \qquad \text{weak-* in} \quad L^\infty(\Omega)$$

for some $u \in L^\infty(\Omega)$ *and*

$$f(u^\epsilon) \rightharpoonup \langle \nu_x, f\rangle \equiv \int_{\mathbb{R}^M} f(\lambda) d\nu_x(\lambda) \qquad \text{weak-* in} \quad L^\infty(\Omega), \tag{12.93}$$

$$\lim_{\epsilon\to 0} \int \varphi(x) f(u^\epsilon(x)) dx = \int \varphi(x)\langle \nu_x, f\rangle dx \; \forall \varphi \in C_0(\Omega). \tag{12.94}$$

Moreover, $u^\epsilon \to u$ *(strongly) in* $L^p(\Omega)$*,* $p < \infty$ *if and only if* $\nu_x = \delta_{\omega(x)}$ *almost everywhere in* Ω*.*

Remark: The weak measurability of the mapping $x \mapsto \nu_x$ means that $x \mapsto \langle \nu_x, f\rangle$ is (Lebesgue) measurable on Ω for all $f \in C(\mathbb{R}^M)$.

In the following examples we use Young measures explicitly.

Example 12.4. Let u be a $\rho-$periodic piecewise continuous function in $\mathbb{R}^1$ defined as follows:

Let $\{\theta_j\}_{j=0}^L$ be a sequence of points in $[0, \rho]$ with $\theta_0 = 0, \theta_j < \theta_{j+1}, \theta_L = \rho$, and let $\{u_j\}_{j=0}^L$ be a corresponding family of numbers. We define

$$u(x) = u_j, \qquad \theta_{j-1} \le x < \theta_j, \qquad j = 1, \ldots, L,$$

and set $u^\epsilon(x) \equiv u[(1/x)\epsilon]$. Then, by Proposition 12.4 for any $f \in C(\mathbb{R}^1)$, $f(u^\epsilon(\cdot)) \rightharpoonup (1/\rho)\int_0^\rho f[u(x)]dx$ weak-* in $L^\infty[0, \rho]$, i.e., for any $\varphi \in L^1[0, \rho]$,

$$\lim_{\epsilon\to 0} \int \varphi(x) f[u^\epsilon(x)] dx = \int \varphi(x) \left\{ \frac{1}{\rho} \int_0^\rho f[u(y)]dy \right\} dx.$$

By the definition of u,

$$\frac{1}{\rho}\int_0^\rho f[u(y)]dy = \sum_{j=1}^L \alpha_j f(u_j) = \left\langle \sum_{j=1}^L \alpha_j \delta u_j, f \right\rangle,$$

i.e.,

$$\lim_{\epsilon\to 0} \int \varphi(x) f[u^\epsilon(x)] dx = \int \varphi(x) \left\langle \sum_{j=1}^L \alpha_j \delta u_j, f \right\rangle \tag{12.95}$$

$$\forall f \in C(\mathbb{R}^1), \qquad \forall \varphi \in L^1[0, \rho].$$

Comparing limit (12.94) with limit (12.95) we obtain the Young measure

$$\nu_x = \sum_{j=1}^L \alpha_j \delta u_j \qquad \text{almost every where} \quad x \in [0, \rho] \tag{12.96}$$

associated with the sequence $\{u^\epsilon\}$.

Example 12.5. Here we generalize the previous example. Let $u(x, y)$ be smooth ρ−periodic in y and bounded function; then from Proposition 12.4 we know for $u^\epsilon(x) = u(x, x/\epsilon)$,

$$u^\epsilon(x) \rightharpoonup \frac{1}{\rho}\int_0^\rho u(x, y)dy \qquad \textit{weak-* in} \quad L^\infty(\mathbb{R}^1)$$

and for any $f \in C(\mathbb{R}^1)$, $\varphi \in L^1(\mathbb{R}^1)$,

$$\lim_{\epsilon \to 0}\int \varphi(x) f[u^\epsilon(x)]dx = \int \varphi(x)\left\{\frac{1}{\rho}\int_0^\rho f[u(x, y)]dy\right\}dx. \tag{12.97}$$

Now recall the following well-known fact from the real analysis.

Fact: Let g: $\Omega \to \mathbb{R}^M$ be a Borel measurable mapping. Then, for any $\mu \in \mathcal{M}(\Omega)$ there exists $\nu \in \mathcal{M}(\mathbb{R}^M)$, induced by the mapping g, such that

$$\nu(B) = \mu[g^{-1}(B)] \ \forall \text{ Borel sets } B \subset \mathbb{R}^M.$$

Moreover, if μ is a probability measure, then ν is also a probability measure.

We apply the above fact by setting $g_x(y) \equiv u(x, y)$, $\Omega = (0, \rho)$, and $\mu = (m/\rho)$, where m is the Lebesgue measure on $(0, \rho)$, to identify the Young measure ν_x defined [from limits (12.97) and (12.94)] by

$$\langle \nu_x, f\rangle = \frac{1}{\rho}\int_0^\rho f[u(x, y)]dy \quad \text{almost everywhere}$$

$$x \in \Omega, \quad \forall f \in C(\mathbb{R}^1). \tag{12.98}$$

The Young measure is

$$\begin{aligned}\nu_x(B) &= \mu\left[g_x^{-1}(B)\right] (= \text{push forward of } \mu \text{ under } g_x)\\ &= \frac{1}{\rho} m\{[\omega(x, \cdot)]^{\rho-1}(B)\} \ \forall \text{ Borel sets } B \subset \mathbb{R}^1.\end{aligned}$$

Example 12.6. Recall the sequence of shear-layer solutions to 3D Euler equations $\{v^\epsilon\}$ considered in Example 12.3:

$$v^\epsilon(x) = \left\{\begin{matrix} v_1\left(x_2, \frac{x_2}{\epsilon}\right) \\ 0 \\ u\left[x_1 - v_1\left(x_2, \frac{x_2}{\epsilon}\right)t, x_2\frac{x_2}{\epsilon}\right]\end{matrix}\right\}.$$

We have

$$v^\epsilon \rightharpoonup \begin{pmatrix} 0 \\ 0 \\ \int_0^1 u(x_1 - v_1(x_2, y)t, x_2, y)dy\end{pmatrix} \qquad \textit{weak-* in} \quad L^\infty(\mathbb{R}^3).$$

Also, by Proposition 12.4, we have for any

$$g \in C(\mathbb{R}^3), \qquad \varphi \in C_0^\infty(\mathbb{R}^3),$$

$$\lim_{\epsilon \to p} \int_0^\infty \int_{\mathbb{R}^3} \varphi(x,t) g[v^\epsilon(x,t)] dx dt = \int_0^\infty \int_{\mathbb{R}^3} \varphi(x,t) \bar{g}(x,t) dx dt, \quad (12.99)$$

where

$$\bar{g}(x,t) = \int_0^1 g\{v_1(s,x_2), 0, u[x_1 - v_1(s,x_2)t, x_2, s]\} ds.$$

We can compute the Young measure $\nu_{x,t}$ associated with the sequence $\{v^\epsilon\}$ by using the fact in Example 12.5. We define the parameterized mapping $v_{x,t}$ from $[0,1]$ into $\mathbb{R}^3$ by

$$v_{x,t}(s) = {}^t\{v_1(s,x_2), 0, u[x_1 - v_1(s,x_2)t, x_2, s]\}; \quad (12.100)$$

then the Young measure $\nu_{x,t}$ is

$$\nu_{x,t}(B) = m\left[v_{x,t}^{-1}(B)\right] \qquad \forall \textit{ Borel sets } B \subset \mathbb{R}^3. \quad (12.101)$$

With this Young measure we can compute the composite weak limits. In particular,

$$\left(v_1^\epsilon v_3^\epsilon\right)(x,t) \rightharpoonup \langle \nu_{x,t}, \lambda_1 \lambda_3 \rangle = \int_0^1 v_1(s,x_2) u[x_1 - v_1(s,x_2)t, x_2, s] ds$$

$$v_3^\epsilon(x,t) \rightharpoonup \langle \nu_{x,t}, \lambda_3 \rangle = \int_0^1 u[x_1 - v_1(s,x_2)t, x_2, s] ds.$$

Thus we have

$$\frac{\partial}{\partial t} \langle \nu_{x,t}, \lambda_3 \rangle + \frac{\partial}{\partial x_1} \langle \nu_{x,t}, \lambda_1 \lambda_3 \rangle = 0. \quad (12.102)$$

Because $v_2^\epsilon \equiv 0$, $\langle \nu_{x,t}, \lambda_2 \rangle = 0$ and $v_1^\epsilon \rightharpoonup 0$, $\langle \nu_{x,t}, \lambda_1 \rangle = 0$ for almost every (x,t), for $\varphi(x,t) \in C_0^\infty(\mathbb{R}^3; \mathbb{R}^3)$, by multiplying Eq. (12.102) by φ_3 and integrating over $\mathbb{R}^3 \times \mathbb{R}_+$ we obtain, after integration by parts,

$$\int_0^\infty \int_{\mathbb{R}^3} (\varphi_3)'_t \langle \nu_{x,t}, \lambda_3 \rangle dx dt + \int_0^\infty \int_{\mathbb{R}^3} \frac{\partial}{\partial x_1} \varphi_3 \langle \nu_{x,t}, \lambda_1 \lambda_3 \rangle dx dt = 0. \quad (12.103)$$

We can rewrite this using $(\langle \nu_{x,t}, \lambda_1 \rangle = \langle \nu_{x,t}, \lambda_2 \rangle = 0)$ as

$$\int_0^\infty \int_{\mathbb{R}^3} \varphi'_t \cdot \langle \nu_{x,t}, \lambda \rangle dx dt + \int_0^\infty \int_{\mathbb{R}^3} \nabla \varphi : \langle \nu_{x,t}, \lambda \otimes \lambda \rangle dx dt = 0. \quad (12.104)$$

Note that Eq. (12.104) is the weak form of Eq. (12.102). Also observe that Eq. (12.104) is the formal integration of the weak form of the Euler equation with respect to the Young measure $d\nu_{x,t}$.

Example 12.6 and in particular Eq. (12.102) [or the weak form, Eq. (12.104)] motivate us to introduce a new concept – a *measure-valued solution* of the fluid equations.

Definition 12.2. *Measure-Valued Solution of the Euler Equation for $L^\infty(\Omega)$ Bound Sequences. Let $v^\epsilon \to v$ weak-* in $L^\infty(\Omega)$. Let $\nu_{x,t}$ be the Young measure associated with the sequence. We call $\nu_{x,t}$ a measure-valued solution of the Euler equation if for any $\varphi \in C_0^\infty(\Omega)$ with $\operatorname{div}\varphi = 0$*

$$\int\int_\Omega \varphi_t' \cdot \langle \nu_{x,t}, \lambda\rangle dxdt + \int\int_\Omega \nabla\varphi : \langle \nu_{x,t}, \lambda\otimes\lambda\rangle dxdt = 0, \tag{12.105}$$

$$\int\int_\Omega \nabla\psi \cdot \langle \nu_{x,t}, \lambda\rangle dxdt = 0 \ \forall \psi \in C_0^\infty(\Omega). \tag{12.106}$$

Note that in the special case of $\nu_{x,t} = \delta_{v(x,t)}$ Eq. (12.105) reduces to the equation for the ordinary weak solution v.

In Example 12.6, Eqs. (12.104) and (12.102) imply that Young measure (12.101) is a (smooth) measure-valued solution.

In Example 12.2, we have by, Proposition 12.4,

$$f(v^\epsilon) \rightharpoonup \int_0^1\int_0^1 f[v(y)]dy_1dy_2 \quad \text{in} \quad \mathcal{M}([0,1]\times[0,1]).$$

Thus

$$\langle \nu_x, f\rangle = \int_0^1\int_0^1 f[v(y)]dy_1dy_2 \qquad \text{almost everywhere} \quad x \in [0,1]\times[0,1].$$

From the fact in Example 12.5, we have that ν_x equals the push forward of the Lebesque measure on $[0,1]\times[0,1]$ under the mapping $v : [0,1]\times[0,1] \to \mathbb{R}^2$.

As we see in the following theorem very weak conditions are enough to guarantee that a Young measure is a measure-valued solution of the Euler equation.

Theorem 12.10. *Let $\{v^\epsilon\}$ be an approximate-solution sequence for the 3D Euler equation such that*

$$|v^\epsilon|_{L^\infty(\Omega)} \le C, \tag{12.107}$$

$$\int\int_\Omega (\varphi_t \cdot v^\epsilon + \nabla\varphi : v^\epsilon \otimes v^\epsilon)dxdt \to 0 \qquad \text{as} \quad \epsilon \to 0 \tag{12.108}$$

for all $\varphi \in C_0^\infty(\Omega)$ with $\operatorname{div}\varphi = 0$.

Then the Young measure $\nu_{x,t}$ associated with the sequence $\{v^\epsilon\}$ is a measure-valued solution of the 3D Euler equation.

Proof of Theorem 12.10. By the definition of Young measure $\nu_{x,t}$ associated with the sequence $\{v^\epsilon\}$ we have for any $\varphi \in C_0^\infty(\Omega)$

$$\int_\Omega\int \varphi_t' \cdot v^\epsilon dxdt \to \int_\Omega\int \varphi_t' \cdot \langle \nu_{x,t}, \lambda\rangle dxdt, \tag{12.109}$$

$$\int_\Omega\int \nabla\varphi : v^\epsilon \otimes v^\epsilon \to \int_\Omega\int \nabla\varphi : \langle \nu_{x,t}, \lambda\otimes\lambda\rangle dxdt, \tag{12.110}$$

as $\epsilon \to 0$.

Thus convergences (12.109) and (12.110) combined with convergence (12.108) implies that $\nu_{x,t}$ satisfies condition (12.105). From the definition of an approximate-solution sequence,

$$\int\int_{\Omega} \nabla\psi \cdot v^{\epsilon}\, dxdt = 0 \ \forall\psi \in C_0^{\infty}(\Omega). \tag{12.111}$$

Taking $\epsilon \to 0$ in Eq. (12.111) we obtain Eq. (12.106) for $\nu_{x,t}$. □

Below we prove the Young measure theorem (Theorem 12.9)

Proof of Theorem 12.9. For each u^{ϵ} there exists a positive measure $\mu^{\epsilon} \in \mathcal{M}(\Omega \times \mathbb{R}^M)$ defined by

$$\langle \mu^{\epsilon}, g(y, \lambda)\rangle = \int_{\Omega} g[x, u^{\epsilon}(x)]dx \ \forall g \in C(\Omega \times \mathbb{R}^M). \tag{12.112}$$

Moreover, from

$$|\langle \mu_{\epsilon}, g(y, \lambda)\rangle| \le m(\Omega) \sup_{\substack{x\in\Omega \\ |\psi|\le R}} |g(x, u)|,$$

we have

$$|\mu^{\epsilon}| \le C, \tag{12.113}$$

where C is a constant independent of $\epsilon > 0$. In particular, for $g(x, u) = \varphi(x)g(u)$,

$$\begin{aligned} |\langle \mu^{\epsilon}, \varphi(y)g(\lambda)\rangle| &= \left| \int_{\Omega} \varphi(x) g(u^{\epsilon}(x))dx \right| \\ &\le \sup_{|u|\le R} |g(u)| \int_{\Omega} |\varphi(x)|dx. \end{aligned} \tag{12.114}$$

Now relation (12.113) implies that there exists a subsequence $\{\mu^{\epsilon}\}$. We use the same notation, and $\mu \in \mathcal{M}(\Omega \times \mathbb{R}^M)$ with $\operatorname{supp}\mu \subseteq \Omega \times B_R$ such that

$$\mu^{\epsilon} \rightharpoonup \mu \qquad \text{(weak-*) in} \quad \mathcal{M}(\Omega \times \mathbb{R}^M),$$

i.e., in particular

$$\lim_{\epsilon\to 0} \int_{\Omega} \varphi(x) g[u^{\epsilon}(x)]dx = \langle \mu, \varphi(y)g(\lambda)\rangle. \tag{12.115}$$

Let $\{g_j(\lambda)\}_{j=1}^{\infty}$ be a countable dense set of bounded continuous functions in $C(B_R)$. Then, for fixed j the mapping from $C_0(\Omega)$ into $\mathbb{R}^1$ defined by

$$\varphi \mapsto \langle \mu, \varphi(y)g(\lambda)\rangle \tag{12.116}$$

is a bounded linear functional on $L^1(\Omega)$. Thus there exists $g_j \in L^{\infty}(\Omega)$ such that

$$\langle \mu, \varphi(y)g(\lambda)\rangle = \int_{\Omega} \varphi(x) g_j(x)dx. \tag{12.117}$$

Moreover, Eq. (12.117) combined with Eq. (12.114) implies that

$$g_j(x) \leq \sup_{|u|\leq R} |g_j(u)| \qquad \text{for almost all } x \in \Omega. \tag{12.118}$$

Also, from Eq. (12.117) with $\varphi \in C_0(\Omega)$, $\varphi \geq 0$, we obtain

$$g_j(x) \geq 0 \qquad \text{for almost all} \quad x \in \Omega \qquad \text{if} \quad g_j(\lambda) \geq 0. \tag{12.119}$$

Equation (12.117) and relation (12.118) combined with relation (12.119) imply that there exists a set N_j with $m(N_j) = 0$ such that the x–parameterized mapping

$$g_j \mapsto g_j(x) \tag{12.120}$$

is a positive linear functional if $x \in \Omega\backslash N_j$. Let $N = \cup_{j=1}^{\infty} N_j$; then $m(\Omega\backslash N) = m(\Omega)$ and for each $x \in \Omega\backslash N$, mapping (12.120) is a positive linear functional for each j. By the Riesz representation theorem, for all $x \in \Omega\backslash N$, there exists a measure $\nu_x \in \mathcal{M}(\mathbb{R}^M)$ such that

$$g_j(x) = \langle \nu_x, g_j \rangle \; \forall j \in N. \tag{12.121}$$

The positivity of mapping (12.120) implies the positivity of the measure ν_x in Eq. (12.121). Combining Eqs. (12.117), (12.121), and (12.115) yields

$$\lim_{\epsilon\to 0} \int_\Omega \varphi(x) g_j[u^\epsilon(x)]dx = \int_\Omega \varphi(x)\langle \nu_x, g_j(\lambda)\rangle dx \tag{12.122}$$

for all $\varphi \in C_0(\Omega)$, for all $j \in N$.

By continuity we can extend Eq. (12.121) [hence Eq. (12.122)] from $\{g_j\}_{j=1}^{\infty}$ to its uniform closure $C(B_R)$.

Setting $g = 1$ in Eq. (12.122), we have $\langle \nu_x, 1\rangle = 1$, i.e., $\nu_x \in \text{prob}\,\mathcal{M}(\mathbb{R}^M)$ for almost every $x \in \Omega$. □

For many applications, including the case in fluid dynamics, the condition of uniform L^∞ boundedness in the Young measure theorem is too strong. In many situations we have only uniformly bounded kinetic energy, i.e., a uniform L^2 bound and the possibility of concentration can occur.

We now extend the notion of a Young measure to that of a sequence with only a uniform L^2 bound.

For a uniformly L^2 bounded sequence, the most general nonlinear function for which we want to know the composite weak limit is one of the type

$$g(v) = \tilde{g}(v)\left(1 + |v|^2\right), \qquad \tilde{g} \in B \subset C(\mathbb{R}). \tag{12.123}$$

However, for many applications, the slightly narrower class of functions of the following type is enough;

$$g(v) = g_0(v)\left(1 + |v|^2\right) + g_H\left(\frac{v}{|v|}\right)|v|^2, \qquad g_0 \in C_0(\mathbb{R}^N), \qquad g_H \in C(S^{N-1}), \tag{12.124}$$

where C_0 denotes the class of continuous functions vanishing at infinity. For example, the terms in the Navier–Stokes or the Euler equations of interest are

$$v \otimes v = g_H\left(\frac{v}{|v|}\right)|v|^2, \qquad g_H(\theta) = \theta \otimes \theta, \quad \theta \in S^{N-1},$$

$$v = g_0(v)\big(1 + |v|^2\big), \qquad g_0(v) = \frac{v}{1+|v|^2} \in C_0(\mathbb{R}^N).$$

Theorem 12.11. *Generalized Young Measure Theorem. Let $\Omega = [0, T] \times B_R$, and let $\{v^\epsilon\}$ be a sequence in $L^2(\Omega)$ such that $\int_\Omega |v^\epsilon|^2 dxdt \le C$ and let $v^\epsilon \rightharpoonup v$ weakly in $L^2(\Omega)$. Then there exists a measure $\mu \in \mathcal{M}(\Omega)$ such that*

$$|v^\epsilon|^2 \rightharpoonup \mu \qquad \textit{weak-* in} \quad \mathcal{M}(\Omega). \tag{12.125}$$

Moreover there exists a weak- μ–measurable mapping $(x, t) \mapsto \{\nu^1_{x,t}, \nu^2_{x,t}\}$ from Ω into $\mathcal{M}^+(\mathbb{R}^M) \oplus \text{prob}\,\mathcal{M}(\mathbb{R}^M)$ such that for functions of the type in Eqs. (12.124) we have*

$$\begin{aligned}\lim_{\epsilon\to 0}\int\!\!\int_\Omega \varphi(x,t) g(v^\epsilon(x,t))dxdt &= \int\!\!\int_\Omega \varphi(x,t)\langle \nu'_{x,t} g_0\rangle(Hf)dxdt \\ &+ \int\!\!\int_\Omega \varphi(x,t)\big\langle \nu^2_{x,t}, g_H\big\rangle d\mu\ \forall \varphi \epsilon C_0(\Omega),\end{aligned} \tag{12.126}$$

where $f \in L^1(\Omega)$ is the Radon–Nikodym derivative of μ with respect to $dxdt$.

Based on this generalized Young measure theorem we can define the more realistic measure-valued solution of Euler equation

Definition 12.3. *Measure-Valued Solution of the Euler Equation for $L^2(\Omega)$Bounded Sequences. Let $\{v^\epsilon\}$ be a sequence in $L^2(\Omega)$ with $|v^\epsilon|_{L^2(\Omega)} \le C$ and $v^\epsilon \rightharpoonup v$ weakly in $L^2(\Omega)$. Let $\mu = (\mu, \nu^1, \nu^2)$ be the generalized Young measure associated with the sequence; then we call $\{\mu, \nu^1_{x,t}, \nu^2_{x,t}\}$ a measure-valued solution of the Euler equation if*

$$\int\!\!\int_\Omega \varphi_t\left\langle \nu^1_{x,t}, \frac{\lambda}{1+|\lambda|^2}\right\rangle(1+f)dxdt + \int\!\!\int_\Omega \nabla\varphi : \left\langle \nu^2_{x,t}, \frac{\lambda\otimes\lambda}{|\lambda|^2}\right\rangle d\mu = 0 \tag{12.127}$$

for all $\varphi \in C_0^\infty(\Omega)$, $\operatorname{div}\varphi = 0$, and

$$\int_\Omega\!\!\int \nabla\psi \cdot \left\langle \nu^1, \frac{\lambda}{1+|\lambda|^2}\right\rangle(1+F)dxdt = 0 \tag{12.128}$$

for all $\psi \in C_0^\infty(\Omega)$. Here, $f = (d\mu/dxdt)$ is the Radon–Nikodym derivative.

A simple but important result of Theorem 12.11 and Definition 12.3 is the following theorem, which is an immediate corollary of the proposition below it.

Theorem 12.12. *Consider a smooth divergence-free velocity field v_0 in $L^2(\mathbb{R}^3)$ and let v^ϵ be any Leray–Hopf weak solution of the Navier–Stokes equations with initial data v_0. Then there exists a subsequence whose limit defines a measure-valued solution of the 3D Euler equation.*

Proposition 12.7. *Assume that $\{v^\epsilon\}$ is a sequence of functions satisfying* $\operatorname{div} v^\epsilon = 0$,

(i) Weak stability: For any $\Omega \subset \mathbb{R}^+ \times \mathbb{R}^M$ there exists a constant $C(\Omega)$ such that

$$\int_\Omega\int |v^\epsilon(x,t)|^2 dxdt \leq C(\Omega). \tag{12.129}$$

(ii) Weak consistency: For all $\varphi \in C_0^\infty(\Omega)$ with $\operatorname{div}\varphi = 0$

$$\lim_{\epsilon\to 0}\int_\Omega\int (\varphi_t \cdot v^\epsilon + \nabla\varphi : v^\epsilon \otimes v^\epsilon)dxdt = 0. \tag{12.130}$$

If $\nu = (\mu, \nu^1, \nu^2)$ is the associated generalized Young measure associated with the sequence $\{v^\epsilon\}$, then ν is a measure-valued solution to the Euler equation on Ω.

Proof of Proposition 12.7. It follows from the definitions that

$$\begin{aligned}
0 &= \lim_{\epsilon\to 0}\int_\Omega\int (\varphi_t \cdot v^\epsilon + \nabla\varphi : v^\epsilon \otimes v^\epsilon)dxdt \\
&= \int_\Omega\int \varphi_t \cdot \left\langle \nu_{x,t}^1, \frac{\lambda}{1+|\lambda|^2}\right\rangle (1+f)dxdt + \int_\Omega\int \nabla\varphi : \left\langle \nu_{x,t}^2, \frac{\lambda\otimes\lambda}{|\lambda|^2}\right\rangle d\mu, \\
0 &= \lim_{\epsilon\to 0}\int_\Omega\int \nabla\psi \cdot v^\epsilon\, dxdt = \int_\Omega\int \nabla\psi \cdot \left\langle \nu_{x,t}^1, \frac{\lambda}{1+|\lambda|^2}\right\rangle dxdt \\
&\quad \forall\varphi \in C_0^\infty(\Omega) \quad \text{with} \quad \operatorname{div}\varphi = 0, \quad \forall\psi \in C_0^\infty(\Omega).
\end{aligned}$$

□

Proof of Theorem 12.11. In virtue of Proposition 12.7 we need to check only the weak stability and weak consistency. For the weak stability, it is well known (Temam, 1986) that for every $v_0 \in L^2(\mathbb{R}^3)$ with $\operatorname{div} v_0 = 0$ in the sense of distributions, there is at least one Leray–Hopf weak solution v^ϵ to the Navier–Stokes equations satisfying the kinetic-energy inequality

$$\max_{0\leq t<\infty}\int_{\mathbb{R}^3} |v^\epsilon(x,t)|^2 dx \leq \int_{\mathbb{R}^3} |v_0(x)|^2 dx. \tag{12.131}$$

Thus, in the high Reynolds number limit, v^ϵ satisfies the weak stability estimate. For weak consistency, multiplying the Navier–Stokes equations by the test function

$\varphi \in C_0^\infty(\Omega)$ with div $\varphi = 0$, we have

$$\int\int_\Omega (\varphi_t \cdot v^\epsilon + \nabla\varphi : v^\epsilon \otimes v^\epsilon) dxdt = \epsilon \int\int_\Omega \Delta\varphi \cdot v^\epsilon \, dxdt. \tag{12.132}$$

Because

$$\int\int_\Omega \Delta\varphi \cdot v^\epsilon \, dxdt| \leq C|v^\epsilon|_{L^2(\Omega)} \leq CT|v_0|_{L^2},$$

identity (12.132) implies that $\{v^\epsilon\}$ is weakly consistent with the Euler equations. □

Because every approximate-solution sequence of 2D Euler equation, by definition, satisfies the weak stability and the weak consistency with Euler, the associated generalized Young measure is a measure-valued solution of the 2D Euler equation.

In the following example we explicitly compute the measure-valued solution of the 2D Euler equation. In this example we reconsider the positive eddies in which concentration–cancellation occurs. We show now how this well-understood concentration phenomenon is cast in the context of the Young measure.

Example 12.7. Recall Example 11.2, which considered the approximate-solution sequence

$$v^\epsilon(x) = \left[\log\left(\frac{1}{\epsilon}\right)\right]^{-\frac{1}{2}} \epsilon^{-1} v\left(\frac{x}{\epsilon}\right), \tag{12.133}$$

where

$$v(x) = \frac{1}{r^2}\begin{pmatrix} -x_2 \\ x_1 \end{pmatrix} \int_0^r s\omega(s)ds,$$
$$\omega \in C_0^\infty(|x| \leq 1), \qquad \omega \geq 0.$$

In this example, $|v^\epsilon|_{L^2(\mathbb{R}^2)} \leq C$ and

$$v^\epsilon \rightharpoonup 0 \quad \text{weakly in} \quad L^2(\mathbb{R}^2), \tag{12.134}$$

$$v^\epsilon \to 0 \quad \text{strongly in} \quad L^p(\mathbb{R}^2), \tag{12.135}$$

for all p with $1 \leq p < 2$, and

$$|v^\epsilon|^2 \rightharpoonup 2C_1\delta_0 \quad \text{in} \quad \mathcal{M}(\mathbb{R}^2), \tag{12.136}$$

$$v^\epsilon \otimes v^\epsilon \rightharpoonup \begin{bmatrix} c_1\delta_0 & 0 \\ 0 & c_1\delta_0 \end{bmatrix} \quad \text{in} \quad [\mathcal{M}(\mathbb{R}^2)]^2. \tag{12.137}$$

From these results we compute (μ, ν_x^1, ν_x^2). First, from *weak convergence* (12.136) we obtain immediately

$$\mu_s = 2C_1\delta_0, \qquad f = \frac{d\mu}{dx} = 0. \tag{12.138}$$

By using $C_0^\infty(\mathbb{R}^2)$ approximations of functions in $C_0(\mathbb{R}^2)$, *strong convergence* (12.135) implies for $g(v) = g_0(v)(1+|v|^2)$, $g_0 \in C_0(\mathbb{R}^2)$, that

$$g[v^\epsilon(\cdot)] \rightharpoonup g(0) \qquad \text{weakly in} \quad \mathcal{M}(\mathbb{R}^2),$$

from which we obtain $\nu_x^1 = \delta_{\bar{0}}$.

The computation of ν_x^2 is more involved and is done in DiPerna and Majda (1987); the result is

$$\langle \nu_x^2, g_H \rangle = \frac{1}{2\pi} \int_{X^1} g_H(\theta) d\theta \qquad \forall g_H \in C(S^1). \tag{12.139}$$

Actually, we can connect this result to *weak convergence* (12.137) by

$$\begin{aligned} v_i v_j \rightharpoonup \langle \nu_x^2, \theta_i \theta_j \rangle d\mu &= \frac{1}{2\pi} \int_{S^1} \theta_i \theta_j \, d\theta d\mu \\ &= C_1 \delta_{ij} \delta_0. \end{aligned}$$

Also, without using *Proposition 12.7*, we can check explicitly that ν_x^2 is a measure-valued solution of the steady 2D Euler equation as follows: We need to check for all $\varphi \in C_0^\infty(\mathbb{R}^2)$ with div $\varphi = 0$ that

$$\int \nabla \varphi : \langle \nu_x^2, g_H \rangle d\mu = 0. \tag{12.140}$$

However, from *weak convergence* (12.137),the left-hand side of Eq. (12.140) is

$$\begin{aligned} \lim_{\epsilon \to 0} \int \nabla \varphi : v^\epsilon \otimes v^\epsilon dx &= C_1 \int \nabla \varphi : \begin{bmatrix} \delta_0 & 0 \\ 0 & \delta_0 \end{bmatrix} \\ &= C_1 \left[\frac{\partial \varphi}{\partial x_1}(0) + \frac{\partial \varphi}{\partial x_2}(0) \right] = 0 \end{aligned}$$

by the condition div $\varphi = 0$.

The following theorem gives us some information on the structure of the generalized Young measures, including the relation to the (classical) Young measure

Theorem 12.13. *Further Structure of Young Measure with L^2 Bounds. Let $\nu = (\mu, \nu^1, \nu^2)$ be a generalized Young measure associated with a sequence $\{v^\epsilon\}$ with $|v^\epsilon|_{L^2(\Omega)} \le C$ and $v^\epsilon \rightharpoonup v$.*

(i) If $|v^\epsilon|_{L^\infty(\Omega)} \le C$, then $\mu_s = 0$ and the classical Young measure $\bar{\nu}$ is recovered from the formula

$$\bar{\nu}_{x,t} = \left(1 + |v|^2\right)^{-1} \nu_{x,t}^1 (1 + f). \tag{12.141}$$

(ii) Weak continuity of the type $g(v^\epsilon) \rightharpoonup g(v)$ holds for all g of the form $g = g_0(1+|v|^2)$, with $g_0 \in C_0(\mathbb{R})$ if and only if ν^1 reduces to a weighted Dirac mass,

$$\nu_{x,t}^1 = \alpha(x,t) \delta_{v(x,t)}, \tag{12.142}$$

where $0 \le \alpha \le 1$ and satisfies $(1+|v|^2) = \alpha(1+f)$.

(iii) Strong continuity holds, i.e., $v^\epsilon \to v$ in $L^2(\Omega)$ if and only if $\mu_s = 0$ and $\alpha = 1$ almost everywhere in Ω.

Remarks: Even if $v^\epsilon \to v$ in $L^p(\Omega)$, $1 \leq p < 2$ [thus, $g(v^\epsilon) \rightharpoonup g(v)$ in $\mathcal{M}(\Omega)$, $g = g_0(1+|v|^2)$]. Part (ii) indicates that $\nu^1_{x,t}$ is not a probability measure in general (i.e., $\alpha \neq 1$ in general). Also, the following example taken from the Greengard–Thomann example in Subsection 11.2.2 shows that, even if $\mu_s = 0$, concentration can happen for $\alpha \neq 1$.

Example 12.8. We again reinterpret a previous example in the context of the Young measure. Recall the Greengard–Thomann example in Subsection 11.2.2. The solution sequence $\{v^m\}$ in this example satisfies

$$v^m \rightharpoonup 0 \qquad \textit{weakly in} \quad L^2(\mathbb{R}^2), \tag{12.143}$$

$$v^m \to 0 \qquad \textit{strongly in} \quad L^p(\mathbb{R}^2), 1 \leq p < 2, \tag{12.144}$$

$$|v^m| \rightharpoonup \mu = \chi_{[0,1]\times[0,1]}dx \qquad \textit{in} \quad \mathcal{M}(\mathbb{R}^2), \tag{12.145}$$

where χ denotes the characteristic function.

As in Example 12.7, strong convergence (12.144) implies that our sequence $\{v^m\}$ corresponds to case (ii) in Theorem 12.13. Thus there exists a function $\alpha(x)$ such that

$$\nu^1_x = \alpha(x)\delta_{v(x)}, \qquad \alpha(x) = [1 + f(x)]^{-1}. \tag{12.146}$$

Weak convergence (12.145) implies that

$$f(x) = \begin{cases} 1 \text{ almost everywhere} & x \in [0,1]\times[0,1] \\ 0 \text{ almost everywhere} & x \in \mathbb{R}^2 \sim [0,1]\times[0,1] \end{cases}.$$

Hence

$$\alpha = \begin{cases} \frac{1}{2} \text{ almost everywhere on} & [0,1]\times[0,1] \\ 1 \text{ almost everywhere on} & \mathbb{R}^2 \sim [0,1]\times[0,1] \end{cases}.$$

Below we compute ν^2_x. First, recall that at level m, for some constant C,

$$4^{-m} \leq 2\pi \int_0^{R_m} |v^m(r)|^2 r\, dr \leq 4^{-m}(1 + C4^{-m}). \tag{12.147}$$

Let

$$R = \left\{ (x,y) \in \mathbb{R}^2 \middle| \frac{k_{11}}{2^{\ell_1}} \leq x \leq \frac{k_{12}}{2^{\ell_1}}, \frac{k_{21}}{2^{\ell_2}} \leq y \leq \frac{k_{22}}{2^{\ell_2}} \right\}$$

be a typical rectangle at the level (ℓ_1, ℓ_2), with $\ell_1, \ell_2 > m$. Then, for any $\bar{g}_H \in C(S^1)$,

$$\int_R\int \bar{g}_H\left(\frac{v^m}{|v^m|}\right)|v^m|^2 = \{\text{number of phantom vortices in } R\}$$
$$\times \int_0^{R_m}\int_{S^1} \bar{g}_H(-\sin\theta, \cos\theta)|v^m(r)|^2 r\, d\theta dr$$
$$= m(R)\cdot\left(\frac{1}{2^m}\right)^{-1}\cdot \bar{g}_H \cdot 2\pi \int_0^{R_m} |v^m(r)|^2 r\, dr,$$

where

$$\bar{g}_H \equiv \frac{1}{2\pi}\int_{S^1} \bar{g}_H(-\sin\theta, \cos\theta)\partial\theta = \frac{1}{2\pi}\int_{S^1} \bar{g}_H(\theta^\perp)d\theta = \frac{1}{2\pi}\int_{S^1} \bar{g}_H(\theta)d\theta.$$

Thus from relation (12.147) we have

$$m(R)\bar{g}_H \le \int_R\int \bar{g}_H\left(\frac{v^m}{|v^m|}\right)|v^m|^2 dx \le m(R)\bar{g}_H(1 + C4^{-m}),$$

from which we obtain

$$\int_r\int \bar{g}_H\left(\frac{v^m}{|v^m|}\right)|v^m|^2 dx \to m(R)\bar{g}_H \qquad \text{as} \quad m \to \infty.$$

Because any Borel set $E \subset [0, 1] \times [0, 1]$ can be approximated by a union of such rectangles R,

$$\lim_{\omega^\epsilon\to\infty}\int_E\int g_H\left(\frac{v^m}{|v^m|}\right)|v^m|^2 dx = m(E)\bar{g}_H.$$

Hence, for any $\varphi \in C_0^\infty([0, 1] \times [0, 1])$,

$$\lim_{m\to\infty}\int_0^1\int_0^1 \varphi(x)\bar{g}_H\left(\frac{v^m}{|v^m|}\right)|v^m|^2 dx = \bar{g}_H\int_0^1\int_0^1 \varphi(x)dx$$
$$= \int_0^1\int_0^1 \varphi(x)\bar{g}_H\, dx = \int_{\mathbb{R}^2}\int \varphi(x)\left\{\int_{S^1} g_H(\theta)\frac{d\theta}{2\pi}\right\}\chi_{[0,1]\times[0,1]}dx.$$

Thus, because $v^m = 0\ \forall m \in N$ on $\mathbb{R}^2 \sim [0, 1] \times [0, 1]$,

$$\lim_{m\to\infty}\int_{\mathbb{R}^2} \varphi(x)\bar{g}_H\left(\frac{v^m}{|v^m|}\right)|v^m|^2 dx$$
$$= \lim_{m\to\infty}\int_{\mathbb{R}^2} \varphi(x)\bar{g}_H\left(\frac{v^m}{|v^m|}\right)|v^m|^2\chi_{[0,1]\times[0,1]}dx$$
$$= \lim_{m\to\infty}\int_0^1\int_0^1 \varphi(x)\bar{g}_H\left(\frac{v^m}{|v^m|}\right)|v^m|^2 dx$$
$$= \int\int_{\mathbb{R}^2} \varphi(x)\left\{\int_{S^1} \bar{g}_H(\theta)\frac{d\theta}{2\pi}\right\}d\mu.$$

Thus ν_x^2 is the uniform probability measure on S^1 (μ almost everywhere) in $\mathbb{R}^2$.

12.4. Measure-Valued Solutions with Oscillations and Concentrations

In this section we use the notion of a measure-valued solution to provide another viewpoint on the phenomena of concentration–cancellation. Investigation of the structure of the measure-valued solution gives us insight into this phenomenon as well as other interaction mechanisms between oscillations and concentrations during the weak limiting procedure. First we observe that because $v^\epsilon \rightharpoonup v$ weakly in $L^2(\Omega)$, in the definition of a measure-valued solution (Definition 12.3), Eqs. (12.127) and (12.128) can be rewritten as

$$\int_\Omega \int \varphi_t \cdot v\, dx dt + \int_\Omega \int \nabla\varphi : \left\langle \nu^2_{x,t}, \frac{\lambda \otimes \lambda}{|\lambda|^2} \right\rangle d\mu = 0 \tag{12.148}$$

for all $\varphi \in C_0^\infty(\Omega)$ with $\operatorname{div} \varphi = 0$, and

$$\int_\Omega \int \nabla\psi \cdot v\, dx dt = 0 \qquad \forall \psi \in C_0^\infty(\Omega) \tag{12.149}$$

(i.e., $\operatorname{div} v = 0$ in the sense of distribution).

On the other hand, the weak limit $v \in L^2(\Omega)$ satisfying Eq. (12.149) is a weak solution of the Euler equation if

$$\int\int_\Omega \varphi_t \cdot v\, dx dt + \int\int_\Omega \nabla\varphi : v \otimes v\, dx dt = 0,$$
$$\forall \varphi \in C_0^\infty(\Omega) \quad \text{with} \quad \operatorname{div} \varphi = 0. \tag{12.150}$$

Let us introduce a matrix Radon measure $V \in \mathcal{M}(\Omega : \mathbb{R}^{N^2})$ defined by

$$dV_{ij} = \langle \nu^2_{x,t}, \omega_i \omega_j \rangle d\mu - v_i v_j\, dx dt$$
$$\text{with} \quad \omega \in S^{N-1}, i, j = 1, \ldots, N. \tag{12.151}$$

Then, simply by subtracting Eq. (12.150) from Eq. (12.148) we have the following proposition.

Proposition 12.8. *Let $\{v^\epsilon\}$ be a sequence in $L^2(\Omega)$ with the associated measure-valued solution (μ, ν^2); then the weak limit v is a weak solution of Euler equation if and only if*

$$\int_\Omega \int \nabla\varphi : dV = 0 \quad \forall \varphi \in C_0^\infty(\Omega) \qquad \text{with} \quad \operatorname{div} \varphi = 0, \tag{12.152}$$

where V is defined as in Eq. (12.151).

This proposition provides us with a general characterization of concentration (and oscillation) cancellation without use of the defect measures, based on the notion of a measure-valued solution. In the typical simple 2D concentration examples in which $v^\epsilon \rightharpoonup 0$ in $L^2(\Omega)$, $\Omega = \{|x| \le 1\}$ and $v^\epsilon \otimes v^\epsilon \rightharpoonup CI\delta_0$ in $\mathcal{M}(\Omega)$ for some constant

$C > 0$, we know that ν_x^2 is a uniform measure on S^1. Thus

$$dV_{ij} = \frac{C}{2\pi}\int_0^{2\pi} \theta_i\theta_j\, d\sigma(\theta)\partial_x = C\delta_{ij}\delta_x. \tag{12.153}$$

On the other hand, in the Greengard–Thomann example, by using the results of computations in the previous chapter we have

$$\begin{aligned} dV_{ij} &= \frac{1}{2\pi}\int_0^{2\pi} \theta_i\theta_j\, d\sigma(\theta)\chi_{[0,1]\times[0,1]}\, dx_1dx_2 \\ &= \partial_{ij}\chi_{[0,1]\times[0,1]}\, dx_1dx_2. \end{aligned} \tag{12.154}$$

By virtue of the condition div $\varphi = 0$ on the test function φ, Eq. (12.152) is trivially satisfied in both cases of Eqs. (12.153) and (12.154). The weak limit $\vec{0}$ is a weak solution of the Euler equation.

Below we restrict ourselves to the special case of two dimensions. For scalar test function $\eta \in C_0^\infty(\Omega)$ denote $\varphi = \nabla^\perp\eta = \binom{-\eta_{x_2}}{\eta_{x_1}} \in C_0^\infty(\Omega)$. Then automatically div $\varphi = 0$, and Eq. (12.152) has the following explicit form:

$$\begin{aligned} 0 = \int_\Omega\int \nabla\varphi : dV &= -\int_\Omega\int \frac{\partial^2\eta}{\partial x_1\partial x_2}(dV_{11} - dV_{22}) \\ &\quad + -\int_\Omega\int \left(\frac{\partial^2\eta}{\partial x_2^2} - \frac{\partial^2\eta}{\partial x_1^2}\right) dV_{12}. \end{aligned}$$

Using the notation

$$\Box = \frac{\partial^2}{\partial x_2^2} - \frac{\partial^2}{\partial x_1^2}, \qquad \Box^\perp = \frac{\partial^2}{\partial x_1\partial x_2},$$

we rewrite Eq. (12.152) as

$$\int\int_\Omega \Box^\perp\eta d(V_{11} - V_{22}) + \int\int_\Omega \Box\eta dV_{1,2} = 0\ \forall\eta \in C_0^\infty(\Omega). \tag{12.155}$$

Thus we obtain the 2D version of Proposition 12.8.

Proposition 12.9. *Let $\{v^\epsilon\}$ be a sequence in $L^2(\Omega)$ associated with the measure-valued solution (μ, ν^2). Then the weak limit v is a weak solution of the 2D Euler equation if and only if*

$$\Box^\perp(V_{11} - V_{22}) + \Box V_{1,2} = 0 \tag{12.156}$$

in the sense of distribution, where V is defined as in Eq. (12.151).

As a corollary of this proposition we obtain Proposition 12.10.

Proposition 12.10. *Assume that*

$$V_{ij} = \sum_{\ell=1}^{L} A_{ij}^\ell(t)\delta_{x_\ell(t)}, \ldots, i, \qquad j = 1, 2, \tag{12.157}$$

where ${}^T A^\ell = A^\ell$ *(symmetric),* $x^{\ell_1} \neq x^{\ell_2}$ *almost everywhere in* $[0, T]$ *for all* $\ell, \ell_1, \ell_2 = 1, \ldots, L$*; then the associated weak limit* v *is a weak solution of the 2D Euler equation if and only if there is equipartition of kinetic energy in the defect matrix* V_{ij}*, i.e.,*

$$A^\ell_{11}(t) = A^\ell_{22}(t), \qquad A^\ell_{12}(t) = 0 \tag{12.158}$$

for almost every $t \in [0, T]$ *for all* $\ell = 1, \ldots, L$.

Proof of Proposition 12.10. Substituting Eq. (12.155) into Eq. (12.154) we obtain

$$0 = \sum_{\ell=1}^{L} \left\{ \left(A^\ell_{11} - A^\ell_{22}\right) \square^\perp \partial_{x^\ell(t)} + A^\ell_{12} \, \square \partial_{x^\ell(t)} \right\}.$$

From the condition $x^{\ell_1}(t) \neq x^{\ell_2}(t)$ if $\ell_1 \neq \ell_2$ we have for each $\ell = 1, \ldots, L$

$$\left(A^\ell_{11} - A^\ell_{22}\right) \qquad \square^\perp \partial_{x^\ell(t)} + A^\ell_{12} \qquad \square \partial_{x^\ell(t)} = 0.$$

Because $\square^\perp \partial_{x^\ell(t)}$ and $\square x \partial_{x^\ell}$ define a pair of linearly independent distributions for each $\ell = 1, \ldots, L$ we obtain Eq. (12.156) as equivalent to Eq. (12.152). □

In all our examples of exact-solution sequences for the 2D Euler equation we have only two cases:

$$\nu^2_x = \delta \frac{v(x)}{|v(x)|}, \tag{12.159}$$

$$\nu^2_x = \text{uniform probability measure on } S^1. \tag{12.160}$$

It is possible that the $O(2)$ covariant structure of the 2D Euler equation restricts the form of the Young measure ν^2_x to the above two types. This suggests the following problem.

Problem: Find an example of an exact-solution sequence for the steady 2D Euler equation with or without L^1 vorticity control so that $\nu^2_x \in \text{prob}\, \mathcal{M}(S^1)$ is neither case (12.159) nor case (12.160).

Below we discuss the oscillation phenomena in 2D fluid flows. Specifically we try to solve the following problem.

Problem: Let $\{v^\epsilon\}$ be an exact-solution sequence of the 2D Euler equation with

$$|v^\epsilon|_{L^\infty(\Omega)} \leq C, \tag{12.161}$$

and *no vorticity control assumed.* Is the weak limit v a weak solution of the 2D Euler equation?

One possible attack to the problem is to show directly that

$$\int\int_\Omega \nabla\varphi : v^\epsilon \otimes v^\epsilon \, dx dt \to \int\int_\Omega \nabla\varphi : v \otimes v \, dx dt \tag{12.162}$$

as $\epsilon \to 0$ $\forall \varphi \in C_0^\infty(\Omega)$ with div $\varphi = 0$. By redefining $\tilde{v}^\epsilon = v^\epsilon - v$ (thus $\tilde{v}^\epsilon \rightharpoonup 0$, div $\tilde{v}^\epsilon = 0$) it is sufficient to show that

$$\lim_{\epsilon \to 0} \int_\Omega \int \nabla \varphi : \tilde{v}^\epsilon \otimes \tilde{v}^\epsilon \, dx dt = 0 \tag{12.163}$$

for all $\varphi \in C_0^\infty(\Omega)$ with div $\varphi = 0$. By writing $\varphi = \nabla^\perp \eta = \binom{-\eta_{x_2}}{\eta_{x_1}}$ for a scalar-valued $\eta \in C_0^\infty(\Omega)$, relation (12.161) becomes

$$\lim_{\epsilon \to 0} \int_\Omega \int \{\Box^\perp \eta [(\tilde{v}_1^\epsilon)^2 - (\tilde{v}_2^\epsilon)^2] + \Box \eta \tilde{v}_1^\epsilon \tilde{v}_2^\epsilon\} dx dt = 0 \; \forall \eta \in C_0^\infty(\Omega). \tag{12.164}$$

The same reduction argument as that in the proof of Theorem 12.2 that uses the Radon transform and the covariant structure of the nonlinear term reduces Eq. (12.164) to

$$\lim_{\epsilon \to 0} \int_\Omega \int \rho(t) h''(x_1) \tilde{v}_1^\epsilon \tilde{v}_2^\epsilon \, dx_1 dx_2 dt = 0$$
$$\forall \rho \in C_0^\infty(0, T), \; \forall h \in C_0^\infty[\pi_{x_1}(\Omega)],$$

where we have separated the time part for the test functions.

To apply Tartar's arguments of compensated compactness (or Gerard's microlocal generalizations of them) we need first to answer to the following question.

Question 2: Is it true that for any exact-solution sequence $\{v^\epsilon\}$ with $|v^\epsilon|_{L^\infty} \le C$ and $v^\epsilon \rightharpoonup 0$ weak-* in $L^\infty(\Omega)$,

$$\lim_{\epsilon \to 0} \int_\Omega \int \rho(t) \varphi(x) v_1 v_2 \, dx dt = 0 \tag{12.165}$$

for all space–time test functions φ and ρ?

The answer to the question is *no*, as we can see in the following example.

Example 12.9. For any smooth radial function f let us define $\{v^\epsilon\}$ with

$$v^\epsilon(x) = f\left(\frac{x \cdot \omega}{\epsilon}\right) \begin{pmatrix} -\omega_2 \\ \omega_1 \end{pmatrix}, \qquad |\omega| = 1. \tag{12.166}$$

Then each v^ϵ is an exact solution of the 2D Euler equation. Furthermore, if we assume that $\int_0^1 f(s) ds = 0$ and f is 1-periodic, then by the averaging lemma Proposition 12.4 we have

$$\lim_{\epsilon \to 0} \int \int \varphi(x_1) v_1 v_2 \, dx = - \lim_{\epsilon \to 0} \int \int \varphi(x_1) \omega_1 \omega_2 f^2 \left(\frac{x \cdot \omega}{\epsilon}\right) dx$$
$$= - \int \int \varphi(x_1) \omega_1 \omega_2 \langle f^2 \rangle dx \neq 0$$

in general.

Notes for Chapter 12

The material in this chapter on reduced Hausdorff dimensions and concentrations comes from the technical paper of DiPerna and Majda (1988), and the material on oscillations and measure-valued solutions first appeared in DiPerna and Majda (1987). The question of the Hausdorff dimension of the space–time concentration set for 2D approximate-solution sequences remains open. Although DiPerna and Majda show it to be less than or equal to 1 for a special cylindrical measure, the best anyone has been able to do for the usual Hausdorff dimension is slightly larger than 2 (Nussenzveig Lopes, 1994, 1997). Furthermore, the connection between concentration–cancellation and the size of the concentration set has been further addressed by Alinhac (1990) with improvements by Zheng (1991) who defined a *strong*-concentration set (larger than the set introduced here) and showed that concentration–cancellation results whenever this set has locally finite 1D Hausdorff measure in space and time. The proof of Theorem 12.1 presented by use of capacity theory can be found in Evans (1990).

We remark briefly about a recent paper by Schochet (1995). He proves that the weak limit of a sequence of approximate-solution sequences of the 2D Euler equations is a solution to the weak form of the vorticity-stream equation provided that the sequence of vorticities concentrate only along a curve $x(t)$ that is Hölder continuous with exponent $\frac{1}{2}$.

The discussion on Hausdorff measure presented here is designed to address only the concerns of the material in this book. For more details on Hausdorff measures see Falconer (1985).

We remark that Lemma 12.1 on approximation by plane waves is closely related to the notion of a Radon transform (see e.g., Ludwig, 1966).

For the proof of Lemma 12.1 see Stein (1970) Theorem 12.5 appears as Theorem 5.1 on p. 79 of DiPerna and Majda (1988).

References for Chapter 12

Alinhac, S., "Un phenomene de concentration evanescente pour des flots nonstationnaires incompressibles en dimension deux," *Commun. Math. Phys.* **172**, 585–596, 1990.

DiPerna, R. J. and Majda, A. J., "Oscillations and concentrations in weak solutions of the incompressible fluid equations," *Commun. Math. Phys.* **108**, 667–689, 1987.

DiPerna, R. J. and Majda, A. J., "Reduced Hausdorff dimension and concentration–cancellation for two-dimensional incompressible flow," *J. Am. Math. Soc.* **1**, 59–95, 1988.

Evans, L. C., *Weak Convergence Methods for Nonlinear Partial Differential Equations*, Number 74 in CBMS Lecture Notes Series, American Mathematical Society, Providence, RI, 1990.

Falconer, K. J., *The Geometry of Fractal Sets*, Cambridge Univ. Press, New York, 1985.

Gilbarg, D. and Trudinger, N. S., *Elliptic Partial Differential Equations of Second Order*. Springer-Verlag, Berlin, 1998, revised 3rd printing.

Greengard, C. and Thomann, E., "On DiPerna–Majda concentration sets for two-dimensional incompressible flow," *Commun. Pure Appl. Math.* **41**, 295–303, 1988.

Ludwig, D., "The Radon transform on Euclidean space," *Commun. Pure Appl. Math.* **19**, 49–81, 1966.

Nussenzveig Lopes, H. J., "An estimate on the Hausdorff dimension of a concentration set for the incompressible 2-D Euler equation," *Indiana Univ. Math. J.* **43**, 521–533, 1994.

Nussenzveig Lopes, H. J., "A refined estimate of the size of concentration sets for 2D incompressible inviscid flow," *Indiana Univ. Math. J.* **46**, 165–182, 1997.

Rudin, W., *Real and Complex Analysis*, 3rd ed., McGraw-Hill, New York, 1987.
Schochet, S., "The weak vorticity formulation of the 2D Euler equations and concentration–cancellation," *Commun. Partial Diff. Eqns.* **20**, 1077–1104, 1995.
Stein, E. M., *Singular Integrals and Differentiability Properties or Functions*, Princeton Univ. Press, Princeton, NJ, 1970.
Temem, R., *Navier–Stokes Equations*, 2nd ed., North-Holland, Amsterdam, 1986.
Zheng, Y., "Concentration–cancellation for the velocity fields in two-dimensional incompressible fluid flows," *Commun. Math Phys.* **135**, 581–594, 1991.

13

The Vlasov–Poisson Equations as an Analogy to the Euler Equations for the Study of Weak Solutions

In Chaps. 9–12 we proved properties of the Euler equation

$$\frac{\partial v}{\partial t} + v \cdot \nabla v = -\nabla p, \qquad \text{div } v = 0, \tag{13.1}$$

for very weak initial data. A large portion of these chapters examined questions concerning 2D solutions with *vortex-sheet initial data* in which the vorticity $\omega = \text{curl } v$ is a Radon measure. We showed in Chap. 11 that such solutions exist as classical weak solutions to the primitive-variable form (13.1) of the Euler equation, provided that the vorticity ω has a *distinguished sign*. We developed the notion of an *approximate-solution sequence* originally formally introduced by DiPerna and Majda (1987a, 1987b, 1988) in Chap. 9 and used some of the elementary results developed there to set the stage for the material in Chap. 10. In Section 11.4 we showed that for any initial vorticity $\omega_0 \geq 0$ a nonnegative Radon measure

$$\omega_0 \in \mathcal{M}(\mathbb{R}^2), \qquad \omega_0 \geq 0, \tag{13.2}$$

and with corresponding initial velocity v_0 with locally finite kinetic energy

$$\int_{|x|<R} |v_0|^2 dx \leq C(R) < \infty \qquad \text{for any} \quad R > 0 \tag{13.3}$$

that there exists a solution to Euler equation (13.1) for all time. Moreover, we can obtain a solution by (1) smoothing the initial data by mollification, $v_0^\epsilon = \mathcal{J}_\epsilon \omega = \rho_\epsilon * v_0 \int \rho = 1$, $\rho \geq 1$, $\rho \in C_0^\infty(\mathbb{R}^2)$, $\rho_\epsilon = \frac{1}{\epsilon^2}\rho(x/\epsilon)$ and (2) using the smoothed initial data v_0^ϵ to solve the Navier–Stokes or the Euler equation

$$\begin{gathered} \frac{\rho v^\epsilon}{\rho t} + v^\epsilon \cdot \nabla v^\epsilon = -p^\epsilon + \nu_\epsilon \Delta v^\epsilon, \\ \text{div } v^\epsilon = 0, \\ v^\epsilon|_{t=0} = v_0^\epsilon, \end{gathered} \tag{13.4}$$

where $\nu_\epsilon \geq 0$, $\nu_\epsilon \to 0$ as $\epsilon \to 0$. Passing to the limit in ϵ produces a vortex-sheet solution to Eq. (13.1). The main issue in passing to the limit in Eq. (13.4) is the behavior of the nonlinear term $v \otimes v$. In particular concentrations occur (see Section 11.2 for some elementary examples) in which the tensor product concentrates as a delta function in the limit. We showed in Section 11.4 that despite a concentration effect, *concentration–cancellation* occurs in which the limiting solution v still satisfies Euler equation (13.1) despite concentration. In Chap. 12 we discussed the case of concentrations in a limiting process and showed that concentration–cancellation always occurs in the case of 2D *steady* approximate-solution sequences.

Many problems involving approximate-solution sequences for the 2D Euler equation still remain unsolved. For example, is there a unique weak solution for given initial data defined by nonnegative measure? Do concentrations occur spontaneously in finite time for vortex-sheet evolution with nonnegative vorticity? How do such concentrations affect the subsequent evolution? Is there an appropriate selection principle to single out a unique weak solution in the high Reynolds number limit of the Navier–Stokes equations? Do different approximate solutions such as can be calculated through various computational algorithms converge to distinct or the same weak solutions?

Zheng and Majda (1994) and Majda et al. (1994a, 1994b) recently addressed essentially all of these issues for a simpler analog problem, the 1D Vlasov–Poisson (V-P) equations for a collisionless plasma of electrons in a uniform background of ions. Our main goal in this chapter is to summarize their results and show how they relate to specific problems discussed in previous chapters of this book. The V-P system has a mathematical structure that is similar to that of the 2D Euler equations. By analogy with Chaps. 8–12 of this book on the weak solution theory of the 2D Euler equations, we introduce definitions of weak solutions for the V-P system, including the *electron patch*, an analogy to the *vortex patch* introduced in Chap. 8, the *electron sheet*, a parallel to the *vortex sheet* introduced in Chap. 9, and a notion of an approximate-solution sequence for the V-P equations following the line of reasoning introduced for the 2D Euler equation in Chaps. 10–12. The simpler structure of the V-P system, because of a less singular Biot–Savart law, presents a more tractable system, yielding answers to some questions that remain unsolved for the case of the 2D Euler equation. For example, in the case of a distinguished-sign one component plasma, nonuniqueness occurs, despite the long time existence of such an 'electron-sheet' solution. The analogous question for the 2D Euler equation, that of uniqueness of solutions with vortex-sheet initial data of distinguished sign, remains an open problem.

13.0.1. The 1D V-P Model

The V-P system we consider here is a model for a collisionless plasma of ions and corresponding electrons. The transport is uni-directional so that we can formulate the problem in one-space dimension. We assume that particle motion is governed solely by induced electrostatic forces, and neglect electromagnetic interactions.

We consider two cases.

The Single-Component Case

In this case the positive ions have enough inertia so that their motion can be neglected. Modeling the positive ions as a neutralizing uniform background field, we let $f(x, v, t)$ denote the density of electrons traveling with speed v at a given position x and time t. We assume here a periodic geometry on $[0, L]$ so that $f(x, v, t) =$

$f(x+L, v, t)$. Without loss of generality, we can normalize the initial data so that $L^{-1} f(x, v)$ is a nonnegative probability measure:

$$\frac{1}{L}\int_0^L f(x, v, t)dxdv = 1.$$

The single-component 1D V-P system is the following active-scalar evolution equation for f:

$$\begin{aligned}\frac{\partial f}{\partial t} + v\frac{\partial f}{\partial x} - E(x, t)\frac{\partial f}{\partial v} &= 0,\\ \frac{\partial^2 \varphi}{\partial x^2} = \rho &= 1 - \int_{-\infty}^{\infty} f(x, v, t)dv,\\ E &= \frac{\partial \varphi}{\partial x}.\end{aligned} \tag{13.5}$$

Here ρ is the aggregate charge density at position x and E is the electric field defined as $E = \partial\varphi/\partial x$, with φ the electric potential.

The Two-Component Case

In the two-component case, the positive ions have small enough inertia that their motion is on a time scale comparable with that of the electrons in the plasma. Denoting as $f_-(x, v, t)$ the local density of electrons traveling with speed v and as $f_+(x, v, t)$ the local density of positive ions traveling with speed v, the particles are transported according to

$$\begin{aligned}\frac{\partial f_\pm}{\partial t} + v\frac{\partial f_\pm}{\partial x} \pm E(x, t)\frac{\partial f_\pm}{\partial v} &= 0,\\ \frac{\partial^2 \varphi}{\partial x^2} = \rho &= \int_{-\infty}^{\infty} [f_+(x, v, t) - f_-(x, v, t)]dv,\\ E &= \frac{\partial \varphi}{\partial x}\end{aligned} \tag{13.6}$$

with initial condition

$$f_\pm(x, v, t)|_{t=0} = f_{\pm 0}(x, v). \tag{13.7}$$

We consider both the problem on the line $x \in \mathbb{R}$ and on a periodic interval $x \in [0, L]$. The net charge in the system is zero, i.e.,

$$\int\int f_+(x, v, t)dxdv = \int\int f_-(x, v, t)dxdv.$$

Our main goal in this chapter is to demonstrate explicitly for the simpler analog problems (13.5) and (13.6), a number of new phenomena conjectured to occur for vortex sheets for the 2D Euler equations (Krasny, 1981; Majda, 1988, 1993).

The analog for Eqs. (13.6) of the high Reynolds number limit in Eq. (13.4) is the (pseudo-) Fokker–Planck limit of solutions $f_\pm^\epsilon$ satisfying

$$\frac{\partial f_\pm^\epsilon}{\partial t} + v\frac{\partial f_\pm^\epsilon}{\partial x} \pm E^\epsilon(x, t)\frac{\partial f_\pm^\epsilon}{\partial v} = \epsilon\frac{\partial^2 f_\pm^\epsilon}{\partial v^2},$$

$$\frac{\partial^2 \varphi^\epsilon}{\partial x^2} = \rho^\epsilon = \int_{-\infty}^{\infty} [f_+^\epsilon(x, v, t)dv - f_-^\epsilon(x, v, t)]dv, \qquad E^\epsilon = \frac{\partial \varphi^\epsilon}{\partial x} \tag{13.8}$$

as the parameter $\epsilon \to 0$. We call this a pseudo-Fokker–Planck limit because we have diffusion in only the v variable instead of in a full x, v diffusion.

In this chapter, we consider several analogy problems to the 2D Euler equations for the V-P system. In Section 13.1 we review the many aspects of how the V-P system parallels the case of 2D incompressible inviscid flows. For example, in Subsection 13.1.1 we describe the physical analogy between the two systems when viewed as nonlinear transport equations. In Subsection 13.1.2 we show that there is an analogous kind of solution to the vortex patch discussed in detail in Chap. 8. We call this solution an *electron patch*. In Subsection 13.1.3 we introduces the concept of an *electron sheet*, the example analogous to that of the vortex sheet introduced in Chap. 9. In Subsection 13.1.4 we introduce the parallel to the point vortex, that of concentrated electron density. In Subsection 13.1.5 we introduce the notion of approximate-solution sequences and concentration, similar to Chaps. 10 and 11. Finally, in Subsection 13.1.7 we show how to construct computational particle methods, similar to those described in Chap. 6 on vortex methods, for numerically simulating solutions of the 1D V-P equations.

In Section 13.2 we discuss the theory for electron-sheet solutions of the one-component system. We show that all of the current open questions concerning vortex-sheet initial data of distinguished sign (namely uniqueness and behavior of solutions achieved through different approximation schemes) are resolved for this simpler analogy problem of the single-component V-P system. In particular, given electron-sheet initial data (see Subsection 13.1.3) for V-P system (13.5), the analog of vortex-sheet initial data (discussed in Chaps. 10 and 11), there are smooth initial distributions along curves so that concentrations in charge develop spontaneously at a critical time (see Section 13.2). For other smooth electron-sheet initial data, singularities in the electron sheet without concentrations develop spontaneously at a critical time. In these instances, solutions computed by means of particle methods always converge to the same weak solution beyond the critical time (see Section 13.2). Furthermore, there are weak solutions in which an electron sheet collapses to a point charge in finite time.

We present analytical examples of nonuniqueness within certain classes of weak solutions to Eqs. (13.5) and (13.6). Careful high-resolution numerical methods establish that for these explicit, nonunique weak solutions different regularizations lead to different weak solutions. For example, the "viscous" limit obtained from Fokker–Planck equations (13.8) as $\epsilon \to 0$ converges to a different weak solution than the limit through other computational regularizations.

Recall from the previous four chapters that vastly more complex behavior occurs in weak solutions of the 2D Euler equations when the vorticity changes sign. One significant open question for vortex-sheet initial data of mixed sign is whether the solution can be continued in time indefinitely as a classical weak solution or whether concentrations develop that necessitate a more general formulation of a weak solution such as a measure-valued solution. In Section 13.3 we address this issue for the two-component V-P equations as an analogy problem. We give explicit examples with more subtle behavior requiring the notion of a measure-valued solution. We also study various computational regularizations of these weak and measure-valued solutions.

13.1. The Analogy Between the 2D Euler Equations and the 1D Vlasov–Poisson Equations

Here we develop the physical and structural analogies between weak solutions of the 1D V-P equations and weak solutions of the 2D Euler equations in Chaps. 7–11.

13.1.1. *The Physical Analogy as Nonlinear Transport Equations*

There is an interesting parallel between the *vorticity-stream form* of the 2D Euler equations (recall, e.g., Proposition 2.1)

$$
\begin{aligned}
&\frac{\partial \omega}{\partial t} + \mathbf{v}\cdot\nabla\omega = 0, \qquad \omega|_{t=0} = \omega_0,\\
&\operatorname{div} \mathbf{v} = 0,\\
&\operatorname{curl} \mathbf{v} = \omega,
\end{aligned}
\tag{13.9}
$$

and V-P system (13.5) and (13.6). Recall from Chap. 2 that the incompressible velocity field $\mathbf{v}$ is determined from ω through the Biot–Savart law by means of a stream function ψ. This stream function satisfies

$$
\Delta\psi = -\omega, \qquad \mathbf{v} = -\nabla^{\perp}\psi = \begin{pmatrix} \frac{\partial\psi}{\partial x_2} \\ -\frac{\partial\psi}{\partial x_1} \end{pmatrix}
\tag{13.10}
$$

such that

$$
v(x) = \int_{\mathbb{R}^2} K(x-y)\omega(y,t)dy
\tag{13.11}
$$

with

$$
K(y) = \nabla^{\perp}\left[\frac{1}{2\pi}\log(|y|)\right].
\tag{13.12}
$$

For single-component V-P equations (13.5), we denote by $\mathbf{u}$ the vector field in the (x, v) plane:

$$
\mathbf{u} = {}^{t}[v, -E(x,t)].
\tag{13.13}
$$

Identifying the coordinates (x, v) for V-P system (13.5) with (x_1, x_2) in the above discussion of the 2D Euler equations, single-component V-P equations (13.5) can be rewritten as

$$
\begin{aligned}
&\frac{\partial f}{\partial t} + \mathbf{u}\cdot\nabla f = 0,\\
&\operatorname{div} \mathbf{u} = 0,\\
&\operatorname{curl} \mathbf{u} = \int_{-\infty}^{\infty} f(x,v,t)dv - 2,
\end{aligned}
\tag{13.14}
$$

because $\operatorname{curl} \mathbf{u} = -E_x - 1$.

System (13.14) describes the nonlinear transport of the scalar function f by means of the divergence-free vector field $\mathbf{u} = (v, -E)$. The electron density f in system

(13.14) corresponds to the vorticity ω in Eqs. (13.9). Equations (13.9) preserve the condition of nonnegative vorticity, $\omega(x,t) \geq 0$ [$\omega(x,t) \leq 0$] provided that, at time zero, $\omega(x,t)|_{t=0} = \omega_0(x)$ is nonnegative (nonpositive). Likewise, the electron density f in Eqs. (13.9), if initially nonnegative, will remain so for all time. Moreover, the third equation in Eqs. (13.9), curl $\mathbf{v} = \omega$, is clearly analogous to the third equation in system (13.14). In this fashion, the physical structure of the single-component 1D V-P equations parallels that for the 2D Euler equations with nonnegative vorticity.

There is also a natural analog for the stream function formulation in Eqs. (13.10) as well as the related potential theory. Define the stream function

$$\Phi = \frac{1}{2}v^2 + \varphi, \tag{13.15}$$

where φ is the electric potential in V-P system (13.5); then as in Eqs. (13.10),

$$\mathbf{u} = -\nabla^{\perp}\Phi = \begin{pmatrix} \frac{\partial \Phi}{\partial v} \\ -\frac{\partial \Phi}{\partial x} \end{pmatrix} = \begin{pmatrix} v \\ -E \end{pmatrix},$$

$$\Phi_{xx} = 1 - \int_{-\infty}^{\infty} f(x,v,t)dv. \tag{13.16}$$

In the periodic setting, the analog of Biot-Savart law (13.11) for the single-component V-P system is the integral equation

$$E(x,t) = \int_0^L K(x,y)\left[1 - \int_{-\infty}^{\infty} f(y,v,t)dv\right]dy \tag{13.17}$$

with

$$K(x,y) = \begin{cases} \frac{y}{L}, & y < x \leq L \\ \frac{y}{L} - 1, & 0 \leq x < y \end{cases}.$$

Note that the kernel K in Eq. (13.17) is anisotropic and bounded but discontinuous; this is a milder singular kernel than occurs in Biot–Savart law (13.11) for the 2D Euler equations. This milder and simpler structure is one of the key features that makes the analysis of the one-component V-P system somewhat simpler than for the 2D Euler equations.

The two-component 1D V-P equations for a collisionless plasma of electrons and positively charged ions are

$$\begin{aligned} &\frac{\partial f_-}{\partial t} + v\frac{\partial f_-}{\partial x} - E(x,t)\frac{\partial f_-}{\partial v} = 0, \\ &\frac{\partial f_+}{\partial t} + v\frac{\partial f_+}{\partial x} + E(x,t)\frac{\partial f_+}{\partial v} = 0, \\ &\frac{\partial E}{\partial x} = \rho(x,t) = \int_{-\infty}^{\infty} (f_+ - f_-)(x,v,t)dv. \end{aligned} \tag{13.18}$$

To see clearly the analogy between the case of mixed-sign vorticity and the two-component V-P equations, we rewrite the vorticity-stream form of the 2D Euler equation by decomposing the vorticity into positive and negative parts:

$$\begin{aligned} \omega &= \omega_+ - \omega_-, & \omega_\pm &\geq 0, \\ \omega_+ &= \max(\omega, 0) & \omega_- &= -\min(\omega, 0). \end{aligned} \tag{13.19}$$

When this decomposition is used, the vorticity-stream form of the 2D Euler equation becomes a system of equations:

$$\begin{aligned} &\frac{\partial \omega_-}{\partial t} + \mathbf{v} \cdot \nabla \omega_- = 0, \\ &\frac{\partial \omega_+}{\partial t} + \mathbf{v} \cdot \nabla \omega_+ = 0, \\ &\mathbf{v}(\mathbf{x}, t) = \int_{\mathbb{R}^2} K(\mathbf{x} - \mathbf{y})(\omega_+ - \omega_-)(\mathbf{y}, t) d\mathbf{y}. \end{aligned} \tag{13.20}$$

Notice that system (13.20) has the same structure as V-P equations (13.18) in which the divergence-free vector field $\mathbf{v}$ in system (13.20) has the same role as the divergence-free $[v, \pm E(x, t)]$. The third equation in Eqs. (13.18) yields a kind of Biot–Savart law for the two-component Vlasov–Poisson system:

$$E(x, t) = \int_0^L \int_{-\infty}^{\infty} K(x, y)(f_+ - f_-)(y, v, t) dv dy, \tag{13.21}$$

where

$$K(x, y) = \begin{cases} y/L, & y < x \leq L \\ y/L - 1, & 0 \leq x < y \end{cases} \tag{13.22}$$

in the periodic setting. Problems (13.20) and and (13.18) have analogous structure as nonlinear transport equations.

13.1.2. Vortex Patches and Electron Patches

In Chap. 8 we discussed the general theory for weak solutions to the 2D Euler equations with initial vorticity $\omega_0 \in L^\infty \cap L^1$. We then concentrated on a special example of such a weak solution, called a vortex patch, in which the vorticity is the characteristic function of a region in the plane,

$$\omega(x, t) = \begin{cases} \omega_0, & x \in \Omega(t) \\ 0, & x \notin \Omega(t) \end{cases},$$

where $\Omega(t)$ is the patch that evolves with the fluid velocity, determined from Biot–Savart law (13.11).

There we showed that the vortex-patch boundary, if initially smooth, would remain smooth for all time.

By analogy, we can construct a weak solution to single-component V-P system (13.5) in which the electron-density function f satisfies $f \in L^\infty\{[0, \infty), L^\infty(S^1)\}$. Specifically we consider the case in which f is constant on bounded open domains of the x, v phase plane. Physically we do not have a "patch of electrons," as the patch is really in a phase plane. However, what this means is that for fixed-position x, the electron-density function f is constant for a range of velocities $v \in [v_-(x), v_+(x)]$ and otherwise zero.

Dziurzynski (1987) showed that the particle paths, defined as solutions of the ODE

$$\frac{dX}{dt} = V(X, t),$$

where $V(x, v, t) = [v, -E(x, t)]$, are Lipschitz continuous in space. He used this to show that an electron patch with an initially Lipschitz continuous boundary would evolve to retain a Lipschitz continuous boundary for all time.

By analogy with CDE (8.56) in Chap. 8, we can derive a contour dynamics formulation of the boundary of an electron patch. In fact, it was the contour dynamics formulation, presented in Berkand Roberts (1970), of the single-component V-P system that motivated the derivation, in Zabusky et al. (1979), of the CDE for the motion of a vortex-patch boundary.

In his Ph.D. dissertation, Dziurzynski shows that an electron-patch layer with an initially C^∞ boundary can develop into a curve that loses C^3 smoothness at a finite time. This is in direct contrast to the vortex patch, which retains all higher-order smoothness for all time (see Subsection 8.3.3).

The focus of this chapter is on weak solutions to the V-P system with electron-sheet initial data (described below – the analogous case to vortex-sheet initial data for the 2D Euler equation). For a detailed discussion of the theory of electron patches, the reader is referred to Dziurzynski (1987).

13.1.3. Vortex Sheets and Electron Sheets

In Chap. 8 we introduced the classical vortex sheet Birkhoff (1962), a more singular kind of weak solution to the 2D Euler equation than the vortex patch, in which the vorticity is concentrated as a measure (delta function) along a curve C. We analyzed the evolution from initial data

$$\omega_{t=0} = \omega_0 = \gamma\delta(|\mathbf{x} - C|), \tag{13.23}$$

where $\gamma(C)$ is a given density chosen so that ω_0 has finite total mass and also induces a velocity with locally finite kinetic energy. The vortex-sheet initial data are nonnegative provided that the surface density, $\gamma(C)$, satisfies $\gamma(C) \geq 0$.

There is a natural analog of nonnegative vortex-sheet initial data for the single-component V-P equations. Consider an initial electron density f defined by a surface measure supported on a smooth curve C in the (x, v) plane:

$$f(x, v, t)|_{t=0} = f_0(x, v) = \gamma\delta(|\mathbf{x} - C|), \tag{13.24}$$

with $\mathbf{x} = (x, v)$ and the density $\gamma(C) \geq 0$ arranged so that

$$\frac{1}{L}\int_0^L \int_{-\infty}^{\infty} df_0(x, v) = 1. \tag{13.25}$$

By analogy, we refer to Eq. (13.24) as *electron-sheet initial data.*

We showed in Section 9.3 that for the 2D Euler equations, planar vortex sheets with nonnegative vorticity quickly develop a rich structure that is due to the *Kelvin–Helmholtz instability* in which the kth Fourier mode of a perturbed flat sheet has a component that grows like $e^{|k|t/2}$. This instability makes even single-sign vortex sheets a classical ill-posed problem (Caflisch and Orellana, 1989). Nevertheless, when the vorticity is nonnegative, numerical computations (Figs. 9.2 and 9.3) and actual experiments (Van Dyke, 1982) develop a complex but significantly more regular structure compared with vortex sheets of mixed sign (see, e.g., Figs. 9.4 and 9.5). As we discussed in detail in Chap. 9, asymptotic analysis and careful numerical simulations predict that singularities form in finite time for vortex sheets with nonnegative vorticity. What we do not know, however, is whether there are nonunique continuations of the sheet past the singularity time.

The single-component 1D V-P equations with electron-sheet initial data (13.24) have a simpler behavior. Dziurzynski (1987) found explicit electron-sheet solutions to V-P system (13.5). Unlike vortex sheets for a 2D fluid flow, these explicit solutions do not exhibit catastrophic linearized instability and hence lend themselves much better to high-resolution numerical computation of electron sheets. Simple examples of electron-sheet motion are known to develop singularities in finite time (Dziurzynski, 1987) and suitable weak solutions have recently been constructed rigorously for all time (Zheng and Majda, 1994). In Section 13.2 we discuss these issues as well as the spontaneous formation of concentrations for the electron-sheet initial data in the single-component case.

13.1.4. Point Vortices for 2D Fluid Flow and Concentrated Electron Densities

Recall from Subsection 9.4.1 that the first attempts at understanding weak solutions to the 2D Euler equations with vortex-sheet initial data considered a superposition of point vortices (Rosenhead, 1932; Chorin and Bernard, 1973):

$$\omega = \sum_{j=1}^{N} \Gamma_j \delta[\mathbf{z} - \mathbf{x}_j(t)], \tag{13.26}$$

where $\mathbf{x}_j(t)$, $1 \leq j \leq N$ satisfy ODEs (9.18). The vortex distribution in Eq. (13.26) is nonnegative provided that the constants Γ_j satisfy

$$\Gamma_j > 0. \tag{13.27}$$

Despite all the qualitative insight that can be gained from such formal point-vortex solutions, as we discussed in Subsection 9.4.1, their theoretical significance as weak solutions to the 2D Euler equations is an unresolved issue because they have infinite local kinetic energy.

In analogy with point vortices, we can consider concentrated electron densities with the form

$$f(x, v, t) = \sum_{j=1}^{N} \alpha_j \delta[\mathbf{x} - \mathbf{x}_j(t)], \qquad \alpha_j > 0, \quad \sum_{j=1}^{N} \alpha_j = 1, \tag{13.28}$$

and $\mathbf{x} = (x, v)$ as candidates for weak solutions of one-component V-P equations (13.5) provided that the $\mathbf{x}_j(t)$ evolve according to the particle-trajectory equations $(d\mathbf{x}/dt) = \mathbf{u}$, where $\mathbf{u}$ is defined by Eq. (13.13). Once again there is simpler but parallel behavior for the analogous solutions in Eq. (13.28) for single-component 1D V-P equations (13.5) compared with the 2D Euler equations with nonnegative vorticity. Concentrated electron densities (13.28) induce rigorous weak solutions of the single-component V-P equations according to the recent theory developed by Zheng and Majda (1994). Furthermore, these weak solutions, in contrast to the point-vortex examples from fluid flow, have locally finite kinetic energy, i.e.,

$$\int_0^L \left[\int_{-\infty}^{\infty} \frac{1}{2} v^2 f(x, v, t) dv + \frac{1}{2} E^2(x, t) \right] dx < \infty.$$

Later in the next Section 13.2, a specific superposition of concentrated electron densities demonstrates explicitly the nonuniqueness of weak solutions.

13.1.5. Concentration, Approximate-Solution Sequences, and the Analogy in Nonlinear Structure

The study (either empirically or analytically) of vortex sheets always involves some kind of smooth approximation because of the very singular nature of the problem. In Chap. 10 we first introduced the notion of an *approximate-solution sequence* for the 2D Euler equation.

Definition 10.2. *Approximate Solution Sequence for the Euler Equation. A sequence of functions $v^\epsilon \in C\{[0, T], L^2_{\text{loc}}(\mathbb{R}^2)\}$ is an approximate-solution sequence for the 2D Euler equation if*

(i) For all $R, T > 0$, $\max_{0 \le t \le T} \int_{|x| \le R} |v^\epsilon(x, t)|^2 dx \le C(R)$, independent of ϵ,
(ii) $\text{div } v^\epsilon = 0$ in the sense of distributions,
(iii) (weak consistency with the Euler equation):

$$\lim_{\epsilon \to 0} \int_0^T \int_{\mathbb{R}^2} (v^\epsilon \cdot \Phi_t + \nabla \Phi : v^\epsilon \otimes v^\epsilon) dx dt = 0$$

for all test function $\Phi \in C_0^\infty[(0, T) \times \mathbb{R}^2]$ with $\text{div } \Phi = 0$,
(iv) L^1 vorticity control $\max_{0 \le t \le T} \int |\omega^\epsilon(x, t)| dx \le C(T)$, $\omega^\epsilon = \text{curl } v^\epsilon$ and a technical condition necessary for assuming the initial data

$$\|\rho v^\epsilon(t_1) - \rho v^\epsilon(t_2)\|_{-L} \le C|t_1 - t_2|,$$

$$0 \le t_1, \quad t_2 \le T \quad \forall L > 0, \quad \forall \rho \in C_0^\infty(\mathbb{R}^N) \tag{13.29}$$

i.e., $\{v^\epsilon\}$ is uniformly bounded in $\text{Lip}\{[0, T], H^{-L}_{\text{loc}}(\mathbb{R}^N)\}$.

Primary examples of approximate-solution sequences for 2D Euler are given by (1) smoothing the initial data (Example 10.1), (2) the zero diffusion limit of Navier–Stokes solutions (Example 10.2), and (3) computational vortex methods (Liu and Xin, 1995). The first two cases were discussed in detail in Chaps. 10 and 11.

Analogously, a sequence of smooth electron densities $f_\epsilon(x, v, t) \geq 0$ together with the corresponding electric fields $E_\epsilon(x, v, t)$ satisfying

$$\frac{\partial}{\partial x}(E_\epsilon) = 1 - \int_{-\infty}^{\infty} f_\epsilon(x, v, t)dv, \tag{13.30}$$

define an approximate-solution sequence for the single-component 1D V-P equation provided that the $f_\epsilon(x, v, t) \geq 0$ satisfy the uniform bounds

$$\int_0^L \int_{-\infty}^{\infty} f_\epsilon(x, v, t)dvdx = 1,$$

$$\max_{0 \leq t \leq T} \int_0^L \int_{-\infty}^{\infty} v^2 f_\epsilon(x, v, t)dvdx \leq C_T \qquad \text{for any} \quad T > 0 \tag{13.31}$$

and weak consistency with 1D V-P equations (13.5) in the limit as $\epsilon \to 0$,

$$\int_0^T \int_0^L \int_{-\infty}^{\infty} [\varphi_t(x, v, t) f_\epsilon + \varphi_\epsilon v f_\epsilon - \varphi_\epsilon E_\epsilon f_\epsilon]dvdxdt \to 0, \tag{13.32}$$

for all smooth test functions φ with compact support. Note that in the case of V-P equations, we use the transport equation (analogous to the vorticity equation for the 2D Euler equtaion) for f to define weak consistency, whereas in the case of an approximate-solution sequence for the 2D Euler equation, we use the primitive-variable form of the equation for weak consistency.

Similarly we can define an approximate-solution sequence for the two-component case. We discuss this in more detail in Section 13.3 (see in particular Definition 13.1).

A natural candidate for an approximate-solution sequence for the single-component V-P equations is the zero noise limit as $\epsilon \to 0$ of Fokker–Planck regularization (13.8). The Fokker–Planck regularization of the single-component V-P equation converges to suitable weak solutions of system (13.5) (Zheng and Majda, 1994). This result is analogous to Theorem 11.4 recently established for Navier–Stokes solutions with nonnegative vorticity and vortex-sheet initial data.

Another type of approximate-solution sequence for the 1D V-P equations uses computational particle methods (see Subsection 13.1.7). In Sections 13.2 and 13.3 we summarize recent computational results for the one-component and the two-component systems, respectively, as empirical evidence for the behavior of approximate-solution sequences of the V-P problem. This is in the same spirit as that of the use of vortex methods for the study of very weak solutions of the 2D Euler equations.

13.1.6. Concentration Effects

Recall from Subsection 11.4.1 that in the proof of existence of weak solutions to the 2D Euler equation with vortex-sheet initial data of distinguished sign, in order for an approximate-solution sequence to converge to a weak solution of the 2D Euler equations, besides the uniform kinetic-energy bound, it is necessary that the off-diagonal terms of the tensor product $\mathbf{v}^\epsilon \otimes \mathbf{v}^\epsilon$ converge weakly to the appropriate limiting values $\mathbf{v} \otimes \mathbf{v}$. In Section 11.2 we showed some explicit examples of steady exact-solution sequences for the 2D fluid equations in which concentration occurs in the limit as $\epsilon \to 0$ and a finite amount of kinetic energy concentrates on a set of measure zero. They included Example 11.1 (Concentration of Phantom Vortices) and Example 11.2 (Concentration in Positive Vorticity).

There is a simple analog of concentration for the single-component V-P equations. From Eqs. (13.30) and (13.31) and the fact that $f_\epsilon \geq 0$, it follows immediately that $E_\epsilon(x, t)$ satisfies the a priori estimate,

$$\max_{\substack{0 \leq t \leq +\infty \\ 0 \leq x < L \\ 0 < \epsilon \leq \epsilon_0}} |E_\epsilon(x, t)| \leq 1 + \frac{L}{2}. \tag{13.33}$$

This estimate and relation (13.31) guarantee that $E_\epsilon f_\epsilon$ is uniformly bounded in L^1:

$$\max_{\substack{0 \leq t < +\infty \\ 0 < \epsilon < \epsilon_0}} \int_0^L \int_{-\infty}^{\infty} |E_\epsilon| f_\epsilon(t) dv dx \leq C. \tag{13.34}$$

Formula (13.32) suggests that the weak convergence of $E_\epsilon f_\epsilon$ to the correct limiting value Ef would define the pair (E, f) as a weak solution for the 1D V-P equation. As long as the limiting value $\bar{E}$ is continuous then the limit is well defined. Charge concentration occurs for the single-component 1D V-P equations precisely when E_ϵ converges to a discontinuous function,

$$\begin{aligned} \rho_\epsilon(x, t) &= 1 - \int_{-\infty}^{\infty} f_\epsilon(x, v, t) dv \\ &\text{as} \quad e \to 0, \\ \rho_\epsilon(x, t) \to \bar{\rho} &= \rho_{\mathrm{AC}}(x, t) + \sum c_j \delta[x - x_j(t)] \end{aligned} \tag{13.35}$$

with $c_j \neq 0$ for some j. In the third line of relation (13.35), $\rho_{\mathrm{AC}}(x, t)$ is a function which is absolutely continuous with respect to Lebesgue measure so that the corresponding contribution to ρ is continuous. In this situation with charge concentration, because $c_j \neq 0$, it is unclear if

$$E_\epsilon f_\epsilon \to \bar{E} f$$

for a suitable definition of $\bar{E}$ because $\bar{E}$ is discontinuous and f contains singular delta functions at exactly the same spatial x locations. This charge-concentration phenomenon for the one-component V-P equations is the analog of the fluid-concentration effect for the 2D Euler equation. The role of the nonlinear product Ef corresponds to the nonlinear terms $\mathbf{v} \otimes \mathbf{v}$, and the duality of L^1 and L^∞ with estimates (13.33) and (13.34) for Ef corresponds to the duality of L^2 and L^2 with uniform local kinetic-energy bound for v. Charge concentration (13.35) leads to similar subtle issues regarding weak solutions as do the concentration effects for 2D Euler. In Section 13.2

we give explicit examples of the spontaneous development of charge concentration in evolving electron sheets for the single-component V-P equations.

13.1.7. Computational Particle Methods for the 1D Vlasov–Poisson and Fokker–Planck Equations

Recall from Chap. 6 the vortex method for computing solutions to the Euler and the Navier–Stokes equations. This method exploits the transport structure of vorticity equation (13.9) by representing the vorticity as a finite collection of blobs.

The computational particle method was first analyzed by Cottet and Raviart (1984) as an analogy to computational vortex methods for the Euler and the Navier–Stokes equations (Beale and Majda, 1982a, 1982b).

The particle approximation of the V-P equations assumes that the electron density is a superposition of a finite number of particles as in electron densities (13.28):

$$f(x, v, t) = \sum_{j=1}^{N} \alpha_j \delta[\mathbf{x} - \mathbf{x}_j(t)], \qquad \alpha_j > 0, \quad \sum_{j=1}^{N} \alpha_j = 1. \tag{13.36}$$

The weights α_j are determined from the initial data and the phase-space coordinates $\mathbf{x}_j$ satisfy the approximate particle-trajectory equations

$$\frac{d\mathbf{x}_j}{dt} = \mathbf{u}_j = [v_j, -\tilde{E}(x_j, t)]^t, \qquad j = 1, \dots, N,$$

with approximate electric field

$$\tilde{E}(x, t) = C \sum_{j=1}^{N^*} K_\delta(x, x_j^*) - \sum_{j=1}^{N} \alpha_j K_\delta[x, x_j(t)].$$

Here $C = \sum_{j=1}^{N} \alpha_j / N^*$ is a constant chosen so that the discrete version of the constraint $\int_0^L E(x, t) dx = 0$ is always satisfied. The kernel $K_\delta(x, y) = K(x, y) * \zeta_\delta(y) = \int_{-\infty}^{\infty} K(x, y - \eta) \zeta_\delta(\eta) d\eta$ is a mollified kernel, where η_δ is an appropriate mollifier, chosen so that K_δ is smooth. One such choice is

$$\zeta = \frac{1}{\delta} \zeta\left(\frac{y}{\delta}\right) \quad \text{with} \quad \zeta = \begin{cases} 1, & -\frac{1}{2} \le y \le \frac{1}{2} \\ 0, & \text{else} \end{cases},$$

which yields the mollified kernel

$$K_\delta(x, y) = \begin{cases} \frac{y}{L}, & y - x \le -\frac{\delta}{2} \\ \frac{y}{L} - \frac{1}{2} - \frac{(y-x)}{\delta}, & -\frac{\delta}{2} \le y - x \le \frac{\delta}{2}. \\ \frac{y}{L} - 1, & \frac{\delta}{2} \le y - x \end{cases} \tag{13.37}$$

The points x_j^*, $j = 1, N^*$ are uniformly spaced on $[0, L]$ and represent fixed particles (ions) of positive charge.

For the two-component case, we approximate the electron and positive-ion densities by superpositions of a finite number of particles:

$$f_+(x, v, t) = \sum_{j=1}^{N^+} \alpha_j^+ \delta[\mathbf{x} - \mathbf{x}_j^+(t)],$$

$$f_-(x, v, t) = \sum_{j=1}^{N^-} \alpha_j^- \delta[\mathbf{x} - \mathbf{x}_j^-(t)],$$

$$\mathbf{x} = (x, v).$$

The weights $\alpha_j^\pm$ satisfy $\alpha_j^\pm > 0$ and the condition of zero total charge:

$$\sum_{j=1}^{N^+} \alpha_j^+ = \sum_{j=1}^{N^-} \alpha_j^-.$$

The phase-space coordinates $\mathbf{x}_j^\pm$ satisfy the approximate particle-trajectory equations

$$\frac{d\mathbf{x}_j^\pm}{dt} = \mathbf{u}_j^\pm = [v_j^\pm, \pm\tilde{E}(x_j, t)]^t, \qquad j = 1, \ldots, N^\pm,$$

with initial data $x_j^\pm(0) = ({}^0x_j^\pm, {}^0v_j^\pm)$ and where $\tilde{E}$ is approximate to the electric field:

$$\tilde{E}(x, t) = \sum_{j=1}^{N^+} \alpha_j^+ K_\delta(x, x_j^+) - \sum_{j=1}^{N_-} \alpha_j^- K_\delta(x, x_j^-).$$

One possible choice of K_δ is the above mollified kernel for the one-component case.

To approximate (pseudo-) Fokker–Planck equations (13.8) we use a random-particle method that combines a first-order operator splitting procedure with a random-walk solution of the diffusion equation in regularization (13.8). The general idea is identical to that of the random-vortex method discussed in detail in Chap. 6. For details on the implementation of these methods, the reader is referred to Majda et al. (1994a, 1994b).

13.1.8. Summary of Analogy between Vlasov–Poisson and 2D Euler Equations

We finish this section on the analogy between the V-P system and the 2D Euler equations by summarizing the connections in Table 13.1.

13.2. The Single-Component 1D Vlasov–Poisson Equation

Recall from V-P system (13.5) that the one-component 1D V-P system for an electron-density function f satisfying

$$\frac{1}{L}\int_0^L f(x, v, t)dxdv = 1.$$

Table 13.1. *Connections between the V-P system and 2D Euler equations*

Property	V-P System	2D Euler Equations
Nonlinear transport	$\partial f/\partial t + \mathbf{u}\cdot\nabla f = 0$ $\operatorname{div}\mathbf{u} = 0$, $\mathbf{u} = (v, -E)$ $\operatorname{curl}\mathbf{u} = \int_{-\infty}^{\infty} f(x, v, t)dv - 2$	$\partial\omega/\partial t + \mathbf{v}\cdot\nabla\omega = 0$ $\operatorname{div}\mathbf{v} = 0$ $\operatorname{curl} v = \omega$
Stream function	$\Phi = \frac{1}{2}v^2 + \varphi$ $\mathbf{u} = -\nabla^{\perp}\Phi$	$\Delta\Psi = \omega$ $\mathbf{v} = \nabla^{\perp}\Phi$
Mixed sign	$\partial f_{\pm}/\partial t + \mathbf{u}_{\pm}\cdot\nabla f_{\pm} = 0$ $\mathbf{u}_{\pm} = (v, \pm E)$	$\partial\omega_{\pm}/\partial t + \mathbf{v}\cdot\nabla\omega_{\pm} = 0$ $\omega = \omega_+ - \omega_-$
Biot–Savart	$E = \int_0^L\int_{-\infty}^{\infty} K(x, y)(f_+ - f_-)(y, v, t)dvdy$ $K(x, y) = y/L\ (y < x \le L),\ y/L - 1$ (else)	$v = K * \omega$ $K(x) = \frac{1}{2\pi}\nabla^{\perp}(\log\lvert x\rvert)$
L^{∞} solution	Electron patch	Vortex patch
δ-fn solution	Electron sheet	Vortex sheet
Numerical method	Particle method	Vortex method
Dual pairing	$E\cdot f$ $L^{\infty}\leftrightarrow L^1$	$\mathbf{v}\otimes\mathbf{v}$ $L^2\leftrightarrow L^2$

is

$$\begin{aligned}\frac{\partial f}{\partial t} + v\frac{\partial f}{\partial x} - E(x, t)\frac{\partial f}{\partial v} &= 0,\\ \frac{\partial^2\varphi}{\partial x^2} = \rho &= 1 - \int_{-\infty}^{\infty} f(x, v, t)dv,\\ E &= \frac{\partial\varphi}{\partial x}.\end{aligned} \tag{13.38}$$

The existence of suitable weak solutions to V-P system (13.38) has recently been established by Zheng and Majda (1994).

Recall that the electron density $f(x, v, t)$ is a nonnegative probability measure with

$$\frac{1}{L}\int_0^L\int_{-\infty}^{\infty} f(x, v, t)dvdx = 1. \tag{13.39}$$

When charge concentration occurs, the nonlinear product Ef can have an ambiguous interpretation because $E(x, t)$ might be discontinuous precisely where the measure $f(x, v, t)$ is singular. For a weak solution of V-P system (13.38), the electric field satisfies

$$E_x = 1 - \int_{-\infty}^{\infty} f\, dv, \qquad \int_0^L E\, dx = 0. \tag{13.40}$$

Because f is a nonnegative measure, the first equation in conditions (13.40) together with estimate (13.33) guarantees that $E(x, t)$ is a bounded function of bounded variation in space–time (see Zheng and Majda, 1994), i.e., $E \in (BV\cap L^{\infty})_{\text{loc}}([0, L]\times\mathbb{R}^+)$.

The weak form of the first equation in system (13.38) is

$$\int_0^{+\infty}\int_0^L\int_{-\infty}^{\infty}(\varphi_t f + \varphi_x v f)dvdxdt$$
$$- \int_0^{+\infty}\int_0^L \bar{E}(x,t)\left[\int_{-\infty}^{\infty}\varphi_v f(dv)\right]dxdt = 0 \qquad (13.41)$$

for all test functions $\varphi \in C_0^\infty(S_L^1 \times \mathbb{R} \times \mathbb{R}^+)$. Although the weak form of the linear terms $\int_0^{+\infty}\int_0^L\int_{-\infty}^{\infty}(\varphi_t f + \varphi_x v f)dvdxdt$ in Eq. (13.41) is obvious, the meaning of the term

$$\int_0^{+\infty}\int_0^L \bar{E}(x,t)\left[\int_{-\infty}^{\infty}\varphi_v f(dv)\right]dxdt \qquad (13.42)$$

requires further interpretation because this involves the nonlinear product Ef. Zheng and Majda (1974) prove that the velocity average

$$\int_{-\infty}^{\infty}\varphi_v f(dv)$$

for any test function ρ is the partial derivative with respect to x of a function $g_\varphi(x,t)$ with finite space–time bounded variation. Thus we interpret Ef as a measure $\bar{E}f$, where $\bar{E}(t,x)$ is Volpert's symmetric average (Volpert, 1967; Volpert and Hudjaev, 1985):

$$\bar{E}(x,t) = \begin{cases} E(x,t), & \text{if } E \text{ is approximately continuous at } (x,t) \\ \frac{1}{2}[E_l(x,t) + E_r(x,t)], & \text{if } E \text{ has a jump at } (x,t) \end{cases}. \qquad (13.43)$$

Here $E_l(x,t)$ and $E_r(x,t)$ denote respectively the left and the right limits of $E(x,t)$ at a discontinuity line – such limits are automatically guaranteed by Volpert's bounded-variation calculus (Volpert, 1967; Volpert and Hudjaev, 1985). The existence of weak solutions of system (13.38) satisfying Eqs. (13.40) and (13.41) in this precise sense for an arbitrary nonnegative initial measure f satisfying Eq. (13.39) is shown in Zheng and Majda (1994). Later in this section we apply this theory of weak solutions for arbitrary measures to the construction of explicit weak solutions defined by concentrated electron densities as described in Eqs. (13.28).

13.2.1. Singularity Formation in Regular Electron Sheets

In this section we present some simple explicit electron-sheet solutions of system (13.38) that can be derived through an exact-solution formula of Dziurzynski (1987). These solutions correspond to velocity perturbations of uniform electron sheets. These solutions demonstrate that even though the initial electron sheet is smooth and perturbed electron sheets do not undergo a Kelvin–Helmholtz instability, smooth electron sheets can still form singularities in a finite time. It is even possible to develop charge concentration [see Eq. (13.35)] in finite time from these weak solutions. We use the computational particle method described in the previous section to examine properties of the solutions past their singularity time.

A uniformly flat electron sheet, the analog of the flat vortex sheet in Eq. (9.12), is a flat curve C in the x, v plane:

$$\begin{aligned} C(\alpha) &= [x(\alpha), v(\alpha)] = (\alpha, 0), \\ f(\mathbf{x}) &= \delta[|\mathbf{x} - C(\alpha)|], \qquad \text{for} \quad 0 < \alpha < 1, \\ E[\mathbf{x}(\alpha)] &\equiv 0. \end{aligned} \tag{13.44}$$

We note that, like the flat vortex sheet in Chap. 8, this elementary weak solution is steady.

Perturbing the velocity field in the uniform electron sheet at time $t = 0$ produces an electron sheet with more interesting dynamics. The initial conditions for this perturbed uniform electron sheet are

$$\begin{aligned} C(\alpha, 0) &= [x(\alpha, 0), v(\alpha, 0)] = [\alpha, g(\alpha)], \\ f(\mathbf{x}, 0) &= \left| \frac{dC(\alpha, 0)}{d\alpha} \right|^{-1} \delta[|\mathbf{x} - C(\alpha, 0)|], \qquad \text{for} \quad 0 \le \alpha \le 1. \end{aligned} \tag{13.45}$$

These initial conditions are the direct analog of a perturbed uniform vortex sheet, Eq. (9.13), which can evolve to form complex structures (see, e.g., Fig. 9.3). The remarkable fact discovered by Dziurzynski (1987) is that problem (13.38) with initial conditions (13.45) has an explicit solution. The derivation of the solution uses a nonlinear differential–integral equation for the curve $C(\alpha, t)$, the analog of Birkhoff–Rott equation (9.11) for vortex sheets. The equation for $C(\alpha, t)$ is linear as long as $C(\alpha, t)$ is the graph of a function in $x - v$ space. The resulting linear equation can be easily solved by Fourier series to obtain the following result.

Theorem 13.1. *(Dziurzynski, 1987.) Exact Weak Solutions for Perturbed Uniform Electron Sheets. Consider electron-sheet initial data (13.45), where* $g(\alpha)$ *is smooth,* $g(0) = g(1) = 0$, *and* $\int_0^1 g(\alpha) d\alpha = 0$. *Then for* $0 < t < \tilde{t}$ *such that* $dx(\alpha, t)/d\alpha > 0$, *there is an exact weak solution of problem (13.38) defined by*

$$\begin{aligned} C(\alpha, 0)[x(\alpha, 0), v(\alpha, 0)] &= [\alpha, g(\alpha)], \\ f(\mathbf{x}, t) &= \left| \frac{dC(\alpha, t)}{d\alpha} \right|^{-1} \delta[|\mathbf{x} - C(\alpha, t)|], \qquad \text{for} \quad 0 \le \alpha \le 1, \\ E[x(\alpha, t)] &= g(\alpha) \sin t. \end{aligned} \tag{13.46}$$

Explicit formulas (13.46) demonstrate that the one-component V-P equation with electron-sheet initial data does not suffer a Kelvin–Helmholtz instability. If $|g'(\alpha)| < 1$ for all $\alpha \in [0, 1]$, then $\tilde{t} = +\infty$, i.e., $dx(\alpha, t)/d\alpha > 0$ for all time, and solution (13.46) is a global weak solution. Also, a simple calculation shows that

$$[x(\alpha, t) - x(\alpha, 0)]^2 + [v(\alpha, t)]^2 = [g(\alpha)]^2,$$

which implies that the particles on the electron sheet are confined to circles in the x–v plane. Thus the perturbed uniform electron sheets are stable with respect to finite-amplitude velocity perturbations $g(\alpha)$, as long as $g(\alpha)$ satisfies $|g'(\alpha)| < 1$ for $\alpha \in [0, 1]$, so perturbed electron sheets do not undergo a Kelvin–Helmholtz instability and the V-P electron-sheet equation is linear well posed.

One might believe that the lack of a Kelvin–Helmholtz instability would imply that solution (13.46) cannot develop spontaneous singularities. However, for larger-amplitude perturbations, these weak solutions can spontaneously develop singularities in the charge at some time $\tilde{t} > 0$. A simple calculation shows that on electron-sheet solution (13.46), the charge is given by

$$\rho[x(\alpha, t)] = 1 - \left[\frac{dx(\alpha, t)}{d\alpha}\right]^{-1}. \tag{13.47}$$

Consequently, if there exists a time $\tilde{t}$ and a value of α, $\tilde{\alpha}$ such that $dx(\tilde{a}, \tilde{t})/d\alpha$ then the charge becomes singular.

We consider two particular choices of $g(\alpha)$ to illustrate ways the charge can become singular. Example 13.1 shows that the charge can form an algebraic singularity as time evolves. Example 13.2 shows that the charge can form an even stronger singularity in time, namely, a Dirac delta distribution that corresponds to the charge-concentration effect described in Eq.(13.35).

Example 13.1. Algebraic Singularity in the Charge. Consider an electron sheet with data (13.45) consisting of a sinusoidally perturbed uniform electron sheet in which

$$g(\alpha) = \epsilon \sin(2\pi j\alpha), \qquad \epsilon > 0, \quad j \in \mathbb{N}, \tag{13.48}$$

where $\epsilon > 0$ and j is a positive integer. If $|2\pi j\epsilon| < 1$, then the solution of V-P system (13.38), defined for all $t > 0$, is

$$\begin{aligned} C(\alpha, t) &= [\alpha + \epsilon \sin(2\pi j\alpha) \sin t, \epsilon \sin(2\pi j\alpha) \cos t], \\ f(\mathbf{x}, t) &= \frac{\delta[|\mathbf{x} - C(\alpha, t)|]}{[1 + 2\pi j\epsilon \cos(2\pi j\alpha) \sin t]^2 + [2\pi j\epsilon \cos(2\pi j\alpha) \cos t]^2}, \\ E[x(\alpha, t)] &= \epsilon \sin(2\pi j\alpha) \sin t \end{aligned} \tag{13.49}$$

for $0 \le \alpha \le 1$. However, if $|2\pi j\epsilon| > 1$, then $\{[dx(\alpha, t)]/d\alpha\} = 1 + 2\pi j\epsilon \cos(2\pi j\alpha)$ $\sin t = 0$ when $\alpha = \tilde{\alpha} = 1/2j$ and $t = \tilde{t}$ with $\sin\tilde{t} = [1/(2\pi j\epsilon)]$. In this case, solution (13.49) is defined for $0 \le t < \tilde{t}$. Furthermore, if we do a Taylor expansion of $\cos(2\pi j\alpha)$ about $\alpha = \tilde{\alpha}$, we find that

$$\frac{dx(\alpha, \tilde{t})}{d\alpha} = (2\pi j)^2(\alpha - \tilde{\alpha})^2 \sin\tilde{t} + \mathcal{O}[(\alpha - \tilde{\alpha})^2]$$

so charge (13.47) forms an algebraic singularity at $\alpha = \tilde{\alpha}$ and $t = \tilde{t}$.

In Example 13.1, at time $\tilde{t}$ we have $dx/d\alpha = 0$ at an isolated point $\tilde{\alpha}$. This leads to an algebraic singularity in the charge. If we can construct an example such that $dx/d\alpha = 0$ for an interval $\alpha = [\alpha_1, \alpha_2]$ with nonzero measure, then it is possible that the charge could form an even stronger singularity. We show that this is possible in the next example.

Example 13.2. Dirac Delta Distribution in the Charge. Consider perturbed electron-sheet initial data (13.45) with $g(\alpha)$ defined in the following way:

For some $\alpha_1 \in (0, 1/4)$ and constant $c > 1$, set

$$g(\alpha) = \begin{cases} c\alpha, & 0 \le \alpha \le \alpha_1 \\ c\alpha_1, & \alpha_1 \le \alpha \le \frac{1}{2} - a_1 \\ c(\frac{1}{2} - \alpha), & \frac{1}{2} - \alpha_1 \le \alpha \le \frac{1}{2} + \alpha_1. \\ -c\alpha_1, & \frac{1}{2} + \alpha_1 \le \alpha \le \frac{1}{2} - \alpha_1 \\ c(\alpha - 1), & 1 - \alpha_1 \le \alpha \le 1 \end{cases}$$

The function $g(\alpha)$ satisfies $g(0) = g(1) = 0$ and $\int_0^1 g(\alpha) d\alpha = 0$. We can use formula (13.46) to write the solution on each subinterval in piecewise fashion:

On $0 \le \alpha \le \alpha_1$,

$$\begin{aligned} C(\alpha, t) &= (\alpha + c\alpha \sin t, c\alpha \cos t), \\ f(\mathbf{x}, t) &= \left[(1 + c \sin t)^2 + (c \cos t)^2\right]^{-1} \delta[|\mathbf{x} - C(\alpha, t)|], \\ E(x(\alpha, t)) &= c\alpha \sin t. \end{aligned}$$

On $\alpha_1 \le \alpha \le \frac{1}{2} - \alpha_1$,

$$\begin{aligned} C(\alpha, t) &= (\alpha + c\alpha_1 \sin t, c\alpha_1 \cos t), \\ f(\mathbf{x}, t) &= \delta[|\mathbf{x} - C(\alpha, t)|], \\ E(x(\alpha, t)) &= c\alpha_1 \sin t. \end{aligned}$$

On $\frac{1}{2} - \alpha_1 \le \alpha \le \frac{1}{2} + \alpha_1$,

$$\begin{aligned} C(\alpha, t) &= \left[\alpha + c\left(\frac{1}{2} - \alpha\right) \sin t, c\left(\frac{1}{2} - \alpha\right) \cos t\right], \\ f(\mathbf{x}, t) &= \left[(1 - c \sin t)^2 + (c \cos t)^2\right]^{-1} \delta[|\mathbf{x} - C(\alpha, t)|], \\ E(x(\alpha, t)) &= c\left(\frac{1}{2} - \alpha\right) \sin t. \end{aligned}$$

On $\frac{1}{2} + \alpha_1 \le \alpha \le \frac{1}{2} - \alpha_1$,

$$\begin{aligned} C(\alpha, t) &= (\alpha - c\alpha_1 \sin t, c\alpha_1 \cos t), \\ f(\mathbf{x}, t) &= \delta[|\mathbf{x} - C(\alpha, t)|], \\ E(x(\alpha, t)) &= -c\alpha_1 \sin t. \end{aligned}$$

On $1 - \alpha_1 \le \alpha \le 1$,

$$\begin{aligned} C(\alpha, t) &= [\alpha + c(\alpha - 1) \sin t, c(\alpha - 1) \cos t], \\ f(\mathbf{x}, t) &= [(1 + c \sin t)^2 + (c \cos t)^2]^{-1} \delta[|\mathbf{x} - C(\alpha, t)|], \\ E(x(\alpha, t)) &= c(\alpha - 1) \sin t. \end{aligned}$$

Consider the middle segment where $x(\alpha, t) = \alpha + c(\frac{1}{2} - \alpha)\sin t$ or $\frac{1}{2} - \alpha_1 \leq \alpha \leq \frac{1}{2} + \alpha_1$. By assumption, $c > 1$, so $dx/d\alpha = 1 - c \sin t = 0$ has a solution $\tilde{t}$ with $\tilde{t} > 0$. Furthermore, for all $\alpha \in [\frac{1}{2} - \alpha_1, \frac{1}{2} + \alpha_1]$

$$C(\alpha, \tilde{t}) = \left[\frac{1}{2}, c\left(\frac{1}{2} - \alpha\right)\cos\tilde{t}\right],$$

$$E(x(\alpha, \tilde{t})) = E\left(\frac{1}{2}, \tilde{t}\right) = c\left(\frac{1}{2} - \alpha\right)\cos\tilde{t}.$$

Thus in this middle segment, the graphs of the curve $C(\alpha, \tilde{t})$ in the $x - v$ plane and the electric field $E(\alpha, t^*)$ versus α are vertical line segments. Because the charge $\rho(x, t)$ is given by E_x, as $t \to \tilde{t}$, $\rho(x, \tilde{t})$ becomes a Dirac delta distribution at $x = \frac{1}{2}$ with strength

$$E\left(\frac{1}{2} + \alpha_1\right) - E\left(\frac{1}{2} - \alpha_1\right) = -2c\alpha_1 \sin\tilde{t}.$$

Thus charge-concentration phenomenon (13.35) occurs explicitly in this example.

The obvious next question regarding Examples 13.1 and 13.2 is the behavior of the weak solutions past the time of initial singularity. The paper by Majda et al. (1994a) addresses this problem by means of numerical simulations of weak solutions by the particle method (Subsection 13.1.7). In one computation, they took initial conditions (13.48) describing the sinusoidally perturbed electron sheet in Example 13.1 with $\epsilon = 2$ and $j = 1$ and found the singularity time to be 0.079. Using the computational particle method of Subsection 13.1.7, they were able to compute past the singularity time to find that the analytic formula for the solution in Example 13.1 does not describe the solution past the singularity time; the singularity time is precisely the time when the curve $C(\alpha, t)$ ceases to be the graph of a single-valued function, and the graph of $C(\alpha, t)$ is multivalued past the singularity time. They showed that the electric field loses smoothness past the singularity time.

They also computed the perturbed electron sheet in Example 13.2 with dynamic charge concentration for the parameters $\alpha = 0.15$ and $c = 2$. For this solution the singularity time is $\tilde{t} = 0.523$. Again, their results show that the analytic formula in Example 13.2 does not describe the solution past the singularity time and the graph of the curve $C(\alpha, t)$ is multivalued past the singularity time. Moreover, it loses smoothness and forms a cusp past the singularity time. Furthermore, at the singularity time the electric field has a vertical section at $x = \frac{1}{2}$, so the charge is a Dirac delta distribution. For details of the numerical simulations, the interested reader is referred to Majda et al. (1994a).

Once again we emphasize the strong analogy between perturbed vortex sheets and perturbed electron sheets. Numerical computations show that initially smooth vortex sheets remain smooth until the formation of a curvature singularity in the vortex-sheet strength. This singularity occurs just before the vortex sheet starts to roll up. Initially smooth electron sheets remain smooth until the formation of a singularity in the charge. The electron sheet starts to fold (becomes multivalued) just after the singularity time. The folded, multivalued electron sheet is the analog of a rolled-up vortex sheet.

13.2.2. Nonuniqueness of Weak Solutions and the Fission Phenomena

In the previous subsection we showed that initially smooth electron sheets could develop finite-time singularities in which the charge exhibits an algebraic or even Dirac delta function at the singularity time. In this subsection we discuss ways in which to continue the solution past singularity time. In order to do this, we must formulate a weak form of the equation that makes sense for such rough initial data.

We consider a special class of initial conditions for the 1D one-component V-P system

$$\frac{\partial f}{\partial t} + v\frac{\partial f}{\partial v} - E(x,t)\frac{\partial f}{\partial v} = 0, \qquad x \in S^1, \quad v \in \mathbb{R}^1, \quad t > 0,$$
$$\frac{\partial E}{\partial x}(x,t) = 1 - \int_{-\infty}^{\infty} f(x,v,t)dv, \tag{13.50}$$

consisting of a single particle with total charge 1:

$$f_0(x,v) = \delta\left(x - \frac{1}{2}\right)\delta(v), \qquad (x,v) \in S^1 \times \mathbb{R}^1. \tag{13.51}$$

Here we take the spatial period to be $L = 1$ for simplicity. Different weak solutions of problems (13.50) and (13.51) correspond to various ways initial particle (13.51) can undergo fission. We present several different examples of weak solutions to problems (13.50) and (13.51). In the first example, the particle does not undergo fission. The second class of solutions corresponds to a discrete fission process in which the initial particle splits into two or three smaller particles at time $t = 0$. The discrete fission process can also occur at any later time with the smaller particles splitting into two or more additional particles. Finally, the initial particle can undergo continuous fission, that is, single particle (13.51) spreads into an interval of uniform charge. In this subsection we give an explicit construction of the first two types of solutions. The continuous-fission solution is constructed in the next section.

We recall the definition of a weak solution from Eq. (13.41), introduced at the beginning of this section, which used the symmetric average for the electric field (13.43). Denoting $\Omega_T = S^1 \times (0,T)$, a pair $\{E, f\}$ of a function $E(x,t)$ and a nonnegative measure $f(x,v,t)$ is called a weak solution to problems (13.50) and (13.51) if $\{E, f\}$ is 1-periodic in x and for some $T > 0$,

(1) $E(x,t)$ is bounded and has a bounded total 2D variation in Ω_T, i.e.,

$$E \in L^\infty(\Omega_T) \cap BV(\Omega_T).$$

(2) $f(x,v,t)$ is a probability measure for each $t > 0$:

$$\int_0^1 \int_{-\infty}^{\infty} f(x,v,t)dvdx = 1.$$

(3) $f(x, v, t)$ satisfies the special property that its velocity average with a test function φ is a special measure, which is the partial derivative with respect to x of a bounded-variation function g_φ, i.e.,

$$\int_{-\infty}^{\infty} \varphi(x, v, t) f(x, v, t) dv = \partial_x g_\varphi$$

as measures for some function $g_\varphi(x, t) \in BV(\Omega_T)$ for all compactly supported infinitely differentiable test functions $\varphi \in C_0^\infty[\mathbb{R}^2 \times (0, T)]$.

(4) $\{E, f\}$ satisfies the Poisson equation in the distributional sense, $E_x = 1 - \int_{-\infty}^{\infty} f dv$, and the normalizing condition compatible with (2), $\int_0^1 E\, dx = 0$.

(5) $\{E, f\}$ satisfies the Vlasov equation in the weak form,

$$\int_0^T \int_{\mathbb{R}^2} (\varphi t_t f + \varphi_x v f) dv dx dt - \int_0^T \int_{\mathbb{R}^2} \bar{E} \left(\int_{\mathbb{R}} \varphi_v f\, dv \right) dx dt = 0, \qquad (13.52)$$

for all test functions $\varphi \in C_0^\infty[\mathbb{R}^2 \times (0, T)]$.

(6) f is Lipschitz continuous from $[0, T]$ to the local negative Sobolev space $H_{\text{loc}}^{-L}(\mathbb{R}^2)$ for some $L > 0$, and $f(x, v, 0) = f_0(x, v)$ in $H_{\text{loc}}^{-L}(\mathbb{R}^2)$.

We recall that special condition (3) is included in the weak solution because we need it to define the product term $\bar{E} f$. Second, we use test functions $\varphi \in C_0^\infty[\mathbb{R}^2 \times (0, T)]$ instead of $\varphi \in C_0^\infty[S^1 \times \mathbb{R} \times (0, T)]$ only for technical notational reasons. Finally, a function or distribution g belongs to the space $H_{\text{loc}}^{-L}(\mathbb{R}^2)$ iff χg belongs to the space $H^{-L}(\mathbb{R}^2)$ for all cutoff functions $\chi \in C_0^\infty(\mathbb{R}^2)$. Condition (6) provides a weak sense in which the initial data are achieved.

We now present the first example of a weak solution.

Example 13.3. The Steady-State Solution without Fission. This steady solution corresponds to isolated particle (13.51), that remains at $x = \frac{1}{2}$, $v = 0$, for all time:

$$f(x, v, t) = f_0(x, v),$$

$$E(x, t) = E_0(x) = \begin{cases} x, & 0 \le x \le \frac{1}{2} \\ 0, & x = \frac{1}{2} \\ x - 1, & \frac{1}{2} < x \le 1 \end{cases}. \qquad (13.53)$$

Intuitively this steady solution is a particle sitting at position $x = \frac{1}{2}$ with no velocity and having unit charge. Verification of the definition of weak solutions for Eqs. (13.53) is left for the reader and can be found as the special case $\alpha = 0$ in the next class of solutions.

Example 13.4. The Discrete-Fission Solutions. Consider an initial condition corresponding to solution (13.53) above. Pick $\alpha \in (0, \frac{1}{2}]$. Imagine that the initial particle splits into three particles, each with charge α, $1 - 2\alpha$, and α. We send the first particle with charge α toward the left and the third particle with equal charge toward the right. As we shall see, the second particle with charge $1 - 2\alpha$ can actually remain stationary.

Formally the particle-trajectory equation for the first particle is

$$\frac{dx}{dt} = v,$$
$$\frac{dv}{dt} = -\bar{E}(x,t) = -\frac{1}{2}(x + x - \alpha), \tag{13.54}$$
$$(x,v)|_{t=0} = \left(\frac{1}{2}, 0\right),$$

where the symmetric mean $\bar{E}$ is used, because the moving charge $\alpha > 0$ induces a discontinuity in E. As the particle moves toward the left, it reduces the electric field $E = x$ by the amount α to $E = x - \alpha$. We therefore find that $\bar{E}(x,t) = \frac{1}{2}(x + x - a)$ for the first particle. The solution of Eq. (13.54) is

$$x_1 = \frac{1-\alpha}{2}\cos t + \frac{\alpha}{2}, \qquad v_1 = -\frac{1-\alpha}{2}\sin t. \tag{13.55}$$

The solution for the third particle is obtained from solution (13.55) by symmetry $v_3 = -v_1$ and periodicity $x_3 = 1 - x_1$, i.e.,

$$x_3 = -\frac{1-\alpha}{2}\cos t + 1 - \frac{\alpha}{2}, \qquad v_3 = \frac{1-\alpha}{2}\sin t. \tag{13.56}$$

We assume that there is no motion for the second particle (because of symmetry). So we have

$$f(x,v,t) = \sum_{j=1,3} \alpha\delta(x - x_j)\delta(v - v_j) + (1 - 2\alpha)\delta\left(x - \frac{1}{2}\right)\delta(v), \tag{13.57}$$

$$E(x,t) = \begin{cases} x, & 0 \le x < x_1(t) \\ x - \alpha, & x_1 < x < \frac{1}{2} \\ x - 1 + \alpha, & \frac{1}{2} < x < x_3(t) \\ x - 1, & x_3 < x < 1 \end{cases}. \tag{13.58}$$

On jumps of E, we use the symmetric mean

$$\bar{E}(x,t) = \begin{cases} x, & 0 \le x < x_1(t) \\ x - \frac{\alpha}{2}, & x = x_1(t) \\ x - \alpha, & x_1 < x < \frac{1}{2} \\ 0, & x = \frac{1}{2} \\ x - 1 + \alpha, & \frac{1}{2} < x < x_3(t) \\ x - 1 + \frac{\alpha}{2}, & x = x_3 \\ x - 1, & x_3 < x < 1 \end{cases}. \tag{13.59}$$

We can verify rigorously that the above construction gives a true weak solution and takes on the correct initial data. For details, the reader is referred to Zheng and Majda (1994).

We can construct a third class of discrete-fission solutions by further splitting the previous three particles at a later time. Here we describe briefly how to generate these solutions without actually calculating them.

Take a solution from the second class with $0 < \alpha < 1$. Let the solution evolve for a short time t_1, then force the particle with charge α moving to the left to split further into two particles with charge β and $\alpha - \beta > 0$. The particle with charge β will continue to move to the left with a bigger acceleration. The particle with charge $\alpha - \beta$ will continue to move to the left also, but with a smaller acceleration. To preserve $\int_0^1 E dx = 0$, we also make the particle with charge α moving to the right split at the same time t, into two particles with charge 0 and $\alpha - \beta$.

Finally, the initial particle can spread into an interval of uniform charge to form a weak solution with a continuous electric field. We discuss this case in Subsection 13.2.3.

13.2.3. The Collapse of an Electron Sheet into a Point

Here we construct yet another solution to problem (13.50), with simple initial data (13.51), which are significantly different from the solutions constructed in the previous section. This solution corresponds to a uniform spreading of initial particle (13.51) into a maximal interval in the x direction. It evolves from a point into a rotating and stretching electron sheet for $t > 0$, and the corresponding electric field is Lipschitz continuous in space and time $t > 0$. Thc location of the sheet $v = V(x, t)$ is a straight segment in the (x, v) plane for each fixed small $t > 0$. The sheet has length $1 - \cos t$ and strength $1/(1 - \cos t)$. When $t \to 0+$, the sheet shrinks to the point $(\frac{1}{2}, 0)$ and its strength goes to infinity, but with total charge remaining unity. Note that the V-P system is time reversible; thus by reversing time, this example demonstrates that an electron sheet can in general collapse dynamically into concentrated point densities [see Eqs. (13.28)]. This weak solution exhibits the property that electron sheets and concentrated point densities can evolve dynamically into each other in finite time. Hence the evolution of electron sheets is not in general more regular than the seemingly more singular concentrated-point-density solutions. It is also interesting to note that although transport equations are known to preserve full-dimensional areas, this example shows that they may not in general preserve lower dimensions.

We now construct the solution formally. Suppose that the initial particle spreads in the x direction uniformly into a maximally possible interval $[x_{\min}, x_{\max}] \subset (0, 1)$ at time $t > 0$. This interval $[x_{\min}, x_{\max}]$ contains (and is symmetric about) $x = \frac{1}{2}$. To find $x_{\min}$ or $x_{\max}$, examine Eq. (13.54) for the particle moving left. The smaller the particle is, the larger the acceleration it has. Hence the fastest particle must correspond to $\alpha = 0$. Taking $\alpha \to 0+$ in the solutions from Eqs. (13.55) and (13.56) yields

$$x_{\min} = \frac{1}{2} \cos t, \qquad x_{\max} = 1 - \frac{1}{2} \cos t. \tag{13.60}$$

To ensure the unit-charge condition, the density of electrons in this interval must be

$1/(1 - \cos t)$. Solving the Poisson equation for E gives

$$E = \begin{cases} x, & x \in [0, x_{\min}] \\ -\frac{\cos t}{1-\cos t}\left(x - \frac{1}{2}\right), & x \in [x_{\min}, x_{\max}]. \\ x - 1, & x \in [x_{\max}, 1] \end{cases} \tag{13.61}$$

The particle trajectories satisfy the following equations:

$$\begin{aligned} \frac{dx}{dt} &= v, \\ \frac{dv}{dt} &= -E, \\ (x, v)|_{t=0} &= \left(\frac{1}{2}, 0\right), \end{aligned} \tag{13.62}$$

where $E(x, t)$ is given in Eq. (13.61). We do not need the symmetric mean as E is continuous.

Because $x_{\min}$ and $x_{\max}$ are special solutions, all the solutions of Eqs. (13.62) are in the form

$$x - \frac{1}{2} = \alpha(1 - \cos t), \qquad \forall \alpha \in \left[-\frac{1}{2}, \frac{1}{2}\right] \quad v = \alpha \sin t, \tag{13.63}$$

i.e.,

$$v = V(t, x) \equiv \frac{\sin t}{1 - \cos t}\left(x - \frac{1}{2}\right), \qquad x \in (x_{\min}, x_{\max}). \tag{13.64}$$

Combining the density and the position gives $f(x, v, t)$ in the nonzero region:

$$f(x, v, t) = \frac{1}{1 - \cos t}\delta[v - V(x, t)], \qquad x \in (x_{\min}, x_{\max}). \tag{13.65}$$

By means of a series of computations and estimates, Zheng and Majda (1994) verify there that $\{E, f\}$ given by Eqs. (13.61) and (13.65) is indeed a weak solution for problems (13.50) and (13.51) for a short time $T > 0$.

13.2.4. Different Regularizations Select Different Weak Solutions

Now that we have demonstrated the existence of multiple distinct weak solutions to the one-component V-P system with Dirac delta function initial data, it is interesting to ask when the different solutions will be realized.

In Majda et al. (1994a), the authors do a numerical study and show that solutions of different regularizations of the V-P system with electron-sheet initial data can converge to different solutions in the limit of vanishing regularization. As in the case of vortex sheets, it is natural to ask if there is a "selection principle" that always chooses a unique solution and whether this selection principle is stable with respect to small changes in the initial data.

Motivated by the vanishing-viscosity–Navier–Stokes regularization of the Euler equation, the authors considered the vanishing-viscosity ($\epsilon \to 0$) limit from the (pseudo-) Fokker–Planck regularization,

$$\frac{\partial f^\epsilon}{\partial t} + v\frac{\partial f^\epsilon}{\partial x} - E^\epsilon(x,t)\frac{\partial f^\epsilon}{\partial v} = \epsilon\frac{\partial^2 f^\epsilon}{\partial v^2}$$

$$\frac{\partial^2 \varphi^\epsilon}{\partial x^2} = \rho^\epsilon = 1 - \int_{-\infty}^{\infty} f^\epsilon(x,v,t)dv, \tag{13.66}$$

as a candidate for a selection principle.

They performed two different regularizations, the computational particle method described in Subsection 13.1.7 and Fokker–Planck regularization (13.66).

They found that the $\delta \to 0$ limit of vanishing kernel regularization of the computational particle method yields a weak solution corresponding to that of steady-state solution (13.53).

For Fokker–Planck regularization (13.66), the numerics indicate that the vanishing-viscosity limit gives the continuous-fission solution from Subsection 13.2.3.

These numerical computations show that the solutions of different regularizations can converge to different solutions of the V-P equation in the limit of vanishing regularization.

Their numerical computations suggest that although the vanishing-viscosity limit is a selection principle for weak solutions with Dirac delta initial data, it is an unstable selection principle for the single-component V-P system. To see this, they showed that there is sensitive order one change in the behavior of the regularized solutions for appropriate perturbations of the initial data. The three-particle discrete-fission solution (from Example 13.4) provides the starting point for this study. They took a perturbed initial condition

$$f(x,v,0) = \alpha\delta\left[x - \left(\frac{1}{2} - d^0\right)\right]\delta(v+v^0) + (1-2\alpha)\delta\left(x - \frac{1}{2}\right)\delta(v)$$
$$+\alpha\delta\left[x - \left(\frac{1}{2} + d^0\right)\right]\delta(v - v^0), \tag{13.67}$$

where $\alpha \in (0, \frac{1}{4}]$ and $d^0, v^0 > 0$ satisfy the relations

$$d^0 = -\frac{1-\alpha}{2}\cos t^* + 1 - \frac{\alpha}{2}, \qquad v^0 = \frac{1-\alpha}{2}\sin t^*, \tag{13.68}$$

for some t^*. We solve equations (13.68) by taking d^0 to be a specified small number and computing the appropriate value of t^*, and then recovering v^0. Initial condition (13.67) becomes an increasingly smaller perturbation of single-particle initial data (13.53) as $d^0 \to 0$ in two ways. The location of the three particles tends to $x = \frac{1}{2}$, $v = 0$, and the L^1 norm of the difference between the electric field for condition (13.67) and the single-particle steady-state field (13.53) tend to zero.

With condition (13.67) as an initial condition, the numerical solution as computed by the particle method closely approximates the three-particle fission solution. Consider now solutions of the Fokker–Planck regularization with perturbed initial condition (13.67). The numerical calculations in Majda et al. (1994a) show that the simultaneous limit of vanishing viscosity and $d^0 \to 0$ in perturbed initial data (13.67) depend on the relationship between ϵ and d^0. For example, if the viscosity is large compared with the particle separation distance for the initial condition then the solutions converge to the continuous-fission solution. However, if the viscosity ϵ is small compared with the separation distance, then the solutions converge to a continuous piecewise linear function with six interior breakpoints as $\epsilon \to 0$.

The interaction of the viscosity with the particles is more easily explained at the level of the electric field. The solution with $\epsilon = 0$ approximates the large-scale motion of particles and viscosity provides a local correction to this solution near the particles. In this local correction, viscosity treats each particle as an individual particle and alters the electric field in the same way it alters the electric field for initial data for a single particle, that is, it produces a local counterclockwise rotation of the electric field. Thus, depending on the size of the perturbation and the strength of the viscosity, it is possible to compute weak solutions that exhibit $\mathcal{O}(1)$ deviations from each other in the limit of small perturbations of the initial data and vanishing viscosity. In this sense, the viscosity selection criterion is not a stable principle.

13.3. The Two-Component Vlasov–Poisson System

In this section we consider the two-component case of the V-P equations in one-space dimension. In the previous sections, we studied the one-component case as an analogy to the 2D vorticity equation for incompressible inviscid fluid flow when the vorticity has distinguished sign. Here we consider the 2D component V-P system as an analogy to the case in which the vorticity has mixed sign. As we saw in Chaps. 8–11, numerical experiments and existing theory suggest that a much more rich behavior can occur in the limiting process for vortex-sheet formation with mixed-sign vorticity. For example, in the case of single-sign vorticity, we showed at the end of Chap. 10 that the zero-viscosity limit of smooth solutions produces a true weak solution to the Euler equation with vortex-sheet initial data. Such a result is not known for the mixed-sign case. Here we are able to show for the simpler analog V-P problem that for the two-component case there is an explicit example with a singular charge concentration. We introduce the concepts of measure-valued and weak solutions and show by these explicit examples that some situations necessarily lead to behavior that is so singular it satisfies the equations in only a measure-valued sense instead of a weak sense. Also we present examples that lead to nonunique weak solutions. We also show that different computational regularizations can lead to different behavior in situations with measure-valued and/or nonunique weak solutions.

The two-component 1D V-P equations for a collisionless plasma of electrons and positively charged ions are

$$
\begin{aligned}
&\frac{\partial f_-}{\partial t} + v\frac{\partial f_-}{\partial x} - E(x,t)\frac{\partial f_-}{\partial v} = 0,\\
&\frac{\partial f_+}{\partial t} + v\frac{\partial f_+}{\partial x} + E(x,t)\frac{\partial f_+}{\partial v} = 0,\\
&\frac{\partial E}{\partial x} = \rho(x,t) = \int_{-\infty}^{\infty} (f_+ - f_-)(x,v,t)dv.
\end{aligned}
\tag{13.69}
$$

In V-P equations (13.69), $f_+(x,v,t)$, $f_-(x,v,t)$ with $f_\pm \geq 0$ denote the density of positively charged ions and electrons, respectively, at location x with velocity v, $E(x,t)$ is the electric field generated by these particles, and $\rho(x,t)$ is the charge density. We consider the initial value problem for Eqs. (13.69) with initial data $f_\pm|_{t=0} = f_\pm^0$, where $f_\pm^0 \geq 0$ are both nonnegative measures satisfying

$$\int\int f_+^0(x,v)dxdv = \int\int f_-^0(x,v)dxdv.$$

The last condition in system (13.6) implies that the plasma has zero net charge:

$$\int \rho(x,t)dx = 0. \tag{13.70}$$

For the two-component case, we take $f_\pm(x,v,t)$ periodic in x with period L for the purpose of numerical computations, in which case the field E has mean zero over one spatial period. However, for the purpose of the analysis, it is simpler to consider the case in which $f_\pm(x,v,t)$ is a function of $x \in \mathbb{R}$, where the last condition means that $E \to 0$ as $|x| \to \infty$.

We now develop concepts of weak solutions and measure-valued solution for Eqs. (13.69) as well as the limiting behavior for certain approximate-solution sequences. A measure-valued solution is a more general concept than weak solution and involves an ensemble average over weak solutions.

As in the case of the single-component V-P equations, we consider two examples of approximate-solution sequences: the zero diffusion limit of the (pseudo-Fokker–Planck) regularization and the high-resolution limit of computational particle methods for system (13.6).

The Fokker–Planck regularization procedure involves the zero diffusion limit $\nu_\pm^\epsilon \to 0$, $\nu_\pm^\epsilon \geq 0$ of solutions of the parabolic (in v) equations

$$\begin{aligned}
&\frac{\partial f_-^\epsilon}{\partial t} + v\frac{\partial f_-^\epsilon}{\partial x} - E^\epsilon(x,t)\frac{\partial f_-^\epsilon}{\partial v} = \nu_-^\epsilon \frac{\partial^2 f_-^\epsilon}{\partial v^2},\\
&\frac{\partial f_+^\epsilon}{\partial t} + v\frac{\partial f_+^\epsilon}{\partial x} + E^\epsilon(x,t)\frac{\partial f_+^\epsilon}{\partial v} = \nu_+^\epsilon \frac{\partial^2 f_+^\epsilon}{\partial v^2},\\
&\frac{\partial E^\epsilon}{\partial x} = \rho^\epsilon(x,t) = \int_{-\infty}^{\infty}(f_+^\epsilon - f_-^\epsilon)dv.
\end{aligned} \tag{13.71}$$

As in the single-component case, we build explicit weak solutions of system (13.6) through superpositions of concentrated point charges:

$$f_\pm(x,v,t) = \sum_{j=1}^N \alpha_j^\pm \delta[\mathbf{x} - \mathbf{x}_j^\pm(t)]$$

$$\text{with} \quad \alpha_j^\pm \geq 0 \quad \text{and} \quad \sum_{j=1}^N \alpha_j^+ = \sum_{j=1}^N \alpha_j^-. \tag{13.72}$$

In Eq. (13.72) we use the notation $\mathbf{x} = (x, v)$ and the locations of the concentrated point-charge solutions; $\{\mathbf{x}_j^\pm(t)\}_{j=1}^N$ satisfy nonlinear ODEs analogous to point-vortex equations (9.18) for 2D Euler equations (13.1).

The explicit weak solutions establish the following facts:

(1) There are limits of explicit weak solutions that converge to an explicit measure-valued solution of Eqs. (13.69) rather than a conventional weak solution with the same initial data. In the language of Chap. 11 for the 2D Euler equation, we say that, in this case, there is no *concentration–cancellation*.

(2) For a large class of examples, there are several nonunique weak solutions arising from the same initial data.

(3) High-resolution numerical simulations of Eqs. (13.69) and (13.71) by means of computational particle methods demonstrate explicitly that, in some cases, different computational regularizations converge to either measure-valued solutions or different weak solutions as the regularization parameters tend to zero.

The behavior of limiting processes for vortex sheets for the 2D Euler equation, as observed in computational vortex methods reveals much more complex behavior when vorticity docs not have a distinguished sign in contrast to the more regular but singular behavior occurring with nonnegative vorticity. Intuitively, when vorticity has two signs, nearby vortices can screen each other by canceling out their effect to high order on more distant velocities, which can drive even further, stronger concentration of vorticity with continued screening effects, as an unstable process. Such a physical mechanism is present in the calculations shown in Figs. 9.4 and 9.5 in Chap. 9.

Recall from Chap. 11 that we introduced several examples of approximate-solution sequences for the 2D Euler equation involving vorticity that changes sign. One such example (Example 11.1) was the sequence of steady-state solutions of "phantom vortices." In this case, we showed that a finite amount of kinetic energy concentrates as a delta function at the origin. Despite this fact, however, the weak limit is still a weak solution to the 2D Euler equation. We called this phenomenon concentration–cancellation. Another such example was due to Greengard and Thomann (Theorem 11.2) and showed that one could use the phantom vorticies to construct a family of solutions whose weak limit has an associated reduced defect measure that is the Lebesgue measure on the unit square. Chapter 12 proved the remarkable result (DiPerna and Majda, 1988) that any *steady* approximate-solution sequences with L^1 vorticity control has concentration–cancellation in the limit, even in the case of *mixed-sign* vorticity. It is not at all clear that the same result holds true for general approximate-solution sequences with mixed-sign vorticity. Indeed, Majda conjectured that measure-valued solutions are needed to describe the limit with a mixed sign (Majda, 1988).

The exact solutions and computations discussed in this section for the two-component V-P system provide unambiguous evidence of the kinds of screening effects that might cause weak solutions to break down for the 2D Euler equation with mixed-sign vorticity. In this section we demonstrate that, for the analogy problem of the 1D V-P system, in the two-component case we can construct approximate-solution sequences that do not converge to a weak solution. A more general class of solution is needed to describe the limit, that of a measure-valued solution. In the two-component V-P equations there is a simple analog of the screening effect in the 2D Euler equation. Simple concentrated point-charge densities

$$f_{\pm} = \alpha\delta(x - x_0) \otimes \delta(v) \tag{13.73}$$

illustrate this screening effect dramatically for solutions of system (13.6). Densities (13.73) define a steady weak solution of system (13.6) with perfect screening because E vanishes identically, i.e., $E \equiv 0$. Later in this section we show by means of explicit examples of time-dependent measure-valued and nonunique weak solutions that screening plays a prominent role in these phenomena.

13.3.1. Weak and Measure-Valued Solutions in the Two-Component Case

In this section we develop three different concepts of solutions to two-component V-P system (13.6): weak, generalized weak, and measure-valued solutions. We show that all three definitions are needed to describe accurately the complex phenomena that can occur in the limiting process of valid approximate-solution sequences for this problem.

For data as general as $f_{\pm}$ a measure, ordinary classical weak solutions are not well defined because of the ambiguity of the product Ef, in that E can be discontinuous whereas f has Dirac delta concentration. Zheng and Majda (1994) overcame this for one-component system (13.5) by introducing the concept of weak solutions by means of the symmetric mean $\bar{E}$ of the electric field E and establishing certain regularity properties of E and f (e.g., $E \in BV$) so that the product $\bar{E}f$ has unambiguous meaning. Unfortunately, the newly defined product $\bar{E}f$ is not weakly continuous with respect to sequences of weak or smooth solutions. Example 13.9 of the next section shows that the limit of a weakly convergent sequence of weak solutions may not be a weak solution; the symmetric mean $\bar{E}$ is too restricitive. Giving up the symmetric mean but requiring E to be only weakly defined and integrable with respect to the measures $f_{\pm}$, we introduce the concept of generalized weak solutions. In fact, the generalized weak solution is also not sufficient to describe all possible cases that can occur from limits of sequences of weak or smooth solutions. Example 13.10 of the next section shows that the limit of a weakly convergent sequence of weak solutions may be double valued on some set of space–time and not be a generalized weak solution. This creates the need for an even more general concept of a solution called a measure-valued solution.

First we define appropriate approximate-solution sequences in a manner that is parallel to that in Chaps. 11–12 for vortex-sheet initial data. Recall that, for the two-component case, we choose to consider solutions on the line $x \in \mathbb{R}$. Let $\mathbb{R}_+$ denote $(0, \infty)$, $\mathbb{R}_+^2 = \mathbb{R} \times \mathbb{R}_+$, and $\mathbb{R}_+^3 = \mathbb{R}^2 \times \mathbb{R}_+$. To simplify notation, in the following discussion, a sequence of functions indexed by ϵ may in fact indicate a sequence indexed by $\epsilon_0 > \epsilon > 0$ for some small positive ϵ_0 or a countable number of those $\epsilon \in (0, \epsilon_0)$ with $\epsilon \to 0$.

Definition 13.1. *A sequence of smooth functions $\{f_{\pm}^{\epsilon}, E^{\epsilon}\}_{\epsilon>0}$ is an approximate-solution sequence for two-component V-P system (13.6) provided that the following conditions hold:*

(i) Conservation of charge: $f_{\pm}^{\epsilon} \geq 0$, $\int\int_{\mathbb{R}^2} f_{+}^{\epsilon} d\,dv = \int\int_{\mathbb{R}^2} f_{-}^{\epsilon} d\,dv \leq M$ for all $\epsilon > 0$, $t > 0$;

(ii) $|E^{\epsilon}(\cdot, \cdot)|_{L^{\infty}} \leq K$, $E^{\epsilon}(\pm\infty, t) = 0$ for all $\epsilon > 0$, $0 < t < T$;

(iii) *Finite energy and decay at infinity: for all* $\epsilon > 0$, $0 < t < T$, *and all* $T > 0$,

$$\int\int_{\mathbb{R}^2} (1 + x^2 + v^2)(f_-^\epsilon + f_+^\epsilon) dx dv + \int_{\mathbb{R}} (E^\epsilon)^2 dx \leq M_T;$$

(iv) *weak consistency with the equations: for all infinitely smooth test functions* $\varphi \in C_0^\infty(\mathbb{R}_+^3)$, $\psi(t, x) \in C_0^\infty(\mathbb{R}_+^2)$ *with compact support,*

$$\lim_{\epsilon \to 0} \int\int\int_{\mathbb{R}_+^3} (\varphi_t f_\pm^\epsilon + v\varphi_x f_\pm^\epsilon \pm \varphi_v E^\epsilon f_\pm^\epsilon) dv dx dt = 0,$$

$$\lim_{\epsilon \to 0} \int\int_{\mathbb{R}_+^2} \left[\psi_x E^\epsilon + \psi \int_{\mathbb{R}} (f_+^\epsilon - f_-^\epsilon) dv\right] dx dt = 0;$$

(v) *The functions* $f_\pm^\epsilon$ *are Lipschitz continuous uniformly in* $\epsilon > 0$ *from* $[0, T)$ *to a negative Sobolev pace* $H_{\text{loc}}^{-L}(\mathbb{R}^2)$ *for some* $L > 0$, *for all* $T > 0$; *and* $f_\pm^\epsilon(0, x, v)$ *converge to* $f_\pm^0(x, v)$ *in* $H_{\text{loc}}^{-L}(\mathbb{R}^2)$.

As in the one-component case, approximate-solution sequences can be generated in various ways. The easiest is by mollifying the initial data $f_\pm^0$:

$$f_\pm^\epsilon(0, x, v) = [\chi_\epsilon f_\pm^0(x, v)] * \rho_\epsilon + g_\pm^\epsilon(x, v). \tag{13.74}$$

Here χ_ϵ is a cutoff function and $\chi_\epsilon = 1$ for $x^2 + v^2 < 1/\epsilon^2$, zero otherwise. The standard mollifier ρ_e, introduced in Chap. 3, Eq. (3.35), scales as $\rho_\epsilon = \frac{1}{\epsilon^2}\rho(x/\epsilon, v\epsilon)$, $\rho \in C_0^\infty(\mathbb{R}^2)$, $\rho \geq 0$, $\int\int \rho \, dx dv = 1$, and the nonnegative compactly supported smooth functions $g_\pm^\epsilon$ are chosen such that $f_\pm^\epsilon(0, x, v)$ satisfy the neutrality condition with $\int\int_{\mathbb{R}^2} g_\pm^\epsilon \, dx dv \to 0$ as $\epsilon \to 0$. The functions $g_\pm^\epsilon$ are needed because the cutoff function χ_ϵ may destroy the neutrality condition of $f_\pm^0$, and the cutoff function χ_ϵ is needed because $f_\pm^\epsilon(0, x, v)$ then have decay $\mathcal{O}(v^2)$ as $|v| \to \infty$ uniformly with respect to $x \in \mathbb{R}$ so that we can apply the existence and uniqueness theory of Iordanskii (1961) to $f_\pm^\epsilon(0, x, v)$.

A physically relevant way to generate an approximate-solution sequence is to use pseudo-Fokker–Planck system (13.71). Smooth solutions of system (13.71) satisfy Definition 13.1 provided they exist. Another way to generate an approximate-solution sequence is by means of a computational particle method.

It is clear that we need to study the limits of products $\{E^\epsilon, f_\pm^\epsilon\}_\epsilon$. Note that these nonlinear terms can be rewritten as

$$E^\epsilon f_+^\epsilon = \left(E^\epsilon \frac{f_+^\epsilon}{f_+^\epsilon + f_-^\epsilon}\right)(f_+^\epsilon + f_-^\epsilon),$$

$$E^\epsilon f_-^\epsilon = \left(E^\epsilon \frac{f_-^\epsilon}{f_+^\epsilon + f_-^\epsilon}\right)(f_+^\epsilon + f_-^\epsilon).$$

Thus both $E^\epsilon f_\pm^\epsilon$ fall into the class of limits

$$g\left(E^\epsilon, \frac{f_+^\epsilon}{f_+^\epsilon + f_-^\epsilon}, \frac{f_-^\epsilon}{f_+^\epsilon + f_-^\epsilon}\right)(f_+^\epsilon + f_-^\epsilon), \tag{13.75}$$

where $g(\lambda_1, \lambda_2, \lambda_3) \in C([-K, K] \times V)$ is continuous. Here K is the constant in (ii) of Definition 13.1 and V is a line segment from the point $(1, 0)$ to the point $(0, 1)$ in $\mathbb{R}^2$:

$$V = \{(\lambda_2, \lambda_3) \in \mathbb{R}^2 | \lambda_2 + \lambda_3 = 1, \lambda_2 \geq 0, \lambda_3 \geq 0\}.$$

We introduce some new notation. Let $\Omega_T = (0, T) \times \mathbb{R}^2$ for $T > 0$. Define $BM^+(\Omega_T)$ as the space of nonnegative finite measures on Ω_T for $T > 0$, and $PM([-K, K] \times V)$ as the space of probability measures. We use the symbol $\mu_\epsilon \rightharpoonup^* \mu$ for weak* convergence of a sequence of bounded measures $\{\mu^\epsilon\} \subset BM^+(\Omega_T)$, i.e., $\int_{\Omega_T} \varphi \, d\mu^\epsilon \to \int_{\Omega_T} \varphi \, d\mu$ for all bounded continuous test functions φ on Ω_T. The symbol $h^\epsilon \rightharpoonup h$ represents weak convergence in L^p, $1 < p < \infty$, i.e., $\int \varphi h^\epsilon \to \int \varphi h$ for all $\varphi \in L^q$, $(1/p)+(1/q) = 1$. Both weak and weak* convergence imply convergence in the sense of distributions for which the test functions are restricted to compactly supported and infinitely differentiable functions. The following lemma identifies the limiting behavior of all continuous functions of the form of class (13.75).

Lemma 13.1. *For every approximate-solution sequence $\{f_\pm^\epsilon, E^\epsilon\}_{\epsilon>0}$ there exists a bounded function $E(t, x)$ satisfying*

$$\lim_{|x|\to\infty} |E(x, t)| = 0, \tag{13.76}$$

a scalar measure $\sigma \in BM^+(\Omega_T)$ for all $T > 0$, and a $\sigma-$measurable map

$$y \to \nu_y$$

from $\mathbb{R}_+^3$ to $PM([-K, K] \times V)$ such that (pass to a subsequence if necessary)

$$E^\epsilon \to E \tag{13.77}$$

in the weak topology of $L^2[(0, T) \times \mathbb{R}]$ for all $T > 0$,

$$(f_+^\epsilon + f_-^\epsilon)dy \rightharpoonup^* \omega \tag{13.78}$$

in the weak topology of $BM^+(\Omega_T)$ for all $T > 0$, and*

$$\lim_{\epsilon\to 0} \int_{\mathbb{R}_+^3} \varphi(y) g\left(E^\epsilon, \frac{f_+^\epsilon}{f_+^\epsilon + f_-^\epsilon}, \frac{f_-^\epsilon}{f_+^\epsilon + f_-^\epsilon}\right)(f_+^\epsilon + f_-^\epsilon)dy = \int_{\mathbb{R}_+^3} \varphi(y)\langle \nu_y, g\rangle d\sigma \tag{13.79}$$

for all $g \in C([-K, K] \times V)$ and all $\varphi \in C_0(\mathbb{R}_+^3)$. Here the angles in Eq. (13.79) denote the expected value,

$$\langle \nu_y, g\rangle = \int_{[-k,k]\times V} g(\lambda_1, \lambda_2, \lambda_3) d\nu_y.$$

Lemma 13.1 is similar to that of Generalized Young Measure Theorem 12.11 for approximate-solution sequences of the 3D Euler equation. The triple $\{E, \sigma, \nu\}$ is called the *generalized Young measure* associated wtih the sequence $\{f_\pm^\epsilon, E^\epsilon\}$. For a proof of Lemma 13.1 see Majda et al. (1994b).

Definition 13.2. *Measure-Valued Solutions. A triplet* $\{E, \sigma, \nu\}$ *of a function* $E(t, x) \in L^\infty(\mathbb{R}_+^2)$, *a scalar measure* $\sigma \in BM^+(\Omega_T)$ *for all* $T > 0$, *and a* σ*−measurable map* $\nu_y \in PM([-K, K] \times V)$ *is a measure-valued solution to V-P system (13.69) in* $\mathbb{R}_+^3$ *provided that E satisfies limit (13.76) and*

(i) for all test functions $\varphi(y) \in C_0^\infty(\mathbb{R}_+^3)$, *and* $\psi(t, x) \in C_0^\infty(\mathbb{R}_+^2)$,

$$\int_{\mathbb{R}_+^3} (\partial_t\varphi \cdot \langle\nu_y, \lambda_3\rangle + \partial_x\varphi v\langle\nu_y, \lambda_3\rangle - \partial_v\varphi\langle\nu_y, \lambda_1\lambda_3\rangle)d\sigma = 0,$$

$$\int_{\mathbb{R}_+^3} (\partial_t\varphi \cdot \langle\nu_y, \lambda_2\rangle + \partial_x\varphi v\langle\nu_y, \lambda_2\rangle - \partial_v\varphi\langle\nu_y, \lambda_1\lambda_2\rangle)d\sigma = 0,$$

$$\int_{\mathbb{R}_+^2} \psi_x E(t, x)dxdt + \int_{\mathbb{R}_+^3} \psi\langle\nu_y, \lambda_2\rangle d\sigma - \int_{\mathbb{R}_+^3} \psi\langle\nu_y, \lambda_3\rangle d\sigma = 0;$$

(ii) both $\langle\nu_y, \lambda_2\rangle\sigma$ *and* $\langle\nu_y, \lambda_3\rangle\sigma$ *belong to the space of Lipschitz continuous functions from* $[0, T)$ *to* $H_{\text{loc}}^{-L}(\mathbb{R}^2)$ *for some* $L > 0$ *for all* $T > 0$, *and both take on the initial data* $f_\pm^0(x, v)$ *in that space.*

Lemma 13.1 combined with the existence of an approximate-solution sequence yields Theorem 13.2.

Theorem 13.2. *For any initial data consisting of measures* $f_\pm^0(x, v)$ *satisfying decay conditions (13.76) there exists a measure-valued solution* $\{E, \sigma, \nu\}$ *to the V-P system. Furthermore, every approximate-solution sequence has a subsequence that converges to a measure-valued solution of V-P system (13.69).*

The proof of Theorem 13.2 follows that of Theorem 12.12. The details can be found in Majda et al. (1994b).

Some measure-valued solutions have enough smoothness to fit the standard definition of a weak solution. Hence we now describe several formulations of a weak solution. The difficulty is to define the nonlinear product $Ef_\pm$ in a distributional setting, where E is typically discontinuous and $f_\pm$ are finite measures. Because of energy assumption (iii) in Definition 13.1, approximate-solution sequences generated by mollifying the initial data or from Fokker–Planck –Poisson system (13.71) (provided smooth solutions exist) actually satisfy two additional estimates:

$$|E^\epsilon(\cdot, \cdot)|_{\text{BV}([0,T]\times\mathbb{R})} \leq C_T \qquad \forall T > 0, \tag{13.80}$$

$$\left|\int_{-\infty}^{x} \int_{-\infty}^{\infty} \varphi(t, y, v) f_\pm^\epsilon(t, y, v)dvdy\right|_{\text{BV}([0,T]\times[-X,X])} \leq C_{T,X,\varphi}, \tag{13.81}$$

for all $T > 0$, $\varphi \in C_0^\infty(\Omega)$, and finite $X > 0$. To verify estimate (13.80) note that from system (13.6)

$$E_t^\epsilon = \int_{\mathbb{R}} v(f_-^\epsilon - f_+^\epsilon)dv.$$

Thus both E_x^ϵ and E_t^ϵ are bounded measures in $(0, T) \times \mathbb{R}$ and hence estimate (13.80) holds. To confirm estimate (13.81), note that

$$\begin{aligned} \partial_t &\left[\int_{-\infty}^{x} \int_{\mathbb{R}} \varphi(t, y, v) f_\pm^\epsilon(t, y, v) dv dy \right] \\ &= - \int_{\mathbb{R}} v \varphi f_\pm^\epsilon f v + \int_{-\infty}^{x} \int_{\mathbb{R}} (\varphi_t + v\varphi_y \pm E^\epsilon \varphi_v) f_\pm^\epsilon \, dv dy \\ &\quad + \nu_\pm^\epsilon \int_{-\infty}^{x} \int_{\mathbb{R}} \varphi_{vv} f_\pm^\epsilon \, dv dy \end{aligned} \tag{13.82}$$

from Fokker–Planck system (13.71). For V-P system (13.6) simply let $\nu_\pm^\epsilon = 0$ in Eq. (13.82). The first term in Eq. (13.82) is a bounded measure in $(0, T) \times \mathbb{R}$. The remaining terms are all bounded functions. This confirms estimate (13.81).

Additional estimates (13.80) and (13.81) provide sufficient regularity to define weak and generalized weak solutions. Passing to a subsequence $\epsilon \to 0$, we obtain by Helly's selection principle a triplet $\{f_+, f_-, E\}$ such that

$$\begin{aligned} &f_\pm^\epsilon(x, v, t) \rightharpoonup^* f_\pm(x, v, t) \in BM^+, \\ &E^\epsilon(t, x) \to E(t, x) \in BV \qquad \text{almost everywhere,} \end{aligned} \tag{13.83}$$

and $\{f_+, f_-, E\}$ satisfy estimates of the form of (i), (ii), and (iv) in Definition 13.1 and estimates (13.80) and (13.81). In particular

$$\begin{aligned} &E \in L^\infty[(0, T) \times \mathbb{R}] \cup \mathrm{BV}[(0, T) \times \mathbb{R}], \\ &\int_{-\infty}^{x} \int_{\mathbb{R}} \varphi f_\pm(t, y, v) dv dy \in \mathrm{BV}_{\mathrm{loc}}[(0, T) \times \mathbb{R}] \end{aligned} \tag{13.84}$$

which guarantees a consistent, measure-theoretic interpretation of the products $Ef_\pm$ in a distributional sense. We can define the product of a bounded $\mathrm{BV}_{\mathrm{loc}}(\mathbb{R}^n)$ function $u(x)$ with a partial derivative of an arbitrary $\mathrm{BV}_{\mathrm{loc}}(\mathbb{R}^n)$ function $w(x)$ by Volpert's symmetric mean (Majda et al., 1994b; Volpert, 1967; Volpert and Hudjaev, 1985). That is, the product $u(\partial w/\partial x_i)$ can be defined as a bounded measure $\bar{u}(\partial w/\partial x_i)$, which means that $\bar{u}$ is integrable with respect to the finite measure $\partial w/\partial x_i$ on any Borel set of $\mathbb{R}^n$. Here the symmetric mean of u is defined as in average (13.43):

$$\bar{u}(x, t) = \begin{cases} u(x, t) & \text{if } u \text{ is approximately continuous at } (x, t) \\ \frac{1}{2}[u_l(x, t) + u_r(x, t)] & \text{if } u \text{ has a jump at } (x, t) \end{cases}. \tag{13.85}$$

Note that in the 2D case, the symmetric mean is not necessarily defined *everywhere* but only almost everywhere defined on any Lipschitz curve in $\mathbb{R}^2$. On the other hand, a partial derivative of a $\mathrm{BV}_{\mathrm{loc}}$ function is a measure that cannot concentrate on any point in two dimensions. Therefore $\bar{u}$ is almost everywhere well defined with respect to $\partial w/\partial x_i$, and $\bar{u}(\partial w/\partial x_i)$ makes sense as a measure multiplied by a "continuous" function.

The only nonlinear terms in system (13.69) involve $\partial_v(Ef_\pm)$. Using relations (13.84) and the symmetric mean, we define them in a weak sense as follows:

$$\langle \partial_v(Ef_\pm), \varphi\rangle = -\int_0^\infty \int_{-\infty}^\infty \bar{E}\partial_x \left[\int_{-\infty}^x \int_{-\infty}^\infty \varphi_v(t,y,v) f_\pm(t,y,v)dvdy\right] dxdt \tag{13.86}$$

for all test functions $\varphi \in C_0(\mathbb{R}^3_+)$. We can therefore define weak solutions as in the one-component case.

Definition 13.3. *Weak Solutions. A triplet $\{f_+, f_-, E\}$ of two measures, $f_\pm$ and a function E, is called a weak solution to two-component V-P system (13.6) if*

(i) it satisfies the basic regularity conditions in (i)–(iii) of Definition 13.1 and BV *control (13.84);*

(ii) it satisfies system (13.6) in the following sense: $\forall \varphi \in C_0^\infty(\mathbb{R}^3_+)$, $\psi(t,x) \in C_0^\infty(\mathbb{R}^2_+)$;

$$\int\int\int_{\mathbb{R}^3_+} (\varphi_t f_\pm + v\varphi_x f_\pm)dvdxdt \pm \int_{\mathbb{R}^2_+} \bar{E}\int_{\mathbb{R}} \varphi_v f_\pm\, dvdxdt = 0, \tag{13.87}$$

$$\int\int_{\mathbb{R}^2_+} -\psi_x E\, dxdt = \int\int_{\mathbb{R}^2_+} \psi \int_R (f_+ - f_-)dvdxdt \tag{13.88}$$

where $\bar{E}$ denotes the symmetric mean of E;

(iii) for some $L > 0$ and for all $T > 0$, both $f_\pm$ belong to the space of Lipschitz continuous functions from $[0, T)$ to $H^{-L}_{\text{loc}}(\mathbb{R}^2)$ and both $f_\pm$ satisfy the initial condition in that space.

Unfortunately the procedure of taking $\epsilon \to 0$ for the one-component case does not work here. In fact, the nonlinear terms $E^\epsilon f^\epsilon_\pm$ are not necessarily weakly continuous. Therefore, determining conditions under which a weak solution exists for the Cauchy problem is an open problem.

The symmetric mean arises naturally in the one-component case. Because the product $\bar{E}^\epsilon f^\epsilon$ is weakly continuous $\bar{E}^\epsilon f^\epsilon \rightharpoonup^* \bar{E} f$ for a sequence of exact solutions, a sequence of solutions of mollified data, or a sequence of the Fokker–Planck–Poisson system with vanishing noise level, existence of weak solutions can be established in the one-component case. The present two-component case is more complex. There is no reason that we should be restricted to the symmetric mean to define the products $Ef_\pm$. In Example 13.9 in the next section, we show that this definition is too restrictive. It is more natural to generalize $\bar{E}$. In fact, Eq. (13.87) is well defined simply because of the key fact that $\bar{E}$ is integrable with respect to all the measures $\int \varphi_v f_\pm\, dv$. Therefore, instead of requiring a number of conditions, such as a symmetric mean, which then guarantee the conditions needed to pass to the limit to a weak solution, we instead construct a version of E, denoted by $\tilde{E}$ for now, that is integrable with respect to all the measures $\varphi_v f_\pm\, dv$. More precisely, we can replace the symmetric mean $\bar{E}$ and the BV conditions of relations (13.84) with $\tilde{E}$, any function satisfying the following conditions:

(1) $\tilde{E}(t,x) = E(t,x)$ almost everywhere with respect to the 2D Lebesgue measure;
(2) $\tilde{E}(t,x)$ is defined almost everywhere and integrable with respect to all measures $\int \varphi_v f_\pm$ for all test functions $\varphi \in C_0^\infty(\mathbb{R}_+^3)$.

A trivial example for $\tilde{E}$ is $\bar{E}$. Another example is to assign zero to $\tilde{E}$ wherever E has a jump. A third example is to assign zero to $\tilde{E}$ also, but on points where $\int \varphi_v f_\pm\, dv$ concentrate. In general it is quite arbitrary to assign values of $\tilde{E}$ on jumps of E or on a countable number of points where $\int \varphi_v f_\pm\, dv$ concentrate. The only requirement is that $\tilde{E}$ be integrable with respect to $\int \varphi_v f_\pm\, dv$. In general we have the following definition.

Definition 13.4. *Generalized Weak Solutions. A triplet $\{f_+, f_-, \tilde{E}\}$ of two measures and a Lebesgue measurable function is called a generalized weak solution to the two-component V-P system if the following conditions are met.*

(i) It satisfies basic regularity conditions (i)–(iii) in Definition 13.1.
(ii) $\tilde{E}(t,x)$ is defined almost everywhere and integrable with respect to both $\int \varphi_v f_\pm\, dv$ for all test functions $\varphi \in C_0^\infty(\mathbb{R}_+^3)$.
(iii) It satisfies system (13.6) in the following sense: For all $\varphi \in C_0^\infty(\mathbb{R}_+^3)$, $\psi(t,x) \in C_0^\infty(\mathbb{R}_+^2)$,

$$\int\int\int_{\mathbb{R}_+^3} (\varphi_t f_\pm + v\varphi_x f_\pm) dv dx dt \pm \int\int_{\mathbb{R}_+^2} \tilde{E} \int_{\mathbb{R}} \varphi_v f_\pm\, dv dx dt = 0, \qquad (13.89)$$

$$\int\int_{\mathbb{R}_+^2} -\psi_x \tilde{E}\, dx dt = \int\int_{\mathbb{R}_+^2} \psi \int_{\mathbb{R}} (f_+ - f_-) dv dx dt; \qquad (13.90)$$

(iv) For some $L > 0$ for all $T > 0$, both $f_\pm$ belong to the space of Lipschitz continuous functions from $[0,T)$ to $H_{\mathrm{loc}}^{-L}(\mathbb{R}^2)$ and both $f_\pm$ take on the initial data $f_\pm^0(x,v)$ in that space.

Although true generalized weak solutions do arise from sequences of exact weak solutions, there are more complex solutions, arising from the limit of certain solution sequences, that can be described only as measure-valued solutions. The notion of a generalized weak solution serves as an intermediate concept between weak and measure-valued solutions.

There is another framework within which to connect the three different types of less smooth solutions. Note that $\{E^\epsilon f_\pm^\epsilon\}_{\epsilon>0}$ are also bounded sequences of measures [from (i) and (ii) in Definition 13.1]. In addition to convergences (13.83), by Helly's selection principle

$$E^\epsilon f_\pm^\epsilon(x,v,t) \rightharpoonup^* \mu_\pm \in BM \qquad (13.91)$$

for some measures $\mu_\pm$. Because the sequence $\{E^\epsilon f_\pm^\epsilon\}_{\epsilon>0}$ is uniformly dominated by the sequence $\{f_\pm^\epsilon\}_{\epsilon>0}$ for any Borel set $S \subset \mathbb{R}_+^3$, it follows that $\mu_\pm$ are absolutely continuous with respect to $f_\pm$. By the Radon–Nikodym theorem (Rudin, 1987), there exist two functions $E_\pm$ measurable with respect to $f_\pm$ so that

$$\mu_\pm = E_\pm f_\pm.$$

If both E_+ and E_- are equal to $\bar{E}$, we have a weak solution. If E_+ and E_- are equal but not equal to $\bar{E}$, we can let $\tilde{E} = E_+ = E_-$ and we have a generalized weak solution. In general, however, E_+ is not equal to E_-. Therefore the E field is double valued for a measure-valued solution.

13.3.2. Examples of Weak Solutions for the Two-Component Vlasov-Poisson System

In this subsection we introduce concrete examples of exact weak solutions to two-component V-P system (13.69). Sequences of these exact weak solutions illustrate the kinds of concentrations that can occur in the limit. In particular, Examples 13.9 and 13.10 in this section show successively the necessity for generalized weak and measure-valued solutions. We also present an example to illustrate the nonuniqueness of weak solutions. To reemphasize, the standard classical concept of a distribution solution does not hold for system (13.6) with measures as initial data. Following the one-component theory, we introduced a definition of weak solution that uses the symmetric mean $\bar{E}$. Viewing this definition as a starting point, we now show that sequences of solutions in this class can have weak limits that are nontrivial generalized weak solutions or even measure-valued solutions.

Example 13.5. Simple Screening Solutions. Consider the initial data

$$f_-(0,x,v) = \alpha\delta(v)\delta(x-x_0), \qquad f_+(0,x,v) = \alpha\delta(v)\delta(x-x_0), \tag{13.92}$$

where $x_0 \in \mathbb{R}$ and α is a positive constant. From the Poisson equation, the corresponding initial E field is identically zero. The two particles screen each other completely. At a later time, the particles remain in the same place and we have a stationary solution

$$f_-(x,v,t) = \alpha\delta(v)\delta(x-x_0), \qquad f_+(x,v,t) = \alpha\delta(v)\delta(x-x_0), \tag{13.93}$$

with $E(t,x) = 0$. Alternatively we can have a translating simple screening solution by considering the initial data

$$f_-(0,x,v) = \rho(x)\delta(v-v_0), \qquad f_+(0,x,v) = \rho(x)\delta(v-v_0) \tag{13.94}$$

where $v_0 \in \mathbb{R}$, $\rho(x) \ge 0$ and $\int \rho\,dx < \infty$. The corresponding E field is also identically zero. We have a translating weak solution

$$\begin{aligned} f_-(x,v,t) &= \rho(x-v_0t)\delta(v-v_0), \\ f_+(x,v,t) &= \rho(x-v_0t)\delta(v-v_0), \\ E(t,0) &= 0. \end{aligned} \tag{13.95}$$

Except for degenerate cases such as $\rho(x) = \alpha\delta(x - x_0)$ solution (13.95) represents a pair of overlapping electron/ion sheets moving with velocity $v = v_0$. More generally, there is a class of screening weak solutions of the form

$$f_\pm(x, v, t) = g(x - vt, v), \qquad E(t, x) = 0, \tag{13.96}$$

where g is any nonnegative finite measure.

Example 13.6. Simple Rotating-Charge Solutions. Consider the initial data

$$f_-(0, x, v) = \alpha\delta(v)\delta(x + a), \qquad f_+(0, x, v) = \alpha\delta(v)\delta(x - a) \tag{13.97}$$

where a and α are positive numbers. The corresponding $\bar{E}$ field is

$$\bar{E}(0, x) = \begin{cases} -\alpha, & |x| < a \\ -\frac{\alpha}{2}, & |x| = a \\ 0, & |x| > a \end{cases}, \tag{13.98}$$

where we use the symmetric mean of E.

The negatively charged particle $\alpha\ominus$ moves according to

$$\frac{dx}{dt} = v, \qquad \frac{dv}{dt} = -\bar{E} = \frac{\alpha}{2} \qquad (x, v)|_{t=0} = (-a, 0).$$

The solution is

$$x = -a + \frac{\alpha}{4}t^2, \qquad v = \frac{\alpha}{2}t. \tag{13.99}$$

The positively charged $\alpha\oplus$ satisfies

$$x = -a - \frac{\alpha}{4}t^2, \qquad v = -\frac{\alpha}{2}t. \tag{13.100}$$

Solutions (13.99) and (13.100) for the negative and the positive particles $\alpha\ominus$ and $\alpha\oplus$ are valid before the particles "meet" at the origin $x = 0$ at time $t^* = 2\sqrt{a/\alpha}$. Shortly after t^*, the particles move apart in time interval $t \in (t^*, 2t^*)$ and move toward each other when $t \in (2t^*, 3t^*)$. They then switch positions and move apart until they return to the initial position at $t = 4t^*$. The solution then continues periodically in time. The E field has compact support so the two particles also screen each other completely. The full solution (Majda et al., 1994b) is, on $0 < t < t^*$,

$$f_- = \alpha\delta\left(v - \frac{1}{2}\alpha t\right)\delta\left(x + \alpha - \frac{1}{4}\alpha t^2\right),$$

$$f_+ = \alpha\delta\left(v + \frac{1}{2}\alpha t\right)\delta\left(x - \alpha + \frac{1}{4}\alpha t^2\right),$$

$$\bar{E} = \begin{cases} 0, & |x| > a - \frac{1}{4}\alpha t^2 \\ -\frac{1}{2}\alpha, & |x| = a - \frac{1}{4}\alpha t^2 \\ -\alpha, & |x| < a - \frac{1}{4}\alpha t^2 \end{cases}, \tag{13.101}$$

whereas at $t = t^*$

$$f_{\pm}(x, v, t^*) = \alpha\delta\left(v \pm \frac{1}{2}\alpha t^*\right)\delta(x), \qquad E(x, t^*) = 0, \tag{13.102}$$

and for $t^* < t < 3t^*$,

$$f_{\pm}(x, v, t) = \alpha\delta[v \pm X'(t)]\delta[x \pm X(t)],$$
$$\bar{E}(x, t) = \begin{cases} 0, & |x| > X(t) \\ \frac{1}{2}\alpha, & |x| = X(t) \\ \alpha, & |x| < X(t) \end{cases}, \tag{13.103}$$

where $X(t) = \frac{1}{2}\alpha t^*(t - t^*) - \frac{1}{4}\alpha(t - t^*)^2$ is the position of the particle $\alpha\ominus$. We can verify (Majda et al., 1994a, 1994b; Zheng and Majda, 1994) that this solution does indeed satisfy Definition 13.3.

Example 13.7. Elementary Weak Solutions. Combining the particle solutions of Examples 13.5 and 13.6 we form more general weak solutions. Consider the initial data

$$f_{-}(0, x, v) = (\alpha + \beta)\delta(v)\delta(x) + \alpha\delta(v)\delta(x + a),$$
$$f_{+}(0, x, v) = (\alpha + \beta)\delta(v)\delta(x) + \alpha\delta(v)\delta(x - a), \tag{13.104}$$

where β is a positive constant. The corresponding $\bar{E}$ field is the same as that of field (13.98). The $\alpha\ominus$ particle at $x = -a$ attracts the $\alpha\oplus$ particle at the origin to form a rotating-charge pair around $x = -a/2$, and the positive ion at $x = a$ attracts the electron at the origin to form another pair around $x = a/2$. What are left at the origin are the $\beta\ominus$ and $\beta\oplus$ particles, which do not move because of the complete screening effect of the α particles. The solution is

$$f_{-}(x, v, t) = \beta\delta(x)\delta(v) + \alpha\delta\left(v - \frac{\alpha}{2}t\right)\delta\left(x + a - \frac{\alpha}{4}t^2\right)$$
$$+\alpha\delta\left(v - \frac{\alpha}{2}t\right)\delta\left(x - \frac{\alpha}{4}t^2\right),$$
$$f_{+}(x, v, t) = \beta\delta(x)\delta(v) + \alpha\delta\left(v + \frac{\alpha}{2}t\right)\delta\left(x + \frac{\alpha}{4}t^2\right)$$
$$+\alpha\delta\left(v + \frac{\alpha}{2}t\right)\delta\left(x - a + \frac{\alpha}{4}t^2\right), \tag{13.105}$$
$$\bar{E}(t, x) = \begin{cases} 0, & |x| > a - \frac{\alpha}{4}t^2 \quad \text{or} \quad |x| < \frac{\alpha}{4}t^2 \\ -\frac{\alpha}{2}, & |x| = a - \frac{\alpha}{4}t^2 \quad \text{or} \quad |x| = \frac{\alpha}{4}t^2 \\ -\alpha, & \left|x - \frac{a}{2}\right| < \frac{a}{2} - \frac{\alpha}{4}t^2 \quad \text{or} \quad \left|x + \frac{a}{2}\right| < \frac{a}{2} - \frac{\alpha}{4}t^2 \end{cases}.$$

Example 13.8. Nonuniqueness of Weak Solutions. We construct another weak solution for the same initial data (13.104) of Example 13.7 for $0 < \beta < \alpha$. This example has all the positively charged particles $(\alpha+\beta)\oplus$ at the origin respond to the attraction of the particle $\alpha\ominus$ at $x=-a$ and all the negatively charged particles $(\alpha+\beta)\ominus$ at the origin to the attraction of the particle $\alpha\oplus$ at $x=a$.

The trajectory of the particle $\alpha\oplus$ to the far right is the same as that of solution (13.100) in Example 13.6. The trajectory of the particle $(\alpha+\beta)\ominus$ is

$$\begin{aligned} \frac{dx}{dt} &= v, \\ \frac{dv}{dt} &= -\bar{E} = -\frac{1}{2}(\beta-\alpha), \\ (x, v)|_{t=0} &= (0, 0), \end{aligned} \tag{13.106}$$

where$\bar{E}$ denotes symmetric mean (13.43). The solution of particle trajectory (13.106) is

$$\begin{aligned} x &= \frac{1}{4}(\alpha-\beta)t^2, \\ v &= \frac{1}{2}(\alpha-\beta)t. \end{aligned} \tag{13.107}$$

The motions of $\alpha\ominus$ and $(\alpha+\beta)\oplus$ are similar. The charge densities and the electric field are, respectively,

$$\begin{aligned} f_-(x, v, t) &= \alpha\delta\left(v-\frac{1}{2}\alpha t\right)\delta\left(x+a-\frac{1}{4}\alpha t^2\right) \\ &\quad + (\alpha+\beta)\delta\left[v-\frac{1}{2}(\alpha-\beta)t\right]\delta\left[x-\frac{1}{4}(\alpha-\beta)t^2\right], \\ f_+(x, v, t) &= \alpha\delta\left(v+\frac{1}{2}\alpha t\right)\delta\left(x-a+\frac{1}{4}\alpha t^2\right) \\ &\quad + (\alpha+\beta)\delta\left[v+\frac{1}{2}(\alpha-\beta)t\right]\delta\left[x+\frac{1}{4}(\alpha-\beta)t^2\right], \end{aligned}$$

$$\bar{E}(x, t) = \begin{cases} 0, & |x| > a-\frac{1}{4}\alpha t^2 \\ -\frac{1}{2}\alpha, & |x| = a-\frac{1}{4}\alpha t^t \\ -\alpha, & |x| < a-\frac{1}{4}\alpha t^2 \\ & \text{and } |x| > \frac{1}{4}(\alpha-\beta)t^2 \\ -\frac{1}{2}(\alpha-\beta), & |x| = \frac{1}{4}(\alpha-\beta)t^2 \\ \beta, & |x| < \frac{1}{4}(\alpha-\beta)t^2 \end{cases}. \tag{13.108}$$

The $\alpha\oplus$ particle and the $(\alpha+\beta)\ominus$ particle collide at a time

$$t^{**} = \sqrt{\frac{4a}{2\alpha-\beta}},$$

where in particular $t^{**} < \frac{1}{2}t^*$. This means that in the time interval $0 \le t \le t^*$ there are two different weak solutions (13.105) and (13.108) corresponding to the same initial data (13.104). Numerical simulations presented in Majda et al. (1994b) indicate that for $\alpha = 1$, $\beta = 0.5$, and $a = 5$, solution (13.108), in which all the delta functions move, is achieved in the limit when the particle method is used to approximate the weak solution. However, with these same parameters, the vanishing-viscosity limit of the pseudo-Fokker–Planck regularization selects solution (13.105). The viscous limit converges to a completely different limit than that of weak solution (13.108), which arises from a time-reversible computational method without viscosity.

Example 13.9. Generalized Weak Solutions. Consider again initial data (13.104) with $\beta = \alpha$. We now construct generalized weak solutions satisfying Definition 13.4, by means of two different limiting processes.

First, we let $\{f_\pm^\beta, \bar{E}^\beta\}$ denote the solution from Example 13.8 and take the limit of $\{f_\pm^\beta, \bar{E}^\beta\}$ as $\beta \to \alpha^-$. The weak* convergence $f_\pm^\beta \rightharpoonup^* f_\pm$, almost everywhere convergent for $\bar{E}^\beta \to E$, combined with symmetric mean (13.43) of E, gives for $0 < t < t^*$ that

$$f_-(x, v, t) = 2\alpha\delta(x)\delta(v) + \alpha\delta\left(v - \frac{1}{2}\alpha t\right)\delta\left(x + a - \frac{1}{4}\alpha t^2\right),$$

$$f_+(x, v, t) = 2\alpha\delta(x)\delta(v) + \alpha\delta\left(v + \frac{1}{2}\alpha t\right)\delta\left(x - a + \frac{1}{4}\alpha t^2\right),$$

$$\bar{E}(x, t) = \begin{cases} 0, & |x| > a - \alpha t^2 \\ -\frac{1}{2}\alpha, & |x| = a - \frac{1}{4}\alpha t^2 \\ -\alpha, & |x| < a - \frac{1}{4}\alpha t^2 \end{cases}. \tag{13.109}$$

Note that Eqs. (13.109) do not satisfy system (13.6) in a sense of distributions, and hence cannot be a true weak solution. The reader can verify (or see Majda et al., 1994b) that, in this example, the particles $2\alpha\oplus$ and $2\alpha\ominus$ at the origin have no motion even though the electric field $E = -\alpha$ is not zero there. However, this limit can be modified to yield a generalized weak solution by defining $\tilde{E}(x, t)$ to be zero at $x = 0$ and otherwise equal to $\bar{E}$. We can easily verify (Majda et al., 1994b) that this modification of $\bar{E}$ produces a generalized weak solution satisfying Definition 13.4.

A second sequence of perturbed initial data is

$$f_-^\epsilon(x, v, 0) = 2\alpha\delta(v)\delta(x - \epsilon) + \alpha\delta(v)\delta(x + a),$$

$$f_+^\epsilon(x, v, 0) = 2\alpha\delta(v)\delta(x + \epsilon) + \alpha\delta(v)\delta(x - a), \tag{13.110}$$

where $\epsilon \to 0+$. Note that all the particles at $x = \pm\epsilon$ may remain motionless because $\bar{E}^\epsilon$ is zero there. We can show that passing to the limit as $\epsilon \to 0$ yields exactly the same generalized weak solution (13.109).

Numerical computations in Majda et al. (1994b) show that for initial data (13.104) with $\beta = \alpha = 1$, the computational particle method converges to generalized weak solution (13.109). However, a second set of computations, simulating the pseudo-Fokker–Planck regularization, reveals that the vanishing-viscosity limit is in fact weak solution (13.105). Thus generalized weak solutions can both be derived analytically and computed numerically. However, they will not typically arise in the zero diffusion limit.

Example 13.10. Measure-Valued Solutions. The initial data

$$
\begin{aligned}
f_-(x, v, 0) &= 8\delta(v)\delta(x) + \delta(v)\delta(x+a), \\
f_+(x, v, 0) &= 4\delta(v)\delta(x) + 5\delta(v)\delta(x-a).
\end{aligned} \tag{13.111}
$$

yield a nontrivial, bona fide measure-valued solution as well as a weak and a generalized weak solution.

The weak solution is

$$
\begin{aligned}
f_-(x, v, t) &= 3\delta(x)\delta(v) + \delta\left(v - \frac{1}{2}t\right)\delta\left(x + a - \frac{1}{4}t^2\right) \\
&\quad + 5\delta\left(v - \frac{5}{2}t\right)\delta\left(x - \frac{5}{4}t^2\right), \\
f_+ &= 3\delta(x)\delta(v) + \delta\left(v + \frac{1}{2}t\right)\delta\left(x + \frac{1}{4}t^2\right) \\
&\quad + 5\delta\left(v + \frac{5}{2}t\right)\delta\left(x - a + \frac{5}{4}t^2\right),
\end{aligned}
$$

$$
\bar{E}(x,t) = \begin{cases} 0, & x < -a + \frac{1}{4}t^2, -\frac{1}{4}t^2 < x < \frac{5}{4}t^2 \\ & x > a - \frac{5}{4}t^2 \\ -1, & -a + \frac{1}{4}t^2 < x < -\frac{1}{4}t^2 \\ -5, & \frac{5}{4}t^2 < x < a - \frac{5}{4}t^2 \\ -\frac{1}{2}, & x = -a + \frac{1}{4}t^2, -\frac{1}{4}t^2 \\ -\frac{5}{2}, & x = \frac{5}{4}t^2, a - \frac{5}{4}t^2 \end{cases} . \tag{13.112}
$$

The generalized weak solution is

$$
\begin{aligned}
f_-(x, v, t) &= 4\delta(x)\delta(v) + \delta\left(v - \frac{1}{2}t\right)\delta\left(x + a - \frac{1}{4}t^2\right) \\
&\quad + 4\delta(v - 3t)\delta\left(x - \frac{3}{2}t^2\right), \\
f_+ &= 4\delta(x)\delta(v) + 5\delta\left(v + \frac{5}{2}t\right)\delta\left(x - a + \frac{5}{4}t^2\right),
\end{aligned}
$$

$$
\tilde{E}(x,t) = \begin{cases} 0, & x < -a + \frac{1}{4}t^2, x > a - \frac{5}{4}t^2 \\ -5, & \frac{3}{2}t^2 < x < a - \frac{5}{4}t^2 \\ -1, & -a + \frac{1}{4}t^2 < x < \frac{3}{2}t^2, x \neq 0 \\ 0, & x = 0 \\ -\frac{1}{2}, & x = -a + \frac{1}{4}t^2 \\ -3, & x = \frac{3}{2}t^2 \\ -\frac{5}{2}, & x = a - \frac{5}{4}t^2 \end{cases} .
$$

Here $\tilde{E}(0,t) = 0$ instead of the natural value -1. This solution is motivated by the solution of Example 13.9; it is not constructed as a limiting process.

To construct the nontrivial measure-valued solution, we perturb the initial data by

$$f_-^\epsilon(x, v, 0) = \delta(v)\delta(x + a) + 8\delta(v)\delta(x - \epsilon),$$
$$f_+^\epsilon(x, v, 0) = 5\delta(v)\delta(x - a) + 4\delta(v)\delta(x - \epsilon),$$

All the particles at $x = \pm\epsilon$ move with the same speed and acceleration because the $\bar{E}^\epsilon$ fields at $x = \pm\epsilon$ are ∓ 1, respectively. It takes some time independent of $\epsilon > 0$ for the particles $\ominus$ and $5\oplus$ at $x = \pm a$ to collide with the particles at $\pm\epsilon$. So for short times, there exists a sequence of solutions

$$\begin{aligned} f_-^\epsilon(x, v, t) &= \delta\left(v - \frac{1}{2}t\right)\delta\left(x + a - \frac{1}{4}t^2\right) \\ &\quad + 8\delta(v - t)\delta\left(x - \epsilon - \frac{1}{2}t^2\right), \\ f_+^\epsilon(x, v, t) &= 5\delta\left(v + \frac{5}{2}ft\right)\delta\left(x - a + \frac{5}{4}t^2\right) \\ &\quad + 4\delta(v - t)\delta\left(x + \epsilon - \frac{1}{2}t^2\right), \end{aligned} \tag{13.113}$$

$$\bar{E}^\epsilon(x,t) = \begin{cases} 0, & x < -a + \frac{1}{4}t^2 \\ -\frac{1}{2}, & x = -a + \frac{1}{4}t^2 \\ -1, & -a + \frac{1}{4}t^2 < x < -\epsilon + \frac{1}{2}t^2 \\ 1, & x = -\epsilon + \frac{1}{2}t^2 \\ 3, & -\epsilon + \frac{1}{2}t^2 < x < \epsilon + \frac{1}{2}t^2 \\ -1, & x = \epsilon + \frac{1}{2}t^2 \\ -5, & \epsilon + \frac{1}{2}t^2 < x < a - \frac{5}{4}t^2 \\ -2.5, & x = a - \frac{5}{4}t^2 \\ 0, & x > a - \frac{5}{4}t^2 \end{cases}.$$

Passing to the limit in ϵ gives

$$\begin{aligned} f_-(x, v, t) &= \delta\left(v - \frac{1}{2}t\right)\delta\left(x + a - \frac{1}{4}t^2\right) \\ &\quad + 8\delta(v - t)\delta\left(x - \frac{1}{2}t^2\right), \\ f_+(x, v, t) &= 5\delta\left(v + \frac{5}{2}ft\right)\delta\left(x - a + \frac{5}{4}t^2\right) \\ &\quad + 4\delta(v - t)\delta\left(x - \frac{1}{2}t^2\right), \end{aligned} \tag{13.114}$$

$$E(x,t) = \begin{cases} 0, & x < -a + \frac{1}{4}t^2 \\ -\frac{1}{2}, & x = -a + \frac{1}{4}t^2 \\ -1, & -a + \frac{1}{4}t^2 < x < \frac{1}{2}t^2 \\ 1, & x = \frac{1}{2}t^2 \\ 3, & \frac{1}{2}t^2 < x < \frac{1}{2}t^2 \\ -1, & x = \frac{1}{2}t^2 \\ -5, & \frac{1}{2}t^2 < x < a - \frac{5}{4}t^2 \\ -2.5, & x = a - \frac{5}{4}t^2 \\ 0, & x > a - \frac{5}{4}t^2 \end{cases}.$$

We use here the weak* limit for $f_{\pm}^{\epsilon}$, almost everywhere convergent for $\bar{E}^{\epsilon}$ and the symmetric mean on E at positions $x = -a + \frac{1}{4}t^2$ and $x = a - \frac{5}{4}t^2$.

The omission of E on the curve $x = \frac{1}{2}t^2$ is intentional; in fact there is no value that can be assigned there to give a generalized weak solution. So we have a sequence of exact weak solutions whose weak limit is not any generalized weak solution. One can check that the limit is a measure-valued solution (Majda et al., 1994b). In another set of computations in Majda et al. (1994b) by use of initial condition (13.111), the computational particle method was shown to converge to measure-valued solution (13.114). Thus measure-valued solutions of the two-component V-P system can be produced analytically and computed numerically. However, the Fokker–Planck vanishing-viscosity limit was shown to converge, for moderate times $0.1 \leq t \leq 0.4$ to weak solution (13.112). For longer times, they observed substantial departure from this solution.

Throughout these examples, screening effects play an important role. Example 13.5 has perfect microscreening so that $E = 0$. In Examples 13.6 and 13.7 each electron–ion pair forms a perfect screen, outside of which the E field is identically zero. In Example 13.9 we find that the concept of a weak solution is insufficient because of a strong local change in concentration of charges; however, the electric field far from the charges remains unaffected because of the microscreening of the particles. Concentration with screening is the mechanism that creates islands or even layers of alternatively charged particles within a microscopic scale in a flow. Finally, in Example 13.10, the same mechanism exhibits itself fully in a dynamical process, resulting in a true measure-valued solution. All three kinds of solutions exist as analytic solutions but also are observed to result from numerical simulations by use of computational particle methods.

Note for Chapter 13

Helly's selection principle can be found in Billingsley (1986, Section 25).

References for Chapter 13

Beale, J. T. and Majda, A., "Vortex methods I: convergence in three dimensions," *Math. Comput.* **39**(159), 1–27, 1982a.

Beale, J. T. and Majda, A., "Vortex methods II: higher order accuracy in two and three dimensions," **39**(159), 29–52, 1982b.

Berk, H. and Roberts, K., "The water-bag model," *Methods Comput. Phys.* **9**, 87–134, 1970.

Billingsley, P., *Probability and Measure*, 2nd ed., Wiley, New York, 1986.

Birkhoff, G., "Helmholtz and Taylor instability," in *Hydrodynamics Instability*, Vol. 13 of the Proceedings of the Symposium on Applied Mathematics Series, American Mathematical Society, Providence, RI, 1962, pp. 55–76.

Caflisch, R. E. and Orellana, O. F., "Singular solutions and ill-posedness for the evolution of vortex sheets," *SIAM J. Math. Anal.* **20**, 293–307, 1989.

Chorin, A. J. and Bernard, P. J., "Discretization of vortex sheet with an example of roll–up," *J. Comput. Phys.* **13**, 423–428, 1973.

Cottet, G. and Raviart, P., "Particle methods for the 1-D Vlasov–Poisson equations," *SIAM J. Numer. Anal.* **21**, 52–76, 1984.

DiPerna, R. J. and Majda, A. J., "Concentrations in regularizations for 2-D incompressible flow," *Commun. Pure Appl. Math.* **40**, 301–345, 1987a.

DiPerna, R. J. and Majda, A. J., "Oscillations and concentrations in weak solutions of the incompressible fluid equations," *Commun. Math. Phys.* **108**, 667–689, 1987b.

DiPerna, R. J. and Majda, A. J., "Reduced Hausdorff dimension and concentration–cancellation for two-dimensional incompressible flow," *J. Am. Math. Soc.* **1**, 59–95, 1988.

Dziurzynski, R. S., "Patches of electrons and electron sheets for the 1-D Vlasov Poisson equation," Ph.D. dissertation, University of California, Berkeley, CA, 1987.

Iordanskii, S. V., "The Cauchy problem for the kinetic equation of plasma," *Tr. Mat. Inst. Steklov* **60**, 181–194, 1961; *Am. Math. Soc. Transl. Ser.* **2**, 351–363, 1964.

Krasny, R., "Computing vortex sheet motion," in *Proceedings of International Congress of Mathematicians, Kyoto 1990*, Springer-Verlag, New York, 1991, Vol. 2, pp. 1573–1583.

Liu, J.-G. and Xin, Z., "Convergence of vortex methods for weak solutions to the 2-D Euler equations with vortex sheet data," *Commun. Pure Appl. Math.* **48**, 611–628, 1995.

Majda, A., "Vortex dynamics: numerical analysis, scientific computing, and mathematical theory," in *ICIAM'87: Proceedings of the First International Conference on Industrial and Applied Mathematics*, Society for Industrial and Applied Mathematics, Philadelphia, 1988, pp. 153–182.

Majda, A. J., "Remarks on weak solutions for vortex sheets with a distinguished sign," *Indiana Univ. Math. J.* **42**, 921–939, 1993.

Majda, A., Majda, G., and Zheng, Y., "Concentrations in the one-dimensional Vlasov–Poisson equations, i: temporal development and nonunique weak solutions in the single component case," *Physica D* **74**, 268–300, 1994a.

Majda, A., Majda, G., and Zheng, Y., "Concentrations in the one-dimensional Vlasov–Poisson equations, ii: screening and the necessity for measure-valued solutions in the two component case," *Physica D* **79**, 41–76, 1994b.

Rosenhead, L., "The point vortex approximation of a vortex sheet or the formation of vortices from a surface of discontinuity-check," *Proc. R. Soc. London A* **134**, 170–192, 1932.

Rudin, W., *Real and Complex Analysis*, 3rd ed., McGraw-Hill, New York, 1987.

Van Dyke, M., *An Album of Fluid Motion*, Parabolic, Stanford, CA, 1982.

Volpert, A. I., "The spaces BV and quasilinear equations," *Math. USSR Sb.* **73**, 255–302, 1967.

Volpert, A. I. and Hudjaev, S. I., *Analysis in Classes of Discontinuous Functions and Equations of Mathematical Physics*. Martinus Nijhoff Publishers, Dordrecht, 1985.

Zabusky, N., Hughes, M. H., and Roberts, K. V., "Contour dynamics for the Euler equations in two dimensions," *J. Comput. Phys.* **30**, 96–106, 1979.

Zheng, Y. and Majda, A., "Existence of global weak solutions to one component Vlasov–Poisson and Fokker–Planck–Poisson systems in one space dimension with initial data of measures," *Commun. Pure Appl. Math.* **47**, 1365–1401, 1994.

Index